Saturn I/IB Rocket

Saturn I/IB Rocket

NASA's First Apollo Launch Vehicle

David Baker

Crécy Publishing Ltd

First published 2022 by Crécy Publishing Ltd.

Layout: Martin Preston

A CIP record of this book is available from the British Library.

ISBN 9781800350281

Printed in Turkey by Pelikan Print

Front cover: Precursor to the grand achievements of the Saturn V and a development of the Saturn I, Saturn IB paved the way for astronauts to fly Apollo, occupy the first re-visited space station and launched the first US-Russia human space flight. *NASA*

Half title page: Saturn IB AS-201 stands ready on Launch Complex 34 at the Kennedy Space Center, in preparation for the first flight into space of the uncrewed Block 1 Apollo Command and Service Module, 26 February 1966. *NASA-KSC*

Title page: Flight configuration of the Saturn C-1 with dummy S-IV and S-V stages, assembled for the first time in the Fabrication and Assembly Engineering Division at Huntsville's Marshall Space Flight Center on 15 February 1961. Put together as a fit-check, the elements would be disassembled for transfer to Cape Canaveral. *MSFC*

Back cover main picture: The first Saturn I (SA-1) lifts off from LC-34 on 27 October 1961. *NASA*
Back cover insets (top to bottom): The S-IV stage is prepared for launch on a Saturn I Block II flight. *(NASA)* The forward section of a Redstone rocket, behind the warhead and instrument section. *(Chrysler Corp)* The first completed S-I stage ready for delivery to Cape Canaveral. (*NASA)*

Crécy Publishing Limited
1a Ringway Trading Estate, Shadowmoss Road, Manchester
M22 5LH

www.crecy.co.uk

Contents

Introduction

The story of the Saturn I and the Saturn IB is an important part of the narrative defining the early years of the space programme, in which launch vehicles determined what was possible. It also has a place on the broader spectrum of military engineering and technology developed by the Germans during the Second World War and built upon by the major powers at the start of the Cold War.

Just as the jet fighter, first deployed operationally by the Luftwaffe in 1944, the modern post-war submarine crafted from the design of the Kriegsmarine's Type XXI, and the Panther tank of the *Deutsches Heer* provided the base upon which modern warfighting machines were designed, so too did the A-4/V2 bequeath possession of one of the most game-changing instruments of conflict – the ballistic missile – to the heavy-lift launchers of the Space Age.

This book is as much about that story as it is about the first big rockets designed and developed by the same team which had, before the end of the war, built and launched several thousand rounds of the world's first ballistic missiles. The linear progression of design, build, test and deployment of those rockets provided a level of experience unique to the German Army. It was the enabling catalyst between the possibilities presented by the laws of physics and chemistry and the dawn of the Space Age, forged by engineers and technicians taken from war-torn Germany to the United States in 1945 and 1946.

In the post-war world, it could be said that the greatest achievement of the early Soviet space programme was the wake-up call it sent to the United States of America and, subsequently, the Moon goal it spawned when, on 25th May 1961, President John F. Kennedy committed NASA to the boldest adventure imaginable: to landing a man on the Moon by the end of the decade. It was achievable only because of the Saturn series of heavy-lift launch vehicles beginning with the Saturn I and progressing to the Saturn V, the only launch vehicle in that series which could do the job in a single flight.

This book is about Saturn I, that lesser-known precursor to the Saturn V which, nevertheless, has a proud place in the annals of space history because it served as the test vehicle for an upper stage of the Saturn V. It launched giant satellites into orbit and the first (unmanned) Lunar Module, the first manned Apollo spacecraft, three teams of astronauts to occupy Skylab – America's first and only national space station – and the last manned Apollo mission, the one which carried three astronauts to dock with the Russian Soyuz spacecraft and its two cosmonauts.

From its first launch in 1961 to the last Apollo flight in 1975, the Saturn I and its successor the Saturn IB were at the forefront of an expanding and rapidly evolving space programme. As such, paradoxically considering its relatively unknown history, it is one of the most important programmes of its age. In just fourteen years it staked a proud claim in those annals and is the subject of this book. But there is much more to the story than the technical description of the rocket itself and the narrative of its 19 launches.

Saturn evolved from a series of rocket programmes which emerged from studies conducted by Wernher von Braun and a group of German scientists and engineers. It was this group that launched America's first satellite and developed the rocket which put the first and the second Americans in space. And while these more famous missions take centre stage in space history, the complex series of associated rocket types studied and evaluated by enthusiasts are frequently side-lined and largely forgotten by all but the most committed space historians. It is that inclusive story this book sets out to tell in a comprehensively researched history embracing captured V2 rockets, the Redstone and Jupiter missiles and the space launch vehicles they matured into. To do that it is divided into four distinct sections.

The first section tells the story of Wernher von Braun, about his previous work in Germany and his work under a new employer from the time he arrived in the United States, through the development of the V2 as a two-stage sounding rocket, to work on the Redstone and Jupiter missiles before they were utilised for space missions as Juno I and Juno II, respectively. It is a story important because all the ingredients for the Saturn I are in there: technical direction, management, engineering evolution of rockets and missiles crucial to the evolution of the Army's rocket programme and a description of the way the designs evolved in a few brief years.

The second section tracks the evolution of new concepts, of Juno III, IV and V, the first two cancelled and the last renamed Super Jupiter before it was finally called Saturn. During this period the personality and charisma of the publicity-savvy von Braun played a major role in turning the eyes of the Army toward bespoke space launchers rather than utilisation of adapted missiles. We explore that through little known aspects of these years to provide background on his influence regarding Army thinking. Here, we examine the various families of the Saturn rocket, designated A, B and C, before describing the C-1 through to the C-5, of which only the first and the last were built and flown, as the Saturn I and the Saturn V respectively.

This frequently tortuous evolution is explained in the description of these rocket types but each is important in its own way, reflecting the prevailing choices of the time and the manner in which various options were considered, cast aside or adopted. We explore the reasons why in each case.

The third section describes in detail the design and structural arrangement of the Saturn I, both in suborbital tests and orbital configuration with the cryogenic S-IV upper stage. It then describes the changes made to the first stage to adapt it into the Saturn IB and explores the design and evolution of the more powerful S-IVB upper stage. In this section can be found the technical description of the Saturn I and the Saturn IB, together with the associated derivatives.

The fourth section of this book chronicles the story of each of the 19 individual Saturn I/IB flights and gives some detail about the missions they launched. It also includes background histories of the White Sands Proving Ground (WSPG) and Cape Canaveral launch sites applicable to the rockets described in this book. But it is essentially about the performance of the rockets themselves, 15 of which resulted in orbital flights after four successful, suborbital test launches. Specific details of weights, event times and performance are provided.

The Saturn I was the first launch vehicle developed specifically for the purpose of placing heavy payloads in orbit and it was the last large rocket programme managed by the US Army, an organisation which has the lasting credit for launching the first American satellite eight months before NASA existed. Integral to its story was the controversy between the Army Jupiter and the Air Force Thor missile during the 1950s and that is told here as it impacted the von Braun group and subsequent work on the Saturn series.

Saturn I has a proud place in the annals of space history and was, from the very beginning to the present day, this writer's most favoured launch vehicle, epitomising as it did the first US rocket to outclass contemporary Russian rockets both in reliability and payload capability. As such, it introduced a new confidence among US engineers and scientists, that they could achieve the goal of landing a man on the Moon and returning him safely to the Earth with the Saturn V – the story of a subsequent book.

David Baker, East Sussex

Above: Saturn I architect Wernher von Braun briefs President Kennedy on the constructional details of Saturn Launch Complex LC-37B at Cape Canaveral, 16 November 1963. Less than a week later, Kennedy would be dead. Author's collection. *MSFC*

Acknowledgements

Any comprehensive history owes much to a large number of people and is never the sole work of one individual. I would like to thank a few who have contributed over the years to providing information which forms the core of this book but there are many more who have influenced the direction taken by this work. Unfortunately, time has taken its toll and a number of these people are no longer with us. Nevertheless, gratitude is extended to their memory and grateful thanks for the contribution they made in allowing me to assemble over many decades a major archive from which the information in this work has been extracted.

I would like to remember the contributions of Joseph M. Jones at the NASA Marshall Space Flight Center, of William J. O'Donnell at NASA Headquarters, and of Richard 'Dick' Young at NASA's Kennedy Space Center. They served as door-openers and providers of original documents, copies of key files in the form of letters from senior managers, technical notes from middle managers and thoughts and musings of many technicians and engineers directly involved with the subjects covered in this book.

I would also like to thank numerous helpers in the NASA history offices who sent box loads of material home to the author's holding address in the UK when he was in the United States. Particularly to Alan S. Wood, Don Bane and Frank Bristow at the Jet Propulsion Laboratory regarding documents covering early work on Army missiles. And to the staff of the history office at the Cape Canaveral Air Force Station who afforded me access and assistance in seeking out the employment contracts for the German scientists who moved to the US in the late 1940s.

I would also like to acknowledge the help and support of Dr. Charles Sheldon, Assistant Director, Select Committee on Astronautics and Space Exploration, and Barbara Luxembourg of the Congressional Research Service who, over many years, provided detailed information on related areas. To Charles F. Ducander, Executive Secretary of the US House of Representatives' Committee on Science and Astronautics, and to George G. Feldman, Chief Counsel and Director, US Senate Select Committee on Aeronautics and Space Exploration, and James J. Gehrig, Senate Committee on Aeronautical and Space Sciences.

In addition, thanks and appreciation to many staff members of the US Office of Management and Budget who provided highly detailed histories of NASA's annual budget process, year in and year out over several decades. Without your help, I would not have in my possession those precious documents which are no longer available – even on the Web – outside agency filing cabinets. Also, my thanks to Bettie Sprigg at the US Department of Defense for similar help over many years and to Joseph Laitin and Alan B. Wade and staff at the Executive Office of the President.

In industry, thanks go to Joyce B. Lincoln, E. F. Hogan and R. W. Martin at Rocketdyne-Canoga Park, to Sue Cometa, A. J. Kearney and David Alter at North American Rockwell and to Roger Koch, William A. Rice, James D. Grafton and Donald B. Brannon at Boeing. It was they who underpinned my research into aerospace companies now under the Boeing banner. Design engineers Robert E. 'Bob' Biggs and Vince Wheelock at Rocketdyne afforded generous and invaluable assistance with some of the nuanced development of engines there and who continued to provide vignettes of information.

I owe special thanks to Norman 'Norm' Ryker and Richard 'Dick' Schwartz who provided valuable insight and under whose leadership at Rocketdyne I spent many happy work hours and some memorable visits to the Santa Susana site to experience first-hand rocket engines on test. The photographs I took will appear in the Saturn V book.

At McDonnell Douglas, Donald R. Stiess and C. W. Birnbaum were especially helpful, together with Roy B. Heath and David C. Jolivette from the Chrysler Corporation. Thomas E. Ross provided outstanding help with Aerojet Liquid Rocket Company.

Special thanks go to Chester M. Lee, Director of Space Transportation System Operations at NASA Headquarters for encouraging me to pursue this history with vigour, for the generosity and personal support of James M. Beggs, NASA Administrator from 1981 to 1986 over many meetings and lunches. I owe him special thanks for his kindness to my students when we visited Washington DC, and to Robert C. Seamans Jr, former Associate Administrator then Deputy Administrator of NASA from 1960 to 1968 who provided valuable insight into major decisions made during the early years.

Ralph C. Perrill at the US Army Materiel Readiness Command at Redstone Arsenal helped with aligning events during the early von Braun years at Huntsville, Capt. Rick. DuCharme at the Department of the Air Force and Robert A. Carlisle at the Department of the Navy provided detail on respective programmes and some unpublished projects.

The images used in this book come from my own archives and files, originating in the assembled collections of government agencies, contractors and individuals who have provided pictures and made them available for publication. The pervasive nature of the Web has allowed many more people to gain access to stunning images in a way they would never have been able to access before and that is a very good thing. But the price has been the rejection of so many more and a high percentage of the images in this book originated as prints, now no longer available except in private archives and certainly not on the Web. There is an especial dearth of historical images and pictures showing specific details of minor activities or hardware which has only passing interest for the general media but rich value for the historian.

All these people have helped over a period of more than 50 years with providing information and documents, allowing access to libraries or archives and/or working with me at various times to compile banks of information pursuant to the subject of this book. Many of them are no longer around but their contribution to this book has been substantial and I thank them all for that. Various sources are used in many publications but I have sought and utilised original documents and reports, many of which are in my possession. That archive could never have been compiled had it not been for these people over many years who gave time and effort to obtain esoteric pieces of information to complete this story in the most accurate way possible.

Two people helped in extraordinary ways. One was the former chief of the Army Ballistic Missile Agency, Maj. Gen. John B. Medaris, who retired in January 1960 on the second anniversary to the day that America launched its first artificial satellite, largely because he had persuaded the Army to unleash Wernher von Braun from constraint when the Vanguard project failed to launch an American satellite on its first attempt. Medaris went on to become an Episcopalian priest and spent many years ministering to his parish, speaking in public at length and helping define and interpret seminal moments in the history of those turbulent and unpredictable times. I felt humbled that he should have afforded me the time he did with fresh perspectives before he passed away in July 1990.

A very special debt of gratitude is owed to Saturn historian Alan B. Lawrie, about whom much could be said and from whose copious and comprehensive research and publications has been derived the engine numbers and individual movement chronologies for Saturn stages published in Section 4 of this book. A rocket engineer himself, Alan has been a major contributor to the universal bank of knowledge about the Saturn launch vehicles and his books on the Saturn I and Saturn V, and on the SACTO facility especially, are essential reading for anyone with a serious interest in the production and test history of these rockets.

In no small measure it was Alan's work that inspired me to prepare this volume on the Saturn I/IB and its companion volume on the Saturn V. In concentrating more on the political, industrial and technical descriptions of the rockets themselves, as well on their flight performance, I have not attempted to duplicate Alan's contribution but rather to add those elements to his existing work on test histories, to which the reader of this book is directed and encouraged to seek out.

Finally, and in no small measure, I thank my publisher for commissioning this and its companion volume on the Saturn V, with special appreciation to Jeremy Pratt, MD of Crécy, and to managing editor Charlotte Smith as well as Martin Preston for putting the whole thing together and Ann Page for proofing, copy-editing and cleaning up a ragged text. Thank you all. No book ever published is without an unintended error or two and I apologise in advance, and accept sole responsibility for, any that may have crept in. *DB*

A Note for Readers

Cognizant of the global reach with this book, measurements are displayed both in Metric and Imperial units, the latter in brackets following the former.

1. To avoid confusion and an interminable exchange of prefixes, while Saturn I missions were prefixed with the letters SA (Saturn Apollo), sources identified Saturn IB flights both as SA and AS (Apollo Saturn). Documents and engineering references from the NASA Marshall Space Flight Center consistently used the SA prefix, while Headquarters and other field centres used AS. Therefore, readers will note that information sources can be found to use either – and sometimes both! We have used AS throughout for Saturn IB and, where mentioned, Saturn V flights, purely to avoid confusion (where both SA and AS could refer to the same vehicle or mission). Perfectionists will know that SA refers to the rocket while AS denotes the mission number but throughout we have used AS for those vehicles.

2. The abbreviation T- indicates launch time (T) minus the time denoted. Where mission times are given they are in elapsed hours, minutes and seconds from the instant of lift-off.

3. Readers should note that the tabular data, timings, events and parameters have been taken from the engineering reports originating at NASA's Marshall Space Flight Center beginning in the 1960s and that these are the basis for the information presented in this book. It will be seen that many equivalent parameters quoted on websites and information available on the internet are sometimes at variance with those figures in this book, for that reason. Moreover, the technical documents themselves are occasionally at variance with each other.

4. Where different values or timings are encountered, the most probable are quoted.

5. Times are primarily pegged to local times at Cape Canaveral with Universal Time in parenthesis and this is so as to align events discussed with the time locally when they occurred.

6. Payload weights are quoted both for the total amount placed in orbit, which includes the S-IV or S-IVB stage, and for the payload component separately, these numbers being frequently interchanged in several published and online sources.

7. Thrust values for all 19 Saturn I/IB flights reflect the nominal sea level thrust at lift-off. Values varied by small amounts on each flight, increasing with altitude as with all rocket motors.

How to Use This Book

The origin and evolution of the Redstone and Jupiter missiles is crucial to a systematic understanding of the Saturn series as laid out in this book. Accordingly, and to define separate elements of the story, as well as helping you find the areas that interest you most, the book is separated into four parts for inclusiveness of content but with seamless continuity. Some readers may care to jump to chapters which interest them most but the book has been written without any perceived breaks between sections so that diehards can start at the beginning and go through to the end:

Section 1: Genesis of the Missile Age

Covers the development work in establishing a team of rocket engineers who became the first employed by a national government to development of large rockets and ballistic missiles. This team underpinned all the developmental achievements of the group which went to the United States and their story is fundamental to that of the first generation of US Army missiles in the 1950s. This section describes the movement of the German team to the United States, the development of the test flights of the A-4/V2 and the Bumper/WAC launches. It covers the history of the Redstone and the Jupiter IRBM and their deployment in Europe.

Section 2: From Jupiter to Saturn

Describes how the Redstone and Jupiter missiles were adapted for satellite launches as the Juno I and Juno II, together with an explanation of how they fitted in to early planning, supporting the early attempts at sending probes to the Moon. It carries the story across to the Juno III, IV and V, the latter named Super Jupiter and consisting of a wide range of baseline rockets with derivative growth options across into cryogenic stages. It follows across to describe those various configurations in the A, B and C categorisations and then explains how the C-series became the progenitor of the C-1 to C-5 options toward the Nova concept. It covers the various hydrocarbon and cryogenic rocket motors in the H-1, F-1, E-1 and RL-10 groups and examines how they fitted the various classes of launcher. It also covers the transfer of the von Braun rocket group to NASA and the all-up concept of 1963.

Section 3: Engineering the Saturn I

Provides a detailed technical description of the Saturn I, the Saturn IB, the S-IV and the S-IVB upper stages, the rocket motors incorporated in each and the different variants proposed during and after flight operations. This part of the book is the technical core and is signed as a reference point for the developmental evolution from the Saturn I to the Saturn IB, each getting separate descriptions and technical detail on the H-1, the RL-10 and the J-2. It is written for the reader without specialist knowledge and does not contain jargon or mathematical calculations which could otherwise act to deter the general reader. Some technical knowledge is an advantage but there are abbreviations and a glossary at the back of the book to facilitate explanations. A separate chapter describes all the derivatives and upgrades considered but never implemented.

Section 4: Saturn Flight Results

Contains an overview of Saturn I and Saturn IB missions, followed by each flight described in detail and in chronological order. No two Saturn vehicles were the same and this section contains information on those detailed differences. The emphasis is on both pre-flight activity and flight operations with the launch vehicle but does not enter into detail about the mission of the payload launched, other than to describe how it fits in to the objectives of the overall programme. In the introduction to the Saturn IB flight section, various mission options for early uses of this launcher are discussed with each change and shift in plan tracked chronologically.

Readers seeking detailed technical information on the Saturn launchers should find that in Section Three while those primarily interested in flights will find that in Section Four. Modellers will find much of interest in the images supporting Section Two while readers fascinated by the background history can read Sections One and Two consecutively. To save space, dimensional data on each individual Saturn launcher can be found in specific launch entries, within the images, or in the captions, which helps keep relevant information in appropriate sections of the book. Finally, we add an Appendix concerning the architect of the Saturn I programme, Wernher von Braun. Readers can consult the contents page for specific identification of individual chapters.

At no point in this book has there been an intention to provide a biography of von Braun. There are enough of those already. But my publisher and I felt it was helpful to incorporate some little-known, and some previously unpublished, aspects of his life and career which together help clarify his early years in the employ of the US Army. In particular, his association and correspondence with members of the British Interplanetary Society after the Second World War. In doing so it is hoped that readers will gain a broader understanding of the important role played by von Braun in the development of Redstone, Jupiter and Saturn.

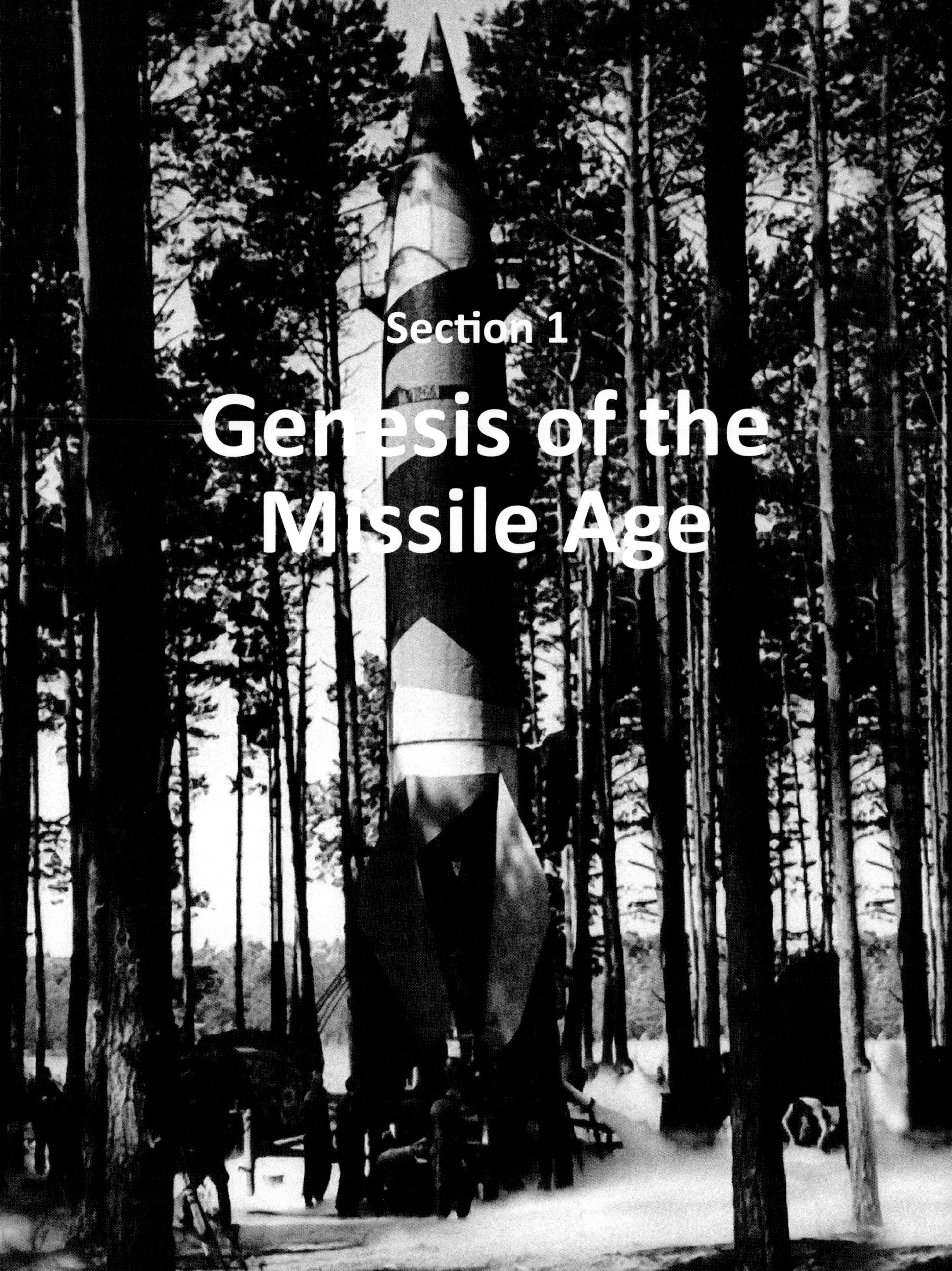

Section 1

Genesis of the Missile Age

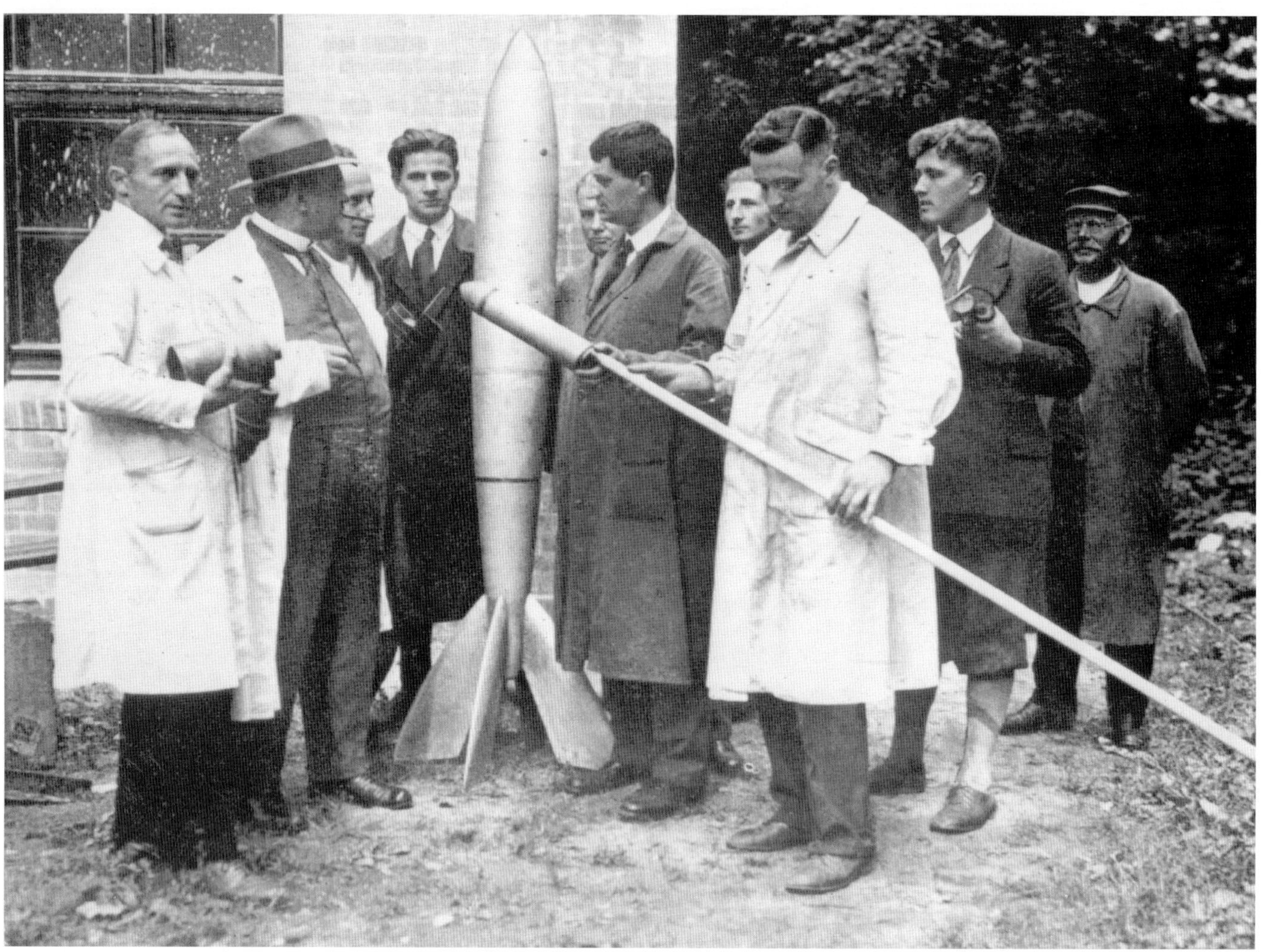

Chapter One

First Tests

The origin of the Saturn series of launch vehicles can be traced back to the immediate aftermath of the Second World War, when a contingent of German rocket scientists, engineers and technicians arrived in the United States to develop a research programme which would presage the dawn of the Space Age. They brought with them knowledge and experience gained through intensive experimentation and development across 15 years, during which time in Germany it had been mobilised into a government programme realising the creative ambitions of a group of civilian enthusiasts who came together to carry out basic research on a very new technology bridging chemistry, physics and engineering. Over the next decade and more it would be they who would put America on the road to the Moon.

Fascination with space travel had erupted in the immediate years after the end of the First World War with an armistice signed in 1919. Fuelled by a surge in advances within the physical sciences, previously by the unification of electricity and magnetism by James Clerk Maxwell in the 19th century and latterly by the development of aviation, science fiction and moving film took up the genre and created a future in which spaceships would ply the voids between worlds. Youth everywhere was inspired to conjure up a world in which such things could become reality. And they had grounds for thinking that could be so. The teachings of the Russian school teacher Konstantin Tsiolkovsky provided the mathematical equations and explained calculations manifestly capable of achieving that, restrained only by the limitations of mechanical engineering and materials technology. When the American engineer, scientist and professor Robert Goddard launched the world's first liquid propellant rocket into a

Above: Members of the *Verein für Raumschiffahrt* (VfR), established in 1927. Left to right: Rudolf Nebel, Franz Ritter, unknown, Kurt Heinisch, unknown, Hermann Oberth, unknown, Klaus Riedel, Wernher von Braun, unknown. *Author's collection*

Left: On 16 March 1926, Robert Goddard prepares to launch the world's first liquid propellant rocket from Auborn, Massachusetts. *Frank H Winter*

cabbage patch, just a few brief footsteps away, at Auborn, Massachusetts, on 16 March 1926, those limitations were lowered still further.

Inspired by this seminal event, and by a surge in popular interest about rockets and space travel, enthusiasts and visionaries in several countries around the world formed groups to study what became known as astronautics, some working the theory, others conducting practical tests with their own rocket designs. Across Europe, in Russia and in the United States, experimenters and enthusiasts formed groups to share information, expand their knowledge and publish literature debating the technical possibilities and discussing the future.

The first of these formal organisations, the *Verein für Raumschiffahrt* (VfR), was set up in Germany by Johannes Winkler on 5 June 1927, less than a year after Goddard's historic achievement. It focused on practical tests and, with more than 500 members, produced a magazine, *Die Rakete* (The Rocket), whose many contributors went on to achieve great acclaim for their theoretical and practical work. Wernher von Braun, a young student with considerable personal means, joined the VfR in 1928 and began to positively influence the direction of experiments there. He was also a great publicist, working with VfR members to guide the Ufa Film Company in its popularisation of rockets and space travel with dramatic and fictional productions imagining a space future. Clearly a born leader of men, von Braun had looks, charm and a bearing that placed him central to any group and this was exploited by some, resented by a few.

Practical tests began in 1930 at the Reinickendorf, a disused ammunition dump outside Berlin named the *Raketenflugplatz* (Rocket Airfield). The most important member at the time was Klaus Riedel, who patented various devices and designed the early test rockets. But he was a keen supporter of human rights and chafed at the politics of the Third Reich when Hitler came to power in 1933, refusing to join the Army team with von Braun. Before that, Riedel was fundamental to the early experiments, which were not aimed at a specific design but rather as concept demonstration tests to prove ideas and theories about the mechanics and the materials associated with rocket engineering. Nobody had any real idea what these would lead to but they were seminal to understanding the fundamental principles of rocket flight.

Early developments began with the Mirak-1 which was powered by liquid oxygen and gasoline, a small stick-rocket with the combustion chamber embedded in the 30.5cm (12in) liquid oxygen (LOX) tank. It was test run in August 1930 but the tank exploded during a second test the following month. Outside the VfR, other experimenters were at work to produce the first successful rocket flight in Europe, that being achieved by Johannes Winkler in association with Hugo A. Huckel on 13 March 1931. Named Huckel-Winkler 1, the

Right: An unambitious school teacher from Siberia, Konstantin Tsiolkovsky (1857-1935) established the principles on which space flight might be possible, including the use of high-energy cryogenic propellants. He is pictured here in 1913 with dirigibles, another of his fascinations with flight. *Author's collection*

rocket ascended to a height of 300m (1,000ft) from a deserted area near Dessau and was followed on 6 October by a second rocket, the Huckel-Winkler 2. That caught fire and crashed from a height of 3m (10ft). Meanwhile, the VfR continued on with Mirak-2 but that too was destroyed on test when the LOX tank burst.

Finally, on 14 May 1931 the VfR's Repulsor-1 made a successful flight to a height of 61m (200ft), the second successful liquid propellant rocket flight in Europe. Repulsor-2 followed on 21 May 1931 to the same altitude but a range of 610m (2,000ft). Named after the Martian spacecraft in Kurd Lasswitz's 1897 science fiction novel *Auf Zwei Planeten* (On Two Planets), Repulsor-series rockets had a water-cooled combustion chamber with the fuel tank mounted above the motor, the assembly flanked by the LOX tank on one side and a nitrogen tank for pressurisation on the other – hence they were called two-stick rockets. The names Mirak and Repulsor were chosen to avoid the use of the term 'rocket', since that was considered to refer to powder rockets such as fireworks.

Later Repulsor rockets achieved an altitude of approximately 1,000m (3,280ft) and, from Repulsor-3 the rockets were fitted with parachutes to prevent destructive impact with the ground, some reaching a height of 1.6km (1ml). These were not inconsequential achievements and made a significant contribution to the knowledge-base but there was a great need for more resources. It was largely at the suggestion of von Braun that the VfR had approached the Army in September 1930, asking for funds and pointing to the military potential of the rocket. It was at the very least premature – the German group had yet to demonstrate it could fire a rocket successfully. The bid was preceded on 23 July 1930 by a static demonstration of the *Kegeldüse* rocket sponsored by the Reich Institute for Chemistry and Technology. It ran for 90 seconds generating a thrust of 66.7N (15lb) on LOX and gasoline. One of those assisting was the 18-year-old Wernher von Braun.

No such military support was in the minds of US enthusiasts, however, when Ed and Lee Pendray formed the American Interplanetary Society (AIS) on 4 April 1930. Uniquely, members were motivated by American pop culture embracing science fiction, space movies and comic books ranging far and wide on the theme of interplanetary travel. Practical experiments concentrated on liquid propellant rockets but in 1934 the AIS changed its name to the American Rocket Society (ARS) to appear more serious. Other groups sprang up, the one most notable for its achievements and place in the history of astronautics being a small group of researchers at the California Institute of Technology (CalTech) which began experiments under the tutelage of Theodor von Kármán. This work would expand during the Second World War with solid propellant and liquid propellant rockets for military use. From this group would emerge the Jet Propulsion Laboratory (JPL), now a part of NASA and famous for its planetary probes. In 1963 the ARS merged with the Institute of the Aerospace Sciences and became the American Institute of Aeronautics and Astronautics (AIAA).

In the Soviet Union, the first formal rocket research organisation was the Society for Studies of Interplanetary Travel set up in 1924 with around 200 members including Tsiolkovsky who earlier had written the equations showing the value of using high-energy hydrogen and oxygen propellants for space rockets. It mirrored the Soviet fascination with technical developments in other countries and a meeting was held on 4 October 1924 to debate a proposal from Robert Goddard that a rocket be sent to the Moon for scientific research – two years before he flew the first liquid propelled rocket.

But the Society was short lived and broke up a year later. It was followed by the *Gruppa izucheniya reaktivnogo*

Above: A Repulsor-4 rocket developed by the VfR in the process of being filled with liquid oxygen. *Frank H Winter*

Left: The American Interplanetary Society launches a test rocket from Staten Island, New York, on 14 May 1933. *Novosti*

Above: On the West Coast of America, rocket experiments developed at the California Institute of Technology (Caltech) would form the foundation of the Jet Propulsion Laboratory. In this image dated 15 November 1936 (seated left to right) are: Rudolph Schott, Apollo Milton Olin Smith, Frank Malina (white shirt, dark trousers), Ed Forman and Jack Parsons (right, foreground). *JPL*

dvizheniya, or GIRD, formed in Moscow during 1931 as an officially approved rocket research organisation which two years later evolved into the *Reaktivnyy nauchno-issledovatel'skiy institute*, the RNII. Central to GIRD's activities had been the guiding influence of aeronautical engineer Sergei Korolev who would mobilise tests and experiments, develop Russia's first intercontinental ballistic missile after the Second World War and manage the Soviet space programme through its early years.

In the UK, the British Interplanetary Society (BIS) was formed in Liverpool during 1933, which along with the American and Russian organisations formed the core of research and development of ideas and possibilities advocating rocketry and space travel. Led initially by Philip E Cleator, the BIS exists today, unbroken since its birth and with headquarters now in London. Noted for its scientific, peer-reviewed *Journal* and by the monthly magazine *SpaceFlight*, as well as the quarterly *Space Chronicle*, the BIS has pioneered many leading concepts. Among which have been ideas and designs for space suits, lunar landers and nuclear propulsion as well as the detailed scientific and engineering design of starships. It continues to serve as a space advocacy organisation. The other societies merged with industry to become professional organisations whereas the BIS was determined from the outset not to spend money on practical experiments to the detriment of its focus on theoretical research and education.

The First Big Ideas

In the early 1930s, enthusiasts at the *Verein für Raumschiffahrt*, received the attention of a small group of ranking officers within the German Army (the *Reichswehr*) who saw in the rocket a means of extending the range and the explosive yield of an artillery barrage, albeit limited by ballistic equations. They regarded rockets as self-propelled shells with the possibility of outperforming the long-range artillery of the First World War. It was this potential which interested the German Army, because limitations on bore size of gun barrels had been restricted by the Versailles Treaty and this was one way of achieving the objective by another means.

In April 1932 Captain Walter Dornberger, an artillery officer, was assigned to the Weapons Department of the *Reichswehr* with instructions to secretly examine the possibilities presented by rockets and to determine the possibility of developing rockets as military projectiles. The technocracy in Army research and development facilities were not convinced something could come of it and to avoid confrontation, the work was conducted under semi-secret conditions for fear its meagre funding would be withdrawn for other and more immediate work.

Dornberger exploited the activities of the small group of amateurs at the *Verein für Raumschiffahrt*. Their work was productive and tests were moderately successful, so much so that it also received the attention of Dornberger's bosses, Captain Ritter von Horstig and Colonel Emil Becker, who agreed to pay for a demonstration launch. The test shot took place on 22 June 1932 but as with most experimental tests at the time, it was unsuccessful. The VfR never did get paid and

Above: Russia's GIRD group fired the first Soviet rocket, known as GIRD-X, on 25 November 1933. With a thrust of 685N (154lb) it reached an altitude of 80m (262ft) before the combustion chamber burned through and it crashed. Sergei Korolev can be seen at left. *Novosti*

it became clear that the activities of the group were ad hoc and uncoordinated.

After Hitler became Chancellor in January 1933 a wide range of military projects were put on hold due to the decision to review policy and to work up plans for significantly expanding the Army and the small air force it was allowed to possess. Contrary to popular belief, Hitler was risk-averse and worried over launching Germany into a military confrontation with its former adversaries before the country was ready for a major conflict. For a brief period, several small-scale development projects were put on hold as the new government adjusted priorities and put resources into public-works programmes, job schemes and employment plans. Only from 1935 when the establishment of the Luftwaffe was declared, openly and in direct violation of the Versailles Treaty, could German arms production and expansion of military forces emerge in plain sight. And that would then include a start on building an ocean-going Navy.

Toward these combined objectives, the rocket team was not constrained by the requirements of a particular concept but directed specifically to development of all kinds and classes of rocket for use by land, sea or air.

Several members of the *Verein für Raumschiffahrt* had been recruited to the Army programme after the Society was dissolved in January 1934, ostensibly because it could not pay its bills, one year after Hitler came to power. Most members were offered work with the Army as civilian scientists and engineers. Some of those objected on political grounds, not wanting to work for the Nazis, but among those who accepted was the young Wernher von Braun, an inspirational devotee of rockets and space travel with apolitical views and a determination to use whatever means he could to develop rockets. When the VfR was shut down all contact with the outside world was cut and previously friendly visits by foreigners, in particular interested members of the American Rocket Society, stopped altogether.

This proved to be a seminal moment in the affairs of the nascent group of German rocketeers, many of whom were appalled at the violent street-politics of the Nazi Party and of the extreme laws enacted when Hitler became Chancellor in January 1933. Several German scientists, philosophers, engineers and social reformers feared the consequences of remaining in Germany and they began to leave while they still could. Among them was Willy Ley, born in 1906 and a founding member of the VfR. Ley was fascinated by scientific matters and saw in the prospect of space travel the next great endeavour for humans. Inspired by the Austro-Hungarian scientist and engineer Hermann Oberth, 17 years his senior, Ley was strongly motivated and wrote extensively on the possibilities inherent in rocket propulsion, enabling travel to other worlds. Disgusted at the new totalitarian Nazi regime, he fled to England in January 1935 with the help of Phil Cleator from the British Interplanetary Society, visited their headquarters in Liverpool on 7 February and then went to the United States. Others remained and hid their feelings and when the war ended, several chose to seek out the Russians and offer their services to the Soviet Union.

But in Germany, before the war began, rocket development was placed under the direction of Dornberger who, in April 1937, moved the project from its initial testing ground on the outskirts of Berlin at Versuchsstelle Kummersdorf-West, to a secret facility built at Peenemünde on the Baltic coast from where test rockets could be fired out to sea undetected. This also constituted the world's first instrumented proving-ground where the flight characteristics of rockets could be monitored and recorded throughout the full duration of flight. Secrecy was everywhere in the Third Reich, the government being particularly sensitive to visits by foreigners and the possibility that they would report unfavourably on developments taking place within the munitions industry in particular. The precise reasons for the closure of the VfR are not totally known today but the absorption of its willing workers was typical of so many organisations across Germany in the 1930s.

Above left: Willy Ley (1906-1969) was a publicist for space exploration, science fiction writer and futurist who would greatly influence early stimulus for rocketry in Germany and across Europe, eventually making his way to the UK and the USA in 1935. *Author's collection*

Above right: One of the early advocates for rocket research and space travel, the Austro-Hungarian-born German Hermann Oberth (1894-1989) inspired many people and became an iconic figure in the origins of astronautics across central Europe in the 1920s and 1930s. *Frank H Winter*

A move, which was to bring some level of inevitable pressure upon the civilian scientists and engineers working on the German rocket programme, required personnel closely associated with the work at Peenemünde to be inducted into the *Nationalsozialistische Deutsche Arbeiterparte* (NSDAP), the Nazi Party. On 12 November 1937, von Braun became party member 5,738,692 and would frequently be photographed wearing the badge on the lapel of his civilian jacket. Later, under pressure to be subsumed into the ranks of the controlling authority, he would join the *Schutzstaffel* (SS) with registration number 185,068. Starting with the rank of Untersturmführer (equivalent to 2nd Lieutenant) he received a series of promotions, finally achieving the rank of *Sturmbannführer* (Major) in June 1943. This brought him some level of authority over both civilian and military personnel although his work largely spoke for itself.

For three years, activity at the Berlin testing ground had focused on a range of projects which since early 1934 had been under the management of the *Heereswaffenamt-Prüfwesen* (Army Ordnance Research and Development Department). In the several years since the German Army became interested in the experiments of the VfR, the technical capabilities of rockets were explored across a wide range of tests, leading to a succession of ideas about how they could be used for military purposes. The origin of the facility on the outskirts of Berlin in 1934 was not to develop long-range ballistic missiles but to first explore the breadth of possibilities inherent with rocket propulsion. A lot of activity centred on jet-assisted take-off (JATO) for propeller-driven combat aircraft and von Braun worked extensively with aircraft manufacturers, particularly Ernst Heinkel, to develop rocket packs for getting heavy aircraft into the air and up to flying speed.

A year after the German Army formally transferred its rocket research programme to the Peenemünde facility which it shared with the Luftwaffe for their Fieseler Fi-103 flying bomb development programme, 1938 proved pivotal to the transition of the German government's aims and objectives. The territorial expansion of the Third Reich had been laid out in the Hossbach Memorandum of 5 November 1937 in which, at a private conference, Hitler made it clear to the senior military and political leadership that Germany was running out of sustainable resources, citing wheat, fuel and steel which could be found in abundance throughout eastern Europe. Military occupation of neighbouring countries south and east of Germany was essential, he said, to obtain raw materials and manufacturing resources. This would be followed by an invasion of the Soviet Union, probably around 1942, prior to settling the future of France and Britain – either through negotiation or by force. However, as history records, events would reverse the direction of German aggression from a series of East-first invasions to West-first attacks, after Britain and France declared war on Germany in September 1939 following the attack on Poland.

But the objectives conveyed to the military leadership set the required standard which ultimately defined the specification for a missile capable of throwing a 1tonne warhead a distance of 322km (200mls) as determined to be the most practical and achievable application of the technology. It was the reason why both the Army and the Air Force ended up with the same explosive warhead and range, the capability stipulated for both the rocket and the cruise missile. From this single decision, high up in the echelons of military-political leadership, came a policy with direct implications for the kind of projects Peenemünde was guided toward. In post-war recollections and autobiographies from key players in the German rocket programme, very little of this appears. So very few in the German military knew of the connection between the geopolitical objectives and the convergence of the two programmes regarding military capabilities.

While considerable effort and time was put on the workers recruited from the VfR, others from across Germany were sought for specialist input in electrical control systems, turbine and turbopump technology, structures, materials science and a wide range of niche research projects. Yet, paradoxically perhaps, there was time to think of the bigger picture and to attempt an audacious bid to begin a research programme aimed at putting the first human on a rocket. Unfortunately, the project had originated in the mind of a Magdeburg businessman, Franz Mengering, who believed the Earth was hollow and the Moon a mere 5,000km (3,100mls) away. Rudolf Nebel, Klaus Riedel, Kurt Heinisch, Herbert Schäfer, Hans Bermüller, Paul Ehmayr and Hemut Zoike were recruited to build the rocket as a precursor step to an ambitious project which aimed at launching a man for initial test flights, that person being Hans Hüter. But first it would be tested unmanned with experimental precursor designs.

A test rig and experimental rocket was set up at Wolmirstedt near Magdeburg and a launch attempt made on 9 June 1933 but it never left the launch tower. A second attempt on 29 June resulted in a flight of around 305m (1,000ft) – but in a horizontal direction. Following a significant redesign, and a launch attempted from Lindwerder Island in Tegeler Lake near Berlin, it reached a height of 914m (3,000ft) before falling back down close to the tower. Further attempts with the 'Magdeburg Rocket', as it became known, were abandoned after similar failures from a boat on Schwielow Lake during August. Perhaps fortunately, private activities such as this were closed down by the Nazi government and nothing more was heard of putting humans on rockets until later in the war

Right: The Magdeburg rocket was the first attempt to develop a programme capable of putting a man into space. Operated by amateurs, it was hopelessly implausible and is best regarded as a curio of its day. *Author's collection*

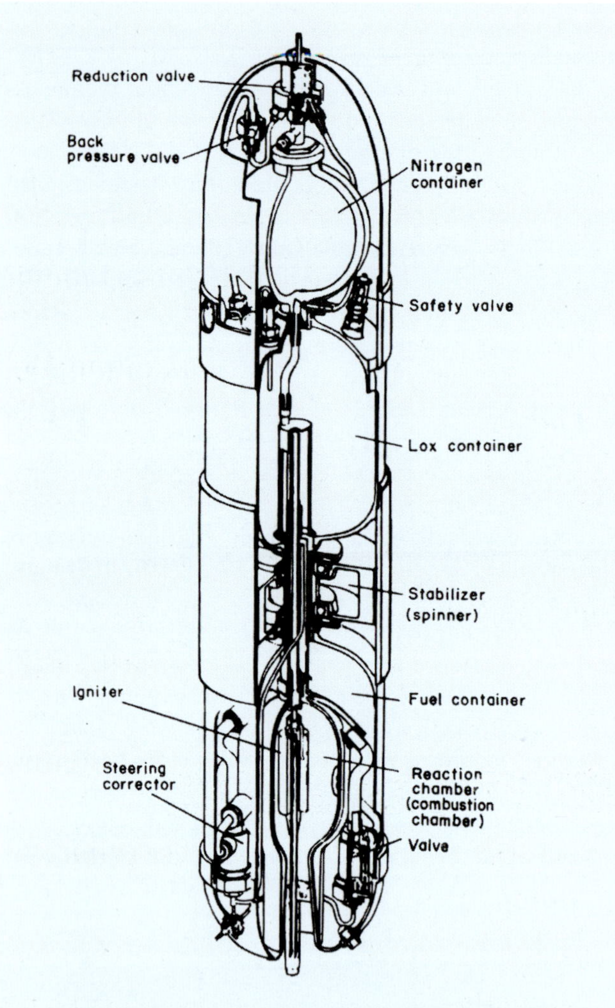

Above: Showing the combustion chamber inside the fuel tank, with the LOX tank on top, a cutaway drawing of the Aggregat-2 (A-2) rocket prepared for test at Kummersdorf. *Frank H Winter*

when such a proposal was mooted for the A-4.

Developments with this and other peripheral concepts were quickly subsumed into the general umbrella of research and development. Much expectation had been raised during the years in which the VfR carried out its early experiments but significant problems were unknown at the time and the challenges were great. Fundamental research, experimentation, test and development were prerequisites for deciding any sort of mission objective and the prospect of putting a man on a rocket seemed to get further away with time.

Set against the background of mixed failures with intermittent successes, by 1938 von Braun had risen high in the management of work at Peenemünde. His leadership and the trust he attracted from the workforce made him as much a valued asset as for any technical input he provided. While at Kummersdorf-West, the rocket team had set forward a development path for the ballistic missile through a series of increasingly more capable Aggregat-series and development work proceeded through a sequence of integrated steps – the evolution of ideas transformed into experimental rockets seeking solutions to technical problems as they emerged. This was evident from the development of the first two rockets at Kummersdorf-West, the A-1 and the A-2.

The A-1 brought together optional technologies on propellants, structural engineering, interior layout and performance targets which were to follow through with successively more advanced designs. It used the preferred liquid oxygen (LOX) and alcohol propellants mixed in a combustion chamber delivering a thrust of 2,936N (660lb) in a projectile with a height of 1.4m (4.6ft) and a diameter of 30.5cm (12in). The A-1 had a fibreglass LOX tank set within the elongated metal fuel tank and a tri-axis gyroscope in the forward part of the rocket for stability during ascent. The rocket could not be set spinning for directional stability due to the inverse centripetal force which would move the propellants away from the bottom to the sides of the tanks, thus preventing delivery to the combustion chamber. After a successful static test, the first attempt at a launch was conducted at Kummersdorf-West on 21 December 1933. It ended in a spectacular explosion, a fateful portent one month before the disbandment of the VfR.

Learning from design flaws in the A-1 design, the A-2 rocket was a rearrangement of the same internal equipment using the A-1 engine but this time buried inside the lower part of the fuel tank for regenerative cooling of the combustion chamber, with the LOX tank separately mounted above. The gyroscope was relocated to the space between the propellant tanks and a pressurisation system using nitrogen for moving propellants into the injector was added. The overall dimensions were the same and two rockets were built, named after characters from the Wilhelm Busch cartoons, Max launched on 19 December 1934 to a height of 2.2km (1.36mls) and Moritz to a height of 3.5km (2.17mls) a day later.

For reasons discussed earlier, greater interest in development of rockets was quick to exploit the evolving intentions of the militaristic state and a considerable sum of money was allocated by Horstig and Becker to build test rigs, provide small locomotives for hauling equipment and buildings for work and test. While still at Kummersdorf-West, the Dornberger team set to work designing the A-3 which was to have a rocket motor with a thrust of 14,678N (3,300lb) and an inertial guidance system, arguably one of the most difficult elements essential for controlled flight along a pre-determined trajectory. It was during tests with this rocket at the refreshed Kummersdorf-West facilities that General Werner von Fritsch was sufficiently impressed to give his support to the programme and to see to it that it received favourable references higher up the chain of command.

The A-3 had all the essential elements of a nascent ballistic missile, becoming the rocket in which it was intended that the fundamental principles of guidance and control were to be met and solved. Design work was underway as early as 1934 but its practical development began when the new test stands at Kummersdorf-West were provided in February 1935. For much of that year work progressed on what was considered

Right: The largest of its type at Kummersdorf, Test Stand 3 was serviced by a tower mounted on rails so that it could be moved back and forth as necessary, here with an A-3 on the firing platform. The concept would influence later test stands in the USA at White Sands and Cape Canaveral. *Author's collection*

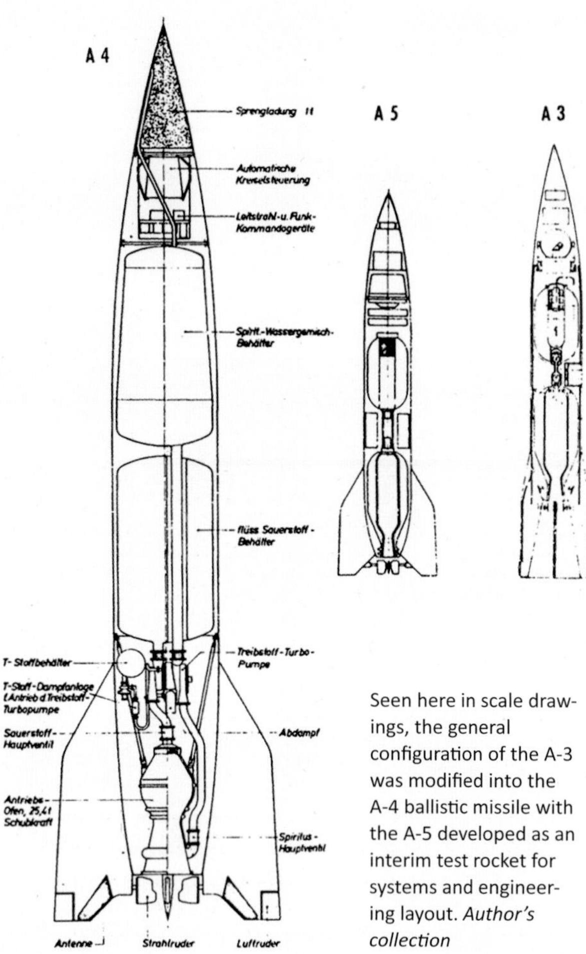

Seen here in scale drawings, the general configuration of the A-3 was modified into the A-4 ballistic missile with the A-5 developed as an interim test rocket for systems and engineering layout. *Author's collection*

the template for any long-range missile, the geometry of the shape and the internal layout of systems aligned with a rocket motor which was designed and built to a target thrust five times the thrust of the A-1/A-2 motors, on LOX/alcohol propellants. It was also to be a testbed for an active inertial guidance system with tri-axis gyroscope and jet vanes in the exhaust manufactured from tungsten, operating on command from the platform for directional and attitude control.

Attitude stability and directional control were the two fundamental elements of a successful missile and both were equally important to its functioning objective. The A-3 was arguably the first step in a design configuration from which the engineers were capable of exploring various technical possibilities. As a research tool it proved invaluable and pointed the way toward the definitive missile, the A-4. The two projects were staggered in conception but quickly ran parallel, with the A-4 design beginning to grow in paper studies during March 1936. Astonishing in retrospect, the A-4 was to be powered by a motor delivering a thrust of 244.6-293.9kN (55,000-66,000lb) and with a design range of at least 281km (175mls) carrying a one tonne warhead. Meanwhile, in the closing months of work at Kummersdorf-West, the A3 was made ready for test flights. The A-3 had a length of 6.7m (22.1ft) and a diameter of 0.68m (2.2ft) with a lift-off weight of 748kg (1,650lb).

The impetus to move quickly toward the detailed design of the A-4 came when General Werner von Fritsch experienced first-hand a test firing of the A-3 engine in March 1936. Development of this rocket itself brought a need to move from the test ground outside Berlin to Peenemünde on the Baltic coast. It had too great a thrust and design range to allow it to be fired from Kummersdorf-West. Much was expected of this rocket, with a stabilisation platform maintained in pitch and yaw by its own gyroscopes. For attitude control of the rocket in pitch, roll and yaw, a stabilisation system known as the SG-33 was fixed to the rocket itself and provided pneumatic servos which received electrical control signals commanding the deflections of the jet vanes protruding into the exhaust. The shape of the rocket itself was copied from the 8mm bullet as some assurance that it would survive supersonic flight.

The flight test programme was named *Operation Leuchtturm* (Operation Lighthouse). Development of the A-3 was delayed by the transfer of equipment, test sheds, engineering structures and associated support equipment during late 1936 and much of 1937. With an agreement that the Luftwaffe should share Peenemünde for development of its Fi-103 flying bomb – essentially a cruise missile – it was left to them to do all the construction work and general erection of buildings and test facilities, as they had experience in working with prefabricated materials. The rocket scientists moved from Peenemünde in April 1937 and almost immediately von Braun sought funds for a wind tunnel capable of creating supersonic conditions. Initially, Dr Rudolf Hermann provided services from his own supersonic wind tunnel at the Institute of Technology at Aachen but a new supersonic tunnel was built at Peenemünde where simulated Mach 4 conditions were provided by 1940.

The first A-3 was not launched from Peenemünde until 4 December 1937 but a parachute designed to lower the rocket to the ground at the end of its propulsive phase deployed prematurely when the engine failed and it crashed close to the launch stand. The second flight occurred on 6 December with a similar result, followed by the third two days later and the

Right: Engineer Kurt Heinisch at the control blockhouse for Test Stand 1 at Peenemünde. *Author's collection*

fourth on 11 December. The last two launches achieved some moderate success and information was obtained from those flights, although the parachutes had to be disabled to prevent a repeat of preceding failures.

Very little was learned and the rocket failed to live up to expectations. Much more was needed and the A-3 had little potential once its limitations were discovered. Several fundamental design and engineering details fell short of requirements and experience with the A-3 emphasised the experimental nature of the programme. There was no adequate way of evaluating design details other than to fly and test. For instance, it was discovered that there was nothing fundamentally wrong with the design of the A-3 other than inadequate understanding of the forces involved. The tri-axis gyroscope, for instance was unable to cope in winds greater than 3.6m/sec (12ft/sec). The only recourse was to retire the programme and insert an interim test rocket to apply lessons acquired from the recent failures. Which is where the A-5 came in.

The fixed requirement for the A-4 long-range ballistic missile was consolidated at the end of 1938 during a visit by Generaloberst Walther von Brauchitsch, *Oberbefehlshaber* (Commander in Chief of the German Army) in which authorisation was provided for a manufacturing plant in which up to 500 A-4 rockets could be produced each year with an initial operational capability of late 1942. Before any of that could achieve reality, what had been expected of the A-3 would be realised by the A-5, from which would emerge the technologies and the operating cycles essential for development of the A-4. The A-5 adopted the engine from the A-3 but incorporated a new control system provided by Siemens. The rocket incorporated a *Brennschluss* receiver enabling a cut-off signal to be sent from the ground and it had improved graphite steering vanes impinging into the exhaust plumes for attitude control and directional stability.

Development of the A-5 was supported by a steadily expanding workforce, growing from 80 people in 1936 to 15,000 by 1943 and names which would underpin the development of rocketry and space exploration two decades later acquired knowledge and gained experience through the pre-war years at Peenemünde. Control mechanisms, jet-vane rudders and gyroscopic systems were under Dr Ernst Steinhoff who managed the flow of materials, instruments and special pieces of equipment from across German industry. Dr Hermann Stauding worked analogue computers, signal processors and the electronic simulators which were increasingly effective in predicting – before testing – the credibility of small and complex engineering devices related to information-processing and flight control. At the core of this effort was the office run by Walter Riedel, the design head for the rocket motors in the Aggregat series, all the way to the A-4 and subsequent paper plans for advanced concepts.

But there was another Riedel, Klaus, who made major contributions to the rocket programme. Klaus had worked for Siemens when members of the VfR took up employment offers with the Army but after initially refusing to join the military programme he joined them in 1937 and worked on the development of ground support and launch equipment. Klaus Riedel was one of those, along with von Braun and fellow rocket engineer Helmut Gröttrup, arrested in March 1944 after allegedly defeatist talk when decrying the fate of

Below: The main test stand at Peenemünde c.1939. *Author's collection*

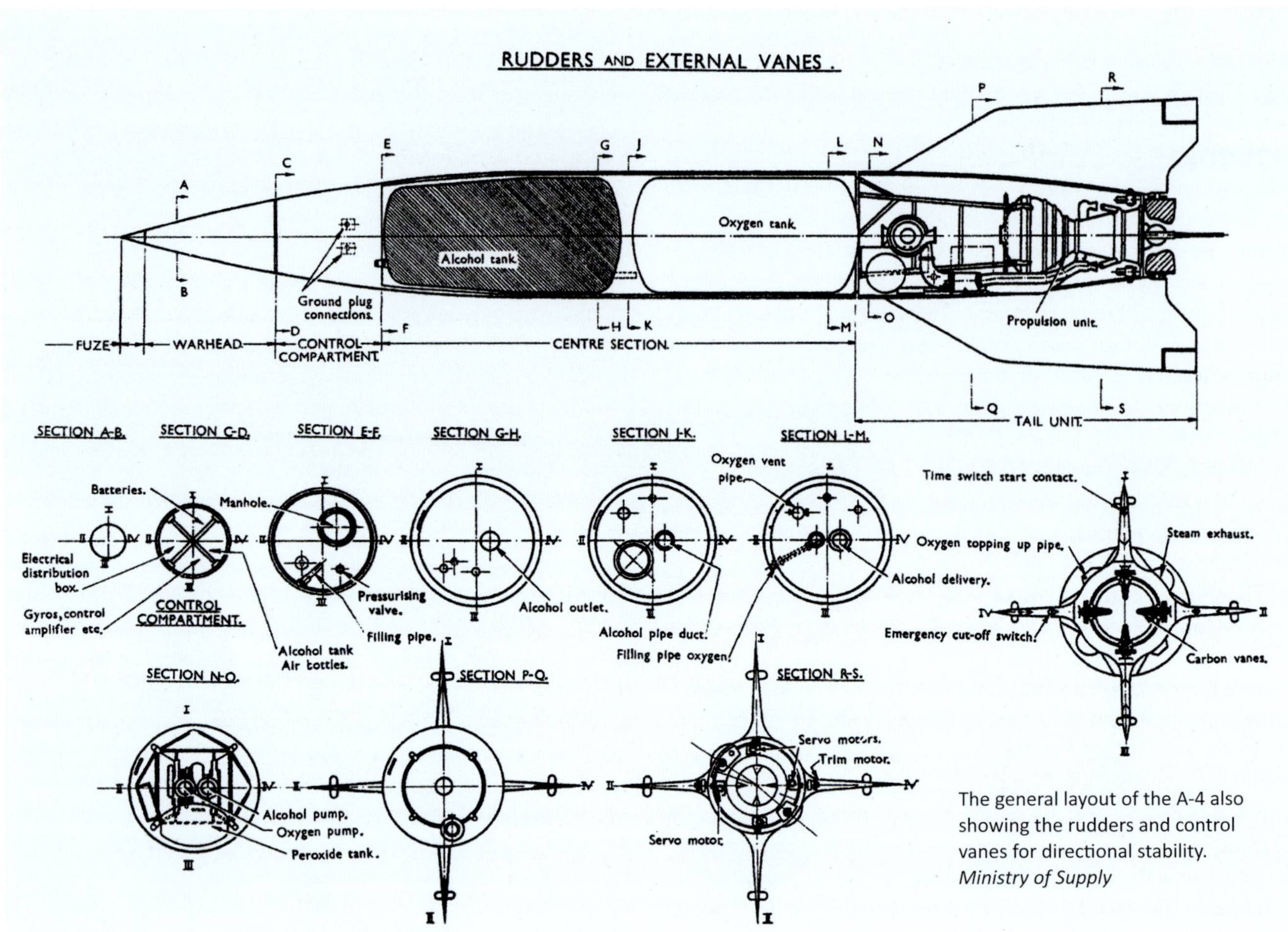

The general layout of the A-4 also showing the rudders and control vanes for directional stability. *Ministry of Supply*

Germany. All three were released when Dornberger intervened on the basis that their work was vital to the war effort. With his liberal views on political issues and an opponent to the Nazi regime, Klaus Riedel was under close observation by Maj. Hans Georg Klamroth, in charge of counterintelligence at Peenemünde, a man who would on more than one occasion save the rocketeers from the clutches of the Gestapo, an organisation ever willing to see intrigue and conspiracy where none existed.

This is no place to add further to the life of Klamroth, except to say that although seemingly an 'ardent Nazi' and a member of the SS, he remained silent over the July 1944 plot to assassinate Hitler and was arrested and hung for that. His position at Peenemünde adds credibility to the claim, made by some German scientists and engineers who went to work in the United States after the war, that there was widespread discussion about the post-war legacy, about the ultimate purpose of their work and about the possibilities opened up in a more peaceful world. Concerns were aired regarding the political regime, about the course of the war and the future of their work, whether it resided in long-range ballistic missiles or space exploration. It would be shared equally.

A member of the team since shortly after von Braun accepted a position with the German Army on 1 October 1932, Arthur Rudolph had very little technical training but was responsible for management of the primary design on the A-1 and A-2 rocket motors and moved to Peenemünde where he worked on building the A-3 test stands. Von Braun placed him in charge of production and much of his work was dedicated to setting up plans for a rocket production facility based on the design configuration of the A-4. After von Brauchitsch approved a major expansion of Peenemünde, slave labour was used to build massive facilities including a liquid-oxygen production plant, work on which began at the end of 1939. After a slow start during the winter of 1940, the resulting building was 70m (230ft) long, 42m (138ft) wide and 20m (65ft) tall, supporting production of 13,000kg (28,665lb) of LOX per day from July 1942. Despite further air raids, LOX production remained here until February 1945.

This was a major task, calling for integration of materials, fluids and propellants to the isolated location on the Baltic coast, plans which were dramatically transformed after the RAF bombed Peenemünde in August 1943. Production of the A-4 shifted to a converted gypsum mine in the Harz Mountains, the infamous *Mittelwerke* facility close to the Nordhausen concentration camp from where the work force was obtained. Rudolph was placed in charge of that activity and this, together with his membership of the Nazi Party and subsequently the SS, did little to help when he faced accusations of complicity in brutalisation of the workforce. It must be noted that slave labour was rife in Nazi Germany and that the pace of the missile production programme had its equal share of these workers, many of them from occupied countries including Eastern Europe as well as France and the Low Countries. Not all slave workers were Jews, gypsies or political dissidents. Much would depend upon the pressed workers, held under atrocious conditions incurring great loss of life.

Challenges and Solutions

Central to activity at Peenemünde, development of the A-5 test and engineering development rocket and the A-4 ballistic missile together with the direction of rocket applications came under the management of Ludwig Roth. Although he controlled several different administrative departments these activities were split between procurement of long-lead items for production and the requirements of emerging technology, including the existing inventory of missile projects. Of which there were many. While famous for the A-4 and the Luftwaffe's Fi-103 flying bomb, other less notable Army projects took place at Peenemünde and all that work carried on after the bombing of August 1943. Arguably the most notable of these was the *Wasserfall Ferngelenkte FlaRakete* (Waterfall Remote-Controlled Air-Air Rocket), essentially a scaled-down A-4 for anti-aircraft defence. Radio-controlled, it was powered by a rocket motor developed by Walter Thiel.

Recruited to the rocket team in late 1936, Thiel remained at Kummersdorf-West until moving to Peenemünde in 1940 conducting engineering design for the A-4 rocket motor, his major contribution. Working with a rapidly expanding team of specialists, Thiel was tasked with speeding development of the engine and for three years he supported that activity until he died, along with his entire family, in the RAF raid of 1943 when the bombs destroyed much of the workers' quarters and accommodation at adjacent Karlshagen. His place was taken by Martin Schilling. Thiel had already left his legacy in the rocket motors he designed as well as in the specific requirement of the Wasserfall missile

Critical to the successful operation of this missile was that it should be held ready for use in the field at very short notice, something that was not possible with the A-4 motor involving LOX propellant. To allow missiles to be fuelled and ready for flight, Thiel used a combination of Visol, a vinyl ether, and RFNA (Red Fuming Nitric Acid), essentially 94 per cent nitric acid and 6 per cent dinitrogen tetroxide. The broad range of experiments and laboratory tests with various combinations of propellant was one of the main legacies of this work and the direction under von Braun represented a more disciplined approach to research in this area than that being implemented in other countries at the time. To their credit the Dornberger leadership allowed the flexibility of basic research and development work, so long as it did not interfere with the A-4.

Much about that missile depended on validation of design trends and verification of engineering with the A-5, from which a great deal was expected. The A-3 lent its shape to the A-5 with a length of 5.8m (19.1ft) and a diameter of 77cm (2.5ft) and a lift-off weight of 900kg (2,000lb). Unlike previous Aggregat rockets, and because of the additional investment in the programme, small models, 1.5m (5ft) long and 20cm (8in) in diameter, were built for drop-tests from Heinkel He-111 bombers. Basic unguided A-5s were fired beginning in October 1939. The first three of the initial four shots

Above: Aerodynamic tests on the A-4 shape were conducted after dropping scale models of the A-5 carried aloft by an adapted Heinkel He-111 bomber. *Frank H Winter*

were made with the SG-52 (*Krieselgeräte*) guidance system which incorporated a tri-axis stabilised platform for attitude together with a separate tilt programme after lift-off in which signals were integrated with rate gyroscopes. The Siemens (*Vertikant*) guidance system was integrated with the A-5 for subsequent flights from 24th April 1940.

Fellow rocket engineer Hellmuth Walter, famous for the work he did on propulsion for the Messerschmitt Me-163 rocket-powered interceptor, produced scale models of the A-5 powered by hydrogen peroxide with potassium permanganate as catalyst to produce experimental results on fin configurations. These were fired from Griefswalder Oie, an island off the coast of Peenemünde from where the A-3 rockets had been fired and from where the precursor models of the A-5 were set off. Numerous tests were conducted with powered A-5 rockets in which stability, guidance and control systems were tested and evaluated for the A-4, then well under development. The A-5 quickly became the experimental testbed for much of the equipment which would fly on the A-4. Between 1939 and 1943, almost 80 A-5 flights took place.

The development and test evolution of the A-4 has been told in many books and need not be repeated here. Suffice to say that at the end of the war the level of experience and quantity and quality of test results was unparalleled. Across all technology applications associated with the A-4 much was learned from the A-5 which helped shorten the missile's development time and bring it to operational status. As noted earlier, the specification for the A-4 evolved from military requirements and from a cross-correlation between the most advanced rocket the team could build and the minimum amount of time necessary to deliver it as a production-line missile. Development which had stalled for some time during the A-3 test failures quickened with A-5 tests

The baseline A-4 engine was designed with a thrust of 244.6kN (55,000lb), the lower end of the performance target laid down at inception, albeit still posing significantly greater challenges than previously experienced with preceding development motors. The A-4 had an overall length of 14m (46ft) with a body diameter of 1.62m (5.3ft), a maximum width across the aerodynamic fins of 3.55m (11.6ft) and a launch weight of 13,000kg (28,665lb). To achieve its demonstrated mean range of 320km (199mls), the A-4 contained 4,900kg (10,800lb) of LOX and 3,710kg (8,179lb) of ethyl alcohol. Propellant consumption was 130kg/sec (286lb/sec) for a burn time of 65sec. After launch the A-4 remained vertical for 4sec after which it began to pitch over attaining a tilt of 49deg at 55sec maintaining that until burnout. The maximum altitude attained was 96km (58mls) after which the inert body of the A-4 arched over and descended to strike the ground at a velocity of 1,100m/sec (3,600ft/sec), or 3,960kph (2,460mph).

A fundamental part of the engine's operation depended upon the turbopump, which was the means by which the propellants would be transferred to the combustion chamber. There was no possibility of pressurising the tanks for a positive expulsion through the delivery lines as that would have added too much weight to the tanks themselves by thickening the walls to resist internal pressure. The turbopump was a steam turbine flanked by LOX and fuel pumps and driven by a mixture of hydrogen peroxide *(T-Stoff)* and potassium permanganate *(Z-Stoff)*. Compressed nitrogen was used to drive the hydrogen peroxide to mix with the potassium permanganate and produce super-heated steam to drive the turbine at 5,000rpm. The two pumps were located inside an aluminium casing with the bearing lubricated with LOX. The delivery lines carried LOX to the combustion chamber through 18 aluminium pipes, three on each of six distributors from the pump, and separate flow control valves. Alcohol was transferred from the fuel pump to two separate distributor outlets, each of which had triple lines for a total of six delivery outlets.

The double-walled combustion chamber was a large spherical, welded chamber with two concentric rings of 'burner' cups, 12 outer and 6 inner, at the top. Each was supplied with a LOX line from the turbopump. Individual burner cups had three rows of oxygen jets and a set of plain openings for the fuel which was sprayed through each orifice. The main supply of alcohol fuel came through a set of six pipes at the base of the venturi tube leading from the combustion chamber to the expansion nozzle. Flowing between the walls of the combustion chamber, it moved upwards to the delivery cups and provided cooling of the chamber and warming of the fuel. Various supplementary routes for fuel provided additional cooling and a more satisfactory propellant spray into the combustion chamber. Ignition was provided by four fireworks configured in the form of a swastika (by sheer coincidence), mounted at the top of the combustion chamber ignited by an electric plug. This produced a cartwheel of sparks which provided the energy to ignite the propellants.

A special plate was situated in the centre of the injector head to control the propellant flow and the thrust of the motor. This allowed the A-4's rocket motor to start operation under gravity feed of the propellants to produce a thrust of 71,181N (16,000lb) allowing for several critical parameters to be observed. If after several seconds the motor was running within tolerances, the turbopumps began, thrust was increased to a maximum and the rocket left its launch platform. There were no restraints holding down the rocket and when the thrust exceeded the weight of the rocket – 13,000kg (28,665lb) – it lifted free and began its ascent. This process allowed the motor to be shut down in the event of instability or some problem which would prevent it flying safely, keeping it on the launch

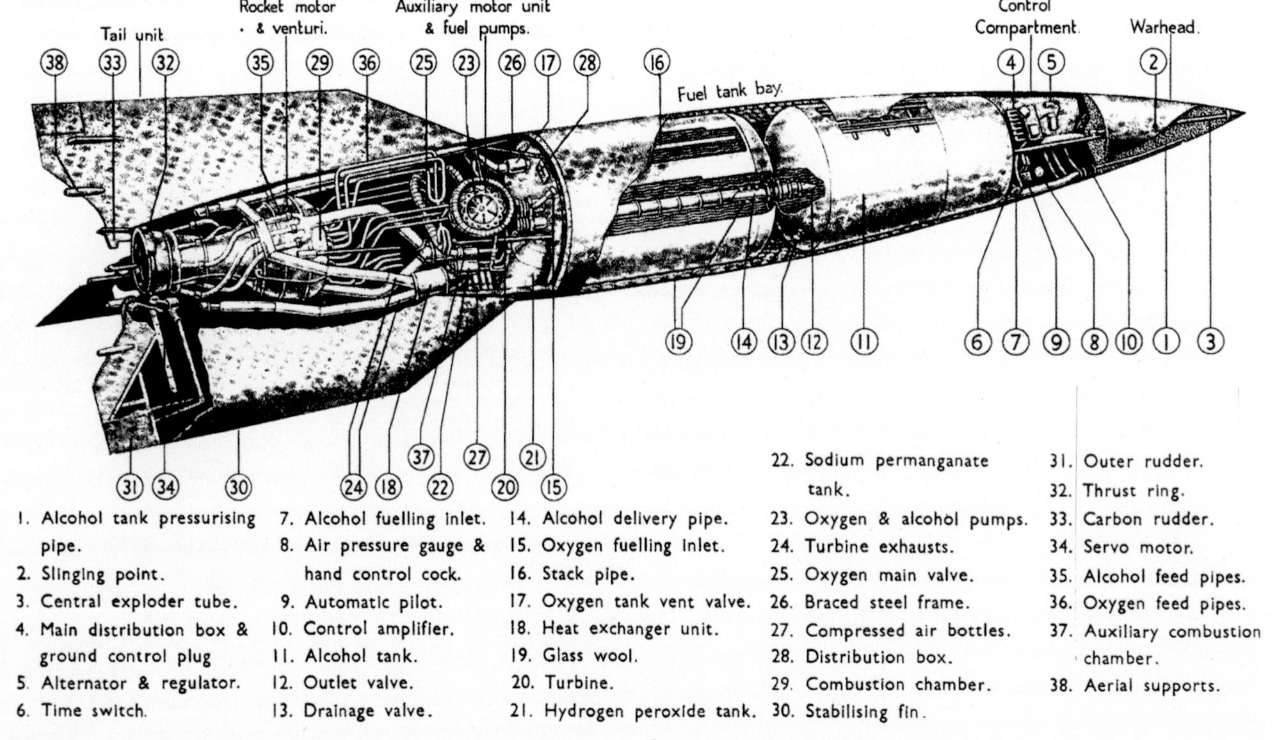

Cutaway diagram showing the internal elements of the A-4 ballistic missile. *Ministry of Supply*

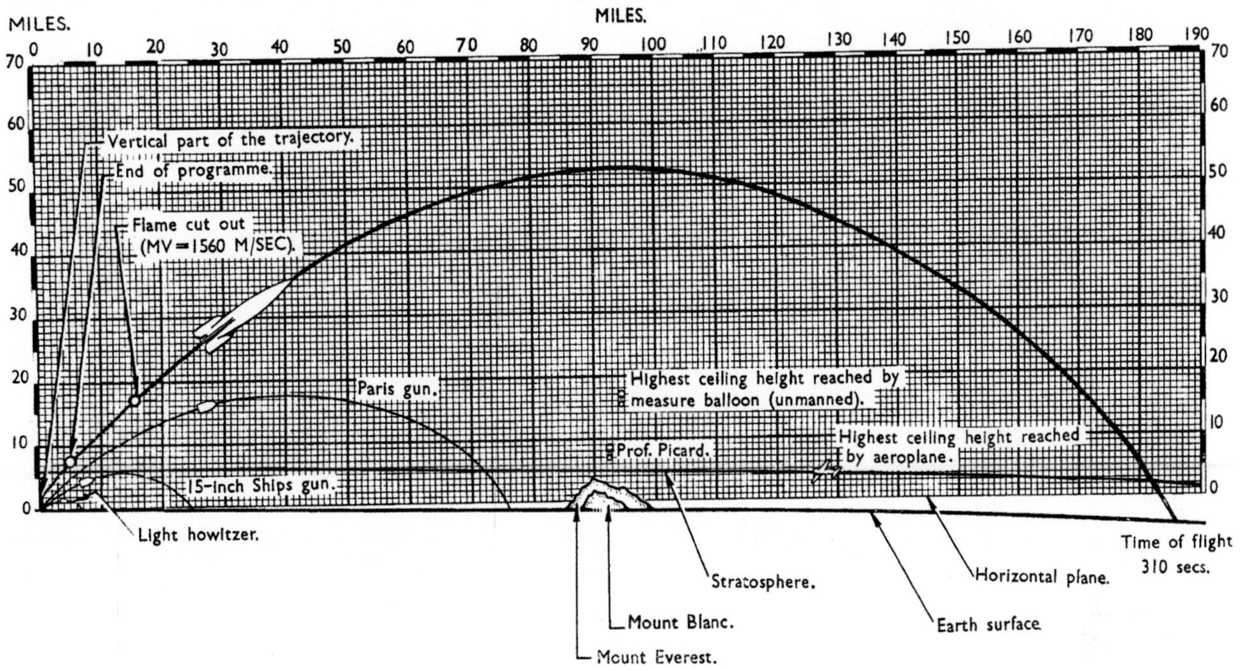

Above: A comparative representation of altitude and range for a variety of projectiles including the A-4 with surface features and heights attained by balloons and aircraft. Note the scale curvature of the Earth's surface over range. *Ministry of Supply*

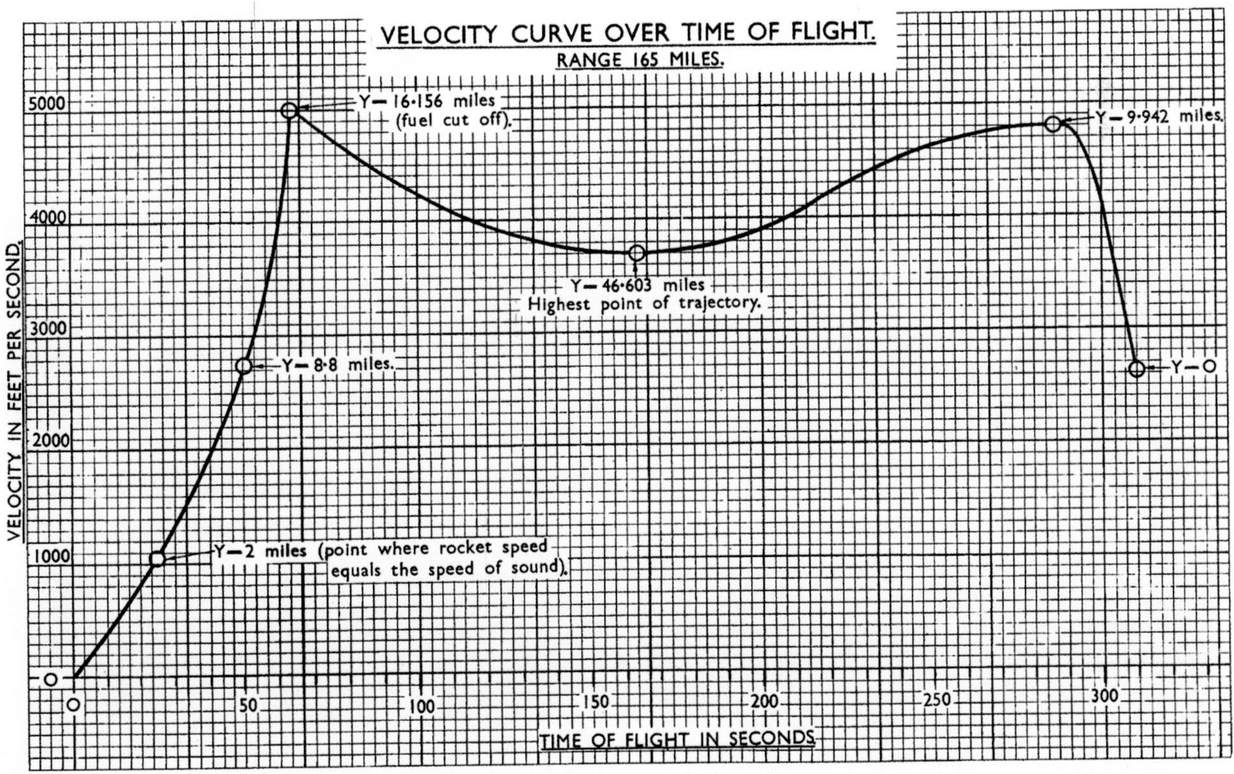

Above: The optimised trajectory of the A-4 rocket as plotted by velocity against time. *Frank H Winter*

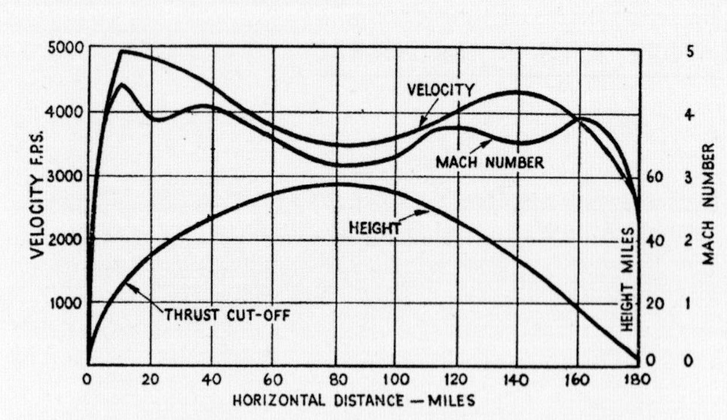

Parameters of height, Mach number, velocity and range for the standard A-4 flight path from launch to target. *Author's collection*

platform and saving the cost of a new rocket! The base end of the rocket was essentially a protective and aerodynamic shroud for the engine assembly and a place on which to locate the guide vanes for directional control and for the aerodynamic rudders mounted to the outer face of the four main fins.

The guidance systems for the A-4 evolved from work conducted by Helmut Hölzer, who met Steinhoff and von Braun while working at the Telefunken factory in 1939. Hölzer had been tutored by the legendary Alwin Walther who was one of the principle engineers behind the emergence of mechanical computing in Germany. Hölzer moved to Peenemünde in 1940 where he developed his radio-guidance system employing two antennae spaced apart. This transmitted alternating signals operating integrally with a vacuum-tube device which was eventually employed for rate measurement in lieu of the rate gyros. He also made an analogue computer for calculating and then measuring the A-4 flight trajectories, a world first and a design which credits him as being the inventor of this device. After the war, Hölzer would find work in the US and play a major role in the use of analogue devices to control aircraft, eventually playing a major role in the development of fly-by-wire (FBW) in military and commercial aircraft.

Hölzer faced many challenges in meeting the requirements of the A-4. Propellant sloshing caused by the physical motion of the large rocket had a far greater effect on the missile's course in flight than had been experienced with the A-5. Wind and other factors would threaten to topple the rocket before it achieved its required trajectory. To understand these problems, a full-size A-4 was mounted on a gimbal and as the engine ran a full duration firing sequence the evolving attitude control systems would fight to keep it stable. But there were too many imponderables and unrealistic perturbations so an electro-mechanical control system was sought, first to simulate the problem and then to test ways to alleviate unrequired excursions in attitude. A simulator was built to replicate the lateral guidance and attitude control. Hölzer devised a *Mischgerät* (mixing computer) to integrate all the variables. This was the world's first fully electronic flight control system from which would evolve the highly advanced analogue flight control computers, successors to which are built in to the most advanced aircraft today.

Of only peripheral regard to the story of the A-4 employing the wide-ranging fields of work involving radio, radar and electronic control equipment, the Junkers design engineer Fritz Haber developed the flight control system for the *Mistel* aircraft, a combination of an unpiloted, twin-engine Ju-88 light bomber stripped of non-essential equipment and loaded with explosives on top of which was a piloted Messerschmitt Bf-109 single-seat fighter attached by struts. The Bf-109 pilot would fly the combination off the ground and to the target, releasing the Ju-88 and control its flight with a small stick in the cockpit. This was the world's first radio-controlled FBW system used in warfare and opened endless possibilities for the remote control of aircraft and, eventually, land vehicles without risking loss of crew.

Problems posed by the A-4 were initially challenged by the need to maintain a stable attitude for the missile in flight by controlling azimuth and for a precise engine cut-off so as to provide some control over the range of the rocket. This latter function was developed by Dr. Friedrich Kirschstein from Siemens, a variation of which was designed by Prof. Wolman in a device which employed a Doppler signal from the ground to measure the velocity. Initially, the radio controlled cut-off (*Brennschluss*) used the radio control method for engine cut-off but later development of a system of onboard integrating accelerometers dispensed with ground sig-

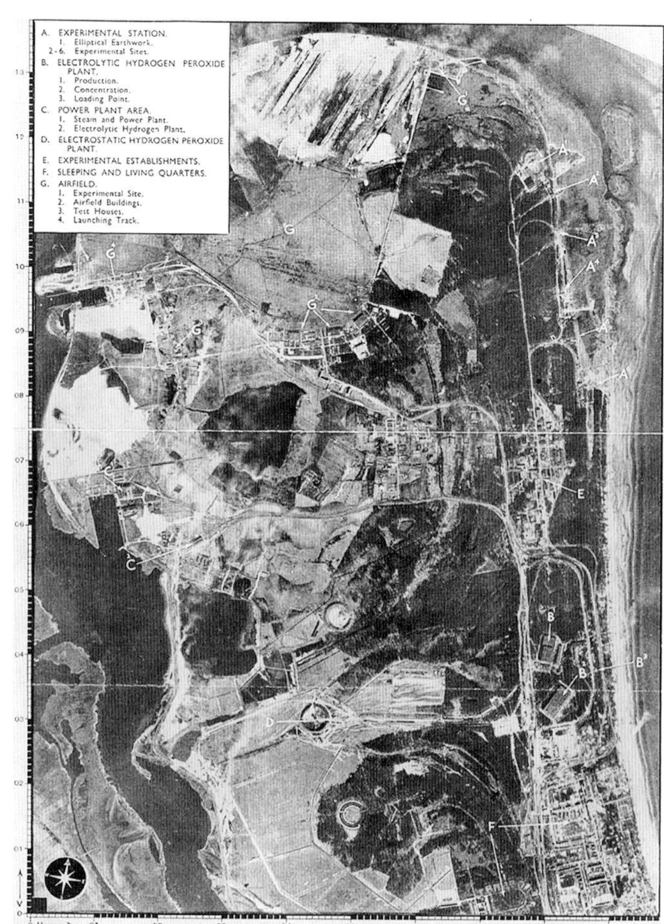

Right: The sprawling facilities of the shared Army-Air Force research station at Peenemünde on the Baltic coast as photographed by an RAF aerial reconnaissance unit. *RAF*

nals. There was concern about Allied signal interference which threatened the ability of the missile to achieve the desired range by precipitating a premature shut-down. Not uncharacteristically, this got the attention of Hitler who pressed for a change to an autonomous control system. Nevertheless, of all the A-4 rockets fired in anger, one in five used the beam-guided system.

The first engineering test model of the A-4 was completed on 25 February 1942 but three launch attempts, made in March, June and August failed before the first successful flight was achieved on 3 October 1942. On that flight, a problem with the pitch command took the rocket to a maximum altitude of 90km (56mls) and a range of 190km (118mls). On that day Dornberger proclaimed that a German rocket had been the first man-made object to reach space and this has been quoted as fact in numerous places since. However, the official, internationally accepted altitude for the beginning of space is 100km (62mls), although the Americans regard that line as 80km (50mls). Not until 20 June 1944 did an A-4 exceed the present internationally accepted height when the rocket launched that day reached an altitude of 176km (109mls) and an object did truly reach space, however arbitrary that line actually is.

Production of the A-4 for operation by the German Army had been long in the planning and initial thoughts had focused on the development of a fixed launch complex on the English Channel coast separating France and the Low Countries from Britain. The first draft of those plans were signed off on 27 March 1942 and the initial location was to be at Watten in the north of France with production from a facility at Peenemünde, at the Zeppelin works at Friedrichshafen and at the Raxwerke works in lower Austria. After Hitler approved large-scale production on 22 December 1942, armaments minister Albert Speer called a conference and on 8 January 1943 met with Dornberger and von Braun to announce that he was taking control of this effort. A committee was formed under Gerhard Degenkolb who pushed for a more industrial approach with responsibility for production taken away from the bureaucracy of Army Ordnance.

Degenkolb was a fanatical Nazi and the senior Peenemünde leadership resisted his efforts but Rudolph visited the Heinkel factory and became convinced that slave labour was a cost-effective way to secure mass production. A detailed study of requirements set the figure at 10-15 labourers from concentration camps for every German engineer, with von Braun in charge of signing off finished components. It was agreed that, initially, between 1,400 and 2,500 workers from SS camps would be set up in a production unit on the lower floor of Building F1 at Peenemünde. This line began production on 16 July 1943 but it was large numbers of these workers that were killed during the RAF raid of 17-18 August. As mentioned previously, production plans changed and shifted to the *Mittelbau* plant in the Harz Mountains but the working prototype for what originated there was set out initially at Peenemünde. On 7 July, Hitler had placed the A-4 rocket on the top priority list opening opportunities for production.

Key to the decisions regarding how the A-4 was tested at Peenemünde was Kurt H. Debus, a member of the SA from

Above: The first successful flight of the A-4 from Peenemünde, 3 October 1942. *Deutche Bundesarchiv*

1933 and the SS from 1934. He was the man in charge of all A-4 testing, particularly at Stand VII where he was in charge of engineering work. Debus specialised in test pad, launch pad and integration of the A-4 and was crucial in developing the plans for fixed launch sites on the French coast. After the war he would become a vital member of the group that went to work for the US Army on launch platforms for ballistic missiles, working up the launch facilities for the Saturn rockets and was eventually appointed director of NASA's Kennedy Space Center. But in 1943 the mode of launch for the A-4 was about to change despite great efforts to build colossal reinforced structures from where these missiles would be launched against England. It is estimated that upwards of 50,000 impressed labourers were involved in unused structures that remain today and which are frequently interpreted by visitors as part of the *Atlantikwall* (Atlantic Wall) of defensive fortifications against a cross-Channel invasion.

The way the A-4 would be operated varied between fixed sites or dispersed locations, the former attracting enormous defensive protection against heavy bombardment while the latter had mobility and ready concealment to favour it. At the meeting on 7 July at the *Wolfsschanze* (Wolf's Lair) in Poland, all the senior leadership at Peenemünde had presented models and plans for the fixed bunker sites at Watten, explaining how the rockets would be prepared under protective cover of reinforced concrete structures, fuelled and their guidance platforms set before moving by rail to concrete

Above: New labour from the Mauthausen-Gusen concentration camp arrives at the Raxwerke facility in Lower Austria where A-4 centre-sections were fabricated. *Author's collection*

hardstands from where they would be launched. Hitler liked this idea and pressed for an additional site or two to expand the launch capability but the Peenemünde group favoured the mobile method. With either method, no more than 900 missiles a month could be produced and launched, quantities limited by the production of LOX and the availability of the ethyl alcohol fuel.

Later that month a meeting was held, over dinner, at which the labour minister, Otto Karl Sauer had stated his intention to push for production of 2,000 missiles each month from the end of 1943. A figure obtained by adding production of 900 a month from each of the three planned factories and rounding up the number. With Hitler's increasing interest in the activities of Peenemünde, other forces came into play, especially that of Reichsführer SS Heinrich Himmler, who made his first visit to the facility in April 1943 and explained how he would expect to have its research activity removed from the Army and adopted by the SS. The Army resisted but Himmler insisted on placing a 'protective' security umbrella over the facility to ensure that its security would remain intact. This was a non-too-subtle move to gain control of the programme and claim success in high places.

Himmler made a second but private visit on 29 June and impressed upon Dornberger the importance now placed on the programme during a lengthy and wide-ranging conversation about the importance placed by Hitler on uniting all the people of Europe against the 'Asiatic horde' which he considered subhuman and useful only for their extraordinary physical resilience and hard work. Himmler said that pursuant to this, Hitler felt 'custodian' of European science and technology and that it was the duty of the SS to manage and control the rocket programme and any other weapon system possessing such enormous advantage. Two months later, Himmler appointed SS Brigadier Dr. Hans Kammler in charge of all building activity associated with production of the A-4. The SS intervention continued until, on 15 March 1944 von Braun, Klaus Riedel and Gröttrup were arrested and charged with treason. Only incidental here, a great deal of negotiation and the strenuous efforts of Maj. Hans Georg Klamroth, secured their release. But it left these men, and those who became aware of the intervention, how fragile was their hold on life itself.

After the Hitler bomb plot of 20 July 1944, SS Generalleutnant Hans Kammler was appointed to seize control over the entire rocket development programme from 8 August, a month before the A-4 was put into operation as the V2 (*Vergeltunsgwaffe Zwei*) – vengeance weapon No 2 – following the operational debut of the V1, the Air Force Fi-103 flying bomb. Propaganda minister Josef Goebbels wanted the people to know that the country was fighting back and talk of 'won-der-weapons' was actively promoted. In Germany, the operational designation for this missile was expressed, appropriately, without a hyphen but later, in the United States the media decided to express it as V-2. This occurred after an error on the part of a US Army typist when writing up a communique from which emerged a press release about the missile. We retain it in its correct presentation from here – V2 and not V-2.

From mid-1943, pressure from Hitler to expand the number of hardened bunker sites for A-4 flight spurred construction to begin at a site close to Wizernes and another near the town of Sottevast, 13km (8mls) from Cherbourg. These launch complexes would cover all of southern and south-east England. Watten would be the largest and require 91,750m^3 (70,152yds^3) of concrete with sufficient capacity for 108 missiles and fuel reserves for three days. Here, it was expected that 36 rockets could be launched each day. So strong were these facilities that they were determined to be impregnable to all but the largest available bombs. But there was disagreement regarding the effectiveness of the bunker method. Dornberger believed that these facilities would attract bombing raids and in a sustained period of air attack between 27 August and 7 September while it was still under construction and vulnerable the Watten site was rendered totally unusable and he was proved correct.

Contrary to the view taken by the scientists and engineers at Peenemünde, who believed that their rockets could only be fired by trained and qualified technicians from prepared sites, Dornberger believed that the missile could be made mobile. He remained convinced that field troops familiar with the equipment and trained in the procedures for firing these weapons would have a far greater chance of survival by continuously moving around in camouflaged vehicles and trailer wagons. By selecting personnel skilled in certain mechanical procedures and training them on a specific and contained part of the operation, he supported the field-mobile concept. While the Watten site was being pounded, up to 50 locations were selected between Cap Gris Nez and the Contentin Peninsular to launch attacks on London, Bristol, Southampton, Portsmouth, Winchester and Aldershot.

As the sites were selected they were prepared with flat concrete sections surrounded by trenches underneath trees to hide the relevant vehicles from air attack. For six months the preparation of these sites went ahead until it was realised that their location could readily be transmitted to England by partisan groups, obviating the advantage of mobility. More detailed examination of the problem found that any flat concrete surface would do and that the vehicles could disperse as opportunistically as possible in appropriate places, securing the site from any pre-preparation or forewarning. These issues of mobility, dispersal, operating procedures, training requirements for operating personnel and the logistical support with propellant and electrical power would underpin similar challenges faced a decade later with the deployment of field missiles by the United States.

Above: Wernher von Braun (fourth from left) and Walter Dornberger (third from left) share social niceties with military officers during presentations of rocket plans in Berlin. *Author's collection*

Below: SS General-leutnant Hans Kammler (centre), who took charge of A-4 production following the attempt on Hitler's life in July 1942. *ZDF*

Deployed in Anger

As control of the A-4/V2 programme slipped away from the Army, the transfer to the SS was completed in a diary entry of 31 August 1944 where Kammler was to take over full control of operations including activities at Peenemünde. The logistical requirements for mobile deployment of the missile were immense, rail lines delivering the missiles and their equipment to an unloading station where they were received by a transport company of the supply detachment, with the warheads being transferred by heavy vehicles. Separate companies also delivered fuel and LOX while the fourth provided the vehicles and units with fuel, oil and grease. Field stores

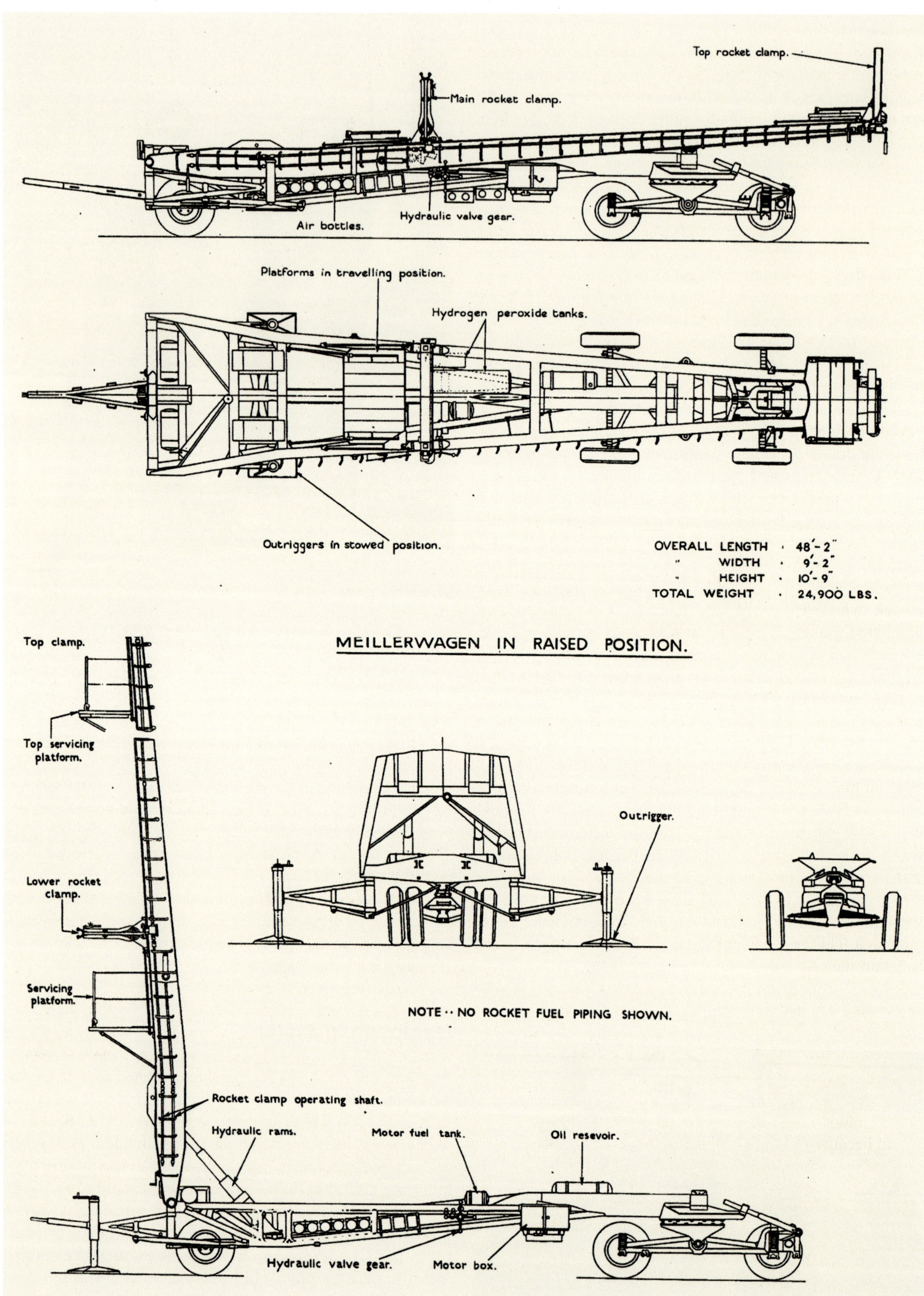
Top rocket clamp.
Main rocket clamp.
Air bottles.
Hydraulic valve gear.
Platforms in travelling position.
Hydrogen peroxide tanks.
Outriggers in stowed position.
OVERALL LENGTH · 48'-2"
" WIDTH · 9'-2"
" HEIGHT · 10'-9"
TOTAL WEIGHT · 24,900 LBS.
MEILLERWAGEN IN RAISED POSITION.
Top clamp.
Top servicing platform.
Outrigger.
Lower rocket clamp.
Servicing platform.
NOTE ·· NO ROCKET FUEL PIPING SHOWN.
Rocket clamp operating shaft.
Hydraulic rams.
Motor fuel tank.
Oil resevoir.
Hydraulic valve gear.
Motor box.

held 30 rockets maintained by the technical detachments and checked and repaired as necessary. The field store would also provide the *Meillerwagen*, the conveying trailer, to which was attached a lifting cradle. The rocket was then moved to the site selected as the mobile launch location in such a manner that after erecting the V2 to a vertical position the empty trailer would leave by a different route, allowing another rocket to be brought in. Usually, three rockets would be launched before the equipment was recovered and another site selected at a very different location.

Normally, a firing battery consisted of three platoons of 39 men divided into separate groups responsible for fire control with their dedicated vehicle, surveying and azimuth alignment, the rocket motor, electrical systems and the handling crew with its *Meillerwagen* and launch table. Preparation time was limited by the local conditions, access, weather and any other local military activities. It required approximately 60 minutes to check out the missile, another 15 minutes to fuel it up, a further 15 minutes to finally prepare the condition of the rocket and its orientation on the launch table and only a brief verification that no aircraft were in the vicinity before the firing command could be given. Useful for hiding the presence of the rocket and associated vehicles and crew, forest clearings provided some protection against low-level winds and the preference was for launching at night to ensure a greater degree of secrecy. Until the launch, when local residents were suddenly made aware of the rocket's proximity!

In preparation for and in support of offensive operations which began on 6 September 1944, more than 300 test launches took place from Peenemünde between June 1942 and the end of the war. In addition, more than 200 test launches took place from Blizna, Poland, beginning on 5 November 1943 where most of the test launches took place after the August 1943 air raid. A further 220 test flights were mounted from the nearby Tuchola forest site beginning 10 September 1944 after the Allies bombed Blizna. Extensive test and development was essential throughout the operational phase. The original war-load mass of one tonne (2,205lb) was achieved only with the use of aluminium and magnesium alloys but their scarcity forced the use of sheet steel of 0.6cm (0.25in) thickness on the nose area. This added weight reduced the size of the warhead to retain the required launch mass and maintain the desired range so that the explosive mass was about 750kg (1,653lb), 25 per cent less than that originally planned.

Upon arriving at its target the scientists had planned to use a proximity fuse for the warhead, imposing maximum lateral blast effect when detonated at a height of 18m (60ft) and considerable effort was made to find a reliable design, which proved impossible. Impact detonation was the only possibility and because of the terminal velocity the specification was demanding and only achieved through much effort. Consistently, the biggest problem during the extensive test phase was premature break-up of the missile during descent. Structural deformation during the tumbling phase as the spent body of the missile descended through the outer layers of the atmosphere placed tremendous stress on the structural integrity of the vehicle and this would cause break-up. Several intermediate measures reduced the failure rate from 30 per cent of launches to near zero by the end of the war.

Left: Designed at Peenemünde and fabricated by the Meiller-Kipper GmbH in Munich, the *Meillerwagen* doubled as a transporter-erector-launcher and as a set of work stands for servicing the rocket. When erect and stabilised by outriggers, the Meillerwagen was able to support technicians at various levels on fixed work platforms as shown here. *Ministry of Supply*

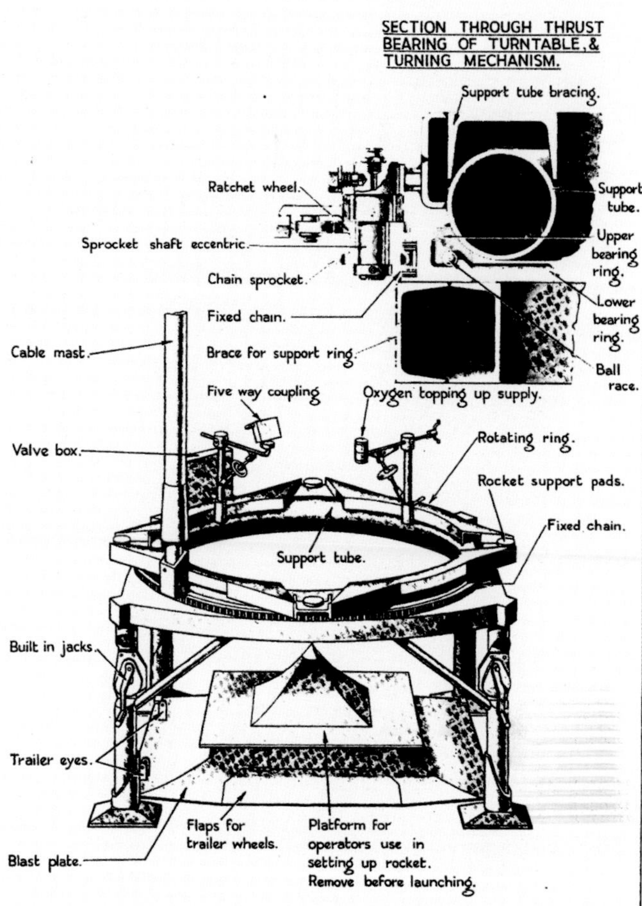

Capable of supporting the A-4 on its four fins and being levelled and rotated to the appropriate launch bearing, the firing table was transportable and could easily be handled by two men. *Ministry of Supply*

The most important remedial step was the addition of a much less sensitive fuse so that break-up would not cause premature detonation of the warhead high in the atmosphere due to high forces of axial acceleration. In a high percentage of cases the rocket body did break up but the warhead remained intact to detonate on impact. Further steps to prevent break-up at all was made with the addition of a short steel collar, or 'sleeve' as it was referred to, on the forward section of the rocket and this did help considerably. But the problem was never fully solved and only served to demonstrate that a lot of research work was required to perfect the operational performance of the A-4.

During the development and test phase and throughout operational deployment, several improvements were continuously being proposed, evaluated and, where justified, placed into further test. Much was learned in practice regarding the way missiles behaved during free flight and it was in this area that the A-4 showed the greatest probability of failure. Little progress was made during the latter stages of the war in addressing in-flight failures other than those outlined here. German rocket scientists and engineers accumulated a wide range of information but the absence of any means of getting direct information from the rocket during its flight hampered any long-term solution of issues observed during operational flights. The dark days of Germany's impending collapse prevented many of these issues being addressed and a great many questions awaited further research in more peaceful times devoid of the pressures of operational demands. Questions which would eventually find answers in another country.

Production of the A-4 accelerated at the facilities described earlier, in total some 5,200 rockets being assembled, about two-thirds after the start of the operational offensive. Of the total 3,172 launched by the end of the war, 1,664 (52 per cent) were fired against targets in Belgium. The majority fell on the port town of Antwerp where, not long after liberation by the British in early September 1944, the Allies were pouring in military supplies along the River Scheldt to consolidate the push north and west toward the Rhine. Until then, supplies had taken the long route around from Cherbourg. Antwerp was a key facility through which military units were replenished. A further 1,402 (44 per cent) were fired against the UK, almost all against London. France received 76 (2 per cent) V2 attacks, the Netherlands 19 (0.6 per cent) while 11 were fired against the Ludendorff Bridge at Remagen in Germany in an attempt to bring it down and stop the Allies crossing the Rhine. That bridge was also subject to attack by German Arado 234 jet bombers.

In this regard, the A-4 was used as both a tactical and a strategic weapon and in the seven months of intensive use a great deal was learned about how to operate field missiles and achieve flexibility and maintainability. The readiness level was greater than that of air combat units of the time and the amount of disruption caused was significant. In terms of overall effectiveness as a weapon of war, the A-4 caused over 9,200 casualties, or 2.9 per rocket fired. For its cost and effectiveness, the V1 Fi-103 flying bomb had equal success, the 8,025 fired against England causing almost 23,000 casualties, 2.8 per missile. For its cost however, the flying bomb was far more effective a weapon and tied up far more military personnel on defence measures than were possible for the supersonic V2.

While production and launch of the V2 required considerable quantities of material and technical know-how, the V1 flying bomb – precursor to today's cruise missile – was capable of being manufactured with fewer exotic materials in less time, involving fewer people and much more cheaply. These comparative evaluations, underway in the UK and the United States before war's end, framed the basis for post-war planning on the potential for rockets and pulse-jet engines respectively. Which is why both were subject to intense development in the United States and which were used as the baseline for new designs. Added to which were the rocket troops who managed the flight programmes, a group of people whose experiences were widely sought. But for all the intensity of work, flight preparation and commitment to producing a working missile, there was time for an artistic representation of iconic images and emblems. This was reflected in the unique rocket culture which built up in the work of scientists and engineers at Peenemünde and continued on into the post-war world.

Responsible for technical designs copied to blueprints and layouts for distribution to design departments, Gerd De Beek wanted to personalise each test rocket, creating motifs and cartoons by which each could be recognised. The early A-5 development programme provided opportunity for nu-

Below: Studies were conducted, and tests carried out, to evaluate the feasibility of launching A-4s from rail trucks containing propellant and equipment necessary for rapid preparation and launch. *Author's collection*

merous flights to be made with these reusable rockets for the first time and personnel became attached to these recognition symbols. More so for the A-4 during its long development. Von Braun himself had high appreciation of the arts, learning as a boy the joy of playing the piano and stringed instruments. De Beek shared this parallel interest in the arts and prepared numerous artistic visualisations of spaceships and rockets of immense size and capability. Soon, all the A-5, and most of the A-4 test rounds, were adorned with illustrations on the aerodynamic fins or the aft body of the missile which allowed some release from the rigours and privations of a barrack-room existence. When von Braun went to America after the war De Beek went with him and played no small part in maintaining morale through his personalised renditions of pin-ups, comic characters and mythological symbols.

Despite the exigencies of war, a wide range of advanced projects were mooted but never really got anywhere, including plans to develop an underwater launch capability for the A-4, one of only several proposed for firing on land targets from beneath the waves but anticipating the emergence of the Submarine Launched Ballistic Missile (SLBM) of the late 1950s. The sea-launch idea originated in 1943 with Otto Lafferenz, head of the *Deutsche Arbeitsfront* (German Labour Front) which had united all workers' unions into one organisation. One proposal was that it be towed by submarine, either to the North Sea for attacking the North of England and Scotland or across the Atlantic as a quick way of attacking US cities on the Eastern Seaboard without waiting for development of a rocket with sufficient range to attack North America from continental Europe – the A-9/A-10 described later. At this stage of the war there was considerable enthusiasm for this but development of the land-launched A-4 prevented serious attention to the concept until late in 1944.

The horizontal container to house the A-4 was designed with a length of 29.8m (98ft), with a total weight of 453 tonnes (500tons) and would be towed across the sea to an appropriate position for launch at which point an aft section would fill with water to raise the container through 90deg. Once in the vertical position the upper end of the container would be released and the missile fired. The container itself was designed with work platforms, propellant tanks and a gyroscopic stabilisation system but the problem of carrying LOX across the Atlantic was arguably the most difficult to solve. Studies showed that the latest Type XXI U-boat could pull three such containers at the same time. It was agreed that conversion of the submarines would be carried out by Blohm und Voss in Hamburg and by the Weser AG in Bremen.

On 19 January 1945 the Elektromechanische Werke GmbH summarised the requirement and defined the specification, emphasising its value for taking the war to areas no longer readily addressed by the Luftwaffe. By this date, however, plans for incorporating an accurate guidance system had been dropped. Early tests were made with 8in (21cm) solid-propellant Rheinmetall-Borsig rockets which were successful and the German Navy expressed some interest, seeing in this technology a new way in which sea power could support general war and maintain a significant role; increasing dissatisfaction in the political leadership with the ability of the Kriegsmarine to defeat the Royal Navy and the increasing effectiveness of anti-submarine warfare placed the German Navy in a critical position; a potential for adding an extra role for the Navy was considered desirable by some, totally irrelevant by the majority of naval commanders.

Above: The service tower for the A-4, in this application set up on a rail car and transported horizontally until required. *Author's collection*

The A-4 application to sea launches was dubbed *Schwimmwest* (Swim Vest) and the Peenemünde facility undertook research into the containers which were known as *Projectprüfstand-12* (Project Test Stand-12) for an Army application but this attracted strong contest with the Kriegsmarine and pressures of mainstream work inhibited further development. Research results were gathered together and stored at Peenemünde but little further work ensued at a time when the Me-264 *Amerikabomber* programme was submitted to the Air Force leadership. Inter-service rivalry played a large part in destabilising any coherent plan to attack the United States – only a few elements in the Army, the Navy and the Air Force vying for that distinction, most considering it fantasy.

Nevertheless, the Air Force believed that in attacking the

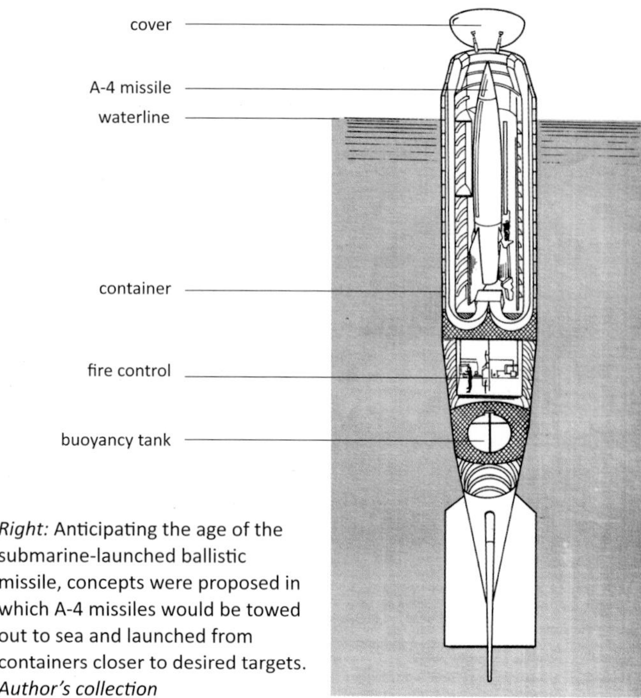

Right: Anticipating the age of the submarine-launched ballistic missile, concepts were proposed in which A-4 missiles would be towed out to sea and launched from containers closer to desired targets. *Author's collection*

United States from the Azores, the *Amerikabomber* raids would prompt a concentration on the home defence of North America to the detriment of the defence of the UK and in so doing open England and Scotland to a more concentrated assault by rocket and glide-bomb. Surprisingly, the German government believed that the presence of US troops in Britain was to consolidate a British Home Defence programme and placed lower emphasis on the invasion plan which had been the intention since the Americans began to arrive in the second half of 1942. In believing that by taking the war to the American continent it would cause a withdrawal of US troops back home for national defence, proponents of the Schwimmwest plan justified their work.

None of these ideas came to anything as the original plan for exploiting the A-4 and the Fi-103 had been delayed by development troubles and came too late for its intended purpose. The D-Day invasion of 6 June 1944 occurred earlier than the Germans had hoped and marginally before operations with the rocket and pulse-jet bomb could begin. Investment made in a wide range of jet aircraft, pulse-jet cruise weapons and ballistic missiles were all just too late to have any serious hope of influencing the outcome of the war and preventing, or at best delaying, the defeat of Nazi Germany. The interrelationship of these separate projects and programmes however is one reason why so many seemingly unrelated weapons were in play at the same time.

Out of This World

Myth has it that the Germans began secret development of rockets because that was one category of weapon denied to them at the Treaty of Versailles, essentially a German surrender ratified by all the belligerent powers except the United States. But this is not really so. It would have been absurd to specify rockets at a time when they were regarded as nothing other than a category of ammunition. Apart from which, the first liquid propellant rocket motor was not demonstrated – in the United States – for a further seven years. The use of powder rockets in combat by the Allied powers during the First World War had been limited to the *Le Prieur*, a firework-type weapon. Attached to the interplane struts of fighters and fired electrically to ignite their powder propellant, it was an early form of anti-aircraft missile. The only German application had been by Rudolf Nebel, a fighter pilot with Jasta 5 who achieved some measure of success when using signal rockets to attack other aircraft. Nebel would eventually join the VfR and become their spokesperson.

Rockets were not included in the Versailles Treaty because someone forgot to put them in, which has been claimed, but rather because in 1919 they had little or no importance. Only with the development of liquid propellant rockets in the 1930s did they become significant as, potentially, independent weapons of war. Moreover, the Germans had not given up on developing a wide range of other weapons specifically banned under the Treaty. Many tank commanders of the Second World War received their training in Russia while fighter pilots were also trained in the Soviet Union or in fascist Italy from the mid-1920s. There is no evidence that the Germans restrained development of any military weapon between the Treaty of Versailles and Hitler coming to power in 1933.

Development of German missiles during the 1939-45 period are well documented elsewhere but it helps to place the technology in context before following the course of the von Braun group as they left Europe and were taken to the United States. Over the period in question, many different types of propulsion and rocket configurations for different purposes had been developed, tested and a few placed in production. Adaptation, upgrades and technical developments were a fundamental part of A-4 production, with block variants in succession. The first 300 off the production line were designated A-4 B, continuing on through successive variants defined by the level of improvement in various systems or pieces of equipment. From December 1943 the second batch of 300 carried the manufacturing designation beyond A-4 B and into C and D variants. Refined manufacturing and assembly processes were continuously fed in to the production line and a wide range of improvements, sometimes with better equipment and less costly materials and sometimes as a result of shortages calling for new and innovative solutions.

But improvements on the production line were also mirrored by work to significantly improve the capabilities of the missile itself, by working the basic design up through an evolution brought about by better understanding of operating procedures and improved technical equipment. Von Braun had masterminded work since the very early days of Peenemünde on a development of the A-4 with much greater range by using lifting surfaces, wings placed on opposing sides of the main body of the rocket, to turn a ballistic missile into a glider. Tests of such a configuration had been carried out in a wind tunnel and later on a specially adapted A-5, seeking to transfer the kinetic energy imparted by the rocket motor to lift by way of decreasing velocity in a glide toward the ground.

The test involved an air-dropped A-5 bearing the type designation A-7. Released from a Heinkel He-111H-4 (NF + AB), the test occurred on 25 October 1942 with the A-7 dropped at a height of 20,000ft (6,100m) and recovered by parachute. The test was a success and plans for further development continued with blueprints for an A-9, essentially a winged A-4. The A-6 was a planned variation of the A-4 but with storable propellants and the A-8 was to have been a winged version of that. Storable propellants would afford greater flexibility in stockpiling missiles in a ready-state for launch, without using cryogenic oxygen as the oxidiser. Much discussion had been held regarding the long-term suitability of a missile which required at least an hour to prepare for launch, not counting the time to prepare the LOX and the transit time of the whole convoy to an appropriate site. All these considerations would be renewed in the 1950s when the United States looked to storable propellants for its long-range ballistic missiles.

Work on the winged A-9 had begun in early 1941. The definitive A-9 would, said von Braun, extend the range of the A-4 from 322km (200mls) to 500km (310mls) or longer if launched from a catapult or an aircraft. A great deal of study went into air-launched ballistic missiles, as there had been by the Air Force on their Fi-103 glide bomb designed to be launched by catapult from an inclined ramp. But the weight of the V2 made air-launch problematic. With a loaded mass of 12,700kg (28,600lb), no existing aircraft could lift an A-4 and there was the further problem of ignition and the stability after separation of a missile dropped horizontally and turned to a vertical position for powered flight. But the winged A-9 launched from the ground promised to significantly extend the range of the missile. Development was slow due to the priority given to the basic A-4 and there were few resources other than the more realistic studies on range-extension which were considered acceptable at this military facility on Peenemünde.

With such strict restrictions on the direction of work, von Braun and colleagues resorted to a subtle but effective subterfuge and requisitioned parts to adapt an A-4 taken off the production line into what it designated an A-4b, indicating it to be a development of the basic missile. But the team working on this rocket had a very different idea. They were already completing preliminary calculations for a massive booster rocket from which to launch the A-9 as the second stage of an intercontinental rocket. In the meantime the A-4b was fitted with an adapted tail section to compensate for the angular momentum induced by the wings, each 6.2m (20.34ft) in length. Larger rudders were also built in to the larger fins. Von Braun and his deputy, Dr. Eberhard Rees, planned to fire 20 A-4b rockets to test various aspects of the design.

It was a legitimate military project but not until early 1945 was the first A-4b ready for flight, the launch on 8 January achieving a height of only 30m (98ft) before the rocket failed and fell back. The second rocket was prepared and successfully launched on 24 January before attaining an altitude of 24km (15mls) and a speed of 4,318kph (2,684mph). One of the wings broke off and the rocket came to ground at a range of only 450km (280mls). But the concept had proven viable, even down to the use of sweptback wings which were another element of the advanced nature of German aeronautics and astronautics – aircraft design and rocketry which, after the war would be avidly sought out and exploited.

Below: Plans to extend the range of the A-4 were matured before the end of the war but Berlin's reluctance to fund new rockets pushed the Peenemünde group to name it the A-4b, simply an adjunct to an existing design. *Frank H Winter*

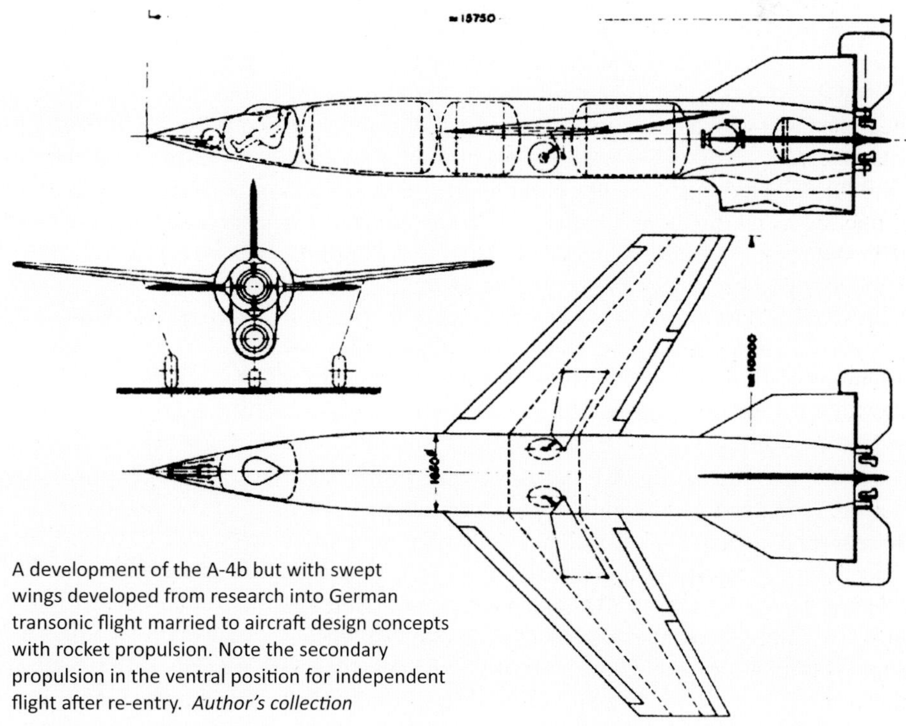

A development of the A-4b but with swept wings developed from research into German transonic flight married to aircraft design concepts with rocket propulsion. Note the secondary propulsion in the ventral position for independent flight after re-entry. *Author's collection*

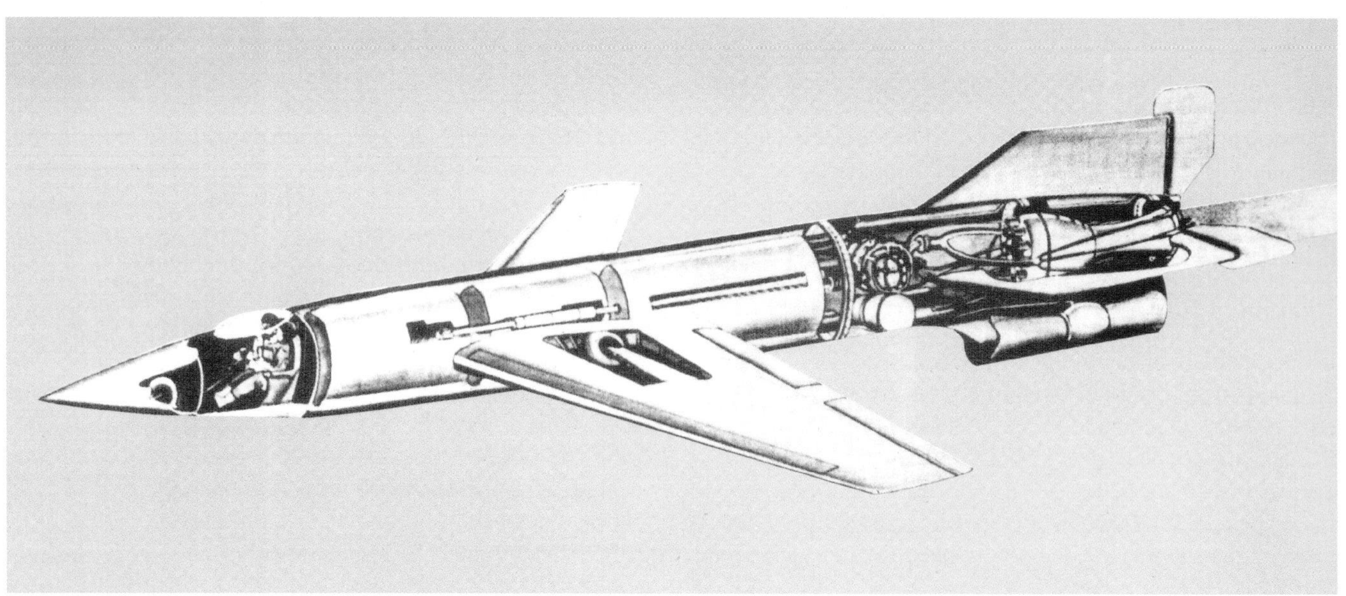

Above: The piloted A-9 would have been launched as the upper element of the two-stage rocket with which the von Braun group hoped to put humans in space. *Author's collection*

The massive booster envisaged by von Braun, Rees and other scientists and engineers at Peenemünde as a first stage to the A-9 was a major progression from the work on the A-4 engine but within the boundaries of the attainable. It had first been proposed by an engineer named Graupe in a communication dated 29 July 1940, seeking discussion of a massive intercontinental rocket for which, on 18 December 1941, Thiel proposed a new engine. This would have a thrust of 269.55kN (60,600lb) arranged in a cluster of six for a total lift-off thrust of 1,617kN (363,600lb). Designated A-10, the idea would be modified over the remaining years of the war and, although far from being a formal programme under development, was succeeded by new concepts for a single rocket motor with approximately the thrust of all six rocket motors combined. This represented a rocket with a thrust level at launch almost seven times the power of the A-4. Yet it was achievable, given the advanced status of the technology known to the Peenemünde scientists and engineers.

Designated A-10, this massive booster would have the appearance of a scaled up A-4, with a length of 20m (65.6ft), a diameter of 4.1m (13.6ft) and a weight of 78,925kg (174,000lb) including 62,100kg (137,000lb) of propellant and operate as the first stage to the A-9. When defined, the A-10 would have had a lift-off thrust of 1,961kN (441,000lb) and a burn duration of 68sec, attaining an altitude of 180km (112mls) and a speed of 4,318kph (2,684mph). Whereupon the A-10 would separate from the A-9 and fall away, deploying a parachute with an area of 2,500m² (2,990yd²) for a slow descent to the sea from where it would be recovered. Meanwhile, producing a vacuum thrust of 274.6kN (61,740lb) for 50sec, the A-9 would attain a maximum speed of 11,000kph (6,835mph), a peak altitude of 350km (217mls) and a range of 5,000km (3,100mls) carrying a warhead with a mass of 975kg (2,150lb).

This was the weapon Hitler dreamed of and even the A-9/A-10 would easily have been capable of striking the Eastern Seaboard of the United States. The transatlantic range of the A-9/A-10 made this the first serious study of an Intercontinental Ballistic Missile (ICBM) and the presentation to Hitler of its potential excited the Führer at a time when he was contemplating bombing the United States. And the German Army was convinced too, although unlikely to apply the necessary resources to progress its development. Hitler again intervened and on 20 October 1943 the *Oberkommando des Heeres* (Supreme High Command of the German Army) assigned to the SS management construction of a massive underground complex with the apt code name *Zement* (cement).

Development and test of the A-9/A-10 was to take place at a special facility inside the mountains close to Gmunden, Lake Trauen in Upper Austria. Two tunnels inside the mountain were to be excavated, ensuring complete protection from any bombing raid other than a nuclear attack. Named *Anlage A* and *Anlage B*, these tunnels would occupy a surface area of 63,998m² (76,543yd²) and house a work force of 3,000. Out in the open and down the slope of the mountain, test stands for the powerful rocket motors would be set up close to launch pads used for development and qualification flights prior to

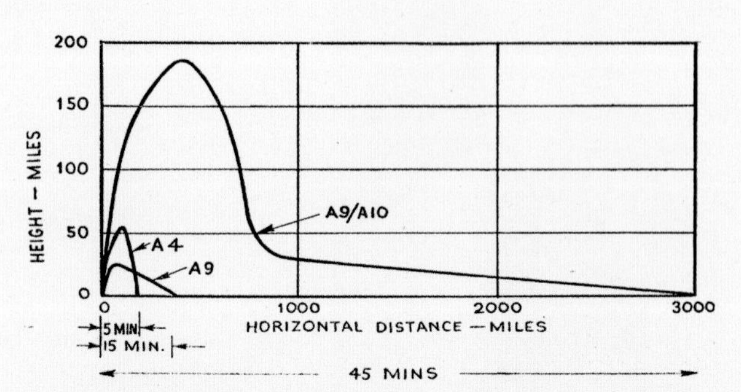

Right: Relative altitude/range capability with the A-4b, the A-9 and the A-9/A-10 combination. The A-9 designation had a variety of sub-variants according to mission objectives. *Author's collection*

operational deployment. Because of the potential range of this two-stage assembly, trajectories for test flights would follow three alignments to the following target zones: Tatra in the Carpathian Mountains; Arlberg, between Vorarlberg and the Tyrol in Austria; and Ortler in the Alpine mountains.

The concept of the A-9/A-10 was highly ambitious, involving a multi-stage rocket assembly with ignition of the upper stage effectively within seconds of the separation of the spent first stage. That had never been done before on a rocket of this size. For operational launches, the A-9/A-10 would be impossible to deploy in a mobile configuration and special hardstands would be required, which would undoubtedly revert in design to the three hardened locations at Watten, Wizernes and Sottevast originally specified for the standard A-4.

There were variations to these concepts too. A proposed piloted version of the A-4b/A-9 would have carried a single occupant over an equivalent range to the basic A-9, equipped with a turbojet engine which would be on the underside of the aft rocket body when the missile was in the horizontal re-entry position. A wheeled landing gear would allow a fast landing, modulated with braking parachutes. The Germans were quite amenable to putting humans on rockets and had several projects of this type. Well known as the first rocket-powered interceptor, the Luftwaffe's Messerschmitt Me-163B became operational late in 1944 but with little success. With swept wings and a single Walter HWK 109-509A rocket motor delivering a variable thrust of 14,678N-980N (3,300lb-220lb) it was powered by a hypergolic mixture of hydrazine hydrate and methanol burned with hydrogen peroxide oxidiser.

Also powered by this engine, the Bachem Ba 349 Natter was developed as a vertically launched interceptor carrying a single occupant and a battery of anti-aircraft missiles in the nose. The occupant was there to direct the missiles into the bombers toward which it would have been launched from an open frame and held on course by stub wings and a controllable tail. The occupant would return to Earth by parachute. Only one attempt was made to launch a man inside, on 1 March 1945 when Luftwaffe pilot Lothar Sieber lost his life shortly after launch when the Natter careened off course and the cockpit canopy came loose.

By the end of the war Germany had demonstrated an eclectic mix of rocket-powered devices and its range of guided and unguided rockets for aircraft and ships was extraordinary and diverse. The enormous effort and expense committed to rockets and missile research and development was unique in the world at that time, presenting a breath-taking array of concepts and ideas. But the most ambitious proposal of all was to develop the A-11, a massive booster stage on top of which would be attached the A-10 and A-9 as second and third stages, respectively, for a weapon of truly colossal potential.

The A-11 would have had a lift-off thrust of 16,673kN (3.75million lb). To place that in perspective, this was twice the thrust of the Saturn IB when it appeared 20 years later and half the thrust of its contemporary, the Saturn V Moon rocket. With this, manned missions into Earth orbit would be feasible and no place on Earth would be immune from attack if used to deliver multiple warheads on several targets from a single launch, dispensing them on a flight path encircling the globe. As a space launch system it could have sent groups of people into orbit and sizeable unmanned payloads to the Moon and the planets.

Few in the German Army and certainly in the political hierarchy could absorb the implications of these ideas and the concepts generated by scientists and engineers who had been working these problems for more than 10 years. Interpreting much of this work as irrelevant, the military leadership was immune to the convictions expressed by plausible argument and short-sighted attention to immediate issues preoccupied those directing the defence of the nation. It was for all these reasons that the rocket scientists and the engineers who expressed their ideas through advanced technology fell into the clutches of suspicious and mischievous state

Below: Luftwaffe rocket-power fighters: a rare photo of the Me 163 in combat (left) and the Bachem Ba 349 (right). *National Air & Space Museum*

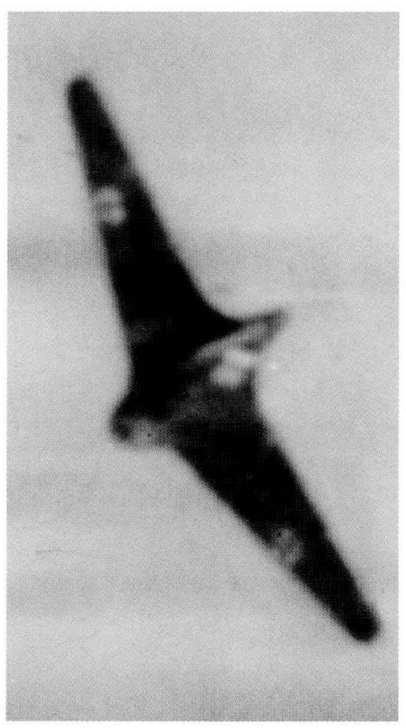

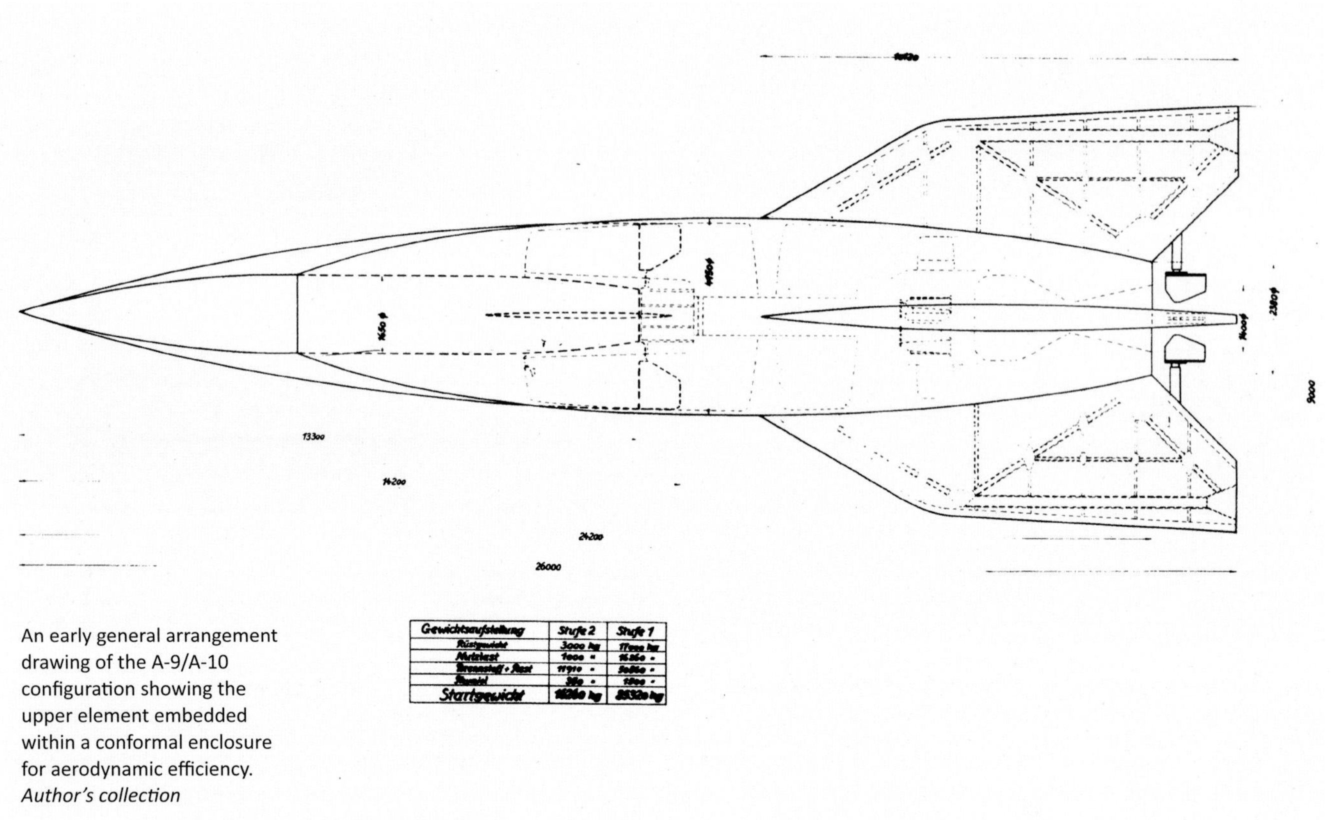

An early general arrangement drawing of the A-9/A-10 configuration showing the upper element embedded within a conformal enclosure for aerodynamic efficiency. *Author's collection*

Right: An artist's depiction of the piloted A-9 with A-10 booster stage displaying scaled-up A-4 design elements. *Spike Rendchen*

police and secret service officials, failing to realise that on these designs could rest the future of Germany, however unlikely that appeared to be.

Yet for all the anachronistic attitudes of these naysayers, only at the very top of political control in Nazi Germany, and among those uniformed administrators living and working with the rocketeers, did it seem to make sense. Very late in the 13 years of his control over Germany did Hitler begin to accept the value of this work, although he had been more enthusiastic than some to bring it all to fruition. And later still, the SS sought control during a bid to absorb as many promising elements of the national capacity for war as they could seize. But this forced a decision on the team under Dornberger, who himself feared the consequences of losing his project and seeing his civilian scientists and engineers fall into their clutches. Aware of the impending chaos as Germany began to collapse, von Braun circled his workforce and gave them free choice in choosing their future – with the Americans to the West or the Russians in the East.

At the core of what they wanted for their collective future was realisation of their dreams – of building very large rockets, opening up the space highways to exploration and sending people to the Moon and Mars, destinations which beckoned them with a will and a determination held by only a very few. The ideas and the initial calculations had matured with the evolution of the A-4 into a spectacular derivative concept. Soon, that concept was being bartered to the Americans for firing in the opposite direction – toward Russia.

Chapter Two

A New Life

When the war ended in Europe in May 1945, the Allies knew they had a potential prize in the horde of hardware, documentation and engineering data on the V2 ballistic missile which, from September 1944, had been falling on London and Antwerp. But as early as the autumn of 1943 British intelligence had known about the V-weapons programme, with its rocket propelled ballistic missile and pulse-jet powered, subsonic cruise weapon, the V1. Captured senior officers had been heard secretly discussing the V-weapons. In addition, the rocket-powered Messerschmitt Me-163 and the jet-powered Me-262 signalled a new era in weapons unlike anything that had gone before and appreciably, in the breadth of its diversity, in advance of anything under development elsewhere. The Americans and the British were as keen to get their hands on this technology every bit as much as the Russians.

As early as late 1944, Wernher von Braun had been on Britain's targeted 'wants' list, a prize to be sought and interrogated when peace had been restored. But it was with incredulity that American units in Bavaria were informed by their captors that they had in their custody the design and management team for the feared missiles. At first nobody believed them, probably due to their insistence on being immediately taken to 'Ike', General Dwight D. Eisenhower, at the time the commander of the Supreme Headquarters Allied Expeditionary Force (SHAEF). They said they wanted to immediately offer their services, show the Americans where all the data on the V2 could be found and go to work at once on building

Above: Peenemünde rocket scientists surrender to the 44th Infantry Division at Reutte in the Tyrol on 2 May 1945. From left: Charles Stewart (counter-intelligence), Dr Herbert Axster, Dieter Huzel, Wernher von Braun (his arm in a sling after a road accident), Magnus von Braun and Harris Lindenberg. Magnus had a bad reputation for theft when he arrived in the US in 1945 and was beaten up by his brother as a punishment. He worked for Chrysler from 1955 but left under a cloud and lived in the UK before returning to the US where he died in 2003 aged 84. *US Army*

rockets for the Allies. Of course the captured Germans were not taken to see Eisenhower and when some of them did get to meet him more than 10 years later he was the President of the United States.

The small group surrounding von Braun surrendered to the US Army late in the day on the 2 May 1945 (officially registered as the day after when formal photographs were taken) but were assertive in proclaiming their pre-eminence in gifting to the Americans the means to rush straight ahead with flights to the Moon and Mars, after developing missiles to keep the Russians at bay. The following month von Braun was moved to the British Interrogation Camp, Schloss Krautzberg, Taunus, where he remained for several weeks before being moved to the US interrogation camp at Garmisch-Partenkirchen. This was the route taken by others within the von Braun group as initial scepticism gave way to a belief that these men really were who they claimed to be.

Some members of the Peenemünde team had already been thinking about the post-war situation. In late 1944 a small group led by Dornberger and von Braun had discussed the impending collapse of Germany and what that would imply. Secretly connecting with the American company General Electric via the German Embassy in Portugal, they had offered their services. Interest had been shown and the Germans were kept waiting for further negotiations into the early months of 1945. Other leading scientists and engineers had also made their own, discrete plans and some of those involved negotiations with the government of Argentina for employment after the war. Most were involved in aircraft and engine design but several leading Wehrmacht officials also tried to negotiate a free pass to a more secure life. At the most senior level, even hardened Nazis were not immune from this dash for a post-war job; no less a person than Heinrich Himmler had already been negotiating through neutral countries to overthrow Hitler and run Germany in his absence, if only the Allies would stop the fighting and allow him to take over.

None of this, however, was known to local Army units and commanders. Initially brought together at Garmisch-Partenkirchen, more than 200 V2 personnel believed to be of special interest were collected from across Germany. Several different Allied military intelligence agencies interviewed the men and learned as much as they could, not only about the programmes in which they had been involved but also about the individuals themselves and how much value they could be if retained under employment. But it was clear from the beginning that the Allies were keen to get their hands on the missiles and to learn as much as they could about operating procedures and how to go about firing the V2.

It took a while for the Allies to realise the full potential of their captives, awareness of which dawned first on the Americans who saw in von Braun's team more than the promise of knowledge and expertise but an entirely new philosophy for the future. On 15 May 1945, less than two weeks after his capture, von Braun had penned a summary of possibilities for the Anglo-American Combined Intelligence Objectives Sub-committee titled 'Survey of Development of Liquid Rockets in Germany and Their Future Prospects'. Most military personnel saw the V2 and the abundance of data and engineering

Above: Mastermind of Operation Paperclip to get German scientists to the US, Maj. Gen. Holger Toftoy played a crucial role in securing the services of brilliant rocket engineers who might otherwise have languished in obscurity or faced war crimes in post-war Germany. *US Army*

Left: Part of the gas turbine for an A-4 rocket recovered at Remagen, against which rockets had been directed in an attempt to inhibit crossing the Rhine at its famous bridge. *US Army*

knowledge captured by the Allies as useful for development of the next step in weaponry. Those with political awareness of encroaching tensions between East and West saw something different. They focused on the last several pages of the report, where they discovered a blueprint for the Space Age with utterly fantastic projections, many of which stunned US interrogators. It was as though they had either stumbled into a world populated by futurists, men of vision and a forthright determination, or occupants of a lunatic asylum!

There had been neither time, manpower nor material resources to develop the ICBM concept, the official workload being completely taken up with making the A-4 a viable weapon as the V2. But the ICBM idea had some traction, a degree of practicality, and the work survived to place it in this vital missive written by von Braun on 15 May, a memorandum which also spoke of multi-stage rockets achieving a velocity of 27,000kph (16,800mph) so that they would orbit the Earth, of space-based reconnaissance platforms and of stations assembled by people dressed in 'divers suits' using rocket packs to move around in weightlessness. Von Braun wrote of space-based mirrors capable of directing sunlight to modify the weather or be turned against surface targets to destroy them with heat rays, ideas which had already been proposed by Hermann Oberth, the German space prophet and futurist.

Below: Tests with the A-4 moved to Blizna in Poland after the British air raid on Peenemünde in August 1943, with additional raids in 1944 and early 1945. *US Army*

Von Braun spoke of flights to the Moon and Mars and of uniting the rocket with nuclear propulsion, something he had undoubtedly discussed on one of the meetings he had with the German nuclear physicist Werner Heisenberg. This was already being promulgated by certain British scientists including Dr. Leslie R. Shepherd who would pursue the idea of nuclear propulsion during the late 1940s. But to intelligence personnel reading von Braun's memo it was heady stuff indeed, although questions were restricted to specific technical matters regarding existing rockets. Later, in 1946 when he was in the United States, von Braun wrote (in German) to Robert Oppenheimer, the coordinator on science for the Manhattan project which developed the first atomic bombs, urging the possibility of developing a nuclear-tipped missile force capable of attacking Russia. In 1946, proposing such a sophisticated capability was absurd and if he wanted support he was unlikely to get it from Oppenheimer.

But the foresight and linear 'futurist' thinking impressed itself upon the senior leadership at the US Army Ordnance Department. Seeing the potential clearly, Col. Holger Toftoy, then in charge of technical intelligence, ordered Majors William Bromley and James Hamill to secure as much of the remaining rocket material as they could. This would come primarily from the production facilities at Nordhausen in the Harz Mountains, close to the concentration camps which had provided slave labour for assembling the rockets. The object was to get this material to the United States from under the prying eyes of the British and to do so before the area around

Above: Much of the production for the A-4 took place at the Mittelwerke facility in the Harz Mountains from where truckloads of parts and components were extracted by the Americans prior to the Russian occupation. *Author's collection*

Nordhausen was handed over to the Russians under the agreement reached on territorial boundaries. An Army detachment extracted a large amount of material, 14 tonnes of documents from the Harz Mountains just ahead of a British Army roadblock being set up as part of the agreed zones. The Russians arrived at Nordhausen on 1st July.

In the month after the war ended there was a frantic race for the US to gather up as much rocket material as possible, take it by road to Antwerp and ship it to America, war booty by a government unwilling to share any of this with its Allies – British, French or Russian. Interpreted more recently as a peremptory move to induce political hostility against Russia, it was in fact a determination by the United States that it alone would reap the rewards of their involvement in the European theatre of operations. This was seized upon as a means by which the USA could gain the benefit of German research and engineering. It was not a *de facto* declaration of political hostility against the Soviet Union, since at this time the British and the French were treated the same as the Russians with regard to securing war booty. In fact, in the months after VE day there was considerable energy expended by the US government to normalise relations with the Soviet Union. The descent into a Cold War was still a little way off.

Recently, revisionist historians have interpreted events as supporting an anti-Soviet stance where in fact there is little evidence that any existed during the summer of 1945. Disinterest in sharing this technical intelligence was more a result of political determination in the United States that having twice gone to the aid of Western Europe within a lifetime, there was a resolve to prevent restoration of military might on that continent. The State Department had, in fact, debated the future direction of Germany, with considerable weight given to the idea that the country should be forever de-industrialised and that there should also be little support for strategically independent states within the European continent (some claiming that Germany should be knocked back to the Stone Age and kept there). Against Britain too, the shutters came down on continuing aid and military cooperation, extending as far as expelling British scientists from the Manhattan project.

The gathering of war booty came at an interesting period in the history of the US military. Long before the war ended, Congress ratified a decision to homologate overall civilian control of the US armed forces through a single Department of Defense which would be administered through the Pentagon, a massive military complex built by Gen. Groves, head of the Manhattan project. But, while the Department of Defense would not be formally instituted until 1947, both Army and Navy were aware of a new force emerging – the soon-to-be independent United States Air Force, a further claimant on dwindling, post-war resources and on a downsized, peacetime budget.

The advent of the atomic bomb muddied the waters, since it had been the Army Air Forces that had delivered the knock-out blow to Japan through the fire-bombing of its cities while the Navy and the Marine Corps had won the tactical and strategic island-hopping battles across the Pacific Ocean. With the Army seemingly without a strategic system, which service would get the principle claim to big budgets for next-generation weapons was a hotly contested question. Among the political elite there was only talk of demobilisation and slashed budgets; US defence spending had increased from 1.4 per cent of GDP in 1940 to 37 per cent in 1945, while GDP increased by 67 per cent over the same period, propelling the dollar total to dizzying heights. Paradoxically, with a paucity of civilian goods (manufacturing had frozen on automobiles and consumer durables), many bought homes with their modest extra income and set up the economy for a post-war boom.

The military saw it very differently. While a substantial number of US politicians favoured a conciliatory relationship with Russia, the armed services were deeply worried about the sheer magnitude of Soviet manpower mobilised for war. It began the shift toward an emphasis in the US on superiority in technical excellence and sophistication as a force-multiplier for the military manpower the US could not sustain. But in a republican democracy, the price paid for war by the average American was equally unsustainable in peacetime. America could not out-man Russia for that reason alone. In the majority of cases the standard of living for most Americans had not grown during the period 1940-1945, taxes had increased to a top level of 90 per cent, the working week had increased from 38 to 45 hours and a six-fold increase in households paying tax. But more were in work, 40 per cent being either in the military or engaged on munitions production for the 'arsenal of democracy'.

It was the tangled convergence of all these disparate factors that created a shift from manpower to machines in the immediate aftermath of the war: demobilisation to build a civilian economy for domestic goods and home consumption while investing in new technologies building unprecedented levels of superiority through new and innovative weapon systems. For the emerging independent Air Force it would mean money for air power; for the Army, it meant a much bigger stake in missiles and rocketry. Starting with less experience and from much lower levels of technical knowledge, the German research and development programmes were a gifted reward. It was from this base that the US Army invested in what it considered to be future of warfare – precursor to the 'push-button' era – and it was to this America that the German rocket scientists and engineers came in 1945.

Exports and Imports

Captured V2 material was shipped from Antwerp and other coastal ports to Las Cruces, New Mexico, in August 1945. Several historical references indicate that the V2s arrived at Las Cruces in almost launch-ready condition but this was not the case. The Germans were themselves insistent on components being stripped, inspected and reassembled from the several hundred rail cars that shipped in the abundance of material; their experience from working on the V2 at the end of the war showing that stored parts were prone to deterioration. But there was another reason for meticulous scrutiny of each part. Manufactured by pressed labour and slaves, frequently worked to death, production components from work facilities manned by inmates from concentration camps were either shoddy or deliberately sabotaged in subtle and discrete ways to create mechanical failure.

Discussion between this writer and General der Jagdflieger Adolph Galland (*Adolph Galland: The Authorised Biography*, Windrow & Greene) testified to the frequent sabotage of most aviation and rocket components assembled under forced labour which left several completed assemblies masking buried flaws rendering them prone to catastrophic failure. For the components secured by the United States there was little option for scrutiny, as the materials came across from Germany without any specific regard for what they were or of their condition. Scrambling to collect up anything and everything connected to the V2 programme, to the untrained eye of regular Army GIs it was impossible to discriminate between valuable components and useless rubbish. Few could make a subjective decision as to the value of the objects they were collecting. A very considerable amount of material arrived in the US damaged in transit or suffering from rough handling.

The rocketeers were longer in migrating from their devastated home country. It had been uncertain what fate they faced. Emotions ran high among the British who, compared to the Americans, had suffered most from almost four years of air attack on their homes and work places. There were calls among the British public for senior military scientists and engineers across a wide spectrum of technological development to be hanged as criminals. Decisions regarding their survival were balanced between the work they could do for national defence and these emotive cries for retribution. Before his death in April 1945, President Roosevelt had pledged not one 'ardent Nazi' be allowed refuge in the United States and the State Department had strict instructions from the Truman government that succeeded him not to change that. On paper, no scientist or engineer who had worked on the German weapons industry would be allowed in and that had already become public knowledge.

But other factors overturned this decision, a lot of which concerned the vast hoard of material arriving in the United States with very little understanding of how to operate, test and evaluate it. In Britain and France, German technicians were already employed identifying and cataloguing sequestered equipment and putting it back in operation. With the bigger items such as aircraft and rockets it needed some specialised advice. And the potential for acquiring detailed technical and scientific knowledge was important. In acquiring these items there was an intention to put them back under test and glean important information about equipment which did not exist anywhere else in the world. For the politicians, any advantage acquired from the backbreaking work of their former enemies was welcome; for the military, who feared the future direction of US-Soviet relations, advantage over a future enemy was essential.

For all these and other reasons, before the end of the war in Europe an Anglo-American organisation known as Technical Force (T-Force) was set up to move rapidly behind the front line to seek out and abduct key individuals and their families for repatriation to Western Europe far from the fighting and the encroaching Russian lines. Once there, their fate could be decided. T-Force was an ad hoc organisation composed of regular Army units and with no special knowledge, only guile, persistence and determination. Tested by results, they trawled a large number of people, vast amounts of information about the location of key locations where documents and research units could be found and where small manufacturing facilities had been moved into forests and remote rural areas to escape the bombing of machine plants and factories.

Left: Insulated with glass wool between the propellant tanks and the walls of the rocket, an A-4 in the hands of the British Army. *Author's collection*

Above: While in England, Wernher von Braun was driven around the streets of London and shown damage from his A-4 rockets – places similar to this which suffered the penultimate rocket strike on the capital. *Author's collection*

As much of the armaments industry as possible had been relocated in the last two years of the war in a largely successful attempt to escape the aerial bombing.

Much of the work sought was of a higher priority than aircraft and rocket projects and included people who had been associated with German attempts to build an atomic bomb and to develop biological weapons. Thorough and extensive, the search embraced a wide range of technical projects from advanced radio and radar equipment to defensive armour for tracked vehicles, from new naval weapons including guided torpedoes and from research on the effects of high altitude flight on the human body to research into new pharmaceutical products and chemical products including synthetic fuels. In all, approximately, 5,000 specific identifiable applications and locations were gathered in with T-Force expanding to more than 2,000 personnel by mid-1945, in excess of 3,000 German scientists and engineers being held in special interrogation and filtering centres across Western Europe.

At this stage the war was not over. Only Germany and its European allies had surrendered and much of the impetus sought to find answers to high-tech weapons insofar as it might inform the fight against Imperial Japan.

There was urgency in the need to get von Braun and his senior personnel out of Germany. On 19 June 1945 he and a select few scientists were moved from Garmisch-Partenkirchen to a holding facility near Munich. Two days later the Russians moved in to Garmisch. From Munich the small group was flown to Nordhausen and from there to Witzenhausen in the American sector. For a brief period von Braun was incarcerated in Kransberg Castle, where a range of high-level officials from the industrial, economic and scientific institutions were being interrogated. The British attempted to get von Braun for detailed interrogation but were refused by the Americans, until Sir Alwyn Crow intervened and arrangements were made for him to visit England.

Along with two others, at the end of August 1945 von Braun was first taken to London where he was kept in a special facility at Wimbledon close to the tennis grounds. It was while he was in London that von Braun met, and had affable discussions, with Sir Alwyn Crow, who had been director of research into ballistic projectiles at Woolwich Arsenal and then in charge of missile projects at the Ministry of Supply, while driving him around to see V2 bomb craters. It was Crow's opinion that von Braun was no different to other scientists or engineers, employed during wartime to build the biggest, best and most effectively destructive weapons possible; there was some talk of putting him on trial for war crimes

Above: Sir Alwyn Crow became pivotal to recommendations before the UK's Committee of Imperial Defence during the mid-1930s for a long-range ballistic missile. *Author's collection*

but in the aftermath of the combined bombing offensive against German cities and the dropping of two atom bombs on Japan, such talk was quickly dropped as being counter-productive and potentially self-defeating.

As early as 1934, the British authorities had been aware of German developments with the von Braun group and the War Office began a series of studies on high-velocity rockets, the first meeting to formally discuss this occurring that December. In the following April, the Research Department at Woolwich Arsenal was tasked with setting up a development programme using cordite as the propelling agent for a wide range of concepts. In July 1936 it reported encouraging results, such that offensive ballistic rockets were discussed by a sub-committee of the Committee of Imperial Defence specialising in air defence research. Four categories were examined: air-aircraft defence, long-range missiles, air-to-air weapons and assisted take-off for heavy aircraft. The first three had been given priority and two types of long-range ballistic missile were considered, with arbitrary ranges of 804km (500mls) and 1,448km (900mls). It may have been coincidental that these were the distances between London and Berlin and London and Rome, respectively.

By 1937 the prospect of war was looming large and it was decided that the effort and investment required to build an offensive system around somewhat poorly defined and nebulous expectations was not as prudent as development utilising a more productive use of government resources. Very little work on liquid propellant rockets had been carried out in the UK and development of a long-range missile was a major challenge. It was at this time that work into an air defence system for Britain was being developed at Bawdsey Manor in Suffolk under the guidance of Robert Watson-Watt, the government deciding that this defensive system was a better use of national resources than the possibility of an atomic bomb as an offensive weapon, also competing for funds.

While less proficient in tests with liquid propellant projectiles, and achieving great success with anti-aircraft, surface-to-surface and air-to-air solid propellant rockets, the more advanced proposals, inspired by intelligence of what the Germans were up to in the mid-1930s, only highlighted the shortfall in knowledge and experimentation. As Sir Alwyn Crow said: 'At the start, practically all that the rocket development group had to offer were faith and hope; we were fortunate in receiving a modicum of charity...'

The implications of this are staggering. That British Intelligence knew about the work of the Germans on ballistic missile development long before the war began has been a closely guarded secret. It took Britain to the very edge of deciding to race the Nazis to a long-range weapon with a more ambitious specification than that envisaged by the Germans with their A-4. Rocket experts in the UK, including some engineers in the British Interplanetary Society were consulted regarding the practicality of such a concept. Their positive judgement was that such a missile was entirely feasible and that if the British chose to, they could build it. Reports continued to leak from the German rocket engineers even after the group moved to Peenemünde in 1937, information passed along by those close to the team under Dornberger. Several anti-Nazis who chose not to work for the German government believed the consequences in not sharing this information could be disastrous, a view expressed quite independently by several scientists who worked with the US Manhattan project building the atomic bomb.

On 20 July 1945 the US set up a secret programme – Operation Overcast – to offer selected German specialists employment in the United States, if they were judged likely to be loyal to their prospective employees and were capable of advancing the aims of the US government and its military objectives. In November, this became Operation Paperclip as a means of allowing high-level German specialists into the United States. It got its name from the process whereby State Department officials would receive a request for immigration

Right: British Army soldiers stand guard on an A-4 being prepared for firing, the first post-war launch of the missile. *Author's collection*

on behalf of the Ordnance Corps with a small paperclip attached to the application. Whereupon the official would re-categorise the individual as being non-Nazi and with no part in any activity for which he could be held for war crimes, without further interrogation. It was what the US military termed, the 'intellectual booty'.

A Transatlantic Migration

Von Braun and six others from the German rocket research station at Peenemünde arrived in the United States on 20 September 1945 as part of the first shipment after a gruelling sequence of flights from London back to Frankfurt, Germany, to Orly Airport, France and from there across the Atlantic to the Newcastle, Delaware army base in a C-54 Skymaster before flying back up to Boston, Mass. Von Braun had signed a six-month contract with the option of a further six months, at an annual salary of 32,000 marks, which at the time was about $9,500. On the evening of 8 October they arrived at El Paso, Texas. Wernher's brother Magnus von Braun arrived in the US on the SS *Argentina* on 16 November. They were destined for Fort Bliss, Texas, where they would prove pivotal to giving the Americans instruction on long-range ballistic missile technology and, more specifically, on preparing the A-4/V2 for flight.

One of the group who would be key in the development of flight operations for the Saturn rockets – and many others along the way – Dr. Kurt H. Debus was part of a small group who had been siphoned off to help with test firings of captured rockets by the British Army. In July 1945, Dr. Debus had been moved to Cuxhaven to assist with Operation Backfire, where three A-4 rockets were prepared for a series of test launches which took place on 2, 4 and 15 October. The first reached an altitude of 69.4km (43.1mls) and a range of 249.4km (154.9mls), the second only partially successful with an altitude of 17.4km (10.8mls) and a range of 24km (14.9mls), while the third achieved 64km (39.7mls) and a range of 233km (144.8mls). These were conducted by the British Special Projectiles Operation Group.

Above: Elements of the A-4 rocket stacked for transportation to the United States, no one single rocket being complete with all elements having to be reassembled on arrival. *Author's collection*

The British had been keen to get their hands on the A4 rocket before the Americans completed collecting up their war booty and shipping it Stateside. There was a general feeling that the Americans would not share their work on the rocket and that this would be the only chance for the British to compile a technical dossier and complete a full assembly, preparation and launch sequence. After all, the British had known about this rocket long before the Americans. But while many in the know wanted to seize the technology for a British equivalent of the A-4, the parlous state of the economy and the lack of political interest in such an idea, rendered Operation Backfire as an exercise in gathering technical information rather

Below: Designated R-1, the Soviet rebuild of the A-4 came quickly after the Russians entered Peenemünde. An expert in guidance, control and telemetry, Helmut Gröttrup led the German team and was shipped, along with 2,500 fellow engineers, to an isolated location 322km (200mls) from Moscow on the night of 21 October 1946. The first Soviet launch of an A-4 occurred at Kapustin Yar on 18 October 1947. *Novosti*

than immediately exploiting it for national defence.

But it was to the Americans that the men and materiel of the Peenemünde team migrated, their chances there providing infinitely greater prospects for carrying on the work begun in Nazi Germany. Departing Cuxhaven in November, Debus was moved to other places in Germany from where he agreed to an offer of employment in the United States, arriving in January 1946. He was given a position with the Guided Missile Branch at the Army Ordnance Research and Development, Redstone Arsenal, supervising the development programmes and responsible directly to the project director from Fort Bliss, Texas.

Walter Dornberger was also taken to Cuxhaven to help with the Backfire tests but he had a different future to that of von Braun and the other civilian rocket scientists. Surrendering to troops of the US 44th Infantry Division on 2 May 1945, along with von Braun he was detained for a month at Oberjoch close to Hinderlang. From there, their paths diverged. After the Backfire flights, on 9 January 1946 Dornberger was transferred to Island Farm Special Camp 11 at Bridgend, Glamorgan in the UK, where he was kept under special guard due to implied threats from other prisoners. The facility had been cleared of its existing German prisoners on 31 March 1945 in preparation for 160 high ranking Wehrmacht officers including Gerd von Rundstedt, Erich von Manstein and Walther von Brauchitsch. While there, Dornberger was the most hated man at the camp; even von Runstedt shunned him.

Dornberger was to have stood trial as a war criminal on the basis of the slave labour employed to produce the A-4/V2 rockets and the targeting of London but he maintained that Hans Kammler had been in charge, not him. His vociferous attitude toward the German High Command made many enemies and his work with the SS was also noted by some as inappropriate to a ranking field officer in the Wehrmacht. Under interrogation, and given the desire to get informed experts and specialists into the US under the Paperclip process, on 9 July 1947 Dornberger was moved to the London District Cage run by MI19 from where he was flown to the United States by military transport.

After working for the US Air Force for three years on several missile projects, he joined Bell Aircraft Corporation where he remained from 1960 to 1965. During this period he fell out with von Braun, recruiting several personnel from Huntsville to work on a number of Air Force projects. While at Bell, Dornberger helped develop several boost-glide projects and for a while he worked on the Air Force Dyna-Soar programme. He retired in 1965 and left the US to live in Mexico before returning to Germany where he died in 1980.

When Debus began his term of employment with the US government he was under a contract initially running until November 1951 with a starting salary of $6,000 ($84,700 in 2021 money), rising to $7,000 at the end of the contracted term. Interestingly his immediate supervisor was Dr. Ernst Steinhoff. His focus was on supervision of about 30 engineers and 60 technicians, mechanics and electricians and to be central to all A-4 firings and subsequent missile development. Both the location and the attitude of locals to these immigrant rocketeers was very different to that which had met von Braun and his group at Peenemünde on the Baltic coast in the late 1930s.

Located on the outskirts of El Paso, Texas, Fort Bliss was about 80km (50mls) south of the White Sands Proving Ground in New Mexico. The northernmost edge of the firing range was 303km (188mls) north of the border and occupied an area 66km (41mls) from east to west. About halfway up the range, off to the side and close to the majestic Sacramento Mountains lay Alamagordo, where the first atom bomb had been detonated on 16 July 1945; on the opposite side of the range were the formidable San Andreas Mountains. Fort Bliss had been established in the mid-19th century to protect locals and the forty-niners against marauding Indian tribes. As a training camp for anti-aircraft artillerymen and their equipment during the Second World War, it was a suitable place from which to launch German rockets.

It was a place familiar with the sound of rocket motors.

Below: A group picture of the 104 German rocket scientists brought to the United States. A number were located at Fort Bliss, Texas, from where they resumed A-4 flights at White Sands, New Mexico. *NASA*

East of White Sands, during the 1930s Roswell had been home to the pioneering rocket scientist Dr. Robert H Goddard who was financed by the philanthropist Daniel Guggenheim. There, he could work on his liquid propellant rockets isolated from the prying attentions of interfering well-wishers and journalists. Goddard had made the world's first flight of a liquid propellant rocket on 16 March 1926, a flight lasting 2.5ec adjacent to a cabbage patch in Auborn, Massachusetts. He had secured international acclaim for his work, continuously so as he developed more capable rockets and patented some innovative technical solutions to problems with guidance stability during flight. His work attracted attention from nascent rocketeers around the world, stimulating scientific and engineering interest across the United States, but with little government interest.

The Germans claimed to have known little of Goddard's work, which is probably not as surprising as it appears. Goddard himself had been reluctant to share his work publicly and chose not to court attention, seeing popular attraction as a distraction from the many whom he thought might steal his designs and copy his rockets. In that, he was not unlike the secretive and suspicious Wright Brothers, in December 1903 the first aspiring aviators to demonstrate heavier-than-air powered flight. Later, Debus would write that in 1950, 'with patent claims against the Army and the Navy, several of us received special clearances and were permitted to examine the Goddard drawings and technical reports. We were astonished to learn that he anticipated some of the fundamental solutions we arrived at in Peenemünde, including the gimballing of engines, employing jet vanes and other basic principles'.

Debus described how conditions at Fort Bliss were far from ideal: 'We needed facilities and we built them. Instrumentation experts and test equipment designers laboured with hammer and saw to construct barracks and never succeeded in completely keeping out the fine sand blown by desert winds. They cut and welded steel, mixed and poured concrete for a static test stand, shovelled earth and laid pipes. When a trainload of supplies arrived, they stripped to the waist to unload, catalogue and store the multiple items. Relieved not only to be alive but to have a paid job in a country

Oberkommando des Heeres
Dienstausweis Nr. 01
v. Braun Wernher
Name Vorname
geboren am: 23.3.1912 zu: Wirsitz
Wohnort: Berlin-Steglitz
Gestalt: groß Gesicht: oval Haar: blond
Augen: blau Bes. Kennzeichen: keine
Berlin, den 15.Februar 1945
Oberkommando des Heeres
I. A.
Dienststempel
Unterschrift des Dienststellenleiters
Zeitstempel
1942 1943
1944 1945
Wernher von Braun

Above: Von Braun's German Army identity card stamped up to date and in 1945 exchanged for commitment to a very different army and a new country. *Author's collection*

where there was a real opportunity to carry on with work that had occupied them for many years. And they wanted equipment deemed essential to their work – even mundane items. One requisition asked for shovels; the Army shipped 1,000!'

The V2 launches from White Sands were to be integrated within the Army's Hermes programme, which had been activated on 15 November 1944. This was essentially a technology development programme which had itself emerged from the Rocket Branch established within the Technical Division of the Office of the Chief of Ordnance in September 1943. On 20 November 1944, the Ordnance Department issued a contract to General Electric for the development of long-range rockets which could be used as surface-to-surface or anti-aircraft missiles. They were not to know that within a few months a treasure trove of V2 rockets and associated ancillary equipment would become available and merged with the objectives of the Hermes programme.

Testing Times

The first V2 rocket static fired in the US occurred on 15 March 1946, strapped into the 100K Static Test Stand. The first commander at White Sands was Lt-Col. Harold Turner and he invited 500 military personnel and a large number of local community leaders to watch what was intended to be a demonstration of its potential. Shortly after ignition the heat and vibration of engine was felt by the spectators as the steel plates bolted down onto the concrete started to glow before the

Left: US pioneer Robert Goddard (far left) and helpers in his New Mexico workshop, working on his P-series rocket c. 1938. *Author's collection*

shock waves sent them hurtling off toward the watching bystanders. As they careened off into the desert setting off several small fires the spectators fled for their lives in all directions. Nobody was hurt in the confusion but it certainly impressed.

This was followed by the first launch of an A-4 on 16 April but here too events unfolded at a frightening pace. After a clean getaway from LC-33 it exhibited erratic tendencies – not at all unusual – and from inside the bunker a technician sent a signal to shut off the fuel flow. After reaching a mere 5.5km (3.4mls) it momentarily stopped before falling back to the ground. Unfortunately, immediately after it had been sent on its way an excited group from inside the blockhouse rushed outside but on seeing it heading straight down again they dashed to get back in only to collide with a second echelon, also trying to get to watch it on its way, creating an impromptu rugby scrum! Fortunately, the missile came down 8.5km (5.3mls) from the launch site, sufficiently far away for it to pose no real danger.

As the *Journal of the American Rocket Society* reported at the time: 'About fifteen seconds after the launching of this first missile, tail surface IV was torn loose from the body and came down separately. The appearance of this piece indicated that it had been repeatedly overstressed, probably by the violent spin, and had finally torn away ripping the spot-welded

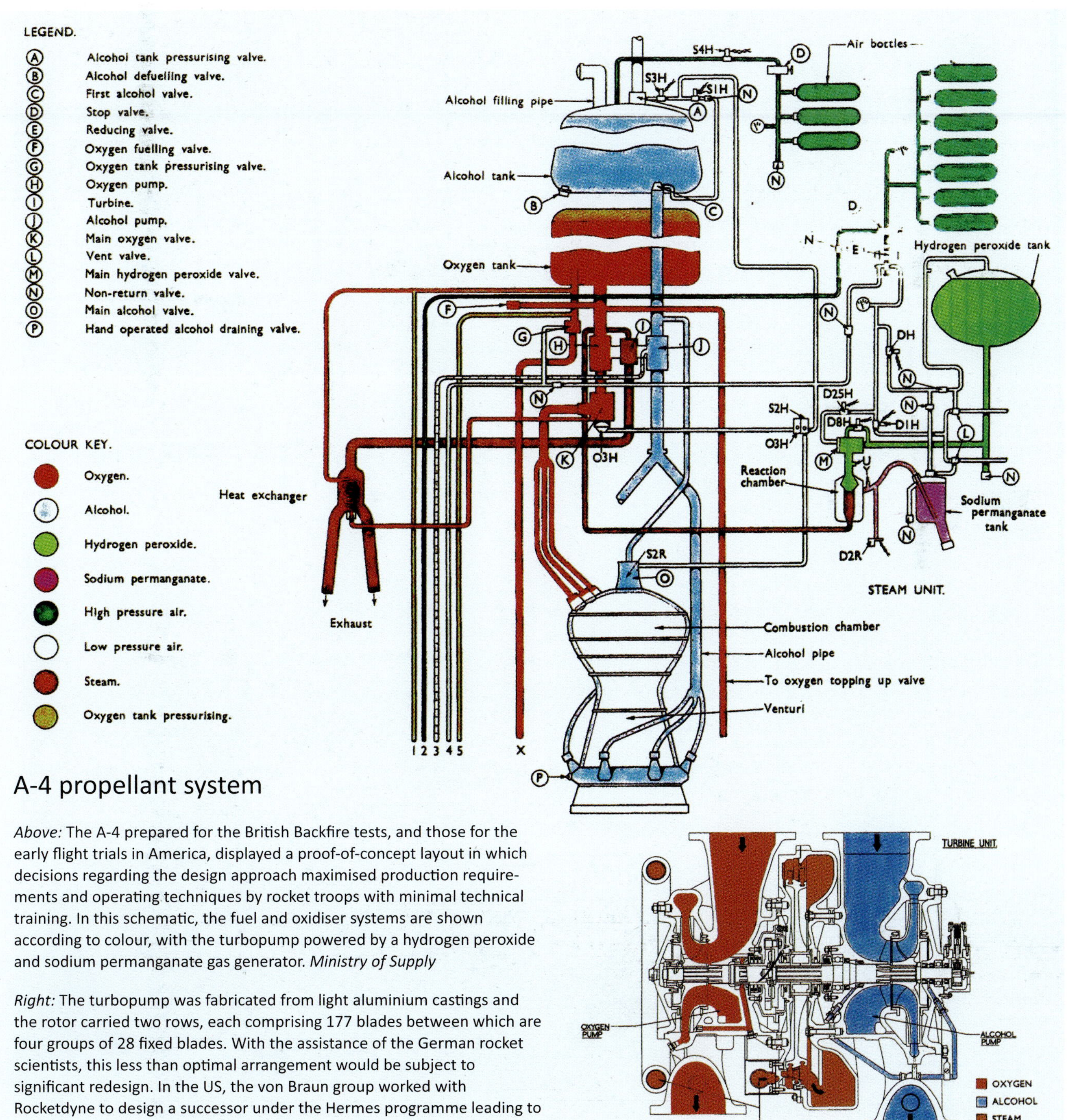

Above: The A-4 prepared for the British Backfire tests, and those for the early flight trials in America, displayed a proof-of-concept layout in which decisions regarding the design approach maximised production requirements and operating techniques by rocket troops with minimal technical training. In this schematic, the fuel and oxidiser systems are shown according to colour, with the turbopump powered by a hydrogen peroxide and sodium permanganate gas generator. *Ministry of Supply*

Right: The turbopump was fabricated from light aluminium castings and the rotor carried two rows, each comprising 177 blades between which are four groups of 28 fixed blades. With the assistance of the German rocket scientists, this less than optimal arrangement would be subject to significant redesign. In the US, the von Braun group worked with Rocketdyne to design a successor under the Hermes programme leading to the Redstone missile. *Ministry of Supply*

skin and breaking the stressed members. This surface carried both the AN/APN 55 S-band beacon receiving antenna and the ARW-17 (40mc) emergency receiver antenna and although both were torn away with the tail surface the emergency receiver operated properly and the beacon signal was strong and steady throughout the flight.' It appeared that this indicated the coaxial cables had 'severed in such a way as to leave appreciable lengths of the center conductors uncovered and ungrounded outside the body of the rocket and that these acted as receiving antennas with sufficient gain to operate the receivers at the relatively short range'.

This initial launch was successful in proving the makeshift facilities and the rocket were capable of launch, paving the way for the first formal firing on 10 May 1946 before officials and military specialists from the Army, Army Air Force and Navy. Preparations had begun at 9.00am local time the previous day. With a launch weight of 12,610kg (27,800lb), at precisely 2.00pm the rocket took off and ascended as expected, propelling the missile under power for 59.2sec to a maximum altitude of 113.9km (70.8mls) but with a slight error of 4deg in the flight path angle from that predicted. Consequently, impact occurred 22.9km (24.5mls) north of the launch site, rather than the 66.7km (41.5mls) predicted. Technicians manning radar stations telephoned in their plot as to its location, with three jeeps and a radio truck setting out across the desert to find the wreckage. Liaison aircraft followed and orbited the spot serving as an indicator to other vehicles bringing the contingent of observers to inspect the crater, 9.1m (30ft) in depth and diameter in the wet sand and gypsum with several large boulders blown around by the impact.

Flight tests would be conducted at White Sands Proving Ground until 19 September 1952 when a total of 77 had been launched but under Spartan conditions and all from Launch Complex 33, which today is rightfully a historical monument. The personnel billeted at Fort Bliss were shipped to White Sands for the tests but accommodation was bleak and there was little attention to comfort. Compared to the elite-status accorded von Braun and his group at Peenemünde, it was a world away and brought some surprise and not a little disappointment in what they had regarded, before arrival, as a land of plenty. Not, it seemed, for the immigrant Germans living strictly under 'Army regulations'.

For a while, especially during the spring of 1946, there was some suspicion of the Germans and they were excluded from any two-way transfer of technical or project information. It was only natural that they should be regarded with distaste – American servicemen were returning home with horrifying tales of concentration camps, extermination facilities, inhuman treatment of Jews, gypsies and political dissidents as well as evidence of atrocities carried out at some prisoner-of-war camps. The Germans were kept well away from local citizenry, such as it was in these dispersed and underpopulated locations. Neither were all US servicemen happy with their charges. Nevertheless, American scientists and engineers, especially those from General Electric, came to question the Germans and glean information, who gave forth the fruits of their experience freely and enthusiastically.

Yet as engineers, for all the respect they received from

A-4 steam generator

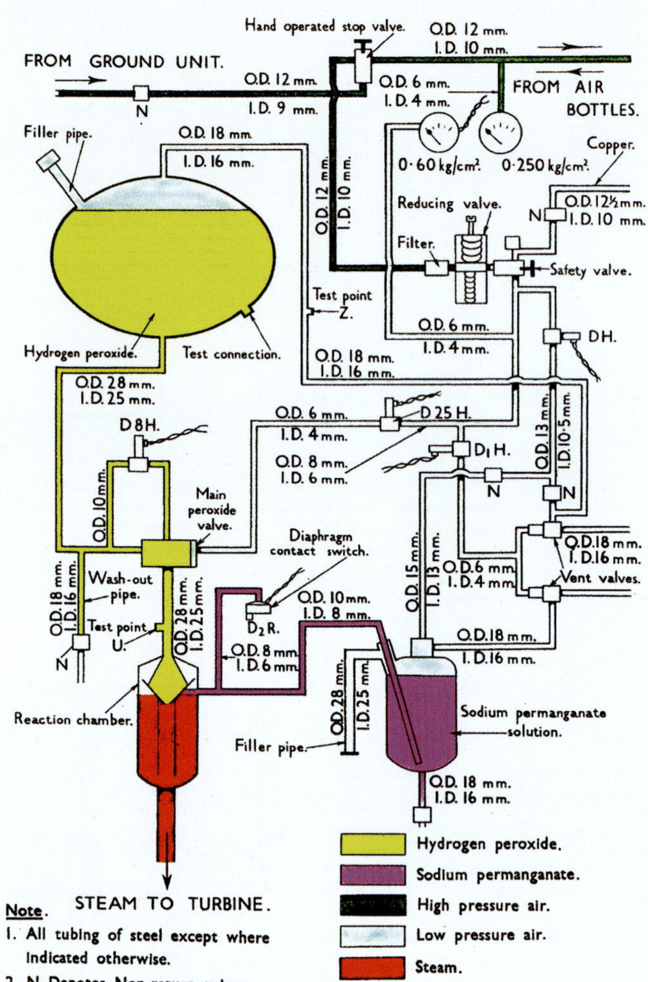

Above: The steam unit for the turbine was located near the turbine and pump assembly and employed air pressure to drive the hydrogen peroxide and sodium permanganate into the reaction chamber. It would be updated and simplified for the Redstone, providing a baseline of design and operating experience which would be taken up by Russia and the United States after the war and form the basis for some technology applications in Britain. *Ministry of Supply*

Right: The turbopump, steam generator and peroxide tank in their original mounting frame (some parts missing or replaced). *National Air & Space Museum*

COMBUSTION CHAMBER

Forged steel cups

Section at X

Section at Y

Section at ZZ

2 rows of 12 feed holes

1 row of 12 } cooling
2 rows of 16 } nozzles

A-4 combustion chamber

Left: The combustion chamber design was the responsibility of Dr. Walter Thiel through the Technical College in Dresden, where Georg Beck and Konrad Dannenberg had worked with Karl Zinner and Hans Lindenberg to bring it to reality. Zinner did most of the work on the 'showerhead' design, which came too late for the A-4 but would later be incorporated in the XLR43, a Rocketdyne development of the original engine. *Ministry of Supply*

Below: A close-up diagram shows the burner cup detail and the relative placement of the fuel and oxidiser orifices. *Author collection*

Bottom: Tagged by the German design team as a compromise between the ideal technical solution and an interim application of known principles, the chamber incorporated 18 burner cups for bringing together the injected fuel and the oxidiser. *Ministry of Supply*

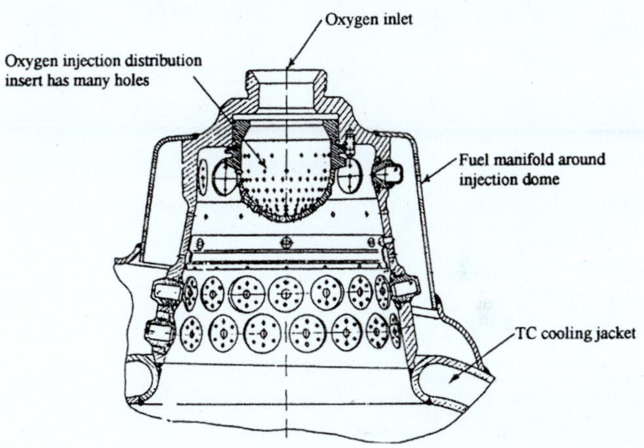

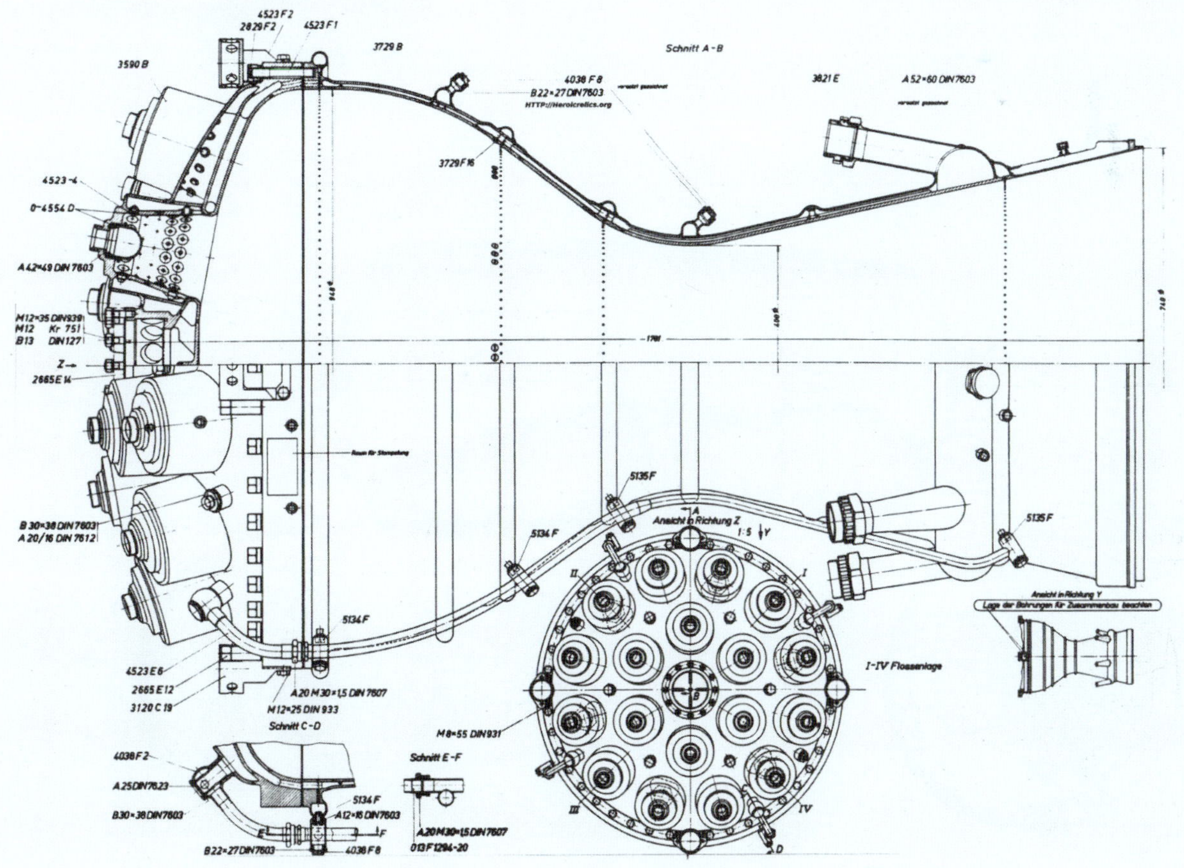

A-4 propellant tanks

Right: In this schematic the upper part of the rocket is at the bottom, showing the external skin of the A-4 consisting of 0.6mm (0.025in) sheet steel strengthened with circumferential ribs spot welded or welded according to the location and the strength requirement. *Ministry of Supply*

Below: The LOX tank contained 5,000kg (11,025lb) of liquid oxygen with a purity of 99 per cent. The bulk requirement for an oxidiser to satisfy the A-4 launch rate required two new dedicated LOX plants as the demand could not be met by existing plants. *Ministry of Supply*

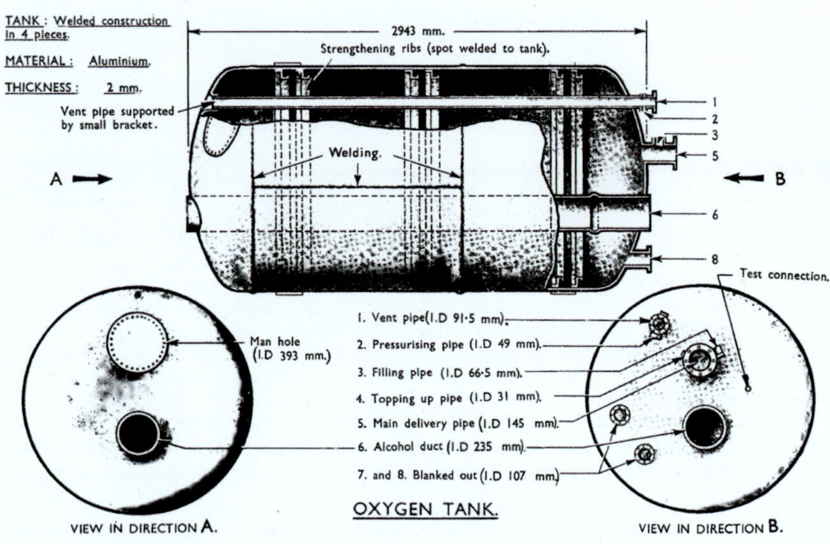

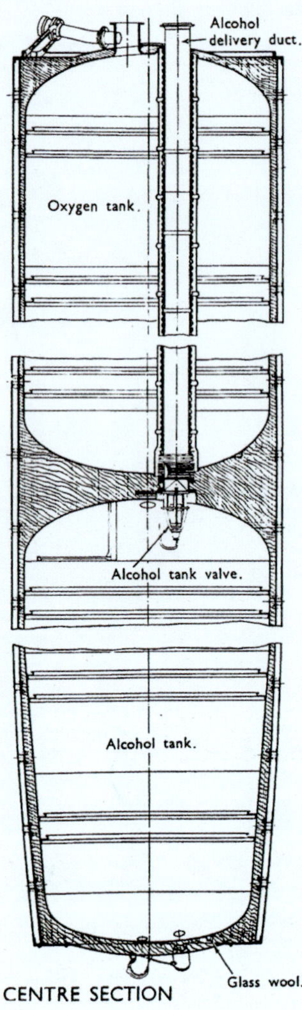

Below: The fuel tank contained 3,700kg (8,158lb) of ethyl alcohol with a specific gravity of 0.86 at 15 per cent with a requirement to eliminate solids or impurities to prevent clogging of the burner cup jets. *Ministry of Supply*

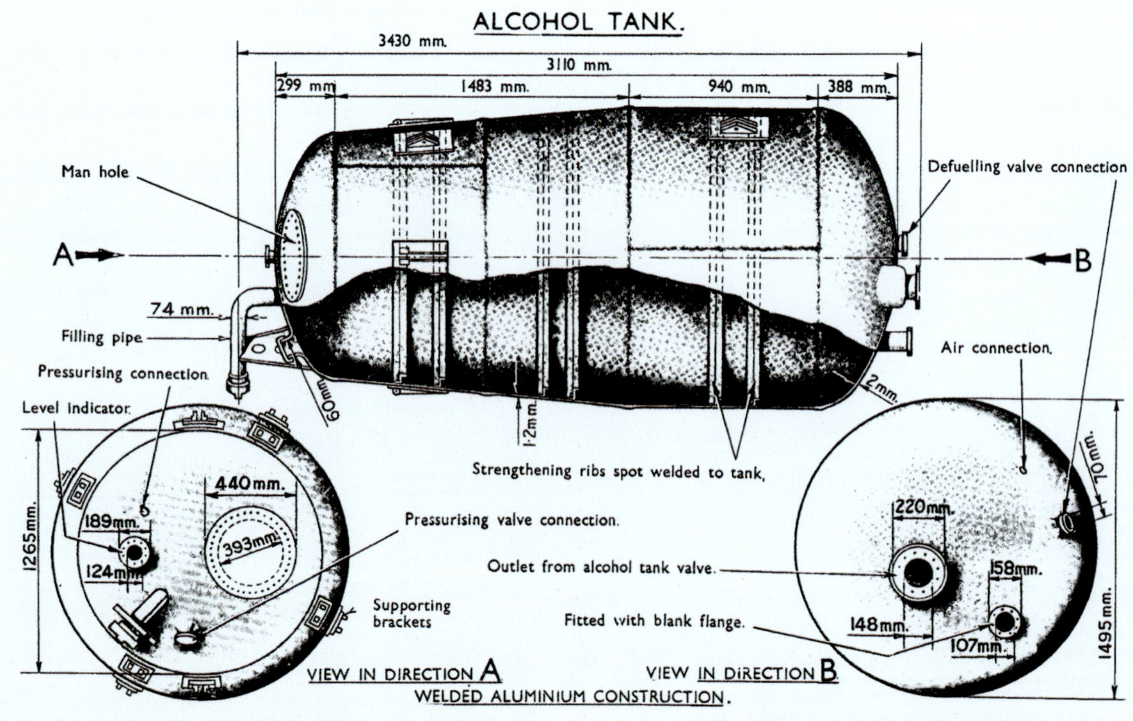

A-4 guidance system

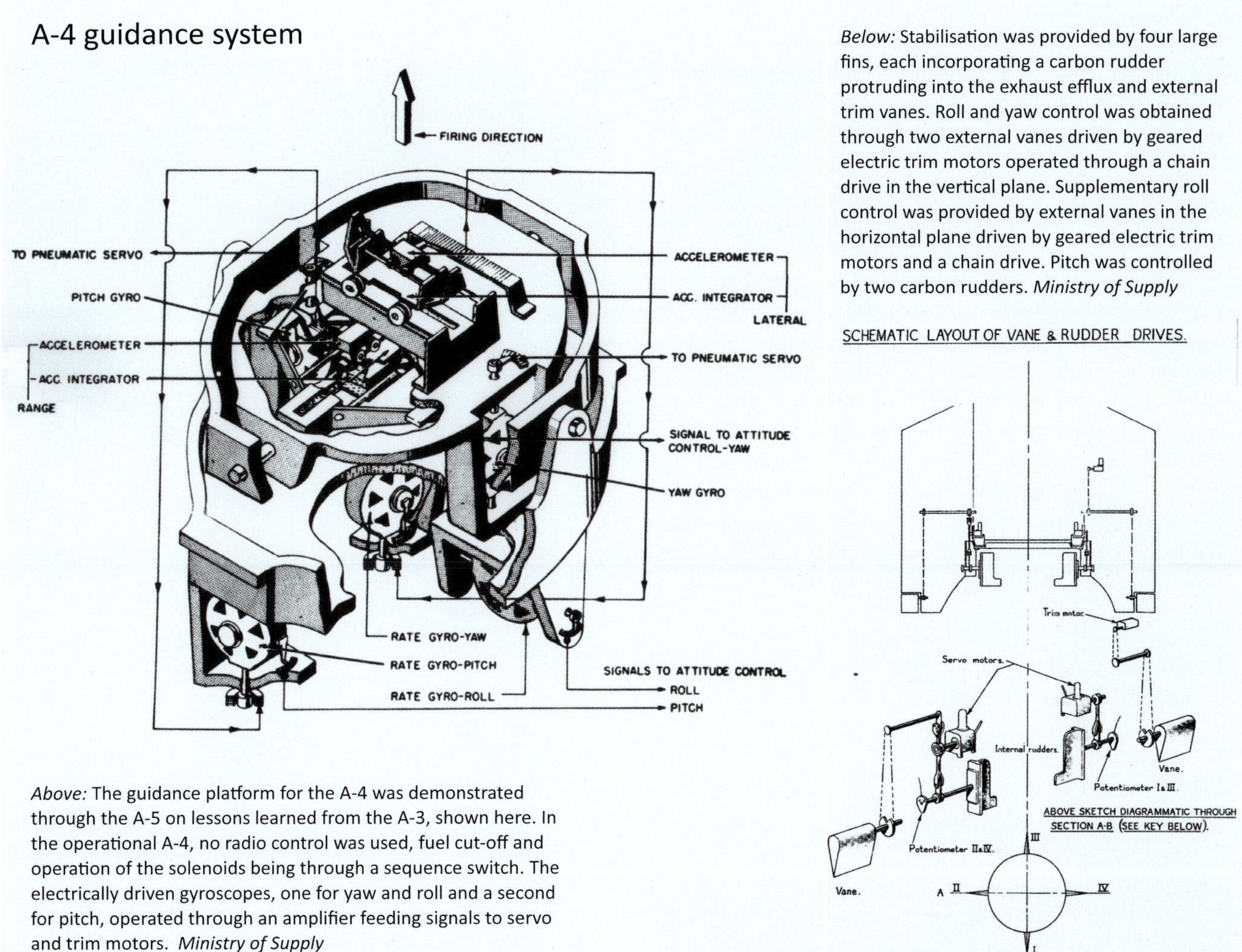

Below: Stabilisation was provided by four large fins, each incorporating a carbon rudder protruding into the exhaust efflux and external trim vanes. Roll and yaw control was obtained through two external vanes driven by geared electric trim motors operated through a chain drive in the vertical plane. Supplementary roll control was provided by external vanes in the horizontal plane driven by geared electric trim motors and a chain drive. Pitch was controlled by two carbon rudders. *Ministry of Supply*

Above: The guidance platform for the A-4 was demonstrated through the A-5 on lessons learned from the A-3, shown here. In the operational A-4, no radio control was used, fuel cut-off and operation of the solenoids being through a sequence switch. The electrically driven gyroscopes, one for yaw and roll and a second for pitch, operated through an amplifier feeding signals to servo and trim motors. *Ministry of Supply*

their opposite numbers, the Germans appeared conservative, risk averse and unwilling to try novel and innovative concepts. Some engineers from General Electric did share a little information about an atmospheric sounding rocket they were designing. It was called Viking and was to have integral propellant tanks in which the tank walls were part of the external structure through which loads would pass, rather than having the propellant tanks installed inside a separately stressed shell, as with the A-4. The von Braun group was also opposed to the suggestion of placing the rocket motor on gimbals so that it could be controlled by an integral guidance system rather than having directionally controlled vanes in the rocket's exhaust, deflecting the plume and controlling the flight path of the rocket, as fitted to the V2. The Germans knew the long, hard road taken in developing the V2 and were cautious about rushing to accept different, as yet unproven, design features. Possessed by a risk-averse culture, this accusation of conservatism would haunt the von Braun group until the early 1960s.

To support scientific research requested by the Army Upper Atmosphere Research Panel, which had been formed on 16 January 1947, eight special development flights took place between 13 May 1948 and 29 July 1950. Known as Bumper, they involved the addition of a second stage to the V2 consisting of a Douglas WAC Corporal, which was the first sounding rocket developed by the United States. It had little to do with the Germans and they were hardly involved at all. But the development was significant and while the technical challenges of adding an upper stage and placing the success of the flight on whether that upper stage fired on command, or not, was problematical, the Peenemünde engineers had long advocated that method for improving the range potential of ballistic missiles.

But the adoption of an upper stage required modifications to the gantry crane because the assembled Bumper configuration had a height of almost 16.6m (54ft) and that was greater than the height of the open structure. To simplify accommodation, platform extensions were added to the south side of the gantry so that it could be manoeuvred up to the assembly rather than over it. With platforms cantilevered directly out from the main structure of the gantry the assembly could be serviced. Later in the programme, an additional launch platform would be provided for the Bumper rocket flights, east of the previous pad with an extension of the perimeter road.

Installed on top of the A-4/V2 in a modified forward section, the liquid propellant WAC Corporal had a thrust of 6.7kN (1,500lb). The prime purpose of the WAC Corporal was to propel a package of instruments to high altitude for a better understanding of the upper atmosphere and to develop technologies essential to the future development of long-range projectiles. Initially separate, the two requirements converged in the WAC Corporal, which emerged from a collaborative venture between the Army and the California Institute of Technology (CalTech), the latter group dating back to the 1930s financed by the Guggenheim Aeronautical Laboratory (GALCIT).

Known as ORDCIT, the Army/Caltech team had begun formal development of its first military missile, named Private, on 24 May 1944. Private was an outgrowth of work conducted by CalTech on solid propellant rocket-assisted-take-off (RATO) packs for conventional propeller-driven aircraft and 41 test launches were made with this winged missile between December 1944 and April 1945. The WAC Corporal rocket employed a hypergolic propellant mixture of red fuming nitric acid (RFNA) and furfuryl alcohol. As a stand-alone sounding rocket it was to have been propelled first by a Tiny Tim booster. It had a parachute system for recovering instruments and a separate nose cone carrying a radiosonde provided by the Army.

Twelve tests with the boosted WAC Corporal took place between 26 September 1945 and 25 October, resulting in some changes in overall dimensions and various modifications before a further 20 launches were made between 7 May 1946 and 12 June 1947, during which a maximum altitude of 28,042m (92,000ft) was achieved. The eight Bumper flights were historic. On 24 February 1949, Bumper 5 set an altitude record of 393km (244mls), achieved after stage separation from the V2 at a height of 32km (20mls). Of greater lasting significance, on 24 July 1950, Bumper 8 became the first rocket launched from Cape Canaveral, at a site designated Pad 3.

Above: Towed behind the fire control vehicle, the A-4 launching table was lowered to the ground and stabilised by a single screw-jack attached to each corner of a support plate. Note that the blast deflector is raised slightly above ground level. *Author's collection*

By 28 June 1951, General Electric had supervised the launch of 67 V2 rockets and two days later the programme was terminated, the last flight being an unspectacular event with an explosion in the tail of the rocket at eight seconds into flight, the 32nd failure in a series of which one-third were due to propulsion and the rest after launch and during flight, of which almost half were guidance problems, or 'steering' as it was termed at the time.

A vital area of research brought challenges in the second half of the 1940s: development of an effective telemetry system for recording vital data between launch and termination of flight. Flight operations with high performance aircraft were limited to subsonic tests where on-board recording devices were sufficient and where wind tunnels could provide a certain level of information regarding the effects of supersonic shapes on aerodynamic performance and the response of materials could be measured. But for rockets and missiles that was impossible. The A-4 had a maximum speed of 5,630kph (3,500mph) with no chance of recovery. It was itself a research tool but largely through experimentation and empirical information. With supersonic research aircraft being developed, the absence of adequate information was be-

Left: A direct line to the first American satellite launcher began with the cooperation between Redstone Arsenal and the Jet Propulsion Laboratory (JPL) when the solid propellant WAC Corporal was developed for application as an upper stage to the A-4/V2 tests at White Sands. JPL co-founder Frank Malina (1912-1981) shows off the WAC Corporal. After founding Aerojet, Malina became dissatisfied with the priority given to rockets as weapons and fell out of favour with the establishment, moving to France and joining the secretariat of the United Nations' UNESCO. *JPL*

coming critical to the next steps in aviation and rocketry.

The Germans had faced the problem of data acquisition with the development of the A-4. When tests with the technology precursor A-5, much smaller than the A-4, began in 1939 it was possible to record nine parameters on board the rocket, including the position of the inertial stabilisation platform (a 3-axis gyroscope), the position of the rudder and the pressure in the combustion chamber both at launch and at motor cut-off. Because the A-5 was recovered, there was no need for a radio-telemetry system but with the long-range A-4 it was impossible to obtain data in any other way.

The big challenge with the A-4 test programme had been producing a radio-telemetry system with accuracy, linearity and reliability, all of which were dubious in the extreme. This was largely brought about because of the low frequency response of 0.2-0.5cps (cycles per second, or Hertz) which only recorded slow-moving events. Much work had been done to provide a more reliable telemetry system and development had been well advanced on a 12-channel system when the war ended.

Recognising the core requirement for data, during the late 1940s the US Naval Research Laboratory (NRL) developed a more capable, 18-channel telemetry system and this was incorporated into the V2 launches and in the Hermes II flights. This was a pulse-code amplitude-modulated (PCM) system with mechanical commutation but the data still had to be ex-

Above: A German A-4 rocket is prepared for launch from LC-33 at White Sands Proving Ground. The first from the United States was sent aloft on 16 April 1946. *Author's collection*

Below: Proclaimed as the 'beginning of a new era in naval weapons' by Rear-Adm D. V. Gallery, Chief of Naval Operations for Guided Missiles, an A-4 rocket was launched from the deck of the carrier USS *Midway* on 6 September 1947 in Operation Sandy. It was immediately followed by sending off aircraft from the same deck to demonstrate ease of dual operation. *Author's collection*

tracted after recovery of the instruments due to the limitation of mechanical needle tracing devices. A modest improvement was achieved with the Cook recorder, which was able to store measurements on film.

A German Influence

Information about the upper atmosphere and the effect of high speed flight on temperatures, configurations, weights and range was lacking. It was a problem facing aircraft designers and rocket engineers alike. The technology of ballistic missiles required substantive data which had to be obtained by operating various rockets and missiles as well as the shapes and sizes of re-entry bodies in the domain for which they were designed to operate. There were no wind tunnels that could simulate that environment; in fact, the wind tunnels were built and configured according to what the flight tests discovered.

Assumptions made by the group at Fort Bliss, which involved a significant number of personnel from previous work on rockets reinforced by the experience of the Germans, was that under present knowledge it would not be possible for a rocket to survive re-entry at a range greater than 1,127km (700mls) because no known material could survive the temperatures. To get data on this flight regime, a ramjet programme was instituted. Called Organ, it would be carried aloft on a modified A-4, separating at an altitude of 20.1km (12.5mls) and a speed of 5,790kph (3,600mph) after 54 seconds. Controlled by its own gyroscopes, jet and air vanes in the ramjet, Organ would provide a thrust of 13.1kN (2,948lb), burning for 400 seconds to achieve a velocity of 3,487kph (2,167mph).

These tests were under the auspices of the Hermes II programme but the first attempt on 29 May 1947 was a flop. Programmed to begin a pitch-over from true vertical and follow a flight profile to the north at eight seconds, it failed to do this and headed south. But a radio command to shut down the rocket motor was not sent for a further 38 sec. After attaining a maximum altitude of 79km (49mls), it fell to Earth and hit the ground near Juarez, Mexico and created what amounted to an international incident resulting in a halt to further tests until safety techniques had been implemented. These involved radar tracking with an electronic continuous impact prediction and a sky-screen for visually judging if the vehicle was deviating from its planned flight path. Analysis showed that the configuration was inherently unstable but the next test did not occur until 13 January 1949 followed by the last two on 6 October 1949 and 11 November 1950, all of which were failures, by which time the programme was retired.

The ramjet concept had been the instigator of Hermes B, a subset of the Hermes II programme and that produced plans for a B-1 test rocket and a B-2 operational tactical missile. The ramjet configuration for the B-1 tests consisted of two wedge-shaped 'wings', one each side of the forward section incorporating the ramjet engine. The real problem was the absence of any aerodynamic specialists in the German team and there were few aerodynamicists anywhere who could confidently design the unusual configuration required.

Above: Built around the German *Wasserfall* anti-aircraft missile, the first Hermes A-1 lifts off from LC-33 at White Sands on 19 May 1950 with weight of 2,993kg (6,600lb) and a thrust of 84.5kN (19,000lb) from its General Electric engine. Control was through an SC-584 radar and inertial gyroscopes operating the electrohydraulic exhaust vanes and four fin rudders. The last of five flights took off on 15 March 1951. *US Army*

Some modifications had been made to the A-4 however, including enlarged fins and some changes made to the front end of the rocket.

Not to be confused with the development path of the A-4, the Hermes A-1 was a copy of the German *Wasserfall* anti-aircraft missile, albeit modelled on the successful shape and overall layout of the A-4 but specifically designed for air defence purposes. It had different propellants. Instead of the nitric-acid/vinyl-isobutyl-ether mix in the *Wasserfall*, GE used a liquid oxygen/water-alcohol mixture. There was some plan to develop the A-1 into an operational anti-aircraft missile but that was passed up in favour of the solid-propellant Nike Ajax rocket. Other derivatives under the Hermes II umbrella also failed to make it out of the test stage and functional roles they were to have performed were taken up by other programmes unrelated to this story.

In April 1948 the inadequacies associated with continued development of A-4 and Hermes launch testing became clear and were defined as such by Col. Toftoy, chief of the Rocket Branch, when he accused the Ordnance Department of failing to provide proper resources and a lack of necessary funds. What was required, he said in a report which was accepted by the Ordnance Department without equivocation, was the establishment of a proper arsenal in which the need for an operational tactical ballistic missile could be met. After looking at a variety of locations and possibilities, on 18 November 1948 it was announced that the Redstone Arsenal at Huntsville, Alabama, was to be the site of the Ordnance Rocket Center. The lack of facilities at Fort Bliss made it imperative to effect a move and the proposal was approved by the Secretary of the Army on 28 October 1949, several months after the Huntsville Arsenal had been deactivated and so made available for the expansion required to adapt its purpose. The Ordnance Guided Missile Center was formally established at Redstone on 15 April 1950. It was none too soon. The world situation was worsening and the Army wanted bigger and better rockets.

On 25 February 1948, communist forces took control of Czechoslovakia in a violent coup and on 18 June a communist insurgency against British forces in Malaysia began. On 24 June 1948 the Soviet Union cut off access by road, rail and canal to the British, American and French sectors of Berlin, deep inside East Germany which the Russians controlled. The only relief possible for Berliners was food and supplies delivered by air – and so began the Berlin Airlift which would not end before the Allied zones were opened again on 12 May 1949. On 29 August the Russians detonated their atomic bomb and by the end of the year the Chinese civil war saw a complete takeover by the communist forces controlled by Mao Zedong. Within months, on 25 June 1950 the communist forces of North Korea invaded territory to the South and the United Nations appealed to world democracies for help. A war began which would last more than three years before an uncertain armistice which remains to this day.

Responding to these events, the decision had already been taken by President Truman to build a thermonuclear weapon – popularly known as the Hydrogen Bomb – and the US military was operating under a new structure which had already realised the independence of the US Air Force on 18 September 1947. No longer part of the US Army, in Washington, DC, the fight was on between the newly formed Air Force and the Navy for the dominant role in the strategic defence of the United States and its allies. That latter commitment had taken on new meaning with the establishment of the North Atlantic Treaty Organisation (NATO) on 4 April 1949, a military organisation harnessing the armed forces of member states in North America and Europe against unpro-

Below: While test launches continued at White Sands, some Bumper launches shifted to Cape Canaveral, inaugurating operations at what would become the Atlantic Missile Range. Here, a WAC Corporal is about to be attached to the nose of an A-4/V2 .*Author's collection*

Above: The first launch from Cape Canaveral as Bumper 8 lifts off on 24 July 1950, an atmospheric research flight programmed on a suppressed trajectory achieving an altitude of 16.1km (10mls) and a range of just over 320km (200mls). *NASA*

voked aggression from the Soviet Union.

It was not this train of events alone which pressed the Army to develop a tactical, theatre-class missile. That had been the intention all along, but it added purpose to the expansion of Army plans for rocket development. Development of such a weapon required better facilities than could ever be provided at Fort Bliss and White Sands, facilities for research and development of a new class of rockets. Von Braun's team moved from Fort Bliss to Redstone Arsenal close to Huntsville, Alabama, in 1950. The inevitable consequence of moving to Huntsville was that the Germans would now live and work among the civilian population and come under closer scrutiny in their everyday lives. Not only that, the former head of the German Army's A-4/V2 programme was now beginning to receive international recognition. Perhaps surprisingly there was acknowledgement from the British too, burdened by painful recollections of the weapon that had claimed more than 9,200 casualties.

On 27 August 1949, Len Carter, the Executive Secretary of the British Interplanetary Society (BIS) had written to von Braun inviting his acceptance of an Honorary Fellowship of that respected organisation. As related earlier, the BIS had been formed in 1933 as a think-tank and advocacy group supporting rocketry and the exploration of space, one of three such organisations with others formed in Germany and the United States during the inter-war years. He received a reply on 29 September accepting the honour and on 2 November that year the noted British rocket engineer Val Cleaver wrote to von Braun with his support for that award.

While being a significant rocket engineer and a talented manager of men, von Braun was a creative thinker and expressed his talent through writing. Prolific since his days at Peenemünde, in 1948-49 he wrote his first book, titled *Mars Project*, and accompanied it with a technical appendix of 120 pages, including contributions from former German rocket engineers. The grand plan he envisaged came straight out of deliberations from the 1930s and owes much to the great thinkers and futurists of that decade, including the great Hermann Oberth. In his imagination, von Braun envisaged ten giant spaceships carrying 70 crew and large winged entry gliders to carry astronauts down to the surface of the Red Planet.

Laid out in the form of a novel, it had sound engineering calculations but lacked realism only in the vast scale of the expedition he envisaged. In a mood of anticipation, during late February 1950 von Braun wrote to Kenneth W Gatland of the British Interplanetary Society and explained that he had sent draft copies to several publishers and that he was convinced that a publication contract was 'just around the corner'. A contract never arrived, commissioning editors finding it totally unsuitable as a matter of fiction but in 1952 von Braun published the technical appendix as *The Mars Project*. That was readily snapped up by those for whom rocketry and the Space Age were just around the corner.

A New Tempo

As commented upon earlier, the Unification Act of 1947 in which the Air Force became independent of the Army and all four armed services (Army, Navy, Air Force and Marine Corps) eventually came under the jurisdiction of the Department of Defense (DoD) housed in the Pentagon building, the functional allocation of roles, responsibilities and missions had polarised rivalry and fomented inter-service battles. This element in the story of the German rocket scientists and their work for the US Army is critical to shaping an academic understanding of the direction of work for the von Braun team, through the development of ballistic missiles to the Saturn launch vehicle. It is important, therefore, to summarise those steps regarding the significance of the broader alignment of defence organisation in the United States.

The significance of the unification and the turbulence it caused within the Army and the Navy is only appreciated when its impact is measured against the traditional roles of these two institutions. Held by them for 149 years since 1798 as immutable pillars of national defence, the uncontested geographic division between land and sea buttressed a mutual and benign coexistence with no contest, as the lines of demarcation between land and sea were obvious. The emergence of air power and the dominant role played by the Army ran into contention when naval aviation entered the medium and brought some conflict over resources for respective air fleets. Upon achieving independence, the US Air Force (USAF) created a tripartite contest which led to unbridled acrimony. And as if to project a neutrality on the subject, when he became President in April 1945, Harry Truman replaced Roosevelt's naval pictures with paintings of early aeroplanes.

Below: President Truman at his desk in the Oval Office, signing the final National Security Act Amendments of 1949. Secretary of Defense Louis A. Johnson looks on to Truman's right. *NASA*

Neutrality was short lived and the fierce confrontation over dwindling defence dollars and priorities over budget allocations and resources quickly spilled over into respective roles, responsibilities and missions. Yet dissent had fostered unanimity in understanding that, as academics Allan Millett and Peter Maslowski so elegantly put it, 'amid the casualties of World War II lay the corpse of American defence policy'. United by the vulnerability the US now faced in the age of mass air bombardment, atomic weapons and ballistic missiles, there was a sense of urgency disconnected from the worsening geopolitical situation. The rush to exploit equal means of defending the 48 continental United States provided furtive ground for redefining those same roles, responsibilities and missions.

The Navy wanted a post-war fleet of 14 battleships and 14 carriers, claiming high presence on the seas to emphasise the role it had for defending the sea lanes that moved its ever burgeoning trade with foreign ports. The Air Force planned a force of long-range jet bombers, fast jet fighters and a general strategy of an intercontinental 'atomic blitz'. What emerged as the National Military Establishment (NME) produced a civilian (a Secretary of Defense, or SecDef) in control of the Pentagon, legislation authorising the establishment of the Joint Chiefs of Staff (JCS), whereby all four armed services were coordinated by their most senior commanding officers in a leadership rotation, and a perpetual struggle for autonomy and real independence. Gradually, power gravitated to the civilian SecDef but not until 1961 would that power become absolute under the Kennedy Administration. But that is not important here. What is of concern is the battle between the services for control over projects and programmes they believed crucial to their operational responsibilities, specifically in regard to rockets and missiles.

While the Air Force had been lukewarm over long-range ballistic missiles, the Army was enthusiastic regarding the potential for its own management of these new rocket concepts. In 1949, the Air Force spent $3billion on development and procurement of manned aircraft and a mere $39million on a range of missile projects. For its part, the Air Force had engaged in a fierce battle with the Navy for possession of the strategic employment of atomic weapons, epitomised primarily through two specific programmes: the six-engine Convair B-36 intercontinental bomber for the Air Force and the proposed *United States*-class super-carrier for the Navy with capacity for carrying up to 24 bombers. The Navy based its case on the fact that no military target was more than 2,800km (1,700mls) from the coast and that it would be better to build many bombers, with a range no greater than that, than a single aircraft of immense range reliant on overseas bases and pre-positioning of a massive support infrastructure in a foreign country.

The nation could not afford both and the new Department of Defense was not about to fund parallel competitors for similar roles and missions. The *United States*-class carrier was cancelled by the SecDef on 23 April 1949, the first assertion of civilian control over a military programme since the overall decision-based structure shifted from the President to his appointee at the Pentagon. But that was not all. Some definition of responsibility was necessary if additional confrontations of role were to be avoided and this included ballistic missiles, one category of weapon which attracted the Army, the Navy and the Air Force. Paradoxically, the Air Force had jumped on the idea of mating an atomic warhead to a missile within hours of the Hiroshima detonation but had been unable to get through the impenetrable security screen surrounding the Manhattan Project responsible for its development, and they lost interest.

Responding in part to the decision in favour of the Air Force regarding the strategic role for delivering the atomic bomb, and the increasing reliance on massive retaliation to any unprovoked attack, the Army sought to break the monopoly in atomic weapons on 24 May 1949 with a proposal to the Joint Chiefs that it should have sole responsibility for all surface-launched ballistic missiles. On paper, the Army held all the negotiating cards. It had the experience of the A-4/V2 personnel, the German scientists and engineers were already preparing to move into a new facility outside Huntsville and this request drew back from direct challenge to the Air Force. The Navy was happy with that arrangement so long as it did not reduce its development of air-to-air and surface-to-air missiles, of which at this time it had three under development. But the Air Force employed a stalling move, expressing the ostensibly logical need to conduct an overall evaluation of potential in the concept before assigning responsibility.

The matter was still left in the air when on 16 January 1950 Secretary of Defense Louis A. Johnson decided to place additional emphasis on four missiles: the Navy Regulus, the Army's Hermes A-3, and the Air Force's air-dropped Rascal missile and Snark cruise missile. Into the fray rode the decision regarding type of atomic weapon to employ and because none of the above could be in service before 1954, the Army got to fast-track development of an atomic artillery shell for the interim, although the Air Force did point out that tactical delivery by fighter-bomber would be a more efficient means of delivery. On 15 March 1950, the Joint Chiefs recommended that the Air Force should have sole and exclusive responsibility for the long-range strategic missile. Johnson confirmed that judgement on 21 March 1950. We will return to this later as the first step toward the Army seeking a different role for very big rockets.

Von Braun left El Paso for Huntsville on 10 April 1950 but it was several months before all the equipment and personnel had been transferred away from the dust and the heat of Texas and New Mexico to the only slightly more balmy Alabama. The move signalled more than anything else the impetus that would now attach to the development of the tactical ballistic missile, which would itself come to play a seminal role in the history of the US space programme. The origins of that resolve can be traced to a report issued much earlier, on 21 May 1946 by the War Department Equipment Board chaired by Lt. Gen. Joseph W. Stilwell. It called for a period of intensive analysis and test leading toward the requirements necessary for developing a theatre missile capable of operational deployment. Immediately after the Second World War, the Air Force had been somewhat dismissive of the long-range missile concept. The Army was not.

In 1948 the Board had convened at Fort Bliss and learned about developments, included among which were the initial Hermes tests. As a result they specified two surface-to-surface missiles for field deployment. The following year the requirement was amended and a range of 804km (500mls) was decided upon as the basic specification. Again, in early 1950 a fresh set of requirements was defined for short, medium and long range missiles, the latter with a range of 1,388km (863mls). These performance variations were homologated into a single requirement defined by the Army Ordnance Department on 10 July 1950, calling for a range of 804km (500mls) carrying a warhead weighing 1,360kg (3,000lb) within a structure with a diameter of 111.7cm (44in) providing a circular error probability (cep) of 914m (3,000ft) – the radius of a circle within which 50 per cent of warheads will fall. Within a matter of weeks the specified maximum weight of the warhead had been lowered to 680kg (1,500lb) and the diameter refined to 81.3cm (32in).

Preparations for this type of missile had been under development as the Hermes C initiative and the formal requirement was embraced by this programme. On 11 September 1950 the contract with GE was changed to embrace the C1 project, at

Left: Launched from Pad 33 on 22 August 1952, TF-1 was the first of five rockets launched by an Army training unit which were also the last A-4/V2 rockets fired. Known as 'Broomstick scientists', they were personnel of the 1st Ordnance Guided Missile Support Battalion, 9393rd Technical Support Unit. *Author's collection*

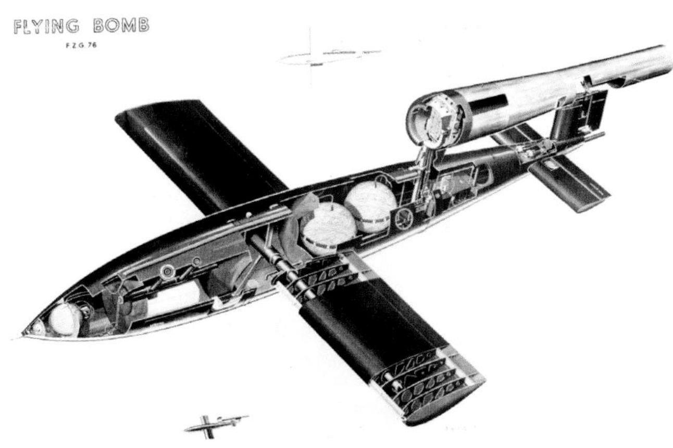

Above: Drawn by the British artist employed by the government to make drawings of German military equipment as it became available from crashed or intact examples, Peter Endsleigh Castle provided one of the first cutaways of the Fieseler flying bomb, dubbed the V1 by Goebbels' propaganda ministry. For a decade after the war, in the US the concept had precedence over ballistic missile proposals. *Author's collection*

which date it lacked a name. But it was more than a bureaucratic designation for a series of test shots. With this new requirement came responsibility for all the design, development, assembly and test of a flight qualified missile. It received the highest priority of any other work at the Guided Missile Center, although the pace of activity was constrained to some degree by the amount of money provided for the project.

Von Braun was appointed the project engineer on the missile and the results of initial studies were presented to the Committee on Guided Missiles on 25 January 1951. These included a wide range of optional configurations: liquid propellant, solid propellant, glide vehicles, ramjet concepts, single stage and two-stage designs were considered. The overriding imperative was to produce an operational missile just as soon as possible – within tight financial constraints. With the onset of the Korean War, activity increased and at this meeting the requirement was modified for a missile with a range of 740-833km (460-517mls) adopting a single-stage design, or a two-stage configuration for greater range. The initial recommendation was for this missile to be powered by a single XLR43-NA-1 rocket motor produced by North American Aviation (NAA) Inc.

North American was an aircraft manufacturer. Having built more than 42,600 airframes during World War Two, the company had to cut its workforce from 91,000 to 5,000 when peace broke out. Famous for the P-51 Mustang single-seat fighter and the B-25 Mitchell bomber, North American was developing a straight-wing jet fighter for the Navy and the swept-wing F-86 for the Army Air Force. Seeing that rockets and guided missiles could well be a part of future opportunities, NAA put William Bollay from the GALCIT team in charge of recruiting people who could get it into that business, quickly securing a small group of brilliant engineers including George P. Sutton, Thomas F. Dixon and Douglas W. Hege.

Key to a lot of the work NAA rocketeers would secure over the next several years, Sutton brought lots of experience from several years at Aerojet Engineering. In more than one

Below: Key to rocket motor development in the post-war period were the requirements for early pilotless cruise missiles such as the MGM-1 Matador, first flown in 1949 and boosted to flying speed by a 240kN (55,000lb) thrust solid propellant rocket from Aerojet General and sustained throughout its cruise phase by a 20kN (4,600lb) thrust Allison J33-A-37 turbojet engine. Successive cruise missile concepts would require much more powerful liquid propellant motors. *US Army*

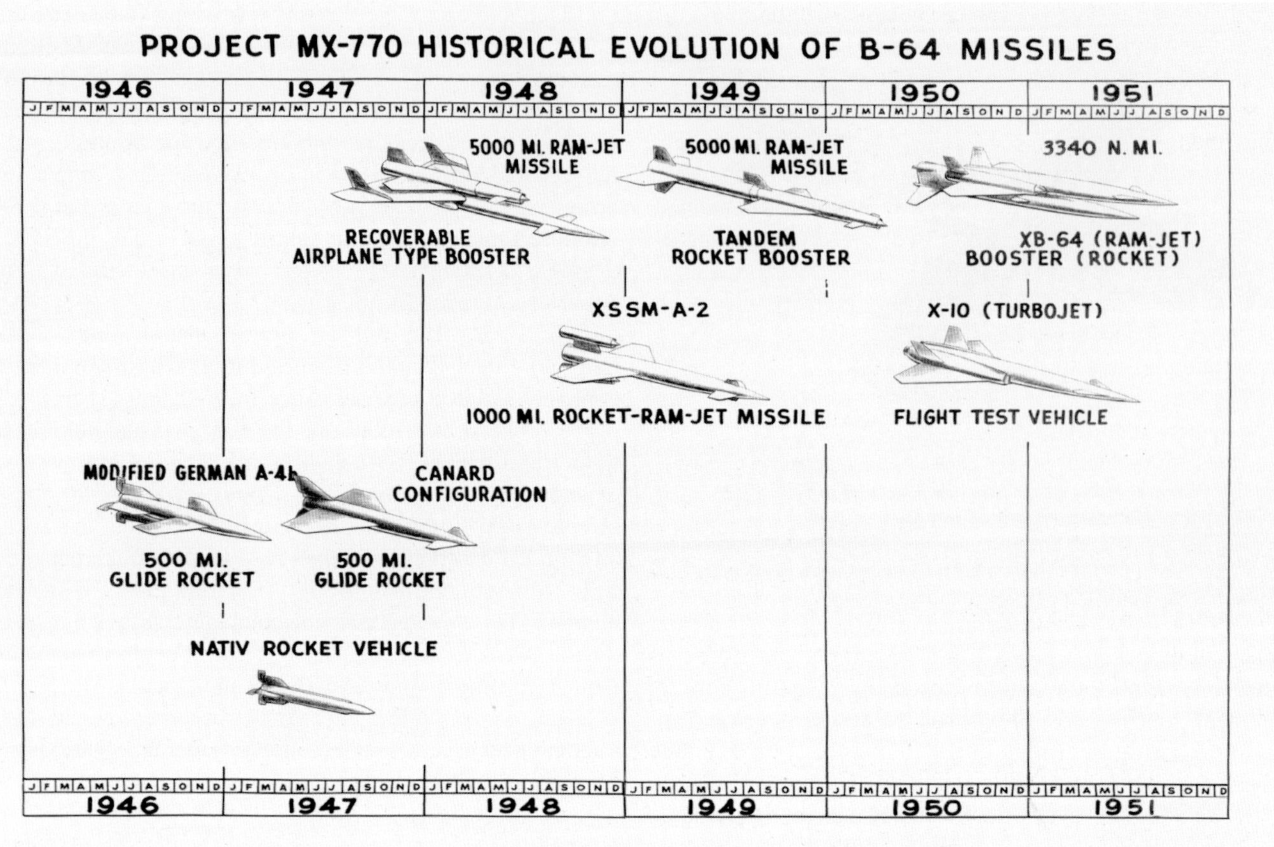

Above: Evolution of the H-1 for the Saturn I began with the MX-770 requirement for the B-64 Navaho supersonic cruise missile. Envisaged initially as a development of the German A-4b, it matured to require a rocket motor produced by Rocketdyne as the XLR41 and on to the XLR43 which would be deemed insufficient for the MX-770 but would be adopted for Redstone. *Author's collection*

respect, Sutton became the 'von Braun' of North American's rocket team and is remembered for his book *Rocket Propulsion Elements*, revised and republished over time and still regarded as the 'bible' for rocket engineers to this day. But the team was bent on producing and developing new rocket motors in advance of that built for the A-4/V2 and set up a test site in the Santa Susana Mountains west of the San Fernando Valley, not far from the Inglewood plant. Sharing rugged landscapes used for Hollywood Westerns, North American sealed a long term lease with the owners of the site, the Dundas family, and set about constructing test rigs.

In November 1945, the Army requested bids on the design of a surface-to-surface missile and Bollay, along with Ray Rice, decided to build a powerful propulsion system based on the A-4/V2 motor. On 2 April 1946 NAA received a contract for the boost-glide missile promising a range of 804km (500mls), designated MX-770 and named Navaho. In consultation with von Braun at Fort Bliss, NAA obtained important information about the design and evolution of the A-4 engine. They also learned about a development which the German rocket scientists had been unable to incorporate in their missile due to production pressures and general chaos toward the end of the war. Technical knowledge would prove crucial to the effort and NAA hired Dieter Huzel to guide them through the design details of the engine which would have powered the A-4 had there been sufficient time and resources to produce the refined motor.

Replacing the 18 injector cups with a 'showerhead' type configuration involving a single circular injector perforated with holes through which propellant would be delivered to the combustion chamber, the B8 injector plate would have been installed in the Model 39a motor with a thrust of 288.7kN (64,902lb). It took many tests to produce a configuration capable of sustained and reliable thrust but the general layout of the concept and the detailed arrangement of the injector would set the standard for all rocket motors thereafter. It was this upon which the NAA engineers focused in developing the engine for the Navaho contract, two captured examples of the A-4 motor having been obtained from the US government at the end of 1946.

Bollay set to work on developing this design, dubbed the Mark I, into an 'Americanised' version, the Mark II for which further assistance was required in the form of Konrad Dannenberg, who had been close to the development of the Model 39a. The A-4 engine had itself pioneered several innovative features, not least being the concept of regenerative cooling where fuel is used to remove heat from the walls of the combustion chamber. In principle, this had been applied in a more direct form by burying the A-2 engine in the fuel tank; bigger engines required a more sophisticated solution and this was perfected by George Sutton in an elegant and simplified concept. But the most fundamental change from the German to the US design was the adoption of a cylindrical combustion chamber rather than the bulbous, spherical shape.

Development testing began early in 1950 at the Santa Susana test site in Vertical Test Stand No 1 (VTS-1). Managed by Jim Broadston, operations came under Bill Cecka while Murray Bebbe, Ed Carlotto and Phil Fons looked after the firings and the instrumentation. Tests with the thrust chamber, fed by propellants from storage tanks, began on 2 March with a select group of invitees, including von Braun, in attendance. In his first assignment, Paul Castenholz was tasked with noting performance for development management. The drama of the event was highlighted by the knowledge that this was the most powerful liquid propellant rocket motor yet fired. And it was certainly dramatic, as the chamber exploded in a ball of fire. The oxidiser dome had been fabricated from mild steel, brittle under high temperatures. Nevertheless, the redesign was sound and the first complete engine was successfully fired for the first time on 15 November 1950.

Dubbed the 75K engine (for the thrust in pounds) it was a measured migration from the German Model 39a, with an increase in chamber pressure from 1,517kPa (220lb/in^2) to 2,193kPa (318lb/in^2) and the XLR43-NA-1 had a thrust of 333kN (74,860lb), and a bi-propellant engine fuelled by liquid oxygen and ethyl alcohol. The throat diameter of the A-4 was retained at 38.86cm (15.30in) and the divergent nozzle held the same 15deg slope angle. The area ratio of the nozzle was increased from 3.4:1 to 3.6:1 with specific impulse (ISP) increased from 203sec to 218sec. The North American engineers had done good work on what they referred to as the NA-704 'Mark III' engine, significantly reducing the weight of the A-4 (Model 39) engine from 1,125kg (2,480lb) to 669kg (1,475lb) while increasing thrust by 34 per cent. Moreover, the overall design of the engine was simplified and made much more reliable. Von Braun had been consulted over the improvements and together the NAA team reduced the overall size of the engine from a height of 4.46m (175in) to 3.33m (131in), reducing the amount of plumbing and increasing the operating efficiency and adding to its overall reliability.

Although it was developed in support of the MX-770 project for the Navaho cruise missile, when the specification for that was increased threefold, the XLR43 was side-lined, only to appear as a readily available powerplant for the Hermes C missile with which it would find application for the Army's first ballistic missile. But there were a lot of additional requirements and next came consideration of the appropriate guidance system, a field of technology in a rapid state of development and in continual flux.

The phase comparison radar came from GE and was operated by continuously measuring azimuth, elevation, range and range-rate, producing extremely accurate values for velocity to within 3cm/sec (+/-0.1ft/sec) and azimuth to within 0.02km (0.012mls). Two stationary antennae less than 6.4m (21ft) apart received a phase difference of returned signals measured in 1/3rd of an electrical degree at 3,000 megacycles. The Consolidated Vultee Corporation put up their Azusa system, which operated on the principle of radio interferometry to give out highly accurate tracking signals when the rocket transmitted a radio signal. Named after the town in southern California where it was developed, Azusa was part of that company's preliminary work on a guidance system for MX-774, a project which would become the Atlas ICBM.

The German scientists preferred the inertial guidance system they had developed for the A-4, in which all the components were incorporated within the instrument section of the rocket. These provided equipment to measure forces acting on the missile and from this it produced information which could send commands to keep it on its desired path by use of tail vanes deflecting the exhaust plume. Three on-board processors would control lateral guidance, range guidance and engine cut-off to fully control all aspects of flight. It had a demonstrated accuracy of 457m (1,500ft), which exceeded

Below: Re-designated SM-64, the developed Navaho was initially designed with the XLR71, a further development of the XLR43, in the booster element with the cruise section (right) mounted on top for sustained flight at Mach 3. *Author's collection*

the requirement and operated with a stabilised platform that provided a space-fixed frame of reference. With added re-entry control during the terminal descent phase, accuracy could be improved to 137m (450ft).

Challenging the Status Quo

From its inception at the beginning of 1951, several significant technical challenges remained to be resolved by decisions relating to the operating profile of the missile itself. To date, all rockets and ballistic missiles contained an explosive warhead which remained with the body of the projectile during flight and to the point of impact, in most cases the spent rocket casing serving to stabilise the trajectory. With the prospect of encountering severe heating conditions during re-entry to the upper atmosphere from near-Earth space, there were difficulties. The kinetic energy of a body is proportional to the mass; the less mass the less of a thermal problem. Moreover, directional stability of a falling body, tumbling through space where there was an absence of aerodynamic control, introduced the need for a separate stabilisation system to maintain orientation and prevent break-up as it encountered the atmosphere. Much of this had been recognised during operations with the V2 toward the end of the war, as recounted earlier.

In the early weeks of 1951, Redstone engineers pondered the dilemma, which the Germans knew well. As we have seen, a significant proportion of V2 rockets broke up as they entered the atmosphere again from a peak altitude of more than 80km (50mls). When speaking to the author, several German engineers working on Redstone believed that this stability problem was the most difficult technical issue in the programme. The notion that the forward section containing the explosive should be designed to separate and return through the atmosphere under aerodynamic control, leaving the main body of the spent rocket to destruction, was problematic. While solving the problem of mass and stability on re-entry, it would introduce complications. But there were advantages: in developing a rocket which effectively became a two-stage missile, an upper element may one day circle the Earth as an independent artificial satellite.

It was a possibility with attractive appeal to a nucleus of engineers who had first discussed such a capability during the development of the A-4. But for a purely engineering solution, it was decided to give the Redstone a separate warhead, an idea that quickly took hold as a standard design practice for any long-range, theatre missile. Later, for the Thor and Jupiter IRBMs and the Atlas and Titan ICBMs that were quick to follow by the end of the decade, separate warheads were designed in from the outset. The pioneering work of the Redstone, and later the Jupiter, missile in this area is fre-

Below: Rocketdyne technicians prepare XLR71 engines for installation in the booster section for the Navaho cruise missile. *Rocketdyne*

Left: With a total 25.12m (82.4ft), the definitive Navaho missile was powered into the air by a booster with a triple-chamber XLR83-NA-1 motor delivering a thrust of 2,048kN (460,444lb) for 90sec. Sustained flight at Mach 3.25 to a claimed range of up to 10,200km (6,340mls) was provided by two Wright XRJ47 ramjet engines, each with a thrust of 890kN (200,000lb). *USAF*

quently overlooked but the physics and the engineering that made it inevitable was developed independently by the National Advisory Committee for Aeronautics (NACA) – but that is a separate story.

The Redstone people were still settling in at Redstone Arsenal during all these developments but a predicted schedule anticipated 20 months between the programme start and the first two missiles being available for launch. The agreement anticipated a total of 20 test missiles for flight trials conducted over a period of 16 months with pilot production beginning 30 months after the formal go-ahead, followed by the first production prototype six months after that. In February 1951 there was a stated requirement for 100 C-1 missiles. This was not unduly optimistic because the rocket motor was already available and tested and several elements, including the guidance package, were in the final stages of detailed design. It was assumed that the production of the missile would be subject to a contract with an industrial manufacturer and this had the advantage of freeing workshops and test facilities.

Significant changes were imposed on the specification in February 1951 when Col. Toftoy informed the Guided Missile Center that the gross weight of the warhead had grown and that the missile should now be capable of carrying a nuclear device weighing 3,130kg (6,900lb), considerably more than twice the weight stipulated the previous month. Carrying this additional weight, the rocket would be incapable of providing the range previously stipulated and this was recognised, with a minimum range of 185km (115mls) now considered acceptable. The diameter of the missile was set at 152cm (60in). The need to carry an atomic warhead overruled range but not reliability as those two factors were considered immutable.

Changes were also made in the test schedule, where now 75 missiles were to be provided for research and development, the first 24 of which would be assembled in-house followed by a production order to an independent contractor. There would be some phasing-in of large scale production, with some of the in-house builds incorporating components provided by the contractor. The Army wanted the first test launch in January 1953 and all 75 missiles delivered by September 1954. Pilot production by a contractor envisaged delivery of two missiles per month, accelerating to 15 a month by August 1954.

The programme was approved on 13 April 1951 but it required further acceleration, with the first 12 test rounds delivered by May 1953. There followed a sequence of names and designations for what had begun as the Hermes C1, initially when all the Army missile programmes were homologated under a single identification system. The C1 became the XSSM-G-14 before that changed to XSSM-A-14. As is wont with any military weapon system, an informal name began to set in when the Redstone personnel referred to it as 'Ursa'; there was a feeling that it should not retain the designation Hermes C1 since it had moved far beyond the original intention of that programme. But the Ordnance Corps decided that it should be called Major, until 8th April 1952 when it was officially called Redstone, after the site where it was developed. Throughout this period a transformation started in the way governments and the military mind addressed the issue of nuclear war and that had great stimulus for the Army rocket programme.

It was of no great moment for the Redstone project, paced as it was by the technical and production capacity of the rocket, but senior Army commanders at the highest level had been encouraged to urge acceleration due to a study known as Project Vista. This was a highly classified consultation and think-tank exercise that threatened to upset the strategic deployment of both Air Force and Army programmes but in a positive way for the latter. It was another reaction to the shock of the Korean War, which began after the communist North invaded South Korea in June 1950. At this date the Air Force's Strategic Air Command (SAC) had authority over atomic attacks on an enemy, specifically defined as the Soviet Union, using bombs of enormous yield and arguably inappropriate

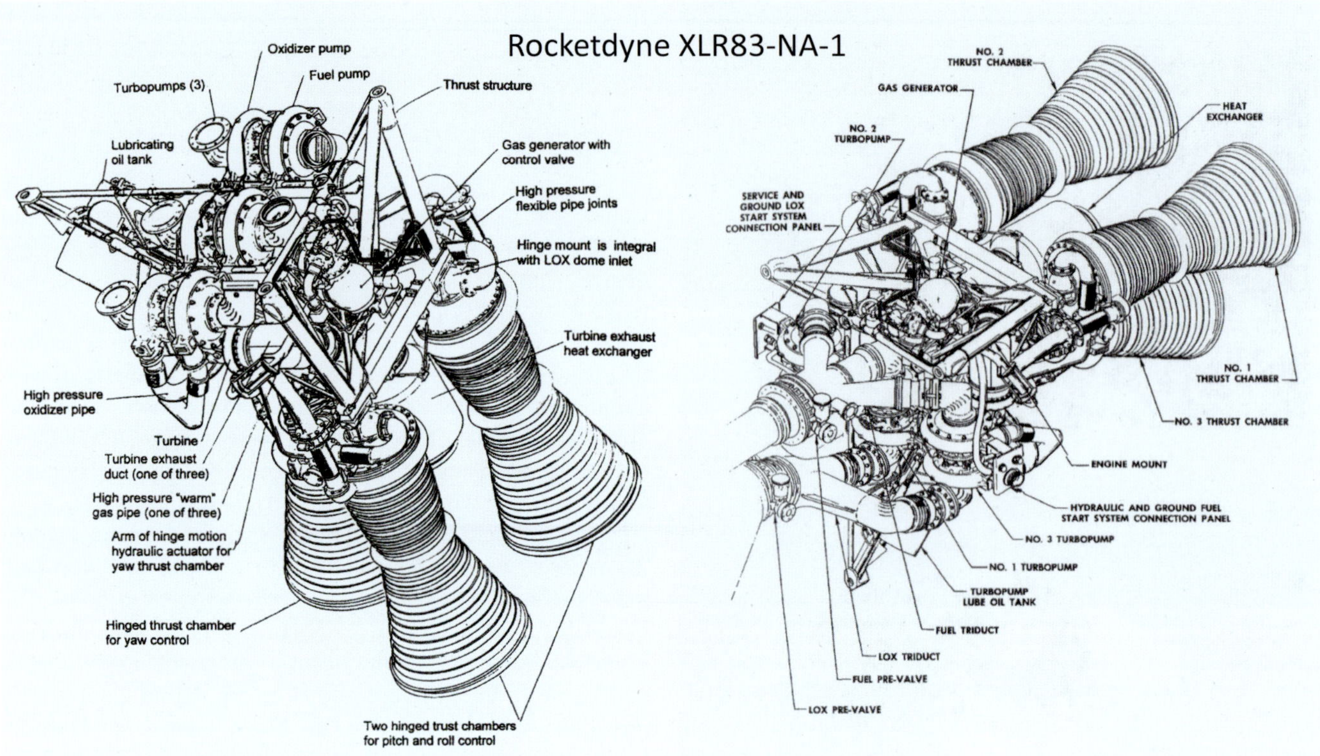

Above: While chasing down the increase in performance demanded of the Navaho missile, valuable experience was acquired in building more powerful liquid propellant rocket motors. The development of the Rocketdyne XLR83-NA-1 being an important step toward the Redstone, Jupiter, Thor and Saturn I engines. Outmoded by the advent of ballistic missiles, Navaho never entered production. *Author's collection*

for limited conflicts or brushfires.

The principal nuclear physicist Charles C. Lauritsen visited the battlefield in his role as adviser and concluded that a stiffer and more robust support for ground troops was essential to modern warfighting of the kind he saw in that region. President Truman had authorised development of thermonuclear weapons – the Hydrogen Bomb – on 31 January 1950 and that promised much more powerful explosive yield with considerably smaller bombs. That in itself rendered them more applicable to a wider range of delivery platforms, not only fighter-bombers but rockets such as Redstone already under development. This broadened their application and potential use. With smaller and more effective bombs, less collateral damage would be suffered. Now even tactical and battlefield use was within reach. It was this that encouraged a closer look at the approach taken by SAC for a total atomic assault on cities, towns and facilities to eradicate vast swathes of an enemy population.

Lauritsen was accompanied for some of his work by Col. James M. Gavin, a highly decorated veteran of the Second World War who would rise to Lt.-Gen. Aware of the encroaching vice tightening around Army applications of rockets for warfare (witness the decision of 21 March 1950 for Air Force supremacy over long-range missiles), Gavin was convinced that by drawing upon German knowledge and expertise, Redstone technology could be applied to bigger missiles. These would have longer range holding prime responsibility for supporting field troops through the use of smaller, tactical nuclear weapons. Aware that UN troops in Korea were hard pressed and lacking the kind of air support desirable, this potential seemed more relevant than ever.

During much of 1951 and in a more formal analysis of strategic and tactical nuclear war planning, more than 100 scientists at the California Institute of Technology conducted the Vista study for the US government, during the course of which it interviewed and received opinion from the country's leading scientists and nuclear weapons engineers. It was outside the influence of the military, although it was held under the auspices of the Army, the Navy and the Air Force. What Lauritsen had seen for himself in South Korea endorsed a prevailing view that dependence on massive aerial thermonuclear attacks on enemy cities and highly populated areas was wrong, in that it was not fighting the battles where victory could be achieved. But what Gavin appreciated was the need for full logistical support and from that grew the fantastic idea of moving large numbers of troops by rocket ship direct to the battlefield, supplanting the use of airborne assault forces. But this later application never made it into the report!

In pulling together its recommendations, Vista incorporated the views of Robert Oppenheimer, who was firmly of the opinion supporting the Army persuasion and opposed to national defence dependency on massive thermonuclear attack. The Vista preliminary report was submitted to the Secretaries of the three services in June 1951 with its firm recommendation for a more flexible response involving both conventional and tactical, battlefield weapons, *de facto* giving the Army greater control over the choice of weapons as deemed necessary by the local commander. At the time this seemed logical but would, in the following decade, be viewed as dangerously devoid of centralised decision-making over whether to 'go nuclear' no matter what the scale. In the early 1950s, nuclear weapons were just another tool in the war bag and widespread public denunciation of an escalating nuclear arms race had yet to emerge.

The Air Force, especially Gen Curtis LeMay in charge of SAC, was vehemently opposed to the idea of dispersing control and responsibility for the use of nuclear weapons away from the Air Force. He vociferously expressed the notion that attack could be held at bay by the threat of enormous and catastrophic destruction to an enemy across both its civil and military centres. But Vista had been triggered by deep concern over the inflexibility of that policy, one of 'massive retaliation' as it became known when officially adopted by the Eisenhower Administration from January 1953 when it assumed office. When the final Vista report was submitted in February 1952 it asserted that a war in Europe involving NATO and Soviet forces could be lost unless the early use of tactical nuclear weapons was made possible.

Several opinion shapers who contributed to the Vista report supported the view that an extended conflict in Western Europe could not be brought to a swift conclusion without the use of tactical nuclear weapons. The distance between Eastern Europe and its communist enclave was far removed from the English Channel and the advancing Red Army would readily outrun its logistical supply lines, much as the Allied armies had in the months after D-Day and before the availability of the port of Antwerp. Unless a battery of ballistic missiles could decapitate the forward edge of the battle area (FEBA) a sustained, costly and casualty-driven erosion of Soviet forces would inevitably follow a stalled front-line for lack of supplies. All of which encouraged the Army leadership to sustain the development of ballistic theatre-based missiles.

Below: SAC chief Gen. Curtis Le May was vehemently opposed to extending ownership of nuclear weapon beyond the Air Force. *US Air Force*

While the report itself was officially rejected, and buried until partially declassified 30 years later, it did have a significant impact, not least in empowering the Army to capitalise on its most valued asset in rocketry, the von Braun team at Redstone. Thus began a major effort to develop significantly more powerful theatre and battlefield rockets which would drive forward the technology for increasingly more capable missiles. The Air Force could continue its push for a strong, strategic and tactical, air-delivered, nuclear capability but without the clear and unassailable lead it had assumed. Compromised by a core command structure loyal to the manned aircraft, the Air Force would not readily absorb the realities of what would, a decade hence, be tagged as 'push-button' warfare but the pressure for the Army to build on the success of the Redstone missile was relentless and had an effect on pushing the Air Force to match it in capabilities.

Redstone Rising

Responsibility for accelerating and formally structuring the Redstone programme fell to a civilian, Kaufman Thuma Keller who in 1950 had been appointed by President Truman to be the part-time director of the Office of Guided Missiles. Keller had made his name as an executive with the Chrysler Corporation, serving as president of the company from 1935-1950. After touring Redstone and imposing his corporate strategies, the missile programme blended this management upscaling with the political support engendered by the shock of the North Korean invasion of the South. What was required, which von Braun did not have the authority to instigate, was a more business-like approach to organisation of the facility. But Keller also insisted on a stringent target accuracy for the missile's nuclear warhead. Terminal guidance was the only way the accuracy of the missile – the cep of the warhead – could be brought down to the requirement demanded by the Army.

To achieve this, von Braun worked with his guidance deputies and Theodor Buchhold, Walter Haeussermann and Fritz Mueller set to work improving the inertial guidance system and its air-bearing gyros rather than contracting out for a new-start approach. Ford Instrument Company would manufacture the improved platform but throughout the Redstone guidance programme, as with the missile itself, there would be many changes and improvements, clearly displaying the missile as a concurrent test bed and production design for operational deployment. In fact concurrency was becoming the hallmark of aviation in the 1940s and 1950s, and now rocket development, and this had its price. Instead of building a series of prototypes for testing, the conventionally sequential steps of detailed design, fabrication, test, modification and production would overlap to the point where detailed design of a production-level hardware set would still be under way when improvements and modifications were being applied to early production rounds.

A lot of administrative arrangements were set in place, creating a foundation upon which all later projects would build, changing temporary work contracts with most of the ex-Peenemünde Germans to full employment contracts with the US government. The very nature of that work required them to take US citizenship. In January 1953 von Braun became civilian head of a division within the Army, the Guided Missile Development Division of the renamed Ordnance Missile Laboratories which would be managed by Col. Toftoy, replacing Hamill who had mired the organisation in lethargic progress through his use of inexperienced junior officers.

Within eighteen months, von Braun was in charge of 950 personnel but by mid-1958 that number would have swelled to 3,925. It was in this role that von Braun excelled, more so than as a development engineer. His management skills were innate and employees would never forget his kindly, cautious and measured approach with high personal regard for the lowliest employee and always presenting an 'open door' policy at his office. From the outset, von Braun mobilised a programme of quality control and high reliability, honing component contractors into a new way of working, demanding microscale perfection with high precision and monitoring. New levels of reliability were set and adhered to and the missile group schooled every level of associated industry in new ways of thinking about production and quality. These interactions were challenging and frequently resulted in strong argument, new standards with which most of the contractors were unfamiliar.

The influence of von Braun's team was seminal in that it brought to the Army a greater level of quality control over devices and equipment which had little or no tolerance and presented unforgiving consequences for shoddy work. This was a new level of precision-engineering to field weapons deployed for service use. Some of the challenges faced by the team of former Peenemünde engineers had, only a few years earlier, faced the same dilemma, balancing the sophistication of the weapon and its high-end performance against rugged construction and simplified operation for battlefield use. Under-appreciated outside the world of rocketry and missile development, design is the easiest part of the process (the possibilities have been worked out mathematically) followed by the more difficult process of prototyping, fly and test, ending with the most difficult of all, that of manufacturing and production for reliability and consistency without compromising performance.

Below: The entrance to the former Redstone Arsenal in Huntsville, which in 1953 became the Guided Missile Development Division of the Army and in 1956 matured into the Army Ballistic Missile Agency (ABMA) under the technical direction of Wernher von Braun. *US Air Force*

In February 1952, Kurt Debus laid out a programme for introducing reliability functions to the topmost levels of management, instilling a rigour and discipline which was lacking in many Army contracts. At the heart of this was to assign an element of the missile – indeed the programme itself – as a 'series' (linear) or a 'parallel' (redundant), part of the whole. A parallel system, element or part, could be allowed to fail because there was a back-up; if an element in series failed it would imperil the entire objective, be it during a flight or in a management flow critical to completing the objective on time. Thus, a flowchart (sometimes known as a 'waterfall' chart) became entrenched within the system at Redstone Arsenal and this would remain in place until the mid-1960s before it was replaced with a more sophisticated management tool (PERT).

Debus demonstrated his concern about the need for change in quality and reliability with a simple illustration. If each component had the standard Army requirement for 99 per cent reliability, a linear arrangement of one hundred elements with that standard would, when assessed overall, have a reliability level of 36. per cent across all test firings. With 300 series elements, or components, reliability would be 5 per cent. A great number of missile and rocket failures to a wide range of programmes and projects during the 1950s and early 1960s were due primarily to poor quality control. Another aspect was the tradition of preparing hardware through a multi-organisational structure: one each for checkout, assembly, test and launch. Debus anticipated that this would lead to a confusion of who had responsibility and he recommended a single unifying organisation for the entire launch phase, resulting in the timely establishment of the Experimental Missile Firing Laboratory over which he had immediate management and control. Accountability throughout the chain would be a key factor in enhancing reliability.

In early 1952, with Hans Gruene, Debus travelled down to Cape Canaveral in Florida where the Air Force Missile Test Range (AFMTR), officially designated as such on 30th June 1951, was being set up and from where Bumper rockets had been fired since 24th July 1950. Previously named the Long Range Proving Ground (LRPG), it was from there that the Redstone would be test fired and that required a considerable amount of time and effort to make it fit for large-scale operations. But there was no other place more suitable and it afforded a clear path extending 8,000km (5,000mls) out into the Atlantic for future development of as yet unfunded intercontinental missiles, with separate tracking stations along the way. These were located at Grand Bahama Island and the islands of Eleuthera, San Salvador, Grand Turk, Antigua and Ascension.

Chapter Three

Birth of the IRBM

After receiving approval for the Redstone missile programme in April 1951, a formal and funded start was made on 1 May, a development programme which would not only see its deployment with the Army but provide a launch system for the first US satellite and put the first and second Americans in space. But there was a long way to go before those achievements. A key element of the Redstone story was the induction of civilian contractors, companies already accomplished and known to the government for reliability and effective work. Several schemes for engineering and manufacture assigned between government and contractor were proposed, changed and reworked over the years, but one of the most enduring legacies of the programme was the way industry was schooled in the new demands from rocket

Above: Developed directly from the work conducted to improve the engine fitted to the German A-4, the XLR43 powered the Redstone, America's first production ballistic missile. *US Army*

and missile engineers and this requirement was built in to the selection process.

Initially, the bidding process saw six contenders submit proposals to the Chief of Ordnance on 18 April 1952, one of whom was Chrysler. But Chrysler was one of three which pulled out, only renewing its interest after a prospective jet engine production deal with the Navy was cancelled, leaving the personnel available for Redstone. Nevertheless, judged as the best qualified for the work, Chrysler received a letter contract on 28 October 1952. Initially, the Army had first approached the aircraft industry for production of the tanks but no interest had been shown. With its long association with the Army, Chrysler was thought to be an ideal choice and had come to know and respect the outstanding work done by von Braun's people in metals and the science of metallurgy.

So universal was the recognition for this work that K. T. Keller, the special assistant for guided missiles under Secretary of Defense George C. Marshall, sent a team from Chrysler to study their techniques. Redstone commander Holger Toftoy pondered the possibility of Chrysler doing the Redstone work and discussed the matter with K. T. Tritchel, a former friend and now executive with the firm. The continuous attention paid by the Army to Chrysler paid off.

Above: Posing with a model of an A-4/V2 rocket, Wernher von Braun settled in at Redstone Arsenal where he would remain for 20 years, working first for the Army and then for NASA. *US Army*

Under the terms agreed in a preceding contract dated 27 March 1951, North American would supply the engines, while in a separate contract dated 14 August 1952 the Ford Instrument Company Division of Sperry Rand Corporation would provide the guidance and navigation section and the Reynolds Metals Company would fabricate the main sectional assemblies.

The first 12 Redstone rockets and missiles 18-29 were built at the Guided Missile Center, with numbers 13-17 and 30-128 fabricated by Chrysler. Of the total, 37 were assigned to the research and development phase, although only 12 of those were a part of the Redstone programme itself. The other 25 were designated as Jupiter A rockets and three were designated Jupiter-C and used as composite re-entry test vehicles, all in support of the Jupiter missile programme. In addition, six were modified for placing satellites in orbit and designat-

US missile and rocket motor evolution

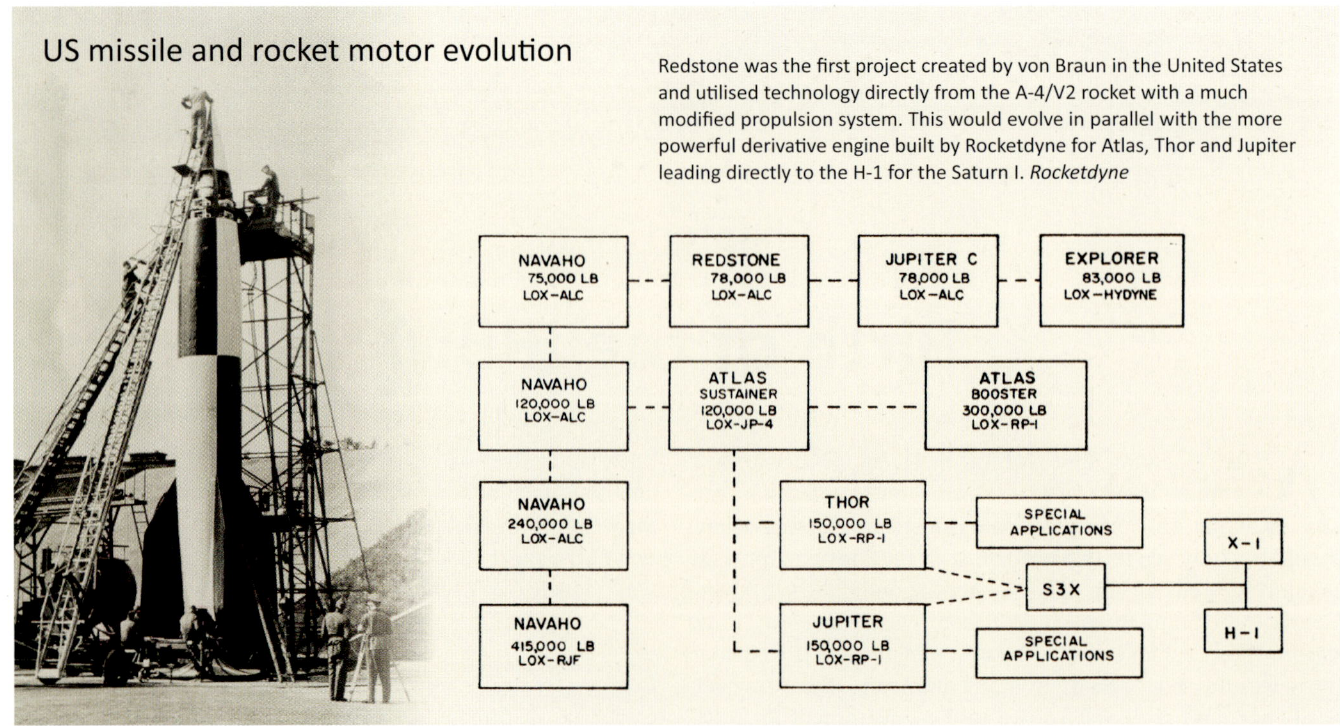

Redstone was the first project created by von Braun in the United States and utilised technology directly from the A-4/V2 rocket with a much modified propulsion system. This would evolve in parallel with the more powerful derivative engine built by Rocketdyne for Atlas, Thor and Jupiter leading directly to the H-1 for the Saturn I. *Rocketdyne*

ed Juno I about which more later on in this chapter. In all, 89 Redstone rockets were launched between August 1953 and October 1963.

In one, somewhat bizarre, proposal, under the Army Missile Transport Program of 1958, Redstone was considered for cargo delivery to the battlefield. Readers will recall that this idea had been mooted by Col. Gavin during the Vista study of 1950-51 and some of that somewhat bizarre idea had found a home. Under the concept, a ballistic trajectory would deliver a logistics container, released during terminal descent slowed by parachute and then by retrorocket to a soft landing. For this, the total assembly would have a height of 20.2m (66.3ft). It would not be the last time that rocket delivery of logistical supplies, even troops, was mooted and would lead in the 1960s to a series of rocket proposals specifically for that function.

Technical development of the missile placed great emphasis on improvements to the XLR43-NA-1 engine and these were considerable over the relatively short test life of the Redstone, beginning from the date of the contract award with a requirement to adopt a modified design, the NAA 75-110. In all, there were seven different engine types across the Redstone flight history, the A-1 being used only on the first two test launches.

The launch of the first Redstone occurred at 9.37am EST on 20 August 1953 from LC-3 at Cape Canaveral, the rocket carrying a 16-channel telemetry system for measuring structural temperatures, vibrations, propulsion, flight mechanics, steering control and take-off and cut-off signals in addition to lateral forces. Lift-off took place 2.7 seconds after ignition and acceleration followed the predicted curve and within planned programming. The missile disappeared into cloud at 42 seconds but eight seconds later it began an unplanned roll and went out of control. The flight safety officer sent the destruct command at 106 seconds with automatic separation of the front end three seconds after that. From the telemetry received, the reason for the failure was evident. In seeking to achieve a tight connection, an engineer had used a screwdriver to tighten a potentiometer which changed the electrically orientated zero position of the rudders. When control commands went in they performed an incorrect alignment, sending the missile off course.

Below: Development of launch facilities at Cape Canaveral for an increasingly powerful family of Air Force and Army missiles required expert help from German rocket engineers and Dr. Kurt Debus was instrumental in developing the launch complex concept which would establish how the facilities at each pad were designed and built. Here, a typical scene in a Redstone blockhouse as launch preparations get under way. *NASA*

Launched on 27 January 1954, the second flight was a success and three more were launched in 1954 followed by six in 1955 and 10 in 1956. The A-2 engine evolution was adopted for the second through fourth launches and featured a liquid oxygen pump inducer to prevent cavitation and the first of these flew on RS-3 launched on 5 May 1954. Subsequent engine types varied only in minor component changes as experience grew and in all cases engine parts were interchangeable. But these were significant. Between the A-1 and the A-7, the number of components in the pneumatic control system was reduced from 31 to 10. With the rigorous test procedures inducted to the entire development programme, each new component was tested both in the laboratory and in static firings before it qualified for production. And because they were built to be operated by field troops in all environmental and weather conditions, they were tested at extreme temperatures and across a wide range of humidity and dust concentrations. It paid off, with engines exhibiting 96 per cent reliability.

Development of the guidance and navigation section continued throughout the programme and several changes were applied. Long lead-time delays in getting the desired ST-80 guidance system resulted in the LEV-3 autopilot standing in but this did permit a start to flight trials where the major concern was to test the rocket motor, the structure and warhead separation. But components of the ST-80 were flown on two of these tests to gain information about their performance without being on the critical path to operations during flight. The first ST-80 for guidance came with the seventh flight (RS-9) on 20 April 1955. Only after the 21st Redstone launch on 18 December 1956 did full guidance and air-vane control come under the command of an ST-80.

Redstone Characteristics

The engineering design of the missile was based around three parts: thrust, guidance and warhead elements. The thrust element included the tail section, four fixed stabilisers each equipped with a moveable rudder and a carbon jet vane, the latter extending into the exhaust stream for control of the rocket until the ascent velocity was high enough to bring the rudder into effect. The main engine provided a thrust of 346.9kN (78,000lb) fed by 11,370kg (25,000lb) of liquid ox-

Above: Handling and transportation of the Redstone rocket was a key requirement in the design of a missile intended for field deployment with rocket troops. *US Army*

ygen and an 8,650kg (19,000lb) fuel load comprising a mix of 75 per cent alcohol and 25 per cent water. From a tank above the oxidiser tank, fuel was delivered to the engine by a pumping system to a turbine operated by hydrogen peroxide decomposing into steam, a similar concept to the A-4.

Essentially a redesigned evolution of the rocket motor built for the A-4, which had a thrust in the region of 245.1kN (55,100lb), in its ultimate variation it delivered an increase of more than 41per cent in output over that motor. But the A-4 engine had been a conservative design hurried to operational status in the interest of the war effort and some of the improvements in design exhibited by the German engineers working on the Redstone motor had already been on the drawing board before the end of the war. However, it was the considerable talent at North American Aviation that did much of the development work, signalling a rapid adaptation within the aircraft industry to new requirements.

As described earlier, the guidance system was subject to several changes, modifications and variations but the ST-80 inertial platform in operational missiles incorporated accelerometers in the stable platform kept orientated by the gyroscopes. The angular position of the missile about its centre of gravity was maintained by the potentiometers, which directed movements between the missile and the stable platform. Tape recorders were able to capture part of the trajectory with the remainder of flight information sent into the guidance computers.

The warhead had a weight of 3,590kg (7,900lb) and carried a W39 weapon in a couple of derivative forms, each with a yield of 3.8MT (millions of tonnes of TNT equivalent). This was the first missile to carry this warhead, a thermonuclear device in the new generation of lightweight weapons made possible by breakthroughs in the technology and engineering of nuclear devices. Of great importance too, the development of ablative and heat-sink nose cones containing the thermonuclear device alone opened the possibility of very long range missiles and would usher in the era of the ICBM, initially through the Air Force Atlas and Titan rockets.

The guidance unit and the warhead were located above the main body of the missile and its propellant tanks and formed an integral part of the complete assembly, some of it flying all the way to the target. The thrust unit was joined to the missile body by six bolts containing explosive charges which were detonated after engine cut-off, the two elements of the rocket separated by air-loaded piston. The base of the guidance section had four air vanes to provide manoeuvring control for the main body of the missile after separation. The forward section of the guidance unit and warhead were strengthened to survive the re-entry temperatures.

The Redstone had an overall length of 21.1m (69.3ft) with a diameter across the main tank section of 1.8m (70in). Redstone had a loaded weight of 28,000kg (61,700lb) and an

Above: The Redstone missile adopted several design characteristics inherited from the A-4, including air rudders and carbon steering vanes in the exhaust efflux. This technology would change on later missiles by making the motor(s) gimbal for directional control. Here a Redstone is made ready for a test firing. *US Army*

empty weight of 7,420kg (16,300lb). It boasted a minimum range of 92.5km (57.5mls) on a lofting trajectory and a maximum range of 323km (201mls) for a depressed trajectory. Comparisons can be made with the A-4/V2, from which so much of the technology for the Redstone emerged, in that while the range was about the same as the German rocket, Redstone carried a warhead almost four times the weight the V2 could lift. The rocket motor had a thrust duration of 96-121sec, depending upon the trajectory flown. Key to its operational effectiveness was its transportability and field deployment for ease of placement and launch.

The start command initiated a sequence beginning with pressurisation of the alcohol and hydrogen peroxide (H_2O_2) tanks. Located just forward of the engine and structurally attached to it, the 235lit (78USgall) H_2O_2 tank was filled after missile erection and then used to pressurise the pneumatic system. When the feed valve opened, the H_2O_2 flowed to the steam generator which hit a bed of potassium permanganate pellets, instantly converting it into steam. Some of that was recovered to maintain required pressure levels in the pneumatic systems and the LOX tank. Thrust levels were maintained by a controller which compared desired level with the actual value, to either increase or decrease the flow of hydrogen peroxide to the steam generator and in so doing control the rate of propellant flow to the engine.

When the alcohol tank was pressurised, LOX tank pressurisation began from the ground supply, a process which took eight seconds from the time the fire switch was depressed. Simultaneously, the igniter squibs were energised with the cartridge starting its 10sec burn duration. The assembly consisted of four electrically fired squibs igniting a pyrotechnic device suspended beneath the injector plate on a thin plastic rod screwed into the top. When the igniter fired up, the main LOX valve opened and liquid oxygen would begin to flow under a combination of gravity and tank pressurisation. Simultaneously, the fuel was forced through the centre ring of the injector head and into the combustion chamber.

Combustion began when an excess of oxygen was present. When the mainstage stick – the ignition sensing device – saw ignition the alcohol valve opened and fuel flowed into the manifold and up through the walls of the combustion chamber, forcing the propellant into the chamber via the injector head. This opened the steam flow to the turbopump which spun it up to its operating speed of 4,800 rpm. As thrust increased so did the propellant flow rate with gaseous oxygen from the heat exchanger directed into the LOX tank to maintain a constant pressure. When thrust exceeded mass, the rocket lifted away from its launch table.

Redstone was not the first battlefield missile deployed by the Army and with the capability of delivering an atomic warhead. When convention had the Army naming missiles after ranks (as noted earlier, Redstone was initially known as Major until someone said 'which major rocket' with the answer 'the one from Redstone'), the liquid propellant Corporal missile was developed in parallel with Redstone and achieved operational deployment in March 1952. But this was different to the WAC Corporal sounding rocket. The product of a collaboration between the Army and the Jet Propulsion Laboratory of the California Institute of Technology, Corporal was developed by the Douglas Aircraft Company and later by Firestone Tire and Rubber Company which was also responsible for fabricating the main body of the missile. Gilfillan Brothers produced the guidance system. Corporal had a range of 48-130km (30-81mls) and 1,101 were built of which 1,046

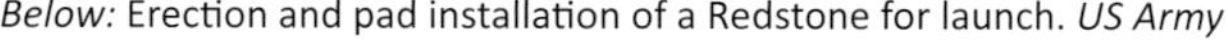

Below: Erection and pad installation of a Redstone for launch. *US Army*

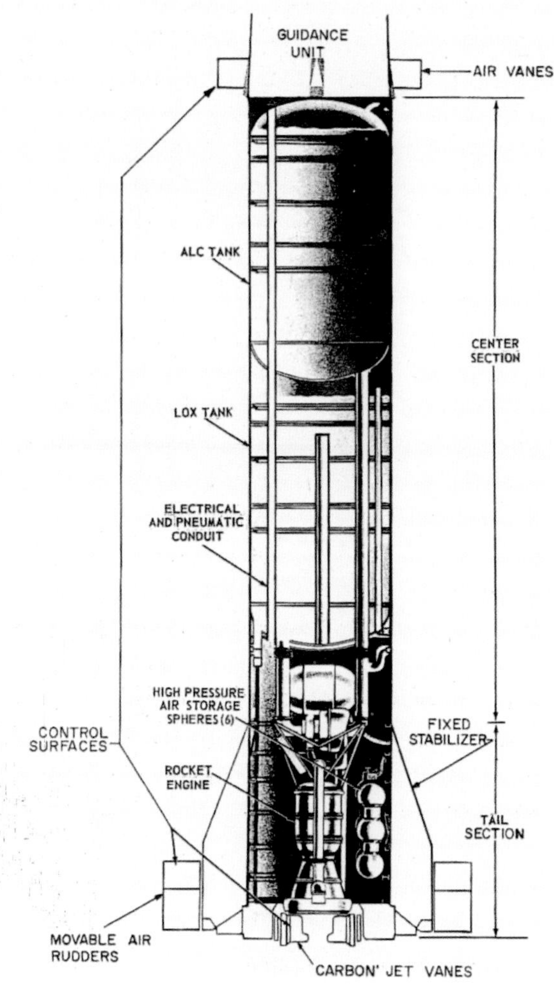

Above: Centre section and tail section containing the propellant and propulsion elements of the Redstone missile. *US Army*

went for operational use.

In mid-1955 the Army began directing personnel to Redstone missile training courses with the 217th Field Artillery Battalion. The first troop launch occurred on 16 May 1958 and two months later the missile was deployed overseas with the 40th Field Artillery Missile Group attached to the Seventh Army in Europe, which hosted the 46th Group establishing a fully operational capability from 25 April 1959. Seen in retrospect, Redstone was an interim step but an essential bridge between the A-4/V2 and the Saturn launch vehicles, albeit one whose role was rapidly superseded by the two-stage, solid propellant Pershing.

Leap to Jupiter

While the development of the rocket motor for the Redstone missile owed much to the A-4 and a development programme at North American Aviation in concert with tests in the Hermes series, the next evolution in missile and motor development was triggered by a robust and long-term effort on the part of the contractors. From the end of the war, that effort within industry began with a conscious decision to invest in the future of rocket motors for increasingly more capable missiles. At the end of World War Two, NAA had set up its astrophysics laboratory for studies on what would be possible and what was required for that need. Development had been divided into six phases, each building on its predecessor and absorbing studies in high speed aerodynamics, heat transfer, propellant chemistry and associated fields.

Phase I had involved a lot of learning from running A-4 engines, followed by Phase II which integrated existing German technology and design trends with new configurations, although that did not involve firing trials. Phase III had resulted in the engine developed for the Navaho cruise missile and which, when the specification for that was significantly upgraded making that motor superfluous, was employed in Redstone. Phase IV was underway in 1951 and was a singular step away from the A-4 heritage line, with a new double-wall chamber, a high-speed turbopump and a bipropellant gas generator fed from the primary propellants. In addition, the 75 per cent alcohol fuel was changed to a 92.5 per cent mix with water. The engine that ensued had a thrust of 533.7kN (120,000lb).

Changes to the specification for the Navaho missile prompted development of the Phase V programme which produced a dual thrust chamber motor fed by a common gas generator and delivering a thrust of 1,076kN (242,000lb), first fired up in September 1953 and imported to the G-26 variant of Navaho. In an engine designated LR71-NA-1, the lightweight construction incorporated brazed tubes to cool the combustion chamber with a more even heat distribution of heat. In January 1956 NAA tested a triple-thrust chamber motor with three turbopumps and a single gas generator which produced a thrust of over 1,800kN (405,000lb) of thrust for the G-38 design iteration of the Navaho cruise missile.

The development cycle for the post-Redstone evolution began in March 1952 with NAA's Rocket Engine Advancement Program (REAP) which aimed to evaluate LOX/hydrocarbon propellants for large rocket motors. Equally, it sought detailed evaluation of high speed, gear-driven turbopumps which incorporated inducer-impeller combinations and improved thrust chambers, instrumentation and starting techniques. It was both a research and development programme as well as a series of analytical evaluations of broad-based concepts with the aim of channelling this information into a development effort which would serve the needs of the Army and the Air Force. Needs which were growing each year and generating requirements above and beyond existing capabilities.

Of only peripheral interest to the story of the origin of the rocket motors which would eventually power the Saturn launch vehicle, it is helpful to note the context in which these developments were made possible. As noted earlier, major decisions were made by the Department of Defense in response to a technological breakthrough with miniaturisation of nuclear warheads during the enhanced geopolitical tensions described previously. While the Eisenhower administration had been keen to constrain military expenditure and sought to cut the defence budget, the Cold War was intensifying and there

Right: This grainy photograph of the first Redstone missile launch on 20 August 1953 marked the beginning of a series of rocket and missile developments that would lead directly to the Saturn I. *US Army*

Redstone A-6 engine

Right: Basic characteristics of the A-series engine developed for Redstone incorporated the showerhead injector assembly for delivering LOX and fuel to the combustion chamber, elongated and smaller than the spherical chamber on the production A-4. These improvements were already in development at Peenemünde when the war ended. *US Army*

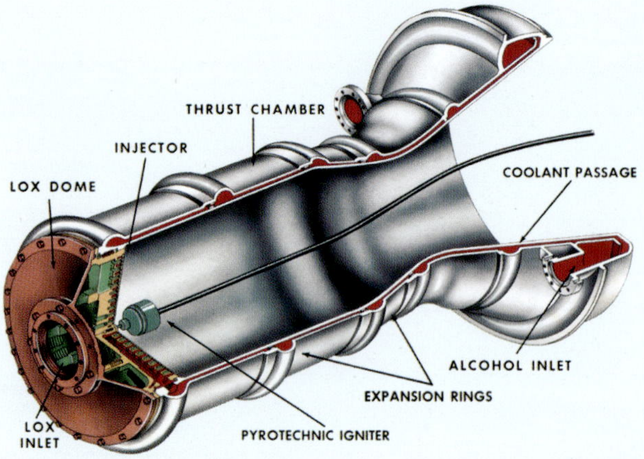

Below: The Redstone A-6 engine surmounted with the single hydrogen peroxide tank for the gas generator driving the turbine and a diagram displaying its primary elements. *Author's collection/US Army*

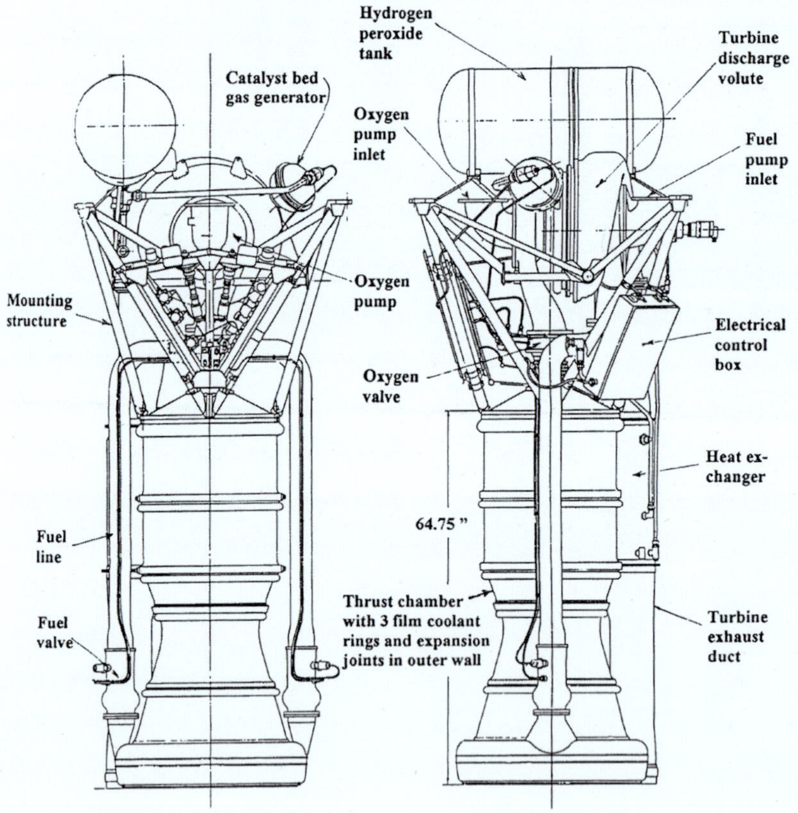

was an urgent need to develop a counter to threatened Soviet hegemony, as perceived by the United States. While missiles and rocket motors could furnish the military with battlefield and theatre weapons, intercontinental weapons were constrained by the magnitude of the thrust output required and by the weight of the early thermonuclear weapons. When those warheads suddenly became lighter and technology development closed the gap on performance needed to match requirements, the ICBM suddenly became feasible.

Following a series of study and development contracts with Convair beginning in the late 1940s, and in response to the detonation of Russia's thermonuclear weapon in 1953, on 14 May 1954 the Atlas rocket was instigated as a maximum effort to build an ICBM. A formal contract was awarded to Convair for Atlas in January 1955. Rocket technology had reached the point where it was possible to throw a thermonuclear warhead, with a mass now down to 1,680kg (3,700lb), from the United States to Russia. In September 1955 this was followed by a contract to The Martin Company for the Titan missile, a back-up to the controversial Atlas, which incorporated unique design characteristics that some believed to be a risk.

The broad development of Army and Air Force missile programmes is a long and tortuous story and is not for this book. However, the fundamental technologies and the management experience so gathered by this general expansion of missile programmes is fundamental to the story of the Saturn rocket and to the rocket motors that were selected for it. It is

Redstone A-7 engine

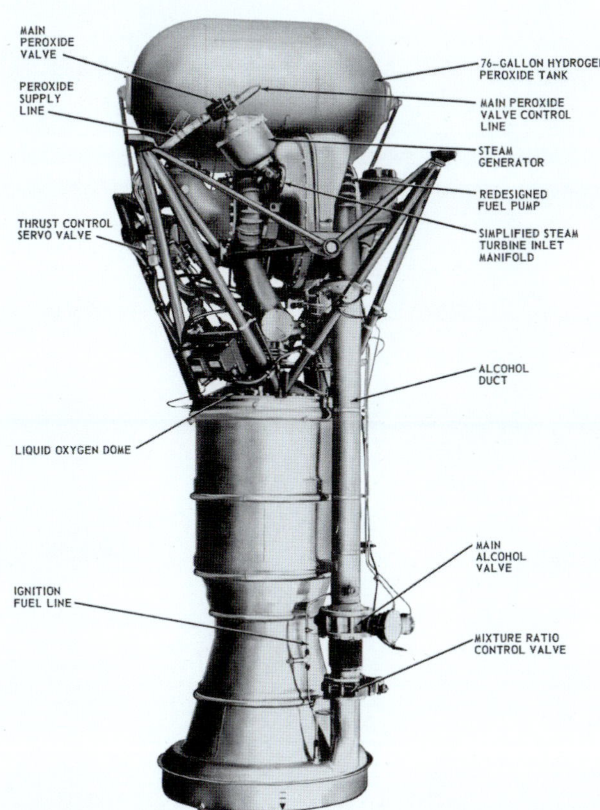

Above: Refinements and improvements brought the A-7 engine into production for the definitive Redstone missile. *Rocketdyne*
Below: The flow schematic for the hydrogen peroxide, gas generator and turbine side. Note the single turbine for the two pumps. *US Army*
Below right: The pneumatic control system operated the valves for all propellant and engine operations including that of the gas generator and its hydrogen peroxide supply. *US Army*

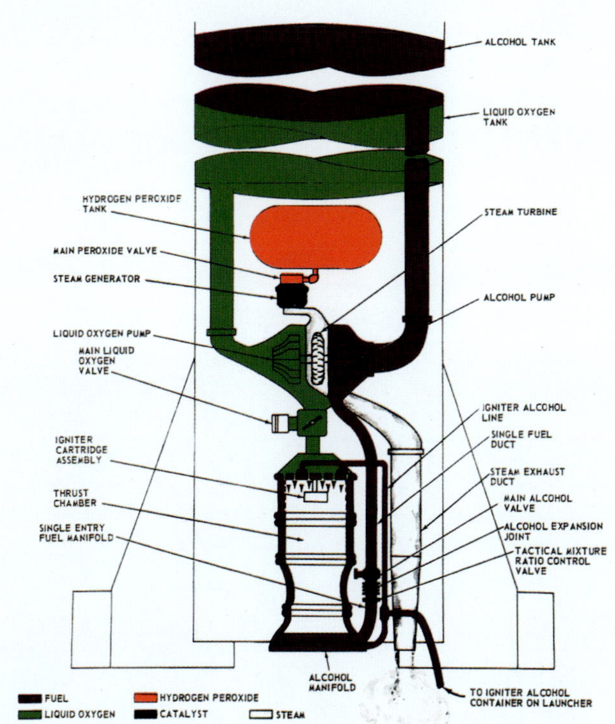

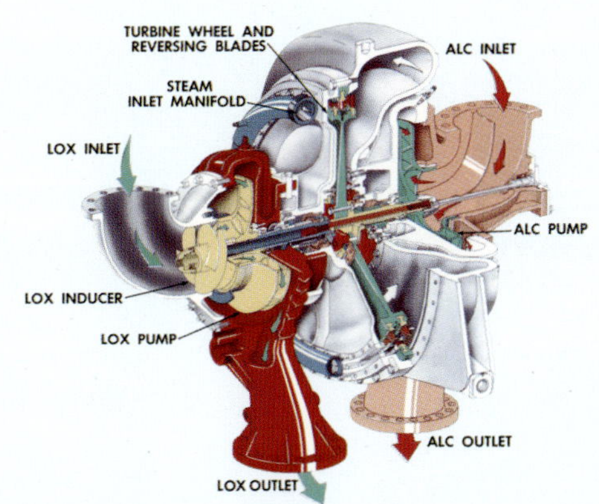

Above: A single turbine wheel for both fuel and oxidiser sides of the propellant delivery system simplified the A-4's duel-turbine set-up.

Below: The steam generator employed hydrogen peroxide to create heat and drive the turbine. *US Army*

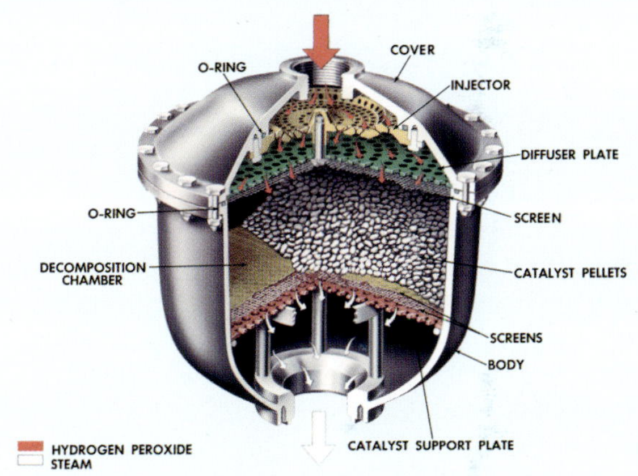

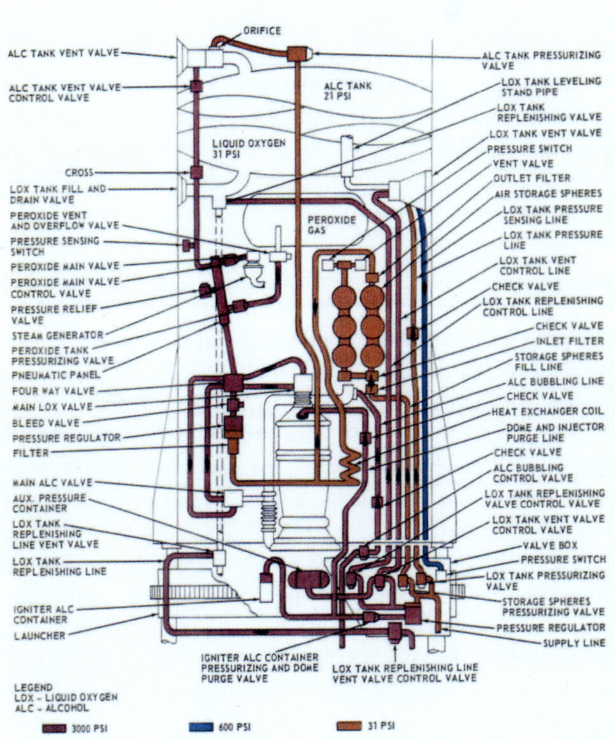

Redstone A-7 engine mounting

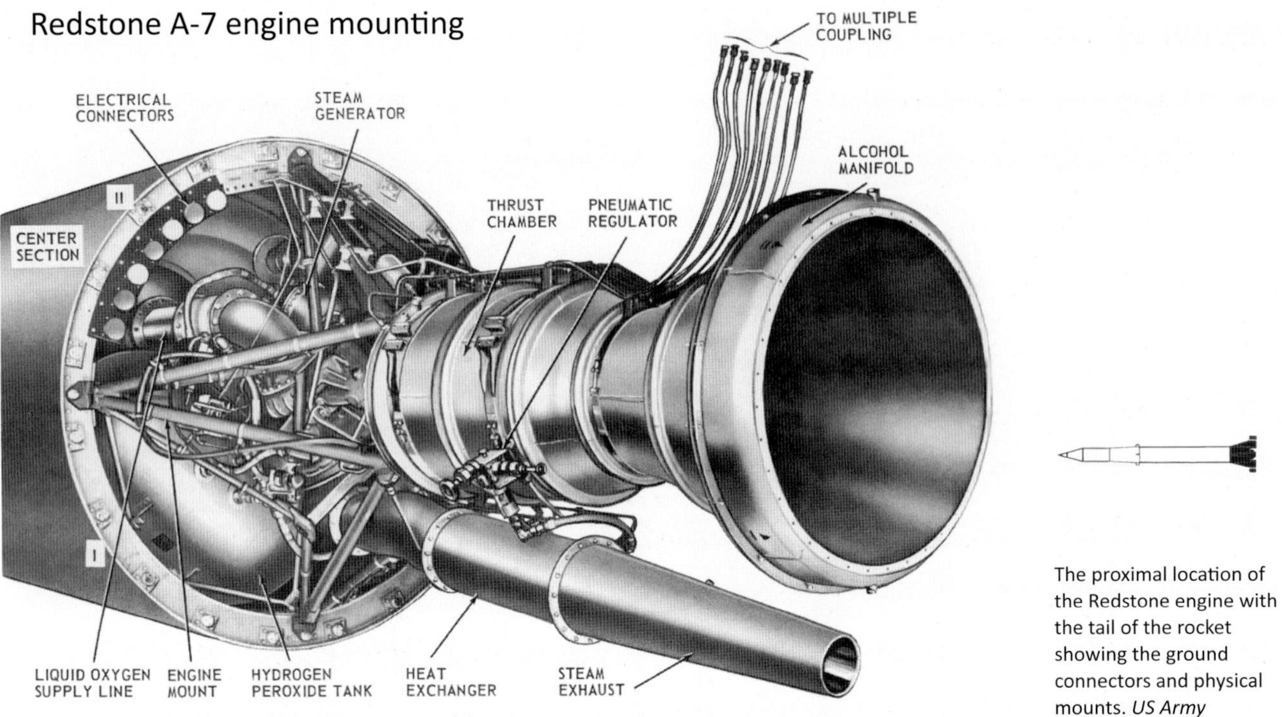

The proximal location of the Redstone engine with the tail of the rocket showing the ground connectors and physical mounts. *US Army*

also crucial to understanding the parallel and complementary, but not always identical goals and aspirations of the von Braun group and the Army. That story spans events spread across the period 1953-1956, arguably the most decisive in the history of military rockets, of the Army's tactical missile programme and of the work of the Germans at Redstone Arsenal.

Nine months after President Eisenhower was inaugurated, in October 1953 Trevor Gardner, the Assistant Secretary of the Air Force, put together the Strategic Missiles Evaluation Committee (SMEC) to examine the state of Army, Navy and Air Force missile projects including the Redstone. Its members included a wide range of the nation's leading experts in the field and, chaired by Dr. John von Neumann, it presented its confidential report on 10 February 1954. Known by its codename Teapot Committee, the main drive within the study leaned toward the Air Force and most of the testimony heard came from that service. The Air Force had an unflinching certainty about its own command and control of long-range missile programmes and disregarded the Army missile efforts under von Braun as peripheral to the strategic and tactical defence of the United States.

In the years since the end of the Second World War, the Air Force had experimented with several cruise missile projects and was sceptical regarding the possibility of a ballistic rocket achieving the required accuracy. But, with the development of thermonuclear warheads possessing a far greater radius of destructive force, blast and heat, plus the demonstrated improvements with inertial guidance, meaningful target damage that could be incurred now came within the predictable capabilities of a long-range missile such as an ICBM.

Moreover, the close working relationship with prime rocket engine manufacturers such as North American Aviation convinced the Air Force that it was possible to accelerate plans for such a capability.

This ran parallel with the desire to outpace the Army's von Braun team and follow a different path, one which brought furious and vociferous disagreement between the two services. As related elsewhere, Convair had been given a contract to develop a precursor test missile under the designation MX-774 and three test flights did take place from White Sands Proving Ground in 1948. But the minimum size of the smallest atomic bomb was greater than the projected lifting capacity of any long-range rocket with the technology then available and the project was quickly cancelled. Convair got the contract for the 1948 missile tests after presenting the Air

Right: Rival rocket man: Convair Technical Director Karel 'Charlie' Bossert, father of the Atlas ICBM. *New Mexico Space Museum*

Force with a rocket design from its rocket engineer Karel Bossart incorporating innovative solutions to the performance problem.

Rather than stage a rocket in sequence, Bossart proposed placing a single 'booster' engine either side of a 'sustainer' engine and igniting all three engines on the pad in parallel. Propellant would be provided from fuel and oxidiser tanks, one above the other, which formed the external skin of the missile. This was so thin that it had to be kept pressurised to prevent it collapsing under its own weight, either filled with propellant or a nitrogen gas. This approach was designed to minimise the dry weight of the missile. At a certain altitude the two booster engines, connected by a thrust harness passing around the central sustainer engine, would be jettisoned, reducing the mass of the ascending rocket still further. It was a solution forced by the early stage of rocket propulsion and engineering design.

Convair received a second contract on 16 January 1951 for MX-1593, a specification which took advantage of advancements in missile engineering and stimulated by progress with the von Braun group as they moved their Redstone project from Fort Bliss and White Sands to Huntsville. MX-1593 envisaged a massive rocket powered by seven engines and dramatically oversized for the job. But the entire Bossart approach was anathema to the German rocket engineers and excited an untypical reaction. Livid at what they considered a totally implausible concept, marketed to a group not wholly in tune with the nuances of rocket engineering, they launched a lengthy and protracted criticism of the Air Force approach being taken. This was not the German way to build rockets and with the availability of increasingly powerful rocket motors on the horizon they saw no need to resort to such radical and high-risk design concepts to produce the necessary capability.

Moreover, in testimony to von Neumann's Teapot Committee they strongly objected to the proposed evolution of types, with the Air Force supporting an initial development of an ICBM before an IRBM. The Air Force claimed, in later testimony before the Senate Committee on Armed Services, that 'a lesser range missile would derive from the ICBM if the ICBM were carried to a conclusion'. They claimed that a more reliable IRBM could be obtained from de-sizing a massive ICBM which would be more than capable of carrying out the role of the less demanding requirement of a shorter-range missile.

Supported by the Army Redstone team, von Braun took vigorous exception to what he considered to be flawed logic, citing that a pressurised, thin-wall tank would never survive the buffeting a missile would receive as it ascended through the upper atmosphere, quoting research results from the WAC Corporal flights in the Bumper programme as evidence. But the issues never went away and the rift imposed by such different approaches would never be bridged. So it was that the von Neumann report created a division of approach which would haunt the von Braun team through the next 20 years and bring a consistently different approach to rocket design and engineering. It was one which would be manifestly effective – the German desire for robust, over-engineered, fail-safe designs providing a high level of reliability and one which

Redstone construction details

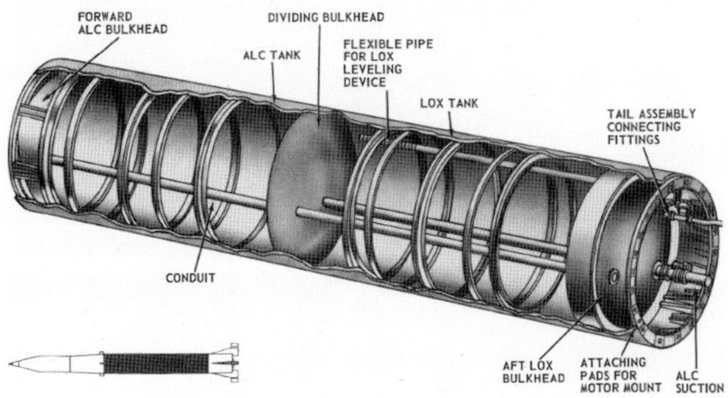

The main body of the missile, or centre-section, contained the separate fuel and oxidiser tanks, the engine mounting physically located into the tail unit, and the propellant transfer lines. *US Army*

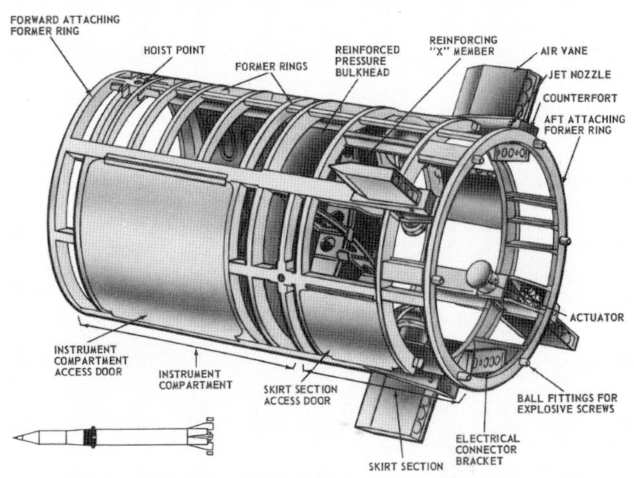

The aft section referenced in the workshop manual, being the forward section of the missile behind the warhead and nose section. *US Army*

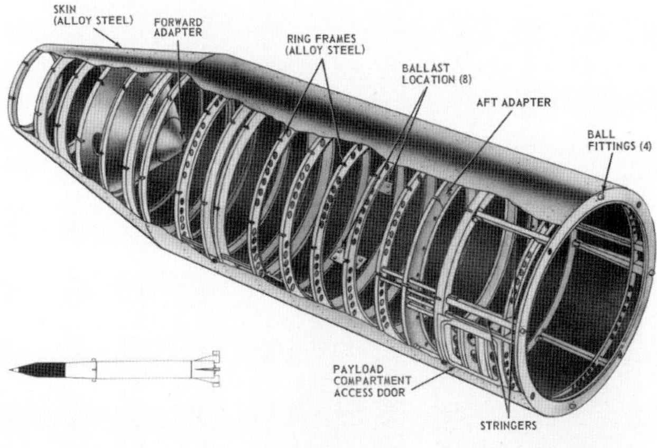

The warhead and forward section of the Redstone missile, showing the location of the frames and stringers. Integral with the design, the separate warhead was configured to detach prior to encountering the atmosphere. This helped with Redstone's adaptation to the space programme and numerous payloads. *US Army*

will become self-evident with the unfolding story of the Saturn launch vehicles.

The von Neumann committee had rightly pointed to the effectiveness of thermonuclear warheads in reducing the requirement for a cep of 460m (1,500ft) at the target. The Operation Castle nuclear tests of March 1954 demonstrate that to be the case, relaxing the required cep to a generous 5.6km (3.5mls) for the MX-1593 missile, soon to be known as Atlas. The findings echoed similar conclusions from a report by the RAND Corporation's Dr Bruno W Augustine, released on 8 February 1954 from data that had also informed the Teapot Committee. Similarly, there was an arguably more influential analysis of the defence posture which was both instigated and delivered to the highest levels of government and that provided opportunity for the Army to mobilise a defence in claiming its own role with an IRBM.

It originated in 1954 when the US State Department became aware of an acceleration in the number of Soviet facilities supporting rocket research and testing of ballistic missiles based on indirect information from the CIA and from the intelligence-gathering assets of the armed services. It came after the detonation of a Soviet thermonuclear bomb on 12 August 1953 and while the expansion of US nuclear arsenal was accelerating with deployment of tactical nuclear weapons carried by fighter-bombers and a range of atomic warheads for tanks and artillery. Responding to this evolving situation and the potential for progressing with a truly long-range, intercontinental missile, on 27 March 1954 an assessment was sought from a 42-member group known as the Technological Capabilities Panel (TCP) chaired by James Rhyne Killian, Jr, President of the Massachusetts Institute of Technology (MIT) since 1948 when he was Editor of MIT's Technology Review.

In its report dated 14 February 1955, the TCP judged that while both superpower states remained vulnerable to surprise attack, the United States would have a superior warfighting capability until the end of the decade, beyond which the USSR would rapidly reach parity and, without further amendments to US strategic and tactical capabilities, possibly overtake the United States. Specific recommendations called for

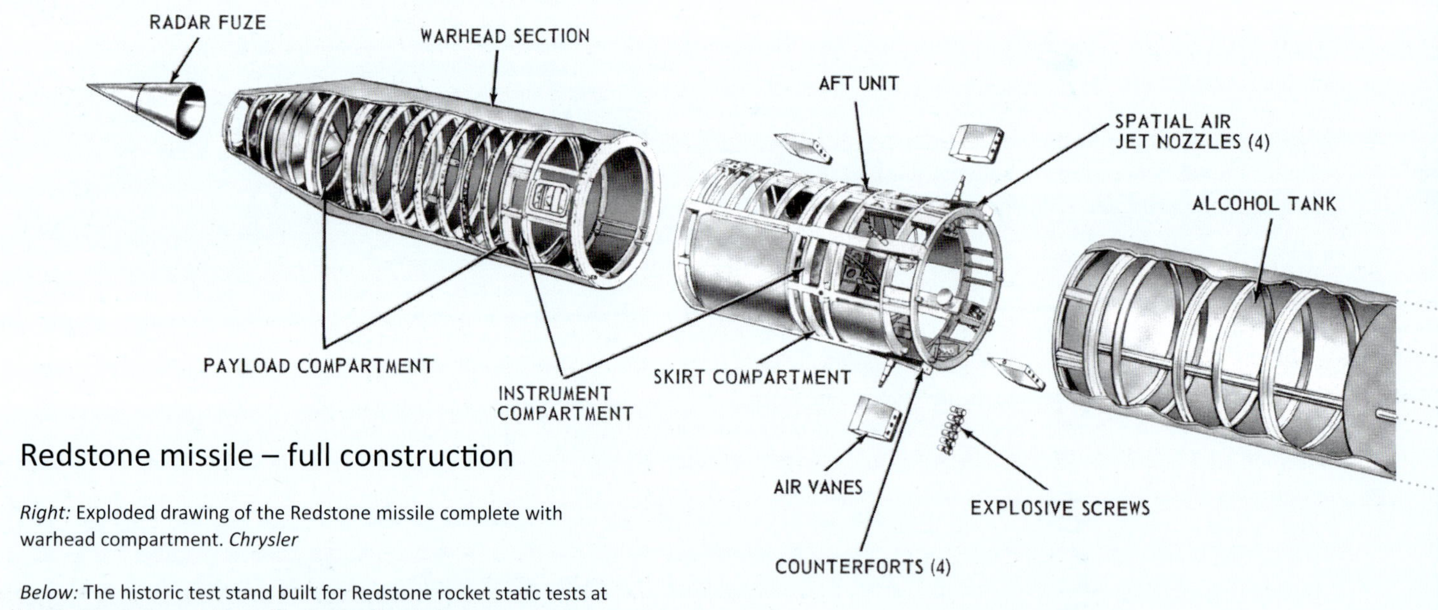

Redstone missile – full construction

Right: Exploded drawing of the Redstone missile complete with warhead compartment. *Chrysler*

Below: The historic test stand built for Redstone rocket static tests at the Redstone Arsenal, Huntsville. *Author's collection.*

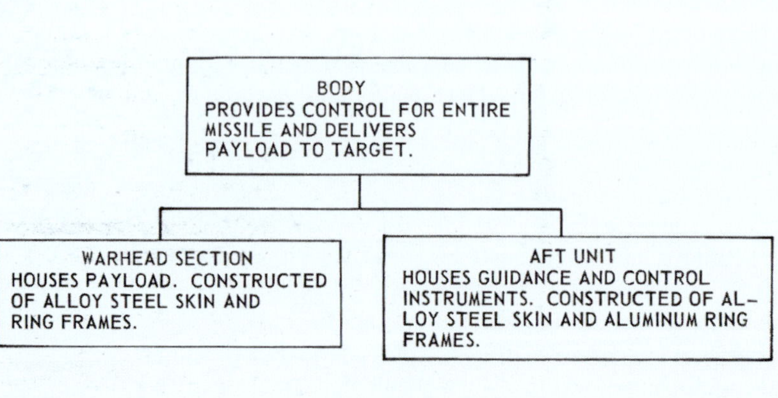

urgent priority on development of an ICBM capability with a range of at least 10,185km (5,330mls) and a ballistic Intermediate Range Ballistic Missile (IRBM) carrying a megatonne-yield warhead and with a range of about 2,780km (1,725mls) for both land and sea-based deployment.

The TCP had taken a broad view of its mandate and provided determinations across a very wide spectrum including the need for a continental defence network. This would stretch right across North America, to Greenland, Iceland and the Faroes. It would be implemented as the Ballistic Missile Early Warning System (BMEWS) and extend down into the UK with a terminal point at Fylingdales, Yorkshire. Many other recommendations were made, most of which would be adopted, but there was wording in the report indicating the value of communications satellites and a section inferring the value of spy satellites and of capabilities extending to the use of military missiles adapted to the role of satellite launcher.

The Redstone Arsenal group had been consulted over many discussions during the compilation of evidence and their input had been crucial to determining the path for rocket and missile technology over the coming decade or two. The TCP had specifically laid out a timetable of potential events – invoking both assets and threats – driving a chronology of capabilities which it recommended should be sought to protect the security of the United States. The Army was making enormous strides in the control and operability of tactical ballistic missiles and the von Braun group was central to that, including the outstanding progress made with inertial guidance already incorporated into the Redstone Short Range Ballistic Missile (SRBM). It needed to progress that and bring it up to the requirements of an IRBM but, as observed, the Air Force challenged that notion and opposed the Amy bid which, it said, unnecessarily duplicated work.

The Killian report added muscle to the argument for an IRBM and in a far more vigorous way than had the von Neumann or RAND reports. The Army upped its chances of convincing the political base through a report prepared mid-1954 by the Department Defense Technical Advisory Committee, chaired by the Deputy Secretary of Defense, Reuben Robertson, no less. This was to be more productively useful for the

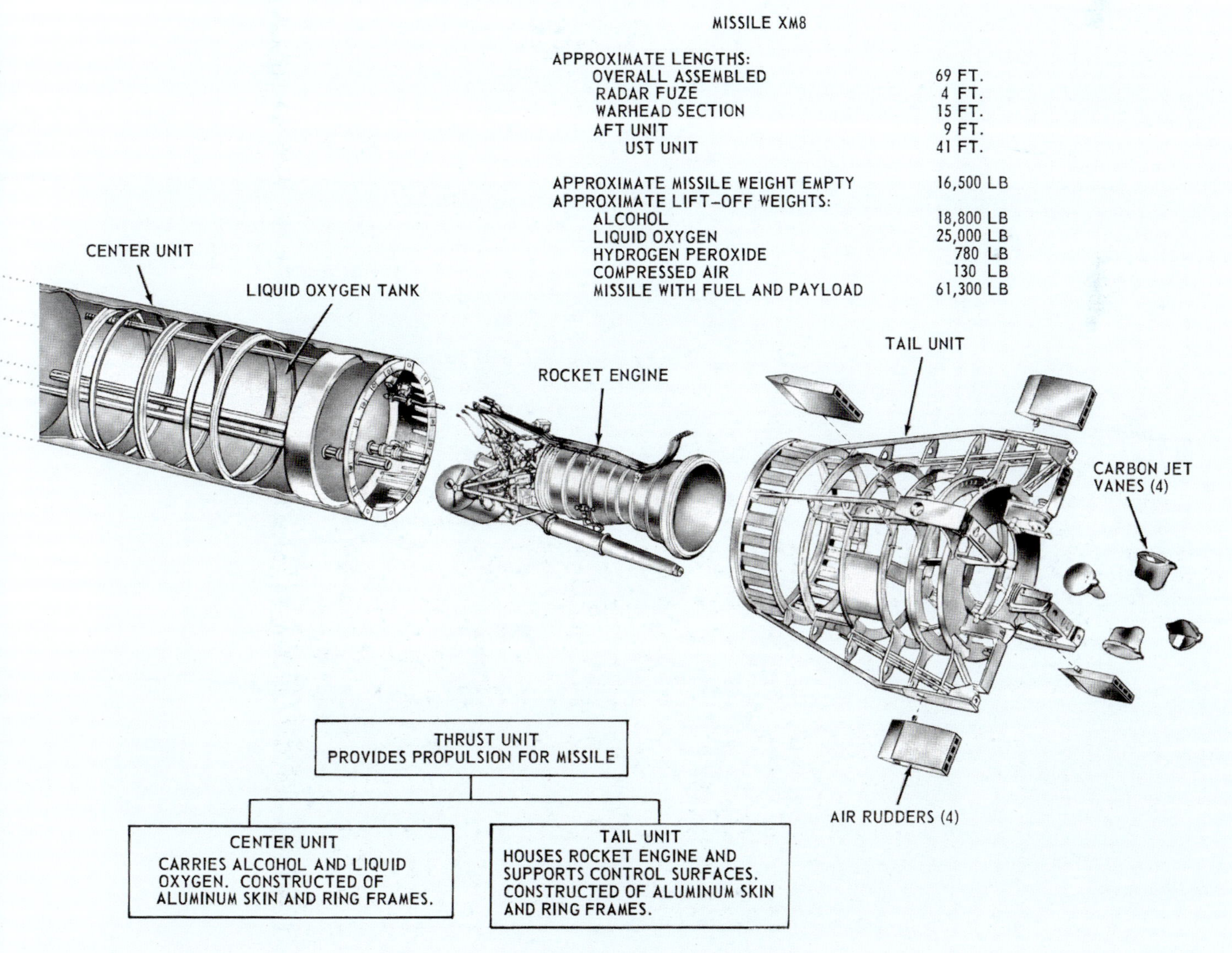

Army-Air Force tussle in that it allowed full airing of the differences between the ways each service would go about providing the nation with an IRBM. The Robertson committee reinforced the idea that both approaches were technically sound and that two, parallel approaches should be encouraged. When it meshed its own conclusions with those of the Killian report, which stressed urgency over technical niceties, it argued well for each service to have its own IRBM programme.

Foreign Affairs

When all these studies and conclusions were sent to the White House it greatly impressed Eisenhower with the common urgency each expressed. Secretary Wilson too concurred and the President decided that timing was of the essence and that 'the political and psychological impact on the world of the early development of a reliable IRBM would be enormous, while its military value would, for the time being, be practically equal to that of the ICBM'. Given the much shorter range of an IRBM, clearly deployment to a NATO country within striking distance of the Soviet Union would be essential to equalising the threat of an ICBM and placing the Soviets on notice that they were in equal danger.

There was also the belief that, by emphasising the presence of an IRBM in Europe and close to their border, the Russians would prioritise weapon systems which could counter the threat on their doorstep rather than placing all their efforts on striking the United States. But that was a nuanced and far too sophisticated reading of the Soviet response for either side to believe the other would desist from developing ICBMs as well. The suspected Soviet missile development received further evidence when during 1955 the National Security Council received intelligence reports that Soviet medium-range tests were already testing rockets with ranges of up to 1,450km (900mls). And, to consolidate Eisenhower's desire to have both Army and Air Force IRBMs, on 1 July 1955 Wilber Brucker became Secretary of the Army, a man with high ambition, tenacity and skilled advocacy for the von Braun team, which the President had already singled out as being far too valuable to risk losing. There were unsubstantiated rumours that the German rocket engineer had been offered a very large amount of money to head up the rocket division of a major aerospace contractor. Eisenhower was adamant that the momentum within the von Braun team must not be lost.

On 26 October 1955, the Army Chief of Staff announced to his subordinates a plan to develop a rocket with a range of 2,780km (1,725mls) which, under the classification of throw-distance, was an Intermediate Range Ballistic Missile

Below: Development of Intercontinental Ballistic Missiles (ICBMs) began with the Atlas, with an innovative way of producing what is referred to as a 1½ stage system in which two booster engines are shed during ascent saving weight and leaving the single sustainer to propel the missile to maximum speed and range. Here in this Convair production facility the paired booster engines can be seen in a row extending back along the line ready for rolling in and mating with the base of the main stage. The SM-65 Atlas made its first flight in June 1957. *Convair*

Left: As a back-up to the innovative Atlas, the SM-68A Titan rocket which had been contracted to the Martin Company in October 1955 incorporated a conventional two-stage configuration. A bold decision when the only proposed multi-stage pairing after the Bumper flights had been the Thor IRBM with Agena upper stage which would not fly until January 1959, one month before the first Titan I. *US Air Force*

(IRBM). It supported the Army concept of a battlefield of considerable depth, which justified possession of an IRBM as a tactical weapon, and the validity of this interpretation was argued before the Joint Chiefs of Staff. But there was disagreement over which service should have responsibility for a missile of such range. As we will see, the Air Force had not wanted an IRBM. Yet, after a coalition of circumstances it had begun development of such a missile in 1954. Named after the Norse god of thunder, Thor was placed in development with the Douglas Aircraft Company as Weapon System 315A (WS-315A) on 27 December 1955, one day before the Air Force was instructed to proceed with the missile.

Worth noting at this point is the length to which the Air Force went in trying to place development of an IRBM after, and not before its ICBM, a sequence noted earlier and which brought such vociferous outrage from von Braun. The Air Force had decided to develop Titan, a hedge against failure with the controversial Atlas configuration. Titan was a two-stage rocket and in July 1955 the Air Force concluded that it could be used separately as an IRBM, calculations showing that it would have a range of 1,482km (920mls), which could, it said be upgraded with little effort to give the desired range of 2,778km (1,726mls). The issue had been turned over to Space Technology Laboratories (STL) – an arm of the Ramo-Wooldridge Corporation – for analysis. STL determined that this was not a good way to go and that the Air Force should design a new missile for the role of IRBM and that it should be air-transportable.

Two features demand note. The Air Force was universally disenchanted with the notion of overseas deployment, having striven hard to find long-range strike aircraft that could reach their targets from the continental United States (at this time Alaska and Hawaii were not States of the Union). It was getting tired of the stresses on its operational readiness, on its resources and of demands on equipment such as a high-capacity transport system to achieve global mobility. It was intrinsically opposed to adding a system designed for thermonuclear war which would only be effective against an enemy by deploying to a foreign country. By developing an IRBM first, a missile it had not wanted in the first place, the Air Force was trapped into basing options involving a continental European country for it to get within striking range of the Soviet Union.

Secondly, the sequential development of missiles proposed by the Army was implicit to that service's longer-term view of supporting the defence of the United States with both an offensive missile system and a network of space-based assets which could be employed defending the United States. The Air Force, in 1955, was heavily committed to the notion of satellites and space systems in support of its offensive programme but with the convincing rationales argued by the von Braun team, much of the senior Army leadership – in both the uniformed and civilian ranks – looked to the Redstone Arsenal to provide a capability which was of little interest to the Air Force. Witness the impending Project Horizon study (about which more later) which put the Army at the core of a space-force with bases on the Moon and a cislunar transportation system using a network of heavy-lift launch vehicles. This had long been a dream of von Braun and his team and it converged with the way the Amy was looking to carve for itself a unique niche in the second half of the century.

In testimony to the Holifield Committee, set up under congressman Chet Holifield to examine civil defence in the

Right: The Air Force Western Development Division opens for business with the Ramo Wooldridge company its chief analytical tool and civilian advisory body. *DoD*

US, Secretary of the Army Wilber Brucker declared 'the fact that the Army pressed on had a great deal to do with propelling the other service', while the Commander of Air Materiel Command, General Clarence Irvine, opined to Representative George Mahon's Appropriation Sub-Committee that 'There is nothing healthier for this country than inter-service rivalry, so called. I would say if anything it helped the Air Force work like hell on Thor, it is the fact that we knew the Army was clawing at our backs'.

In receiving the instruction to develop Thor, the Air Force had been made aware that time was of the essence, increasingly worrying intelligence reports made it plain that Soviet missile development was accelerating and would soon have operational missiles capable of threatening all the cities in Western Europe. Conventional practice dictated development of the flight hardware first (be it aircraft or missile) followed by development of all the ground support equipment and operating hardware when the definitive flight vehicle had been declared ready. But if applied to the IRBM that would add five years or more to the programme and to save time the Air Force instituted the concept of 'concurrency' in which both elements went in parallel and not in sequence, something the Army had already introduced with Redstone.

In addition, relying on the proven performance of Douglas, the Air Force had the first Thor missile fabricated on production tooling and thus avoiding the 'prototyping' stage. Both Douglas and the Air Force funded a static test facility at Sacramento, California on facilities leased from Aerojet General Corporation so that the various elements of the propulsion elements could be tested without involving a missile. As with von Braun's team, engineers working the Thor programme were required to 'invent' technology the specification for which had not been written when the programme started and to do so on a pre-set timeline. It also required development of the missile and its ground handling equipment compatible with air transportation using C-124 Globemaster II and C-133 Cargomaster, existing transport aircraft built by Douglas.

This approach saved large amounts of time, money and development effort and had been roundly pushed by (the then) Brig. Gen. Bernard Schriever himself, commander of the Western Development Division (WDD), which had been set up on 1 July 1954 out of Air Research Development Command (ARDC). It had been formed specifically to manage the slowly evolving ICBM programme and the fast-paced IRBM Thor programme. It was re-designated as the Air Force Ballistic Missile Division (AFBMD) of Air Materiel Command (AMC) on 1 June 1957 by which time it was in full authority with the Titan I ICBM in addition to Atlas and Thor. As the AFBMD grew its space-related programmes, a potential conflict was resolved on 1 April 1961 when the organization was split into two: the Ballistic Systems Division (BSD) and the Space Systems Division (SSD). By this date Schriever was commander of the ARDC. A consolidation of the two brought unification on 1 July 1967 with the formation of the Space and Missile Systems Organization (SAMSO). On 1 October 1979, SAMSO was granted two headquarters, one for the Ballistic Missile Division and the other for the Space Division.

Left: Battery A of the 217th Field Artillery Missile Battalion, 40th Artillery Group prepares to launch a Redstone missile from LC-5 at Cape Canaveral, flown on 17 May 1958. *US Army*

A key element in the decisions made as a result of the 1954-55 studies into national defence issues, thermonuclear weapons and missiles was the consultation provided by the Ramo-Wooldridge Corporation, which had formed as recently as September 1953. Both Simon Ramo and Dean Wooldridge had worked for Hughes Aircraft and left to form a consultancy specialising in defence, missiles and space with strong connections to the Air Force. Respected aerospace engineers in their own right, both were members of the von Neumann Teapot Committee. The Air Force placed them under contract to part-manage the technical side of the ICBM and IRBM programmes. Rapidly expanding their portfolio, in 1958 they merged with Thompson Products with whom they had been working, to form TRW, which grew quickly to build satellites and space probes. In some respects, albeit loosely, in the early days Ramo-Wooldridge provided for the Air Force some elements of the technical base which von Braun and his team provided for the Army. By that implication, the Army-Navy divide became formalised in 1955.

The formal instruction to proceed with development of an IRBM was issued by Secretary of Defense Charles Wilson to the secretaries of the Army, the Air Force and the Navy on 8th November 1955, While verifying that both the Army and the Air Force should each have an IRBM it mirrored the commitment made by the DoD to the Air Force having two ICBMs (Atlas and Titan). All four programmes were to receive the highest national priority but on strict instruction that the two IRBM projects should not hinder work on the strategic ICBMs. Implicit within the instruction was a clear declaration that the prime weapon would be Thor while the Army Jupiter missile would be secondary and, as such, potentially vulnerable to cancellation.

But there was a stipulation that the Army was to work with the Navy in producing a missile which could be used by either service, the Navy being desirous, albeit tentatively, of a missile which it could launch from the deck of a rolling ship. The Army was reassured by this as it believed it would be more difficult politically to cancel a weapon which was the only missile in play for the Navy. As events would determine, it was incorrect in thinking this. Development would be coordinated through a Joint Army-Navy Ballistic Missile Committee (JANBMC), but chaired by the Secretary of the Navy. Instrumental in expanding the internal and administrative structure to command all these programmes, announced on 22 November 1955 and effective from 1 February 1956, the Army Ballistic Missile Agency (ABMA) was set up under General John B. Medaris, with von Braun as its technical director.

However, the Navy was worried about aspects of the Army's IRBM in that the liquid propellant concept was not suited to shipboard operations; the slow lift-off of a liquid rocket following a few seconds delay for thrust stabilization (bringing enormous heat to bear on the deck), plus the greater length of a liquid rocket over a solid of equivalent capability, raised persistent concerns. Consequently, the Navy began talks with the Lockheed Missile Division for a solid-propellant solution to their requirement, a missile with a range equivalent to that of an IRBM but with a completely different concept. One which could achieve lift-off at the instant of ignition and one with storable propellants ready for immediate launch.

There was difficulty here because the efficiency of a solid propellant is much lower than that of a liquid propellant motor and the level of development of solid propellant missiles in 1955 was insufficient to provide the performance capability sought by the numerous committees which had looked intensively at the IRBM. There was a weight issue too, but that argued against the Army missile. Potentially, a solid propellant missile would weigh less than one-third that of a compromised, solid propellant version of the Army Jupiter IRBM. As to throw-weight, physicist Edward Teller convinced the Navy that a 1MT warhead could be developed for a new and lightweight solid propellant missile.

What was being fought for here was the efficacy of the Army position in a rapidly changing political and industry environment. During this shift from manned aircraft to push-button warfare, implied by the SRBMs, IRBMs and impending ICBMs, the public perception of the Army was one of 'slug-it-out-slowly' warfare rather than a quick-and-clean decisiveness of the all-pervading influence of the atomic bomb and thermonuclear weapons. Moreover, with its 'arsenal' approach, in which new weapons were designed and produced in Army facilities by specialised engineers and technicians who then went out to industry for manufacturing and assembly on production lines, there was no room for design bids. Skills of aircraft builders were threatened by cuts in defence dollars and new rationales for producing weapons.

The Army received increasing criticism from the aviation industry fearing for orders and jobs. Increasingly, defence dollars were going to missile projects and the expertise of the Army 'arsenal' system was feeding new technology companies emerging from the expanding electronics industry. The classic example being Ramo-Wooldridge and Space Technology Labs. The Air Force had no such problem because it wrote the requirements and decided the concept, leaving it to industry to come up with prototypes and projects. When the Army discussed a Jupiter production contract with Chrysler the issue was inflamed and the Aircraft Industries Association complained bitterly that work should have been put to competitive tender with aircraft manufacturers and not a car builder. The very fact that production of Redstone and Jupiter tanks was a simple production job for any steel-bender missed the fact that a shift was taking place which was unique to the Army and the way it worked.

All this was in plain sight for von Braun, who persistently lobbied for a greater degree of salesmanship in the Army's position within the political and the public arena: 'The Army has got to play the same game as the Air Force and the Navy', he said. Which had validity. The Air Force was proudly boasting of the capabilities of its new jet force and worked with Hollywood to project its fighter and bomber forces with support for big screen films and TV series and the Navy was boasting about its new ships and nuclear-powered submarines with equally unabashed pride. But the Army could find no way of turning tanks and rockets into vehicles conveying excitement and drama.

With the decision that priority for IRBM development had already gone to Thor, the Army was expecting to be squeezed out of the missile field altogether. It showed in the allocation of defence expenditure. In 1955, the top four listed defence contractors were Boeing, North American, General Dynamics and United Aircraft, representing more than 19 per cent of Pentagon dollars. Seven of the top 10, 11 of the top 15 and 13 of the top 20 defence contractors were aircraft firms with deep ties to the Air Force. Chrysler ranked sixth largest defence supplier during the Korean War but by 1955 it ranked 94th. It was therefore inevitable that the future of the long-range missile would probably be in the control of the Air Force. It had the support of industry, the Pentagon brass, the Chairman of the Joint Chiefs (Adm. Arthur W. Radford) and numerous politicians.

During this period significant advances were made in the specific impulse of solid propellants – the measure of how much thrust is obtainable from a given amount of propellant – and this challenged the initial design of a sea-launched Jupiter, designated Jupiter-S. This would have replaced the engine and propellant tank with a cluster of solid propellant rockets but was expected to weigh approximately 72,576kg (160,000lb) which was too heavy and unacceptable to the

Right: The Air Force took over operational deployment of the Jupiter missile and in January 1958 stood up the 864th Strategic Missile Squadron (SMS) at the ABMA, followed by the 865th and 866th later that year. Hopes that Jupiter could be the sole IRBM for the Air Force were dashed when Sputnik brought an urgent need to acquire the more troublesome SM-75 Thor. *DoD*

Navy. Tests performed by Keith Rumbel and Charles Henderson at the Atlantic Research Corporation, and further verification from Werner Kirchner at Aerojet-General Corporation, provided a solution. It emerged that a different solid propellant missile would provide the required performance within the size limitations of an internal launch tube of a nuclear-powered submarine.

The Navy asked the JANBMC for permission to develop its own missile and that was granted on 12 March 1956, followed on 11 April by a systems development contract awarded to Lockheed and Aerojet-General. All this was going on while Redstone Arsenal proceeded to refine the Jupiter-S concept, the independent Navy missile development considered by the Army rocket engineers to be a very high risk venture. In reality there was little development work on Jupiter-S apart from two test ships, the *Pass Island* and *Observation Island*, which had been fitted for test flights. But these would be reconfigured for tests in the Navy Polaris programme.

On 8 December 1956 the Secretary of Defense Charles F. Wilson acquiesced to pressure and authorised the Navy to delete the liquid propellant missile from its planning and to proceed with development of the solid propellant Lockheed missile, later named Polaris. In a stroke, this opened the possibility of deploying a solid-propellant missile in a submarine. Instead of being launched off the deck of a rolling ship, Polaris would be ejected from the launch tube of a submerged submarine by pressurised gas, its engine ignited only when the projectile broke the surface.

Great progress had been made with nuclear-powered submarines; now the submarine would get to support a third leg of the strategic triad after bombers and land-based missiles operated by the Air Force. This immediately brought dissolution of the JANBMC and the formation of the Army Ballistic Missile Committee (ABMC), chaired by the Secretary of the Army. With these decisions the die was cast: the Air Force would develop two ICBMs and a single IRBM (Thor) designed for operation from fixed base launch locations while the Army would field its semi-mobile IRBM. Meanwhile, during all these shifts in structural planning, the Army was progressing well with Jupiter.

Throughout the spring and summer of 1956 the Army began to suspect that there would be only one IRBM and that Jupiter would be cancelled. The alignment of targeting for both Jupiter and Thor was firmly in the hands of the Air Force – it was they, after all, who would have done the reconnaissance and intelligence-gathering to know what targets to hit so far behind the enemy border with Western Europe. But nothing is ever that simple. But the issue reached a discussion point when President Eisenhower, who earlier had championed the work of the von Braun team at Redstone Arsenal proclaimed that this requirement for target information 'puts you right square in the Air Force business'. But this was only part of inter-service rivalry for control of Jupiter, the President getting increasingly agitated by the sparring.

As regaled later, to bolster assurance of their broad capabilities, in 1955 von Braun had proposed a satellite launching capability using the Jupiter-C but that had been rejected; given the existing Project Vanguard programme supporting a national satellite programme for the International Geophysical Year of 1957-58. Proposals for the resuscitation of Project Orbiter began in the spring of 1956 when Army Secretary Brucker tried to get it aired in the Pentagon. Prudence prevailed and pressure at this sensitive time in Air Force-Army dialogue restrained the effort. Nevertheless, on 7 September, a modified Jupiter-C lobbed a Jupiter re-entry vehicle to an altitude of 1,100km (683mls) and a distance of 5,300km (3,295mls) downrange from Cape Canaveral. The news was deliberately suppressed to prevent an upset!

Above: Chairman of the Joint Chiefs Admiral Arthur W Radford favoured the propriety given to Air Force missile programmes such as Thor IRBM and Atlas and Titan ICBMs, raising concerns within the Army that they would receive scant attention for funds from Congress. *DoD*

Four days after the Jupiter-C flight, the Army issued two regulations asserting that their missiles were an integral part of their weapon inventory in a continuing bid to impress the Department of Defense and the White House with the value of the missiles to their bid for concurrency in high-tech weaponry. In seeking to wrap up business before he retired, Defense Secretary Wilson made a definitive decision to bring the matter of missile authority to a neat and tidy conclusion. On 26 November 1956, he decreed that the Army would not be allowed to develop missiles with a range greater than 322km (200mls), resolving the development issue. Although developed by the Army, the Air Force would take over responsibility for deployment of all IRBMs and from 30 June 1957 take over all subsequent funding.

There was one more issue to resolve. The Army had designed the Jupiter for mobile operation at the instigation of von Braun, who convinced the ABMA of that concept's enhanced survivability over a fixed-base launch site. It was especially sensitive to attack given the proximity to enemy airfields in the event of a war where bombers or enemy missiles could readily destroy a known site long before the missile

Rocketdyne S-3D / Jupiter IRBM

Above: A variant of the Rocketdyne LR79, the engine was badged as the S-3D for the Jupiter missile and while initial versions delivered a thrust of only 533.76kPa (120,000lb), the engine fitted to production missiles delivered a thrust of 667.2kN (150,000lb). *Chrysler/Heroicrelics.org*

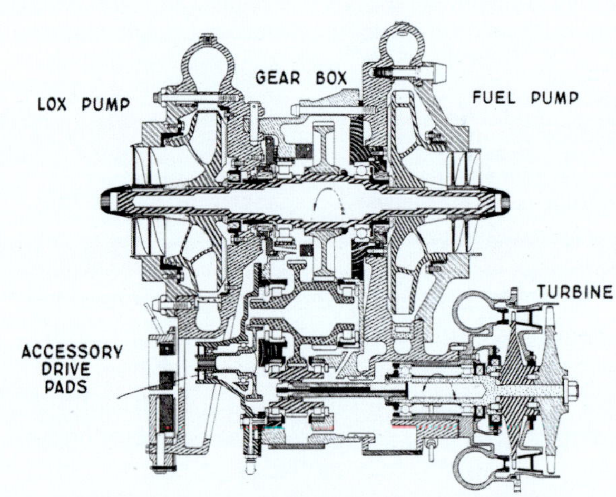

Right: A significant improvement over the turbine for Redstone, the two-stage turbine Mk III for the S-3D was powered by a spherical gas generator rather than the system utilising decomposed potassium peroxide in the Redstone. *Author's collection*

Below: Layout of the SM-78 Jupiter missile had been compromised by successive design iterations and by the changing user base but would retain the basic requirement as an IRBM along with the Air Force Thor. *US Army*

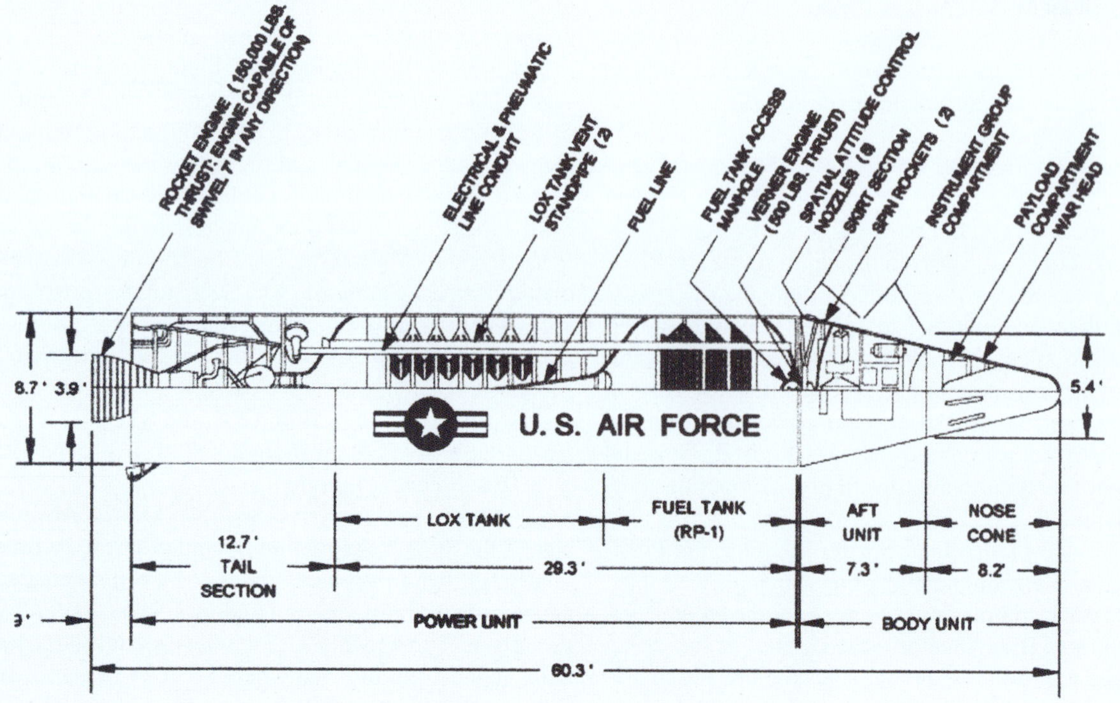

could be launched. But the Air Force disliked that – perhaps because it would require Army field troops to be responsible for its operation.

More specifically, during 1957 the British and American governments concluded a range of secret deals for the UK to gain access to the technology in the Atlas rocket motors for their Blue Streak IRBM, in reciprocation for agreement to deploy 60 Thor missiles on RAF bases in Britain. Blue Streak was to have had a range of 4,022km (2,500mls), deployed in silos spaced so that no single 20MT thermonuclear warhead could disable more than one missile and its launch centre. In accepting Thor, the RAF disliked the mobile launcher concept and urged the US to adopt fixed-base operations. In November 1958 the Air Staff asked Redstone Arsenal to redesign Jupiter for a fixed-base deployment as was the case already with the Thor IRBM.

However, unrelated to the development path toward the Saturn series and therefore not within the remit of this book, when the range limitation was imposed by the Wilson Memorandum, the ABMA was already working on an Army requirement for a medium-range missile for tactical deployment based on solid propulsion. In what was initially known as Redstone-S (for solid propellant), a successor to Redstone, development proceeded in cooperation with the Martin Company when the range limitation was rescinded on 7 January 1958 to a range of 1,287km (800mls). Nine days later it was renamed Pershing, a missile which would enter service in 1962, evolve through three generations and be retired in 1991.

The pace of the Thor development programme was fast and effective, but only because the Air Force had already invested heavily in the development of the Atlas ICBM, which was already in detailed design. After the award of the contract it took Douglas less than one year to deliver the first Thor missile to Cape Canaveral, due largely to the adoption of the concurrency development strategy discussed earlier and the use of critical elements from Atlas. These included the LR79 engine, the twin vernier motors, the gimballed directional control, the re-entry warhead and the inertial guidance system which had initially been developed by AC Spark Plug.

Jupiter Engine Technology

As proposed by the Redstone team, the essential requirement for the Jupiter missile was to throw a 907kg (2,000lb) warhead to a distance of approximately 2,400km (1,500mls) and for that an existing engine was proposed based on the North American LR89 already under development for the Atlas ICBM, in which it operated in pairs to provide the booster element. This engine would also be used on the Thor missile, for which it was designated LR79-NA-9. But the version for these IRBMs would be different to that for Atlas because of the longer burn time required. Much of the development was down to Samuel K. Hoffman, brought in to manage the rocket motor development operation for North American Aviation. It was a vital development for the evolution to the Saturn launch vehicle and as such is worth noting here.

On 7 November 1955 the Propulsion Center was renamed Rocketdyne as a separate arm of NAA with Hoffman as president of that division, along with the establishment of three new and independent divisions: Autonetics; Missile Development; and Atomics International. Most people understood what a rocket was but when asked what 'dyne' meant, the general public was told that it was the Greek word for 'power'. To the modern engineer the dyne is an outdated SI unit replaced by the newton, which signifies the force acting on a mass of 1gm that accelerates it by 1cm/sec^2. The Air Force pressed Hoffman to switch rocket motor propellant from the alcohol/water mixture to JP-4, a coarse-grade kerosene (paraffin) for jet aircraft, effectively a hydrocarbon fuel mentioned above. Under the general umbrella of the REAP research and development work, special standards and chemical specifications were set for kerosene identified for rocket motors and this was known as RP-1. The switch proved productive with many innovative design concepts and engine configurations studied.

Uniquely badged for the Jupiter IRBM as the S-3D, this variant of the LR79 was crucial to the eventual development of the H-1 for the Saturn rocket and was on the critical path which made that engine possible. In the late 1950s, there were no computer models of combustion chamber pressures and fluid flow patterns with density curves in different parts of the conduits. Instead, empirical design led to the countless successes which were achieved with several small groups appointed to work out solutions, test, possibly fail, sometimes succeed, but always to press on until a workable solution was invented. Engineering, like scientific experimentation, learns equally from failure and success. Because the main engine variants for Thor and Jupiter were developed from the LR79 there was some degree of subsidy which allowed the IRBM motors to emerge as a branch line from the primary development path for the Atlas motors.

In the configuration for Jupiter missiles, the S-3D had a thrust output of 667.2kN (150,000lb) with a burn duration of 2min 58sec. In addition to its increase in thrust output, it was different in two significant ways: it had a fuel-burning gas generator and it had gimbal capability for directional control of the missile. The first engine of this type was delivered by Rocketdyne in June 1956 for static testing but that had a thrust of only 600.5kN (135,000lb). Firing trials would take place at Rocketdyne's Santa Susana site, at a Douglas facility in Sacramento and at the Leuhman Ridge, near the dry lakebed site at Edwards Air Force Base in the Mojave Desert. It used a single thrust chamber and a single turbopump controlled through an electrical circuit and operated with pneumatic and hydraulic systems. There was no thrust control and a combination electronic-hydraulic system controlled thrust chamber pressure, which held it steady through a closed-loop servo system.

The thrust chamber was fabricated from formed nickel tubes which were brazed together and supported by external steel bands and rings. The injector and chamber pressure was 3,620kPa (525psi) with RP-1 as fuel and a regeneratively cooled thrust chamber, the engine displaying a specific impulse of 247.5sec versus 252sec for the MB-3, the Thor rocket motor. The multi-ring, flat-plate injector with a common set of orifices fed a central spray disc for a proportionately symmetric ignition flame before mainstage ignition. A pyrotechnic igniter screwed into the centre of the injector spray disc provided ignition from the independent feed.

Thrust vector control was achieved by gimballing through a 14deg conic angle by means of a cross-bearing block mounted on top of the injector dome. This was a significant departure from the Redstone system, which used exhaust vanes and air rudders on the tail fins, much like the A-4/V2. To achieve gimballing, flexible stainless-steel sections in both LOX and RP-1 feed ducts were necessary, with hydraulic actuator arms located between outriggers on the thrust chamber and jibs on the fixed thrust frame. Pneumatically operated butterfly main propellant valves were installed in the main ducts adjacent to the chamber inlet ports.

The two-stage turbine was driven by fuel-rich gases produced in a spherical gas generator, differing from the system on the Redstone engine which utilised decomposition of potassium peroxide to create steam. The combustor for the gas generator incorporated a simple injector and an extra fuel spray nozzle at the centre, with mixing assisted by a cylindrical basket inside the uncooled combustor where gases were heated to 649°C (1,200°F). The two centrifugal pumps were geared down to a ratio of 4.88:1 at the turbine rpm with two accessory drive pads used for the hydraulic pump and for actuation of the gimbal drive. The lubrication system employed pneumatic pressurisation to feed the bearings and gear-cooling jets.

The pneumatic system consisted of nitrogen held at 20,685kPa (3,000psi) in a separate tank, reduced to a normal working pressure of 5,240kPa (760psi). The gaseous nitrogen-operated valves provided purging and pressurised the lubrication system and the turbopump gearbox. A four-way solenoid valve operated the main LOX and ignition-stage RP-1 fuel valves and a second solenoid valve operated both main fuel and gas generator valves, both attached to a single manifold.

The flow of propellant began by opening a double-blade, single-pneumatically activated valve which was an integral part of the combustor, flow rate being controlled on the LOX side by a throttle valve operating to maintain a constant pressure in the main combustion chamber. Fuel flow was controlled

Below: Jupiter East and West Static Test Stand at Redstone Arsenal for the engine qualification firings. *Author's collection*

Redstone–Thor–Jupiter engine evolution

The refined and simplified design configurations for the Thor and Jupiter rocket motors as an evolution from the Redstone motor which was itself a development of the German A-4 motor. *Rocketdyne*

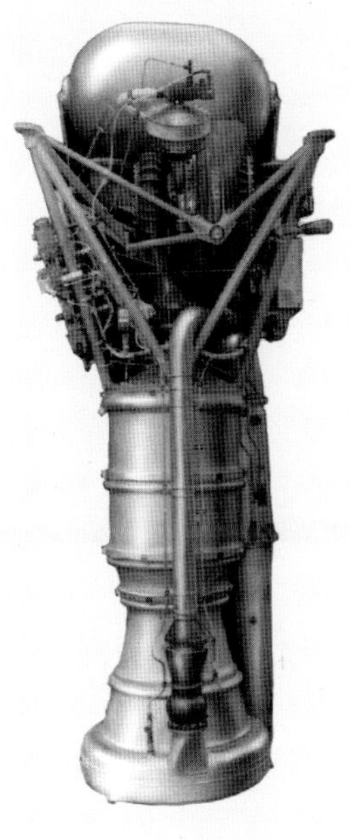

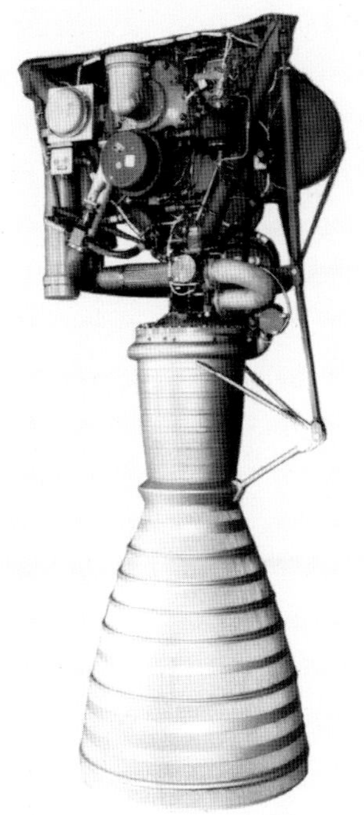

by a fixed, calibrated orifice. Ignition was effected by three pyrotechnic charges mounted on the combustor, the gas generator being started by propellant flow from two spherical, pneumatically-pressurised, ground-mounted tanks which also fed the main engine combustion chamber with igniter fuel. As the chamber pressure increased to 80 per cent of steady-state operation the gas generator became self-sustaining. But the generator served a secondary function, providing roll control during powered ascent by swivelling the turbine exhaust 25deg either side of the longitudinal axis of the missile as required, removing the need for vernier engines to counter roll tendency.

The hydraulic system maintained thrust chamber pressure at a predetermined level within a tolerance of +/- 1 per cent, control achieved by sensing chamber pressure and comparing it to the magnitude of a predetermined value and controlling LOX flow to the gas generator. The combustion chamber pressure was monitored by a pressure transducer which produced an electrical signal proportionate to the magnitude of the pressure. The output signal was summed with a reference voltage proportional to the predetermined pressure level. Errors were amplified by a servo amplifier producing an output proportional to the magnitude and the direction of the error signal. This would reposition the LOX valve by means of the hydraulic servo valve, channelling the LOX flow to the gas generator in an appropriate direction to reduce the error to zero. This closed-loop control mechanism was made possible because combustion chamber pressure was an implicit function of turbine power which was, in turn, a function of the LOX flow to the gas generator.

Electrical power for engine operation consisted of a 28-volt dc supply with all signals coming via an integrated missile tail distribution box through a harness to operate valves and to receive position indications from micro-switches on the main valves. This provided several pre-ignition and pre-launch safety and sequencing functions. A 115-volt, 60Hz ground power supply operated some heaters to prevent components freezing due to their proximity to the LOX at -252ºC (-423ºF). The 115-volt 400Hz power supplied to the thrust control computer amplifier was tapped directly from the electrical supply system on the missile.

Great attention had been paid to making engine operation and reliability as secure as possible and the start sequence was engineered into the S-3D via a step-ladder system whereby one successful operation activated the next. This eliminated a complex process of sequential commands unsuited to field operation by rocket troops and differed from the Redstone, and the A-4/V2 preceding it, in that regard too.

It began with pressurisation of the missile's propellant tanks to 276kPa (40psi), at which point a signal was given to pressurise the ground-mounted RP-1 and LOX start tanks. On receiving that command a circuit closed which fired a pyrotechnic igniter in the main chamber. Burn-through links in the igniter itself signalled the main LOX valve and the igniter RP-1 valve to open and the resulting pilot flame signalled the gas generator igniters to fire. That burned through a link wire stretched across the thrust chamber nozzle which signalled the gas generator igniters to fire. Similar burn-through links signalled the opening of the main RP-1 valve and the blade valve in the gas generator. This triggered the flow of fuel and oxidiser from the ground start tanks which began to flow under a pressure of 295kg (650lb) into the gas generator where it burned, starting the turbopump spinning. The pump speed accelerated rapidly and as the RP-1 filled the main thrust chamber cooling jacket, the main flow of fuel arrived at the injector at a high pump speed.

Meanwhile, the flow of liquid oxygen had also increased

and the transition to full thrust took place under high flow rates and rising pressure in the main combustion chamber, while a portion of the propellant was diverted from the high-pressure ducts to the gas generator. As the pressure of the propellants increased to exceed that of the pressure from the ground-start system, the engine overrode that supply to begin self-sustaining combustion, a process referred to by rocket engineers as 'bootstrapping'. As with Redstone, the engine would induce lift-off when the thrust exceeded the mass of the Jupiter missile and the process became self-sustaining all the way to cut-off.

Shut-down was activated by simultaneously removing power from the solenoid control valves. In a programmed sequence, this employed pneumatic restrictors in the vent ports on the control side of their respective solenoids to close the LOX valve shortly after the gas generator valve with a slight time lag before closing the main fuel valve. The time taken for the LOX valve to close was the prime determinant in control of the cut-off impulse, the later closing of the fuel valve having little effect. The speed in which the LOX valve closed was determined by the maximum hydraulic surges withstood by the high-pressure and low-pressure ducts.

Deployment

Operationally, while there were similarities in the range and performance requirement between the Air Force Thor and the Army's Jupiter IRBM, the two services each saw the function of their respective missiles very differently. The Air Force accepted that Thor would have to operate from fixed bases, targeting enemy airfields to minimise the reciprocal threat to their own bases and to degrade the operability of enemy air forces by destroying the facilities from which they operated. In this way, the Air Force saw Thor as a supplementary support system which aimed to react within a 15-minute warning window before the fixed bases could themselves be targeted.

This placed great emphasis on the detailed design of the entire system and, as we have seen, the S-3D was a product of the MB-3 (Thor) engine which had itself derived directly from the dual-mounted LR79 engine for Atlas. But it was the rapid-reaction requirement that had driven a completely new philosophy in missile operations; instead of proceeding through a series of separate steps between ignition and lift-off inherited from the V2 days – with ignition, pre-stage and mainstage, pausing at every step for an operator to give the OK to the next – there would be a need to put the steps into a naturally evolving sequence, a continuous flow. This had enormous impact on transforming rocket motor design and played a major role in preparing the S-3D for the job it would have to perform with Saturn.

As with most missile programmes, the development of Jupiter was fraught with changes induced by military and political shifts in the balance of priorities and roles, in addition to the technical and engineering learning curve that naturally attaches to these complex requirements. The tortuous story of Jupiter as a weapon system – and Thor for that matter – is not for this book but technical issues with the propulsion system certainly are. Initially, the S-3D produced a

Above: Deployment of Jupiter, together with Thor, was a consequence of Cold War response to Sputnik and the acceleration of Soviet missile development, without which these outdated missiles might never have been sent to the UK, Italy and Turkey. *USAF*

thrust of only 533.76kPa (120,000lb) which was raised to 578.2kPa (130,000lb) before achieving the specified requirement of 667.2kPa (150,000lb). With the Air Force in the lead with their MB-3 motor, they got to push hard for improvements but that story also is too long and fractious to relate here!

Each step brought its own bag of problems and technical challenges, where supposed improvements proved a step backward. One example of which was the larger bell-shaped expansion skirt, or nozzle, which was less rigid than the previous conical design and required additional stiffening rings in the expansion section. With the increased power load, turbine blades began to fail and their redesign was essential, incorporating interlocking tip shrouds which solved the main cause of the problem and also improved efficiency by several per cent. Cooperation between Rocketdyne and the ABMA engineers solved several areas of potential weakness, with critical bearing retainers being redesigned to preclude bearing walk, where a 'creep' sets in to destabilise the geometry of the mountings. In other areas, a quill shaft connecting the turbine to the gearbox was redesigned and a non-foaming oil was investigated which also improved the lubrication of bearings. But one highly significant change was to pressurise the gearbox to 34.47kPa (5psi) so that the vacuum of space would not change the chemical nature of the lubricant. Each step was occasioned by tests, failures, redesign and retest – a whole lot of learning.

There were other changes including the propellant feed system and associated ducting which were improved by simply changing supplier and using larger ducts. But there were subtle changes on a small scale too. The fuel pump exhibited a frequency output of 500cps, which seemed to be amplified by the mitred shape of the vaned elbows in the high-pressure duct, which was then retransmitted to the bell-shaped thrust chamber. Tests with data from accelerometers attached to the dome of the thrust chamber registered 40g at 500cps and this was demonstrated on high-speed film where the flickering flame could be seen. All it needed were three extra stiffening rings on the nozzle to shift the spectrum of the frequency and solve the problem. But there were operating limitations driven by the missile itself. Structural limitations imposed a 10g acceleration boundary which meant that there had to be some way of throttling the engine as propellant was consumed and the mass went down.

A hot gas valve in the duct connecting the gas generator to the turbine was thought to be the answer. But that was no good. It failed. Then a thrust control, closed-loop servo system was designed which would have been capable of limiting the thrust to a predetermined level. But that produced complications and the engine became less reliable. The sensitivity to g forces had been driven by the delicate mechanisms in the guidance and navigation system but they proved to be far stronger than had been anticipated and there was no need to alter the acceleration level after all. But it had led from a desire to more fully understand the dynamic stresses on the vehicle as it ascended and that proved of great value, creating models which proved beneficial in later programmes, including the Saturn launch vehicle.

The guidance platform was an evolution from the ST-90 developed by ABMA with industry assistance from the ST-80 on Redstone. As refined for Jupiter, the platform was aligned on the ground by two gyroscopic pendulums in the guidance section which then aligned the inertial platform. Three additional pendulous accelerometers sensed motion through signals providing information on altitude, range and cross-range. These signals controlled pitch and yaw through the gimbal actuators moving the engine, roll control being managed by the moveable vernier described earlier. Overall, Jupiter's guidance and navigation section weighed 136kg (300lb) compared to nearly 317kg (700lb) on the Thor missile. Moreover, where Thor had a cep of 3.7km (2.3mls), Jupiter demonstrated a cep of 1.48km (0.92mls). Much of the development work on the Jupiter guidance and navigation system had been managed by Fritz Mueller at ABMA.

During its evolution, the length of Jupiter had changed from its original configuration at 28m (92ft) to accommodate the Navy requirement for a shorter, fatter missile – in fact the Navy wanted a missile with a length of only 15.2m (50ft). When the Navy pulled out, Jupiter retained much the same overall configuration of the compromised layout, the definitive operational missile having a total length of 18.29m (60ft) and a tank diameter of 2.66m (105in) with a dry mass of 4,860kg (10,715lb) and a weight at ignition of 49,353kg (108,804lb). The rocket carried 30,683kg (67,645lb) of LOX and 13,703kg (30,209lb of RP-1. The S-3D had a burn duration on a nominal trajectory of 2min 53.8sec, a maximum altitude of 660km (410mls), 9min 10sec after lift-off and a total flight time to impact of 16min 57sec. Tests demonstrated an operational range of 2,844km (1,767mls), considerably better than the brochure advertised!

As indicated earlier, the operational function of the missile emphasised a high level of mobility but during a particularly untidy series of debates with the Army, after the Air Force was given control of Jupiter deployment it assumed a fixed base like Thor. But the requirement to get the missile launched within 15 minutes of an alert was retained and deployment assigned 15 missiles to a US Air Force squadron with the requirement to fire them all in a salvo within that time. The missiles would be dispersed to five outlying locations with each launch position having three missiles, three launchers and a single triple launch control trailer and this was known as the 'three-by-five' deployment configuration. The missiles would be emplaced without the ability to move them under alert conditions and for this reason were considered highly vulnerable.

Missile testing began on 1 March 1957 from the Atlantic Missile Range (AMR). Leaving the pad at 16.51hr EST, it reached an altitude of 14,630m (48,000ft) before it broke up 74sec after launch due to overheating in the tail section. The second launch occurred on 26 April in a test of the airframe and the rocket engine but this was terminated at 93sec because of propellant sloshing in the tanks. Interestingly, the Thor missile never experienced this and it was found that the dynamic motion of the fluids were constrained by Thor's slimmer diameter of just 2.44m (8ft). To correct this, slosh baffles were built in to the Jupiter tanks to break up the motion of the fluids as the vehicle flew its ascent trajectory. The third flight, on 31 May 1957, was a total success, although it did fall a little short of its planned range.

After several more test flights, on 18 May 1958 the first IRBM tactical nose cone was flown downrange and success-

Right: A Jupiter IRBM with its 'petal' cover open. The cover allowed crews to service key components of the missile in all weathers. *US Air Force*

fully recovered, the first of its type. On 17 July, the next launch carried the complete ST-90 inertial guidance system and demonstrated a miss distance of 1.8km (1.15mls) over a range of 2,297km (1,428mls). This flight also demonstrated the type of nuclear warhead designed for the missile and for the nose cone and warhead combination which would carry it to the target.

On 13 December 1958, missile AM-13 (the 12th Jupiter launched) carried a South American Squirrel Monkey on a ballistic trajectory which, although being considered highly successful, resulted in the loss of the monkey when the retarding parachute failed. This was the first in a series of science missions carried out by the Jupiter IRBM independent of its use in that role by NASA. After several more engineering tests of the rocket, at 02.35hr EST on 28 May 1959 a Rhesus Monkey (Able) and a Squirrel Monkey (Baker) together with biomedical samples of yeast, corn, mustard seed, fruit-fly larvae, human blood, mould spore and fish eggs were successfully carried on a full range flight and safely brought back to Earth. Despite not reaching orbit, it marked the first recovery of living creatures from a flight through space.

More test flights followed, some carrying special equipment for measuring the relative accuracy and drift values of the ST-90 guidance and navigation system, with the first Chrysler-built rocket launched on 21 October 1959. The following flight, also a Chrysler missile, was the first launched without a preceding static test of the rocket motor. This was followed by a demonstration of a shorter countdown sequence reducing the duration of pre-launch activity to four hours – half the normal time allocated. The last of 29 test launches left the pad at 19.19hr EST on 4 February 1960, all test objectives completed. One further test on 20 October that year was fired under simulated tactical conditions using standard 'government supplied equipment' (GSE). Following that, the first of four combat training launches opened the sequence on 22 April 1961, the last taking place on 18 April 1962.

Above: Squirrel monkey Miss Baker poses with a model of the rocket in which, with rhesus monkey Able, she was launched on a ballistic trajectory, becoming the first living creatures safely returned from a flight through space. *NASA*

Below: Jupiter AM-18 shortly before launch on 28 May 1959 in which Able and Baker would experience space travel and a brief period of weightlessness during the 15min flight. *NASA*

Operational deployment of the Jupiter missile had begun in July 1960, when 30 missiles were set up in Italy in a move completed in July 1961. Initially, France had been mooted as a deployment location but the French government refused to accept US IRBMs on its soil. An additional 30 missiles were deployed to Turkey from November 1961 under the IBRAHIM 2 programme and completed in March 1962. The US kept control of the warheads, which were held separate to the missiles under the control of respective NATO forces. This was complementary to the deployment of Thor to the UK, where 60 IRBMs were under the control of the RAF and on full deployment from April 1960, the warheads retained by the US separately.

Jupiter may be considered merely a side-note in the annals of Cold War history but it was important nevertheless for the part it played in pushing Soviet Premier Nikita Khrushchev to deploy nuclear-tipped missiles on the island of Cuba in October 1962, triggering the infamous Cuba Missile Crisis which brought the US and the USSR to the brink of war. That inducement lay in the decision to send Jupiter to Europe and deploy it within range of Moscow. In a secret protocol to the settlement with the Russians, the US began withdrawal of Jupiter missiles from Europe in April 1963 and completed withdrawal of Thor from the UK by September 1963.

Section 2

From Jupiter to Saturn

Chapter Four

Ushering in the Space Age

Redstone and Jupiter were vital precursor projects to the development of the Saturn launch vehicle. Understanding them in some detail, as we have done here, is vital to fully appreciating the great strides made in facing the enormous challenges yet to come. There was a linear progression from one to the other and without these engineering steps, leading to Saturn, the launch vehicles vital to the future of the US space programme as viewed in the second half of the 1950s would not have been possible in the time frame in which they were achieved.

But there were other advantages which underpinned the next major phase in rocket design: the unique nature of the guided missile and the need for automated countdown, launch and flight control. The rapid development of the Saturn launch vehicle was made possible only because these hard-won lessons had been acquired and, had it not been for the pioneering work of the Redstone and Jupiter programmes, the ABMA would not have been able to accelerate development of a very different kind of rocket. While the Army pulled away from the development of long-range ballistic missiles, it more than made up for that vacuum through its availability with space launch vehicles which would sustain US prestige – such as it was – after Sputnik 1 in 1957.

Redstone is well known to space aficionados as the basic missile from which variants were used to launch satellites and space vehicles into orbit. That work evolved via the development of Redstone into the Jupiter-C (Jupiter Composite). To provide the necessary velocity for accelerating the simulated warhead designed for the Jupiter IRBM to re-entry velocities anticipated for the bigger Jupiter missile, a cluster of eleven

Above: Redstone in assembly. The missile had capacity for improvement and adaptation which would see it evolve into the Jupiter-C, a misnomer since the only relationship to a Jupiter IRBM was to facilitate tests with the re-entry warhead for that missile with upper stages which would make it into the Juno I space launch vehicle. *Chrysler*

solid propellant Sergeant rockets, descaled in thrust to produce an average vacuum output of 89.5kN (20,130lb), were situated on top of the stretched first stage. In the centre of this cluster were three similarly scaled-back Sergeant rockets with a total average vacuum thrust of 24.4kN (5,490lb) operating as a third stage by separating from the barrel cluster on burnout at 6.5sec and firing for a similar duration to push the small re-entry simulator (Re-entry Test Vehicle, or RTV) to terminal velocity.

Each rocket had a length of 119.4cm (47in) with a diameter of 15.2cm (6in) with the second stage cluster in a ring with their axes parallel to the line of flight, the axis of symmetry being 33cm (13in) from the centreline to the middle of each rocket, thus creating a space in the centre to which was attached the three motors of the third stage. All motors were held by three transverse beams for each cluster. For high-altitude RTV tests, a fourth stage, slightly shorter than the other rockets, was positioned on top of the third stage cluster. The solids were adapted from the Sergeant rocket, a name given to an early development which should not be confused with the MGM-29 Sergeant, developed by JPL and which followed Private and Corporal as a short-range, surface-to-surface missile with a range of 139km (86mls) and a W52, 200KT nuclear warhead.

Several changes in Redstone missile length were instituted for different applications in military and non-military roles. Developed as a re-entry test vehicle for Jupiter missile warheads, Jupiter-C had an extended-length main body section between the thrust and the guidance and navigation sections, increasing its total length from 9.8m (32.08ft) to 11.4m (37.5ft) and the total length of the rocket from 16.6m (54.5ft) to 19.8m (65.2ft). The extended tank section was retained for

Below: ABMA commander Gen. Medaris (left) and von Braun pose for the cameras with a model of the Redstone rocket, 20 January 1956. *Author's collection*

Above: Outside the headquarters of the Army Ballistic Missile Agency, Wernher von Braun with his 1957 Mercedes-Benz Type 220S Ponton coupé. At the time, von Braun was also a director of Mercedes-Benz North America. The car was acquired by Dr. Donald R Jacobs in 1960 who in 1991 restored it to its original condition. *Dr. Donald R. Jacobs*

the Mercury-Redstone launchers of 1960/61, adapted from their military role for supporting NASA's first manned spacecraft. Later Redstone rockets switched the fuel to Hydyne, a 60 per cent unsymmetrical dimethylhydrazine (UDMH) and 40 per cent diethylenertriamine (DETA) mix which increased thrust to 370 kN (83,000lb). Mercury-Redstone rockets reverted to the original propellants of the liquid oxygen/alcohol-water combination.

Multi-stage versions of Redstone provided a viable reason for von Braun to consider such a configuration, with little modification, capable of placing a satellite in orbit long before Jupiter-C became available. Under the name Project Orbiter, on 25 June 1954 he had proposed a four-stage Redstone to the Ad Hoc Committee on Special Capabilities. All that was required, he said, was the clustering of Loki II-A solid propellant rockets. If instrumented, that final solid propellant fourth stage would become an artificial satellite providing data about the near-Earth space environment, a 2.27kg (5lb) object orbiting at an altitude of almost 500km (310mls). The proposal was submitted to the Secretary of Defense on 20 January 1955 but just five days later President Eisenhower announced that a 'civilian' satellite programme supporting the International Geophysical year (1957-1958) would be based on a new launcher – Vanguard – specifically designed for scientific research out of the existing Viking sounding rocket. Project Orbiter was officially rejected on 3 August 1955, the decision having already been taken to develop Project Vanguard.

Under the management of the Naval Research Laboratory (NRL), Vanguard would be far removed from any association with the development of military missiles, a factor encouraged by President Eisenhower. This alone invalidated Project Orbiter for fear it would raise international tension by seeming to assign exploration of space to the military services. Paradoxically, the Air Force already had a highly classified satellite project underway which would use adapted Thor missiles to place them in space. The spy satellite was under the codename

Corona and consisted of a rocket stage to the forward end of which was fitted a camera system, film transport mechanism and a recoverable pod designed to survive re-entry and deliver the film to analysts. Named Agena, over time this stage would be developed into a versatile upper stage for many satellites and spacecraft. Built by Lockheed, it was powered by a Bell rocket motor, initially with a thrust of 69kN (15,000lb).

Very few people knew about the Corona project and responsibility for America's satellite project was generally believed to lie exclusively with the NRL and its Vanguard project. When flights began in 1959, Corona would be publicly declared as a scientific research project under the name Discoverer. But all this was buried deep inside the black world of classified military intelligence while the 'civilian' Vanguard satellites were expected to demonstrate an essentially peaceful objective for US space operations. Much of this was of no immediate concern to von Braun and the ABMA, who knew little or nothing about Corona and the spy satellite programme.

With the Redstone available for research and technology development purposes in support of the Jupiter IRBM programme, the first variant was designated Jupiter-A, a limited series of Redstone rockets set aside with modifications for technology development applicable to the IRBM of that name. The first Jupiter-A was launched on 14 March 1956, a mere six weeks after the formation of the ABMA. Jupiter-B never matured but was mooted as a solid propellant alternative to the liquid-propellant missile. As a test vehicle for Jupiter warheads, the Redstone variant Jupiter-C made its first launch that year and its availability and performance endorsed confident assertions from von Braun that his team could send up a satellite within weeks of the go-ahead.

This first Jupiter-C (RS-27) was a proof-test shot for the nose cone development flights which were to follow and the flight began at 01:47am EST on 20 September 1956. It was powered by the A-5 engine for a test of the separation and ignition sequence of the two upper stages, an evaluation of missile structures and to test the increased duration of motor burn. It propelled a 14kg (30lb) test mock-up to an altitude of 1,100km (682mls) and a downrange distance of 5,300km (3,295mls), conducting the first deep penetration of space. In achieving a maximum speed of 25,745km/hr (16,000mph) it pushed the terminal stage to more than 90 per cent of orbital velocity. Had the payload been lighter and the trajectory appropriately shaped, it could have put a small satellite in orbit on that very first flight. And there was a nagging persistence in the ABMA's desire to demonstrate that it could do that. Von Braun had recruited Dr. Ernst Stuhlinger to engage with the scientific community and garner support for such a satellite, launched by the ABMA. There was a feeling that the Van-

Below: Jupiter-C with a variety of test cones for evaluating the aerodynamic properties of different re-entry shapes and a range of different refractory and heat-sink materials. Note the Jupiter tank sections with larger diameter to the right of the picture. *Chrysler*

guard programme was languishing and that it would not deliver as promised.

One of the elite group of Germans who came to the United States with the von Braun group, Stuhlinger had a remarkable life. Educated at Tübingen University, where he received a doctorate in physics, in 1941 he was sent to the Russian front as an infantry soldier before surviving the battle for Stalingrad. In 1943 he became involved with missile tracking for the V2 scientists and was one of 20 hiding out near Weimar when the war ended. Stuhlinger was appointed to lead the Research Projects Office at Redstone Arsenal and left to take up a position at Princeton University, where he engaged with a wide range of scientific minds while maintaining contact with the rocket engineers. Thus did this physicist become the interlocutor in gathering support for the proposition during 1953-54, notably from Dr. James A. Van Allen from the University of Iowa who was interested in researching the Earth's upper atmosphere.

The second Jupiter-C (RS-34) launch took place at 02:55am EST on 15 May 1957 and was the first to carry the scaled-down Jupiter nose cone and was to have performed the first operational nose cone recovery test. The flight proceeded normally until 6 seconds before cut-off of the Redstone booster when the attitude of the missile began to deviate due to a loss of pressure in the instrument compartment, disabling the pitch gyro. The resulting trajectory was 190km (153mls) higher than the predicted 465km (289mls) and 761km (473mls) short of the planned range of 2,037km (1,266mls). Radio contact was maintained for 795 seconds, 149 seconds short of impact with the surface of the ocean where the capsule came down. It was not recovered.

Launched on 8 August 1957, the third flight (RS-40) was the second flight fitted with a one-third scale nose cone designed for the Jupiter IRBM, carrying it to an altitude of 481km (299mls) and propelling it to a downrange distance of 2,150km (1,336mls) in a three-stage configuration. Some anomalies occurred during this flight, including a more rapid spin rate of 463rpm instead of the planned 380rpm. The nose cone failed to separate from the third stage as planned but released itself during descent when the magnesium separation joint melted. The deceleration switch operated as planned and the parachute was released at an altitude of 3,377m (11,080ft) and a deceleration of 3g. The nose cone was picked up by the Diver class rescue and salvage ship USS *Escape* some 104km (65mls) further downrange than expected. Despite the overshoot due to problems separating the re-entry nose, the trajectory of the Jupiter-C had been sound, lateral dispersion being a mere 2.5km (1.56mls) off nominal.

This was the first recovery from space of a nose cone designed for a nuclear-tipped missile. It is frequently claimed that the first successful recovery of a man-made object from space was the Discoverer XIII recovery capsule brought back intact on 10 August 1960. This was not so, the claim frequently failing to qualify that this was the first recovery of an object from orbit. The third Jupiter-C launch was indeed the first recovery of an object from space in a programme which was itself shrouded in secrecy, instrumental in shrouding public awareness of this accomplishment. The United States was

Above: Jupiter C RS-27 was the first vehicle fired under the re-entry test programme and got off the ground on 20 September 1956 powered by a Rocketdyne A-5 series engine for a successful flight of 5,302km (3,295mls). *Author's collection*

ahead of the Soviet Union in warhead technology and in the range of materials it was developing for protecting them from the heat produced by kinetic energy as they re-entered the atmosphere. Jupiter-C was crucial to that early work before the more powerful Atlas rocket could take over for that research.

Redstone and the von Braun people were crucial in getting an early start with an existing missile to evaluate, and to compare, the four primary methods of protecting the aeroshell: heat-sink, radiation, transpiration and ablation. Plastics, fibres, ceramics and hybrid materials involving sophisticated composites were tried and tested. The Air Force preferred the heat-sink concept – it had done a lot of theoretical work on that for hypersonic spaceplanes – while the Army liked ablation. It took a while but the Air Force came around to the Army approach.

That third flight of Jupiter-C in August 1957 was also the first to carry the fourth stage configuration, fully capable of putting a satellite in orbit. Only the previous month, Dr. Stuhlinger had revealed the satellite-launching capability of this rocket in a speech at the Army Science Symposium at the US Military Academy, West Point, New York. Hailing its value as a research tool, he described a new computer programme developed at the ABMA capable of calculating the oblateness of the Earth, the density of the upper atmosphere and high altitude ionisation. His enthusiasm took hold and rumours began

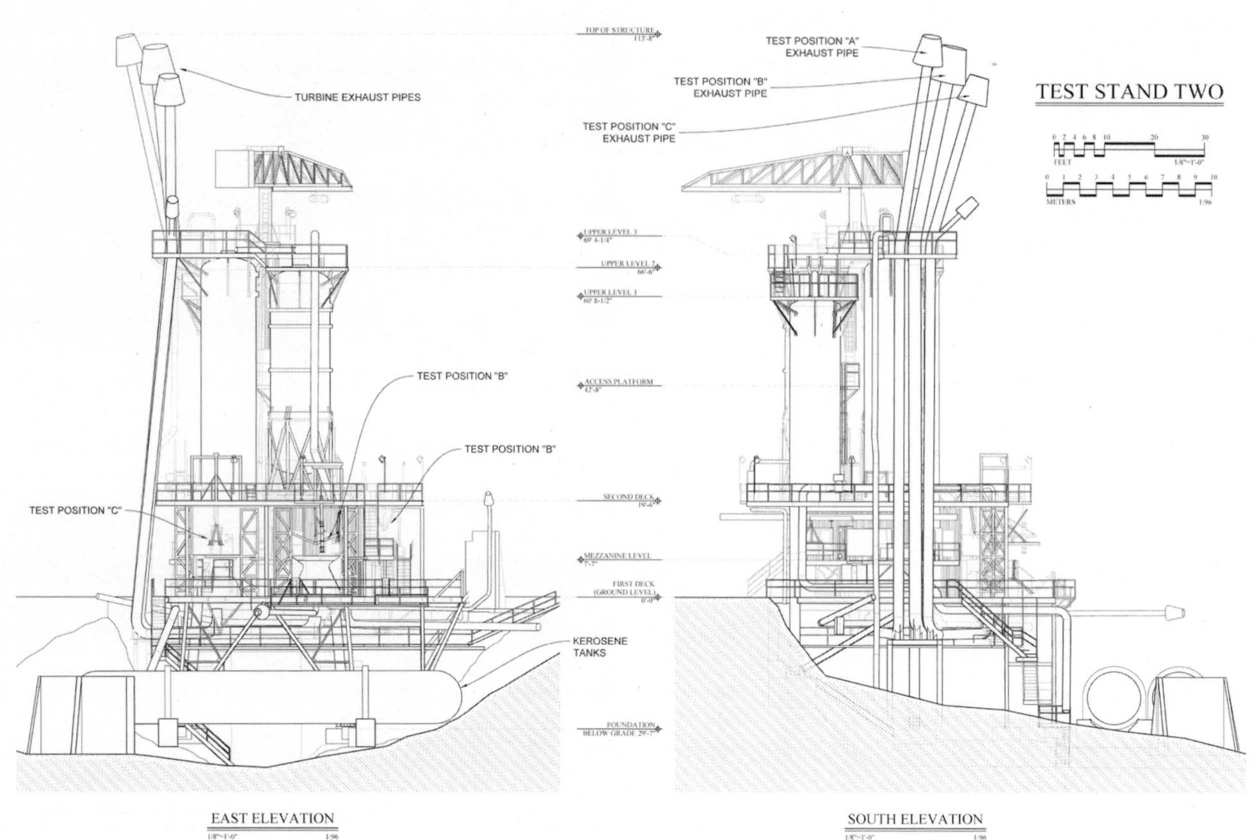

Above: Test Stand B at the Santa Susana test site. The original Bravo complex was designed by the Los Angeles architectural engineering firm of Daniel, Mann, Johnson & Mendenhall (DMJM) with the aid of German engineer Walter Riedel. A portion of the site was based on a World War II German rocket engine test site. *Author's collection*

to spread that the Army was about to get the jump on the Vanguard programme. Alarm bells rang and the Secretary of Defense – Charles Wilson – ordered the Department of the Army to silence any further talk of this nature.

But events took hold of the situation. On 4 October 1957, the Russians put up Sputnik I. In the Army, however, ABMA commander Gen. Medaris had already given the quiet word to von Braun's people that the call may be coming for them to wheel out a Jupiter-C and lob a satellite into space, such was the lack of confidence in the Vanguard programme. Five days after the Russians launched Sputnik 2 and the dog Laika into orbit on 3 November 1957, Medaris gave the word to silently begin preparations for the Army to launch a satellite. When Vanguard collapsed in flames on its own launch pad at Cape Canaveral on 6 December that became an official order and von Braun's Project Orbiter was resurrected on the approval of the new Secretary of Defense, Neil H. McElroy.

Below left: The first Jupiter in the Santa Susana Test Stand, January 1957. *Author's collection*

Built around von Braun's upgraded Project Orbiter concept, it envisaged a three phase programme. By replacing the nose cone on a Jupiter-C with a fourth stage instrumented for transmitting scientific information, it was proposed that two rockets of this type would launch a satellite apiece, followed by five Jupiter-C rockets sending up satellites with TV facilities, with a Jupiter-C carrying a 136kg (300lb) surveillance satellite. On 6 December, the first attempt at a Vanguard satellite launch failed when the rocket blew up on the launch pad at Cape Canaveral in full view of the media. Two days later the Secretary of Defense ordered the ABMA to prepare two Jupiter-Cs for satellite launches and to get the first attempt off the pad by March 1958.

The addition of the fourth stage to the Jupiter-C extended the length of the missile from 19.8m (65.2ft) for the composite RTV shots to 20.7m (68ft). The fourth stage weighed 34kg (75lb) versus the 158kg (350lb) of the RTV section. Because weight was at a premium there would be no active atti-

Left: In the late 1950s, great national pride attached to the rise of an impending space age as the armed services canvassed support for defence programmes. Rockets and missiles could be seen placed in strategic transport hubs to impress the public, as with this Redstone on display at New York's Grand Central Station on 7 July 1957. *Author's collection*

tude-control system, rather spin-stabilisation induced by two General Electric dc electric motors to keep the upper solids bore-sighted along their flight path at a rotation rate of 600-750rpm. Calculations indicated that the first stage would propel the configuration to a velocity of 2,390m/sec (7,842ft/sec), the second stage would provide a delta velocity of 1,571m/sec (5,154ft/sec), stage 3 a velocity of 1,534m/sec (5,033ft/sec) and stage 4 some 2,189m/sec (7,182ft/sec). Thus would it achieve orbital velocity of 7,684m/sec (25,211ftsec – rounded to nearest whole numbers).

Designated Juno I, to avoid the inference that the US was switching to the militarisation of space, the first Jupiter-C and its instrumented fourth stage were prepared for a flight which would have taken place on 28 January 1958 had it not been for an issue with strong upper winds. Nevertheless, and just 84 days after authorisation to launch, at 10:48:16pm on 31 January 1958 local time the first US satellite was sent into orbit by RS-29 from LC-26A at Cape Canaveral. In the internationally recognised log of satellite launch times, it is recorded as having taken place at 03:48am UTC on 1 February. Instrumented by Dr. Van Allen, the satellite was designated Explorer I and was placed in orbit at 10:55:05pm local time.

The orbited component had a launch mass of 13.9kg (30.8lb) and a total length of 203.2cm (80in) of which more than half consisted of the inert fourth stage which remained attached to the instrument section. It was coated with vertical stripes of aluminium oxide known as Rokide-A for thermal control. Two redundant telemetering systems provided data transmission to Earth at 50mW RF power on a frequency of 108.03mc. A lower-power transmitter delivered 10mW at 108.00mc.

But failure was a more common fate for satellite launches during 1958, the Russians losing four out of five attempts while the Americans lost 17 in 23 attempts. Of those failed attempts, six were tiny objects, each weighing 1.05kg (2.3lb) launched by rockets released from a converted Douglas F-4D1 Skyray, a US Navy fighter, during July and August. Developed for the military Pilot programme, they were prototypes of a series of planned weather satellites the operational versions of which were intended for launch on demand. They were never a success and the programme was abandoned, eventually in favour of the Defense Meteorology Satellite Program (DMSP) the first test launches of which were made in 1962.

Of the conventional US launch attempts in 1958, three involved the Thor-DM18 Able-1, the first adaptation of the Thor IRBM as a satellite launcher paired with the Able-1 upper stage. All three were attempts at an ambitious plan to launch Pioneer space probes to the vicinity of the Moon. Vanguard experienced a second failed launch attempt on 5 February, three days after the successful launch of Explorer I, before its first success on 17 March when it placed Vanguard 1 in orbit on its third attempt. The first successful flight of a Juno II (see later), a satellite launcher derived from the Jupiter IRBM, occurred on 6 December, followed by a thinned-down Atlas placing itself in orbit on 18th December carrying a tape-recorded Christmas message from President Eisenhower which

Right: Installed and awaiting launch, Explorer 1 as the instrumented forward section of the fourth stage and the tub of 14 solid propellant rocket motors in the second and third stages. *NASA*

was broadcast to the world.

Of them all, Juno I was the more successful during this first year of the Space Age. Explorer II (RS-26) was launched from LC-26A at 13:27:59hrs local time on 5 March 1958. The ascent proceeded as expected until the time of ignition for the fourth stage, which failed to fire and the assembly, including the expended stage 3 cluster, impacted the ocean about 3,700km (2,300mls) downrange. Switching to LC-5 at Cape Canaveral – for this and the remaining three Juno I flights – Explorer III (RS-24) was launched on 26 March 1958 and achieved the planned orbit from where it provided comprehensive data on cosmic ray intensity. Equipped with four radiation monitors, Explorer IV was launched on the fourth Juno I (RS-44) on 26 July to an orbit with high eccentricity to carry it 2,570km (1,597mls) from Earth, charting the radiation belt as it went along its elliptical path.

The last two Juno I launches were unsuccessful, with Explorer V delayed first by urgent modifications to adjust the payload specification to accommodate results from Explorer IV and then by a leaking fuel valve. It lifted off on 24 August but the main stage collided with the upper clusters, tipping the flight path off course from which it never recovered. This satellite was instrumented to support Project Argus, a series of three nuclear detonations high in the atmosphere to study the effects of a presumed radiation belt created in that region by electrons which would inhibit the arming mechanisms of any incoming Soviet warheads. The results were ambiguous.

Launched on 23 October 1958, the last Juno I flight (RS-49) carried the Beacon satellite which was to have been placed in high Earth orbit using a fifth 'kick stage' to achieve a circular path 750km (466mls) above Earth. The idea was to place the fourth stage and equipment section in an elliptical path and then fire the fifth stage at apogee to raise perigee and achieve a circular orbit. Developed by NASA's Langley facility, the Beacon payload was an inflatable balloon which would have been used for highly accurate tracking of an orbiting object. It failed when the fourth stage and payload broke away from the rest of the stack just 1min 51.5sec after lift-off.

In its brief role as a satellite launcher, Juno I had done much to maintain the pace of US space activity, meagre as it was in this first full year of the Space Age.

Game Changes

The shock of Sputnik was timely for several decisions regarding the future of Jupiter versus Thor; the Army was bracing itself for cancellation of its IRBM, especially so since the Air Force had been granted governance over its deployment and was now bearing the full cost of that programme. In reality,

Below: Vanguard on LC-18A at Cape Canaveral. There was little haste with the Vanguard satellite programme and the Eisenhower Administration was determined to maintain a dignified pace. This, despite warnings from US Secretary of Defense Donald Quarles in 1953 that because the USSR had been behind the USA in the development of atomic and thermonuclear warheads there was a danger the Russians would attempt to put up a satellite first. *Author's collection*

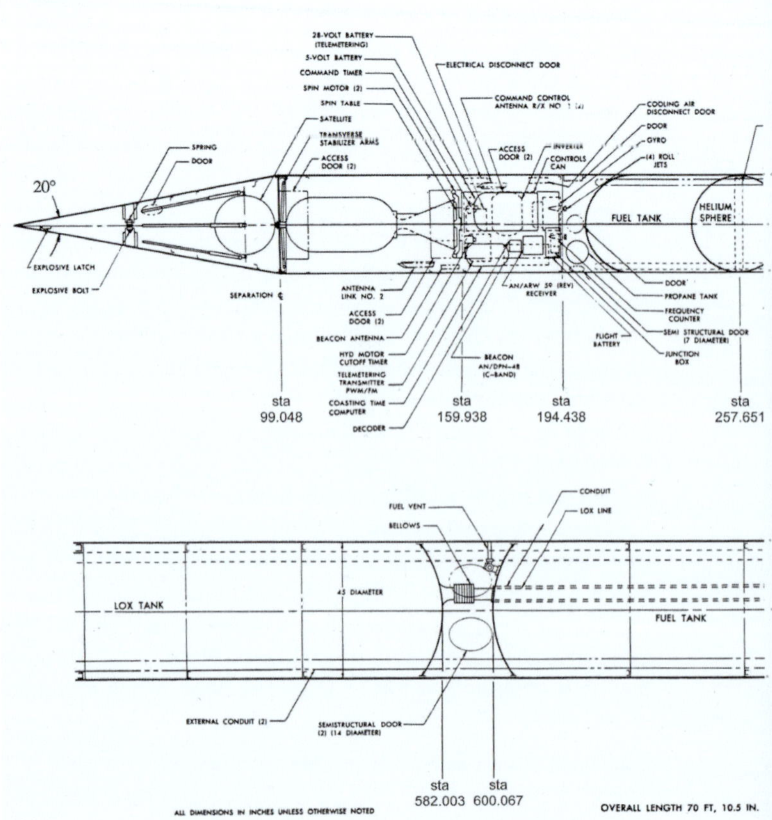

Sputnik was not so much of a shock to the White House, to the intelligence community and to the military as has been so frequently claimed. In June 1957, Deputy Secretary of Defense Donald A. Quarles publicly spoke of the imminent deployment of a Soviet ICBM and the intelligence community had expected that to occur during the second half of the year. Spy flights with the Lockheed U-2 had provided ample evidence to that effect.

The use of an ICBM to place the first satellite in orbit served notice on the US that the Russians were integrating both military and scientific research. It always astounded them that there was this division in the West between military and civilian space projects. They reflected on that in Moscow, to this writer several decades later, on how that appeared quaint and somewhat strange to them, even self-defeating for the US in that it lessened the domestic pressure for financial resources in the one area where they could brandish American technological capabilities.

For the Department of Defense, balancing the decision over Jupiter and Thor – whether to buy one or both – was an expediency built around another balancing act: that between speed and economy; the speed being approximately proportional to the effort expended and the economy being the necessity to eliminate duplication or repetitious financial resources. Prior to Sputnik and the reaction in Congress and around the country, Jupiter appeared poised for cancellation in a decision built around economy. Thor could have been produced at a per-unit cost less than that of an equal number of each since tooling for one could be eliminated, duplication replaced with fiscal efficiency. Sputnik, however, brought speed over economy and two missiles produced at the same time would double the number available.

These were political and not military decisions and had the analysis been advanced further it is likely that the resources on the IRBMs would have gone to the two ICBMs under development. But that would have been illogical, given the state of development for both Atlas and Titan, which no amount of money could have accelerated at that late stage of their development; Atlas was already entering flight trials and those would begin on 11 June 1957, although a successful flight would only occur with the third flight, on 17 December 1957. But that was with a test missile (SM-65A) equipped only with booster engines. The first full range demonstration flight would not take place before 29 November 1958. The Atlas D would go on full operational alert in September 1959.

With inertia within the Atlas and Titan programmes, the Defense Department announced on 27 November 1957 that both Thor and Jupiter would be deployed, those having been described in the preceding chapter. Ironically, the decision was partly justified on economic grounds, in that initial outlays for tooling had already been spent and would inflate the

Below: With a total length of 21.6m (80.87ft), Vanguard evolved from the Viking sounding rocket and comprised two liquid propellant lower stages and a solid propellant upper stage. The first stage had a thrust of 134.7kN (30,303lb) and a launch mass of 10,050kg (22,156lb). The satellite it was intended to launch had a diameter of 16.35cm (6.43in). *NASA*

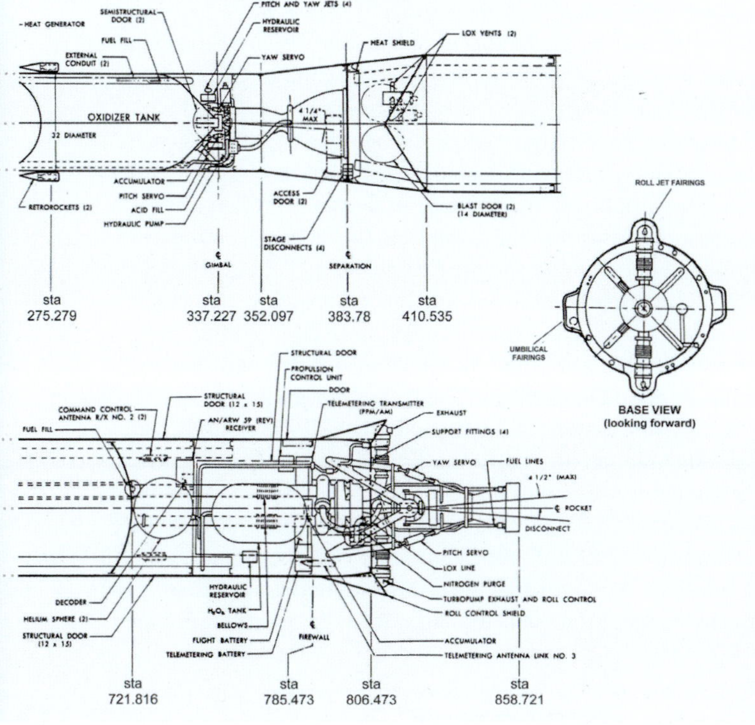

Below: The first attempt to send up a US satellite ended in disaster on 6 December 1957 when Vanguard stalled. It fell back on the launch pad destroying itself and the TV-3 satellite it was carrying in front of officials and several journalists. *NASA*

cost of one programme if all the sunk costs of the two missiles were levied on the head of the selected contender! But McElroy was himself swept into a vortex of decisions over numerous projects and programmes which swamped his desk in the post-Sputnik weeks and months.

In the wake of Sputnik, with public confidence shattered, surveys from US embassies around the world suggested the USSR was now favoured by most people surveyed as the world leader in technology. While the Pentagon was unlikely to concede an unfavourable balance of deterrence capabilities, there was a highly visible absence in the balance of confidence in the Western alliance. It was this latter fact that urged deployment to Europe – of Thor to the UK and Jupiter to Italy and Turkey – where confidence could be regained in NATO and US resolve. And it was that decision which spurred confidence in the von Braun team at Redstone Arsenal that further developments were opportune. But there was more to the deployment than appeared immediately obvious.

As recorded, Sputnik had a significant effect on defence decisions and on the consolidation of the NATO alliance. In the greatest gathering of governments since the 1919 Treaty of Versailles, the semi-annual NATO meeting was held in Paris on 16-19 December 1957. On the first day Secretary of State John Foster Dulles offered the IRBM to NATO allies and presented deployment options which had already been thought through. In early 1956 the Air Force had proposed 18 squadrons of 15 Thor IRBMs in the UK, as that country was now the third member of the nuclear club. The decision to offer the Thor to Britain had been put on hold over the Suez crisis of October-November. Detailed discussions resumed in December after a mending of the rift between the US and the UK. In March 1957 these negotiations concluded with an agreement for deployment of four squadrons, each with 15 missiles.

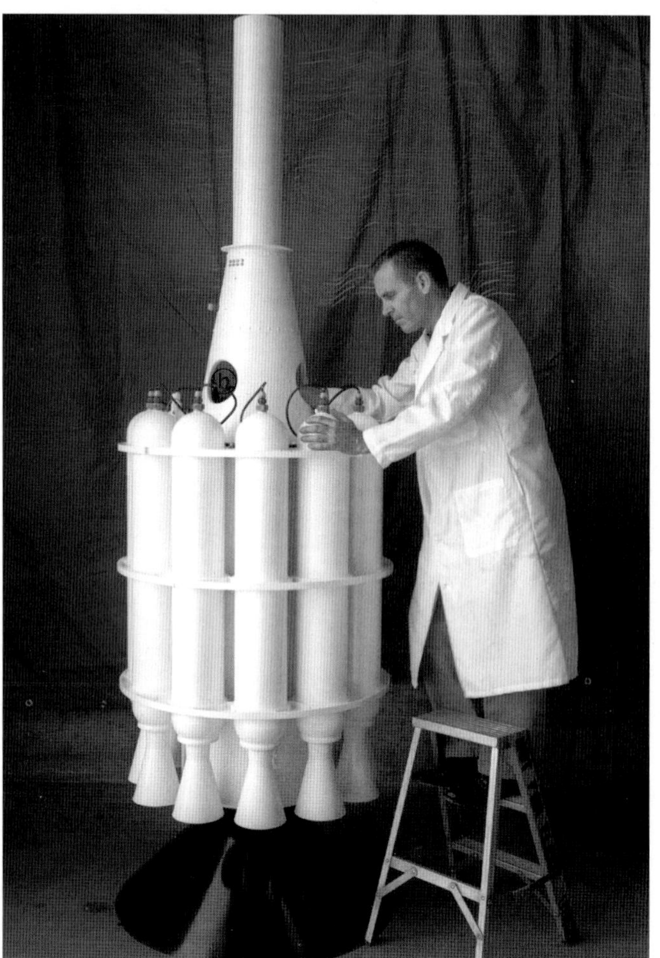

Left: An engineer poses for scale alongside a cluster of Juno I upper stage solid propellant rockets. *Author's collection*

Sputnik had trumped unstated expectations that the Thor and Jupiter missiles would be consigned to history before deployment; many military officials were nervous about basing IRBMs in foreign countries as they had to be to make them viable. But the uncertainties created by Russia's ICBM and the compelling influence of its artificial satellites catapulted Thor and Jupiter into the foreground as tools of propaganda rather than instruments of military value. However, strong messages could be sent to a potential aggressor by the very presence of the IRBMs based within striking distance of critical targets in Russia. On 7 November 1957 a report by the Gaither committee pressed for an increase from 60 to 240 in the number of IRBMs deployed. Named after its chairman, H. Rowan Gaither, the so-called Gaither report was robust in its recommendations for bigger budgets on missile development and larger force deployments.

Although in large part ignored by Eisenhower, the Gaither report defined uncomfortable realities in that it highlighted a fault-line in US strategy: that Eisenhower's satisfaction with his own classified intelligence assuaged fears held by the public regarding America's true capabilities. It was to restore faith in a public deeply concerned about the Soviet threat that the decision of 27 November to develop both Thor and Jupiter was made. But, de facto, to be credible and effective as instruments of reassurance to a domestic electorate and a very real threat to the enemy, due to their range they had to be deployed in Europe.

Thor deployment had been sealed in an agreement between President Eisenhower and the UK Prime Minister Harold Macmillan in Bermuda during 21-24 March 1957. But that left the deployment of Jupiter to be decided, an offer made at the semi-annual NATO meeting that December. Central to these discussions was Gen. Lauris Norstad of the US Air Force and Supreme Allied Commander Europe (SACEUR) between 1956 and 1963, who coordinated and chaired discussions embracing both political and military imperatives. It had always been assumed that France would be the prime location for continental deployment and the French, being the third strongest NATO partner after the US and the UK, had indicated their enthusiasm for that. However, to get the IRBMs within range of Soviet targets, after the UK for Thor, the JCS had selected Turkey, Alaska, Okinawa and France, in that order.

During discussions in the first quarter of 1958 politics prevailed and France moved to the top of the list after the UK after concerns were raised about Soviet reaction to deployment in Turkey, close to the NATO border with the Warsaw Pact countries. But difficulties emerged with the French, who had issues over command and control and they linked deployment to cooperation with the Americans in producing their

Above: Prime Minister Macmillan and President Eisenhower meet in Bermuda, March 1957, where it was agreed to deploy Thor IRBMS in the UK. *US Library of Congress*

own IRBM, similar to the deal between the US and the UK over the British nuclear capability. That trans-Atlantic cooperation involved development of Blue Streak, the UK's Intermediate Range Ballistic Missile powered by two RZ2 engines developed by Rolls-Royce out of the Rocketdyne S-3D. Norstad rejected any such linkage and after the collapse of the last government in the Fourth Republic the incoming regime of Charles de Gaulle in late 1958 vacillated and while France's military had favoured deployment of Jupiter the increasingly hostile, anti-American stance of the Fifth Republic caused that optimism to evaporate.

Norstad had focused too long on France and turned next to Italy. Approached first in February 1958 in a failed attempt to rattle the French into a decision, Italy had no plans for an independent nuclear programme and therefore placed no conditions on a deal to host Jupiter but party politics delayed a final decision before agreement on 26 March 1959. Late the previous year, West Germany had emerged as a candidate when the government sought deployment which ran counter to public and partisan opinion. Moreover, with Germany divided, both Washington and Moscow had learned just how sensitive were issues over nuclear weapons either west or east of the ideological divide and even Norstad never seriously considered deployment in the Federal Republic.

In early 1959, Norstad looked at Greece for potential deployment but internal politics swung between the conservative Karamanlis government and the left which favoured a neutral stance with Russia and that disallowed the opportunity. By mid-1959, however, despite Eisenhower's dismissive views to the contrary, Turkey emerged as the most logical second choice for continental deployment. While the agreement for deployment was a forgone conclusion it was only on 26 October 1959 that the government formally acknowledged the decision, when it announced that before the Grand National Assembly of the Turkish government. It had been signed on 19 September.

And so it was that Italy would receive 30 missiles while Turkey received 15; a plan to deploy a further 15 IRBMs to

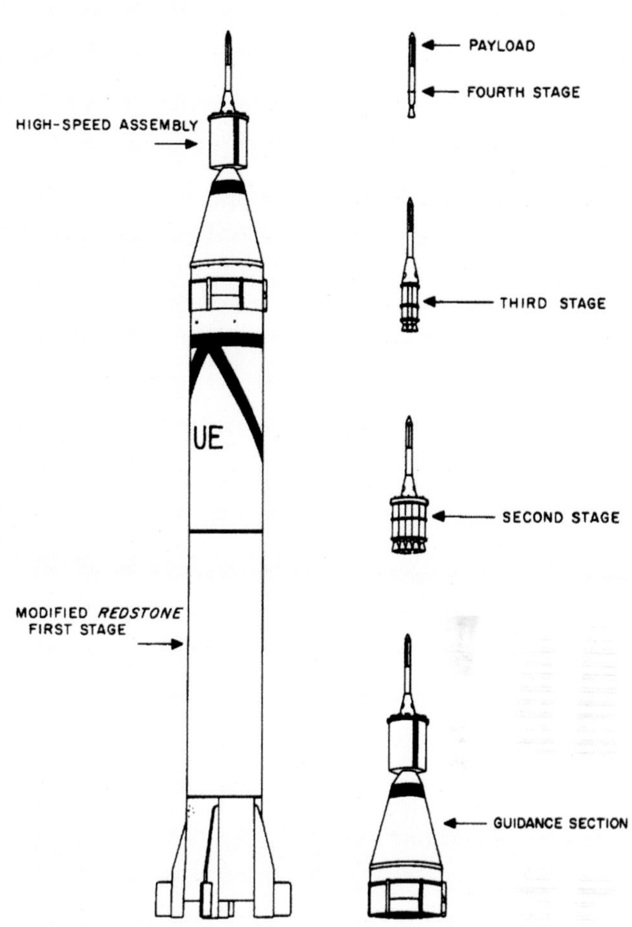

Above: The inherent adaptability built in to the Redstone rocket had expanded its value as a research tool verifying warhead design for re-entry vehicles assigned to the Jupiter IRBM programme, hence its designation Jupiter-C (Composite) consisting of groups of solid propellant upper stages. The addition of a fourth stage with an instrumented front section constituted the first US satellite, Explorer I. Jupiter-C had Hydyne propellant, the A-5 engine and longer propellant tanks. *Author's collection*

Below: Quickly re-designated Juno I for space launches, the upgraded Jupiter-C incorporated 11 solids for the second stage, three for the third stage and one for the fourth stage. As shown, the fourth stage is suspended by the thrust ring which attaches to the adapter on top of the second and third stage cluster.

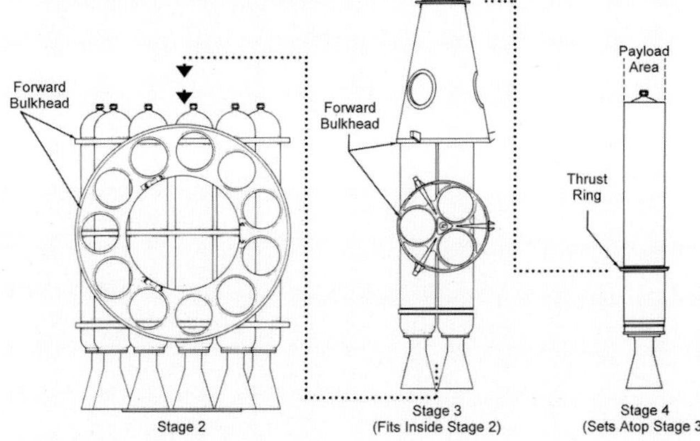

Turkey was shelved. On several occasions, lightning strikes on the Turkish missiles partially armed them and came close to detonating the warhead, after which steps were taken to provide lightning protection.

Largely unrecorded in almost all popular histories covering development of the IRBM and deployment decisions, the impact of Sputnik on keeping the Jupiter programme alive was crucial to how events unfolded over the next several years. Equally unrecorded, the decision to place IRBMs so close to the Soviet Union was the trigger for deployment of Soviet IRBMs on Cuba in 1962. The fear that this would be a consequence had been expressed widely at the time but seemingly disregarded by Eisenhower's successor, John F Kennedy, when expressing such outrage at Castro receiving IRBMs from Moscow. During Khrushchev's visit to the US in September 1959 the Soviet Premier tossed a side comment to labour leader Victor Reuther: 'How would you feel if there were Soviet military bases in Mexico and Canada?' To which Reuther replied: 'Who is keeping you from having them? Set them up!'

At no other point did Khrushchev raise the IRBM deployment either formally or through informal comments. But US officials were aware of the risk of reciprocity. In the written summary of a conference on 16 June 1959, the reporting memorandum noted that: 'The President said one thing... bothering him a great deal (which was) the plan to put IRBMs (in Europe). If...Cuba had been penetrated by the Communists, and then began getting arms and missiles from them, we would...look on such developments with the gravest concern and...it would be imperative for us to take positive action, even offensive military action...He wondered if we were not simply being provocative.'

It was that prescient comment bordering on expectation that underpinned the deployment of SS-4 and SS-5 missiles to Cuba in 1962, resulting in the tension created by and defined as the Cuban Missile Crisis. It was due to a secret protocol to the agreement with the Russians that the Jupiter IRBMs were withdrawn from Italy and Turkey, albeit of a redundant and operationally defunct system which had only ever been deployed as a political reassurance to a concerned American public.

Change in Direction

A clear and distinct advantage attached to Army operations at Redstone was the commanding hold they had on the design, engineering, technology and fabrication of rocket engines, rocket vehicles and missiles, together with launch operations support from the pad to completion of development firings. The Air Force did not have this in-house and acquired that experience through connections with industry, from which, as we have seen, they obtained their knowledge through industry consultants and emerging companies specialising in systems engineering, aerospace design and electronic equipment. There is no better example of that than in the selection of rocket engines, utilising the designs from Rocketdyne and Aerojet-General.

The Atlas engine package of double-booster motors and single sustainer, was known as MA-1 but successive developments led to the MA-2 and MA-3 evolutions, each with a developed set of elements and greater thrust. Aerojet-General developed the LR87 for the more conventional, two-stage Titan I which consisted of two rocket motor thrust chambers in the first stage and a single LR91 with a thrust of 355.8kN (80,000lb) in the second stage. From these engineering developments grew a broad and highly successful series of rocket motors. In this way, the Air Force acquired propulsion expertise and sought design configurations from Convair/General Dynamics for the Atlas and from the Martin Company for Titan. And all the while they honed their expertise with operational deployment and management of the deterrent.

Above: A Thor IRBM on the launch pad. One of the effects of Sputnik was to persuade the US DoD to double down on its investment in IRBMs pending the arrival of the Atlas ICBM. *USAF*

Immediate repercussions for Redstone Arsenal were positive in that there was a determination from the Pentagon for the team to continue with the solid propellant Pershing missile as well as the Nike-Zeus anti-ballistic missile system with a programme office set up there. Overdue in some respects, in that there was no central authority in the Pentagon with overall command and control over military technology developments, on 7 February 1958 a departmental directive established the Advanced Research Projects Agency (ARPA) as a direct consequence of the controversy over Thor and Jupiter. This would have increasing influence in further Army projects at Redstone Arsenal and come to play a major role in determining future rocket and missile policy.

Charged with avoiding wasteful duplication of roles and projects, its first director, Roy W. Johnson, channelled ARPA's efforts into productive, efficient and cost-saving avenues which dispensed with service in-fighting insofar as it related to decisions over competing capabilities. In one example of how this worked, when the President's Science Advisory Committee asked ARPA to push solid-propellant chemistry and provide a breakthrough for missiles technology, Johnson sought bids from 40 chemical companies. Then he parcelled out analytical research and evaluation of those bids to successfully reduce the feasible ones to just four. Johnson saw ARPA as an enabling agency in an authoritative position to ask the Army, Air Force and Navy to work together when previously it had been impossible for the Pentagon to ask any one of those to work with the others.

For the Army, rocket and missile projects would first be channelled through ARPA and its tech-savvy specialists, rather than taking ideas up to the Pentagon and lobbying for them within groups not totally au fait with the science or the engineering. The Army was confidently reassured that it would receive a cooperative support for the Huntsville team and that Johnson had no intention of replacing that group or of seeing its role diminished. This alone brought rumbles of dissatisfaction from the Air Force and the Navy but ARPA had the support of the President and retained its authority. ARPA would play a significant role in the story of later Juno proposals and in particular the Saturn launch vehicle. But there were further concerns in the immediate aftermath of Sputnik and while ARPA held responsibility for initial space projects from both the Army and the Air Force, Congress wanted more and so did the President.

It has been said that Eisenhower was firm in his conviction that space exploration should be in the hands of a civilian arm of the government, leaving the military to proceed in great secrecy with the clandestine Corona spy satellite programme. That is not quite true. Between November 1957 and January 1958, hearings before the Military Preparedness Subcommittee of the Senate Committee on Armed Services struggled to define the difference between ballistic missiles and Earth orbiting satellites, seeing the latter as very different to the former. This helped prepare the way for the Eisenhower administration to adopt the same approach to the challenge of Sputnik as it had to the call for an IGY satellite – to keep it within the civilian sector so as not to militarise the next frontier before it had been settled. But it had not been Eisenhower's first inclination. Only when Vice President Richard Nixon and James R. Killian intervened did he acquiesce and agree to it being an open, civilian agency.

On 25 November 1957 Senator Lyndon B. Johnson, Chairman of the Senate Committee on Armed Services, opened a series of hearings over 20 days at which more than 70 witnesses, most from the DoD, produced more than 2,300 pages of testimony on what to do next. There were so many different points of view that a flood of bills were put before Congress, including one that sought to give the Atomic Energy Commission the major share in a future space programme. In rejecting that and the other bills, the White House felt incumbent to come up with its own proposition. On 5 March 1958 the President approved the recommendations of his Advisory Committee on Government Organization that leadership in space research should rest with the National Advisory Committee for Aeronautics (NACA), endorsing a direction in which Killian and the President's Science Advisory Committee (PSAC) had been moving for some time.

But it would need restructuring. The NACA had been formed on 3 March 1915 with 12 unpaid members and an annual budget of $15,000. Not until 1917 did the NACA get its first facility – the Langley Memorial Aeronautical Laboratory in Hampton, Virginia – from where it could begin to conduct original research. This was followed by the Ames Aeronautical Laboratory at Moffett Field, California in 1939, the Aircraft Engine Research Laboratory at Brook Field, Ohio in 1942 and the Muroc Flight Test Unit, California, in 1945. By the end of the Second World War it had 6,800 paid employees. In 1958 it boasted a Main Committee of 17 members: five from the DoD, five from other government departments and seven from non-government bodies. Five technical committees and their 23 subcommittees had 450 members.

While the NACA had a very high proportion of its work dedicated to space research, engineering and technology conducted by other more appropriate departments of the government, there was constructive opposition to its tilt away from pure aeronautics. Nevertheless, many NACA engineers and scientists had been drawn to work on the missile programmes, solving problems regarding propulsion, aero-thermal challenges of re-entry and control of vehicles at the edge of space such as the X-15 and the Vanguard programme. Now, the Sputnik challenge swung many more engineers toward the challenges of space exploration, their position strengthened by events.

Convinced that these rapidly evolving challenges would engage the NACA in a big way, NACA Director Dryden and Chairman Jimmy Doolittle organised a dinner at the Hotel Statler for 18 December 1957. Several 'third echelon' employees were invited to attend as the potential leadership which would have to construct a national space programme for the United States. After reviewing various courses of action there was overwhelming support for Dryden's view that space would become a dominant function for the NACA. On 16 January 1958 the Main Committee passed a resolution recommending an increase in staff from 8,000 to 17,000 over a three-year period

Right: A Jupiter IRBM emplacement. Early plans to deploy the missiles in France were scotched by the increasingly anti-American government of President Charles de Gaulle. *US Army*

with a corresponding increase in its budget from $80million to $180million a year to prepare it for its new role.

Congress had already set about organising itself for the coming wave of legislative activity. The Senate had created the Special Committee on Space and Astronautics on 6 February 1958, chaired by Senate majority leader Lyndon Johnson, while the House of Representatives set up the Select Committee on Astronautics and Space Exploration effective from 5 March under House majority leader John W. McCormack. One member was a future US President: Gerald R. Ford. The first of 17 days of House hearings involving statements from 51 individuals began on 15 April resulting in testimony running to 1,541 pages in the public record. The metamorphosis of the NACA would be complete when it became the National Aeronautics and Space Administration (NASA) effective from 1 October 1958. It would be headed by T. Keith Glennan with Dryden as his deputy.

When the draft legislation was sent to Congress on 2 April, President Eisenhower took a close personal interest in requiring the NACA and the DoD to 'jointly review the pertinent (space) programs currently under way within or planned by the (DoD, to recommend)…which of these programs should be placed under the direction of the new Agency'. Talks to achieve that were held by Dryden and Deputy Secretary of Defense Quarles. NASA would get ARPA space programmes, including Vanguard and some lunar probes then in the planning stage. Differences of opinion surrounded the Air Force MISS (Man In Space Soonest) project, a plan to send pilots into space in a capsule launched on top of an adapted missile, about which more later. Initially, the NACA and the DoD had agreed to joint management of MISS but the Bureau of the Budget (BoB) objected on the grounds that it would cost too much to have this duality and it was agreed to transfer that too. NASA would rename it Mercury, America's first manned space programme.

Juno II Space Flights

The Redstone and Jupiter missiles had bent the learning curve toward much larger, purpose-built rockets which would underpin America's desire to out-perform the Soviet Union in payload weights and lifting capacity. What the engineers did not know was that there was a still more aggressive learning curve to come. But the reinforcement of an independent rocket design and engineering team at Redstone, occasioned by persisting alarm over Sputnik, the failure of the first Vanguard flight attempt which propelled von Braun's Jupiter-C as America's first satellite launcher, the formation of ARPA and the establishment of NASA, transformed the fortunes of the Army rocket team.

Employees testify to the excitement and sense of opportunity which now unfolded and to the optimism for new concepts to gain ground and gather momentum. Recollections from former employees testify to a restlessness, renewed belief in what a few years previously would have been considered outlandish ideas. For von Braun there was an intellectual harnessing of possibilities, a view that with astute management of long-term objectives, the intermediate steps would not lurch along as they had in the past, with military requirements dictating projects and programmes, rather a defined path with interim objectives toward a final goal. In the clearest sense of what that meant he began to formalise the grand idea – what scholars would come to term the 'von Braun paradigm' – with routine human flights into orbit preceding construction of a permanently populated space base prior to expeditions to the lunar surface. Only then would the goal – a mission to Mars be possible, he concluded.

Above: Formed in March 1915 in response to great strides across the UK and Europe in the technical development of aircraft, the National Advisory Committee for Aeronautics (NACA) was remodelled in 1958 into the National Aeronautics and Space Administration (NASA) with a charter to assume responsibility for collecting up all non-military space activity. *NASA*

But NASA itself was considered a threat. There was extraordinary loyalty and exceptional devotion to the Army and a robust engagement with the gritty conflicts in the Pentagon. NASA was viewed not as a potential enabler opening doors to an expanding space programme but rather as another claimant on limited national funds, however enlarged (on one scale) they were by the political and prestige pressure applied because of Sputnik. Feedback from the new NASA made it clear that Glennan had his sights on the Huntsville team and its capacity for thinking big and delivering on visionary concepts and innovative techniques. The von Braun people liked the discipline and the forthright determination of the Army and distrusted free-wheeling scientists, politicians and bureaucrats. They saw in NASA an overblown version of the Vanguard programme, centrally managed but with too many intervening organisations and elements. But the writing was on the wall. As NASA began to absorb existing space projects spread across various groups and organisations, it had to believe that its future direction was still uncertain.

At the ABMA, Gen. Medaris was concerned at the lack of support for some of the advanced, post-Jupiter design concepts being mooted and von Braun was persistent in wanting to retain the operating principles of Redstone Arsenal. When responding to a question during testimony before the House Select Committee on Astronautics and Space Exploration on 15 April 1958, von Braun asserted: 'We do things in-house, whenever there is a need for a quick improvised solution and where results are needed fast, where we cannot wait to evaluate lots of contractors' proposals and so forth. But our philosophy is to do in-house as little as possible and to contract out what we can just as well get from a contractor. In our experience this combined arsenal-industry method is the quickest way of getting results.'

During that same testimony, an indication of a forward-thinking approach to sequential growth came when von Braun described how he envisaged the first manned flight into space, on a ballistic trajectory fired by a converted Redstone rocket: 'We have worked out a detailed proposal to send a man with a Redstone missile (to an altitude of 241km/150mls). I am convinced, in fact I know, that this could be done sooner than the X-15...We propose to separate the nose section with the man from the rest of the missile prior to re-entry in the atmosphere. The man himself will be in a pressurised capsule which is inserted into the nose section. The latter is equipped with controllable air brakes which retard the fall...There is also a parachute for the last portion of the descent to land him safely and smoothly on the ground.'

Congressional hearings during the first half of 1958 aired a very wide range of ideas and projections, both on specific goals and on possible directions for a national space programme. ABMA already had its hands full with existing small-scale projects and would exploit the opportunity to conduct space flights using both Juno I and the Jupiter, adapted as a space launch vehicle. Stripped of a military role for larger rockets, the Huntsville rocket engineers readily set to work exploiting their new-found role as pioneers in the dawn of the Space Age, using both Redstone and Jupiter to develop a new series of launch vehicles.

Although useless in their acclaimed role as weapons of deterrence due to their vulnerability and the lengthy time required to fill them with LOX and hold in readiness, both Thor and Jupiter would play a major role in the evolution of the space programme. But, while Thor would go on to become the mainstay of the medium-lift launcher inventory under its own name and that of the Delta derivatives, Jupiter too would play a vital part, albeit somewhat ingloriously at first in launching the US space programme. Following the Army test with monkeys referred to earlier, NASA utilised Jupiter as a launch vehicle which the ABMA named Juno II, following in sequence to the Redstone space launches with Juno I.

Surprisingly, the stimulus to give Juno II a prominent place in the history books came exactly three weeks after the launch of Sputnik 1 and more than three months before the launch of America's first satellite – on a Jupiter-C later re-designated Juno I. On 25 October 1957, JPL pushed forward with a set of proposals for establishing a national space exploration pro-

Below: Stored in horizontal 'coffins' to protect them against pre-emptive attack and erected vertically for launch, the Atlas ICBM was a contemporary missile development with Juno I and Juno II but would soon acquire a new role as a satellite launcher. *USAF*

Above: Launched on 18 December 1958, Project SCORE (Signal Communications by Orbiting Relay Equipment) involved an Atlas B rocket placing itself in orbit, relaying a taped Christmas message to the world. A blatant piece of propaganda, the urgency to raise the stakes in the emerging Space Race prompted some bizarre attempts at reclaiming public approval on the global stage, a reputation shattered by the Soviet 'firsts'. *USAF*

gramme. The director, William Pickering, had great hopes of putting his laboratory at the forefront of space research. Having already carved a niche in provision of solid propellant upper stages for the Jupiter-C and Juno I launchers, Pickering now wanted to fill the void in extending the scientific exploration of space to other worlds. To do that, he proposed the Red Socks programme in which nine small 7.75kg (14.9lb) probes would be launched to the Moon for gathering science data and photographing its surface, including the unseen far side always facing away from Earth.

Pickering had a radical view of the dangers in leaving the lead in space exploration to the Russians, believing that 'it is essential for the United States to initiate an immediate programme for the scientific exploration of the Moon…national interest appears to require that…US science likewise has this capability'. Pickering's plan was first to use the Jupiter-C to orbit the Moon in June 1958 followed by a flight to take photographs of the far side during the second half of 1958 and return them to Earth in a re-entry capsule much like the warheads tested by Jupiter-C. The remaining seven flights would have been spread over 1959 and 1960, sending 54kg (119lb) probes orbiting the Moon. Pickering did propose the somewhat fantastic notion of detonating a nuclear bomb on the lunar surface to cause expulsion of material at escape velocity, fragments which would hurtle earthward, 'producing beneficial psychological results'.

Absenting the nuclear detonation, this plan was put to Lee DuBridge, president of CalTech, who supported the idea and helped Pickering take it to the Pentagon. First to the Army where Lt. Gen. Gavin became a firm supporter and then to Deputy Secretary of Defense Donald Quarles, who preferred to have the Air Force involved too. JPL's association with the von Braun team made the Army route the logical path but once inside the Pentagon, broader issues framed choices. Quarles encouraged mutual support for a national space programme and knew that the Air Force would want a piece of the action. In the general unravelling of national interests and the scientific exploration of space which ensued after the shock of Sputnik, proposals such as Red Socks were placed on hold as control moved first to ARPA and then to NASA.

Cooperation between the Army and JPL had been cemented through the Jupiter-C and Juno I programmes and while von Braun had initially assumed that he would put together the satellite that became Explorer I, Medaris had been taken aside by JPL's William Pickering and convinced otherwise. It would be more effective, said Pickering, if JPL worked up the payloads and left the AMBA to develop the rockets to send them on their way. Beyond that, JPL had already laid plans for a rework of the Red Socks programme and this would eventually see it lead America – and the world – in planetary and deep-space exploration. Initially, however, the adaptation of the Jupiter IRBM as a space launch system was built around its ability to send heavier payloads into space than possible with the Redstone derivative – Juno I.

The programme to which Juno II contributed at first was

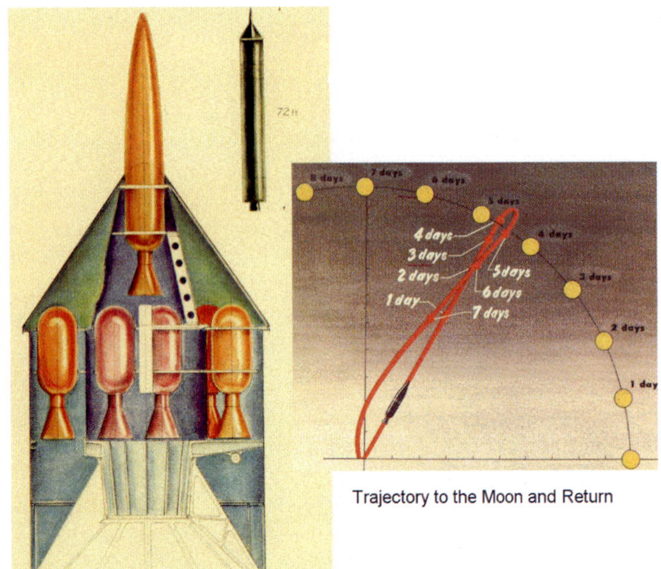

Left: Conceived by William Pickering and physicist Lee A. DuBridge in response to Sputnik, the Red Socks lunar fly-by proposal was brought to the Army for application with the Juno II launcher. It got nowhere until the ARPA received an official go-ahead from Secretary of Defense Neil McElroy on 27 March 1958, resulting in the Army and Air Force Pioneer probes. *Author's collection*

the two Army Moon probes, simultaneous with the Air Force utilising the Thor with an Able upper stage for the three probes which were managed by the AFBMD. Able was, in effect, a readily available adaptation of a Vanguard upper stage solid. The Air Force had already worked with Space Technology Laboratories to propose a series of lunar missions. After ARPA took over control of non-classified space missions, on 27 March 1958 it had authorised three Thor-Able Air Force missions and two Juno II launches from the Army. An important and lasting contribution from the Army was the construction of a 26m (86ft) diameter space tracking dish at Goldstone, California which was begun in April 1958 and completed seven months later. This was too late for the three Air Force Moon probes but not for the two Army probes. It would form the first leg of what would become the Deep Space Network.

The Air Force probes were launched from Thor pad LC-17A at Cape Canaveral, the first attempt being on 17 August 1958 which terminated 1min 16sec later when the Thor stage exploded. The second attempt on 11 October put Pioneer 1 to an altitude of 115,365km (71,700mls) after failing to achieve escape velocity. The third flight on 8 November 1958 with Pioneer 2 lasted 42min after the third stage motor failed to ignite. Pioneer 1 had been the first successful space probe to reach deep space and after NASA took over the Pioneer programmes from its establishment on 1 October 1958, only the two Army flights remained for launch.

To adapt the Jupiter into Juno II, the ABMA added a 0.9m (3ft) section to the main body of the rocket, increased the quantity of propellant by 6,214kg (13,700lb) and extended the burn time to 2min 58sec. The main stage consisted of the same NAA S-3D rocket motor as the IRBM with a thrust of 667.2kN (150,000lb) and incorporating retrorockets in the aft section of the main body to decelerate after stage separation. An instrument unit was attached to the top with clusters of solid propellant rockets in a barrel of three upper stages in a similar configuration to that used to convert a Jupiter-C into a Juno I satellite launcher.

The 331kg (730lb) second stage consisted of 11 Baby Sergeant rocket motors located in an annular ring about a centre tube and held in position by three transverse bulkheads. It produced a thrust of 89.5kN (20,130lb) and had a burn duration of 5.52sec. The 84.8kg (187lb) third stage comprised three rocket motors also carried on three transverse bulkheads for a stage thrust of 24.4kN (5,490lb) for 5.52sec. A conical support on top held the single 29.7kg (65.4lb) fourth stage, with a thrust of 8.14kN (1,830lb) for 5.5sec, which supported the payload above.

Each of the scaled-back Sergeant motors contained 22.7kg (50lb) of T17-E2 solid propellant composed of a polysulphide fuel with an ammonium perchlorate oxidiser. Unlike Juno I, however, the solid rocket upper stages were contained within an aerodynamic shroud, with a diameter of 122.9cm (48.4in), which also included the payload. Referred to as JPL's 'high-speed stages', the upper three stages were spun up to 400rpm by an electric motor until stage 2 ignition. With all these changes the overall height of the rocket increased to 23.16m (76ft). The payload capability was a modest 41kg (90lb) to a low Earth orbit or a payload of 6kg (13lb) delivered to escape velocity. After staging of the Jupiter core and separation, the shroud separated 12.5sec later followed by a coast phase lasting anywhere between about 48sec and 348sec before ignition of the three solid stages in rapid succession, ignition of each at 9sec intervals.

Development by the ABMA of the Juno II shifted from the Army to ARPA on 1 May 1958 and the first Juno II launch occurred just over two months after NASA acquired the programme. That mission, AM-11 was launched at 05:45:12UTC on 6 December 1958 carrying a tiny payload, the 6.67kg (14.7lb) Juno II-A, later named Pioneer 3 by NASA. A cone-shaped object, 58cm (22.8in) tall and with a base diameter of 25cm (10in), the object was intended to pass by the Moon almost 34hrs after launch. Initial plans had been for the probe to carry a small camera to photograph the lunar sphere but after results from the Explorer I satellite discovered the Van Allen radiation belts, a Geiger counter replaced it. A de-spin mechanism was to have reduced the rotation rate from 400rpm to 6rpm using small weights on the ends of two wires, 1.5m (4.9ft) long unspooled to their full length from stowed reels.

Initially, the launch went as expected but the main stage shut down 3.7sec early due to the failure of a propellant depletion sensor and the flight path angle was greater than required which, when the de-spin system failed, prevented mission objectives being achieved. The probe reached a maximum distance of 102,360km (63,606mls) before falling back and burning up in the atmosphere 38hr 6min after launch. It did, however, discover a second radiation belt around Earth which added greatly to the expanding field of knowledge regarding the planet's geomagnetic environment.

The second Moon probe, Pioneer 4 followed on the heels

Right: The first mission for Juno II, Pioneer 3 being prepared for stacking on the clustered second and third stages of AM-11 at LC-5, Cape Canaveral, for an intended flight to the Moon. It ended up in a single, highly elliptical, orbit of the Earth after launch on 6 December 1958. *NASA*

of Russia's Luna 1 spacecraft, which launched on 2 January 1959 to become the first man-made object placed in heliocentric orbit after passing the Moon at a distance of 6,400km (3,977mls) just 34hrs after leaving Earth. Pioneer 4 was almost identical to Pioneer 3 and carried a small photo-sensitive cell with which it was hoped to trigger pictures when the bright light of the Moon came into view. Launch occurred at 05:10:56UTC on 3 March 1959 from LC-5, last of the five ARPA deep-space science probes taken over by NASA. The first stage lifted the assembly to a height of 113km (70mls) before staging coasting for 5sec. Separation of the fairing 8sec later followed by axial alignment and the successful firing of the 11 solids in three stages beginning 37sec after that.

It was vital for the de-spin system to work so that a slow rotation of the probe would allow the photocells to respond to the increased sunlight reflected from the Moon as it passed. The de-spin weights were released at 11hrs into the mission and in 0.25sec reduced the rate from 420rpm (after being de-spun from 600rpm at launch due to the operation of the solids) to the desired 11rpm. As the probe sped through the space it detected what was at the time believed to be a third radiation belt encircling the Earth. Measurements of this region were of special interest to engineers planning manned flights to distant places. As with the previous Pioneers, several ground tracking stations were involved in maintaining contact with the probes, including the Jodrell Bank radio telescope near Manchester, Cheshire.

The closest approach to the Moon had been targeted for approximately 32,000km (19,885mls) but slight perturbations in the trajectory from the solid stages increased that to 58,983km (36,652mls) and a speed of 7,230kph (4,493mph), 41hr 13min after launch. This was insufficiently close to trigger the photocell. Periodic tracking continued with the last session 82hrs into the mission at a distance of 654,860km (407,000mls). Tracking from the FM/PM link was through a 275mw signal and could have endured to a distance of 1,126,000km (700,000mls) had the batteries lasted longer. Tracking stations were located at Cape Canaveral, Florida, Mayaguez, Puerto Rico, and the new station at Goldstone, California. A tiny, man-made planet in the Solar System, Pioneer 4 remains in orbit around the Sun, approximately at the radius of Earth but unlikely ever to collide with it.

The limited success of the Pioneer probes brought great cheer to the von Braun team and the Huntsville engineers at a time when, as we shall see, ambitious plans were being made for growth in launch vehicle capabilities. Celebrations were only temporary as attention was focused on new rockets and increasing numbers of engineers were drained out of the Jupiter programme to work on those. They did afford some degree of celebration, more so now that they were part of a new civilian space strategy only then being forged, but that was tempered by fears that the organisation might be swallowed whole and lose its independence. But there were still several more Juno II launches booked which would keep the ABMA busy on operational flights for a further two years, next up being the 41.5kg (91lb) Explorer S-1 satellite, an investigation into the Earth's radiation bands and cosmic rays. Work which had started under V2/Bumper sounding rocket flights was now being extended through the use of artificial satellites.

Launch occurred at 17:37UTC on 16 July 1959 from LC-5 when AM-16 took off on a flight lasting only 5.5sec before the range safety officer sent the destruct signal. A short circuit in the power supply regulator cut off power to the engine gimbals on the main stage, incurring a hard-over which sent the rocket into one of the most dramatic cartwheels seen at Cape Canaveral. Over at LC-26B preparations were soon underway for a very different type of satellite, packaged into a tiny compartment on top of a two-stage solid cluster.

That launch, designated AM-19B, was in support of an inflatable sphere, Beacon 2, which would have assumed a maximum diameter of 3.6m (12ft) when pressurised from a container with a length of 80cm (31.5in) and a diameter of 17.8cm (7in). Launch of the beacon satellite occurred at 00:31UTC from LC-26B on 15th August 1959, the first Juno II from this pad, but the vehicle failed when ejectable flares added to the guidance compartment for this flight caused a fire which took hold and put the upper elements in the wrong angle preventing the payload reaching orbit. This was the second Beacon launched, the first by Juno I on 24 October 1958 which failed when the upper stages broke away from the ascending vehicle.

On 16 September 1959, a standard Jupiter missile test rocket (AM-23) was fired from LC-6 but shortly after lift-off the rocket encountered difficulties which required its detonation by range safety, causing damage to Juno II AM-19A on the adjacent pad, LC-5. The launch was held pending a decision on its suitability for flight and that was given on 28 September proceeding to launch on 13 October at 15.30UTC. The flight was a success, placing the 41.5kg (91.5lb) Explorer VII satellite in orbit where it operated for two years providing valuable information on solar x-rays, radiation levels at various distances from Earth, cosmic rays and with a secondary role of recording the Earth's heat balance, essentially a repeat mission of the failed Explorer S-1 launch.

Next up was AM19C, the launch of Explorer S-46 which occurred at 13:35UTC from LC-26B on 23 March 1960. It failed to achieve orbit when one of the solids in the second stage failed to fire sending the upper elements back down through the atmosphere. Explorer 8 followed off LC-26B when AM-19D was launched at 05:23UTC on 3 November 1960. With a weight of 40.9kg (91lb), the satellite was delivered safely to orbit where it reported electron density levels and ion concentrations in near-Earth space. A battery failure on 27 December 1960 shortened its planned life but, although inert, it remained in space until 28 March 2012 when it decayed out of orbit.

Not so successful was AM-19F launched from LC-26B at 00:13UTC on 25 February 1961 carrying the satellite Explorer S-45 which failed to reach orbit after a malfunction to the spin stages. From the same pad, the 37.2kg (82lb) Explorer 11 was launched by AM-19E at 14:16UTC on 27 April 1961 which was the first gamma-ray telescope sent into space. It was followed by the final Juno II launch at 19:48UTC on 24 May 1961 carrying Explorer S-45A from LC-26B but the second stage solids failed to fire and it fell back through the atmosphere and into the Atlantic Ocean. In 10 flights between 6 December 1958 and 24 May 1961, Juno II achieved only a 40 per cent success rate, six failing to achieve their objectives. In several respects, Juno II represented the culmination of attempts to use the Redstone and Jupiter missiles as satellite launchers with significant steps being taken during that period to raise the lifting capacity and the reliability of the launch vehicles.

Juno II, III and IV

In the early days of the US space programme there were only four credible rockets for application as space launchers, albeit with appropriate upper stages. These were, in increasing lift capacity: Jupiter from the Army, Thor built by Douglas, Atlas from Convair and Titan produced by The Martin Company. Each would comprise the core stage of a multi-stage launch system capable of supporting early satellites and space probes. Only Jupiter, as Juno II, would not remain in the inventory because of its limited lift capacity. The rest were Air Force projects and each would enjoy a long and varied future equipped with increasingly more powerful and effective solid and liquid propellant upper stages. The pressure for a more substantial, heavy-lift capability came from both political and scientific lobbies. The Eisenhower Administration was keen to accelerate available of lifting capacity and sought to assuage public concern over Soviet success in the weight-lifting contest. Still in its infancy, the space science community sought lifting capacity to fulfil their plans.

As we have seen with the Red Socks plan proposed by JPL, the early weeks and months of the post-Sputnik 1 era saw a wide range of ambitious plans and proposals. In November 1957, galvanised by the prospect of launching America's first satellite, the von Braun team proposed a new launch vehicle to begin the process of transitioning to an evolving space programme based on highly ambitious goals. To achieve those it needed a bigger rocket and a series of intensive meetings were held at which optional development strategies were considered.

Some engineers wanted to start with a clean slate and pushed for a very powerful rocket built around a combination of high-thrust liquid propellant engines and optional upper stages. A few wanted to start immediate development of a cryogenic upper stage for exploiting the lift capacity of a powerful first stage. Others pressed for caution and a gradually evolving capability based on the existing Juno II. Von Braun opted for an evolutionary and conservatively managed series of increasingly more powerful rockets based firmly on demonstrated principles of rocket engineering.

The initial step was, logically, designated Juno III. This was a Jupiter rocket first stage equipped with more powerful solid motors from Grand Central Rocket Company which it was calculated could lift 64.4kg (120lb) into low Earth orbit – a payload growth of about 20 per cent over Juno II. One proposal was for Juno III to launch a small spacecraft to photograph the far side of the Moon. Nothing came of Juno III because the performance gain was too little for the effort required to bring it to fruition. The limitations of solids was apparent in calculations of potential thrust levels, the mass of possible upper stage configurations and the velocity required for orbital or escape trajectories.

The Juno IV concept originated on 20 March 1958 when the ABMA's Gen. Medaris asked JPL to examine the possibility of replacing the solid upper stages with an arrangement of liquid propellant second and third stages for a new configuration, albeit one based on Jupiter. Juno III was not included in the list of candidate launchers which doomed that design. Instead of further use of solid propellant clusters as with Juno I and Juno II, which had a much lower specific impulse, ARPA too wanted the Army to do with new launchers what the Air Force was doing in developing liquid propellant upper stages for Thor and Atlas. Mass for mass, liquid propellant motors provided greater thrust. Seven days later, ARPA announced Operation Mona which aimed to put satellites out to the Moon, the three Thor-Able and two Juno II launches cited earlier.

The initial Juno IV design study was completed by JPL in early April 1958 and ABMA was briefed on the results. Hypergolic (self-igniting on contact) propellants were proposed for the pressure-fed upper stages with the proposed design capable of lifting up to 20 times the payload of a Juno II or III. Propellants would be nitrogen tetroxide (N_2O_4) and hydrazine (N_2H_4). Two new engines would be installed for the upper stages and helium as the pressurising gas would be employed through a unique heated-hybrid concept, the hydrazine tank being pressurised by heated products from the gas generator. This initial Juno IV proposal envisaged a ve-

Below: The dimensional proportions of the Juno II launch vehicles, a modest improvement on performance over the Juno I but rapidly outclassed by derivative variants of the Thor and Atlas which had USAF Agena upper stage elements. *JPL*

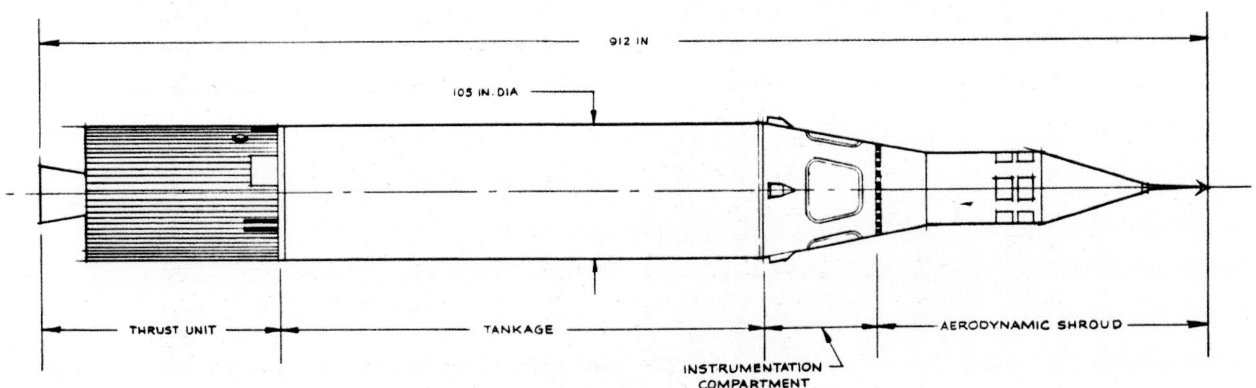

hicle with guided upper stages which could be developed in time for a first flight in March 1959.

During the next few months the design studies continued with JPL concentrating on lunar and deep-space missions and ABMA concentrating on Earth satellite missions. It was found that the optimum configuration for all Juno IV missions could be obtained by merely altering the tank sizes in the various stages. Thrust levels for the upper stages were selected on the basis of a compromise between performance and the technical challenges which could threaten timely availability of the launch vehicle. The design vacuum thrust for the second and third stages were to be 200.16kN (45,000lb) and 26.69kN (6,000lb) – known as the '45K' and the '6K', respectively.

For a time, JPL wanted to employ General Electric to upgrade its 405H engine which delivered a thrust of 146.8kN (33,000lb) but the budget would not allow that so JPL proceeded alone with developing the second stage motor. The 405H was a development of the GE 405 operating on a mixture of LOX and kerosene producing a thrust of 120kN (27,000lb) and the first engine to employ turbine exhaust for roll control. It had been developed as the first stage rocket motor for the Vanguard launcher and formed the basis for several proposed evolutions. It was also assigned as the motor for the Vega upper stage proposed for the Atlas rocket but this was eventually cancelled in favour of the more efficient Centaur cryogenic upper stage which had great success.

As work on Juno IV progressed it became apparent that the time needed for development could not match the requirement for a flight in March 1959, so it was proposed that a vehicle with only two stages (Juno IVA) be prepared for early missions, consisting of the basic Jupiter and the initially proposed third stage. Such a vehicle would have respectable Earth satellite capability but no lunar or planetary performance. After the formal go-ahead in August 1958, a programme was put in place for three two-stage and three three-stage (Juno IVB) missions and JPL set about designing satellites for those. With slightly enlarged tanks for the smaller upper stage, Juno IVA could lift 494kg (1,090lb) to a 400km (250mls) orbit. It was further determined that the proposed missions could be made without the more powerful second stage, a kick-stage being employed instead.

There was some repetition here. Although the Juno IV payload capability was about three times that of the Air Force Thor-Able combination, it was only about equal to the Thor-Agena A with which it would be a contemporary. But Agena A was a highly classified project and NASA only learned about it at the end of 1959 so this could not be factored in to launch vehicle availability planning. Earlier, JPL's director William Pickering had asked his organisation to study a Mars fly-by mission with a spacecraft weighing 159kg (350lb). When NASA opened for business on 1 October 1958 it inherited the Juno programme from ARPA and on the 9th it met with ARPA officials to debate Juno IV and decided against proceeding, even after a second meeting the following day to clarify some technical points. The Juno IV programme was formally cancelled on 17 October and all funding was diverted to what had emerged as the next project – known initially as Super Jupiter. It is interesting to note that JPL's plan for space probes, intended for Juno IV led quickly to the Ranger Moon and Mariner Mars/Venus spacecraft, all of which flew on Atlas Agena launchers.

There was a lot of overlap here. As noted earlier, between January 1958 and October 1959, the Army and NASA flew six Juno I launches and, between December 1958 and May 1961, Juno II made 10 scientific flights for NASA as a 'civilian' vehicle sending satellites and space probes on their way. In addition, the Redstone had been adapted to fly the Mercury spacecraft on unmanned launches and test flights supporting NASA's first human space flight programme. Five Mercury-Redstone launches took place between December 1960 and July 1961, the last two each carrying an astronaut on a suborbital flight and putting Americans into space for the first time.

For a time, Jupiter was considered by NASA as a test rocket for the manned Mercury spacecraft in addition to Redstone for ballistic flights, Jupiter being used for high-velocity re-entry profiles before committing the spacecraft to orbital flights with Atlas. Adapting Jupiter for Mercury was difficult in that the overall mass was slightly greater than the capacity of Jupiter, even for a ballistic trajectory, and the diameter of the nose section would have required an engineered adapter. It was all considered too expensive and time consuming and the idea was dropped.

But this was not only the end of the Juno I and Juno II programmes but the demise of Juno III and Juno IV as well, bringing to a halt derivative applications with the Redstone and Jupiter rockets. Cancellation also led to the end of JPL's solid propellant rocket development programmes but that also encouraged them to press ahead with the '6K' rocket motor.

Left: Launched 13 October 1959, Juno II AM-19A successfully deploys the Explorer 7 Earth-orbiting satellite. *NASA*

Left: Juno II AM-16 stands on LC-5. ready to launch the 42kg (92.6lb) Explorer S-1 satellite into Earth orbit – the third flight of a Jupiter space launcher derivative.
Below left: AM-16 goes awry seconds after lift-off 16 July 1959. *NASA*

Along with GE's 405H, it was considered for the Atlas-Vega launcher but that entire programme was cancelled by NASA on 7 December 1959 when the decision was made to use Agena-B as upper stage to Atlas. The whole Agena programme was classified because it related directly to Top Secret military programmes but the stage became available to the civilian agency. Thus began a long association with a series of upper stages for the civilian space programme bearing the Agena name.

Devoid of propulsion work, JPL turned to the deep-space missions which it had already been studying and so began an enduring commitment to building spacecraft and managing missions for planetary exploration. In fact, the deep-space missions studied by JPL for allocation to the Juno IV launcher formed the basis upon which the Ranger and Mariner spacecraft evolved for flights to the Moon and to Mars and Venus respectively – using Agena B and later Agena-D upper stages on Atlas rockets. But this was a prescient move. None of the NACA facilities and laboratories which NASA would inherit were tuned to work on interplanetary probes and even satellites and space science missions had no natural home. That would soon change with the construction of the Goddard Space Flight Center, Maryland. On 1 January 1959 the majority of JPL projects came under the jurisdiction of NASA, leaving only a few residual military projects such as the Sergeant missile. Hereafter, JPL would enjoy the same autonomy of project management as that it had under its work with the ABMA.

While JPL went its own unique way and the use of JPL-developed solids for Juno launchers no longer applied, except in support of remaining Juno II flights, the ABMA was already busy with its Super-Jupiter, or Juno V. That had a functional origin in the ABMA's early 1958 outline for a national programme of space exploration. Guided by von Braun, the ABMA supported a highly ambitious set of objectives at a time when a great deal of jockeying was taking place among organisations and military services anxious to stake their claim on the new frontier. Immediately after the successful launch of Explorer 1, and maximising his new place in the limelight, von Braun put his weight behind a concept originating at the ABMA for sending the first human into space. In two tests during which a balloon carried a gondola to an altitude of 30km (18.6mls), the Air Force had already tested the effect of high-altitude ascent on a man during the Manhigh programme. A third and final test would be made in late 1958.

Von Braun wanted the Army to go one step further and in the spring of 1958 had managed to get the cooperation of scientists working for the Air Force in aerospace medicine. Distinguishing it from the Air Force effort as Man Very High, the proposed ride involved two conical warhead re-entry bodies positioned base upon base to form a diamond shape, placed on top of a Juno I launch vehicle without the upper stage solids. A pilot would be inserted into the lower conical section and propelled to an altitude of 240km (149mls) before the upper conical section would separate and the lower section parachute to a splashdown 250km (155mls) off the coast of Cape Canaveral. It would provide six minutes of weightlessness during which biophysical data could be obtained.

The ABMA rebadged it as Project Adam (or 'First Man') and took it to the Army on 13 May and thence to the NACA. At a Congressional hearing, NACA Director Hugh Dryden dismissed it as ridiculous, with 'about the same technical value as the circus stunt of shooting a young lady from a cannon'. Ironically, the NACA was working with the Air Force on Man In Space Soonest (MISS) which had much the same idea, a fact which may have prejudiced Dryden's judgement, although there is no direct evidence to support that. ARPA boss Roy Johnson told the Army that Adam had no traction and rejected it on 11 July but the project was submitted to the Secretary of Defense on 8 August, whereupon it was dropped. Significantly, politically charged up due to Sputnik launches, Dryden's opportunity to head up the new space agency was dismissed by the politicians for a lack of boldness and imagi-

JUNO IV-A

JUNO IV-B

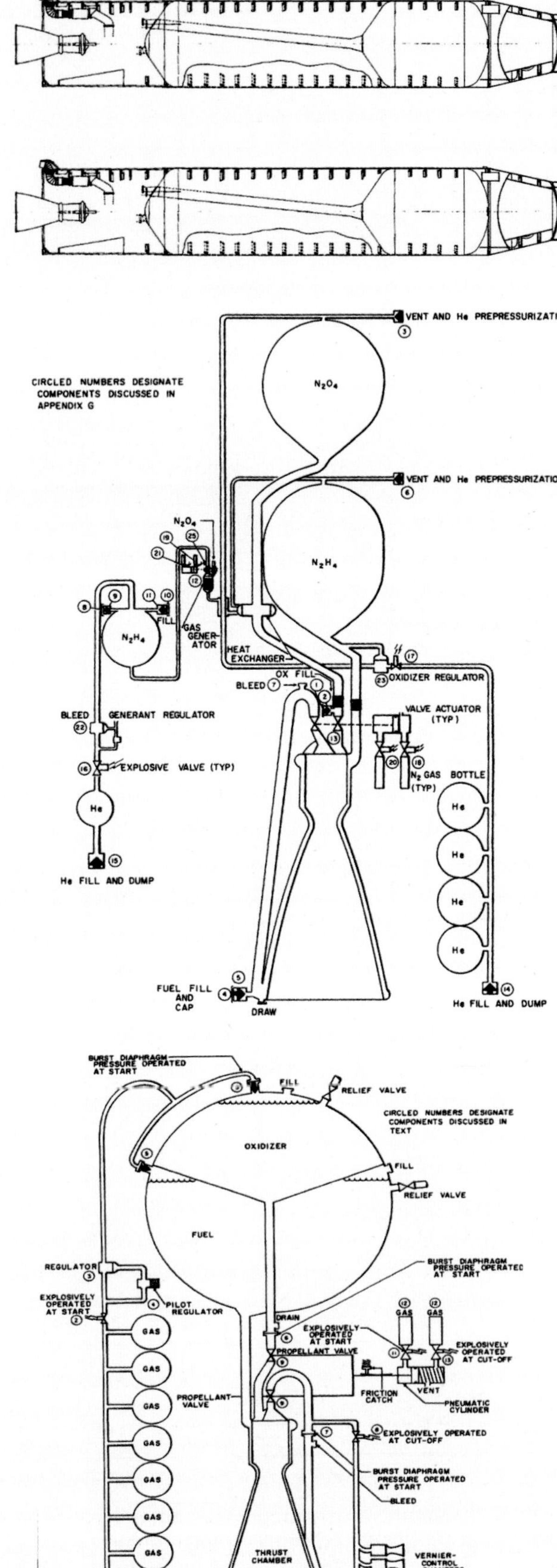

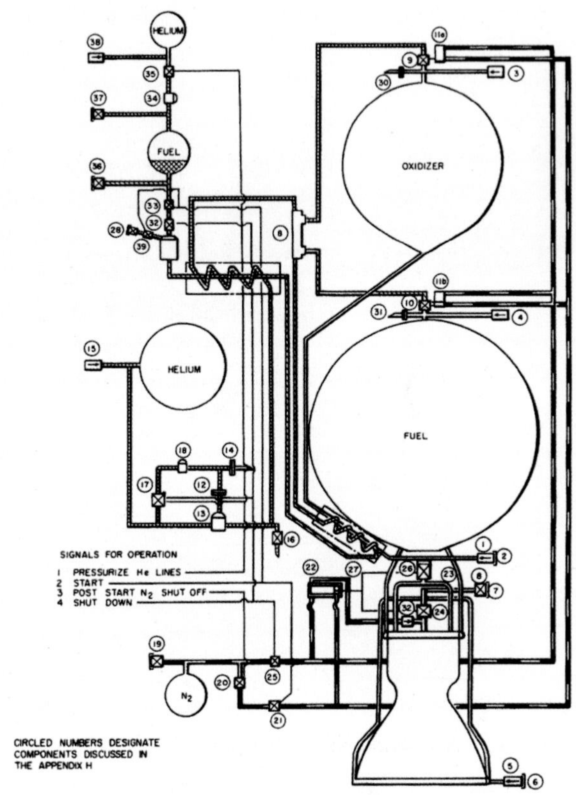

Juno IV rocket

Top: Juno IV-A and IV-B were proposed as a development of Juno I and II but despite presenting a workable solution to the limited weight-lifting capacity of Juno II, it was insufficient to warrant development, the 'Super Jupiter' powered by E-1 or F-1 engines being favoured. *JPL*

Above left: The hybrid Juno IV propulsion system employed a nitrogen tetroxide mono-propellant gas generator to pressurise the fuel tank and helium heated by exchange with the products of the generator to pressurise the oxidiser tank. A gimballed thrust chamber was used for yaw and pitch control. *JPL*

Above right: A schematic of the Juno IV flow control process of the gimbal motor control and heated helium pressurisation system. *JPL*

Left: The design of Juno IV provided the opportunity to develop a more sophisticated upper stage system and this included a vernier motor helium pressurisation system for the 26.7kN (6,000lb) thrust high altitude stage, as displayed here. The verniers would have been used as ullage motors for stage separation. *JPL*

nation the nation craved. He was passed up for T. Keith Glennan, albeit retaining a post as his deputy.

When NASA became effective from October it adopted MISS from the Air Force and renamed it Project Mercury with the added purpose of orbital flight. Von Braun had seen Adam as one step on the way to development of large rocket-powered troop transports, an outgrowth of an earlier idea for using rockets to deliver logistical supplies to the battlefield. As a double irony, a Redstone (in the lengthened tank version) was used by NASA for the two Mercury suborbital flights in 1961, precisely as von Braun had proposed three years earlier, putting the first two Americans in space in a spacecraft which would individually place the first four NASA astronauts in orbit during 1962 and 1963.

But this was only one programme proposal engineered by von Braun with the full backing of the ABMA. In early 1958 it had prepared a draught programme which envisaged a small space station launched in 1962 supporting four occupants, followed by a manned expedition to the Moon in 1966, a permanently manned base on the lunar surface in 1973 and the start of a manned expedition to Mars and – if deemed hospitable – Venus in 1977. This was broad-brush politicking, trying to display a sense of awareness in national needs and aspirations in the post-Sputnik world and a stake-holding claim to defining future programme policies. In the general turmoil of 1958 it did not survive long but the enabling sentiments informed what would emerge the following years as Project Horizon, about which more later.

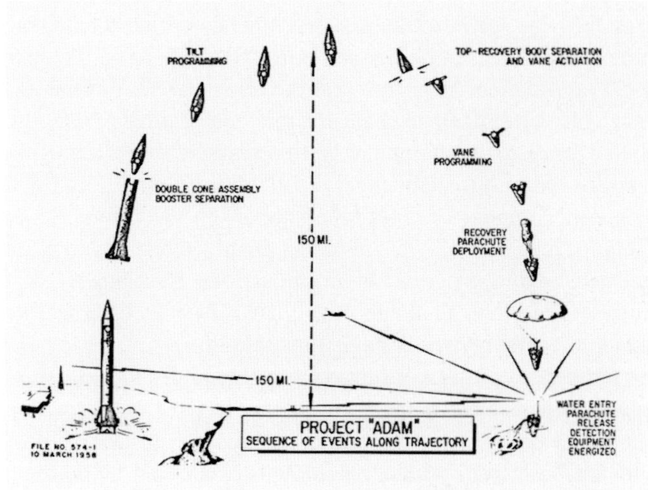

Above: Competing with the Air Force's 'Man-In-Space-Soonest' (MISS) for a ballistic shot into space as a pilot trial for Dyna-Soar, in February 1958 von Braun proposed using a Redstone to lob an astronaut into space and back by the end of 1959. The Air Force objected and the idea was dropped. *ABMA*

Super Jupiter

While propelling the prestige of von Braun to new heights, despite support required for the remaining launches from the Redstone and Jupiter programmes, for the ABMA the cancellation of Juno IV on 17 October 1958 was a worrying precedent, potentially threatening the very existence of the organisation itself. The Air Force was able to offer NASA, now hungry for satellite launchers, variants of its Thor, Atlas and Titan rockets at a price heavily subsidised by the large scale production of military missiles. The Army had nothing with which to compete. But there were concepts in development, if not yet on the drawing board, polarised around a set of options all based on the premise that a very powerful first stage would enable basic payload capabilities offered through a series of optional upper stages.

The bold idea – and it was an extravagant departure from the preceding philosophy of a step-by-step process – was known within the Development Board as Super Jupiter. And 'super' it was. The story of that concept began eighteen months before the cancellation of Juno IV. The urgency for a Super Jupiter, to maintain the core Braun team intact and to keep the Army in the rocket business, stemmed from groups within the ABMA encouraging the von Braun people to think in a very different way to the preferred, evolutionary path, to a capability which would leap above and beyond the payload possibilities offered by the Air Force with Atlas and Titan, each of which offered an evolving variety of upper stages to choose from for a specific requirement. The pressure to come up with something radical was engendered by the lack of cooperation from the Air Force in the wake of Secretary Wilson's decision for the Army to desist from further development of ballistic missiles with a range in excess of 322km (200mls) and that the Air Force should take over operational deployment of the Jupiter IRBM.

But these internal motivations at the ABMA were peripheral to a series of studies during 1957 of future needs, a list of prospective missions examined by the Department of Defense and covering the interests of all four military services (Army, Navy, Air Force and Marine Corps). In addition to a wide range of scientific requirements for defining the nature of the space environment and the conditions under which engineering solutions would need to be found, in this pre-Space Age year there was a profound awareness of a much-needed learning curve. Essential requirements focused on weather and communication satellites with an eye to more advanced missions in the future.

Moreover, the DoD had previously supported spaceplane and hypersonic research programmes known as Hywards, Brass Bell and Robo, anticipating the early development of an operational piloted strike and reconnaissance system. This would later emerge as Dyna-Soar and it would need a very powerful launch vehicle. In 1961 Dyna-Soar would be relegated to an experimental programme known as X-20 and then cancelled but it did play a role in the evolving story of the Super Jupiter. The initial specification for that was developed by Heinz Hermann Koelle in his function as the chief of the Preliminary Design Branch and kept quiet for a little time because he feared that von Braun would balk at such a radical proposal.

Supported by internal analyses, in April 1957 the ABMA decided that future lift requirements for all anticipated payloads were in the range of 9,070-18,140kg (20,000-40,000lb) for Low Earth Orbit (LEO) satellites and space vehicles and 2,720-5,440kg (6,000-12,000lb) for payloads designed for deep space and other missions requiring escape velocity. This is testament to just how fast the US military space programme

Above: Thirteen years junior to von Braun and never a member of the Peenemünde rocket team, Hermann Koelle was a pilot during the war and formed space and rocket advocacy groups in the late 1940s before becoming editor of a space magazine. Eyed by von Braun as a visionary thinker with technically innovative inspiration, Koelle was invited in to become a member of the Huntsville team and is largely credited with the Jupiter-C concept. *Author's collection*

was evolving, prior to the national urgency which would significantly increase momentum after Sputnik 1 in October 1957, and belies the notion that the Eisenhower Administration was 'sitting on its hands' as it 'sleepwalked' into the Space Age – to quote popular press opinion after the first Russian satellites were launched. As a stalking horse, the open 'civilian' Vanguard programme supporting the International Geophysical Year of 1957-58 was arguably at too slow a pitch and its languid progress would induce the Russians to rush ahead where they could so easily have been second, factors which the ABMA was at pains to point out.

This heavy-lift requirement, an assembly of potential space projects already under discussion in 1957 but for which there was at the time no feasible launch system planned, provided a base from which a suitable rocket could be defined. From these payload masses the rocket would require a lift-off thrust of 6,672kN (1.5million lb), almost four times the thrust of the Titan ICBM, at the time the most powerful US rocket under development. A wide range of upper stage configurations was contemplated but the sheer size of the proposed rocket was a design and development engineer's nightmare. Aware that existing ICBM projects Atlas and Titan with appropriate upper stage were each capable of launching satellites, von Braun calculated that even with high-energy cryogenic engines their payload capability would not exceed 4,536kg (10,000lb). This would accommodate less than half the potential payloads which even in 1957 were being considered as serving potential military space requirements for weather forecasting, communications and reconnaissance.

In an initial draft of the design, Koelle and his analytics expert Dr. H. G. L. Krause used IBM computers to run some calculations, finding that four E-1 engines under development at Rocketdyne would provide a total thrust of 5,871kN (1.32million lb) which was deemed capable of meeting the heavy payloads envisaged by the DoD that it would justify development. Koelle was so enamoured with the E-1 that he proclaimed it to be the ultimate rocket motor – that no bigger motor would ever be required. He loved the idea of clustering them to provide a truly heavy-lift capacity. The E-1 had a target thrust of 1,334-1,779kN (300,000-400,000lb) but in designing the E-1, Rocketdyne found that there was, theoretically, no limit to the output of a liquid propellant rocket motor. It was to exploit this possibility that designers at Rocketdyne started preliminary work on a much bigger motor, one with a thrust of 4,448kN (1million lb). This was designated F-1, for which initial work focused on the turbopump because it was with this element that the greatest stress, and the toughest engineering challenge, was believed to lie.

But it was the Air Force that liked the idea of such a powerful motor, giving an instruction for it to be developed. Rocketdyne placed the F-1 under the management of Dave Aldrich, recently from the REAP programme. Dom Sanchini was selected as his deputy, with Bob Linse picked out as the senior project engineer. Quite soon, Ted Benham was added to the team as development project engineer and he eventually succeeded Linse, Benham's position filled by Stuart Mulliken when he moved up to fill that slot.

Although the E-1 was static tested by Rocketdyne, it was

Left: Mercury Redstone awaits launch, incorporating the A-7 engine and LOX/alcohol propellants with the enlarged tanks of Jupiter-C. An additional auxiliary hydrogen peroxide tank was also mounted adjacent to the main tank at the top of the engine. It also incorporated a sophisticated failure detection apparatus integrated with the Mercury spacecraft Launch Escape System. The ST-80 guidance system was replaced with the less complex and more reliable LEV-3. *NASA*

Mercury-Redstone rocket

Right: After the absorption of the ABMA Development Operations Division and the Air Force MISS programme into NASA, von Braun's idea was resurrected and the Mercury Redstone programme began as a test vehicle for putting humans into space on short-duration ballistic shots before orbital flights on an Atlas. *NASA*

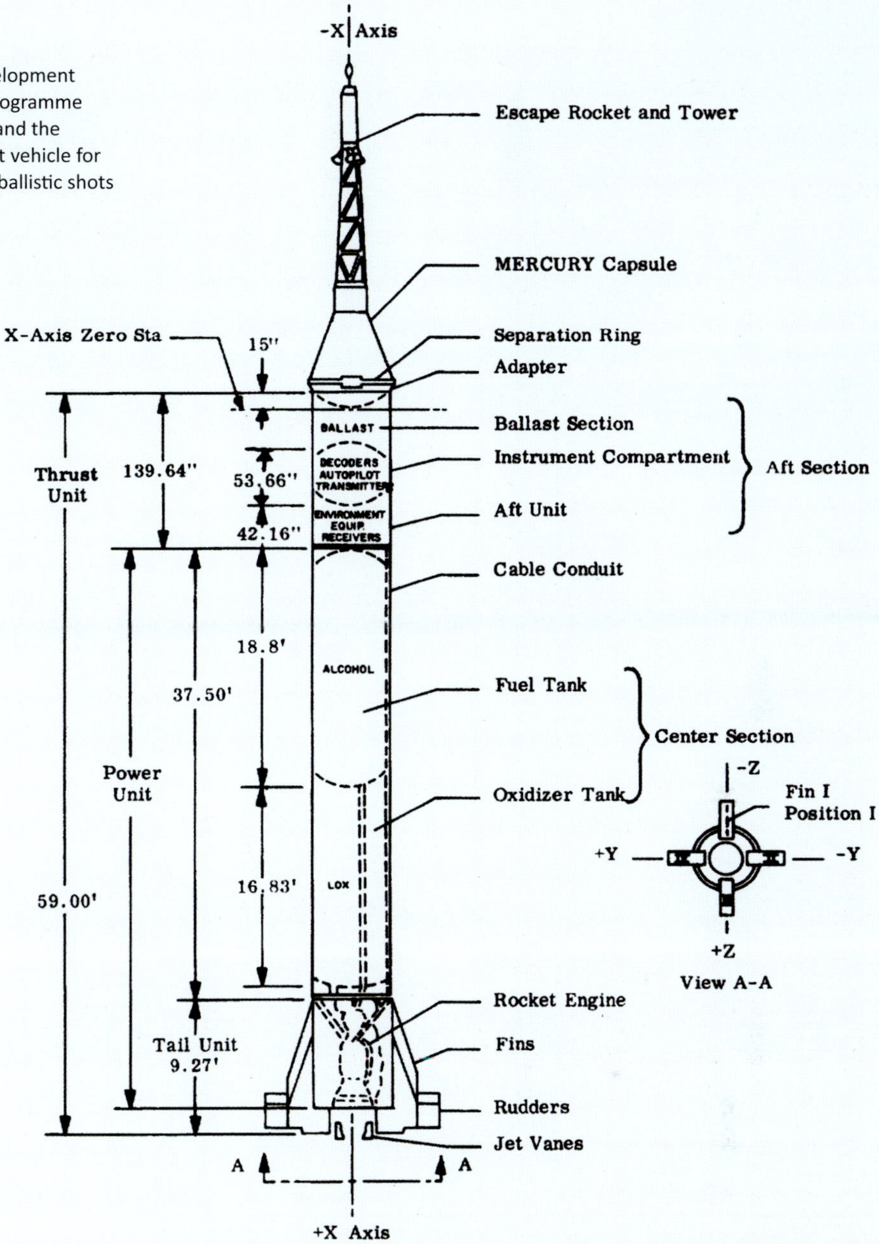

Above: Mercury Redstone represented the peak of development for this rocket, which found greater value in the space programme than it did in military deployment, launching America's first satellite in 1958 and the first US astronaut in 1961. *NASA*

seen to be too small a step beyond the engine already in production for the Atlas ICBM/space launcher and so was passed over for the F-1. As detailed testing of the turbopump got under way, a boilerplate thrust chamber was fired up, but only for short duration runs because it was not cooled. The sheer size of the combustion chamber was something altogether unknown to the existing cadre of rocket engineers and was soon dubbed 'King Kong'! The story of the F-1 is best told in our companion volume on the Saturn V because it only has a role in the story of the Super Jupiter.

Several vital parameters were factored in to the analysis of different concepts and configurations. Performance was clearly the paramount aim in the overall design, with the demand for much higher levels of reliability than were seen at the time with ballistic missiles of the Atlas and Titan generation. From the outset, this heavy-lifter was considered the first true workhorse for big space projects and as such would also be required to lift humans into space, beyond the limited potential of the one-man Mercury spacecraft. In this regard, Juno V was subject to a far greater degree of reliability and safety analysis than had been required for the IRBMs and ICBMs that came before it. And there was also the consideration that launch frequency would play a significant role in determining production and launch parameters, which led some to consider booster recovery as an essential element of any long-term space programme.

By December 1957 both ARPA and the ABMA were agreed that clustering rocket motors in the first stage was the only practical solution to the need for a powerful satellite launcher and that recommendation was forwarded to the Department of Defense by the ABMA in a document titled 'A National Integrated Missile and Space Vehicle Development Program'. The E-1 had been dismissed as a candidate for pro-

SPARTA / WRESAT

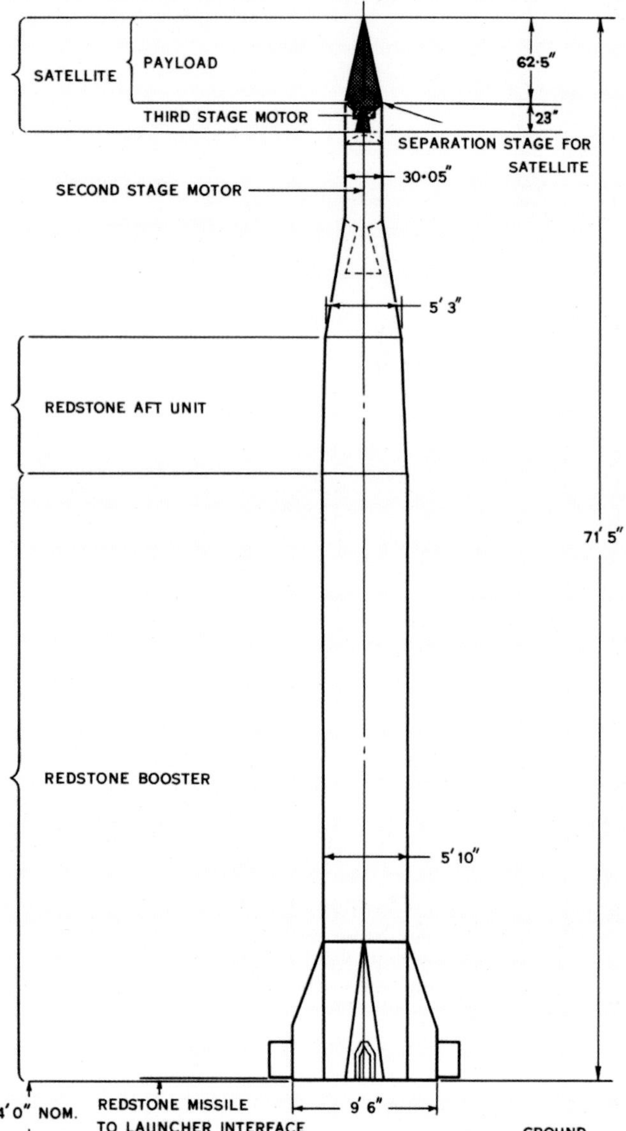

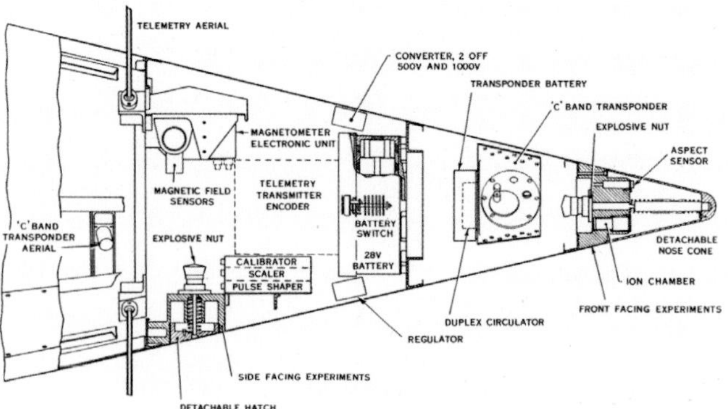

Left: The 10 SPARTA Redstone rockets were refashioned by von Braun at Redstone Arsenal as re-entry test vehicles and produced by Chrysler from August 1966 forlaunch from the Woomera Test Range in Australia. In re-entry tests the B-3 stage would fire to drive the simulated warhead down into the atmosphere at 21,600kph (13,400mph), faster than the British Black Knight used previously. The first successful test shot occurred on 28 November 1966.
Above: The WRESAT payload which, along with the third stage, went into orbit. From there it would gather information on the way the solar radiation interacted with the upper atmosphere.
Below: Launched by the last SPARTA rocket, WRESAT entered an orbit of 1,295km (804mls) by 193km (120mls) at an orbital inclination of 83.2deg. It operated for five days until the battery ran down after 73 orbits of the Earth. *Author's collection*

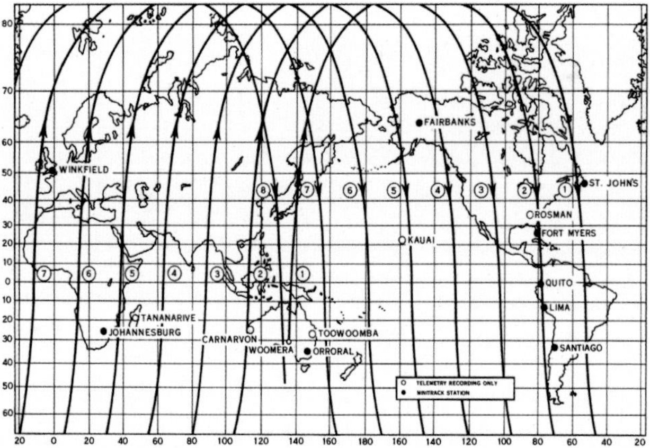

pulsion due to the amount of time required to build, test and qualify this new engine and the F-1 was even further down the road. In these early years of rocket motor development uncertainties about what was possible inspired innovative thinking and a lot of testing. The sheer size of the F-1 was unprecedented and did not find favour with many due to the potential for a catastrophic disaster if used on its own for a first stage.

In July 1958, when ARPA engineers David A. Young and Richard B. Canright visited ABMA they learned about the Super Jupiter and the E-1 engine plan but under the need to move fast with a heavy booster capability, they suggested using eight, modified S-3D engines in a cluster instead of four E-1s. Readily available and becoming operational with Thor and Jupiter, by adopting these proven designs an estimated $60million and two years of development time could be saved. Moreover, the von Braun team knew these motors well and understood their various nuances.

The suggestion was challenged by von Braun, who believed that two or three rocket motors were the most that could satisfy the balance between reliability and efficiency. He based his opposition on two factors: the more engines there are the more there is to go wrong; and, the more engines the heavier the inert mass of the spent stage and the lower the payload fraction of the fuelled weight. He also believed that there was a higher degree of potential failure with increased numbers of motors essential for lift-off.

That in itself could be countered, said Young and Canright, by reverting to the V2 principle of running the engines first to verify integrity of the combustion process before ramping up to full thrust and committing to lift-off. Moreover, with an increased number of motors in the first stage, an inflight failure could be countered if interconnected propellant supply allowed surviving engines to burn longer, consuming the propellant not used by the failed engines. To achieve this, there

Right: As a postscript to the Redstone story, the last rocket of this type launched was in support of Australia's WRESAT (Weapons Research Establishment Satellite, pronounced ree-sat) under the SPARTA (Special Anti-missile Research Tests) programme. Carrying an Antares-2 second stage and a BE-3 third stage, on the last of 10 Redstone flights, SPARTA launched the 45kg (99lb) WRESAT satellite on 29 November, 1967 making Australia the fourth country to launch its own satellite after the USSR, the USA and France. *Author's collection*

would have to be an acceptable margin of thrust over mass to allow some fail-safe accommodation should an engine failure, or two, to be overridden by the reduced thrust of the remaining eight engines. This became the classic equation of the engine-out capability which would be written in to the design of Juno V candidate configurations.

And there were other advantages too. The decision to use a cluster of eight existing, albeit modified, S-3D motors allowed the first stage to be shorter than would have been the case with a single F-1 type motor, which was still under consideration, and it was judged that by switching to an eight-engine configuration, testing could itself begin three or four years earlier. It also reduced challenges for directional control during ascent since a single, large rocket motor would be required to exert greater structural forces than would be the case with eight smaller motors, each of which could play an incremental role in tweaking pitch and yaw. Lastly, transportation would be easier for getting the required motors to the launch pad.

In the overall analysis, there were other advantages to clustering. Reliability was a real problem, not helped by the acceptance that military weapon systems – of which Thor, Atlas and Titan were a part – had a tolerable level of failure. Part of the estimation for the number of IRBM/ICBM rounds required took account of the probability that only two-thirds of them had any chance of reaching their targets. It was that way with artillery rounds, free-fall bombs and torpedoes launched from submarines. The military mind-set was such that 'tolerable failures' became an acceptable reality and this failure rate could only be overcome with a surfeit of numbers, the failures being compensated by the 'overkill' inventory making up for losses.

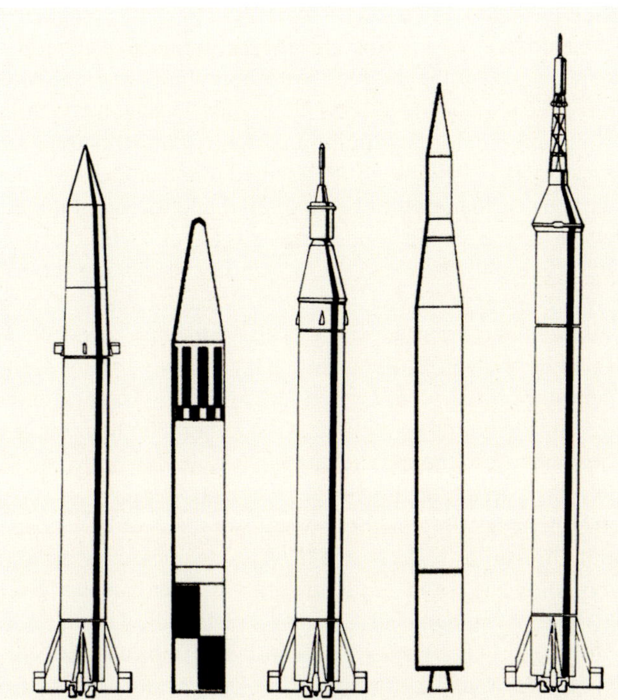

The evolution of the Redstone. In addition to launching the first US satellite into orbit, on 21 July 1958 a Redstone became the first rocket to detonate a thermonuclear device in space, at an altitude of 76km (47.3mls) with a yield of 3.8MT. A second test took place on 11 August 1958 with a detonation of 3.8MT at 43km (27mls), both under Operation Hardtack out of Johnston Island in the Pacific Ocean. *Chrysler*

But the space programme was emerging as something completely different. Which is why, when adapted for launching satellites and spacecraft, Thor, Atlas and Titan received higher levels of fidelity in reliability assurance through quality control and redundancy. Such was the case with the Atlas and Titan rockets adapted for lifting Mercury and Gemini spacecraft, respectively. Considerable resources were applied to making them safer and much more reliable, to the point that, of the 14 Atlas and Titan launched in putting people into space, there were no failures. This approach could be built in to the development of Juno V, because it was an inherent part of the concept from the very beginning and not a post-development adaptation to an existing rocket. Which is why, with only the hint of criticism, it would be said of the von Braun team that they over-engineered their rockets.

This philosophy was attractive to ARPA and the safety factor achieved through clustering provided inherent compensation for a single failure. The von Braun team built in, through the propulsion and guidance equipment, the ability of their Juno V to continue to achieve performance objectives should one or two engines fail after lift-off. By using eight existing engines, with an increasing reliability record, if any one should fail the other seven could continue to operate and get the payload into orbit. That this became a part of the flight test objectives was a requirement demonstrated in flight, proving that the design of the rocket could achieve a 'fail-operational' performance. This was taken to an advanced level in the design of the Saturn V, where on two occasions, engine failures did not prevent the launch vehicle from achieving its objective.

When briefed on the proposal for the big booster in April 1958 ARPA showed interest and defined its early availability as a priority with favoured status being the use of eight clustered Jupiter engines. Full authorisation was given in ARPA directive 14-59 bearing the date 15 August with a full-scale captive dynamic firing by the end of May 1959. Added to this on 23 September was authorisation for four test launches, the last two carrying a live second stage the exact specification for which was awaited. ARPA officials visited the ABMA on 4 November to evaluate the proposed programme and discuss the possibility of assigning initial payloads, in particular placing a heavy communications satellite in geostationary orbit.

In its briefings to ARPA, ABMA emphasised the flexibility of the programme, asserting that what would begin with the Juno V launcher would allow successive levels of development to support space missions anticipated for the following two decades. They emphasised the flexibility of the core stage as a potential test-bed for different engine configurations, by example for instance the replacement of the four inner engines with a single F-1, or another rocket motor of similar thrust. The potential for conveying troops and cargo between two locations on Earth was asserted for this booster as it had been in previous proposals. But reliability was considered a prerequisite for effective operations and the enormous cost of each rocket was addressed in proposals that spent stages be recovered by parachute and used again, cost savings of 50 per cent being claimed.

Below: On 23 March 1960, Juno II AM-19C carrying Explorer S-46 was launched from LC-26B at Cape Canaveral but would fail when one of the eleven second stage solids refused to fire, causing asymmetric thrust vector and the total loss of the vehicle. *NASA*

Throughout 1958 and 1959 the concept of 'safety through simplicity' became the yardstick by which the Juno V was measured. An engine-out situation was directly built in to the overall design. Should one engine fail a fraction of a second after lift-off, a second could stop firing 60sec later without any impact on the success of the ascent, delivering the payload to the pre-determined location. But there was a caveat. While it was decided that the inner four engines would be fixed, the outer engines would gimbal for directional control in pitch and yaw axes and three of those would have to be operable throughout the ascent. Nevertheless, as originally speculated, should one or two engines fail, the others would burn longer consuming the propellant unused by the failed engines.

In this regard, the complement of vehicle configurations, quality and safety-assurance criteria and fail-operational philosophy, as America's first heavy-lifter and the first launcher specifically designed to carry humans into space, set precedents. It became the hallmark for space engineering and safety concepts which exist today, albeit in modified and restructured form. This seminal year in which decisions were made that would rewrite the way space launchers were developed and to what standards, is today largely under-appreciated. But it cannot be emphasised too highly that the success of manned space operations in particular was ensured through the work conducted by ARPA and the ABMA in 1958.

Paradoxically, a reduction in performance accuracy at booster cut-off followed the rule of simplicity and enhanced safety. The parameters at booster engine shut-down and thrust termination would be within much wider ballparks than was the case with existing missiles adapted to launching satellites. This allowed the stage to be simplified and to put all the dispersion mitigation into the upper stages, meaning that they would be required to have the capability of making increasingly precise shut-down parameters so that the terminal stage would achieve great accuracy. This innovative concept enshrined by the von Braun team meant that the driving design criteria for orbital accuracy – essential for any launch vehicle – would ride with the terminal stage rather than be built in to every stage from the ground up.

In engineering the first stage, von Braun imposed a 20 per cent structural safety factor above the criteria which had been applied to earlier rockets and missiles such as the Redstone, Thor, Atlas and Titan. In anticipation of this launcher being used to carry humans into orbit, this placed the Juno V in the same category as commercial aircraft and improved the reliability of the system as a whole. Thus, the Juno V was the first 'man-rated' launch vehicle designed for significantly greater levels of reliability driven by the mandatory preservation of life in the event of a major failure. It also originated the appellation that the Juno V would be the 'workhorse' of space, the DC-3 commercial aircraft carrier of the rocket inventory, words resurrected for the characterisation of the Shuttle programme merely a decade later.

Chapter Five

From Juno V to Saturn I

In early 1958, ABMA examined several different booster configurations of which four appeared most promising. The first envisaged a single, large propellant tank with a diameter of 5.486m (18ft) divided between the LOX and RP-1 tanks by a common bulkhead. The main advantages here were minimum overall dimensions, minimum plumbing and the use of existing knowledge with the design experience on conventional tankage in rockets. Negative aspects emerged when handling became an issue, since the tank could not be broken down into separate elements for transportation. A stage of this size would be much bigger than anything manufactured to date. It would require new tooling for fabricating the structures, fuel feed lines would have to extend down through the LOX tank and an insulated bulkhead together with heavier anti-slosh baffles would be required.

The second option would hark back to the early days at Kummersdorf in the mid-1930s, a concentric arrangement of two tanks, one inside the other. The outer cylinder would contain the fuel while the inner cylinder would contain the LOX, a highly unusual but practical solution. It would eliminate the need for fuel lines through the LOX tank and it would all but eliminate the slosh problem. However, this arrangement would weigh about 20 per cent more than the conventional arrangement and that was unacceptable for a rocket of this size.

The third option consisted of eight tanks carrying the same dimensions as a Redstone rocket (177.8cm/70in) clustered around a central tank with the diameter of a Jupiter rocket (266.7cm/105in). These diameters were chosen so that the same machine tools used to fabricate the Redstone and Jupiter tanks could be used to fabricate the tanks for this configuration. Handling and transportation would be greatly simplified as each tank could be moved around and shipped separately. The use of existing materials and tools would greatly reduce the cost, while bulkheads and penetrating propellant delivery lines would not be necessary and the existing Jupiter anti-slosh system could be used. The only disadvantage was the greater diameter of clustered tanks than with a single monolithic tank as defined by the first two options.

The fourth contender was a variation of the third, with eight Redstone-size tanks in a circular cluster with the same diameter as the third option (6.5m/21.33ft), each tank being,

Above: The Saturn weight test simulator on the way to its destination, displaying the unprecedented scale of the new rocket. *Author's collection*

in essence, a Redstone missile with LOX and RP-1 tanks separated by an insulated bulkhead. Longer tanks would be required to provide the same burn duration required but by omitting the centre tank propellant delivery lines from the vertically stacked segments, delivery lines could be brought down the centre tunnel to the rocket motors below without penetrating the LOX tanks.

Clearly, the third option was the most satisfactory and was the one selected for Juno V but that left two options for how to operate the stage. The conventional method would have been to burn the propellants to depletion in all tanks and separate from any or all upper stages following shut-down. But there was another way. The author is uncertain as to whether a concept of parallel staging was inspired by the Atlas system whereby the two boosters, each flanking the fixed sustainer engine, were discarded during ascent to reduce weight but it was an intriguing proposition. In this, all eight engines would be ignited for lift-off with engines and tanks dropped off as the stage requirements were fulfilled, the remaining tanks and engines continuing as the next stage. The propellants used during the first-stage burn phase would be supplied from the tanks dropped at first stage separation.

Several disadvantages are obvious with this concept. Because some of the engines would fire all the way from the launch pad to orbit, the expansion nozzles could not be optimised for altitude operation and the last stage would be heavier than otherwise due to the additional valves and plumbing plus the thrust frame to which all tanks would be attached at launch. Additional development time was also necessary, delaying the introduction of the parallel stage concept and this would delay the introduction of Juno V. ABMA did conclude that close observation of the Atlas missile development may provide lessons for such a concept.

But there were several advantages including greater flexibility in burn times for individual missions which would offer greater flexibility for specific tasks. It also eliminated the problem of ignition of upper stages at altitude which, in 1958 was considered a highly risky business and likely to increase the failure rate of the overall system. Yet a smaller number of engines would be required to perform the same mission and would be better utilised since the centre engines burn longer. With all engines burning, a shorter total firing time was required. Smoother acceleration would also ensue and this was felt desirable for future manned missions. Moreover, it would result in a shorter, more compact vehicle which would reduce assembly, launch and handling challenges.

The very concept of clustering brought unique problems: The aft end of the rocket could be a source of several problems for acoustics, vibration, thermal energy management, plume impingement, back-flow stage-creep and other such unwanted issues, each a potential disaster waiting to bring the rocket down. Acoustics created by the operation of the rocket motor are a major parameter in any design but when compounded with a new range of frequencies coupled in parallel, the frequency spectrum is a major consideration in the design of the payload. They also play an important part in launch vehicles designed to carry humans, an issue also relating to vibration and the wave and amplitude as measured in cycles per second. The human body can only withstand certain levels of acoustic and vibration spectra.

Thermal management is different with a cluster than with single or paired rocket motors. This, connected to plume impingement, can impact the placement and location of hydraulic, pneumatic and propellant feed lines. What is satisfactory for a single engine is a problem compounded when engines are clustered. Radiated and conducted thermal energy loads drive the selection of materials, lubricants and seals and existing engines, grouped in clusters, and frequently require subtle changes in the materials selected for use. Plume management in a close-clustered group of rocket motors brings unique problems. Clustering patterns must consider any void into which a lower pressure region above the engine expansion chambers (nozzles) can suck up hot gases which could impinge upon the flat base of the thrust structure and burn through. This is also a specific design issue with thermal back-flow in inter-stage adaptors for multi-stage launch vehicles.

Balancing all these factors, it was decided to proceed with further definition of the Juno V concept on the basis of a conventional staging sequence, with additional upper stages taking over after first stage shut-down. But this was solely to fast-track initial flight tests and the option of parallel staging for later Juno V flights was kept open. The emphasis on upper stage choices was, at this point, on existing equipment rather than anything new, so as to get an early flight test and move quickly to an operational capability. There was a belief among von Braun's group that some adaptation of Juno V for supporting extensions to the hypersonic X-15 programme or Dyna-Soar may call upon that rocket for launching these vehicles. But there was also the urge to make choices because the booster stage itself could not attain orbital velocity. Yet just because of this, lofting X-15 or Dyna-Soar vehicles on suborbital trajectories was an advantage for a single-stage Juno V since it would not waste an over capacity, multi-stage vehicle where a boost was all that was required to get the vehicle beyond the atmosphere, which is where both aero-vehicles would operate.

With a generalised agreement between ARPA and the von Braun people regarding clustering and staging, the decision over upper stages was more complicated, with a variety of factors being relevant. Here, as said, existing upper stages were preferable due to the time urgency of getting a heavy-lift capability. That urgency was in large part a response to the launch of Sputniks I, II and III which appeared to show a capability gap embarrassing to the political leadership in the United States. But while existing upper stages using LOX/RP-1 were clearly available and in development, the promise of high-energy, cryogenic propellants was highly attractive due to the greater payload capacity they promised. But so massive was the Juno V core booster stage that existing ICBM stages could, it was felt, be utilised for an early, heavy-lift capability.

Accordingly, the ARPA/ABMA activity focused on two generations of Juno V: the first utilising conventional propellants with alternate, high-energy upper stages and a simple booster recovery programme with an operational capability sometime between 1960 and 1970; the second, a higher-thrust 8,896kN (2,000,000lb) first stage with a sophisticated recovery system for reuse and high-energy or storable-propellant upper

A key turning point in the further development of the Juno series was the suggestion that existing engines could be clustered together to create a powerful first stage, with upper stages being either existing rockets or new ones. The determining factor was the configuration of the propellant tanks and there were several options including conventional arrangement of fuel and oxidiser tanks in tandem (1), concentrically arranged tanks with the fuel tank buried inside the oxidiser tank (2), or a central tank and four outer oxidiser tanks, the latter spaced by four fuel tanks (3 and 4). *MSFC*

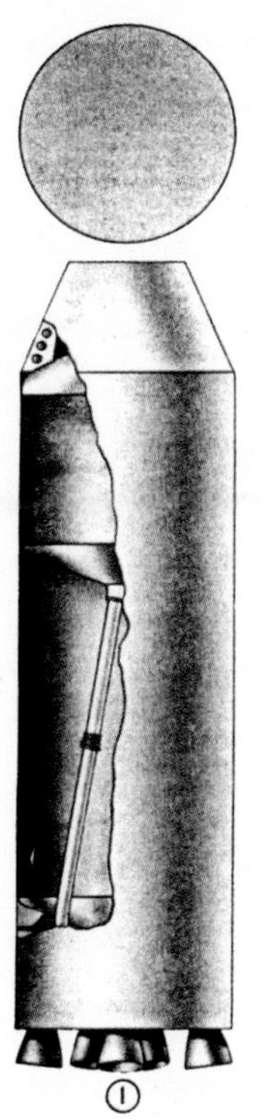

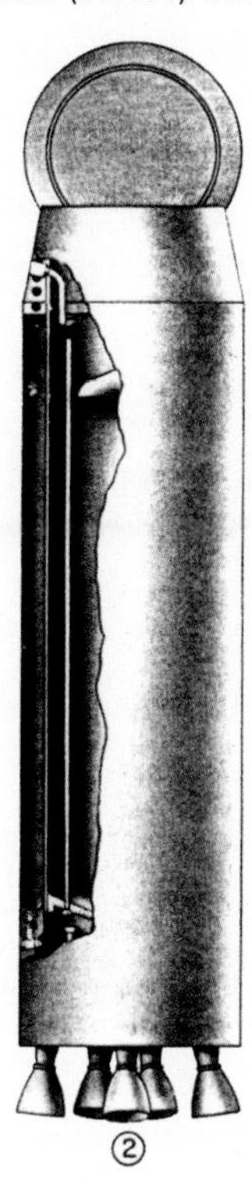

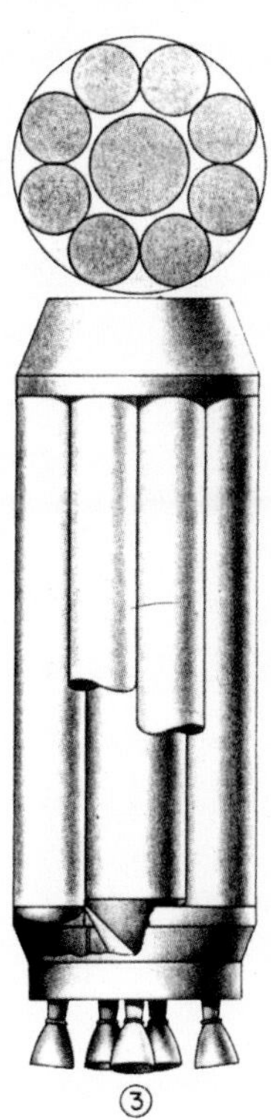

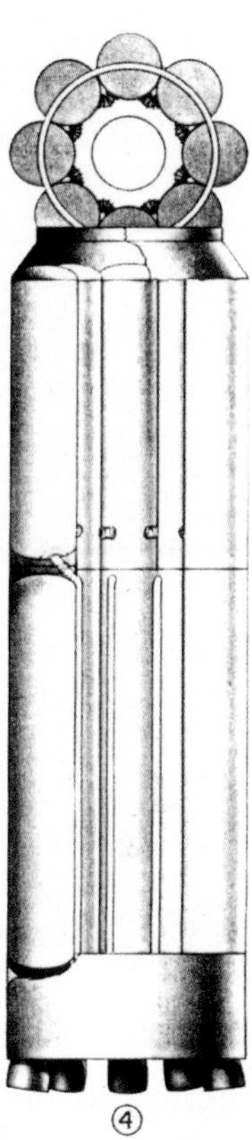

stages for specific missions. The first generation would comprise six building blocks. The first would utilise the booster stage with the eight improved S-3D motors with Atlas or Titan upper stage structure producing an optimum thrust of 890kN (200,000lb) but converting later to a cryogenic motor. Depending upon the selection for the second stage, a cryogenic third stage with a thrust of 133.4kN (30,000lb) would ensure total payload and mission flexibility. There was consideration of a small fourth (kick) stage with a thrust of 26.7kN (6,000lb), a legacy from the now defunct Juno IV configuration.

Safety and reusability were introduced as prime requirements for the design of the Juno V at all stages and in all configurations. The safety aspect included a fire mitigation approach where each of the eight engines had a fire wall and a unique fire extinguisher, enabling that failed system to be shut down and made safe while the vehicle itself continued to operate. But there were other considerations unique to Juno V. At first, reusability was to be a key factor in pushing the idea of a giant super-booster and a considerable amount of time and effort was put into the best way to recover the stage intact, reuse it on a later launch and save production costs. Ideas about how to effect that safe recovery were varied, from a trailed ballute or a solid speed brake to parachute recovery and retrorockets.

This latter concept appealed most and envisaged some form of decelerator, slowing the stage to Mach 1.5-1 from its peak velocity of Mach 4-5, upon which a single ribbon parachute would pop out and reduce the speed to Mach 0.8-0.7. That would be jettisoned and three 32m (105ft) diameter main parachutes would be deployed, slowing the descent of the booster to perhaps 24m/sec (80fps) or 88kph (54.5mph). Eight solid propellant rockets situated around the base of the thrust structure just above the rocket motors would be fired on the signal from a contact switch trailing on a 30m (100ft) wire hanging from the base of the stage. Bringing the stage to a complete halt by the time it reached the surface of the water, the stage would settle gently and be recovered from the splashdown point, some 322km (200mls) from the launch site at Cape Canaveral. Floating on the buoyancy of its empty propellant tanks, the stage would be recovered to a barge by crane on a Navy Landing Ship-Dock alongside.

Transported back to Redstone Arsenal, the stage would be disassembled and cleaned to prevent corrosion with engine parts returned to the inventory and reused in new engines off the production line. Initial water recovery tests on used engines showed that the majority of parts could be used again. Other tests on the thrust structure and the spaceframe itself encouraged confidence that they too could be recycled. Early on it was decided to use 5456 aluminium alloy which welds and repairs easily, at the cost of a slightly lower strength-to-weight ratio than other alloys which could more easily corrode.

Several methods of returning stages to Redstone Arsenal were examined. With the stage disassembled, elements could be airlifted in 11 Douglas C-124 Globemaster II cargo aircraft with nine carrying the tanks and two the motors but this would

Above: The high costs of manufacturing rockets meant that reusability was an important early consideration for the Juno V launch vehicle. In this contemporary visualisation, a combination of parachutes and retrorockets slow the burnt-out first stage for a soft landing at sea (left) and recovery by ocean-going barge (right). *MSFC*

cost $200,000, not counting the need for bespoke facilities at each end, compared with shipping the stage for $110,000. The decision was clearly in favour of shipping, 1,370km (852mls) from Cape Canaveral around the southern tip of Florida, into the Gulf of Mexico and 2,722km (1,690mls) up the Mississippi, Ohio and Tennessee rivers. A journey lasting 13-21 days by special barge. Only a 12.8km (8mls) section of new road was necessary to truck the stages to Redstone Arsenal.

Rocket Motors Galore

Things began to happen in a hurry. As said earlier, on 15 August 1958, ARPA issued Order 14-59, authorising the ABMA to proceed with funded development of Juno V, firmly endorsing a clustered-tank booster with eight clustered rocket motors from the S-3D developed for Thor and Jupiter. With time the urgent driver, long-lead elements needed to be in contract as soon as possible and on 11 September 1958 Rocketdyne received a contract to develop what would be referred to as the H-1. The following month, ARPA redefined 14-59 by requiring the ABMA to study all potential upper stage configurations and to recommend a shortlist leading to the further development of Juno V beyond the original requirement. ABMA was also to study a complete vehicle system which would be operated from the Atlantic Missile Range, Cape Canaveral, and on 11 December, Order 47-59 gave the go-ahead for the Army Ordnance Missile Command (AOMC) to design launch facilities

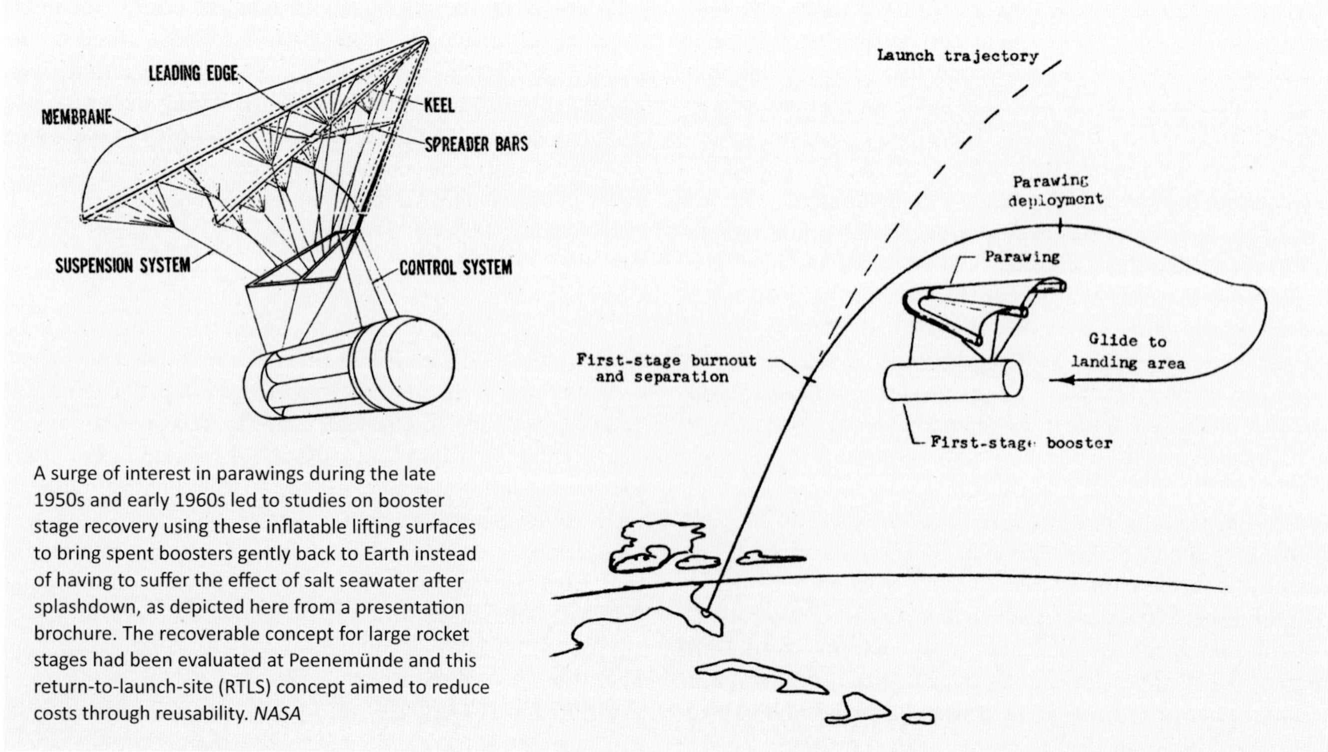

A surge of interest in parawings during the late 1950s and early 1960s led to studies on booster stage recovery using these inflatable lifting surfaces to bring spent boosters gently back to Earth instead of having to suffer the effect of salt seawater after splashdown, as depicted here from a presentation brochure. The recoverable concept for large rocket stages had been evaluated at Peenemünde and this return-to-launch-site (RTLS) concept aimed to reduce costs through reusability. *NASA*

and to work up the design of a static tower supporting Juno V stage testing.

The contract for the H-1 included a requirement to increase the thrust of each motor to 836.224kN (188,000lb), thereby achieving the total required stage thrust of 6,672kN (1.5million lb). But this would take time and a phased evolution of the H-1 would not hinder early flight testing, which would be different to strategies used previously with other rockets and missiles. The only other large multi-stage rocket in development was the two-stage Titan I ICBM. With Titan I flight tests, the first four launches would carry dummy second stages before full, two-stage launches. But the Juno V programme was different in that the upper stage was not yet firmly decided upon and would probably involve several different upper stages, yet to be selected. What was certain, however, was the evolution of a hydrocarbon-fuelled launch vehicle to a hybrid – one in which only the first stage would use RP-1, upper stages using cryogenic LOX/LH_2 for greater efficiency per thrust level. To achieve that, two rocket motors would evolve as fundamental to the story of the Juno V, the RL-10 and the J-2.

For a long time, von Braun had favoured cryogenic propellants for upper stages but, with the time-consuming process of fuelling and launching a rocket stage with non-storable propellants, they had little application for the military, which required storable and not cryogenic propellants for instantaneous launch. Both Atlas and Titan I would use liquid oxygen but Atlas would be phased out as the solid propellant Minuteman was introduced and, with storable propellants Titan II would quickly replace Titan I. Only Juno V was afforded the freedom to employ high-energy, cryogenic propellants. Because it was a space launcher, whereby longer preparation time posed no constraint on that vehicle's usefulness, it was operationally suitable in a way in which it had little application for weapon systems.

The advantages of cryogenic propellants had been known for a very long time. It was a matter of chemistry. The most abundant element in the universe, hydrogen, was a game-changer for space launch systems, significantly increasing the specific impulse (I_{sp}) of the propellant combination. I_{sp} is defined as

Above: Redstone Arsenal worked closely with Rocketdyne to develop the S-3D, which had evolved for the Thor and Jupiter programmes, into the H-1 which it was believed could be clustered to achieve the necessary thrust levels for placing heavy military payloads in Earth orbit. *Rocketdyne*

the thrust integrated over time per unit weight of the propellant, 'weight' here being regarded as that produced by the mass under the acceleration of 1g. It is measured in seconds, simply being the number of seconds a propellant mass of 1kg will produce a thrust of 1kg. Clearly, the higher the Isp the greater the efficiency of the motor. A rocket engine operating within the Earth's atmosphere has less thrust (exhaust velocity) than it would have in a vacuum. This is due to the pressure of that atmosphere on the exhausted gases. Further, solid propellants have less specific impulse than liquid rocket motors and among the latter, cryogenic propellants have a much greater Isp than rocket motors using a hydrocarbon propellant as fuel.

For instance, LOX/RP-1 propellant will have an Isp of around 305sec while a LOX/LH_2 motor will show around 450sec. It would, therefore, be a great advantage to use cryogenic engines in upper stages but the technology was difficult to master and required unusually sophisticated handling techniques, which had not yet been applied to rocket motors. Ironically, for all its lack of heavy-lift capacity, a product to some

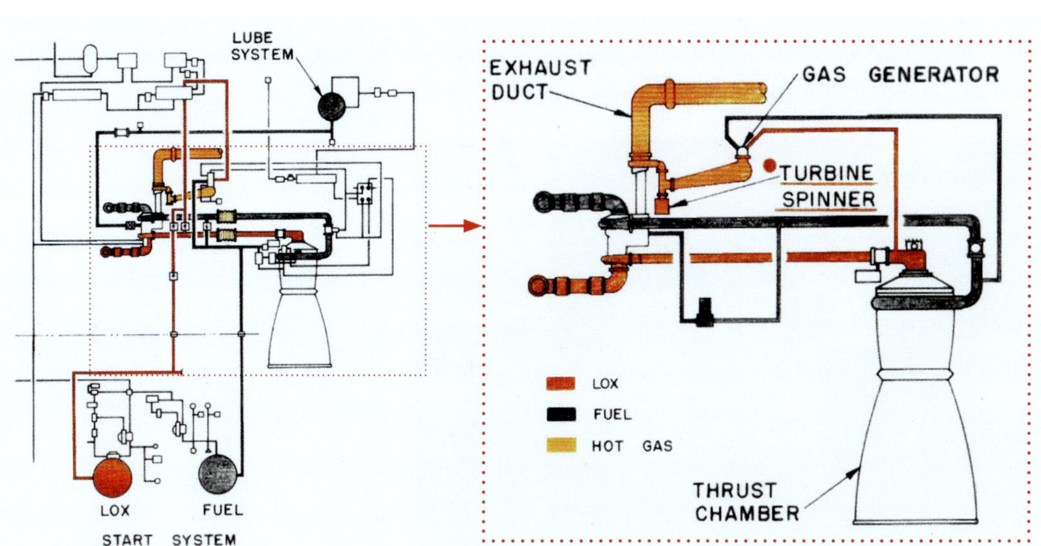

Schematics comparing the S-3D (left) with the later H-1 (right), which would go on to power 19 Saturn rockets. The derivative H-1 was essentially a completely refined design incorporating many changes and modifications introduced in the knowledge that reliability had to be considerably greater than that for unmanned launchers. *Rocketdyne*

extent of a lack of innovative thinking, the United States would become the world leader in cryogenic rocket motors for the next several decades. But there were profound challenges in handling the propellant, not least because of the low temperatures required to maintain hydrogen in the liquid state, -252.8°C (-423°F) compared with liquid oxygen at -182.9°C (-297.3°F).

Moreover, there was a further challenge in the design of a rocket stage, where LOX (stored at a density of 1.42kg/m^3 (0.088lb/ft^3) has a liquid/gas density ratio of 1:860 compared with 1:851 for liquid hydrogen. It would be impossible to design a rocket with an internal structural volume more than 860 times that of a rocket containing gaseous rather than liquid propellants. In designing the interface between the two cryogenic tanks however, given the extreme difference in temperature between the two liquids, the second law of thermodynamics demands that the tanks containing these propellants will have conductivity and boil-off features unique to each. With conventional rocket stages the two liquids associated with, say RP-1 at room temperature and liquid oxygen, would be designed to handle a single temperature gradient, that from ambient temperatures to liquid oxygen, whereas a LOX/LH_2 tanked vehicle would handle three gradients (ambient/LOX/LH_2). There is one other insidious impediment to conventional engineering: a hydrogen molecule is very small and will require much tighter mechanical joints to seal the tanks.

The development of cryogenic propulsion in the United States, and specifically for the Saturn programme, originated in the work conducted by British engineer Randolph Rae and his influence on Lockheed's innovative designer Clarence 'Kelly' Johnson, soon to achieve legendary status for the U-2 and SR-71 spy-planes. Before these emerged, in the early 1950s Rae briefed specialists in the Development Projects office at the Wright Air Development Center (WADC) on a unique design he had for an integrated O_2/H_2 fuelled turbine which would power an aircraft capable of flying at altitudes above those at which conventional gas turbines could operate. This became embraced within the Suntan programme, which envisaged just such a system and led to work at Pratt & Whitney (P&W), the aero-engine builder, under its chief engineer Perry Pratt. The Suntan design was not adopted but it prompted Johnson, who became aware of the concept, to design a Mach 2.5, hydrogen/oxygen powered spy-plane able to operate at 30,300m (99,400ft). That too was rejected, prompting Johnson to propose Aquatone, entering service as the U-2.

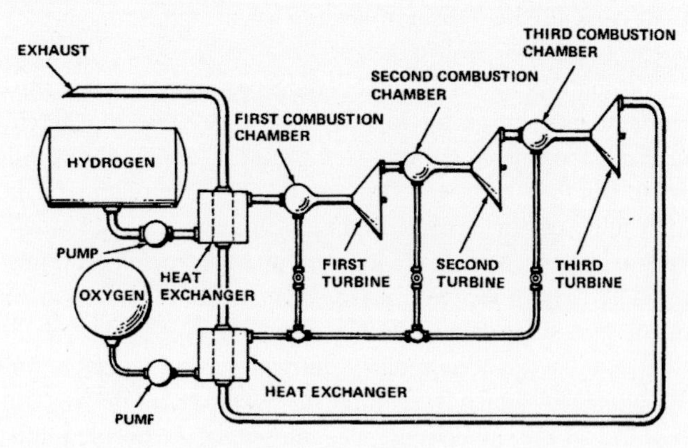

Above: The principle behind Rae's REX I engine is displayed in this graphic, whereby liquid hydrogen and liquid oxygen is warmed to a gaseous state through heat exchangers and delivered to three small combustion chambers, each driving a turbine before recirculation back to drive the heat exchangers. *Author's collection*

But investigation into hydrogen as a fuel, then regarded in most laboratories as a useless curiosity, continued unabated and there were factions within the Air Force that believed in the use of this element as a fuel of the future for truly exotic aircraft. If such confidence was ever to be realised, with hydrogen-fuelled aircraft a part of the inventory, someone had to find out about production of this exotic propellant. Johnson worked his research out of an old converted bombproof revetment and a cryostat he installed there produced 9 litres (2.3USgall) of liquid hydrogen per hour. The Bureau of Standards, meanwhile, got in on the act and set up a Cryogenic Laboratory at Boulder, Colorado, providing a facility which could produce 2,860 litres (757USgall) a day.

This activity, while never a convincing solution to the requirements of the Air Force, produced a vast depository of technical research which provided the backdrop to a broader confidence in cryogenic rocket motors: the theory was sound, the engineering requirements were well known and the practical production of large amounts of hydrogen had now been demonstrated. In fact, exceeded. The demands of the rocketeers for sufficient hydrogen to fuel their upper stages would be far less than squadrons of front-line aircraft would require. This work, conducted during the early to mid-1950s, reduced the development time for cryogenic upper stages by at least five years.

Inspired by the secret development work on the theoretical possibility of hydrogen-fuelled aircraft, and being already in the vanguard of rocket engine development, under a small team led by John Tormey, Rocketdyne had been working on cryogenic motors of this type since early 1957. In a state of competitive challenge, Rocketdyne's Tom Dixon, Vice-President for Development, had convinced key elements in the Pentagon that cryogenic motors of the performance improvement

Left: Seen here in 1955 during a hydrogen tank test, the English propulsion engineer Randolph Samuel Rae was an early convert to hydrogen-powered aircraft and conducted many tests to prove the validity of his radical designs. It was this work that would propel the US far ahead in the design of cryogenic propulsion systems. *Author's collection*

they accorded. They should, he said, be a top priority for the value they would have in upper-stage applications for space launch vehicles developed from the Thor, Atlas and Titan missiles. Embarking on a personal crusade, Dixon took Robert S. Kraemer with him for technical back-up.

The involvement of the Air Force was crucial to starting work on a cryogenic engine – they had the money and the inclination to take advantage of this new technology, while the Army had only limited funds for this kind of research. But while enthusiastic for cryogenic propellants, von Braun and his team remained sceptical about the practicality of the cryogenic motor. While enthusiastic over the possibility of such a payload-lifting advantage, he favoured the use of conventional fuels for the Juno V, influenced perhaps by some bad experiments with liquid hydrogen fuels in Germany, during development of the rocket research programme there in the early 1940s, souring his view.

Again, the paradoxical nature of von Braun's decisions and commitments raises questions about the balance between boldness and conservatism in his conclusions regarding challenging technology applications from established engineering concepts. Nevertheless, von Braun would be forced to fall in line under pressure from consensus views and import the cryogenic rocket motor to upper stage planning. This was another turn of opinion, similar to that he had already made over the clustered rocket concept and as he would again over the best way to send men to the surface of the Moon. But the cryogenic engine would transform Juno V into a flexible launch system and von Braun eventually became an ardent supporter.

The origin of Rocketdyne's work on a cryogenic stage can be traced back even further, to the British rocket engineers Val Cleaver and Leslie Shepherd in the late 1940s, as discussed earlier. The technical papers published in the *Journal of the British Interplanetary Society* in 1948-49 outlined the possibilities for a nuclear upper stage using hydrogen fuel and this stimulated work on further studies at the Oak Ridge National Laboratory, citing the British studies. Unfortunately, this activity was highly classified and failed to get the circulation it deserved but through a protracted sequence of connections, development of a nuclear stage was consolidated at Los Alamos. Eventually known as the Nuclear Engine for Rocket Vehicle Application (NERVA), it required extensive work on a hydrogen fuelled rocket motor. Rocketdyne was asked to design a turbopump capable of moving 37,850 litres (10,000USgall) per minute at 10.342kPa (1,500psi) and to build high-expansion ratio nozzles, initially water cooled and then regeneratively cooled.

During this period, P&W received the advantage of work carried out at the NACA's Lewis Flight Propulsion Laboratory, a facility dedicated to aircraft engines and research into their advanced and exotic engineering solutions. It would be renamed the Lewis Research Center when NASA emerged out of the NACA in October 1958 and in 1999 it became the John H. Glenn Research Center after the first American to orbit the Earth. In the mid-1950s, however, it had already carried out a considerable amount of research into cryogenic rocket motors. P&W received detailed information on their tests with regeneratively cooled cryogenic engines and this put them a step ahead of Rocketdyne in the race to be pre-eminent in this field.

In June 1958, ARPA ordered Pratt & Whitney to produce a cryogenic rocket motor based solely on their earlier work on Suntan, despite that company having never built a single rocket motor. Four months later, the Air Force Special Projects Office directed P&W to design a cryogenic rocket motor with a thrust of 66.72kN (15,000lb). The Air Force wanted this engine to be throttleable for its lunar lander in a proposed Air Force manned Moon programme called Lunex, which was then cancelled. Designated RL-10, the Air Force knew it as the LR115 and the motor would eventually become one of the most noted in the history of rocketry, surviving today as the primary motor for several launchers and powering the upper stage for NASA's Space Launch System.

Several design innovations were introduced for the RL-10 and these will be explained in the next section but the company had already taken steps to fit itself for new and challenging work. In 1956, under the incoming chairmanship of its owner, United Aircraft Corporation (UAC)'s H. Mansfield Horner, P&W acquired a large tract of land in Palm Beach County, Florida. This was a long way from its facility at Hartford, Connecticut and the new general manager of the satellite facility, Charles Roelke had to hire local engineers to prevent time-wasting travel between the two locations. Engineering operations had begun there in November 1956 with authorisation for a hydrogen plant completed in late 1957. Thus it was that P&W, already heavily involved with the Air Force through its aero-engine work, set up the facilities, research projects and manpower for development of the world's first cryogenic rocket motor.

Meanwhile, in 1955-56, Rocketdyne had been given an early introduction to the concept of a cryogenic rocket motor when Krafft Ehricke visited the company and discussed his

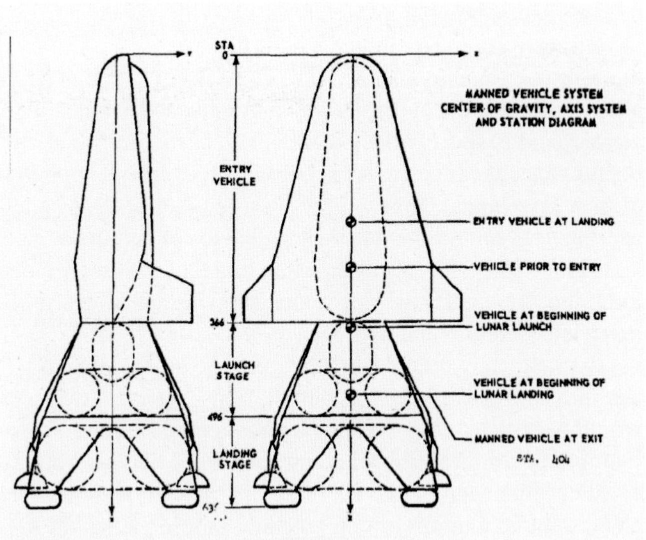

Right: By the late 1950s the US Air Force was contesting responsibility for the future of US space exploration with its Lunex Moon programme and an initial manned landing launched by an undefined vehicle in 1967 followed by permanent lunar bases. Like the Army Project Horizon proposal of 1959, Lunex was overtaken by the formation of NASA and the decision to fund a Moon programme through that agency. *Author's collection*

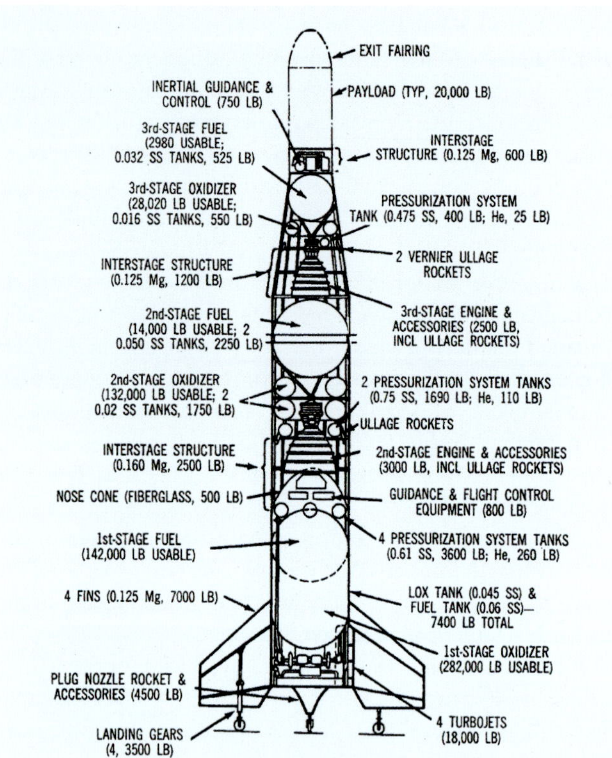

Left: Capitalising on uncertainties over the future of the Juno V/Saturn series, and seeking to begin a commercial launch vehicle market, in 1959 Rocketdyne engineers Kramer and Beyers proposed a three-stage vehicle capable of placing 9,072kg (20,000lb) into low Earth orbit with a recoverable first stage, matching Juno V. When Saturn went to NASA the project was dropped. *Author's collection*

idea for an upper stage to the Atlas ICBM. Atlas was under accelerated development and as a former member of von Braun's Peenemünde team, Ehricke had strong ideas about how to build a sustainable space programme. Ehricke had joined Convair in 1954 shortly before they got the contract to build Atlas. He simplified his engine concept, using pressurised propellant tanks rather than the standard turbopump but Rocketdyne dispensed with the pressurised tanks, put the turbopump in and changed the concept back to a conventional layout. Ehricke called this stage Centaur. When mated to the P&W RL-10, it would be a legendary contribution to the space programme through to the present day.

Surprised that P&W had got the contract to build the LR115 cryogenic motor, Rocketdyne's effort was placed under the direction of Stanley Gunn, which, despite his experience with rocket motors, brought new challenges. With low density and low viscosity, liquid hydrogen was difficult to handle and even more troublesome to operate with. They discovered that some alloys were brittle under its chilling influence while others were found to be unaffected. Adapted from the Mk 9 turbopump designed for the Atlas booster, Rocketdyne applied it to their cryogenic engine and began to build a series of engineering tests and design details intended to convert the cynical von Braun. They did this by showing him the leak-free welded joint and the compilation of alloys showing no affects from being exposed to liquid hydrogen.

Rocketdyne would design their cryogenic engine to a target thrust of 889.6kN (200,000lb). This placed it well, as would be realised later with its application in the Juno V programme. In terms of a chronology, the choice of power output was not influenced by the P&W RL-10, since it had been decided to build a motor of that potential before the RL-10 was placed under contract. Rocketdyne would name their cryogenic motor J-2, because it followed the H-1 in development, leapfrogging the letter 'I' as it could be confused with '1'. Rocketdyne had started out with the intention of using a series of letters beginning with 'A' for the Redstone engine (developments of which followed a numerical sequence) but that system fell into disuse until the 'E' engine described earlier and the mighty 'F' engine which would power Saturn V. The 'G' designation was used for an advanced fluorine/hydrazine system which was never developed.

Although the decision to put Rocketdyne under contract to build the J-2 would not be made until 1st June 1960, as we will see both the RL-10 and the J-2 would play a significant role in the development of these super-boosters and give Juno V tremendous growth capability with numerous permutations and options for different stage configurations. By late 1958, von Braun was slowly coming round to the added value of a cryogenic stage and began to integrate that option into detailed studies of what the Juno V could lead to.

A Challenge of Ownership

In November 1958 the von Braun team had a draft outline for Juno V tests, for which they scheduled 16 launches, starting with just the first stage and progressing to full qualification of a four-stage stack. The first two flights were slated for October 1960 and January 1961 and the initial configuration of eight H-1 engines delivering a total thrust of 5,337kN (1,200,000lb) with a total propellant load of 294,840kg (650,000lb) but with dummy second and third stages. These were primarily structural test shots, an evaluation of the launch processing and checkout procedures and of the booster recovery system. The next two flights were scheduled for June and October 1961, with a usable propellant load of 340,200kg (750,000lb), a dummy second stage but an active LOX/RP-1 third stage delivering a thrust of 355.8kN (80,000lb) and an attempt at delivering a test payload to orbit. As of November 1958, these were the only launch tests funded.

Subject to approval, and following these initial verification launches, three flights, in December 1961 and February and April 1962, would have the benefit of the upgraded H-1 motors for a lift-off thrust of 5,871kN (1,320,000lb), a dummy second stage but the active third stage and the 26.7kN (6,000lb) fourth, or kick, stage targeting a payload into a low Earth orbit. Confidently, flights 8-13 were scheduled for a five month period between June and November 1962 and were to incorporate the 890kN (200,000lb) live second stage together with the third and fourth stages from previous launches. The last three test shots (14-16), scheduled for January-March 1963 would be the same as the preceding series but with the final upgrade of the H-1 motors producing a lift-off thrust of 6,690kN (1,504,000lb).

These 16 test shots were considered to be only a preliminary programme and the von Braun team pushed for a 32-flight test regime, asserting that only then could they demonstrate a reliability level of 75 per cent – a dramatic understatement both of the number required to qualify the Juno V for space operations and the level of reliability they would be able to achieve. But it is interesting that the von Braun team were again thinking of using Juno V for payload delivery on surface-to-surface missions. Calculations showed that as a single-stage booster, Juno V could deliver a 136-234tonnes (300,000-450,000lb) payload across a distance of 322-482km (200-300mls) and a two-stage version a payload of 9tonnes (20,000lb) over a transatlantic distance. All this in a matter of minutes. The preoccupation with delivering very large cargo loads across great distances by rocket would endure well into the late 1960s.

For orbital flights, calculations showed that the Juno V equipped with conventional upper stage propellants could deliver 9,980kg (22,000lb) or 11,340kg (25,000lb) with cryogenic, high-energy propellants. Of equal value to the DoD, it could deliver 2,132kg (4,700lb) into a geostationary orbit where the military wanted to park their communications satellites, but that anticipated a launch site on the equator so that it would fly directly into the equatorial plane. A launch out of Cape Canaveral at a latitude of 28.5°N would require a dog-leg manoeuvre to get into the equatorial plane and that would significantly degrade the geostationary orbit capability. Payload performance was crucial and Juno V offered capabilities far beyond any launcher, or missile, available, at least on the drawing boards of US companies and agencies. Which is why there was significant talk of utilising a nuclear upper stage to gain even greater payload capability.

The ABMA expected the payload capability to grow and for Juno V to be a family of rockets rather than one single design, eventually powered by much more powerful motors, such as the F-1. In this regard, the von Braun team wrote into their documents a predicted capability for launching 90,700kg (200,000lb) to low Earth orbit or sending 9,070kg (20,000lb) to the Moon with the ultimate development of the Juno V. With this, the von Braun team confidently asserted that this capacity would satisfy all the requirements of the US space programme to the end of the 1970s. But from where to launch the giant rockets?

Extensive consideration was given to the test requirements and to a suitable facility for launching Juno V. Here too, the time factor was a prominent criterion and documents of the period consistently maintain that as a major consideration in any decision. Thus, it was recommended that existing facilities should be used wherever possible and that the Cape Canaveral site should be used to build a suitable launch complex from where test shots could be fired down the Atlantic Missile Range. None of the existing pads were suitable so decisions had to be made as to precisely where a new complex could be sited. That depended on the blast effect should a catastrophic failure result in a pad explosion.

Below: Convair's Krafft Ehricke. A former colleague of von Braun's at Peenemünde, he was instrumental in the early development of cryogenic engines of the type eventually employed on Saturn's upper stages. *NASM*

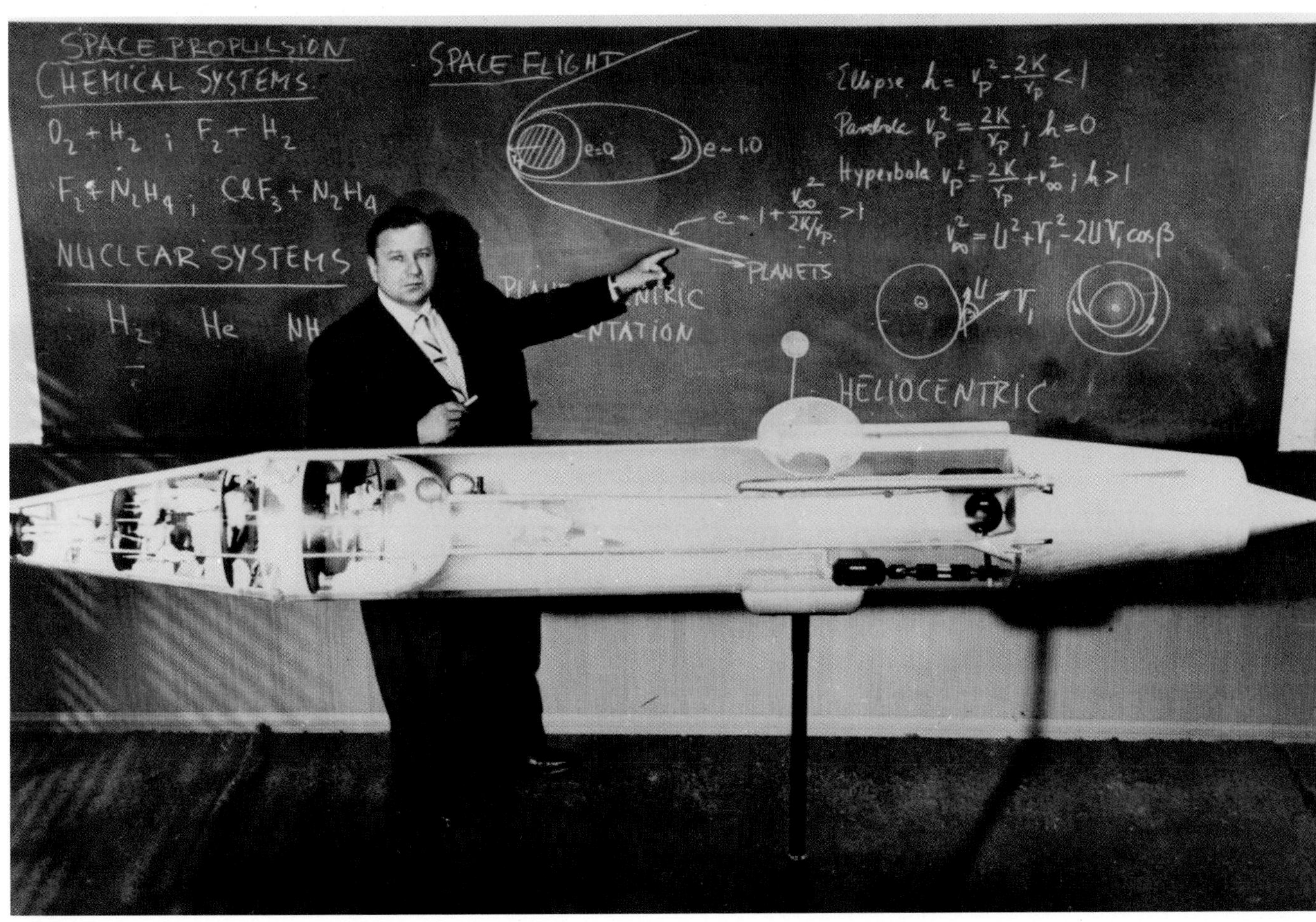

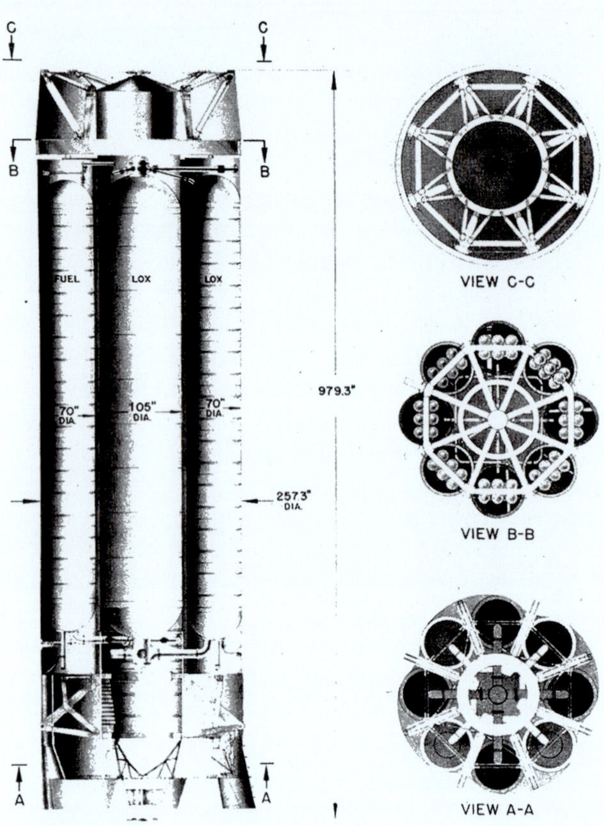

Juno V first stage configuration

Above: With uncertainty surrounding the future market for such a large launch vehicle, the decision over how to provide propellant for the cluster of rocket motors was settled in favour of an arrangement whereby a central tank surrounded by four outer LOX tanks, the latter interspersed with four fuel tanks, was decided. These were not 'left-over' Redstone and Jupiter tanks as is often said but individually designed and machined tanks utilising Chrysler jigs and tools from those previous programmes. A subtle difference perhaps but one of historical importance. *Author's collection*

Right: Initially, Juno V was envisaged as a booster stage capable of accepting a range of upper stage configurations, some with hydrocarbon and some with cryogenic fuel, including existing ICBMs such as Atlas and Titan to provide heavy lift capability as this drawing from a presentation brochure intimates. *Author's collection.*

The explosive fireball and shock wave would have the force equivalent to 50 per cent of the propellant mass being equivalent in explosive yield to the same weight in TNT (trinitrotoluene). That set the safety perimeter 1,649m (5,410ft) from the pad, which, it was judged, should be a structure 70m (230ft) in diameter and have a blast resistant area 49m (160ft) in radius. The blockhouse from where all commands would be sent to the rocket was to be of reinforced concrete, located 320m (1,050ft) from the pad and equipped with facilities for 130 personnel, judged to be the number required to conduct the countdown and launch.

Looking further down the development path, ABMA proposed an equatorial launch site for geostationary payloads and named Christmas Island as their choice. Situated only 10deg south of the equator, it would take out much of the disadvantage in launching to that orbital location from Cape Canaveral. For other military missions, Cape Canaveral would do, or a heavily defended site on the continental United States. In 1958, missions anticipated for the military space programme included a manned space base in equatorial orbit, the launch of surface-to-surface transport vehicles, a global surveillance system and a space defence programme. Clearly, in the official documents there was a declared urgency for a manned space station and this was judged to be a driving factor in selecting the sequence of test shots and the requisite upper stages. A hint of von Braun's influence is seen in the potential application of Juno V for what were as yet undefined missions for the rocket. Summary reports of late 1958 assert that if a full-scale programme could be funded in 1959, it would be possible to use this rocket to make a manned circumnavigation of the Moon in 1964.

But in the small print of the proposed uses for the Juno V was its ability to launch a very heavy cluster of thermonuclear warheads, sending a punishingly destructive volley to targets in the Soviet Union. The limitations imposed by the use of liquid oxygen meant that this could be possible only when planned ahead and executed at a premeditated time rather than in response to an attack. Or, to prepare such a strike to the heartland of the aggressor in the event of an overwhelming land attack by conventional enemy forces. Nothing was to come of this but it was clearly on the capabilities list drawn up by ARPA, the ABMA and the von Braun group. None of these military applications survived for long and there was one good reason – ownership of Juno V and the decision over its operational use was soon to be taken from the hands of ARPA and the ABMA.

Of tremendous moment in the history of US space activities, as a result of the almost endless series of hearings and recommendations, held and offered since Sputnik 1 in October 1957, the Eisenhower administration preferred a civilian organisation running the nation's response, reserving for the military only those classified programmes about which there was to be no public debate. Little has been made of the fact that this response was typical of Eisenhower's desire for detailed intelligence information about the Soviet Union obtained by clandestine means while directing public attention to innocuous, civilian activities. Just as Vanguard served as a foil to the much more advanced Corona spy satellite programme, so too would the civilian space agency – NASA – attract the attention that might surround less publicly declared activities.

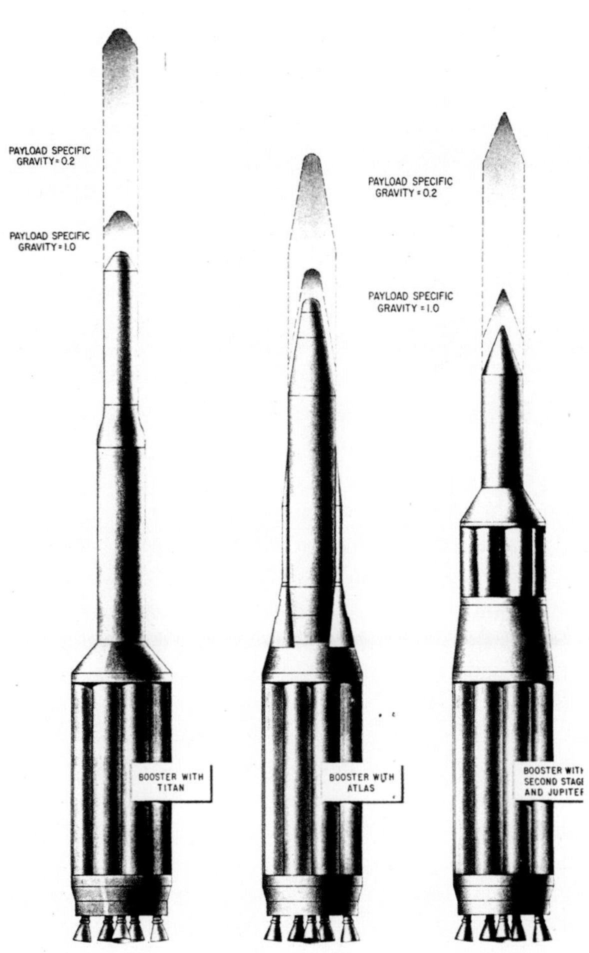

Juno V first stage options

Left: Optional upper stage and payload configurations proposed in 1959 as the ABMA considered two and three-stage configurations for a 13,600kg (30,000lb) payload to low Earth orbit. From left, Juno V with a two-stage Titan I ICBM for second and third upper stages, Juno V with an Atlas and Juno V with a new second stage and a Juno II as third stage. Replete throughout all of von Braun's proposals was a repurposing of existing rocket and launchers whose development costs had already been paid for. *Author's collection*

Below: Juno V embraced a wide range of configurations for meeting certain requirements including lunar circumnavigation, a Dyna-Soar boost-glide vehicle and a geostationary communications satellite for the military in a sequential series of more advanced capabilities. The numbers in brackets refer to the number of stages required to satisfy respective objectives. *Author's collection.*

	*DRY GROSS PAYLOAD, POUNDS (AMR LAUNCH)		
MISSION	LUNAR CIRCUMNAVIGATION	DYNA SOAR-TYPE (2 STAGE)	24 HR. SATELLITE
A-1	6,800 (3)	12,500 (2)	3,800 (3)
A-2	7,000 (3)	8,500 (2)	3,500 (3)
B-1	11,000 (3) 15,000 (4)	12,600 (2)	5,000 (3) 9,000 (4)
C-1	9,000 (3)	24,000 (2)	5,500 (3)
C-2	15,500 (3)	40,000 (2)	9,000 (3)
C-3	25,000 (3) 34,000 (4)	54,000 (2)	12,000 (3) 21,800 (4)

So it was that the National Advisory Committee for Aeronautics (NACA) metamorphosed into the National Aeronautics and Space Administration, effective 1st October 1958. A date selected because it signalled the start of fiscal year 1959, making that a tidy bureaucratic entry on the government expenditure sheet. Very quickly, with T. Keith Glennan as Administrator and former NACA boss Hugh Dryden as his deputy, all non-military programmes were transferred from the Air Force, the Navy and ARPA to NASA. By this date, ARPA had (in August) formally signed off on development of the Juno V and a month later Rocketdyne had been placed under contract to deliver the H-1 and attention focused on the big Juno V work at Redstone.

A NACA Working Group on Space Technology, more generally known as the Stever Committee under its chairman Guyford Stever, was formed on 12 January 1958 and reported preliminary findings on 18 July. These formed the basis of initial NASA planning. Essentially the von Braun report from the ABMA, it identified a five-step capability in launch systems beginning with Vanguard – the satellite launcher for the International Geophysical Year – followed by the Juno I, Juno II and Thor launchers. Next up were Atlas and Titan ICBM rockets for satellite capabilities, with Juno V supporting heavy lift. The final category was a theoretical booster powered by two or four F-1 engines producing up to 25,000kN (6millionlb), although there had been no detailed study on the latter.

From this the Stever Committee laid out a broad-brush programme including a manned lunar landing as early as 1965, a 50-man space station two years later, advanced lunar expeditions from 1972 followed shortly after that by a permanent Moon base and manned flights to Venus and Mars in 1977, essentially what the ABMA had already formulated as a national space strategy. Mindful by this time that there was to be a civilian space agency as a part of government, the committee urged development of the F-1 concurrent with the Juno V, predicting a total cost of $17.21billion for this programme, involving 1,823 launches of all types of vehicle over the next 20 years. It also endorsed booster recovery as a significant step toward cost reduction.

Von Braun was noted for having pressed the priority of developing a capability rather than a tightly controlled set of mission objectives, preferring to build an infrastructure first and then to define the roadmap of space goals. All of which stressed the need for a portfolio of upper stages to turn the Juno V boosters into effective launch vehicles. During the formative processes being defined in mid-1958, the von Braun team began to distil options and provide a set of sequential capabilities within the Juno V programme, mirroring the wide range of lifting capacity across the aforementioned spectrum of different vehicles beginning with Vanguard.

These matters were pressing with some urgency on the new space agency and on 2 January 1959, Glennan received a memorandum written by his assistant, W L Hjornevik, defining the dilemma: whether to prepare a fixed sequence of Juno V launchers or prepare a set of missions to which the requirement could drive launcher development. This was heady stuff and would require a national commitment. Six days later, a joint ARPA/NASA committee on 'Large Clustered Booster Capabilities' met to consider a series of two-hour presentations from engine builders Aerojet and Rocketdyne and from airframe manufacturers Convair, Douglas and Martin which had been held during the preceding week. It noted that if starting with a clean slate the cluster concept would have been better served by bringing three Atlas rockets together as the first stage and a cluster of three 3m (10ft) diameter stages as the upper stage, first with conventional propellants and then with a cryogenic engine. But it judged that the existing work already conducted by the ABMA would give it a head start of six to nine months, citing as a further defining factor, the safety feature of an engine-out capability designed in to Juno V.

On 27 January 1959, a report was drafted for President Eisenhower defining a national launch programme, already signed off by Abraham Hyatt, the new Chief of Launch Vehicles at NASA. The prime architect of the report was Milton Rosen, who defined the hasty preparation of existing launchers – Juno I, Juno II and Thor (with the Able upper stage) – in response to the Russian challenge.

Rosen saw three successive launcher programmes as forming the stable of future workhorse capabilities. First would be a version of the single-stage Atlas for space launches followed by an Atlas with a cryogenic Centaur upper stage, also serving as a technological proof-of-concept for LOX/LH_2 engines utilised in advanced Juno V launchers. The second category was Juno V itself with increasingly more powerful upper stages. Juno V-A would produce a three-stage vehicle by placing a two-stage Titan I rocket on top of the clustered booster and Juno V-B would replace the second stage of Titan with a cryogenic stage which could be Centaur or a new design. The third category was named Nova, envisaging four F-1 engines clustered in the first stage, a single F-1 in the second stage and cryogenic third and fourth stages developed from the Juno V programme. Only with this could NASA hope to achieve a manned landing on the Moon, said Rosen. Nova would grow over the next three years but never achieve development status.

The sheer magnitude of the tasks expected of the new space agency was daunting, with several programmes coming under agency aegis and the prospect, soon, of Redstone being transferred to NASA. Even before NASA formally opened for business, on 18 September 1958, Glennan and Dryden visited the ABMA in Huntsville and met with General Medaris and both expressed surprise at the level of effort on Juno V, asking bluntly: 'What on Earth is the Army doing in this business?' Glennan grasped the vital role the ABMA would play in the options for space programmes open to NASA and wanted to secure the services of that organisation as well as the Juno V launcher.

At this point, NASA would inherit the Langley, Lewis and Ames facilities of the old NACA and would have a new base, named the Goddard Space Flight Center, from where it would carry out development of science and applications satellites. At this date, not everybody believed that NASA would be the dominant practitioner for development of a US space programme and the Air Force, with its massive ICBM and IRBM missile programmes was thought to have prime hold on launch services, stage development and since there were to be a lot of military satellites, on launch services too. The NACA had been an advisory committee, not an operating system such as the Air Force, engaging with industry to design, manufacture and assemble equipment it would deploy operationally.

The Army, on the other hand, did things in a very different way and this was all too clear to Glennan and Dryden when

Left: NACA's Special Committee on Space Technology, called the Stever Committee meets on 26 May 1958. Left to right, Edward R. Sharp, Director of Lewis Laboratory; Colonel Norman C. Appold, USAF, Assistant to the Deputy Commander for Weapons Systems, ARDC: Abraham Hyatt, Research and Analysis Officer, Bureau of Aeronautics, Department of the Navy; Hendrik W. Bode, Director of Research Physical Sciences, Bell Telephone Laboratories; W. Randolph Lovelace II, Lovelace Foundation for Medication Education and Research; S. K. Hoffman, General Manager, Rocketdyne; Milton U. Clauser, Director, ARDC, The Ramo-Wooldridge Corporation; H. Julian Allen, Chief, High Speed Flight Research, NACA Ames; Robert R. Gilruth, Assistant Director, NACA Langley; J. R. Dempsey, Manager, Convair-Astronautics; Carl B. Palmer, Secretary to Committee, NACA Headquarters; H. Guyford Stever, Chairman, Associate Dean of Engineering, MIT; Hugh L Dryden (ex officio), Director, NACA; Dale R. Corson, Department of Physics, Cornell University; Abe Silverstein, Associate Director, NACA Lewis; Wernher von Braun, Director, Development Operations Division, ABMA. *NASA*

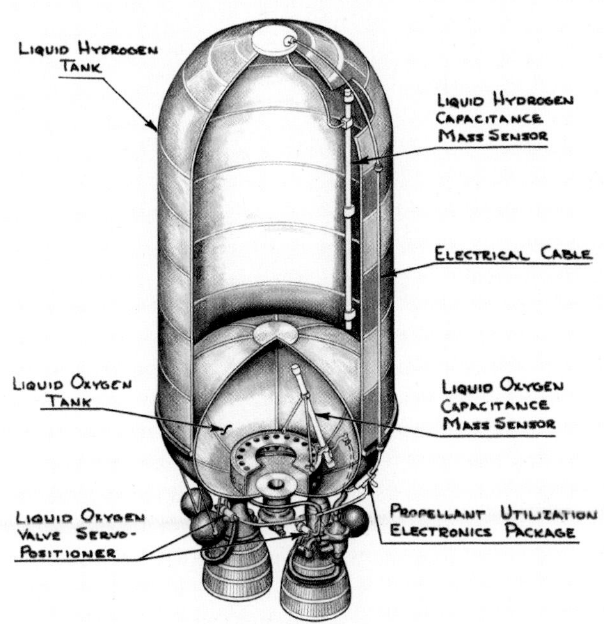

An early drawing of the Centaur upper stage, which it was agreed would pioneer NASA's use of crygenic propellants. *Convair*

they visited Redstone, not for nothing being called an Arsenal. Under the arsenal concept, the Army manufactured its own equipment and did not have the same relationship with industry, bringing companies in to do specified work rather than develop designs and concepts for approval. The von Braun people at the ABMA, while employing Chrysler to build Redstone and Jupiter missiles, had little or no experience at negotiating contracts. Everything they asked industry to do originated at Redstone, except for the engines which were held with not a little suspicion, to the extent that when delivered they would be completely disassembled and put back together.

Clearly, Glennan wanted von Braun and on 10 October 1958 he went to the Deputy Secretary of Defense, Donald Quarles, who promptly asked him to go to his boss, Secretary of the Army Wilber Brucker. It was there that Glennan put forward the idea from his chief of space flight operations, Dr. Abraham Silverstein, that NASA should take the von Braun team out of the ABMA, about 2,000 people in all. He received a short, sharp lesson in just how much the Army wanted to hold on to von Braun – in fact, the whole ABMA organisation, which at this time had 3,925 employees in addition to 1,081 other personnel. With confident plans of great things looming, the ABMA, von Braun and Juno V were the Army's investment in 'big ticket space'.

Von Braun was opposed to breaking the organisation in half and fuel was added to the fire when General Medaris leaked that to the press in hope of rallying opposition. But the decision to shift all non-military work to NASA had the signature of the President and on 3 December a compromise was reached whereby NASA got the ABMA 'under contract', a half-bitter pill sweetened when NASA's bid to absorb JPL was approved on the same day. This did end the partnership JPL had with the ABMA but it allowed it to concentrate on unmanned lunar and planetary exploration. But it was just the beginning of the politics.

The ABMA team had chafed under the NACA rejection of its Project Adam as a suborbital 'hop' for putting people above the atmosphere, albeit on only a ballistic space flight. As recorded earlier, when NASA inherited the Air Force MISS programme and rebadged it 'Mercury', it took a different view to Dryden's NACA and accepted the Redstone rocket as a means of performing suborbital tests. The Air Force had scheduled the use of Thor for high-velocity tests of what was now the Mercury spacecraft but von Braun persuaded NASA to accept the Jupiter rocket instead, even going so far as to commit the ABMA to funding any difference between the two! In shaping its programme to a tight budget, NASA dropped the idea of the Jupiter-Mercury missions. But through all this, with NASA having decided to use a modified Atlas for placing Mercury astronauts in orbit, the agency knew that it would have to rely on Juno V for launching its successor, whatever that might be.

The question of ownership hung over the agency like a pall. Juno V was an Army rocket and NASA was a civilian operation. If this clustered booster was to father a generation of heavy-lifters, the new agency wanted to declare ownership by giving it a new name. Invariably, aircraft evolving from a pedigree are referred to as modified variants of an existing type, a ploy to allay fears among politicians of a surge in funding requests, which usually attaches to a completely new system. That was, after all, why the Army chose a progressive, numerical sequence for what were, in effect, a polyglot mix of very different conceptions all under the Juno label. But sometimes, it does help to break with a linear 'family' and boldly assert a completely new generation, for which there is greater support for the gain by taking a step forward than there is for investing in something from the past. So it was with the Juno family. On 3 February 1959, in a proposal from von Braun himself, Juno V was renamed Saturn in an ARPA memorandum which severed it from the earlier proposals. It was all very logical – Saturn was the next planet out past Jupiter.

Threats and Choices

The vexed issue over who controlled the ABMA was a thorny matter which would not go away. But that was rendered irrelevant when Herbert York, the director of the Department of Defense Research and Engineering, declared to ARPA boss Roy Johnson in a memorandum dated 9 June that he was cancelling the Saturn programme. It was, he asserted, a wasteful duplication and developments by the Air Force with Atlas and Titan various for their application as space launch vehicles meant that Saturn was unnecessary for launching military payloads. There was a degree of sound reasoning in this and were it not for NASA, the matter might have ended right there. Both Atlas and Titan held long-term promise for many of the missions envisioned by the military and this was the period during which the Dyna-Soar spaceplane was well on the way toward approval for development. Atlas or Titan could be used for early development flights although Saturn was recognised as equally applicable.

The Air Force had a plan to attach very large solid propellant boosters to the Titan first stage, upgrading it to what be-

Above: In negotiation: Gen. John Medaris (left), Wernher von Braun (centre) and NASA administrator T. Keith Glennan in Huntsville. *US Army/ABMA*

came known as Titan-C and this would have almost the lifting potential of Saturn but at much less cost. Moreover, it was an Air Force launcher for an Air Force spaceplane. And that was another reason for supporting cancellation of Saturn, which had only been assigned one real-world payload, a large Army communications satellite for which the Air Force had no use.

The impending cancellation of Saturn mobilised NASA in its first active 'turf-war' with another government agency and the fundamental role played by the Saturn rocket made its survival imperative to the nascent agency. But NASA could not afford to pay for the entire ABMA organisation and that rendered a dilemma: take complete ownership of the organisation or look elsewhere for a heavy-lifter. The fact that there was none, and that the abandonment of Saturn would destroy the core NASA long-range plan, left only one place to go – the President himself. Before that, on 3 September, Glennan went to Redstone to mobilise a survival strategy at a facility where the personnel were wedded to the super-lifter and were as fearful of its cancellation as NASA. Perhaps playing the propaganda card (we will never know for sure), on 14 September Brucker spoke to Medaris and told him that von Braun threatened to leave if Saturn was abandoned and, in a bitter twist of pique, that he would go the Air Force instead. Such was the ascendancy of his power and influence, arguably at its peak.

The critical engagement came on 16-18 September 1959 when York and Dryden chaired meetings across those three days at which the Titan-C, Saturn and Nova were discussed and hotly debated. Faced with threats from von Braun, edicts from the Department of Defense to allocate all military space matters to the Air Force, and a lack of justified payload manifests for any rocket of the potential of Saturn, only NASA had any plausible reason for taking ownership of a concept only then coming into the hardware stage and still two years away from first launch. In a letter to his parents dated 4 October, von Braun waxed poetical in his summary of the situation. He had, he said, found himself 'in the position of a beautiful young girl who has two suitors and must be on alert that she marries honourably and is not dishonourably led astray'.

In a surprising *volte face*, which possibly was in response to a nod from the highest office in the land, Glennan suddenly switched tone and agreed to fund the ABMA if Medaris agreed to hand it over. There was, after all, nowhere else for it to go and the agency badly needed Saturn to fulfil its aspirations. On 21 October the announcement was made that the ABMA would go to NASA; disillusioned, Medaris planned to retire but held that announcement back so as not to link the two and sour the sweetness for NASA. On 7 November, all parties, including Eisenhower's science adviser George Kistiakowsky, sat down to work out the logistics of the transfer moving von Braun's Development Operations Division to NASA. In a somewhat maturing if not reflective mood, touching on his lifelong desire, von Braun wrote to Kistiakowsky favouring this next move in ownership to 'an organisation that pursues space flight for space flight's sake, and not because of some temporary military objectives'.

At the Department of Defense, York felt that Saturn was too big for any military objective and that it rightly belonged within the civilian agency. With it went connections to ARPA, who had done so much to bring it to fruition through all the formative years of Juno V. The details of the transfer were defined over the next several months and were eased through the services of McKinsey and Company, a management consultancy which had been so useful when NASA emerged overnight from the NACA. All administrative ties were completed by 16 March and the Saturn programme had the NASA badge on its office doors. There was even greater purpose now that a separate NASA committee had sealed the future for the agency by defining the Moon as its first major goal. Those recommendations, which came from the Goett Committee would define the Saturn programme.

Six months to the day after opening for business, on 1 April 1959, NASA formed a Research Steering Committee chaired by Harry Goett from the Ames facility and who would later become director of the Goddard Space Flight Center. They met first on 25 May and heard from Rosen as well as Maxime A. Faget and Alfred J. Eggers, Jr, that the immediate step beyond Mercury should be a new spacecraft accommodating two pilots for one week in orbit to gain experience for longer trips to the Moon and back. Bruce Lundin's future missions committee wanted a wide range of objectives and stressed the need to treat a Moon goal as merely a first step toward grander expeditions into the solar system. Further meetings were held, the last on 9 December, to formulate recommendations and to decide on an immediate goal of sending two men on a circumlunar flight lasting two weeks. But, to a point, the recommendations were conservative and, with hindsight, appear to have been framed in the conscious knowledge of Eisenhower's reluctance to launch an all-out space race.

As Harry Goett would say of the decision to place the Moon landing as an ultimate goal rather than a stepping stone:

'A primary reason for this choice was the fact that it represented a truly end objective which was self-justifying and did not have to be supported on the basis that it led to a subsequent more useful end.' Nothing could have more clearly defined the way the political elite in Congress would treat the Moon landing objective as a single-shot race than this limitation. By implication an activity which was little more perhaps than an adventure, a supreme effort to outfly the Russians with little or no long-term plan. Of course NASA rapidly coalesced around Saturn, seeing it as the goose with a golden egg but for an offspring as yet not fully defined. That definition would hinge on possibilities structured by the development plan. As this would be defined by the upper stage configurations selected, it was time to put von Braun's preference for a 'capabilities plan' to the test.

The challenge was enormous, as defined by a couple of engineers at Huntsville who summed it up admirably: 'The state of the art at this time classified the Saturn booster as almost impossibly complex.' But opting to use cryogenic propulsion for the upper stages was a big gamble, one made on the basis that there would have been several successful flights with the Centaur and its twin RL-10 engines by the time the Saturn flew cryogenic stages. Development of the H-1 engine was on track, the first production engine (H-1001) delivered to the ABMA on 28 April 1959 with the first test firing completed on 26 May. Elsewhere, on 27 July the last Jupiter IRBM airframe was completed at Redstone Arsenal and the shops began re-tooling to produce the tanks for Saturn. In recognition of these steps, on 1st August ARPA authorised the ABMA to proceed toward a captive live test of Saturn.

The second half of 1959 proved fruitful for the studies being undertaken on various upper-stage options which the ABMA and NASA worked up for selection. The essential mission requirements, set by technical studies conducted by ARPA, the ABMA and NASA, were broadly separated into three classes: 4,500kg (10,000lb) for escape missions to lunar and deep-space destinations; 2,268kg (5,000lb) for payloads destined for geostationary or geosynchronous orbits; and 4,500kg (10,000lb) for manned missions to low Earth orbit. The teams also examined again the options for alternative upper stages using conventional and high energy propellants and divided individual classes into options A, B and C. Of significance too was the diameter of existing candidate upper stages and how they would stack on the Saturn first stage, which would have a diameter of 660cm (260in).

The first configuration in the initial series of upper stage options, the A-1 launcher would add a variant of the Titan I first stage with a cryogenic Centaur as the third stage, each having a diameter of 304.8cm (120in). This option conformed to an urgent requirement to use existing hardware ahead of new and undeveloped designs, the Centaur being considered as a stage well advanced and therefore potentially proven technology – which was optimistic at best. The slender upper elements of the A-1 launcher would provide marginal structural integrity and the teams did look at a new second stage with a larger diameter but that would contravene the need for current hardware. An unlikely configuration, an A-2 vehicle would have replaced the second stage of the A-1 with a cluster of IRBMs, unfavourable for the same reason and because it would not fulfil the requirement for the three mission models.

The B-1 option met all the requirements but did require development of a new second stage with a diameter of 558.8cm (220in), powered by four H-1 engines and delivering a stage thrust of 3,558.4kN (800,000lb). This was greater than the sea-level output of the H-1 engines because they would be operating in a vacuum. The third stage would be a new LOX/LH2 cryogenic stage of the same diameter as the second stage but with four RL-10 engines for a stage thrust of 266.88kN (60,000lb). The fourth stage would be the Centaur, the third stages on the A-1 and A-2 variants. Here too, the cost and time consumed in development outflanked the advantage.

These initial options clearly relied on cryogenic upper stages to match performance needs for the mission models. It became clear that if risk and some element of time-consumption was accepted for bringing a truly capable vehicle to the pad, it should be acceptable to dispense with conventional engines in the second stage and go straight for a cryogenic stage there too. It was at this time that thoughts began to coalesce around a more powerful cryogenic engine, one at the thrust level Rocketdyne had been working on – on paper at least. This led directly to the Saturn C series, of which there were initially three increasingly more capable variants.

The Saturn C series would use all-cryogenic upper stages, sitting atop the S-I stage with eight conventional H-1 motors. The third stage of the proposed B-1 would be adopted as the (S-IV) second stage of the C-1, and a Centaur as the (S-V) third stage. The C-2 would have a powerful (S-III) second stage employing two cryogenic J-2 class engines for a total thrust of 1,779kN (400,000lb) and the third stage would inherit the second stage from the C-1, both with a diameter of 558.8cm (220in), a Centaur as the S-IV fourth stage topping out the stack. This configuration had a more forgiving engineering design with fewer stress points due to the three decreasing stage diameters: first stage, second and third stages and fourth stage. That made it more forgiving in high wind and would expose it to less shear effect.

The C-3 was different in that it responded to the need to

Above: The mighty H-1, the first production example of which was delivered in April 1959. *Rocketdyne*

Early Juno V / Saturn configurations

Right: The A and B series adopted conventional propellants in the first two stages while the three C-series configurations made use of cryogenic upper stages. Note the use of Titan and Jupiter upper stages, the increasing reliance on Centaur, and how the C-3 increases the capability of the first stage. *Author's collection*

STAGE	1	2	3	4
A-1	LOX/RP EIGHT H-1 ENGINE CLUSTER	LOX/RP TITAN 120" DIA.	CENTAUR 120" DIA. TWO 15K ENGINES	
A-2		CLUSTER OF IRBM'S	CENTAUR 120" DIA. TWO 15K ENGINES	
B-1		LOX/RP * 220" DIA. FOUR H-1 TYPE ENGINES	LOX/LH * 220" DIA. FOUR 15-20K ENGINES	CENTAUR 120" DIA. TWO 15K ENGINES
C-1		LOX/LH * 220" DIA. FOUR 15-20K ENGINES	CENTAUR 120" DIA. TWO - 15K ENGINES	
C-2		LOX/LH * 220" DIA. TWO 150-200K ENGINES	LOX/LH * 220" DIA. FOUR 15-20K ENGINES	CENTAUR 120" DIA. TWO 15K ENGINES
C-3	LOX/RP 2.0 MILLION POUND THRUST ENGINE CLUSTER	LOX/LH * 220" DIA. FOUR 150-200K ENGINES	LOX/LH * 220" DIA. TWO 150-200K ENGINES	LOX/LH * 220" DIA. FOUR 15-20K ENGINES

*Nominal tank diameter
H-1 Engine - 165,000 to 188,000 lb. thrust

Below: The B series filtered out into four possible configurations, their physical properties as shown in the upper table. *Author's collection*

Below right: Acceptability for specific missions shows that the capacity to lift a 4,536kg (10,000lb) geosynchronous communications satellite into a 24-hr orbit was only possible with the B-1. Since the launch of the Advent communication satellite was critical to the Juno V, the rest of the A and B series were eliminated from further consideration. *Author's collection*

COMPARISON DATA FOR EARLY SATURN VEHICLES

Vehicle	Stage	Propulsion System (Thrust) lb	Specific Impulse, sec	Approximate Propellant Loading, lb	Diameter, in.	Vehicle Length, ft	Remarks
SATURN B (Minimum)	1 2 3	8 H-1 = 1,500K 2 × 180K = 360K 2 × 15K = 30K	257 289 412	742,000 168,000 26,000	257 120 120	195 to 210	Marginal Performanc for all Missions
SATURN B (Near Minimum)	1 2 3	8 H-1 = 1,500K 2 × 180K = 360K 2 × 15K = 30K	257 299 412	742,000 168,000 50,000	257 120 120	230 to 240	Very Marginal from Contro Viewpoint
SATURN B (Interim)	1 2 3	8 H-1 = 1,500K 2 × 180K = 360K 2 × 15K = 30K	257 299 412	697,000 218,000 47,000	257 160 120	195 to 210	Acceptable, but no Growt Potential
SATURN B-1	1 2 3 4*	8 H-1 = 1,500K 4 × 180 = 720(880) 4 × 20 = 80K 2 × 20K = 40K	257** 299(312) 420 420	600,000 300,000 75,000 25,000	257 220 220 120	O 201	Optimum Con- figuration

*Optional.
**Varies with mission.

PAYLOAD CAPABILITIES OF EARLY SATURN CONFIGURATIONS

Vehicle	Stages	Weight, lb	96-Minute (307-Nautical Mile) Orbit	Escape Mission	24-Hour Orbit Equatorial Dogleg AMR	Soft Lunar Landing	Remarks
SATURN B (Minimum)	3	Net	19,000	4,200	2,400	700*/ 1050	Very Poor Performance
		Gross	23,000	7,300	4,600	900*/ 1350	
SATURN B (Near Minimum)	3	Net	23,000**	8,400	5,000	2600	Very Marginal Bending Frequency
		Gross	27,000**	12,000	8,100	3100	
SATURN B (Interim)	3	Net	27,000	8,400	5,000	2650	No Growth Potential
		Gross	31,500	12,000	8,100	3100	
SATURN B-1	3	Net	35,000	10,250	5,200	3400	Orbital Refueling Capability
		Gross	40,000	14,000	8,800	3900	
SATURN B-1	4	Net	Not	11,900	7,800	4000	Only for High Speed Missions (Optional)
		Gross	Feasible	15,500	10,200	4550	

*300 seconds specific impulse.
**Heavier structure to carry larger payload. Required only for low orbit mission.

SATURN DEVELOPMENT PLAN & LAUNCHING SCHEDULE
(BASED ON ORIGINAL SCHEDULE)

	VEHICLE / CY	1960–1969	TOTAL
SINGLE STAGE VEHICLE	"B-I", R & D 8 X 165K BOOSTER DUMMY UPPER STAGES NO PAYLOAD		4
2 STAGE VEHICLE	8 X 188K BOOSTER 4 X 220K SECOND STAGE DUMMY THIRD STAGE OR NOMINAL PAYLOAD		2
3 STAGE VEHICLE	8 X 188K BOOSTER 4 X 220K SECOND STAGE 4 X 20K THIRD STAGE PAYLOAD		2
4 STAGE VEHICLE	8 X 188K BOOSTER 4 X 220K SECOND STAGE 4 X 20K THIRD STAGE 2 X 20K FOURTH STAGE PAYLOAD		2
	"B-I", OPERATIONAL THREE-STAGE VEHICLE FOUR-STAGE VEHICLE		15 15
	PROGRAM REQUIREMENTS SECOND STAGE	START ENGINE R & D START VEHICLE R & D FIRST DELIVERY TO ABMA	
	THIRD STAGE	START R & D FOR UPRATING CENTAUR ENGINE START VEHICLE R & D FIRST DELIVERY TO ABMA	
	FOURTH STAGE	INITIATE PROCUREMENT OF STAGE FIRST DELIVERY TO ABMA	

Above: Of the four B series options, the ABMA selected the B-1 as optimum and in October 1959 presented a schedule that called for 10 development flights before a first operational flight from late 1964. Within the same structure 10 Saturn C-1 development flights were planned after all other options had been eliminated. *Author's collection*

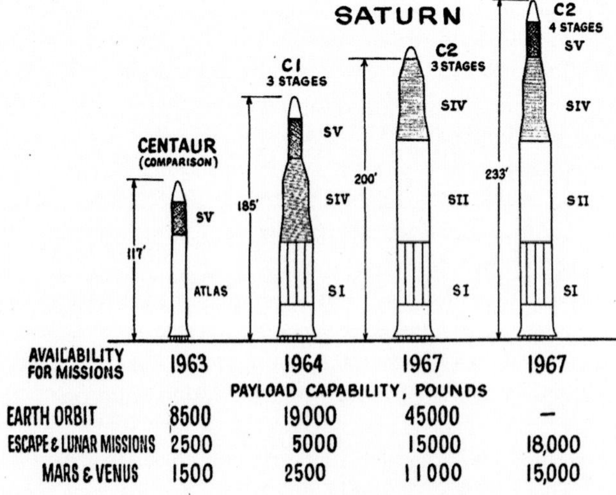

Above: The C series Juno V launcher proposals leaned on the Centaur and its RL-10 engines to provide cryogenic upper stages for greater payload potential. *Author's collection.*

increase the capability of the first stage to maximise the effectiveness of the highly efficient upper stages. The vehicle would have a more powerful cluster of first-stage rocket motors, possibly replacing (as considered earlier) the inner four H-1 engines with a single F-1, increasing the thrust level to more than 8,896kN (2millionlb) or by uprating the existing H-1 set. For the C-3, the second (S-II) stage would have four J-2 class rocket motors delivering a total thrust of 3,558.4kN (800,000lb). The second and third stages of the C-2 would comprise the third and fourth stages of the C-3. The recommendations were clear and precise: develop a new and more powerful cryogenic stage using four RL-10 engines uprated to 88.96kN (20,000lb) each, develop a new cryogenic engine for the S-II and S-III stages and conduct a design study for the stage itself as to how it would be configured and applied.

This work on stages and configurations brought together activity which had been continuing throughout the year. A formal analysis was predicated on Eisenhower's decision on 2 November 1959 to transfer the ABMA's Development Operations Division to NASA and on 17 November NASA's Director of Space Flight Development, Abraham Silverstein, formed the Saturn Vehicle Team – known as the Silverstein Committee – with the concise aim of drawing up a development plan. That framed these recommendations which were presented to Administrator Glennan on 15 December. Out were adaptations of Titan or Atlas and clearly the A and B series were limited in capability; to go down that route would take resources and time, delaying the induction of cryogenic upper stages so eloquently justified through the three optional C series configurations. The development of stages assigned Roman numerals indicated the sequential introduction through the building block concept, starting with the basic S-I first stage and progressing rapidly over time with increasingly advanced capabilities. The C-3 represented the peak of projected development as defined by the end of 1959, with a capacity for lifting a maximum 24,495kg (54,000lb) to low Earth orbit in a two-stage configuration, or sending 15,422kg (34,000lb) around the Moon. At this date, S-I static tests were to be conducted by the end of 1961, with three S-I stage ballistic flights and dummy upper stages. Two-stage flights were to commence in 1963, three-stage launches in 1964 and enhanced capability the following year.

Armed with this sequence of Saturn launchers, a major future missions planning conference was hosted by the Director of Lunar Vehicles, Donald R. Ostrander, in January 1960. Reminding attendees that to this date the prime, long-range goal was a landing on the Moon during the early 1970s after a circumlunar flight by astronauts by the end of the 1960s, he emphasised the need to plan across a broader spectrum and to aim for a diverse range of capabilities. Enshrined within the National Aeronautics and Space Act that resulted in the formation of NASA was a requirement to make the United States 'pre-eminent in space'. That underpinned much of what the senior leadership sought, along with members of Congress and the administration.

Mobilised by a determined effort not to fall short of launch requirements, a variation of the C-3 emerged in early 1960, one in which the S-I stage was upgraded, replacing the eight H-1 engines, or the hybrid H-1/F-1 mix proposed by Silverstein, with two F-1 engines for a total lift-off thrust of 13,344kN (3millionlb). The view at MSFC was to provide redundancy for expected requirements and with thoughts of a future Moon landing underway, that argued for a series of launches to assemble in Earth orbit the hardware required to land astronauts on the surface. The C-3, with a modified first stage identified as an S-IB and S-II and S-IV upper stages, was believed to provide that guarantee of performance. It would lift 36,288kg (80,000lb) to low Earth orbit or send 13,560kg (29,900lb) to the Moon. By mid-1961 the C-3 had been cleared to fly the Apollo B missions – those destined for deep-space lunar missions (the Apollo A configuration was for Earth orbit only).

However, a new configuration of planned stages defining the Saturn C-4 as an even more powerful launch vehicle was proposed at a series of meetings in 1960, still with the Earth Orbit Rendezvous (EOR) method in mind for assembling an expedition to the Moon. By coincidence it identified a modified S-IB stage powered by four F-1 engines producing a lift-off thrust of 27,000kN (6millionlb), four J-2 engines in the S-II second stage and a single J-2 in the third stage, designated S-IVB. It would be capable of lifting 99tonnes (218,000lb) to low Earth orbit or sending 32tonnes (70,000lb) to the vicinity of the Moon. This was a significant increase over the C-3, indicative of the doubling of first stage thrust. Its very existence, on paper at least, appeared to challenge the logic of its numerical predecessor; who would need a C-3 with the C-4 around?

Mission Modelling

The decision over upper stages, and the strategy for the future which was laid out as a result, received acclaim from the President on down the political chain, to a largely agreeable

Right: Committee man – NASA's Director of Space Flight Development, Abe Silverstein was charged with steering Saturn's development. *NASA*

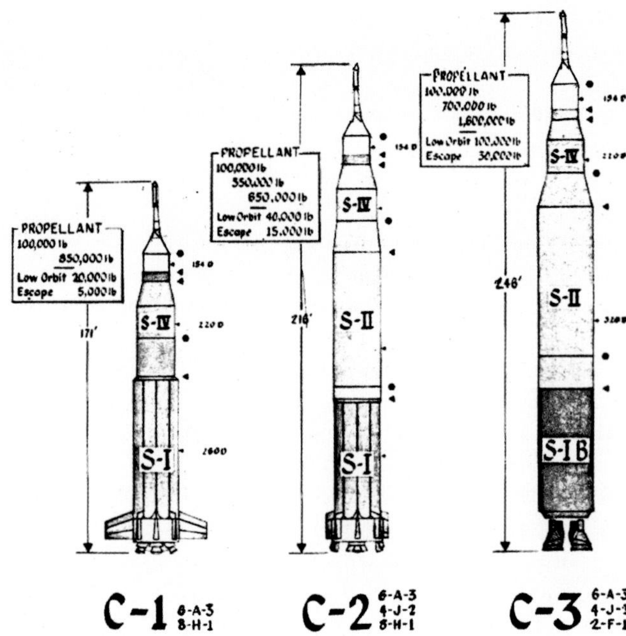

The block stage concept allowed any permutation of these five elements with the C-2 adopting the J-2 engine for the S-II cryogenic stage. For this reason, the stage designations for Saturn I and Saturn V were inherited from progressively more powerful stage designs from the late 1950s. Introducing the F-1 engine with the C-3 began the evolutionary shift toward the configuration that would become the C-5 by way of the C-4. Note that the first stage of the C-3 was initially designated the S-IB but when that was adopted for the Saturn IB it was re-designated S-IC by which it was known for the Saturn V. *Author's collection*

APPROX DIM.	STAGE	THRUST	PROPELLANT	NO. ENGINES	FIRST FLIGHT
80' × 21½'	SI	1,500,000	RP1-O2	8	1961
30' × 10'	SV	35,000	H_2-O_2	2	1961
40' × 18½'	SIV	70,000	H_2-O_2	4	1963
60' × 21½'	SII	800,000	H_2-O_2	4	1965

Congress united on support for getting the big booster ready quickly. Eisenhower wrote to Glennan on 14 January 1960 supporting the Saturn programme and approving funding requests to maximise the effort, assigning four days later the DX rating of highest national priority for work bringing it to fruition. A mock-up of the giant rocket had already been assembled in the ABMA test stand to check fitment and servicing methods and would soon be replaced by a complete test booster and, as the entire programme was being transferred to NASA, the first industrial contract was let for upper stages. After a review of proposals from contenders received during February, on 26 April 1960 Douglas Aircraft Corporation received a contract to build the S-IV stage.

The extraordinary growth in rocketry and the sheer scale of launch vehicle planning stretched resources and available manpower to the point where legacy programmes were suffering. A survivor of the mid-1950s, the Juno II programme was continuing to send payloads into space, yet by March 1960 it had failed four times in six attempts. The only two successes had been Pioneer 4 on 3 March 1959 and Explorer VII on 1 October 1959. Concerned at the low level of success, a review committee was established on 25 May 1960 on the order of Gen Don R. Ostrander, NASA's Director of Launch Vehicle Programs. It visited the ABMA on 1 June to hear testimony on the Jupiter core stage and JPL the following week to discuss the solid propellant upper stages. What they discovered was testimony to how fast the pace of development, and state-of-the-art in rocketry and space operations, had been since the first flight of a Juno II just nineteen months earlier.

The minutes of those meetings records the sorry catalogue of a redundant configuration, a lack of adequate testing and disinterest in the wake of new and greater possibilities, those being constrained by dependency on an outmoded vehicle sapping resources and manpower. It concluded that 'Juno II is an old, closed-end, long-delayed, overextended project', citing the cluster of solid propellant upper stage rockets to be 'a 1955 design with many built-in compromises and well-recognised limitations. For these reasons JPL was uninterested in the Juno II project after the Pioneer 4 launch and tried to be disassociated from it'.

In discussions with the committee, ABMA declared the over-ambitious payload requirement to have been the cause of delays amounting to twelve months and that this has 'led to a reduction in enthusiasm for the project, a need to store vehicle subassemblies, and the exposure to interference from the high-priority Saturn projects', asserting that storing booster engines 'can be a serious matter because of ageing and deterioration'. To the committee there was a marked level of frustration at the flagging Juno II programme, its limited capabilities paired with unrealistic payload requirements and the distraction it imposed on the Saturn programmes.

In summary, the committee was reconciled to the terminal nature of the Juno programme, prophesying correctly that of the four Juno II launches remaining, only two had a likelihood of success. But its conclusions were that the programme was plagued by inadequate testing, lack of development pace, decided disinterest from JPL as it turned its attention to planetary probes for launch on bigger rockets and on the allocation of work required for the Saturn programme, sapping Juno II of personnel and the most talented engineers. Of the four Juno II launches remaining, two were indeed a success, closing a programme with the last launch on 24 May 1961, barely five months before the launch of the first Saturn I.

There were lessons here. The committee exposed the fallacy of placing core stage, upper stages and payload development and integration with three separate bodies: ABMA, JPL and NASA, respectively. They were convinced that only integration would solve problems haunting the continued use of Juno II, inferring that as each organisation had outlived, or outgrown, its original place within the Juno II programme, it had become more distanced from the other two. ABMA was focusing all its efforts on the Saturn series; JPL was preoccupied with developing planetary probes; NASA was allocating science payloads driving significant changes to the operating envelope of the launcher, which brought that demand back to ABMA. More work there meant time and effort subtracted from the Saturn programme and the urgency it demanded.

In the early months of 1960, with NASA still formulating future plans, extravagant expectations for great things in space had fuelled confidence at the ABMA, grandiose plans supported by senior Army officers and not a few in the Department of Defense. Much of that thinking evaporated as NASA absorbed the Development Operations Division but a lot of the planning which would inform the space agency originated when von Braun was employed under an Army pay cheque and after some of the advanced planning for very large deep-space expeditions had been completed.

On 20 March 1959, Lt. Gen. Arthur G. Trudeau, Army chief of research and development, issued a formal request, highly secret, for the ABMA to conduct a study for a lunar outpost under the aforementioned Project Horizon. Clearly in an attempt to seize the high ground of space exploration, this fast-paced study had restricted circulation, adamantly insisting that there would be no 'contacts with agencies outside the Army' until the results had been presented to the Department of Defense. The preliminary investigation was to be delivered by 15 May but much of the work had already been done.

Lost today in the escalating successes of NASA, it is pertinent to recall the suspicion from the military that surrounded the formation of this civilian agency. Believing it to lack the muscle for managing giant projects and denied the heft of big budgets, there was a distinct belief from both the Air Force and the Army that NASA was weak, would be ineffective and could suck out resources best assigned to them. Project Horizon was not only a specific attempt to avoid the nation suffering from another embarrassment to follow Sputnik and to move quickly to get Americans first to the Moon and then to Mars; it was an attempt to seize authority over 'big ticket' operations.

When it was delivered on 9 June 1959, the report anticipated that under Project Horizon, the first manned lunar landings would take place in the spring of 1965 followed by an operational outpost, permanently manned, late the following year. It projected a robust and rapid use of Saturn, fully operational by 1963, and of a developed and more powerful version with cryogenic upper stages, by 1964. By the end of that year, it was expected that a total of 72 Saturn rockets would have been flown with cargo deliveries to the Moon beginning in January 1965 building to a 12- man task force in place on the lunar surface by November 1966. This build-up phase would require 61 Saturns and 88 advanced Saturns by this date, with an average launch rate of 5.3 per month, during which time some 222 tonnes of cargo would have been transported to the Moon. Between December 1966 and the end of 1967, it further envisaged 64 Saturn launches required to lift an additional 121 tonnes.

In almost every respect this was a fine-tuned successor to von Braun's more informal mission plans of the early 1950s, which doubtless played a major part in rapidly delivering the Horizon report. But the figures were staggering, considering the pace and state of the US space programme at the time. Nevertheless, some interesting aspects of this very early phase in serious planning for lunar Missions and, potentially, flights to Mars, are revealed from the formal documents emanating from Horizon. Of certain interest to readers in the UK, attached to the final project report were several bibliographical references to the work on lunar mapping by a certain Patrick Moore, to published work by Kenneth Gatland (eventually a President of the BIS) and to several papers in the *Journal of the British*

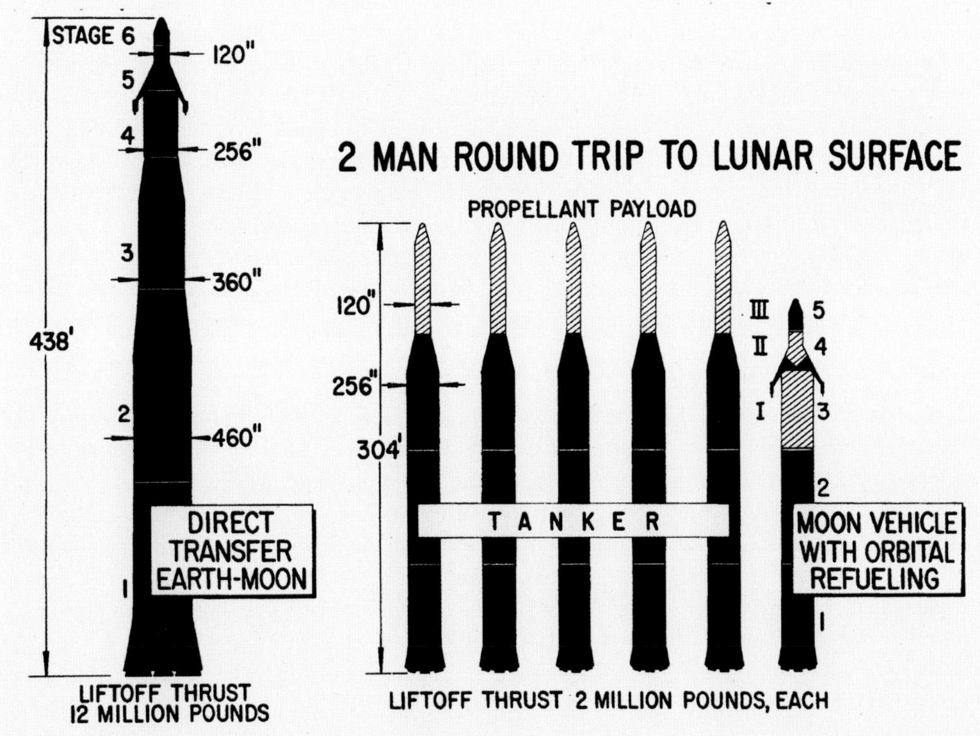

Seizing the initiative and proposing the Juno V series for a wide range of space missions, under the tutelage of von Braun the ABMA crafted an architecture involving circumlunar and lunar landing missions. In 1959 it proposed a series of launchers and flights supporting long-term objectives, making it clear that manned missions to the Moon were clearly possible with Juno V, soon to be renamed Saturn. *Author's collection*

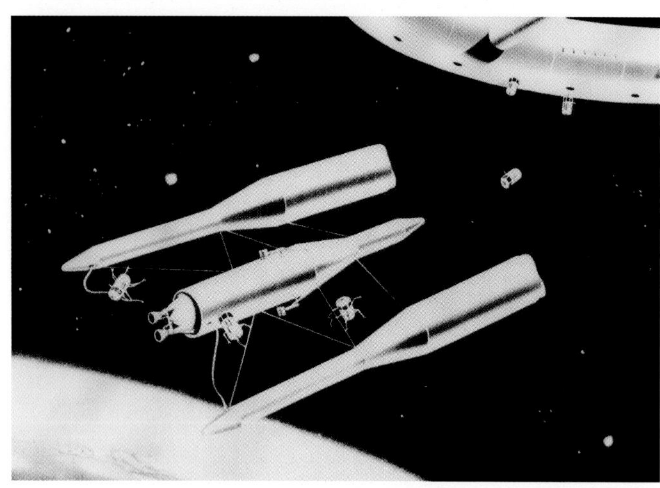

Above: Tanker vehicles supporting a rotating space station for expeditions to the Moon and Mars departing from the orbiting facility. The ABMA and its Development Operations Division under von Braun helped push the US toward a greatly expanded vision for a space future in which permanent bases on the Moon and Mars were considered logical. *Author's collection*

Interplanetary Society on landing sites.

Project Horizon never stood a realistic chance of approval. The new civilian space agency was picking up the slack and crafting a national programme of exploration and development. By early 1960, with America still a year away from putting its first human in space on a ballistic Mercury hop lifted by a converted Redstone, plans were already being drawn up for the next step in human flight. This took the form of a set of objectives defined by Robert O. Piland from the Space Task Group, set up from the Langley Aeronautical Laboratory to manage the Mercury programme and examine future directions of travel. Overall, NASA wanted to send men around the Moon by the end of the decade, and land humans on its surface sometime in the early 1970s.

In a series of briefings conducted by Piland across a number of NASA facilities during April 1960, the future manned space vehicle was being designed for compatibility with the Saturn C-1 and C-2 on circumlunar trips, a multi-mission vehicle weighing a maximum 6,800kg (15,000lb) and capable of supporting three people for up to 14 days. It was to be compatible with space stations and orbiting laboratories and to conduct extensive reconnaissance of the lunar surface from an altitude of approximately 80km (50mls).

Over a two-day event opening on 28 July, NASA held an industry planning conference at which potential contractors and interested bodies were briefed on the broad outline of requirements, technology hurdles in need of detailed analysis and on the general operational use of a vehicle which would, said George Low, NASA's head of manned space flight activity, achieve circumlunar flight by the end of the decade leading to a lunar landing and a permanently manned space station during the 1970s. NASA's Hugh Dryden said it was to be called Apollo. All this was energised by the selection of the Saturn C-series launchers, around which there was much expectation.

At this point, MSFC was working to fly 10 Saturn C-1 rockets to prove the system, the first three of which would carry dummy upper stages before lifting the S-IV for orbiting payloads. The launcher was aligned immediately with the planned Apollo test phase but its ability to lift such a spacecraft into orbit was questioned when it appeared that the mass of the crew module alone could grow from 2,268kg (5,000lb) to 3,630kg (8,000lb) but that this would be within the capability of the C-2 to support circumlunar missions. Mapping the Moon from low lunar orbit was considered a priority, an essential prerequisite to a Moon landing. Very little definition had been conducted about the launcher capable of supporting a lunar landing – that still appeared to be more than a decade away – but cursory regard was made to what NASA called the Nova, a massive rocket, now envisaged with eight F-1 engines in the first stage producing a lift-off thrust of 53,300kN (12millionlb). Demonstration of the F-1 was reserved for the C-3 at the earliest (with one F-1 replacing the four inner H-1 engines) and the engineering challenge of Nova was substantial.

Welded to a future manned (Apollo) vehicle by the Silverstein Committee, MSFC devised the flight vehicle number system beginning with single digits. Development Saturn C-1 flights would be known in the Saturn-Apollo sequence as SA-1 through SA-10 and would be followed by a succession of operational flights carrying manned Apollo spacecraft starting with SA-111. In a planning schedule dated 12 March, 1960, before the advanced manned vehicle was named Apollo, MSFC envisaged the first C-1 flight of an unmanned Apollo A configuration (Earth orbit only) in March 1963, a flight to verify the integrity of the heat shield on a fast re-entry at lunar return velocity. By April it was envisaged that the first manned Apollo flight on a C-1 would take place in February 1966. By the end of September that plan was under strenuous revision, a determination being made that the C-2 launcher would be vital for manned circumlunar flights and that the S-II stage should be placed under contract immediately for a first flight in 1965.

Implicated in the evolution of Apollo support testing were flights of the Atlas-Agena and Atlas-Centaur for high-velocity re-entry tests to evaluate the Apollo heat shield and for tests of the launch escape system but these had disappeared from the schedule by September 1960. However, the schedule persistently retained manned C-1 flights with the first scheduled for 1965 followed by the first manned C-2 launch in early 1968 supporting a lunar fly-by. It predicted a lunar orbit flight in December 1968 (which by pure coincidence, did actually happen in that month but by a Saturn V and not a C-3!). Three more lunar fly-bys were scheduled by mid-1969.

In much the same way Redstone, and at one point Jupiter, were proposed for early Mercury spacecraft development flights, so too were Atlas-Agena launches mooted as unmanned test flights for various elements of the Apollo programme. In late January 1961, a total of 21 such flights were deemed necessary to qualify a wide range of technologies developed for the Apollo spacecraft. At this date, MSFC planned for the first fully configured flight of an unmanned Apollo spacecraft on SA-10 in mid-1965 with the first manned flight on a C-1 a year later. Between then and 1968, NASA planned to fly 12 manned Apollo missions, overlapping the first unmanned Apollo C-2 flight in late 1966, with a second in mid-1967.

The first manned C-2 mission was to have been a low Earth orbit flight in late 1967 followed by eight more by early 1969. Piloted circumlunar missions would start with the

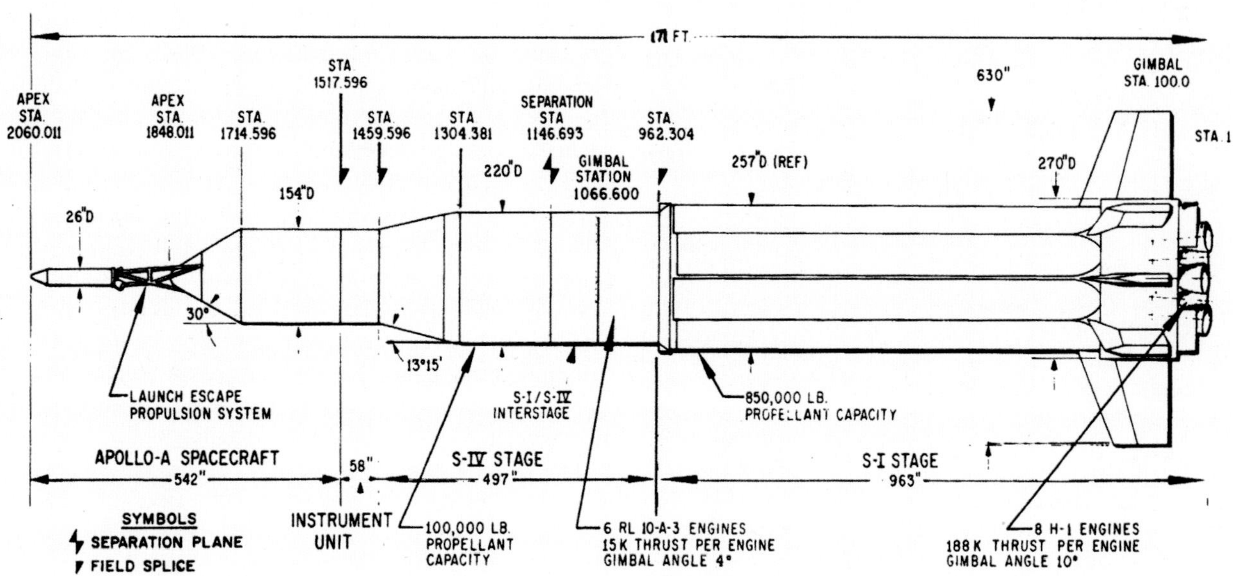

Saturn C-1 configuration

Above: The definitive configuration of the Block I C-1 (Saturn I) assigned to Apollo-. At this early stage, later Block II C-1s were to have carried manned spacecraft. *ABMA*
Right: The predicted flight profile of a Saturn C-1 with cryogenic S-IV upper stage carrying a payload to low Earth orbit at 185km (115mls/100nm). *ABMA*

Below: An evolution of the C-1 and C-2 over the period from 1959 to 1961. The 1959 Saturn C-1 is in Block I configuration while the 1961 drawing is the Block II. Note that the S-III cryogenic stage with two J-2 engines only appears in the original (1959) C-2, the S-II stage with four J-2 engines replacing it on the later (1961) vehicle. Both this stage and the S-IC for Saturn C-4 would get a fifth engine before the drawings were released for manufacturing purposes, converting it to the C-5. *Author's collection*

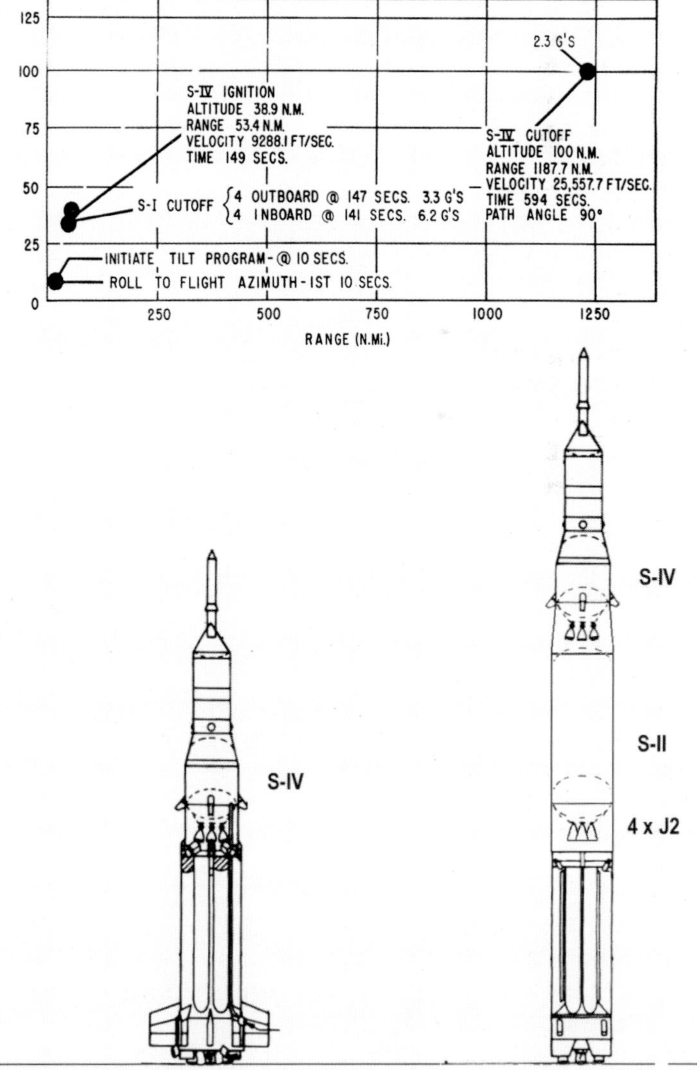

S-V
S-IV
S-IV
S-III
2 x J2

SATURN C-1
1959

SATURN C-2
1959

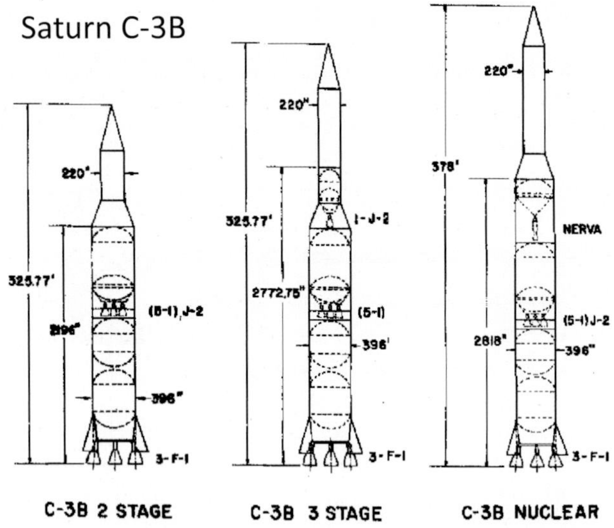

Above: By December 1961 the definition of the C-3 was nearing finality with the C-3B offering two or three conventional and nuclear stages envisaged. As well as three F-1 engines in the first stage, there were now five J-2s in the second stage, plus a single J-2 in the third stage (soon to become the S-IVB). The NERVA was in the nuclear stage.

Below: A variant of the C-3/Nuclear launch vehicle for a lunar landing employing two F-1 engines in the first stage and cryogenic propellants for the lander and the six J-2-engined second stage. The standard, all-chemical C-3 had four J-2 engines in the S-II stage. Its payload capability to low Earth orbit is 45 tonnes (100,000lb), with a minimum escape capability of 13,150kg (29,000lb). *MSFC*

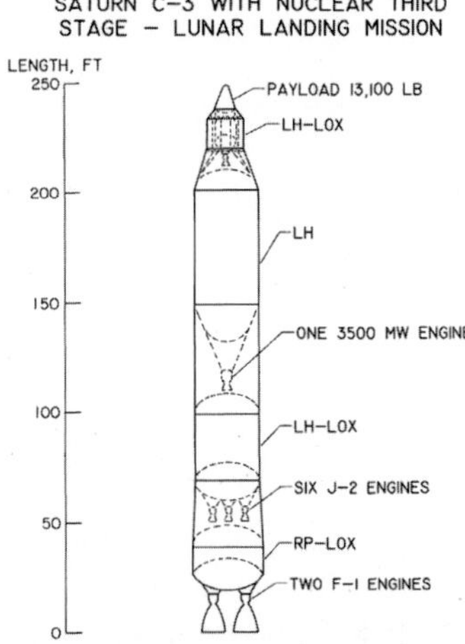

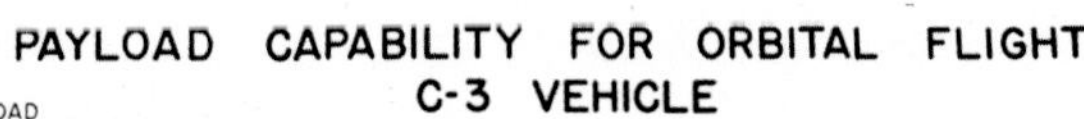

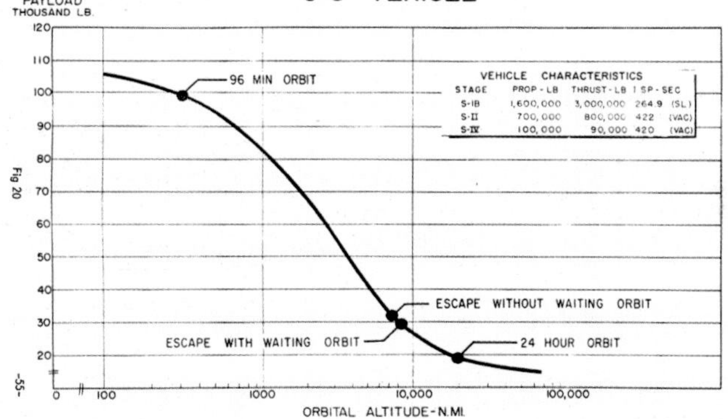

launch of a C-2 in mid-1969, two more fly-bys later that year and four lunar orbit missions in 1970. Designations ran to the 100-series for manned Saturn C-1 flights and 200-series for the C-2. Of supporting Atlas flights, nine would have been dedicated to development of propulsion systems for lunar landing, although there was no mention at all of any such capability on the late January 1961 manifest. This schedule and all the planning that had gone on before, was to get humans to the vicinity of the Moon, and quickly thereafter to a series of lunar orbit reconnaissance flights.

Although not centrally associated with the story of the early Saturn launch vehicles, this was an implicit part of the burgeoning aspiration for deep-space exploration that pushed expectations for a long-term goal of landing on the Moon, expectations beginning at least two years before President Kennedy announced political backing for that objective on 25 May 1961. By mid-1960 there was inertia in the move to expand the national space programme, from one inherited out of existing military activities to a broad-based expansion of science, technology and applications; one of human exploration, supported by robotic and automated missions.

These maturing plans threatened to grow beyond the limits on public expenditure set by President Eisenhower, who never did see Sputnik as a technological and scientific challenge but rather as a dangerous psychological obstacle to retaining pro-American world opinion. And there is no evidence to suppose that at this early stage the USSR thought that either – at least for the two years after Sputnik 1. But it was equally incredulous for anyone to believe that a race was not already underway. While the clandestine intelligence-gathering world, managed at times on a direct level by Eisenhower himself, convinced the government that the Russians were somewhat behind the Americans in overall military strength and atomic weapons, the propaganda war was one he chose to fight indirectly through various government agencies. These included the US Information Agency (USIA) which distributed positive, life-style stories to Americans at home and foreigners abroad.

The higher echelons of government knew full well about the impending launch of Sputnik. When the idea of launching satellites for the International Geophysical Year had been put to Eisenhower by the US National Committee for the IGY in October 1954 it was agreed that the US should do so but it was not considered a priority. Only on 16 April 1955 when the USSR announced with much fanfare that it was setting up a space flight commission did it inject a sense of urgency. This prompted Joseph Kaplan from the US National Academy of Sciences to warn the government that it was in grave danger of suffering a propaganda disaster and he quoted the *Washington Post* when it reported 'The critical shortage of time…(which) cannot be over-emphasised'. A National Security Council note dated 20 May 1955 warned that 'Considerable prestige and psychological benefits will accrue to the nation which first is successful in launching a satellite'.

The connections between the ability to place a satellite in

Right: Adjustments to the projected sequence of increasingly more capable Saturn launchers as the clustered engine/tank configuration formed the basis for B and C series options between 1958 and January 1960. *NASA*

orbit and the demonstration of an ICBM capability were not necessarily tied together, the US for instance put its first satellite in orbit with a modified IRBM. But the sheer weight of the Sputnik satellites had not been foreseen, although that same NSC note in early 1955 warned explicitly that: 'The inference of such a demonstration of advanced technology and its unmistakable relationship to intercontinental ballistic missile technology might have important repercussions on the political determination of free world countries to resist communist threats, especially if the USSR were to be the first to establish a satellite.'

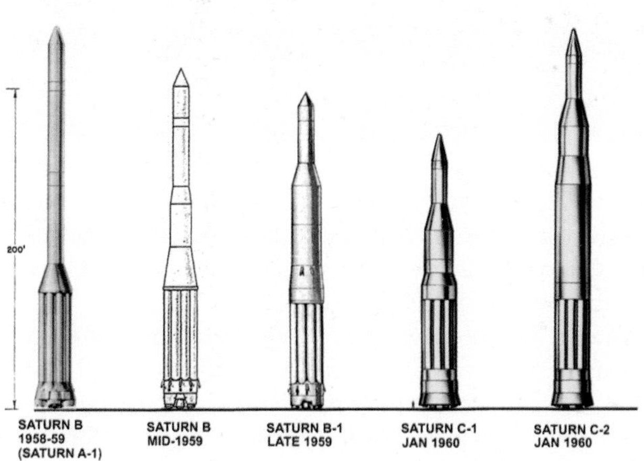

Above: The driving imperative for the Juno V/Saturn series had been the Advent military communications satellite for which there had been an urgent need serving Army, Navy and Air Force global commitments. *ABMA*

It was this stark NSC warning that accelerated development of the Atlas and Titan ICBMs and drove the Eisenhower Administration to deploy IRBMs to Italy and Turkey, placing IRBMs in the UK as an interim message to the Kremlin. But in the 1960s, the availability of the Saturn launcher, implicitly addressing the national concern over lifting power and the perceived advantage to Russia, also hardened Eisenhower's will against investing in Apollo, NASA's proposed successor to the manned Mercury capsule. To Eisenhower the ICBM and Saturn rockets had solved both ends of the equation: the perceived missile gap and the decrement in lifting capacity for US space missions.

Thus it was that nobody in government was really very surprised at all that the Russians had achieved international acclaim and were now regarded as a potential world leader in science and technology. The US had been slow to react and the consequences had been predicted both inside and outside the government. But this view was not universal among those groups responsible to the electorate who received an avalanche of concern about national security and the swing in world power status. Plans for a series of ever more powerful Saturn and Nova rockets assumed a momentum out of synchronism with the White House; while the legislature had chased after big

goals and grand objectives, the executive held its nerve and refused to join the rush to the 'superiority' aim enshrined within the Space Act by Congress for NASA to fulfil.

Paradoxically, it was precisely because of the formation of NASA that the Saturn programme survived at all. Had NASA not emerged from the NACA and given the rocket a reason for existence the Pentagon would have cancelled it, as the Army leadership had proposed already. But while endorsing the public clamour for big rockets – mistakenly interpreting that as a measure of who was in front in the race for technology supremacy (which was not in reality the real issue) – Eisenhower was opposed to the advanced, ambitious and potentially expensive programmes it enabled.

It was for that reason that in the final year of office he rejected plans to bring Apollo in as a funded goal for the United States, claiming that with no man having yet gone into space it was preferable to await results from those flights before paying up front for a line of increasingly advanced manned vehicles.

Left: On a visit to the new Marshall Space Flight Center, President Eisenhower is shown a scale model of the Saturn I. In the background is the S-I-1 stage, minus the expansion skirts for the eight H-1 engines. *MSFC*

As someone in the White House would say later to this writer: 'It was, thought "Ike", akin to the Wright Brothers building a jet airliner before they made the first hop in the Flyer.'

But that was just a political excuse, invoking fiscal prudence as a fig-leaf for conservative limits on runaway spending, however laudable that might be. Notwithstanding a call for constraint from the White House however, the space goals kept getting bigger. In several ways, these issues became political footballs in an election year where the new president would be selected. A year in which politically motivated polarisation grew ever more influential. Eisenhower's dismissal of NASA's request to put Apollo square and central to its future goal only sparked further campaigning by politicians, industry and astute observers among the electorate. It brought divisions among ardent supporters of an expanded national space programme but it also spawned frustration in that, having come this far since Sputnik, the future possibilities would be cut short by apathy in the executive.

NASA had generally assumed that just as the proposed C-2 and C-3 series had made the tentative transition from clustered H-1 engines in the first stage to one or two F-1 engines for a greater payload capability, Nova would be the next evolution for which eight F-1s were proposed, as described earlier. At this date, nobody had assessed the multiple use of Saturn C-2 or C-3 rockets for assembling in Earth orbit a lunar landing spacecraft, simply because the entire focus had been on circumlunar and lunar orbit capability with a manned vehicle, not the much bigger challenge of taking astronauts down to the surface.

Nova was effectively deferred in the operational flight planning stage but over the next several years that name would grow in the imagination of rocket designers to embrace a wide range of possibilities, including solid propellant versions, thrust levels growing to phenomenal and quite impractical levels. Which is probably very fitting: a nova is a star which suddenly brightens with impressive intensity before fading away to obscurity. But even the C-3 lent itself to a wide range of configurations, with stretched upper stages or different numbers of rocket motors in each. But all this was about to change as the new US President, John F. Kennedy, up-ended the sequence and instead of supporting a lunar-orbit-first strategy, succumbed to events which conspired to leapfrog that initial plan and go straight for a Moon landing goal.

In response to the flight of Yuri Gagarin on 12 April 1961, upstaging NASA's plan to put the first human in space – albeit on only a ballistic hop aboard a Mercury-Redstone – President Kennedy ordered an immediate evaluation of potential goals which stood a real chance of achieving a momentous 'first' in space. Tasking Vice President Lyndon Johnson with reporting to him on possibilities, on learning of a shortlist designed to outdo Russian capabilities, Kennedy selected a Moon landing as the potential objective in which the US could prevail.

Never convinced that space exploration had anything other than propaganda value, Kennedy had been indecisive on space matters to this point, choosing to maintain Eisenhower's grip on further plans for an Apollo programme to follow Mercury. In an interim budget review shortly after reaching office Kennedy did increase the funding for big boosters, but only a little and he still denied the agency funds for Apollo. NASA had been relieved that Eisenhower's decidedly negative response to NASA's desire to press ahead with a permanent human space flight programme was toned up a tad by the Kennedy administration. But neither was there much interest in it before the Gagarin flight. Now, on 25 May, Kennedy publicly charged NASA with taking a select few American astronauts to the Moon before the end of that decade, an end date little more than eight years and seven months ahead.

As though to amplify the magnitude of the challenge, much has been written about NASA's unpreparedness to respond to Kennedy's mandate. For more than three years, NASA had been looking at a phased evolution from circumlunar to lunar landing but as two distinct programmes. Even two separate sets of hardware, with an original Apollo spacecraft intended for Earth orbital and lunar orbital operations launched on a Saturn C-1, C-2 or C-3. Following that, a Nova launch vehicle to launch a significantly changed Apollo spacecraft with a very large landing stage to lower the entire spacecraft to the surface, support surface activity and return the crew to Earth. In that it was not so very different from the approach which would be taken in the Soviet programme, which itself had two completely separate endeavours: circumlunar flights with Chelomei's Proton launcher and a lander programme with Korolev's N-1.

But in dashing for the Moon, as a result of the new decision, NASA's circumlunar objective was reworked into a preliminary test phase with its primary objective of evaluating hardware for a lunar landing and, as such, demoting the Saturn C-1 itself to qualifying hardware for the landing. But even there, the new Apollo spacecraft requirement inherited the Block A/Block B approach when it was redefined for Block I (Earth orbit) and Block II (lunar), which were essentially the roles originally defined for the two generations. In so doing, Block I would be a cut-back version of the Block II, without docking equipment or S-band communications equipment, the latter only required for deep-space missions. And it would afford early access to orbital Apollo flights and hardware qualification to demonstrate advanced technology for the Block II.

Following the Moon decision, several long-term evaluations of rocket performance were instituted. In May 1961 changes to the C-1 were approved which eliminated the immediate need for the S-V stage, effectively side-lining that for all but exceptionally unusual missions. It was during this month too that the effectiveness of the C-2 was re-examined, which was felt to be underpowered for the restructured goals and it was cancelled on 23 June, all engineering development work focusing on the C-3 while studies into the Nova concept continued. Success with the C-1 had been largely trouble-free and it remained on schedule, NASA's jewel in the large-booster race for high payload capability. The first flight of the SA-1 vehicle occurred on 27 October, restoring some degree of US pride in beginning to reverse the significant lag in launch capacity compared to Russia.

Even as the first Saturn I was being prepared for its inaugu-

Left: Key political players in the formulation of NASA and the lunar landing objective, President John F. Kennedy (left) and Vice-President Lyndon B. Johnson, each inaugurated on 20 January 1961. *The White House*

ral flight, NASA was in an intense period of contract negotiations on a wide range of equipment required for the accelerated Apollo programme. Several contractors held back on what they considered less lucrative contracts in hope of getting better ones. NASA had a policy of not awarding more than two contracts to the same firm and it also had a geographical (for which read political!) spread across the United States which factored strongly in source selection.

At this date there were preliminary paper plans for the Nova vehicle and some work was well underway on possible motors for the upper stages, the first stage being powered with eight F-1 engines for a first stage thrust of 53,376kN (12millionlb). Baselined as supporting eight J-2 engines in the second stage and one in the S-IVB third stage, the configuration was in a great state of flux. Aerojet-General had developed its Y-1 cryogenic concept engine producing 4,448kN-6,672kN (1-1.5millionlb) of vacuum thrust as a candidate for the Nova second stage, two replacing the eight J-2s. It was designated AJ-1000 by Aerojet-General. This had evolved from the M-1 cryogenic motor which began development in 1960 as a potential Air Force upper stage for space launchers of the future and only latterly found a potential home in NASA's post-Apollo era. Rocketdyne too had a cryogenic candidate for the Y-1 concept, a development of the F-1 designated F-1H. None of these would make it off the drawing board although some technology and engineering development did occur.

On 26 September 1961 NASA held an industry conference at New Orleans regarding S-I stage fabrication and assembly and a total of 37 companies were invited to bid. Only seven submitted proposals, including Avco, Boeing, Ford's Aeroneutronic Division, Chance Vought, Chrysler's Missile Division, Northrop and Lockheed Missile and Space Company. Chrysler got the contract on 17th November for 20 stages. Fabrication of both S-I and S-IB stages would be at the Michoud Ordnance Plant, operated by the industry contractor under the management of MSFC. At the time of contract award there were missions for 27 Saturn I booster stages with MSFC planning to build the first 11 and Chrysler the remainder.

Little more than a week after the first flight of the C-1, on 6 November MSFC directed North American Aviation to redesign the S-II stage to carry five J-2 engines bringing total stage thrust to a nominal 4,448kN (1millionlb). Here again, MSFC had been studying the capabilities of the C-4 and, by essentially retaining the same first stage structure, albeit modified, it proposed a C-5 vehicle which would locate a fifth F-1 engine in the middle of the four-square arrangement on the C-4. By placing a fifth engine in the centre, base heating problems were overcome to the advantage of the overall performance of the launcher. The contractor for the powerful S-IC stage with its five F-1 engines was provisionally selected on 15 December when Boeing was named as the winning bidder, pending a formal contract for 24 flight stages and one ground stage. Later, incremental production contracts pegged the total number of flight stages at 15.

The decision to develop the three-stage C-5 was made public on 25 January 1962 and on 11 July 1962, the day that NASA announced it had decided on the Lunar Orbit Rendezvous mode of lunar landing, a new Saturn vehicle was also declared. Designated C-1B, it would incorporate upgraded H-1 engines and adopt the S-IVB stage powered by a single J-2 cryogenic engine. This would be used for early flights of the Apollo spacecraft, both unmanned and manned, and the unmanned Lunar Module before the C-5 could begin flight tests. In fact, the Saturn C-1B was every bit an interim test launcher allowing the Apollo programme to get an early start

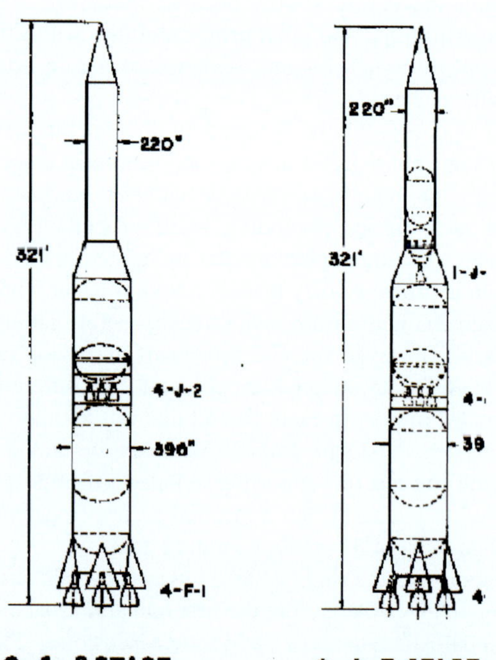

Above: By 1961 the urgency for a super-heavy launch vehicle had been triggered by the selection of the Moon landing goal as a national objective, building on the C-3 and creating the two-stage and three-stage C-4, each with four F-1 motors in the first stage, four J-2s in the second stage and the latter with a single J-2 in the third stage. One step further would produce the C-5 by adding a fifth engine to the first and second stages. *NASA*

Left: On 5 May 1961, Alan B. Shepard is congratulated by President Kennedy having becomer the first US astronaut in space aboard Mercury-Redstone MR-3 (far left). The suborbital ballistic shot cleared the way for a Moon decision, articulated by Johnson, submitted by Kennedy and approved by Congress following a national message delivered by the President on 25 May. *NASA/The White House*

on flying the S-IVB stage.

The lunar landing decision was responsible for the reshaping of the launcher inventory and refining the conclusions of the Silverstein Committee, which had made its upper stage selection prior to any hurry-up from the White House. Scrambling to readjust requirements for this singularly focused goal, in May MSFC directed Douglas to increase the diameter of the S-IVB stage to 660cm (260in) for a more optimum size and volume and to make it more compatible with the diameter of the S-I stage, and also to provide a better fit with the C-5. At the same time, Boeing was told to decrease the length of the S-IC stage from 42.9m (141ft) to 42m (138ft) and MSFC decided to increase the length of the S-II stage from 22.9m (75ft) to 24.8m (81.5ft).

The Apollo programme was itself dragging in a lot of expansion in both resources and hardware; the decision to build an interim, two-man spacecraft called Gemini was made in December 1961 in response to a requirement for an early start evaluating 14-day flights, rendezvous and docking and spacewalking – EVA, or extravehicular activity. It would be built by McDonnell as a production follow-on to Mercury. There was no competitive bid because it was a successor to an existing spacecraft and there was no logical purpose to wasting public funds on a completely new start-up, especially when McDonnell had been working on an expanded, two-man version of Mercury since early that year.

The big challenge for MSFC was to come up with a programme plan and a lunar landing support package using existing, or studied, Saturn launchers. It had been assumed, specifically by von Braun, that the lunar landing would be achieved with assembly of the spacecraft in Earth orbit before heading off for the Moon. In 1959, Project Horizon had been a precedent for this but that was when attention focused on the C-1. Over the preceding three years a new generation of C-class Saturn rockets had appeared, closing the gap between the payload mass required to get to the Moon and the lift capacity available with the bigger Saturn launch vehicles. The determination to move along to a manned landing on the Moon found initial expression with the formation of the Manned Lunar Landing Task Group on 6th January, three months before the Gagarin flight and analysis of how to get there rapidly coalesced around the use of the more powerful C-series rockets, as outlined above.

At that time, it was assumed that multiple launches of C-2 rockets would need to be launched for assembling the Moonship in Earth Orbit Rendezvous (EOR) but additional modes of Lunar Orbit Rendezvous (LOR) and Direct Ascent (DA) were discussed. LOR came out of the Langley Research Center and proposed a separate Lunar Module to go to the surface from lunar orbit, to which the Apollo spacecraft had carried the crew. The DA method would carry two astronauts launched on a Nova, or C-8 (for the eight F-1s in the first stage) as it was being known, to a direct descent without intermediate lunar orbit. The DA method had strong support and various landing configurations evolved. By the end of 1961, with the decision still pending, the lander stage had legs at the base for a descent right down to the surface but by early 1962 the landing legs were up on the base of the existing Apollo design, with a deceleration stage beneath for slowing the whole assembly before jettisoning itself, leaving Apollo to back down to the surface.

There were advantages and disadvantages to each mode and readers will be familiar with those. The EOR mode required multiple launches, rendezvous and docking, fuel transfer and then a flight to the Moon and direct descent to the surface – although the option for an intermediate lunar orbit remained. The LOR method dispensed with multiple launches and could be carried out with two C-4 launches, while the DA mode required development of a new and more powerful launch vehicle with new facilities for stage testing and a launch pad greater than anything presently available or planned. With the LOR mode selected in mid-1962, the final decisions had been made and despite setbacks and a single catastrophe with the planned launch of the first manned Apollo, that goal was achieved just seven years later.

All-up Policy

When the ABMA Development Operations Division was officially transferred to NASA in 1960, that component of the Redstone facility near Huntsville, Alabama, became the George C. Marshall Space Flight Center (MSFC). Honed on the 'arsenal' principle alluded to earlier in this chapter, MSFC would undergo considerable change and move toward a more contractor-based focus, managing overall design, test and evaluation but also retaining the laboratories that distinguished this place from other facilities which gave it a more in-depth access to problems than was often the case at other facilities. The core philosophy at MSFC was inherited from the German experience, all the way from Peenemünde to White Sands and on to Huntsville. It was to dive deep into the work of the contractors and differed from standard military practice in the degree of monitoring, sometimes intervention and always a robust verification of what was going on at the industrial facilities doing work for MSFC. This had enormous benefits for the wider space programme.

Over time this would come to fuel a major expansion and as work on the Saturn rockets progressed the traditional ways were replaced by a new sense of MSFC being at the hub of a giant managerial challenge rather than being fundamentally a set of laboratories, machine shops and test facilities. Between 1961 and 1963 the budget, and the procurement funding they managed, increased threefold and this forced change masterminded by von Braun himself. He transferred control of the various laboratories into project offices which could reach across several facilities at MSFC and control and coordinate to more effectively tighten management of production requirements and timelines. Somewhat outside the story of the Saturn I, but greatly indicative of the new managerial style only marginally preceding a NASA-wide shift the way programmes and projects were managed, von Braun ensured that the laboratories had priority when it came to decisions.

That the organisation now managed by von Braun as Director of MSFC from 1960 would stand out within the NASA structure was due directly to the 30 years of experience in building, testing and developing rockets of increasing size and capability. The commitment within the group was seminal and explained by von Braun himself: 'Ever since the days of the young Raketenflugplatz Reinickendorf in the outskirts of Berlin in 1930, we have been obsessed by a passionate desire to make this dream come true... We have had only one long-range objective: The continuous evolution of space flight.'

A convincing argument for defining the unique nature of the organisation that now had responsibility for the Saturn family was the consistency of personnel and the extensive experience brought by the German rocket engineers. Von Braun would occasionally reflect on the single assistant he had when working for the German Army from 1932, the 80 or so within the group when they went to Peenemünde in 1937 and the 120 that came with him to the United States after the war, to the 1,600 personnel at the ABMA that swelled to approximately 6,000 when the ABMA transferred to NASA. Throughout, von Braun was proud of the 'dirty hands' philosophy, as he called a management style in which the managers were never very far from the engineers preparing materials, forming assemblies or testing complete rocket motors.

Von Braun's successor as Director of MSFC when he retired in 1970, Eberhard Rees maintained that the NASA management structure was incomplete and without an efficient structural apparatus that could survive the accelerated growth and expansion it experienced in the early years, especially after Kennedy's Moon decision. Rees felt that no single monolithic management structure could possibly survive the transformation it experienced. In the initial two years of the programme at MSFC, the Saturn Systems Office (SSO) ran the project with three offices: those of the S-I Vehicle Project Manager; the S-I Stage Project Manager; and the S-IV/S-V Stages Project Manager. Relatively small, with only 154 people by spring 1963, the SSO was modelled on the example of other administrative support offices and the layout of line divisions, nine technical departments or laboratories in each of which several hundred personnel were employed.

Below left: The Saturn engine cluster mock-up and weight test simulator ready for installation. *Author's collection*

Below: Clustering of the Saturn first stage underway at MSFC. The circular objects in the foreground are fuel pressurisation spheres. *NASA-MSFC*

Von Braun ran meetings at which all nine division heads were present to discuss technical issues and to reach consensus either through discussion or by specific presentations. This had been the model preferred by von Braun from the beginning of his work at Redstone Arsenal in 1950 and it survived as the core framework for maintaining close links up and down the management chain. Inclusivity was key to keeping people involved at various levels in projecting a universal integrated 'team' ethic and providing a sense of continuity that employees could see directly in the way they worked with others up and down the line. The consensus approach involved von Braun at many levels and inspired the work force at different stages of their tasks.

Much of this was integrated within NASA through frequent visits by the managers and von Braun himself to headquarters in Washington DC. Two specific dates defined the changes which were to occur and transform the way the existing practice was modified by separate circumstances: the decision in 1962 to develop the Saturn IB and the Saturn V and the strategic change in overall NASA management strategies in mid-1963. A shift toward a more management-central approach was predicated by the introduction of the cryogenic upper stages for the Block II Saturn C-I and the Saturn C-IB, which was progressively introduced during the period 1961-1962 and this would expand into the development of the Saturn C-5 and its big cryogenic S-II stage. But it was the decision to build the Saturn IB and the Saturn V which had the greatest impact.

Enlargement of the SSO was crucial to success with the two new vehicles and this brought challenges in that MSFC believed that these projects were far beyond the internal capacity of the facility to manage in-house. Accordingly, von Braun set MSFC on a progressive series of adaptive steps incorporating major industrial contractors to handle work which it would never have outsourced in the time of Redstone and Jupiter rockets. Added pressures from budget restraints, tight schedules and the integrated operating structure involved with several prime and hundreds of secondary contractors brought a need for reorganisation.

Largely as a result of these the separate Saturn C-I and C-IB offices were brought together as a single Saturn/Apollo Integration office, although the technical divisions remained largely the same and reported directly to von Braun. These covered areas involving electronics, mechanical engineering, flight mechanics, etc. Each division director was encouraged to not only manage a high level of expertise in his organisation but also to maintain links with other government agencies, monitor theoretical research in his field and keep a broad-based awareness of peripheral developments which may impact operations. For instance, von Braun was aware of the unionisation of many organisations and companies in the subcontractor supply chain and wanted his managers to be constantly aware of the impact problems in that sector could produce. In this way, it was felt that industry contractors, frequently at a distance from government agencies they worked for, would respect that here was a level of expertise that in some cases was greater than their own skills and knowledge.

Von Braun attracted a wide range of opinions and sought different views, thoughts and propositions, encouraging counter-arguments and new opportunities via debate. He encouraged a work ethic that embraced personal respect and trust and sought to consolidate that through extra-mural activities which he organised. For instance, there was a strong connection between the German rocket engineers and music, most of them being proficient in at least one instrument. He found it useful to encourage interaction between people at different levels of management and employees in the workplace to assemble a little orchestra at MSFC where they all played an intrinsically important part of the whole by holding concerts, even taking such performances out to the local communities. On several occasions he was heard expressing the view that as music was a universal language it helped weld together people from two very different cultures and that it did much to level out ranks and pay grades.

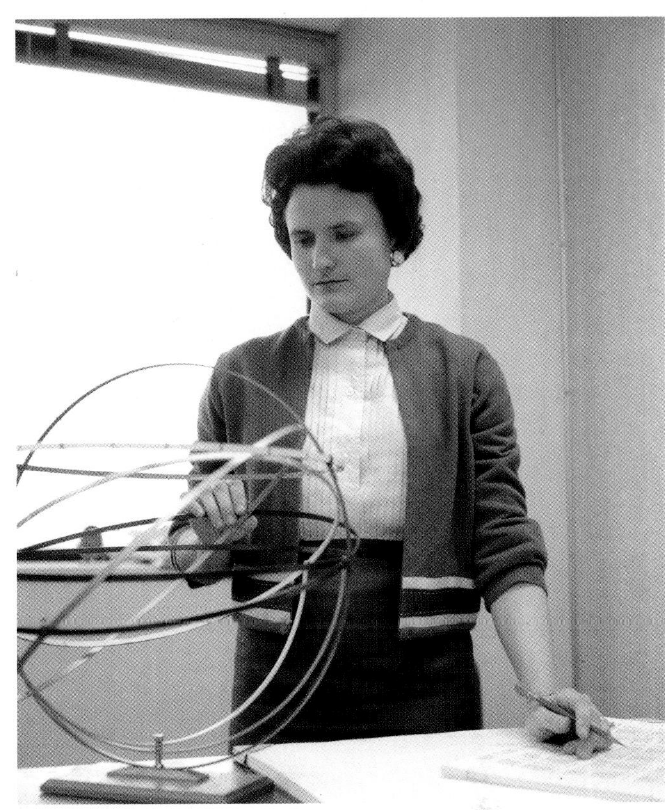

Above: Despite being encouraged by her father to 'get married, settle down and be a good housewife', in a policy decision by von Braun to give women opportunities for work in the space industry, Ethel Bauer joined MSFC and did so much to help the space programme, here posing for a picture from her routine job calculating trajectories. *Author's collection*

In requiring his managers to adopt a systems management approach, von Braun inculcated a sense of control in the Project Director, where the structure of a particular set of objectives were held to account rigidly and only adapted to problem solving and solutions for obstacles. Von Braun required his managers to set up the programme and take control of it, asserting personal responsibility – and liability – for how it turned out. If there were insurmountable obstacles 'there is usually something wrong with either the engineering concept or with the ratio between the implementation plan and the resources committed to it', he would say. Many changes were driven by the introduction of the Saturn C-5 programme and that is the subject of another book but it was the reorganisation of mid-

Left: President Kennedy made his third visit to Cape Canaveral on 16 September 1963 and in the LC-37 blockhouse is given a presentation by manned space flight boss George Mueller with models (from right) of Mercury-Redstone, Mercury-Atlas, Gemini-Titan, Saturn I Block II and Saturn IB. Just out of sight at left is a Saturn V model on the Crawler Transporter. *NASA*

1963 that involved all three launchers then under development at MSFC.

From 1st September 1963, the Center Director's office grew two operational segments which reported direct to von Braun: Research and Development Operations (R&DO) and Industrial Operations (IO). The latter had a direct line of communication with NASA headquarters due to the contractual and budget implications. The R&DO came directly out of the separate technical divisions and the IO grew from the SSO with three programme offices assigned to Saturn C-1/C-1B, Saturn C-5 and engines. The engines office extracted responsibility for the H-1, RL-10, F-1 and J-2 motors from the laboratories which smoothed management control and monitoring.

The chain of communication from HQ ran from George E. Mueller, the Associate Administrator for Manned Space Flight to the MSFC Director and thence to the programme manager with an informal channel reporting through to Mueller's GEM boxes which, signified by his initials, were introduced when he joined NASA in the role. These changes involving NASA headquarters were strategic in their implications and were identified as an important part of his early actions when Mueller commented on why he had implemented that: 'I put together this concept of a program office structure, geographically dispersed, but tied with a set of functional staff elements that had intra-communications between program offices that were below Center level and below the program office level so as to get some depth of communications.'

A specially important aspect of overall project management and control was in the relationship of MSFC with the industrial contractors and in more than a few instances the contractor felt that he should be left alone to do the work. It was at times a delicate balancing act and would come to a head when failures and fatalities resulted from a lack of corporate transparency regarding problems and technical challenges which defied solution but appeared not to stand in the way of schedules. Several German scientists from the original von Braun group wanted to maintain a strict control over the contractor, insisting on close scrutiny and monitoring. Eberhard Rees was especially concerned about loosening the reins, asserting that 'it became clear that close and continuous surveillance of the contractor operation was required on an almost day-to-day basis'. Key to that was the Resident Manager's Office (RMO) located at each contractor which mirrored the project manager's office back at MSFC. This produced a concerted shift in contractor attitudes and did much to improve the performance of the industry, which by the mid-1960s was considerable.

In development of the S-IVB for both classes of Saturn rocket, the stage contractor Douglas was late in configuring its management structure and strong intervention from NASA was required to sort it out. The problem originated with the Sacramento Test Operations (SACTO) facility which had management connections to both the engineering division at Santa Monica and the manufacturing division at Huntington Beach. Both sections of the company were pulling strings at SACTO until NASA made Douglas restructure SACTO so that it was responsible to the highest levels of corporate management and could be operated separately to the manufacturing base of the organisation. NASA also insisted on some significant management changes and went through more than 90 applications for the occupant at the RMO before settling on a man who could maintain that ethic while placing Douglas' deputy director of its Saturn programme to run SACTO. The new RMO took a substantial drop in salary just to get the job.

Management issues aside, refinement of the many possible configurations of stages in the Juno V A-, B- and C-series launcher families into three types funded for development reflected the entrenched link with the past inherent in the classification of successively larger Saturn rockets. The prefix 'C' was abandoned on 7 February 1963 when a management decision was passed to all centres that henceforth the C-1 would be known as Saturn I, the C-IB as the Saturn IB and the C-5 as

Left: The 62m (204ft) tall Saturn dynamic test stand at the Marshal Space Flight Center showing the installation of the dummy S-V and the completed stack suspended within the steel structure for bending loads, temperature effects of propellant loading and simulated flight vibrations. *MSFC*

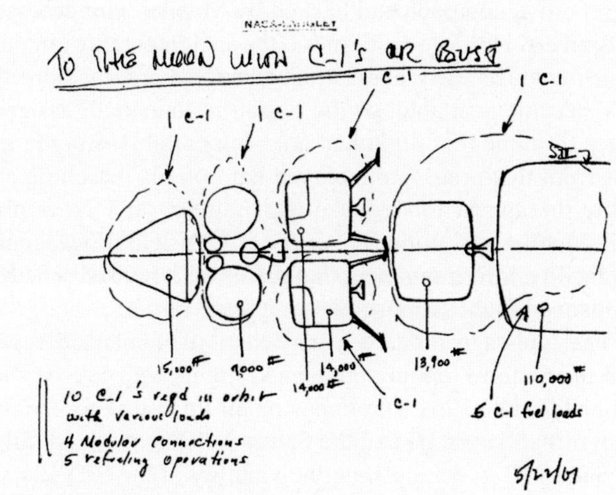

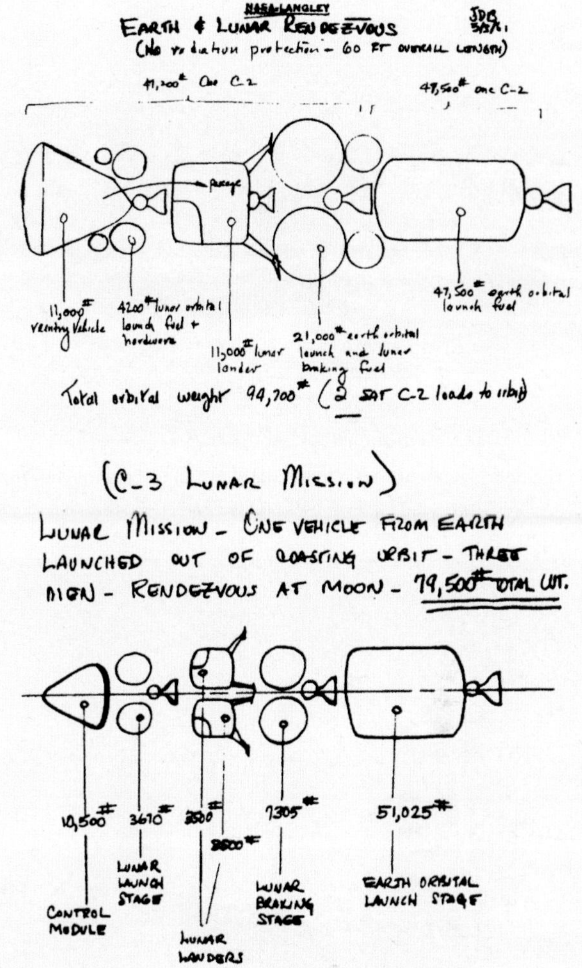

Shooting for the Moon

Above: In May 1961, calculations of the possibilities for lunar missions with the Saturn C-1 concluded that 10 launches would be sufficient to place two landers on the surface. Five of the launches would be propellant tankers and the mission profile called for no less than four rendezvous and docking operations. *Author's collection*

Above right: With two C-2 rockets, each carrying a payload of just over 21,320kg (47,000lb), it was believed that a single lander could send a spacecraft to the lunar surface following a rendezvous in Earth orbit and return the crew safely to Earth. *Author's collection*

Right: Studies also highlighted the possibility of using a single C-3 to place humans on the Moon. However, the weight assumptions for critical elements were grossly over-optimistic and badly underestimated the total mass of equipment required for a landing. *Author's collection*

Saturn V. In so many ways, 1963 would see the greatest change and challenge to the Apollo goal and see the shift away from the German legacy to a very American method of managing big projects and controlling giant programmes, of which this was a seminal example.

In late 1963 the new management style was invoked throughout the entire agency and von Braun acquired about a dozen US Air Force officers highly skilled in the systems management approach, in budgeting and in administration. General Edmund O'Connor was appointed the director of industrial operations and Arthur Rudolph, bringing experience from managing production of the V2 rocket in Germany, ran the Saturn V programme office. He festooned the Project Control Center with 279m^2 (3,000ft^2) of wall with charts defining every element of the programme, its manufacturing flow, acceptance and test routines and pre-flight preparation. Saturn was a giant management challenge and the effectiveness of the latter would engage with the former to deliver on time. These changes took effect from 1 September 1963, but there was one further effect, one again predicated by events around the life of John F. Kennedy.

A little over 900 days after declaring the Apollo Moon goal, on 11 November 1963, President Kennedy was assassinated in Dallas, Texas. He had lived to view for himself the mighty Saturn I at Cape Canaveral and to be impressed by its awesome size and scale. For him, the need for the US to possess a powerful launch vehicle was paramount. As recorded in greater detail previously, on taking office at the White House his initial amendment to the Eisenhower budget proposal was to increase funds for development of Saturn I while remaining ambivalent over the initial objectives of Apollo, preferring to await the outcome of the Mercury project but very quickly finding, in the collective availability of assets, the political opportunity of setting America on its way to the Moon.

As Kennedy's successor from November 1963, Lyndon Johnson was pressured by the Congressional critics of high spending that NASA's burgeoning budget had to be kept under tight controls. That filtered down to NASA boss Jim Webb who sought solutions for cuts to the fiscal year 1965 spending plan, due to be announced in early 1964 for the 12 months beginning 1 July that year. The new head of manned space flight operations, George Mueller had already been instrumental in moving to an all-up testing plan for the Titan II and Minuteman missiles when he had been responsible for technical operations in a range of Air Force programmes.

What he saw in the Saturn I programme had appalled him, its wastage of time and money in sequential testing of incremental stages before a final full configuration launch, appeared counter-intuitive. The sequential flight operations plan had been developed at ABMA, conforming to von Braun's

Above: The two Sacramento Test Operations (SACTO) test stands with their respective liquid hydrogen tanks at right. *Douglas Aircraft Co.*

Above: A battleship tank in the SACTO test stand ready for a static test of the S-IV propulsion system. Note the five (of six) ejectors for releasing steam from two large accumulators fed into ejectors to create a vacuum around the S-IV engines. Once ignited, the engine exhaust created sufficient vacuum for the simulated altitude. The spherical tank at right contained liquid hydrogen. *Douglas Aircraft Co.*

Above: A SACTO engineer operates the valves on one ofthe liquid hydrogen tanks supporting stage tests with the S-IV, a facility also used for the S-IVB. *Douglas Aircraft Co.*

conservative approach and backed by Mueller's predecessor, D. Brainerd Holmes. In fairness, the incremental testing approach enabled early evaluation of the S-I stage before the S-IV became available, so the option to launch all assigned stages from the first flight had not been possible. But the lessons from that programme played hard on his reasoning and before the appeal for reduced expenditure, on 1 November 1963 Mueller sent notices to every NASA field office, centre and facility, informing them that there was a revised schedule for manned flight support.

That message ordered a cancellation of manned flights with the Saturn I, ending unmanned flights with SA-10, then planned for late 1964, and imposing all-up testing for the first flight of the Saturn IB and the Saturn V. Thus it would fall to the Saturn IB to demonstrate the viability of the S-IVB stage and to Saturn V to demonstrate the efficacy of the S-IC and the S-II stages. A very bold move but one which would pay dividends in time saved and costs reduced. Von Braun was in determined opposition and conducted what some historians have described as 'lively debate' between the Director of MSFC and his boss at headquarters. But Mueller was adamant amid almost universal opposition to what many regarded as a reckless and foolhardy decision.

Development of cryogenic propulsion systems had not been easy and the Saturn IB was now to demonstrate a satisfactory re-design of the S-I stage and the inaugural operation of the S-IVB, so very different to the S-IV, all on the same flight. Diplomatic to a fault, on 8 November von Braun gave a formal response, putting his reputation behind the idea and pledging that shift on behalf of MSFC, despite so many of his staff having grave misgivings. Less than a month later, when Mueller visited MSFC he was approached by Arthur Rudolph who quietly drew his attention to the very big difference between a relatively solid propellant rocket like Minuteman and the vastly more sophisticated technology of the liquid-propellant Saturn V. Mueller was circumspect and summarily dismissed the relevance of the comparison.

The all-up concept dramatically changed the Saturn I and IB programmes, effectively speeding up the schedule not only by eliminating unnecessary initial flights but also by saving time and money on re-configuring the launch pad after every sequential launch. It also sharpened the preparation cycle and greatly simplified flight planning, introducing both the Saturn IB and Saturn V to a single set of flight activities (because all stages would be active from the outset) but also by eliminating unnecessary planning operations for the different assembly of stages. Just as Saturn IB was a pioneering test vehicle for the S-IVB, and for the Apollo spacecraft itself, it would now test Mueller's all-up approach. Under expectations of late 1963, there was less than two years to the first flight of that rocket.

This was a critical period in the integrated history of the Saturn I, the Saturn IB and the Saturn V. The S-I stage had already proven itself to be a robust and reliable booster on four suborbital flights but in late 1963 it had still to fly the cryogenic S-IV stage and to demonstrate an orbital capability. Cryogenic operations were key to the efficiency of all three Saturn launchers and the technology was crucial to a Moon landing during the 1960s.

Section 3

Engineering Saturn I

Chapter Six

Stage Design and Systems

The Saturn I played an operationally significant role because of the technology it advanced, in the creation of an efficient and highly effective management style and in proof-testing engineering designs and operational capabilities that were enablers in the bid to get to the Moon before the Russians. It is probably not true to claim that NASA would not have got to build the Saturn V were it not for the Saturn I, but something very close to that is borne out by the record: that development of the use of cryogenics and particularly in the use of the S-IVB could not have occurred as rapidly as it did were it not for the Saturn I, where that generation of stages flew almost four years before it got to ride on a Saturn V. The engineering of the Saturn I was crucial to many processes and technologies vital for the Moon effort.

In engineering applications, as well as operational demonstrations, the Saturn I programme was divided into two phases, each with unique elements and operating capabilities: the Block I, which consisted of an S-1 first stage without fins, dummy S-IV and S-V stages and a research and development payload for ballistic research and development flights of the clustered first stage in addition to modal measurements on the full stack; and Block II, which had a live S-I stage with base fins, a live S-IV stage, an instrument unit and an orbital capability. Block I flew four flights (SA-1 to SA-4) and Block II flew six missions (SA-5 to SA-10). The original plan had been to use these as precursor flights to a series of six manned Apollo missions (AS-111 to AS-116), later reduced to four before being completely removed from the flight lists in 1963 with manned flights transferred to the Saturn IB, initially known as the Uprated Saturn I.

The design of the Saturn I evolved between 1957 and 1959 with a series of cryogenic upper stages for the C-series adopted as the baseline for detailed planning. ARPA had approved development of the cluster concept on 15 August 1958, with four test launches sanctioned a month later. A mock-up of the boost-

Above: A complete Saturn I, including dummy S-IV and S-V upper stage, takes shape at MSFC for engineering tests. *NASA-MSFC*

er's tail section was placed in the ABMA Static Test Tower on 4 January 1960 followed by the S-I-T test stage on 28 February, this being the first full-size representation of the first stage.

Assembly of the first Saturn flight vehicle began on 26 May before formal transfer of the programme, along with the Development Operations Division, from ABMA to NASA on 1 July. On 26 July NASA signed a contract with Douglas for design, development and fabrication of the S-IV, which at this date consisted of four RL-10 engines delivering a stage thrust of 355.84kN (80,000lb).

A study contract for the S-V was awarded to Convair on 21 October 1960. Static firing tests of the S-I began on 2 December with a two-second firing of the eight-engine cluster with booster tests concluded on 20 December. In late January 1961, von Braun recommended abandoning plans to move to a three-stage configuration and the S-V stage was cancelled during May, although a dummy stage would continue to fly on early Block I launches.

With missions planned for 27 Saturn I rockets, NASA invited manufacturing and assembly bids for 20 S-I stages from 37 companies but only seven submitted proposals, largely because others wanted to go for more lucrative agreements supporting the upcoming Apollo work. Interested parties included Avco, Boeing, Ford Motor Co, Chance Vought, Chrysler, Northrop and Lockheed Missile & Space Co. NASA wanted to build the first 11 S-I stages at MSFC, the rest by a contractor, proclaiming that the life of the launcher would run to six or seven years. Other Apollo contracts coming up included a new Y-1 cryogenic engine with a thrust of 4,448kN (1million lb) to 6,672kN (1.5million lb) which would have propelled the second stage of the Nova rocket. Eight F-1 engines would power the first stage with a cluster of four J-2 engines in the third stage.

On 17 November 1961, NASA announced that Chrysler had been selected for manufacture of the S-I stages at the Michoud Assembly Facility. Outsourcing was becoming the order of the day. The last of the Saturn 177.8cm (70in) tanks manufactured by MSFC was completed during the week of 4 December, subsequent tanks being manufactured by Chance Vought in Dallas, Texas, prior to their assembly into stages at Michoud. The contract was signed in January 1962 but on 6

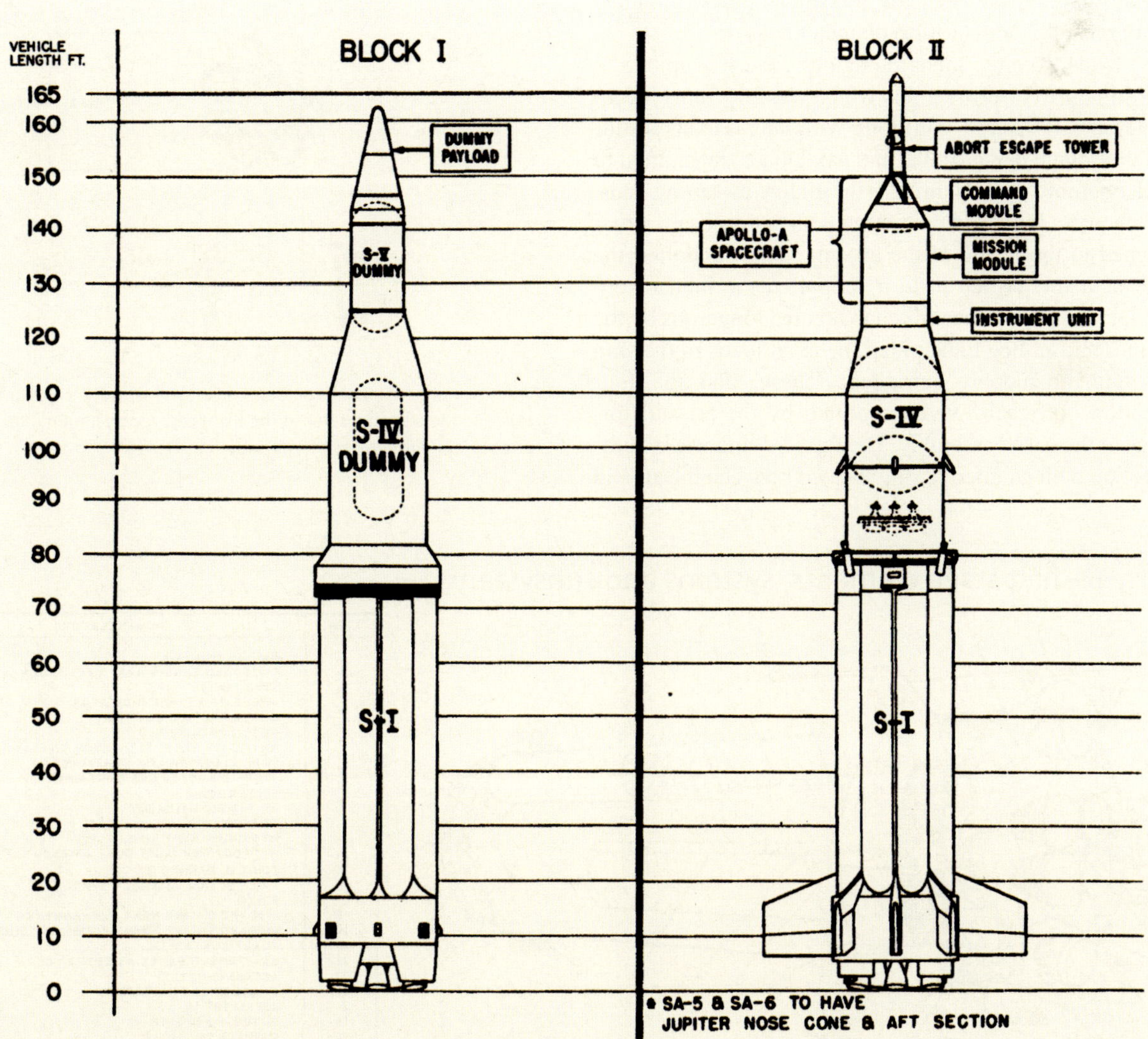

Principal characteristics of the four Block I suborbital and six Block II orbital versions of the Saturn I rocket, the latter incorporating several changes to the original design with upgrades to the Instrument Unit for the more complex missions of the Block II variant. *MSFC*

August NASA amended that to 21 S-I stages to be delivered between 1964 and early 1966. Work on the first industry booster began on 4 October (S-I-8 flown on SA-8).

S-1 Engineering Design

Overall, the S-I had a length of 2,444.5cm/24.44m (962in /80.2ft) and a diameter of 652.8cm/6.53m (257in/21.4ft) across the tanks and 696cm/6.96m (274in/22.8ft) across the thrust structure. The opposing fins on the Block II had a maximum span of 1,239.5cm/12.4m (488in/40.7ft) across the four larger fins of the eight (including four stub fins) attached.

The tail section supported the eight H-1 engines and transmitted loads to the five LOX containers while supporting the four fuel containers in addition to protecting the engines and associated installations from aerodynamic loads and engine heating. Thrust loads were transmitted to the LOX containers through the aluminium-alloy thrust structure. Equally spaced on the 162.5cm (64in) diameter aft ring, the four inboard engines were mounted in a fixed position, canted 3deg from the vehicle centreline. The four outboard engines gimballed and were equally spaced between the inboard engines, mounted equidistant on a 482.6cm (190in) diameter ring.

Thrust loads from the inboard engines were transmitted to the thrust structure barrel assembly which was 266.7cm (105in) in diameter and 190.5cm (75in) long with lateral loads resulting from the engine being canted and axial loads transmitted to the barrel assembly aft ring through the engine mounting pads. The aft ring was a built-up box section with a crossbeam structure attached to the inside of the aft ring. This supported the fixed-link actuators which in turn supported the inboard engines. Axial loads were transmitted to tapered longerons by the aft ring and in turn they transmitted the axial loads to the skin and the four fin-support outriggers. These, and the four thrust-support outriggers were supported by the aft and forward rings of the barrel assembly. The forward ring was a built-in box section with an internal ring located between the aft and

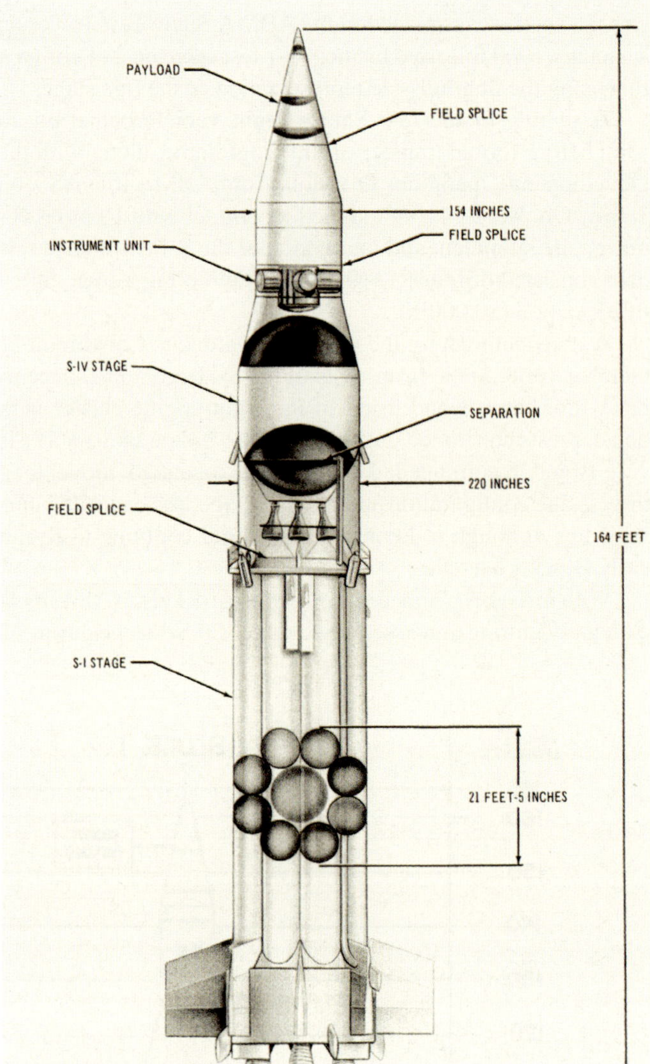

*R*eference dimensions of the Block II Saturn I with the cryogenic S-IV stage. Readers should refer to the flight test reports in Chapter 10 for accurate dimensions of individual vehicles (SA-5 to SA-10). *MSFC*

Saturn 1 – first stage elements, systems and subsystems

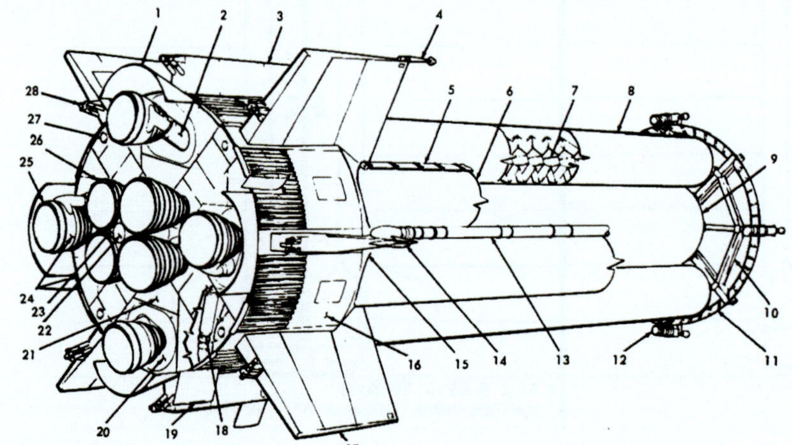

ITEM	NOMENCLATURE
1	ENGINE SKIRT (4)
2	OUTBOARD ENGINE HEAT EXCHANGER (4)
3	STUB FIN (4)
4	ANGLE OF ATTACK INDICATOR (2; SA-7, 9 & 10)
5	CABLE ROUTING DUCT (4)
6	FUEL CONTAINER (4)
7	ANTISLOSH BAFFLES (IN EACH CONTAINER)
8	70-IN. DIAMETER LOX CONTAINER (4)
9	105-IN. DIAMETER LOX CONTAINER
10	SPIDER BEAM
11	45° SHROUD ASSEMBLY
12	RETROMOTOR (4)
13	HYDROGEN VENT LINE (3)
14	HYDROGEN EXHAUST DUCT ASSEMBLY (3)
15	CONICAL FAIRING (8)
16	TAIL SECTION ASSEMBLY SHROUD
17	FIN (4)
18	INBOARD ENGINE HEAT EXCHANGER (4)
19	INBOARD ENGINE TURBINE EXHAUST DUCT (4)
20	FLAME CURTAIN (4)
21	HEAT SHIELD PANEL ASSEMBLY (4)
22	ACCESS CHUTE
23	FLAME SHIELD ASSEMBLY
24	ASPIRATOR (4)
25	OUTBOARD ENGINE (4)
26	INBOARD ENGINE (4)
27	WATER QUENCH DISCONNECT (4)
28	SUPPORT AND HOLDDOWN POINT (8)

Saturn I first stage elements, systems and subsystems. *Author's collection*

Saturn 1 – Block II flight configuration and dimensions

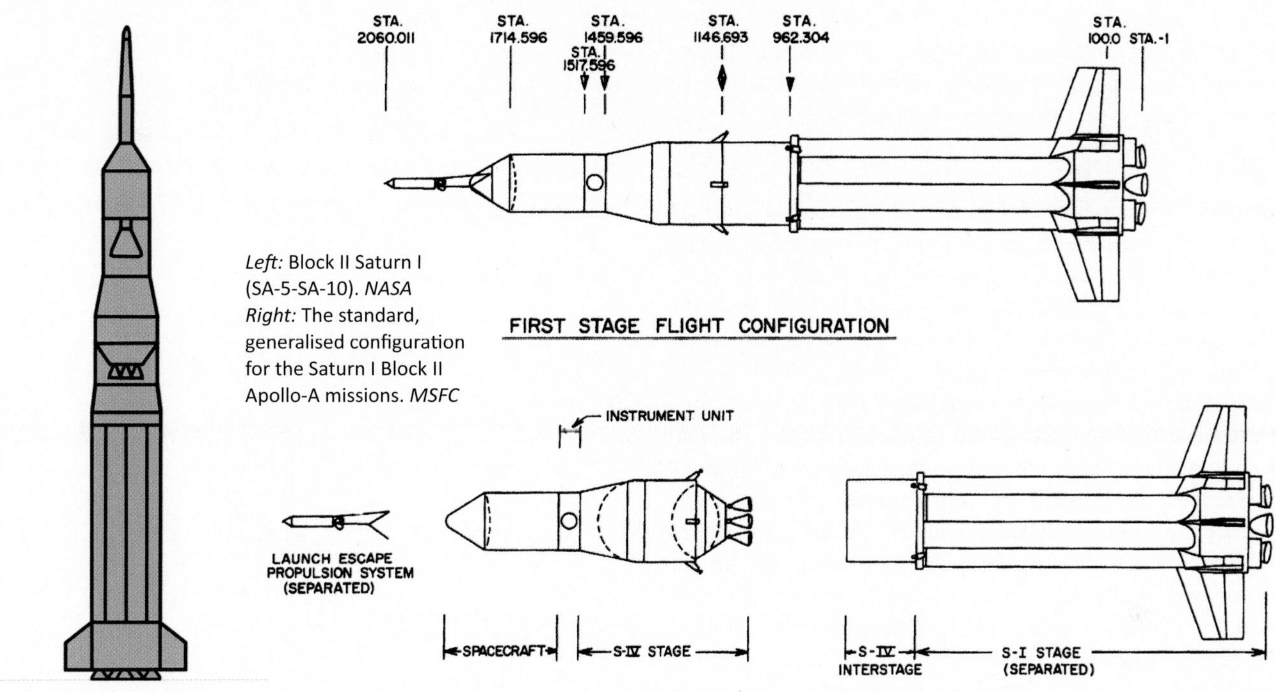

Left: Block II Saturn I (SA-5-SA-10). *NASA*
Right: The standard, generalised configuration for the Saturn I Block II Apollo-A missions. *MSFC*

Saturn 1 – station measurements and fins

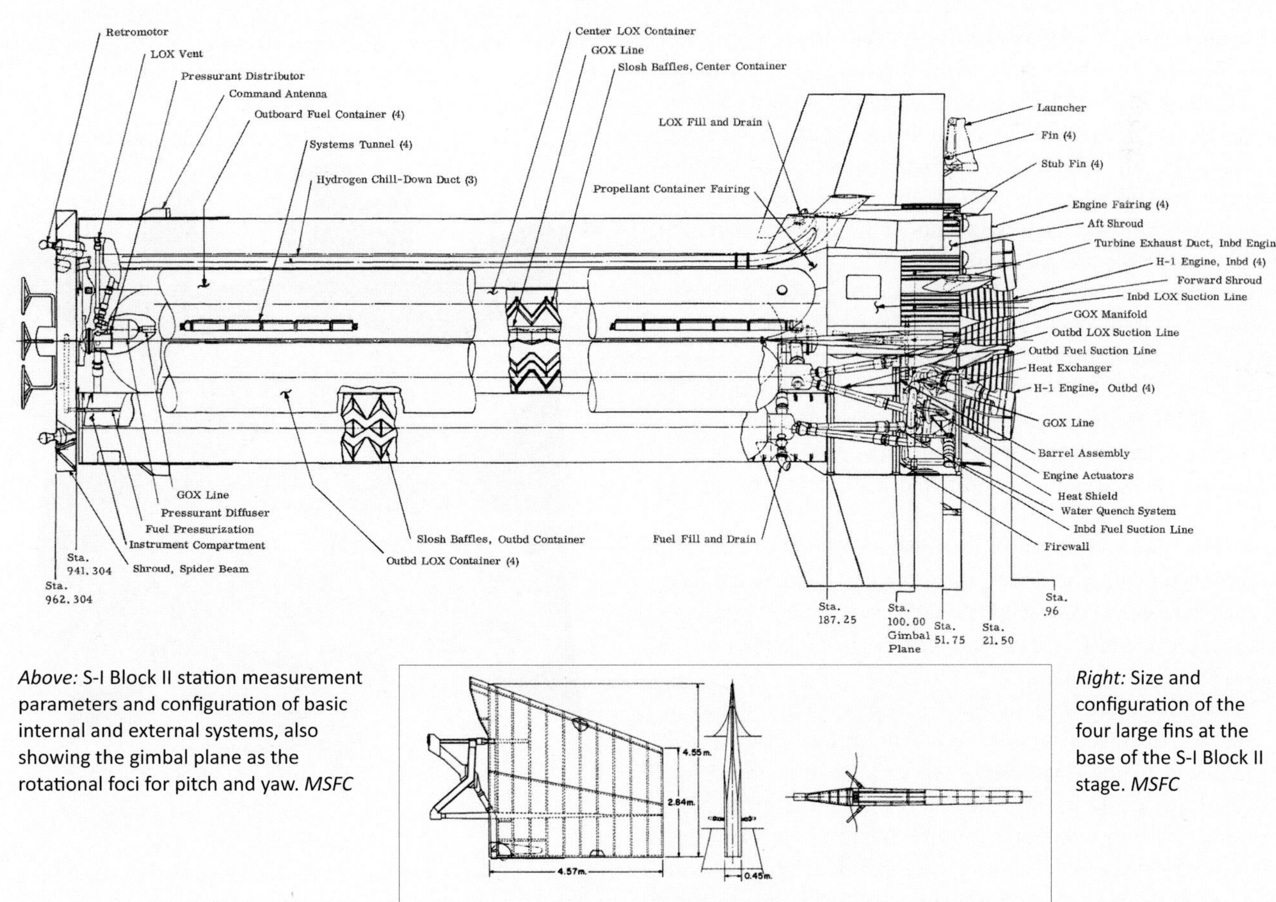

Above: S-I Block II station measurement parameters and configuration of basic internal and external systems, also showing the gimbal plane as the rotational foci for pitch and yaw. *MSFC*

Right: Size and configuration of the four large fins at the base of the S-I Block II stage. *MSFC*

forward rings which supported the barrel skin.

The forward ring of the barrel assembly was attached to the centre LOX tank and part of the thrust load from the four inboard engines was transmitted there with the remainder transmitted to the four fin-support outriggers. The fin-support and thrust-support outriggers were attached to the barrel assembly, from which the four fin-support outriggers received thrust loads. The thrust-support outriggers supported the outboard engines. Two mounting points on each of the outriggers supported the outboard propellant containers which were on a 475cm (187in) diameter ring with each outrigger carrying a support point for a fuel and a LOX container. Thrust loads were transmitted from the outriggers to the outboard LOX containers, the fuel tanks not taking any of the thrust load. All support points were capable of carrying lateral loads.

Each outrigger consisted of two plates stiffened with horizontal and vertical members with thrust loads from the outboard engines transmitted to the plates through thrust beams, located between the plates and which served to back up the outboard engine mounting pads. Actuators for the outboard engines were attached to a beam assembly mounted on the thrust-support outriggers, with upper and lower ring segments, each with a radius of 343cm (135in), attaching the outboard ends of the outriggers. Eight forward shroud panels were attached to the ring segment shroud support plates. The panels protected the compartment between the propellant tanks and the engines from aerodynamic pressure and thermal loads, each panel stiffened with internal longitudinal and circumferential members with a door for access to the compartment.

Located at the aft end of the thrust structure were firewall panels attached to the aft ends of the outriggers and the lower ring segments. These panels formed a fire barrier between the forward (propellant tank) compartment and the aft engine compartment. The aft compartment was protected from aerodynamic pressure and thermal loads by the aft shroud which was attached to the lower ring segments. This shroud, 685.8cm (270in) in diameter and 152.4cm (60in) in length formed a continuous corrugation supported by internal rings. The corrugated skin exposed the greatest amount of surface area to the engine compartment for maximum heat dissipation.

The power end of the aft compartment was closed out by the heat shield, designed to provide protection from the engine heat. Constructed of stainless steel stiffened panels, it was covered on the aft face with an additional insulation. The panels were supported by a complex arrangement of cross beams which were attached to the aft end of the aft shroud with cut-outs provided in the shield for gimballing the outboard engines. The cut-outs were sealed with flexible curtains, fabricated from fibreglass cloth and Refrasil – a silica cloth – attached to the engines and the heat shield.

The flame shield was supported from the heat shield by the conical frustum access chute which, at the forward end, was attached to the heat shield star assembly at the centre portion of the shield itself. This flame shield was located between the four inboard engines at the thrust chamber outlets and was constructed of stainless steel and attached to the inboard centre engine thrust chambers with steel bands insulated with fibreglass cloth. The four engine skirts attached to the

Saturn I – thrust structure details

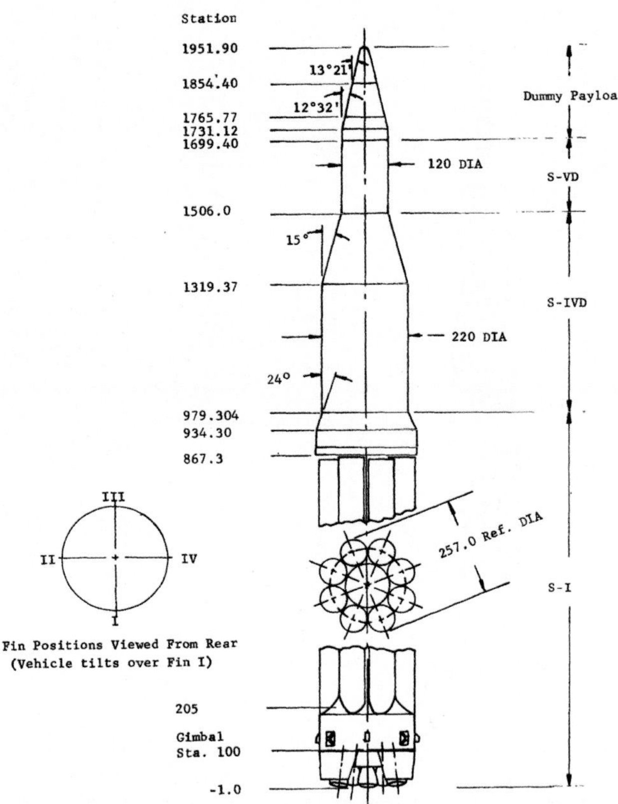

Above: Principal characteristics of the four Block I suborbital and six Block II orbital versions of the Saturn I rocket, the latter incorporating several changes to the original design with upgrades to the Instrument Unit for the more complex missions of the Block II variant. *MSFC*

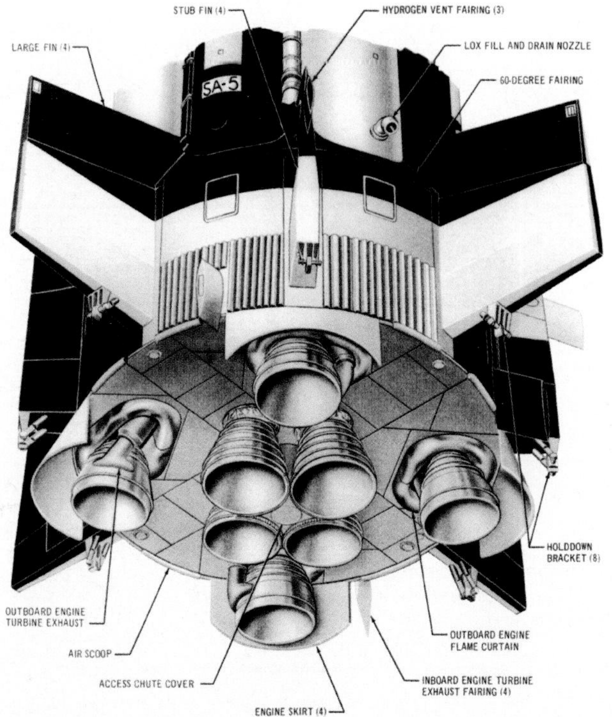

Above: The thrust structure of the Block II stage with modifications, upgrades and some subtle design changes. *MSFC*

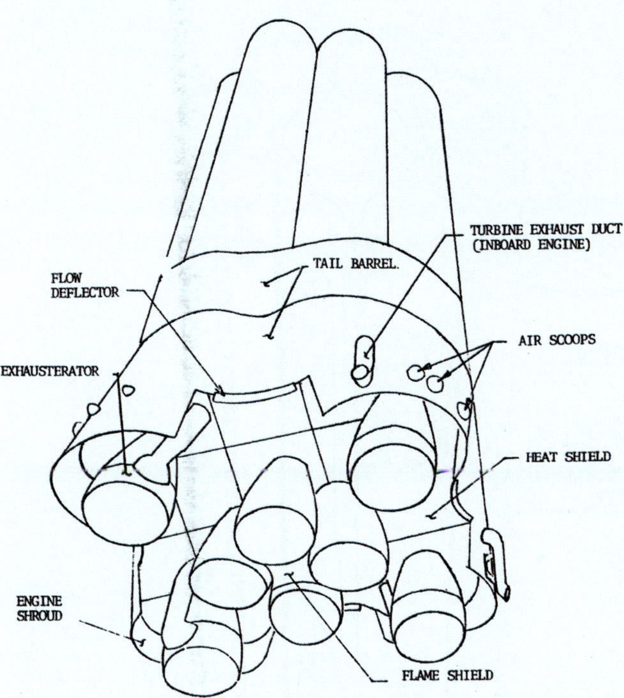

Above: The general configuration of the Saturn I placed emphasis on legacy hardware and use of improved lineal engines dating back to development motors for the Navaho cruise missile. The clustering concept opened opportunities for America's first heavy-lift launcher which would emerge in 1964 with the Block II. *MSFC*

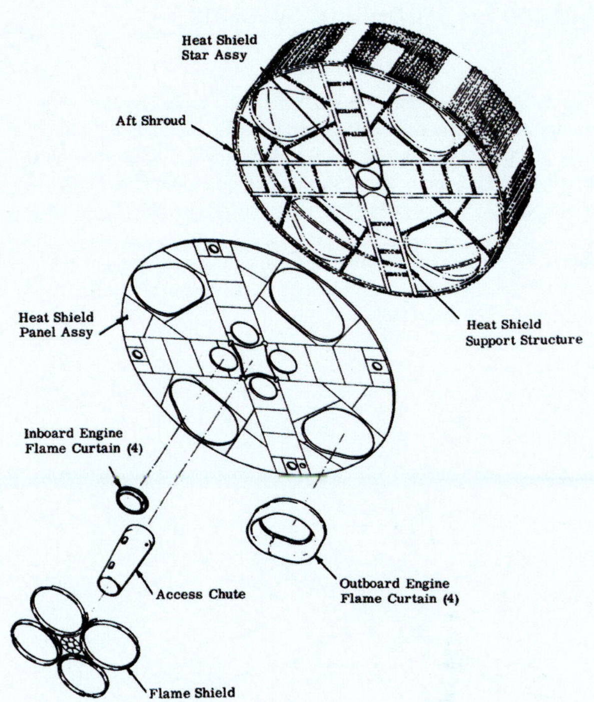

Above: The arrangement of the heat shield and the flame curtain attached to the thrust structure. MSFC

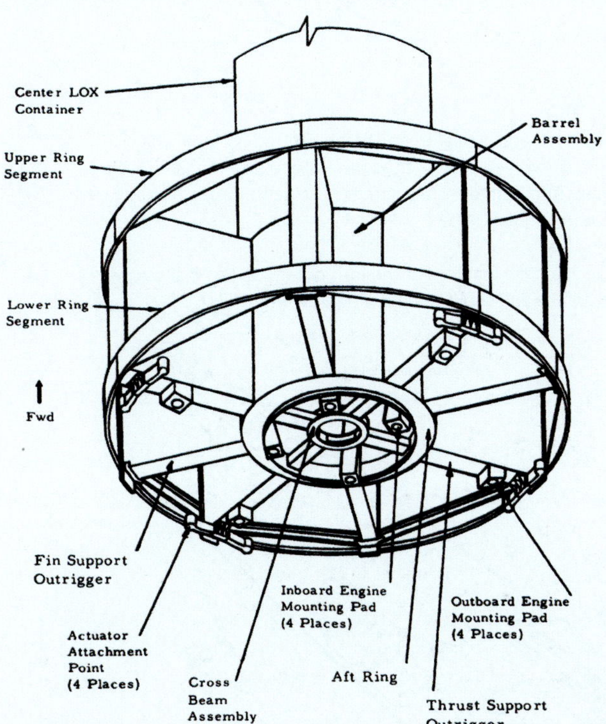

Above: The essential structural elements of the thrust structure displaying the loads path from the four inboard engines transmitting through the aft ring, with the four outboard engines carrying loads through to the lower ring segment directly to the barrel assembly upon which was mounted the central LOX tank. *MSFC*

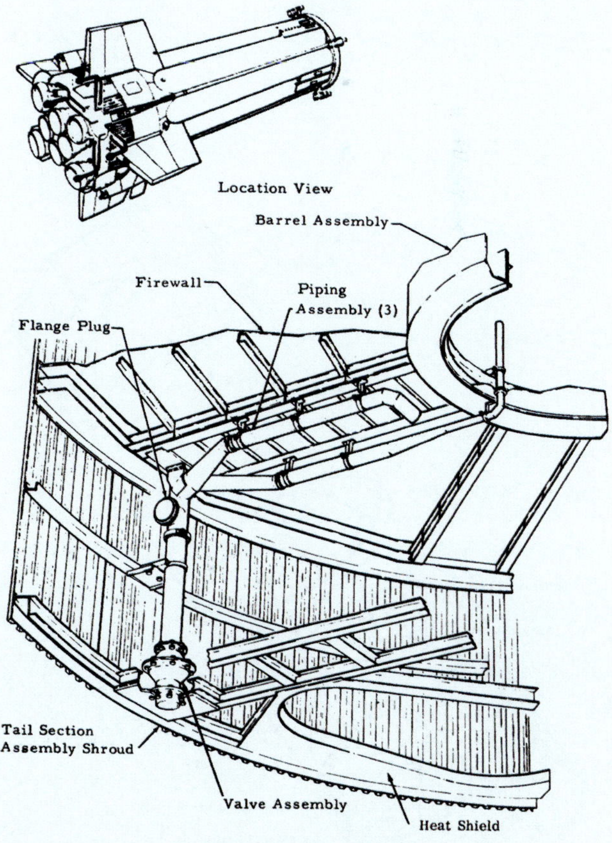

Above: Relevant to the S-IB in this particular illustration, the water quench system was also attached to the S-I stage. MSFC

Saturn I – propellant systems

Right: The configuration of the LOX and RP-1 tanks together with nominal arrangement of baffles and external conduit – left out on a lot of completed models! See the individual flight reports where changes and upgrades to the slosh baffles are identified. *MSFC*

Below: Geometric and numbering arrangement of the eight H-1 engines as viewed looking aft with the S-1 stage in the horizontal position for transport. *MSFC*

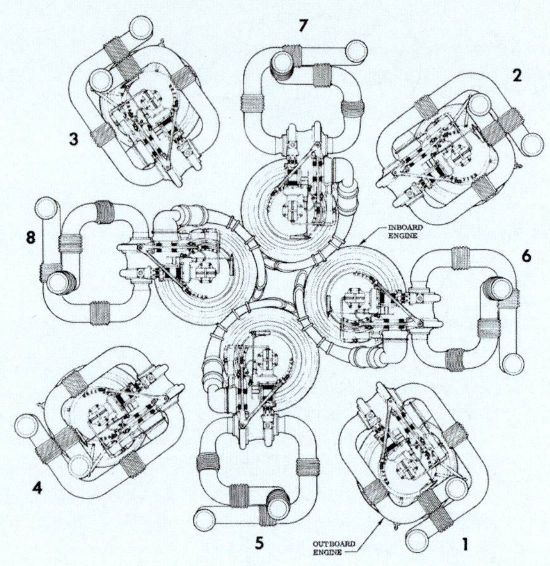

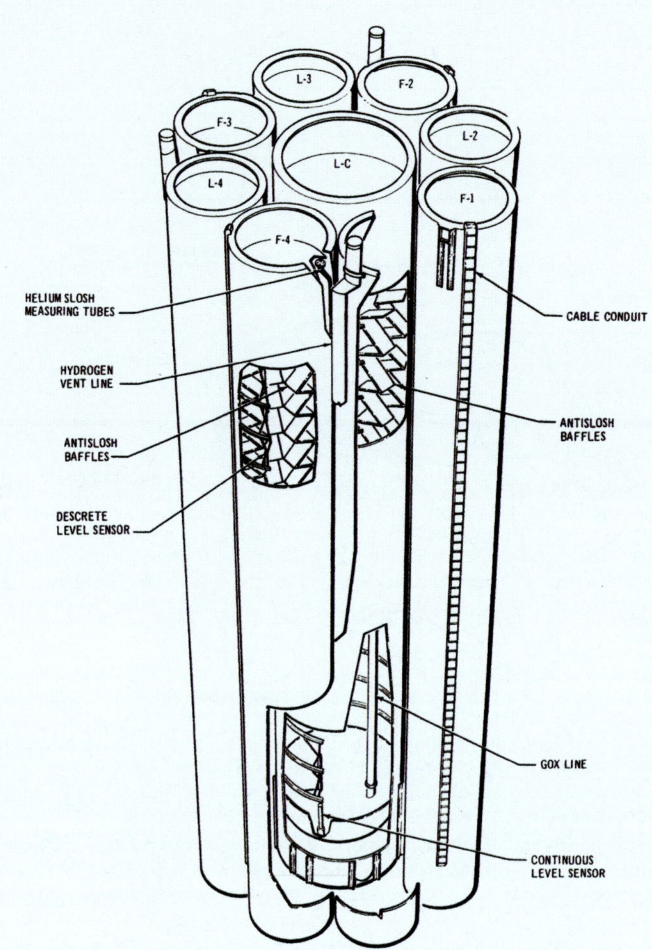

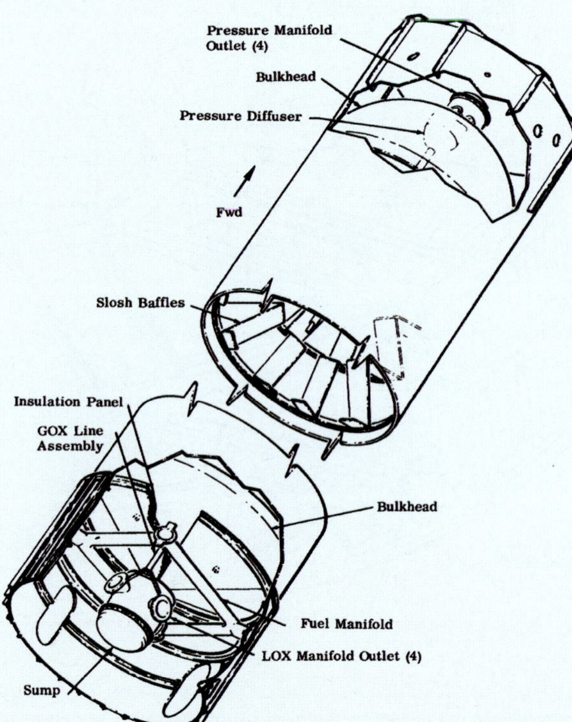

Left: The general arrangement of the centre LOX tank which had the same diameter as the Jupiter IRBM and Juno II satellite launcher and employed the same construction using Alcoa aluminium barrels. *MSFC*

Below: Principle features of the four outer LOX tanks showing slosh baffles and supplementary equipment. *MSFC*

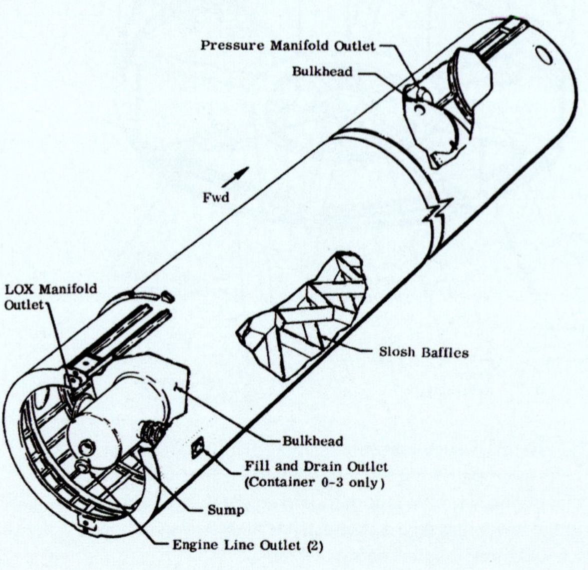

heat shield protected the engines from aerodynamic forces that could produce excessive loads on the control actuators. The skirts were conical segments, 81.3cm (32in) long, with the inside surface below the heat shield and protected from engine heat by ablative insulation.

The Block II (SA-5 to SA-10 vehicles) S-I stage carried four large and four stub fins to aid in aerodynamic stability with the fins also doubling as hold-down and launch pad support points for the entire stack, transmitted to the thrust structure outriggers. The support points were located on the aft face of the fins on an 873.7cm (344in) diameter circle. The stub fins were located at the outboard engine positions, with fins II, III and IV covering liquid hydrogen vents. The large fins were spaced equally between them, both types having trapezoidal planforms. The large fins had an area of 11.89m² (128ft²), the stub fins having an area of 4.83m² (52ft²), with steel leading edges swept back 20°. The main fin structure was composed of aluminium alloy with an ablative insulation applied to the exterior surface. Initially, four fins were selected of 11.32m² (121.9ft²) but these were abandoned.

Propellant Tanks

The S-I stage consisted of a cluster of nine cylindrical containers, eight 177.8cm (70in) diameter tanks circumferentially around a 266.7cm (105in) central tank joined structurally fore and aft. The two types of tank were fabricated from the same tooling utilised for the Redstone and Jupiter missiles respectively and were not left over from production of those missiles, as some texts have implied.

The aluminium cylindrical sections were contour milled to save weight with the sheets rolled and joined by automatic fusion welding. Each tank was completed by spot-welding the internal frames, perforated to form anti-slosh baffles with hemispherical end dome bulkheads added. The end domes were formed from a single, chemically milled disc in a combination shear/spin/machining operation and joined by internal fillet welds. The tanks were kept dimensionally consistent by routing the skin sheets to the precise length needed and by parallel trimming of all the tank sections.

Approximately 36 per cent of the LOX for the S-I stage was contained in the centre tank, a cylinder with a diameter of 266.7cm (105in) and a length of 1,722cm/17.2m (678in/56.5ft) and tori-spherical end domes producing a total length of 1,902.5cm/19m (749in/62.4ft), designed and fabricated to carry flight pressurisation and propellant loads induced by acceleration. Fabricated from 5456 aluminium alloy, the centre tank also transmitted part of the thrust load from the thrust structure to the second stage adaptor. Assembled from 5086 aluminium alloy, the tori-spherical bulkheads were joined to the cylinder by circumferential welds, the aft bulkhead containing a sump with four outlets for connection to the LOX manifold.

The forward bulkhead had four outlets connected to the pressure manifold and three connections to vent lines. A pressure diffuser was attached to the forward bulkhead. Longitudinal stringers were attached to the cylindrical skin in the area above and below the tank forward and aft container skirts. These distributed loads received at the tank support points with cut-outs for pressurisation and vent lines located in the skin forward of the container. Circular rings welded to the interior of the cylindrical section supported slosh baffles arranged in eight vertical rows spaced equally around the periphery of the tank interior.

Approximately 16 per cent of the LOX carried in the S-I was in each of the four outboard oxidiser tanks, each with hemispherical bulkheads, a diameter of 177.8cm/1.77m (70in /5.83ft) and a cylinder length of 1,722cm/17.2m (678in/56.5ft). With hemispherical bulkheads, total tank length was 1,894.8cm/18.9m (746in/62.16ft). Designed to carry flight pressurisation and propellant loads from acceleration, each LOX tank contained transmitted thrust loads from the tail section to the second stage adaptor. Each container was supported at the aft end by the thrust structure outriggers and at the forward end by the spider beam in the adaptor. Two diametrically

Below: The upper parts of the S-1 stage with the relative locations of the propellant tanks, spider beam, retrorockets, Instrument Unit and associated propellant fittings, pressure manifold and fairings. MSFC

Below right: Distribution of helium pressurisation and nitrogen control pressure system spheres located on top of the spider beam assembly. *MSFC*

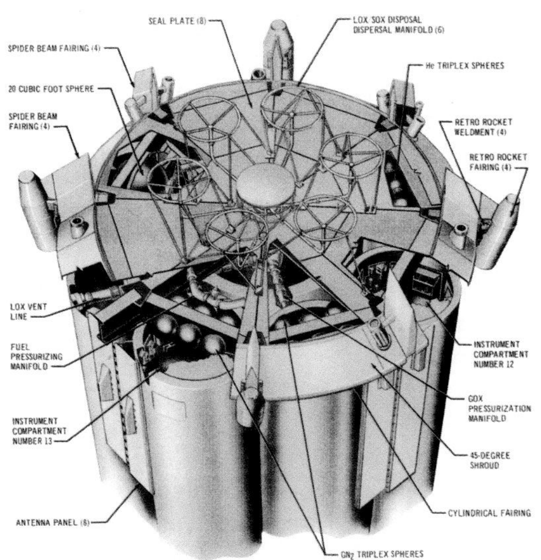

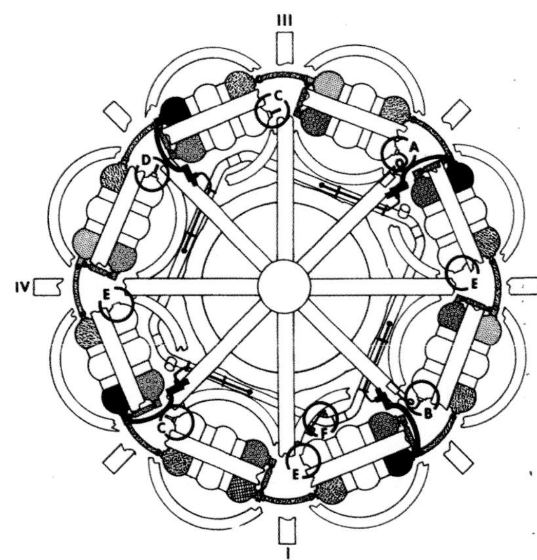

opposed supports points on the outriggers were located for each tank to transfer axial and lateral loads. The spider beam also carried two diametrically opposed support points, each consisting of an adjustable mounting stud for transmitting axial and lateral loads.

Each LOX tank was fabricated from 5406 aluminium alloy, with recessed hemispherical bulkheads from 5086 alloy joined to the main cylinder by circumferential welds. The aft bulkhead had a sump with three outlets, two for the engine lines and another for the LOX manifold interconnect line. Tank O-3 (tank 3 of the LOX group) had an additional outlet used for fill and drain and the forward bulkhead had an outlet for a pressure manifold connection. In the area above and below the forward and aft tank skirts there were longitudinal stringers attached to the skin, these distributing the concentrated loads received at the two tank support points. The skin above and below the tank had cut-outs for lines connecting the various outlets. Circular rings welded to the interior of the cylindrical section supported the slosh baffles arranged in six vertical rows spaced equally around the cylinder.

Approximately 25 per cent of the fuel for the S-I stage was contained in each of the four fuel containers, cylinders with hemispherical aft bulkheads and tori-spherical forward bulkheads. The containers had a diameter of 177.8cm (70in) and a length of 1,656cm/16.5m (652in/54.3ft). The containers were designed to carry flight pressurisation and propellant loads from acceleration. The containers were supported at the aft end by the thrust structure outriggers and at the forward end by the spider beam in the second stage adapter. On the outriggers there were two diametrically opposed support points for each container. Each support point transferred axial and lateral loads. On the adapter spider beam there were also two diametrically opposed support points for each container, each consisting of a sliding pin joint. The pin joint resisted lateral loading but allowed for differential expansion between the fuel and LOX containers in the longitudinal direction.

Fabricated from 5486 aluminium alloy, the overall tank was 1,887cm/18.8m (743in/61.9ft) long. Recessed into the forward and aft ends of the cylinder were bulkheads fabricated from 5086 aluminium alloy. The bulkheads were joined to the cylinder by circumferential welds. The aft bulkhead had three outlets, two for engine lines and another for the fuel manifold. Container F-1 (tank 1 of the four fuel tanks) had an additional outlet for fill and drain while the forward bulkhead had a pressure manifold outlet. In the area above and below the container (forward and aft container skirts), longitudinal stringers were attached to the cylindrical skin. The stringers distributed the loads received at the two container support points. The skin above and below the tank had cut-outs for lines connecting various outlets. Above tanks F-1 and F-2 were compartments for mounting electronic equipment. Baffles identical to those for the LOX tanks were installed inside the fuel tanks.

A systems tunnel was attached externally to each of the four fuel containers. Each tunnel joined the tail section and second stage adapter. Three of the tunnels shielded electrical cables; the other was for routing tubing. The tunnels were constructed in sections to allow easy removal for maintenance and repair. A conical-shaped fillet extended forward from the aft ends of the propellant tanks, fairing in the area between the tanks and the 685.8cm (270in) diameter forward shroud panels, with the exterior of the fairing coated with ablative insulation. Three LH_2 chilldown vent lines were located on the exterior of the S-1, connecting to equivalent vent lines on the aft interstage. The lines ran aft and were ducted through three of the four stub fins.

Loads were transmitted to the second stage through the second stage adapter composed of a spider beam, seal plate panels, a 45deg shroud assembly and a cylindrical fairing. The spider beam supported the propellant tanks at the forward end. Fabricated from 7075 aluminium alloy, the spider beam was composed of an octagonal ring and eight radial beams which extended inward from the points of the octagon and were joined at the centre with plate gussets. The octagonal ring and radial beams were 50.8cm (20in) deep I-sections. To absorb vertical loads, the radial beams were stiffened at the propellant container support points.

The spider beam was bolted to the S-IV stage at eight points at vehicle station 962. Mounted on the forward side of the spider beam were seal plate panels of honeycomb sandwich construction. Sections of the seal plate could be removed for access to the forward propellant container area. The 45deg shroud assembly was attached to the periphery of the seal plates and to the ends of the radial beams. Attached to the lower end of the shroud was a cylindrical fairing. The shroud and fairing protected the forward container area from aerodynamic loads. Helium spheres were mounted on the spider beam. To insure separation, four Aerojet MB-I retrorockets were also mounted on the spider beam, each producing a thrust of 164.5kN (37,000lb) for 2.15 seconds.

Attitude Control

The Saturn I attitude control and stabilisation function maintained a stable vehicle motion (through the engine gimballing system) and adjusted this motion in accordance with guidance commands. During the ascent phase this function directed the vehicle orientation about its axes, maintained the angular rate of vehicle movement within allowable limits and dampened any first bending mode oscillation of the vehicle structure. The attitude control and stabilisation function performance was limited by various factors. During S-I stage flight, the high aerodynamic pressures encountered by the launch vehicle resulted in structural constraints and related control challenges.

The basic launch vehicle was aerodynamically unstable, therefore, a minimum angle-of-attack flight prevented excessive structural loading from aerodynamic forces and large gimbal deflections of the control engines. A constraint existed because of the natural bending of the vehicle structure. During S-I stage powered flight, any oscillations occurring in the first bending mode of the structure had to be actively damped by thrust vectoring. The Saturn vehicle was required to maintain the launch orientation for several seconds after lift-off, permitting it to rise above the launch facilities to gain manoeuvring clearance. The size and complexity of the launch vehicle and launch facilities constrained the launch vehicle to a specific orientation.

Saturn I – propellant systems (contd.)

Right: The fuel tanks were essentially the same as the LOX tanks but with unique arrangement for the RP-1 propellant. *MSFC*

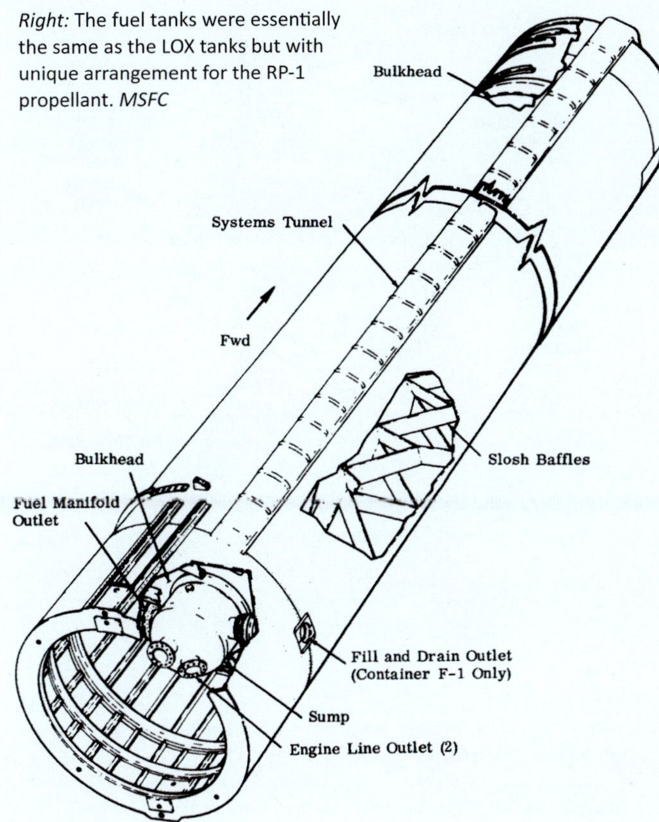

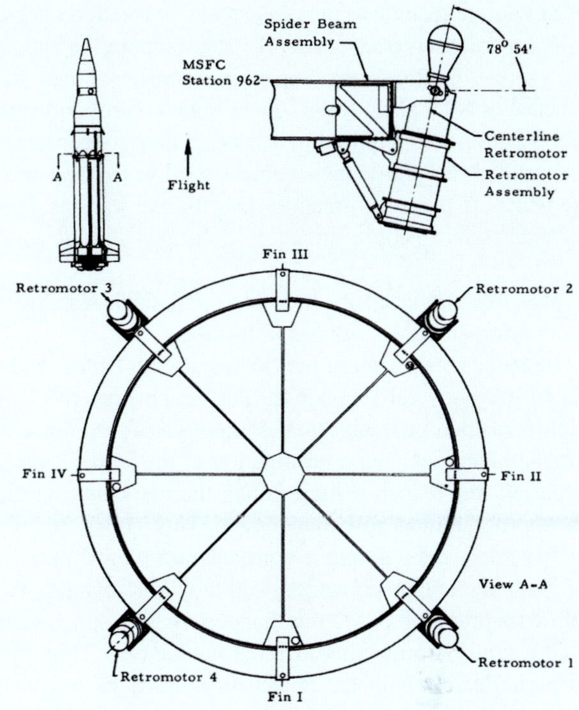

Above: The location and angular pointing alignment of the four retro-rockets attached to the forward part of the S-1 stage, used to decelerate the stage and prevent contact with the upper elements of the launch vehicle after separation. *MSFC*

Below: The attachment strut and bridge wire assembly for firing the retrorockets. *MSFC*

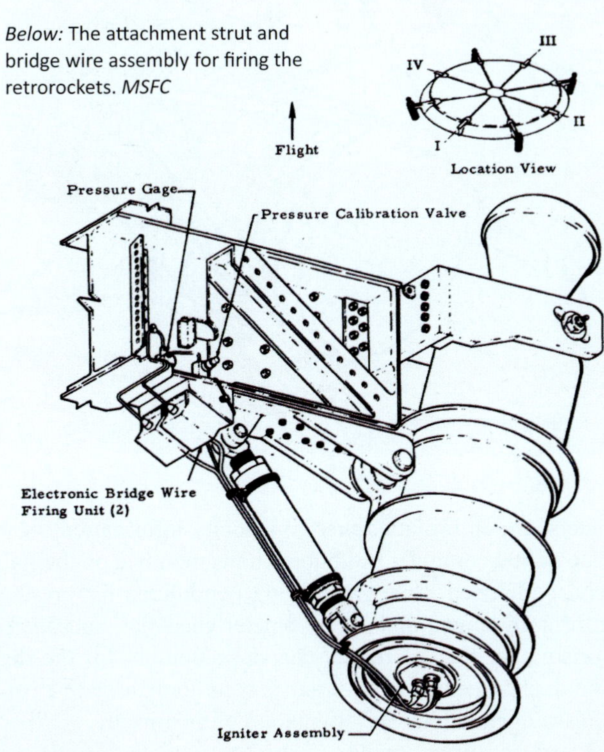

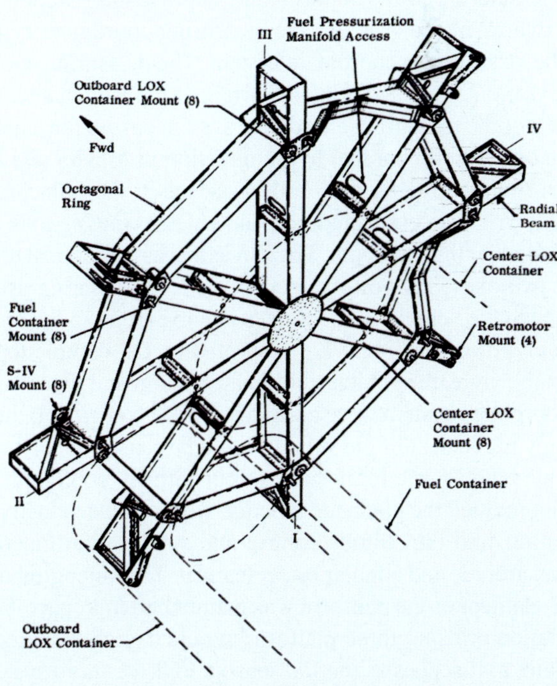

Above: The spider beam assembly formed the securing structure for the nine propellant tanks and the forward mounting positions for the upper structure. *MSFC*

Immediately prior to vehicle staging, the attitude control and stabilisation function had to restrain the launch vehicle to a constant pointing angle to prevent excessive rotation rates during the separation process. After S-I stage separation and S-IV stage engine ignition, any separation transients had to be damped. For S-IV stage flight, the attitude control and stabilisation function was required to accept guidance-steering commands and direct the launch vehicle motion to meet the requirements of these commands. Due to the various launch vehicle and control constraints, a programmed attitude control, without active guidance, was used for S-I stage powered flight. This was accomplished in three phases: launch stabilisation, manoeuvring and pre-staging stabilisation.

The launch stabilisation period began with lift-off and terminated after several seconds during which time the launch vehicle ascended vertically to attain a physical clearance with the launch facilities. Upon termination of the launch stabilisation period, the launch vehicle began the manoeuvring phase with a programmed roll manoeuvre. This consisted of the launch vehicle maintaining a constant rate of roll until such time as its pitch plane coincided with the flight azimuth. Coincident with initiation of the roll manoeuvre, the launch vehicle started a gravity-turn, time-tilt pitch manoeuvre. This rotated the longitudinal axis of the launch vehicle in the pitch plane toward the flight azimuth. A few seconds prior to vehicle staging, the time-tilt manoeuvre was terminated. The pre-staging stabilisation phase began at that point, with S-I stage outboard engine cut-off. The pre-staging stabilisation maintained the launch vehicle in a fixed attitude orientation. During S-IV stage flight, the attitude control and stabilisation function maintained a stable vehicle motion and oriented this motion as directed by guidance steering commands.

The Saturn I attitude control and stabilisation function utilised two reference systems, the measuring coordinate system and the vehicle axes coordinate system. The measuring coordinate system (X_m, Y_m, Z_m) had its origin at the launch site. The Y_m axis of the measuring coordinate system passed through the centre of the Earth parallel to the direction of gravity and was positive outward from the Earth's surface at the launch site. The X_m axis was orientated perpendicular to the Y_m axis and lay along the flight azimuth. The Zm axis was orthogonal to the other two axes. The roll manoeuvre performed during ascent orientated the vehicle X and Z axes to correspond to the measuring coordinates X_m and Z_m, respectively. On completion of this manoeuvre, the vehicle coordinate system and the measuring coordinate system were considered to be coincident, therefore any movement of the vehicle could be sensed against that.

The ST-124 platform had a four-gimbal configuration, which provided the guidance function with vehicle velocity information, and the attitude control and stabilisation function, with an attitude and angular rate reference. The inner gimbal or stable element of the platform was maintained in a space-fixed orientation utilising three platform mounted, single-degree of freedom, gyroscopes as inertial sensors to drive servo systems which positioned the platform yaw, pitch and roll gimbals. The power to orientate the gimbals was provided by dc direct drive servo motors attached to the gimbals. The stable element also carried three orthogonally-mounted integrating gyro accelerometers, which provided inertial velocity information for use by the digital computer. Additional units mounted on the stable element included three gas-bearing pendulums for pre-flight platform alignment and accelerometer checkout, and one poro-prism synchro and digital encoder assembly for pre-flight azimuth alignment. A poro-prism is commonly used in an optical instrument to alter the orientation of the image.

Platform gimbal angles relative to vehicle attitude were measured by four pancake type resolvers, electrically connected in a chain with three command resolvers located in the

Saturn I – attitude control system

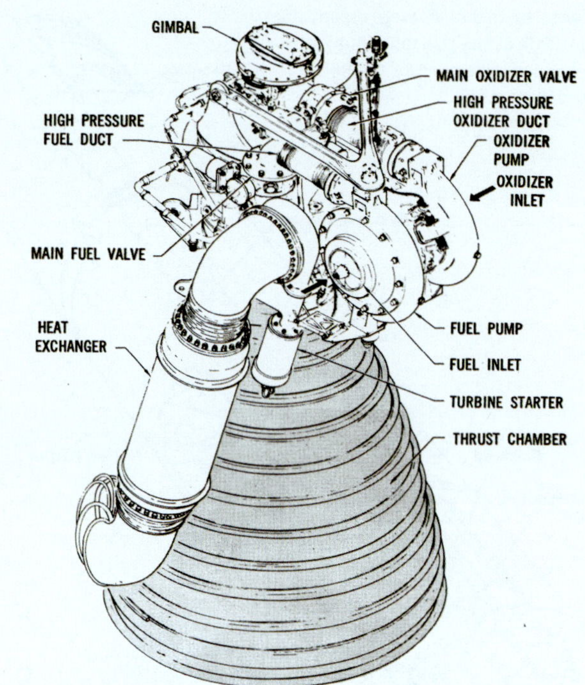

Above: Principle features of the H-1 engine, showing the gimbal mounting. *Rocketdyne*

Below: Two gimbal actuator assemblies were provided for each outboard engine with dedicated main hydraulic pumps for all engines.

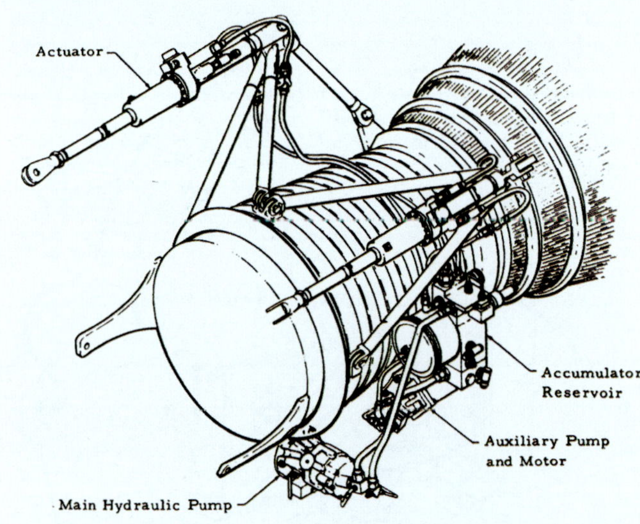

Saturn I – attitude control system (contd.)

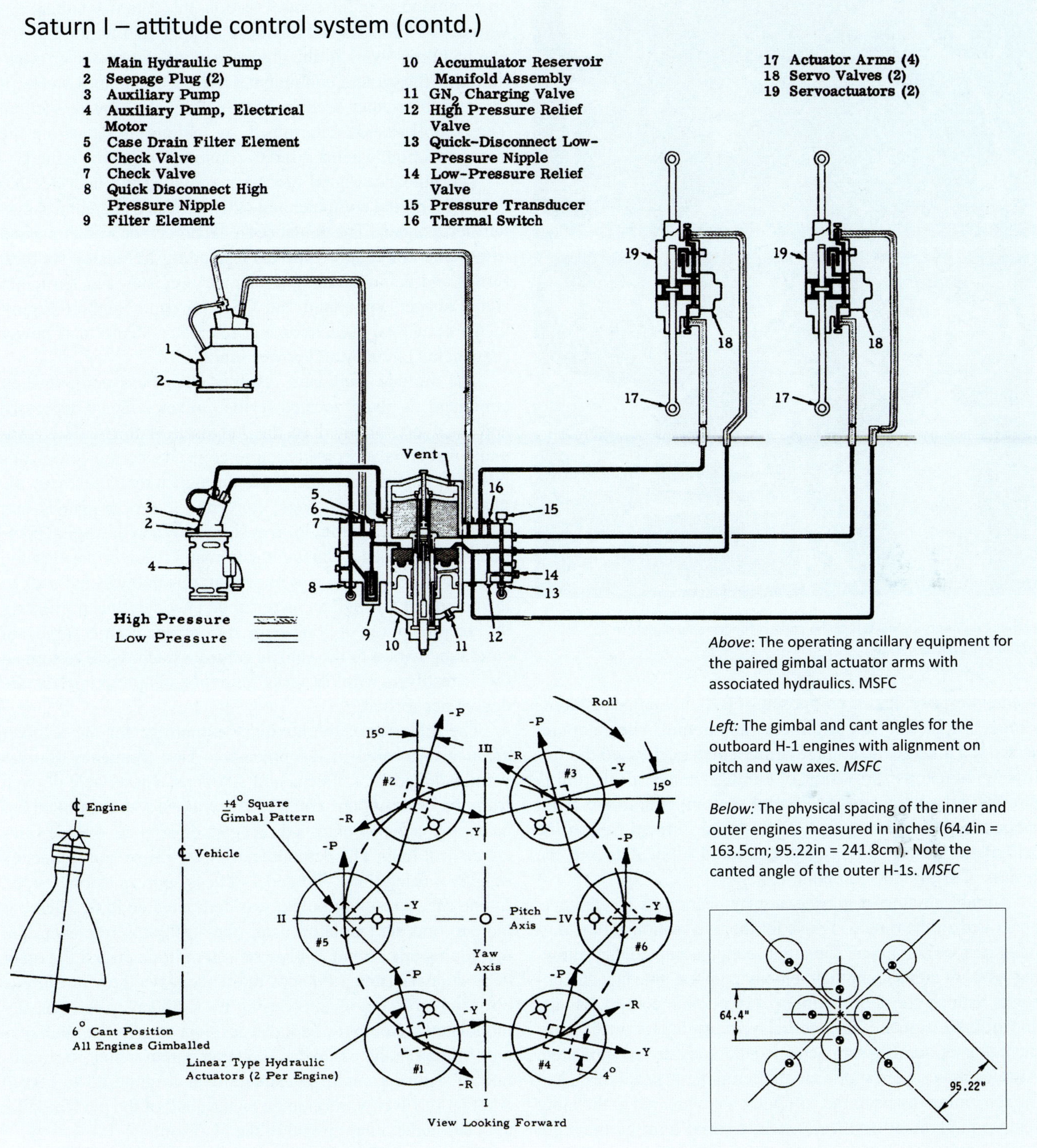

Above: The operating ancillary equipment for the paired gimbal actuator arms with associated hydraulics. MSFC

Left: The gimbal and cant angles for the outboard H-1 engines with alignment on pitch and yaw axes. *MSFC*

Below: The physical spacing of the inner and outer engines measured in inches (64.4in = 163.5cm; 95.22in = 241.8cm). Note the canted angle of the outer H-1s. *MSFC*

guidance signal processor. This resolver chain converted space reference guidance commands to vehicle referenced steering commands, which were applied to the flight control computer. During the alignment, the guidance and control system operated with an RCA-110 computer on the ground to accomplish platform alignment in the azimuth plane. The platform checkout module was contained in an electronics box which also included the circuitry essential for operation of the stabilised platform.

The ST-124 stabilised platform system was sealed so that operation in a near vacuum was possible. Auxiliary heaters in the platform system external cover provided preheating and inflight temperature control. The gas bearings for platform gyroscopes and accelerometers were supplied with nitrogen by the gas bearing supply, and this supply conditioned the gas by controlling temperature, pressure and impurities.

During S-I stage flight the computer provided the source of programmed attitude changes for the attitude control and stabilisation function and generated guidance steering signals during S-IV stage flight. The computer was a serial, binary,

Above: Engineers at MSFC test the correct functioning of Saturn's gimballing system. *MSFC*

special purpose device composed of five functional sections: storage; input; control; arithmetic; and output. The computer contained a magnetic storage drum which performed the following functions: provision of timing signals to all timing circuits; instructions to the control section; data to the arithmetic, control and output sections which formed part of the arithmetic section shift registers; stored input data and the results of arithmetic computations.

Timing circuits, using the storage section as a reference, generated all the timing signals for the five sections of the digital guidance computer. The input section accepted the following types of signals: inertial velocity components; launch constants; launch constant modifiers; programme control signals; and discrete signals. The inertial velocity inputs were incremental, continuously sampled and automatically accumulated. Attitude inputs were applied to the computer in serial form, and the remaining inputs to the computer were applied to the input section in parallel form, converted to serial form by the input section, and stored in the storage section by a command from the control section.

The arithmetic section received data from either the input section or the storage section, and after performing any mathematical operation defined by the control section, stored the result in the storage section. The arithmetic section performed five mathematical operations: addition of one number to another; subtraction of one number from another; comparison of one number with another; multiplication of one number by another; and conversion of grey code inputs (vehicle rates) to binary numbers.

The output section received data from the storage section on command from the control section; this data was either converted to a proportional analogue voltage and applied to the command resolvers in the guidance signal processor or transmitted to the ground equipment. Discrete commands were applied to the output section directly from the control section. The control section determined, by monitoring data from the storage section, when a discrete command was to be issued.

The guidance signal processor provided the interface between the digital computer and other guidance and control system components. The guidance signal processor was composed of attitude command resolvers (including frequency sources, servos and demodulators), a telemetry register, accelerometer signal shaper, command and GSE (Ground Support Equipment) networks, accelerometer telemetry shaper and power sequencing circuitry and power supply.

The attitude command resolver chain was comprised of command resolvers located in the guidance signal processor and resolvers mounted on the stabilised platform. The command resolvers accepted space-referenced steering commands from the digital computer and, through interaction with the platform mounted resolvers, converted these commands into vehicle referenced attitude error signals. When the digital computer commanded a change in command resolver positioning, an analogue of the resolver rotor shaft position was fed back to the computer through an incremental encoder, thus preventing the accumulation of long-term rate errors. Large surges of command values to the vehicle control system were restrained by the resolvers within a speed limitation of approximately one degree per second.

The majority of the auxiliary equipment for the resolver chain was located in the processor. Two frequency sources were included, at 1,500cps and 1,800cps. These were derived from the basic 400cps voltage. The voltage was controlled because any error becomes a direct gain error in the overall vehicle control loop. The demodulators were phase and frequency sensitive, using the 1,500 and 1,800cps sources as references. In one case a resolver output was demodulated in two demodulators: one demodulator, using the 1,500 cps reference, demodulated this output to give the roll attitude errors; the other demodulator, using 1,800 cps reference, gave the yaw attitude error. A third demodulator, using the 1,500cps reference, demodulated the output of another resolver to give the pitch output. All demodulators had a 3volt/deg output which was accurate to within a small percentage over a range of +/-15deg. Another resolver was mounted on the shaft of the pitch module to position the outer gimbal of the platform.

Telemetry of guidance functions was performed with the telemetry register. In the Saturn I launch vehicle, computer words were buffered and fed at 100 words/second to the processor. A command would be applied to the telemeter each time one of the desired words passed through the register, and when the telemeter command reached the processor, the telemeter gate opened and the next word from the accumulator entered the shift register. The data was then available in parallel form to the PCM telemeter system during flight, and the GSE during prelaunch. The accelerometer signal shapers converted sine-wave acceleration information received from optisyns into square-wave information for sampling by the digital guidance computer. Optisyns generate electrical square-wave signals indicating the rotation of the accelerometer shafts.

The platform-mounted signal outputs from the accelerometer encoders were sine-wave and cosine-wave signals which were applied to the accelerometer signal shapers. The signal shapers conditioned the signals into square waves of voltage displaced 90deg. The square waves were applied to the digital computer which processed the information contained in these signals to obtain steering signals. The switching network selected the GSE or the command system as a source for loading the computer, and this provided the capability of loading either while on the ground or in a coast condition. In addition, it allowed the ground or command system to control various modes of computer operation. The accelerometer telemetry shapers received signals from the accelerometer encoder shaper and conditioned them for telemetry. The square wave from the accelerometer shapers were given specific dc levels, added together and sent to the telemetry system as discrete levels between 0 and +5 volts.

The guidance signal processor power supplied all power required in the processor and power for the encoders on the platform. In addition, power to the computer passed through the processor. Since the drum of the computer utilised 400-cps, two-phase power, it was necessary to convert the three-phase power available from the vehicle inverter to two-phase. This was done by a Scott connected transformer or similar device in the processor. Approximately 70W of 81.5 (+/-2.5) volt, 400-cps, two-phase power was required by the drum. In addition,

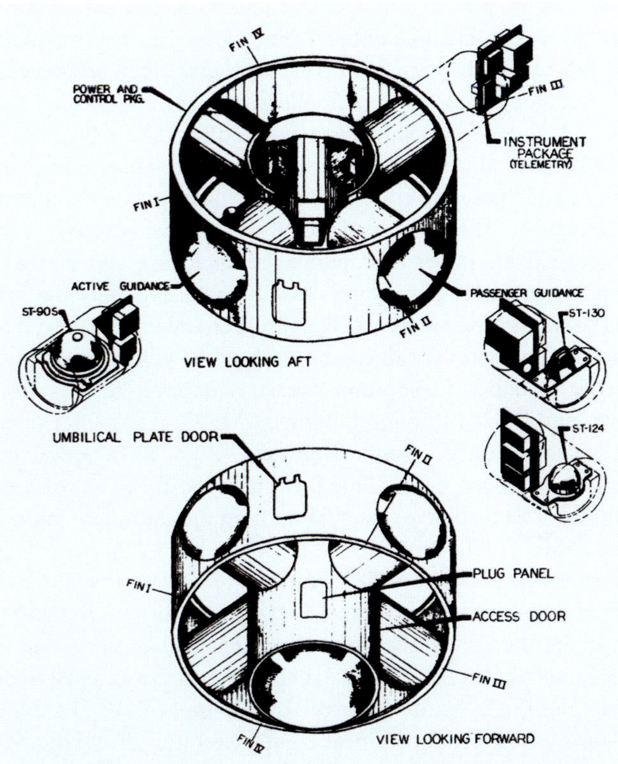

The Instrument Unit for the Saturn I consisted of a cylindrical section above the first stage and comprised four instrument package points slotted in at 90deg intervals around the circumference, housing ST-90S, ST-10, ST-130 or ST-124 equipment. *MSFC*

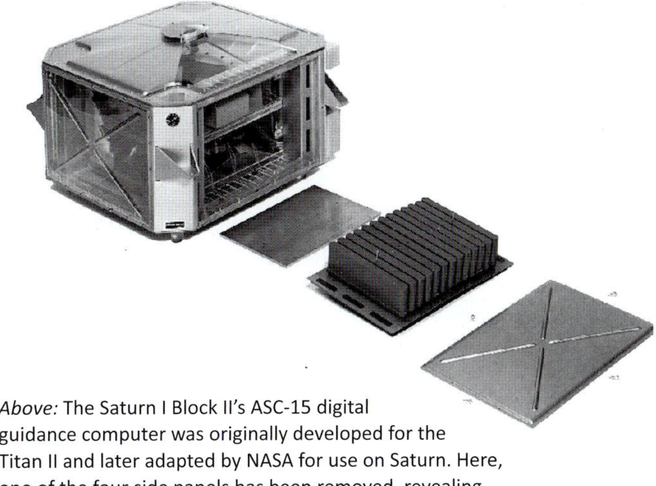

Above: The Saturn I Block II's ASC-15 digital guidance computer was originally developed for the Titan II and later adapted by NASA for use on Saturn. Here, one of the four side panels has been removed, revealing 13 of the 52 logic sticks. *Edgar Durbin*

approximately 240W of 28 (+/-2.0) volt dc passed through the processor to the dc-to-dc converter in the computer where the various levels necessary for computer operation were developed and regulated. The guidance signal processor required approximately 65W of 115-volt, 400cps, single-phase power and 215W of 28V dc power.

The analogue flight control computer accepted signals from the stabilised platform, control accelerometers and actuator position feedback potentiometers. After performing signal filtering, shaping and mixing, the computer provided steering and control signals to the engine gimbal actuators. The major modules of the flight control computer were the servo amplifiers, filtering and shaping networks and the gain programmers. The servo amplifier was a magnetic amplifier plug-in module used for signal mixing, scaling and polarity selection. The signal filtering and shaping networks provided signal conditioning based on the dynamic qualities of the vehicle. The gain programmer was a motor-driven cam which positioned a potentiometer to adjust the gain in each channel.

Two control accelerometers were used in the launch vehicle to measure lateral acceleration (perpendicular to the longitudinal axis) in the vehicle pitch and yaw planes. The outputs of the instruments were used by the control system to reduce structural loading and engine gimbal angle. The control accelerometer was a spring mass, fluid-damped accelerometer with an inductive pick-off. The range of the instrument was +10m/sec2. Two linear, double-acting, equal-area, electro-hydraulic servo actuators gimballed the engine in response to commands from the flight control computer. A feedback transducer mounted on each actuator transmitted an electrical signal to the flight control computer which was proportional to the actuator position.

Electrical System

The two stages of the Saturn I launch vehicle and the instrument unit were electrically independent. Each contained a complete electrical system which supplied all of its power requirements. The Saturn I electrical systems were active throughout all mission phases. During the prelaunch phase and the majority of the launch phase, primary power (28-volt dc) for the systems was supplied by generators located at the Automatic Ground Control Station (AGCS). The generator source of primary power was maintained until near the end of the launch phase (approximately T minus 35 seconds) at which time the primary power was switched without interruption from the ground generator source to stage batteries by a network of relays.

At launch, explosive switches, connected in parallel with the relay contacts, were fired to permanently lock the power transfer functions, thus preventing power interruptions that could occur due to relay failure or contact bounce. Throughout the mission, the ac power 115 volt (three-phase 400cps) for each electrical system was supplied by stage electrical distributors and networks.

The operation of each electrical system was similar. The S-I stage electrical system was comprised of two 28-volt batteries, dc power supplies, a dc-to-ac inverter, distributors, a flight sequencer, a slave unit, and several types of J-boxes. Inflight power for the stage was supplied by two 28-volt batteries, the battery cells being constructed of zinc-silver oxide using potassium hydroxide as the electrolyte. Each was rated at 1,650 ampere-minutes and was provided with taps which were used to adjust the output voltage to 28-volt dc (nominal) under load. A 450-volt-ampere solid-state inverter was used to convert 28-volt battery power to 115-volt, 400 cps, three-phase power. The output voltage was used to power various components in the measuring system.

The master measuring voltage supply was a solid-state dc-to-dc converter. It converted 28-volt dc inputs into 5-volt dc outputs (one amp capacity), controlled to within 0.25 per cent. The reference voltage for measurement transducers and signal conditioners was supplied by this unit. The distributors were the switching and distribution centres for all of the electrical circuits in the stage, containing relays, buses, and current limiting components. The stage switching and distribution functions were assembled into groups of identical or similar functions (measuring, power distribution, etc.) and a distributor was furnished for each group.

Prior to primary power transfer, the stage primary power was supplied to the power distributor from ground generators. After the power transfer and during vehicle flight, primary power was supplied to the distributor from the 28-volt batteries. The power distributor contained two separate dc output buses, one for steady loads, and one for varying loads. The steady-load bus supplied power to the measurement system components while the varying-load bus supplied power to relays, valves, and other control equipment. A third bus system, supplied from the inverter, served all the ac power loads.

Vehicle functions that were initiated or controlled by the flight sequencer were handled by the main distributor unit. The main distributor dc power was supplied by the power distributor. The propulsion system distributor received 28-volt dc power from the power distributor and distributed it to the circuits that control the engine functions. Thrust-OK pressure switches, and relays used for the operation of fuel and LOX fill and drain and replenishing valves, were contained in this unit.

The flight sequencer, a relay device, distributed 28-volt dc power to stage relays and control devices. The capacity of the basic unit was a 10-step programme. Each step was expandable in multiples of 10 steps by the addition of slave units. The timing pulses for driving the flight sequencer originated in the guidance computer (part of the guidance and control system). The J-box was a standard connector with outer terminals soldered together to form junction points, or used to connect simple circuit elements into the circuits of the distributors. In this way, the J-box functioned as a small remote distributor and signal conditioner.

H-1 Engineering

Two models of the H-1 rocket engine were used: H-1C for the four fixed, inboard engines; H-1D for the four gimballed, outboard engines. Basically, the physical characteristics of the two models were identical, with the exception of the exhaust system and vehicle attach hardware. The major exhaust system difference was the exhaust gas exit, H-1C engines containing a curved exhaust duct while H-1D engines utilised an aspirator. Each outer engine was attached to the vehicle structure by a gimbal assembly. The inboard engines were fixed in their positions by struts attached to the thrust chamber stabilising lugs. The outboard engines had gimbal actuators attached to thrust chamber outriggers (described earlier) permitting the outboard engines to gimbal for vehicle directional control.

The H-1 rocket engine thrust chamber chamber was gimbal-mounted, regeneratively cooled with fuel and had an exhaust-nozzle expansion ratio of 8:1. The propellants, liquid oxygen and RP-1 fuel, were supplied to the thrust chamber by a turbopump, powered by a gas generator using the same propellants as the thrust chamber. Re-servicing of the propellants and fuel additive and replacing the solid propellant gas generator, initiators, igniters, hypergolic cartridge, and main LOX valve closing control valve was required for restart. The thrust chamber and gimbal assembly included an oxidiser dome, injector and hypergolic container, thrust chamber body and gimbal assembly. The purpose of the thrust chamber was to receive the propellants under turbopump pressure, mix and burn the propellants, and impart a high velocity to the expelled combustion gases to produce thrust. The thrust chamber also served as a mount or support for all engine, and certain vehicle hardware.

LOX, under turbopump pressure entered the oxidiser dome and flowed through the injector LOX ring orifices into the thrust chamber combustion area. Fuel, also under turbopump pressure, entered the fuel inlet manifold and was distributed. The turbine exhaust system on the outboard engines consisted of a turbine exhaust hood, heat exchanger, heat shield, heat exchanger LOX supply line and aspirator to distribute and direct the exit flow of gas generator exhaust gases. On inboard engines, the turbine exhaust system consisted of a turbine exhaust hood, heat exchanger, turbine exhaust duct, heat shield, and heat exchanger LOX supply line. The combustion gases, produced by the gas generator, drove the turbine and were vented through the turbine exhaust hood into the heat exchanger, where the heat of the gases was used to convert liquid oxygen into gaseous oxygen (GOX) for vehicle systems. The fuel-rich exhaust gases then entered the thrust chamber aspirator or the exhaust duct to be directed

1 MAIN LOX VALVE
2 FUEL HIGH-PRESSURE DUCT
3 FUEL PUMP
4 TURBINE ASSEMBLY
5 SOLID PROPELLANT GAS GENERATOR ATTACHMENT POINT
6 GAS GENERATOR COMBUSTOR BODY
7 FUEL BOOTSTRAP LINE
8 GAS GENERATOR CONTROL VALVE
9 LOX BOOTSTRAP LINE
10 HYPERGOL CONTAINER
11 GIMBAL ASSEMBLY
12 MAIN FUEL VALVE

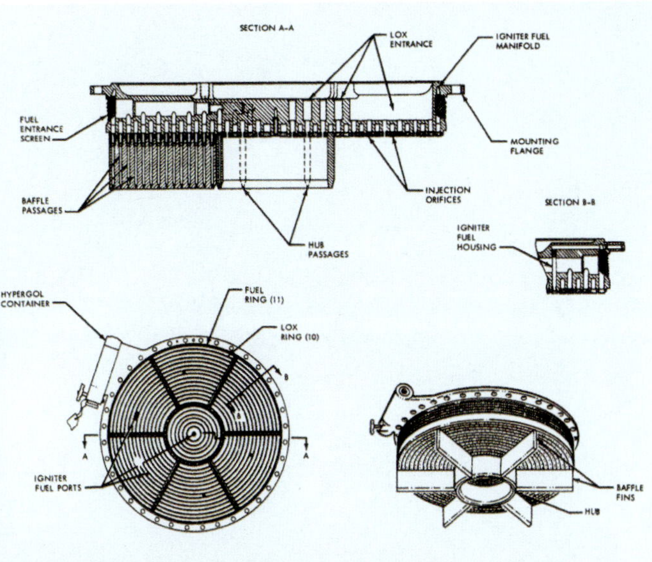

Rocketdyne H-I engine

Above: Injector for the H-1 combustion chamber showing baffle passages with LOX ring and fuel ring and igniter ports. MSFC

Left: The configuration of H-1 engine elements as a compact, efficient package developed for minimal space and conservative design, and aiming for maximum reliability over absolute performance. *Rocketdyne*

Below: A 'breadboard' diagram of the operating cycle of the H-1 engine clearly showing the turbine and turbopump for LOX and RP-1, with gaseous oxygen (GOX) lines and ground services connections. *MSFC*

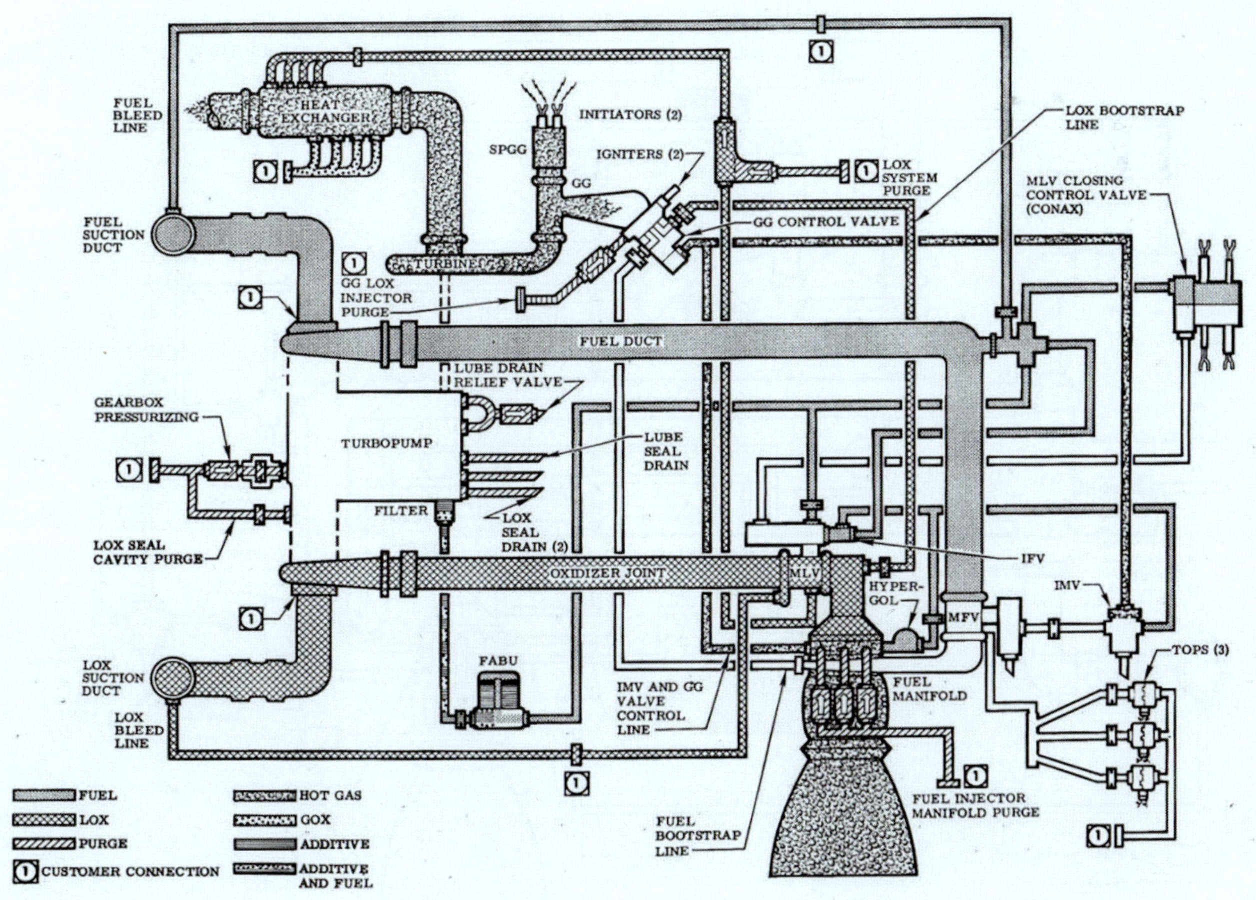

Above: A Rocketdyne H-1D engine awaiting delivery. *Rocketdyne*

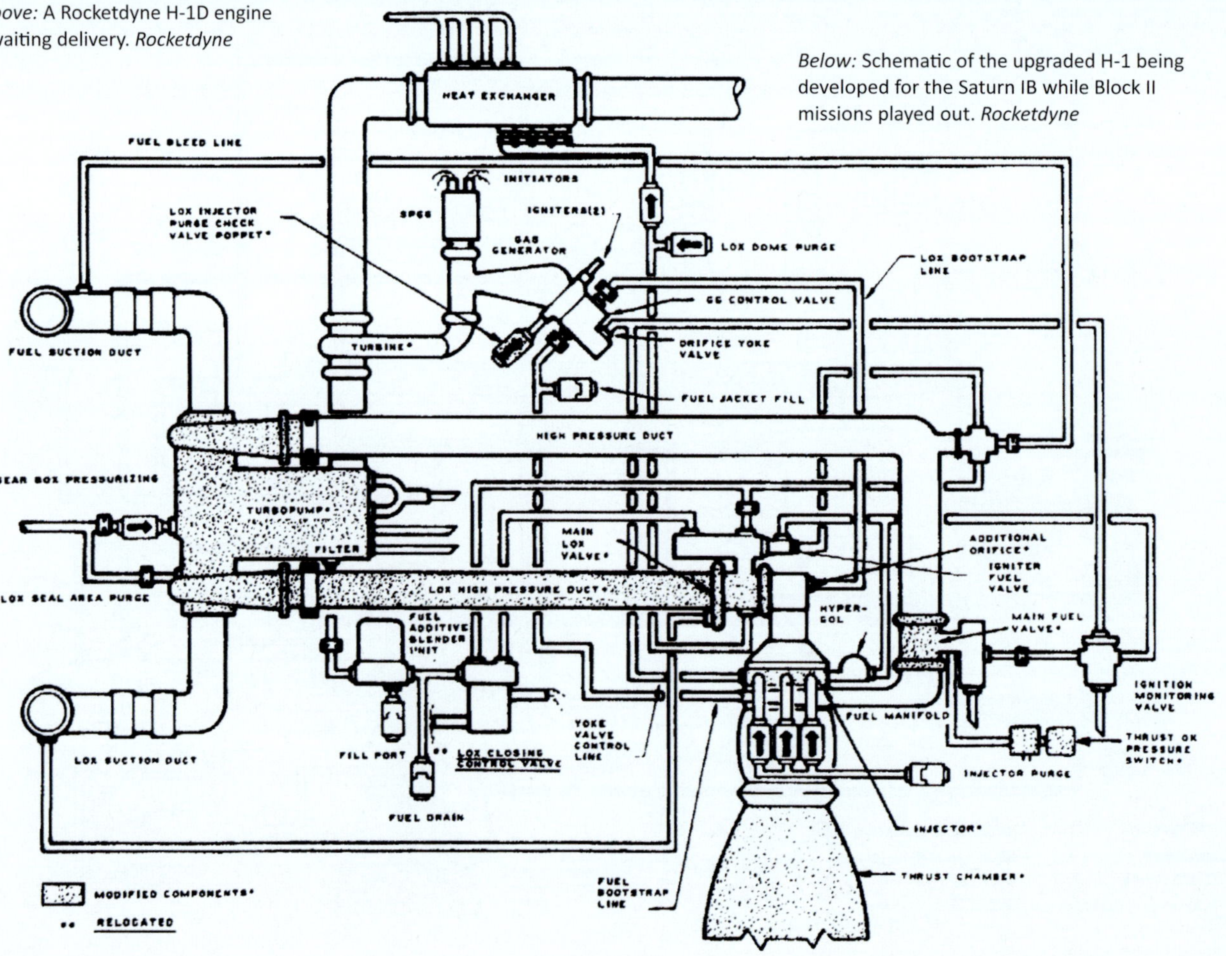

Below: Schematic of the upgraded H-1 being developed for the Saturn IB while Block II missions played out. *Rocketdyne*

into the thrust chamber exit flow stream.

The turbine exhaust hood was a stainless-steel, welded elbow assembly incorporating two mating flanges, two doubler rings, a bellows section and an integral liner to protect it. The bellows permitted the degree of movement required by the system. The heat exchanger was a welded, stainless-steel assembly consisting of an outer shell, inlet and outlet flanges, a helix-wound, four-coil system with inlet and outlet manifolds. Turbine exhaust gases passing through the shell heated the coils. Liquid oxygen, under turbopump pressure, entered three coils of the four-coil system, to be converted to gaseous oxygen for vehicle system use. The fourth coil was blanked off and not utilised. On development of H-1C and H-1D engines for evolving engine variants in the Saturn IB, a clamp was installed on the two top coils of both the inside and outside coils opposite the GOX outlet. The clamp prevented coil movement and possible damage by holding coils firmly against the coil displacement support assembly. The turbine exhaust duct was a curved, stainless-steel assembly, mounted on inboard engine thrust chambers, which consisted of a mating flange, forward support bracket, bellows section, curved duct, and aft support bracket.

The exhaust gas aspirator was a welded, Hastelloy C shell assembly installed over, and extending beyond, the thrust chamber exit on all outboard engines. It was welded to the thrust chamber forward channel band, located approximately 50.8cm (20in) forward of the thrust chamber exit. The aft end of the aspirator was not secured to the thrust chamber. This provided a 1.11cm (0.44in) clearance between the thrust chamber fuel return manifold and the aspirator, through which the gas generator exhaust gases could escape.

The gas generator and control system consisted of a liquid propellant gas generator, ignition monitor valve, purge check valve, orifices, bootstrap lines, thrust-OK pressure switches and the hose and line assemblies which made up the series control line. The gas generator and control system controlled engine start sequencing and supplied the power to drive the turbopump. Initial power required to spin the turbopump and increase the control pressures necessary for an engine start was provided by the firing of the solid propellant gas generator. The control pressure (fuel) was utilised to open or actuate the main LOX valve and the fuel additive blender unit, and to rupture the hypergolic cartridge burst diaphragms. Rupture of the diaphragms allowed pyrophoric fluid, followed by igniter fuel, to enter the thrust chamber for ignition. If a satisfactory ignition was achieved, thrust chamber fuel injector manifold pressure actuated the ignition monitor valve which directed control pressure to actuate the main fuel valve. Main propellant ignition pressures opened the gas generator control valve initiating gas generator operation and the high-pressure gases thus produced powered the turbopump during mainstage operation.

The gas generator incorporated a solid-propellant cartridge, bolted to the liquid- propellant gas generator flange, which supplied power to the turbine for starting the engine. It was a disposable unit that could not be reloaded or reused. The mounting end was closed out by a burst diaphragm and an orificed retaining plate. Threaded bosses were provided for two initiators which ignited the solid-propellant gas generator grains. The engine start signal energised both initiators. As the solid-propellant grains started to burn, the burst diaphragm ruptured (pressure was approximately 4,481kPa [650psi]), releasing a gas flow rate of approximately 2.12kg/sec (4.68lb/sec). This constant flow rate was maintained for approximately one second. These gases spun the turbine which, in turn, drove the LOX and fuel pumps until fuel control pressure opened the liquid-propellant gas generator control valve. This began to receive bootstrap propellants from the turbopump propellant discharge ducts as solid-propellant grains were consumed. Ignition of liquid-propellant for the gas generator was accomplished by the solid-propellant grains, which burned approximately 100-200 milliseconds after LOX and fuel entered the combustor. Ignition of liquid propellants was ensured by the use of two auto-igniters.

The ignition monitor valve was a three-way, pressure-actuated valve which physically sensed satisfactory thrust chamber ignition before directing control (fuel) pressure to open the main fuel valve. The monitor valve was mounted on the main fuel valve and connected to the main fuel valve opening port by a close-coupled orifice fitting. During engine start, the main LOX valve opened the igniter fuel valve which directed igniter fuel to the hypergolic container and fuel pressure to the ignition monitor valve inlet port, for subsequent main fuel valve actuation. When satisfactory thrust chamber ignition had been achieved, the pressure build-up sensed at the thrust chamber fuel injector manifold opened the ignition monitor valve. If ignition was not established, ignition monitor valve actuation pressure would not be available and the main fuel valve would not open. During engine shut-down, decreasing thrust chamber pressures allowed the main fuel valve actuator spring to close the valve, and the ignition monitor valve to close and vent.

Engine shut-down was achieved by an electrical cut-off

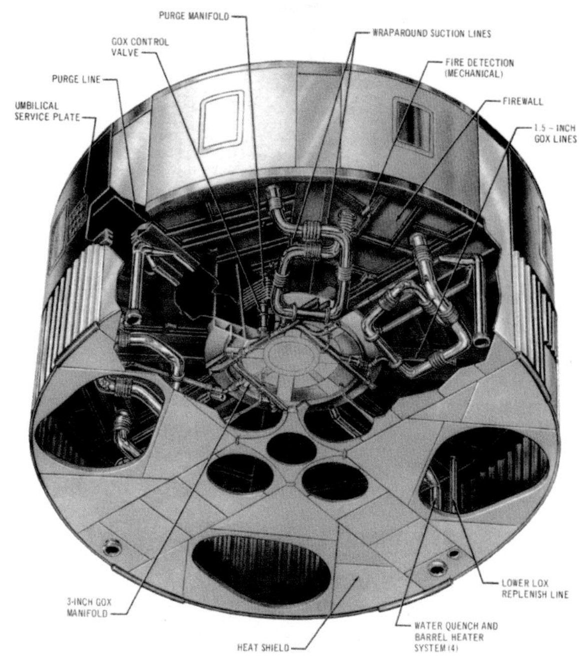

Right: A cutaway illustration of the base region of the boattail assembly showing some of the interior plumbing supporting the eight engines and the pressurisation system. *MSFC*

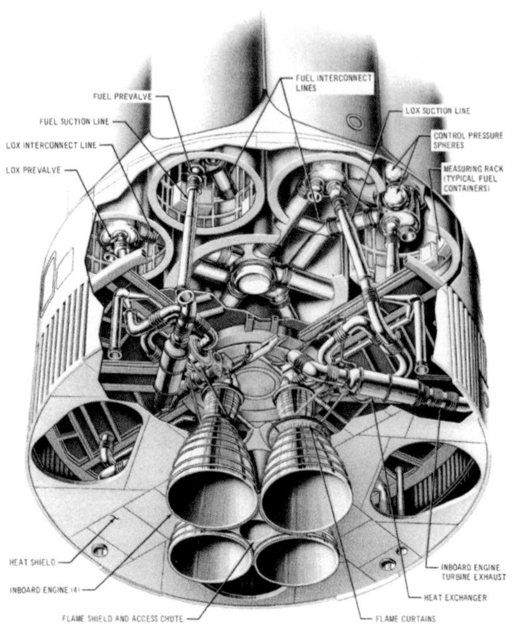

Above: With the four moveable outboard engines deleted, the four fixed inboard H-1s show the relative disposition of associated hardware in the boattail end of the Saturn I. *MSFC*

signal which fired a pyrotechnic-actuated main LOX valve closing control valve. Opening of this valve allowed fuel pump discharge pressure to enter the closing port of the main LOX valve actuator, equalising the fuel pressure on both sides of the valve actuation piston. The main LOX valve then closed by the fuel pressure acting on the larger piston surface area on the closing side, plus spring force, cutting off LOX flow to the thrust chamber and gas generator, resulting in turbopump speed and propellant pressure decay. Closing the main LOX valve allowed the igniter fuel valve to close, shutting off fuel pressure to the hypergolic container inlet, ignition monitor valve inlet and the main fuel valve opening port. When the main fuel valve actuation pressure decayed to approximately 1,379kPa (200psig), the main fuel valve closed under spring force, completing engine shut-down approximately one second after accepting the cut-off signal. A fuel-rich shut-down was provided to prevent a temperature spike in the liquid propellant gas generator. A natural, fuel-rich shut-down occurred in the thrust chamber.

Development of the H-1 began in 1957 and after receiving the contract to build the H-1 engine on 11 September 1958, Rocketdyne set to work design at the Canoga Park, California, facility with test and qualification work assigned to the Santa Susana Field Laboratory (SSFL). Tests began in the Canyon Area with the first main stage firing on 31 December, the initial prototype (H-1001) tested on 6 March 1959 before delivery to Huntsville where it was fired from the Power Plant Test Stand on 26 May. The engine was of the initial series delivering a thrust of 836,224N (188,000lb) but for the initial flight series the engines were down rated to 733,920N (165,000lb) The ABMA accepted the first flight engine on 14 October 1959.

Rocketdyne delivered the first engine rated at 889,600N (200,000lb) on 31 March 1964 with qualification signed off on 7 April 1965. The programme to certify this engine extended to 17 firings with an accumulated duration of 2,325sec and these were first employed in the AS-201 flight vehicle, the first Saturn IB. NASA received the first engine rated for a thrust of 911,840N (205,000lb) on 21 October 1965 and its qualification programme was completed on 16 June 1966. The first flight verification testing was completed on 20 October 1966.

Cryogenic Challenges

One of the greatest achievements of the early Saturn programme was to introduce the use of cryogenic propellants to large launch vehicles and thus achieve a 30 per cent increase in efficiency compared to conventional LOX/RP-1. By using the brute power of LOX/RP-1 engines in the first stage, the more important role was held by the upper stages. The S-IV provided approximately 85 per cent of the payload kinetic energy placing 9,980kg (22,000lb) in low Earth orbit while the S-IVB provided 90 per cent initially inserting 16,330kg (36,000lb) into the same path, rising to 18,144kg (40,000lb). When applied to the Saturn V, the S-IVB contributed 60 per cent of the payload kinetic energy, 50 per cent of which was in firing a second time to put the payload on a cislunar trajectory. But the S-IV for Saturn I and the S-IVB for Saturn IB provided the energy which transformed the potential of each and provided a solid engineering base for the cryogenics in the Saturn V.

As discussed earlier, cryogenic research in the United States began in the early 1950s and was far more advanced than in any other country. The success of the Saturn series of launch vehicles was directly attributable to that initial research, where liquid hydrogen was thought to have great promise for propulsion. The US National Bureau of Standards conducted much ground-breaking work on how to handle liquid hydrogen, with much of the technology for liquid oxygen being found highly relevant. Key areas for development included valve design, propellant transfer techniques, purging and fill procedures, venting techniques, safety and instrumentation. The very design and spatial layout of ground installations was configured according to handling criteria.

In some respects it was the ground installations and the equipment for filling the Saturn propellant tanks that posed the biggest challenges. There was very little problem with valve design, the technology for LOX valves being found equivalent to the requirements for handling liquid hydrogen, with the major difference being in the LOX packing glands being located sufficiently distant from the liquid hydrogen to keep shaft and seal temperatures equivalent to that of the LOX servicing lines. The much smaller hydrogen molecule does pose severe leakage problems however, necessitating much tighter shaft and seal tolerances.

For engine and stage tests, the large storage vessels were located a greater distance away than were the LOX storage vessels, with consequent increase in the length of the transfer lines, jacketed to minimise boil-off. Welded flanges ensured a leak-free join but the large liquid hydrogen tanks required purge techniques to avoid accumulations of oxygen, moisture, liquid or frozen air, frozen nitrogen and other contaminants. Initially, the liquid hydrogen tank was filled with nitrogen for leak checks, which was then purged with helium until the nitro-

Left: A cutaway display model of the S-IV stage displaying the common bulkhead between the two cryogenic tanks and the six RL-10 engines. *Author's collection*

gen content fell below 1 per cent by volume. The helium too had to be moisture-free and of the highest grade of purity obtainable. Low fill rates were used until the helium was chilled to near working temperatures to prevent inversion of the tank wall through geysering, which could cause rapid cooling of the liquid hydrogen. With the liquid hydrogen operating temperature at a low level, the tank was filled rapidly to the 99 per cent level, topping up the additional 1 per cent being conducted at a slow rate. The LOX was also subject to the same loading cycle.

The loading process imposed severe thermal stress on the common bulkhead and to minimise that the LOX tank was filled first, placing the upper face adjacent to the LH_2 tank under compression. These stresses were relieved by the introduction of the cold hydrogen. LH_2 has a wide range of flammability in air, from 4 per cent to 75 per cent concentration by volume but whenever it was handled, leaks were possible and when vented or dumped a static discharge could ignite the gas. Venting through a closed system to a remote location from the test stand helped prevent this and Douglas had a burn pond where hydrogen could be consumed under control conditions. Experience showed that there was little risk of fire from low pressure hydrogen, even in large quantities, but whenever high-pressure leaks of 1,379-2,068kPa (200-300psi) occurred, fires invariably resulted. Because hydrogen burns with an almost invisible flame, burn wires were strung at various locations in the engine compartment and near the tank fill and engine feed lines and valves, automatically triggering an alarm if broken.

Special materials were required in addition to the challenging handling issues. The AISI 300 series of stainless steel, defined as having 18 per chromium and 8 cent nickel alloy, was adopted for ground installation due to its low temperature characteristics and good welding tolerance. The specific alloys used were the low carbon 304L, the titanium stabilised 321,

S-IV stage – structural elements

Cutaway illustration showing the principal structural elements of the Douglas S-IV stage with systems layout. Development of the cryogenic stage provided learning experience for other stages using liquid hydrogen as a fuel. *Douglas Aircraft Co.*

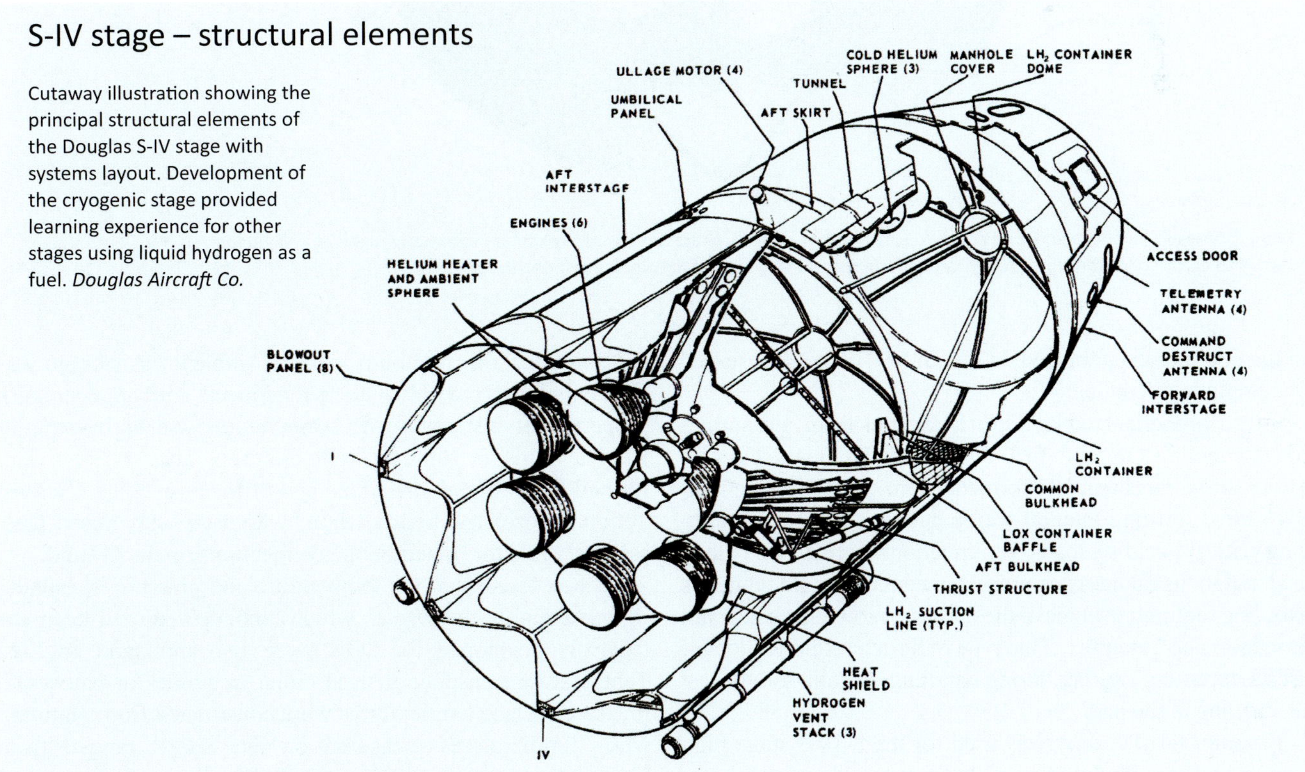

S-IV stage – structural elements (contd.)

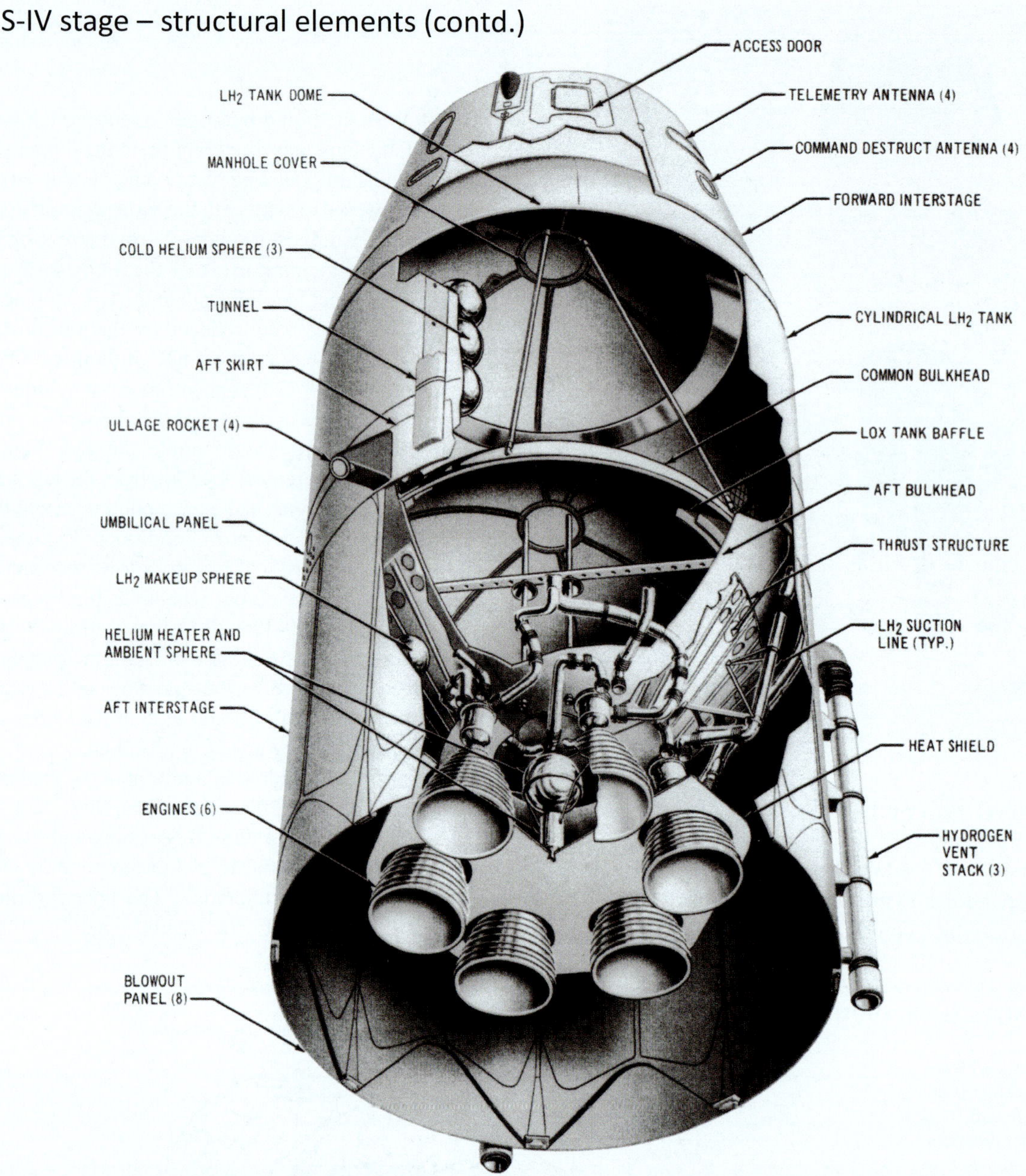

From a different perspective the S-IV stage displays design aspects characteristic of the unique common bulkhead utilised for the S-IVB and also for the North American S-II, the second stage of the Saturn V. *Douglas Aircraft Co*

and the columbium stabilised 347, with SO4L or 321 reserved for welding applications.

Saturn propellant tanks were fabricated from aluminium alloy for both LOX and LH_2 tanks with Douglas selecting 2014T6 alloy, mechanically formed, chemically and mechanically milled forming integral waffle patterns in areas encountering critical buckling loads. Tank sections were welded using metal and tungsten inert gas processes which brought demands on tooling required to achieve the necessary tolerances in detail fabrication and assembly. This type of structure did lend itself to self-supporting rigidity during handling, installation for test and stacking at the pad.

Titanium 6414V alloy was used for the S-IVB stage high pressure bottles with the 56cm (22in) spheres immersed in liquid hydrogen for minimum weight. Ambient gas storage was effected with 61cm (24in) diameter spheres with good strength properties at low cryogenic temperatures and high strength/weight ratios at ambient temperatures. The spheres were formed from forged, machined and diffusion-welded sections. Teflon, fluorinated hydrocarbon, was used with raised-face, serrated seals for bolted flange connections on the ground.

It was found that the temperature and pressure transducer instrumentation techniques which had evolved, and been extensively developed for LOX were also applicable to LH_2. Temperature transducers, used either as probes or patches to measure surface temperatures were constructed from platinum wire. Temperatures measured on the S-IVB ranged from -262°C to 982°C (-440°F to 1,800°F). Square temperature patches from 0.25-2.54cm (0.1-1in) were bonded to the vehicle structure, the black boxes and the propellant lines.

Pressure transducers on the cryogenic stages comprised strain gauge and potentiometer types. The strain gauge consisted of a stainless steel diaphragm with the gauge circuit attached. Pressure deflected the diaphragm with elastically stretched wire. The potentiometer transducers handled low pressures in moderate environments and point level sensors indicated liquid levels by measuring the change in capacitance of four stainless steel concentric rings, the level indicated by the difference in dielectric between the liquid and the gas.

S-IV Engineering Design

The decision to incorporate the S-IV was made by the Silverstein Committee and presented to the ABMA and NASA in December 1959, based on the LR115 (RL-10) from Pratt & Whitney, which had been under development since late in 1958. For stage construction, a bidders' conference was held at Huntsville on 26-27 January 1960 and 11 proposals were received by the closing date on 29 February. On 26 April 1961, NASA announced that Douglas Aircraft Corporation had received the contract for the S-IV stage. Initially, consideration was being given to using six LR115 engines, rather than four of the more powerful LR119s, which promised a unit thrust of 77.84kN (17,500lb), but development problems argued against that. Instead, approval for a six-engine configuration was granted on 29 March 1961 using the RL-10-A1 instead. This was later changed to the RL-10-A3 with compatibility fit checks during May 1961.

Because of the low density of the hydrogen/oxygen propellant combination, a tank of minimum volume was required to reduce overall inert weight. The forward chamber contained the liquid hydrogen and the smaller aft chamber the liquid oxygen. This required development of a common bulkhead which, for ease of manufacturing and tooling, was the same shape as the forward and aft domes. The bulkhead itself formed a barrier to avoid solidification of the LOX, which would occur if the bulkhead on the LOX side reached the temperature of the LH_2. The hemispherical segment of fibreglass hexagonal honeycomb was bonded to the 2014 T-6 aluminium face with more than 122m (400ft) of welding required. Tests revealed that this was not the best solution, the combination of acoustic frequencies from the engines and thermal stresses causing frequent de-bonding. Surprisingly, there was no effect on stage performance. The welded aluminium, dome-ended propellant tank had a diameter of 558.8cm (220in) divided into the two (LOX and LH_2) chambers by the welded-in and insulated double-walled common bulkhead.

The forward and aft interstage elements consisted of aluminium honeycomb panels attached to an aluminium alloy reinforcing structure. Located between the aft bulkhead of the LOX tank and the exhaust nozzles of the six engines, the heat shield served to protect the tanks, engines and control equipment from radiated and convective heat transfer. The engines were attached to a thrust frame which was connected to the aft bulkhead of the LOX container. The aft skirt and the aft interstage not only protected the engine assembly but also carried the loads to the interface with the first stage. The aft interstage contained blow-out panels for relief of the chilldown gases. The conical forward interstage connected the instrument unit with the S-IV stage and the payload. Bonded aluminium honeycomb was used for both skirts and interstages.

The stage was designed to handle flight loads of 7g in longitudinal acceleration and 1g in the lateral axis and the aft interstage structure was designed for a maximum aerodynamic pressure of 89.6kPa (13psi). The S-IV had a nominal dry mass of 5,307kg (11,700lb) of which the propulsion system and ancillary equipment accounted for 1,586kg (3,498lb). The S-IV stage structure was 1,262.4cm/12.6m (497in/41.4ft) long and 558.8cm/5.5m (220in/18.3ft) in diameter. An aft interstage, an aft skirt, a thrust structure, a base heat shield, two propellant containers, and a forward skirt were structurally joined to make up the stage. Loads from the first stage were transmitted to the S-IV stage through the aft (S-I/S-IV) interstage.

The interstage, a cylinder approximately 467cm/4.6m (184in/15.3ft) long, was constructed of eight 45deg cylindrical segment panels joined by longitudinal splices. The panels were of honeycomb sandwich construction consisting of 7075 aluminium alloy faces bonded to a 5052 aluminium alloy core. Loads were introduced to the interstage at eight points through a field splice with the S-I stage at station 962. The loads were carried forward by tapered longerons which sheared the concentrated loads into the sandwich panels. The panels distributed the loads uniformly to the aft skirt. Between the longerons, at the aft end of the structure, were triangular vent ports covered with fabric blowout panels. The panels were removable to permit servicing of equipment within the structure. Three hydrogen chilldown vent lines were mounted on the exterior of the structure.

Loads from the S-I stage were transmitted to the S-IV LH_2 tank through the aft skirt, which was 121.9cm/1.2m (48in/4ft) long and also constructed of eight 45deg cylindrical segment panels joined by longitudinal splices. The skirt and aft interstage were attached by explosive bolts. When fired, the bolts allowed the S-IV stage to separate from the first stage at station 1147. The skirt was welded to the LH_2 container at the tangent point of the aft bulkhead. Four ullage motors and fairings were mounted on the exterior of the aft skirt. Cut-outs were provided in the aft skirt for the umbilical plate, propellant fill and topping lines, oxygen vent line and ground air conditioning line.

The thrust structure transmitted engine thrust loads to the LOX container. This 7075 aluminium alloy structure was a conical frustum with an aft diameter of 248.9cm/2.49m (98in /8.1ft), a forward diameter of 431.8cm/4.3m (170in/14.1ft) and a length of 152.4cm (60in/5ft). The skin slope was tangent to the LOX container aft bulkhead at the interface. The six engines, mounted on a circle 233.6cm (92in) in diameter, were canted 6deg from the vehicle centreline. Two control actuators for each engine were also supported by the thrust structure. Lateral loads (resulting from engine gimballing and cant angle) and axial loads were transmitted from the gimbal bearing joints to the LOX container aft bulkhead through the thrust structure beams, skins and stringers, which were supported by an aft ring, two internal intermediate rings and a forward ring. Lateral loads were sheared by the aft ring into the thrust structure skin. Axial loads were transmitted from the aft ring through the thrust beams and external longitudinal hat-section stringers to the forward ring, which was attached to a milled land on the LOX container aft bulkhead. Loads transmitted from the forward ring were distributed to the LOX container aft bulkhead.

The base heat shield protected the forward propulsion area from engine heat. The heat shield was located approximately 121.9cm (48in) aft of the engine gimbal plane and was supported from the thrust structure. The heat shield was an insulated honeycomb sandwich panel. Cut-outs in the panel, sealed with flexible curtains attached to the engines and heat shield, provided clearance for the engine gimballing action.

The LOX for the S-IV stage was contained in a 2014 aluminium alloy container. Two bulkheads, an aft and a common, were attached through two rings to form the container. The aft bulkhead, a hemisphere with a spherical radius of 279.4cm (110in), was constructed of six gores and a circular centre piece welded together. It was designed to support flight pressurisation and propellant loads resulting from acceleration. The other bulkhead, termed a common bulkhead because it was common to both the LOX and LH_2 containers, was a spherical segment also with a radius of 279.4cm (110in). The common bulkhead was of honeycomb sandwich construction consisting of 2014 aluminium alloy faces bonded to a fibreglass core and it had sufficient insulating properties to prevent the LOX from freezing during a 12-hour ground-hold period. Two compression rings were welded to the periphery of the bulkhead. These rings were attached to the aft bulkhead by welds and mechanical fasteners. A milled land on the aft bulkhead provided a mounting surface for the engine thrust structure. Engine thrust loads were transmitted through the land to the aft bulkhead, then carried into the LH_2 container cylindrical section.

Ring baffles of aluminium alloy were installed in the container to prevent sloshing of the LOX. The baffles were supported by a sheet metal conical frustum which was attached to the aft bulkhead at the common bulkhead joint. A manhole in the centre of the aft bulkhead provided access to the container. Outlet fittings in the sump at the bottom of the bulkhead were provided for six LOX engine lines and for two vent lines. A screen in the aft bulkhead over the engine line outlets retarded formation of vortices during draining.

Below: The S-IV-9 stage for SA-9 in Hangar AF at Cape Canaveral. *NASA*

The LH_2 for the S-IV stage was contained in a 2014 aluminium alloy container 652.8cm/6.5m (257in/21.4ft) long. The container was composed of a cylindrical section closed at the forward end by a hemispherical bulkhead and closed at the aft end by the LOX container (discussed above). The forward bulkhead and LOX container aft bulkhead were welded to the cylindrical section. Designed to support flight pressurisation loads, the forward bulkhead was constructed of six gores and a circular centre piece welded together to form a hemisphere. The bulkhead had a spherical radius of 279.4cm (110in). Three openings were provided in the bulkhead; one for container access and two for hydrogen vent lines. The LH_2 cylindrical section was designed to carry pressurisation, propellant loads due to acceleration, and external flight loads. It was composed of three 120deg cylindrical segments each 279.4cm (110in) long. Each segment was machine milled on the internal surface to a square waffle pattern with a 45deg skew angle. The segments were welded into a cylinder. The waffle stiffeners provided sufficient buckling strength to give the structure a free-standing capability when the container was unpressurised.

First stage loads were introduced into the LH_2 container through a weld joint connecting the container to the aft skirt. The LH_2 container transmitted loads to the forward skirt through a weld joint on the forward bulkhead. Six LH_2 engine line-outlet fittings covered with anti-vortex screens were located just forward of the aft bulkhead/common bulkhead joint. With the exception of the common bulkhead, all inside surfaces of the liquid hydrogen container were insulated with polyurethane foam. Bonded to the container walls, the insulation limited hydrogen boil-off during launch operations and flight.

The forward skirt (forward interstage) transmitted the loads from the LH_2 container to the instrument unit. The skirt was a conical frustum approximately 330.2cm/3.3m (130in/10.8ft) long with an aft diameter of approximately 543.5cm/5.4m (214in/17.8ft), and a forward diameter of 391.2cm/3.9m (154in /12.8ft). The slope of the forward skirt was tangent to the LH_2 container bulkhead at the aft interface.

The skirt was constructed of eight 45deg conical segment panels joined by longitudinal splices. The panels were of honeycomb sandwich construction consisting of 7075 aluminium alloy faces bonded to a 5052 aluminium alloy core. Loads were transmitted to the panels through a weld joint at the LH_2 forward bulkhead. From these panels, the loads were transmitted to the forward ring which provided an interchangeable mating face for the attachment of the instrument unit (a field splice at station 1460). A door in the forward skirt provided access to the equipment installations and cut-outs were provided for the hydrogen vent line, telemetry antennas and range safety antennas. Mounting provisions for two retromotors were located on the forward skirt.

The systems tunnel, designed to accommodate cables and tubing, was located externally on the S-IV stage body and extended from the aft skirt to the forward skirt. The fairings were designed to carry aerodynamic pressure and thermal loads. The instrument unit structure transmitted loads from the S-IV stage to the payload. The aluminium-alloy structure was 391.2cm/3.9m (154 in/12.8ft) in diameter and 86.4cm (34in)

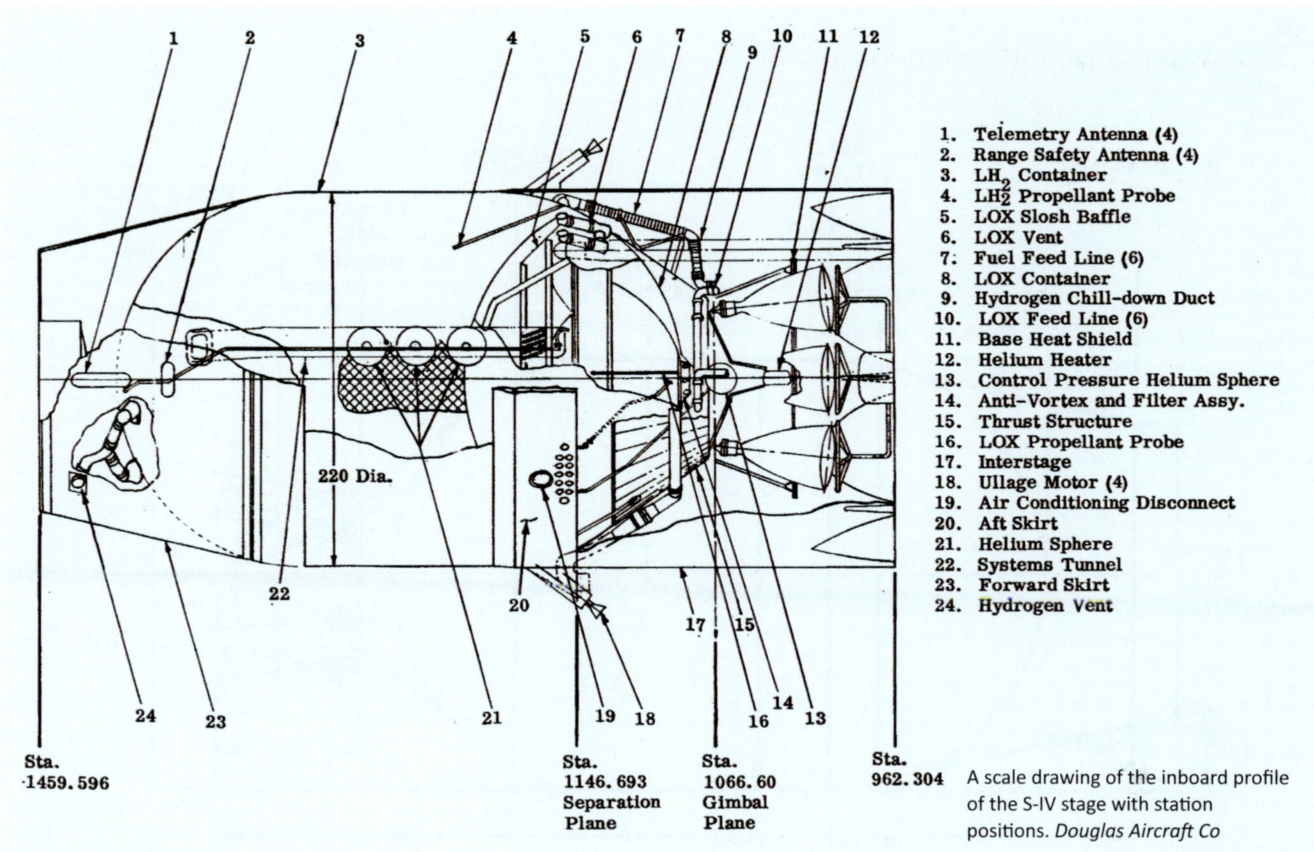

A scale drawing of the inboard profile of the S-IV stage with station positions. *Douglas Aircraft Co*

long. Axial loads and bending moment were carried by internal longitudinal hat-section stringers and the shear load was carried by the skin. The aft and forward rings provided mating faces for attachment to the adjacent structures. Loads were transmitted to the aft ring by the S-IV stage through a field splice at station 1460. The aft ring transmitted axial and shear load to the stringers and skin. Loads were transmitted by the forward ring to the payload at station 1494. Internal longitudinal stringers, attached to the skin and rings, provided support for the equipment mounting plates. Access to the instrument unit was through the S-IV stage forward skirt. Cut-outs were provided in the skin for the umbilical plate, stabilised platform window, antennas and vents. Four equally spaced vents, located at the forward end of the instrument unit, provided a common environment for the S-IV forward skirt, the instrument unit and the spacecraft adaptor.

Four solid-propellant ullage motors were used to accelerate the S-IV stage providing propellant positioning and sufficient turbopump inlet pressure for starting the engines. Each ullage motor was mounted in a fairing bolted to the aft skirt of the S-IV stage at two points using frangible nuts. The motors, located 90deg apart around the skirt, were canted 35deg from the vehicle centreline to minimise the effect of exhaust gases on the vehicle hardware. Each ullage motor provided a nominal average thrust of 15.39kN (3,460lb) at 21°C (70°F) to position S-IV propellants for RL-10A-3 engine ignition and to aid in separation during S-I/S-IV staging. After the motors were expended, the four fairings were jettisoned by breaking the frangible nuts. This occurred approximately 20 seconds after the separation signal was initiated. Retrorockets were not required on the S-IV stage for separation of the S-IV stage and instrument unit from the Apollo payload. However, the vehicle was designed with a capability for inclusion of two TX-280 solid-propellant retrorockets on that stage.

The Instrument Unit on Block II Saturn I vehicles contained four horizontal cylindrical compartments, essentially 1.02m (40in) diameter tubes extending radially from a 1.78m (70in) diameter central, vertical cylinder as the fifth compartment. The overall diameter was 3.9m (12.8ft) with a height of 2.31m (7.58ft).

Compartment I carried an active ST-90S guidance package which provided inertial guidance systems control for the vehicle. The ST-90S maintained programming through a magnetic tape actuator. Compartment II held an ST-124 passenger guidance package, which was a digital computer guidance system, and the gyro stabilisation electronic box. Compartment III contained telemetry equipment to transmit signals for approximately 335 measurements of pressure, temperature, strain, vibration, flight path and other data from within the instrument unit. Installed in compartment IV was the AZUSA transponder and the power supplies for the instrument unit and other components throughout the vehicle. The guidance signal processors, flight control computers and sequencers, power distributors, the UDOP transponder, and MISTRAM transponder were contained in the centre cylinder (compartment V).

The five compartments were air conditioned and pressurised by gaseous nitrogen from two high-pressure spheres mounted on the lower outer wall of the central cylinder. The cooling system maintained a stable acceptable temperature within the instrument unit during vehicle pre-flight and flight.

S-IV stage – structural elements (contd.)

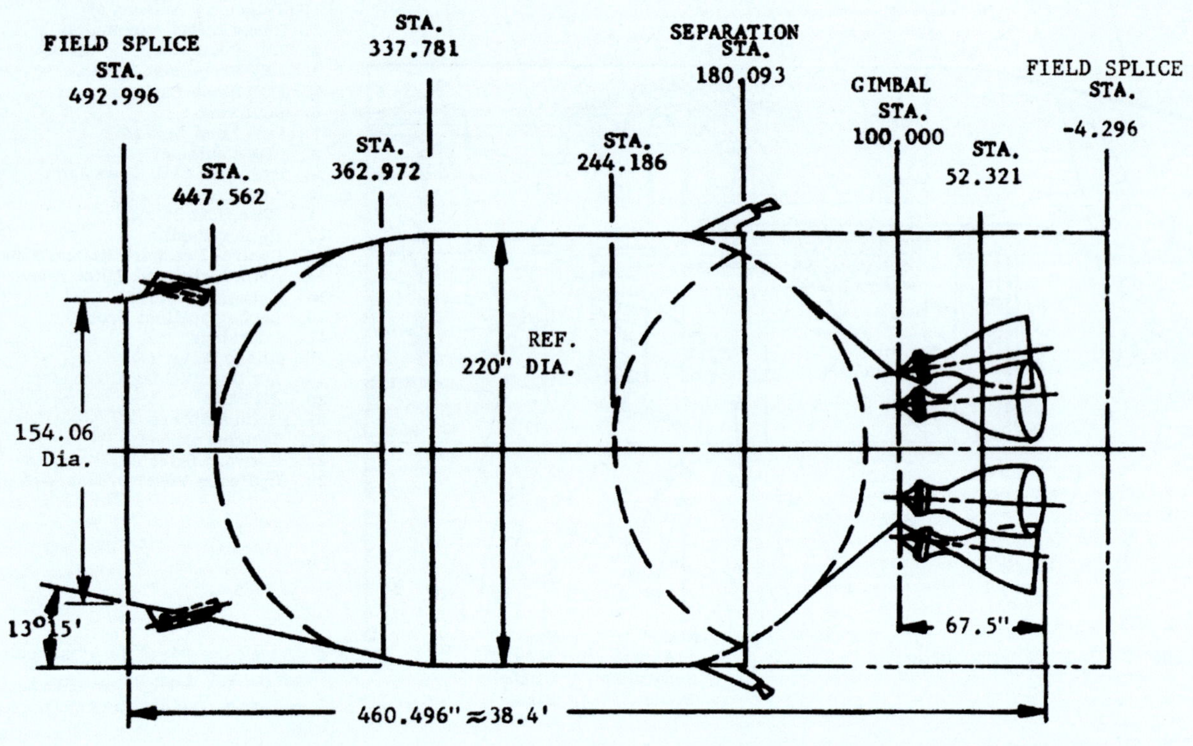

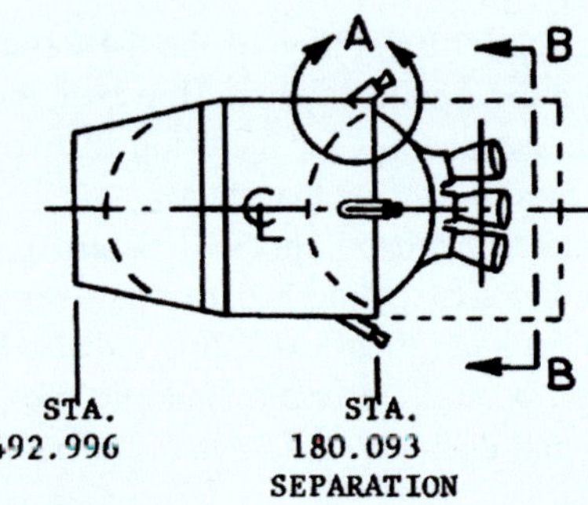

Above: Principal dimensions and sizing parameters for the S-IV. The stage had undergone some changes to the overall size with adjustments for elements made proportionate to the overall size of the stage. The diameter was limited by the size of the S-I stage while the length was controlled by the quantity of propellant required fo consumption by the cryogenic engines.

Left & below: The S-IV ullage rockets showing relative location, thrust vector and alignment. *MSFC*

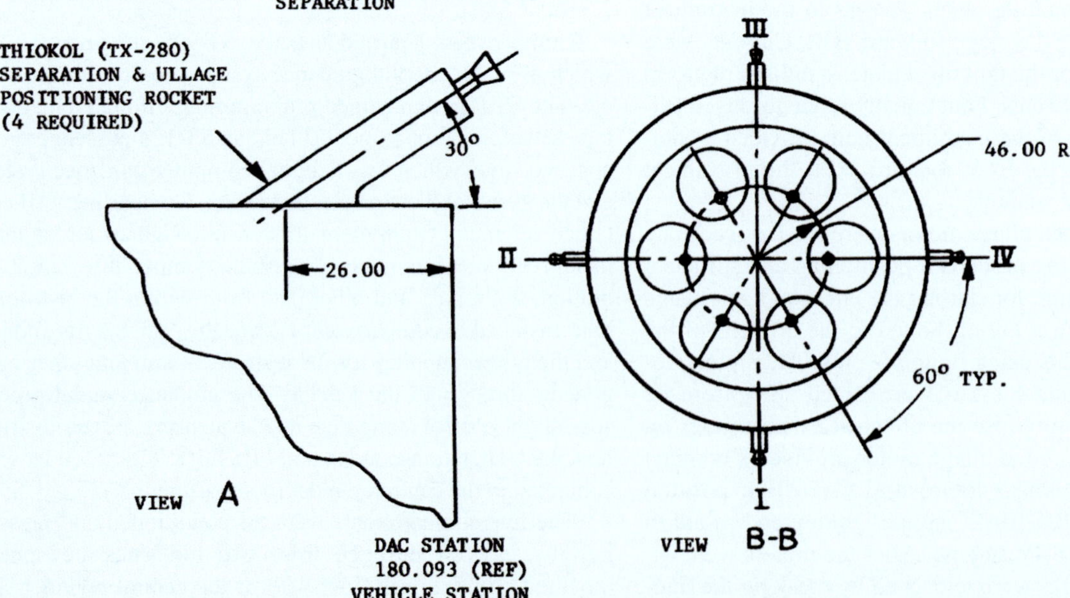

The Instrument Unit began to function before lift-off in order to sequence the S-I engine start and control. Programming and flight control from the unit continued throughout the operational flight of the S-I stage. The Instrument Unit initiated stage separation sequencing at the termination of the S-I stage operation as well as engine start for the S-IV stage. At the time of separation, control was switched from the S-I stage to the S-IV stage by the Instrument Unit. Instrument programming and flight control of the S-IV continued during the operation of the S-IV engines. Three telemetering sets were also controlled by the instrument unit.

A significant change occurred with SA-9, the first to fly with a completely redesigned Instrument Unit in which the components were attached to a circular wall with a diameter of 3.91m (12.8ft) and a height of 86.4cm (54in), the total weighing 1,179kg (2,600lb) versus the 2,449kg (5,400lb) of the previous arrangement. This was also flown on SA-8 and SA-10 and began the adaptation to the Instrument Unit design for the S-IVB.

RL-10 Engineering

The S-IV stage was powered by six RL10A-3 liquid-propellant rocket engines. Each engine incorporated a regeneratively cooled thrust chamber and a turbopump-fed propellant system. Heat absorbed by the fuel in cooling the thrust chamber provided power for a hydrogen turbine that drove the propellant pumps.

A nominal propellant consumption rate of 15.9kg/sec (35.2lb/sec) at a 5:1 LOX/fuel ratio, enabled each engine to develop a thrust of 66.72kN (15,000lb) at a 60,960m (200,000ft) altitude rating at a nominal specific impulse of 427 seconds and an absolute thrust chamber pressure of 2,068kPa (300psi). The firing duration of each engine was 470 seconds.

Each engine had a dry weight of approximately 131.5kg (290lb). Arranged in a circular pattern, the engines were gimbal mounted to provide a +/-4deg thrust vector in a square pattern for vehicle attitude control. All six engines were used for pitch and yaw control with engines 1, 2, 3 and 4 providing roll control through 12 actuators mounted in a star pattern. The cryogenic upper stage for the Saturn I Block II provided approximately 45,000kg (100,000lb) of LOX/LH_2 for a maximum burn duration of 470 seconds.

The thrust chamber consisted of a body, a propellant injector, and a spark igniter. The thrust-chamber body was a brazed assembly consisting of an inlet manifold, 180 short single-tapered tubes, turnaround or rear manifold, 180 full-length double-tapered tubes, exit or front manifold, and external stiffeners. The full-length tubes led axially rearward from the hydrogen exit manifold and, for the full periphey of the combustion chamber, the throat and the forward part of the expansion chamber. The short tubes led rearward from the hydrogen inlet manifold and interweaved between the full-length tubes to form the remainder of the expansion chamber. The turnaround manifold at the aft end of the expansion chamber nozzle interconnected with the short tubes to the long tubes. Brazing between the tubes served mainly as a seal. Inlet and exit manifolds provided entrance of unheated fuel and exit of the regeneratively heated fuel, respectively. The chamber hoop loads were carried by reinforcing rings. The thrust-chamber body, designed with a 40:1 expansion ratio, employed a truncated nozzle to minimize weight.

The propellant injector, located on the thrust chamber, atomised and mixed the LH_2 and LOX to provide the correct conditions for ignition and efficient combustion. The propellant injector consisted of 216 elements arranged in eight equally spaced concentric circles. Each element was composed of a LOX nozzle and a concentric fuel annulus. With the exception of those in the inner and outer rows, all nozzles used swirlers to produce efficient mixing. The LOX nozzles were fed from a conical LOX chamber, within which was a conical fuel chamber that fed the fuel annulus. The fuel chamber wall facing the combustion chamber was formed of porous welded steel mesh to provide transpiration cooling of the injector face. This cooling was accomplished with a fuel flow of 0.25kg/sec (0.56lb/sec) or 10.4 per cent of the total fuel flow. The spark igniter was a recessed centre electrode, air-gap type that ignited the propellants by a high-voltage capacitor discharge at a rate of 20 sparks per second. The igniter was recessed in the injector face to form a chamber that kept the combustible mixture near the spark. Because of the spark concentration in the vacuum conditions, the proximity of the propellant mixture to the spark was critical.

The turbopump assembly consisted of a two-stage hydrogen turbine, gear box, two-stage fuel pump and single-stage LOX pump. The turbopump was an integral unit which pumped pressurised propellants from the vehicle containers to the engine thrust chamber. The two-stage, partial-admission impulse-type turbine was driven by expanding hydrogen gas flowing from the jacket and through a venturi. Both blade stages were mounted on a single rotor and were fully shrouded to minimize blade tip leakage. A rated turbine speed of 28,400rpm developed 441.45kW (592hp) from a hydrogen flow rate of 2.52kg/sec (5.56lb/sec), approximately 95 per cent of the total rated flow, working between inlet conditions of 89.2ºC (192.5ºF) 4,475kPa (649psi) total pressure and exit conditions of 99.8ºC (211.6ºF) and 3,006kPa (436psi).

The turbopump gearbox transmitted power from the main turbine drive shaft to the LOX pump shaft through a 2.5:1 reduction gear train. Gearbox and oxidiser shaft cooling was provided by a 0.0045kg/sec (0.01lb/sec) LH_2 coolant flow from the first-stage pump volute. The main drive shaft provided LH_2 coolant bleed flow from the second-stage fuel pump inlet to the support bearings at the turbine drive end. Gearbox pressurisation was maintained at 124-172kPa (18-25psi) above ambient, and excess gas was vented into a cooldown vent manifold.

The fuel pump consisted of two stages mounted back-to-back to minimise axial thrust. A common shaft drove the fuel pumps directly from the turbine. The pump had a constant velocity collecting volute for equal circumferential pressure distribution, and a straight-tangential nozzle diffuser for velocity-head recover. A power requirement of 379.5kPa (509hp) was necessary to drive the fuel pump at a rated operating speed of 28,400rpm and a flow rate of 2.65kg/sec (5.85lb/sec). The first-stage fuel pump was preceded by a three-bladed axial flow inducer which operated at the same speed as the aluminium-alloy impeller. A 50deg exit angle backswept blade design was incorporated into the back shrouded impeller to provide a suitable low-flow allowable stress characteristic. The second-stage fuel pump impeller was also of aluminium alloy and incorporated a back-shrouded radial blade design with a 90deg exit angle.

The LOX pump was mounted on the turbopump gearbox beside the fuel pump and was driven through the 2.5:1 reduction gear train located within the gearbox. A three-bladed axial flow fully-shrouded stainless steel inducer increased impeller inlet pressure above the vehicle's supply pressure to prevent impeller cavitation. The centrifugal pump had a single-stage fully shrouded stainless steel impeller. A constant velocity collecting volute designed for equal circumferential pressure distribution and a straight tangential discharge nozzle diffuser for velocity-head recovery were employed within the oxidiser pump housing. An accessory drive pad, located on the aft end of the oxidiser pump shaft, provided a mounting for the main hydraulic pump. The oxidiser pump operated at a nominal speed of 11,350rpm, with a nominal flow rate of 13.29kg/sec (29.3lb/sec) when operating at inlet and discharge pressures of 334kPa (48.5psi) and 3,199kPa (464psi), respectively. A pump efficiency of 59.7 per cent at the rated conditions resulted in a power requirement of 58.3kW (78.2hp).

The fuel pump and oxidiser pump inlet shut-off valves controlled the flow of the propellant form the vehicle containers to the engine pumps. Both valves were similar and normally closed, two-position rotating ball-type valves. The valves were opened by a 3,102kPa (450psi) control helium actuator piston and were spring closed. The fuel pump inlet shut-off valve moved from closed to fully open in approximately 30 milliseconds, moving to fully closed in approximately 389 milliseconds. The oxidiser pump inlet shut-off valve moved from closed to fully open in approximately 17 milliseconds, and moved from fully open to closed in approximately 158 milliseconds.

The prestart and start solenoid valves controlled the flow of helium pressure from the stage storage tank to the engine system propellant-control valves. The prestart and start solenoid valves were identical in design, operation and construction. When energised, the solenoids operated a two-way poppet. That in turn controlled the flow at 3,102kPa (450psi) helium pressure to the propellant control valves. In this manner the helium actuator flow was controlled. The prestart solenoid valves controlled the helium pressure which opened the fuel and oxidiser pump inlet shut-off valves. The prestart solenoid valves remained in the open position as long as the solenoid remained energised. The prestart solenoid valves were closed by a spring at engine shut-down. The start solenoid valve controlled the helium pressure which initiated the opening of the main fuel pump inlet shut-off valve and the closing of the interstage and downstream cooldown and bleed valves. The start solenoid valve was opened by a start signal, which occurred 41.6 seconds after the prestart signal and remained energised throughout engine operation, holding the start valve in the open position. The valve was closed by a spring when the engine was cut off.

Two independent prestart, or cooling sequences, one for the LH_2 system and the other for the LOX system, were initiated by electrical signals from the vehicle. The first energised the fuel prestart solenoid valve, which permitted control helium at 3,068kPa (445psi) to pressurise, and open, the actuator of the fuel pump inlet shut-off valve. Liquid hydrogen flowed through the two pump stages and was discharged overboard through the cooldown valves. A signal, approximately 32 seconds later, energised the oxidiser prestart solenoid valve which admitted control helium at 3,137kPa (455psi) to the actuator of the oxidiser pump inlet shut-off valve. Liquid oxygen then flowed through the LOX pump, discharging through the propellant injector. At the end of the prestart sequence, the propellant pumps would have cooled down to a temperature which would prevent cavitation during pump acceleration.

The six engines had to pass through a prestart cooldown sequence because of the low temperature characteristics of their propellants. The engines were started in unison a minimum of 41.6 seconds after the prestart signal had been initiated which was activated by an electrical signal from the vehicle energising the start solenoid valve. An interval of at least 20 seconds between the first prestart signal and the start signal was required. The start signal also energised the ignition system. Pressurised helium flowing through the energised start solenoid valve opened the main fuel pump inlet shut-off valve and partially closed the fuel pump cooldown and bleed valve. This permitted fuel from the pump discharge to flow through the thrust chamber tubes, absorb heat providing the energy for the turbine to overcome static friction and start turbopump rotation. The partially closed fuel pump cooldown bleed and pressure relief valve acted as a bleed during acceleration to provide the fuel pump with transient stability.

In the start position, the oxidiser flow control valve managed LOX flow as a function of inlet pressure. When a combustible mixture was developed in the thrust chamber, the propellants were ignited by the spark igniters and the engine accelerated to rated thrust. The fuel pump cooldown bleed and pressure relief valve closed as the fuel pump discharge pressure increased. The oxidiser flow control valve opened as a function of rising LOX pump pressure to provide the proper mixture ratio for engine acceleration. During steady-state operation the metering orifice area in the oxidiser flow control valve was varied for propellant utilisation control. Thrust was controlled by the thrust control valve which regulated turbine bypass flow as a function of chamber pressure.

Termination of the electrical signal from the stage sequence initiated shut-down. The solenoids returned to their normally closed positions, shutting off the helium supply and venting helium from all valve actuators. The fuel pump cooldown, bleed and pressure relief valves opened, draining fuel from the system to prevent a pressure build-up caused by closing the main fuel pump inlet shut-off valve. This stopped the flow through the turbine, thus halting pump rotation and causing the system to come to rest. The oxidiser pump inlet shut-off valve closed, stopping the LOX flow into the engine. The remaining oxidiser in the engine vented through the injector into the thrust chamber. The fuel pump inlet shut-off valve closed, preventing fuel from entering the system.

Due to its high heat capacity, the liquid hydrogen was very

RL-10 engine – primary elements

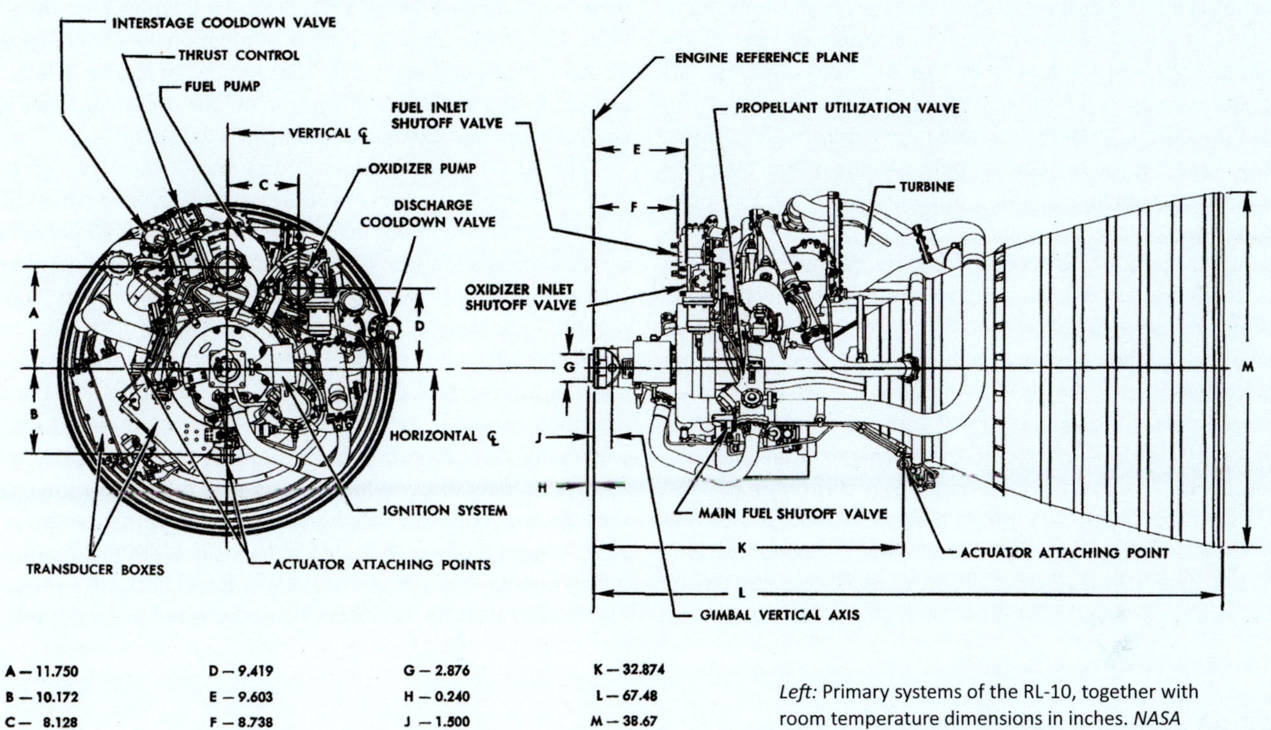

Left: Primary systems of the RL-10, together with room temperature dimensions in inches. *NASA*

RL-10 engine – turbopump

Below: Cutaway diagram showing the main component of the RL-10 turbopump. *MSFC*

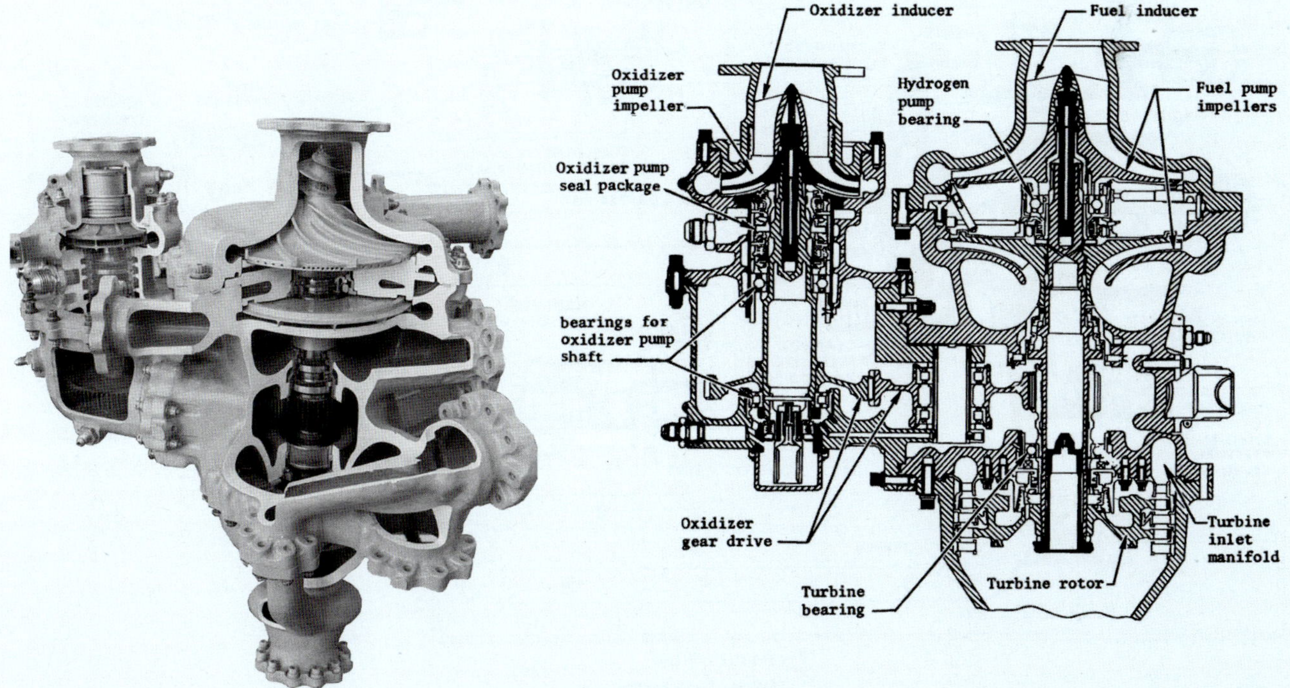

Above: The integrated turbopump for the RL-10 engine with a common housing reflecting the maturity of the design. *MSFC*

effective at cooling the thrust chamber. While passing through the thrust chamber tubes, the hydrogen picked up heat and was expanded in a two-stage turbine to drive a single geared turbopump. The fuel was then injected into the combustion chamber. This 'topping' cycle provided a performance gain of approximately 0.5-1 per cent over that of a conventional gas generator type cycle. Oxidiser was pumped directly to the propellant injector through the mixture ratio control valve. Thrust control was achieved by regulating the amount of fuel bypassed around the turbine as a function of combustion chamber pressure in order to vary turbopump speed and thereby control engine thrust. Ignition was accomplished by means of an electric spark-torch igniter recessed in the propellant injector face. Starting and stopping were controlled by pneumatic valves which received their supply of helium through electrically operated valves.

As described above, the RL-10 evolved from the Pratt & Whitney 304 turbine engine and an amalgamation with elements of the J57 hydrocarbon-burning aircraft engine. Most RL-10s for the Saturn S-IV programme were built at the company's Florida Research and Development Center (FRDC) with some tests taking place at NASA's Lewis Research Center and from a single stand at MSFC. Some 20 pre-flight tests were conducted on FX-121-9 and completed in April 1960 with the initial intent being to install four LR119 engines in the S-IV under the contract issued on 10 August that year. As we have seen, that changed to six LR115/RL-10 engines and this was fired 230 times prior to further evaluation in a twin-engine Centaur configuration which began on 6 November 1960. An explosion during the second test on the following day destroyed them and extensively damaged the stand.

The RL-10 test programme was one of the mist rigorous ever conducted on a rocket engine, some 712 development firings accumulating more than 19hrs of burn time. Initial development orientation was for the Centaur stage in a twin-engine configuration and the first was test fired in March 1962 but as recorded previously, the first Centaur launch on 8 May was unsuccessful and ended in disaster. Many historical sources concentrate on the Centaur application for the RL-10 but the development put in by MSFC for the S-IV was highly useful in eventually bringing the motor to operational status, the S-IV RL-10A-3 operating on higher pressures than the original A-1 for Centaur. With increased demand, the Saturn I programme alone would require 36 flight engines, an additional manufacturing facility was opened at P&W's East Hartford, Connecticut, facility with the first from there delivered in July 1962.

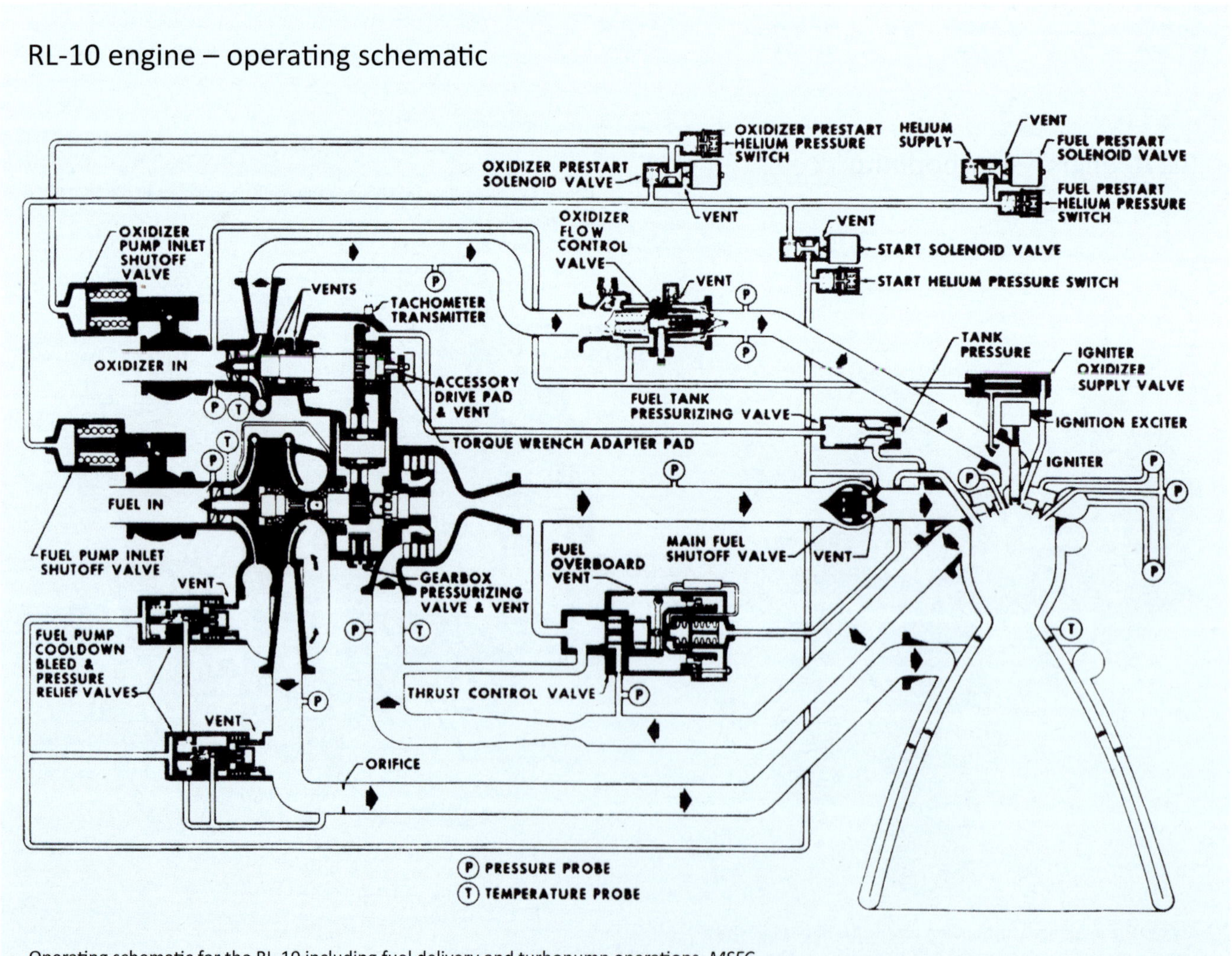

Operating schematic for the RL-10 including fuel delivery and turbopump operations. *MSFC*

Chapter Seven

Saturn IB Evolution and Design

Formal authorisation for development of the two-stage Saturn IB was announced on 11 July 1962, on the same day that NASA declared the Apollo lunar landings would be conducted using the Lunar Orbit Rendezvous mode. Both Saturn IB and Saturn V would use the powerful S-IVB stage with a single J-2 cryogenic engine, the development and evolution of the upper stage coming straight out of design and fabrication lessons from the S-IV.

At the time NASA planned eight Saturn IB missions in support of the Apollo programme, the first two being re-entry tests of the Apollo heat shield; the third would conduct fluid motion evaluation on the liquid hydrogen in the S-IVB stage. The Saturn IB was still considered the launcher of choice for whatever followed Apollo and soon this vehicle would have responsibility for many assignments in the Apollo Applications Program (AAP), later survived in greatly reduced form as the Skylab programme.

With completion of SA-10 and the end of the Saturn I test phase, NASA was still converting LC-34 and LC-37 to support Saturn IB operations and, eventually, manned Apollo missions. Complex 34 was assigned AS-201 and -202 with Complex 37B assigned the AS-203 mission. AS-204 and AS-205 would carry Apollo spacecraft on Earth orbit tests, AS-206 flying an unmanned Lunar Module for full systems evaluation. AS-207 would fly an Apollo spacecraft and AS-208 a Lunar Module for rendezvous and docking tests on the same integrated mission.

Under a revised contract, Rocketdyne was to increase total S-I stage thrust from AS-206 by 177.9kN (40,000lb) through a series of minor improvements, raising the thrust of

Above: Ready for clustering, tanks for the first S-IB stage which began to come together in 1964 while the first three Block II Saturn I launch vehicles were proving out the S-IV stage. *MSFC*

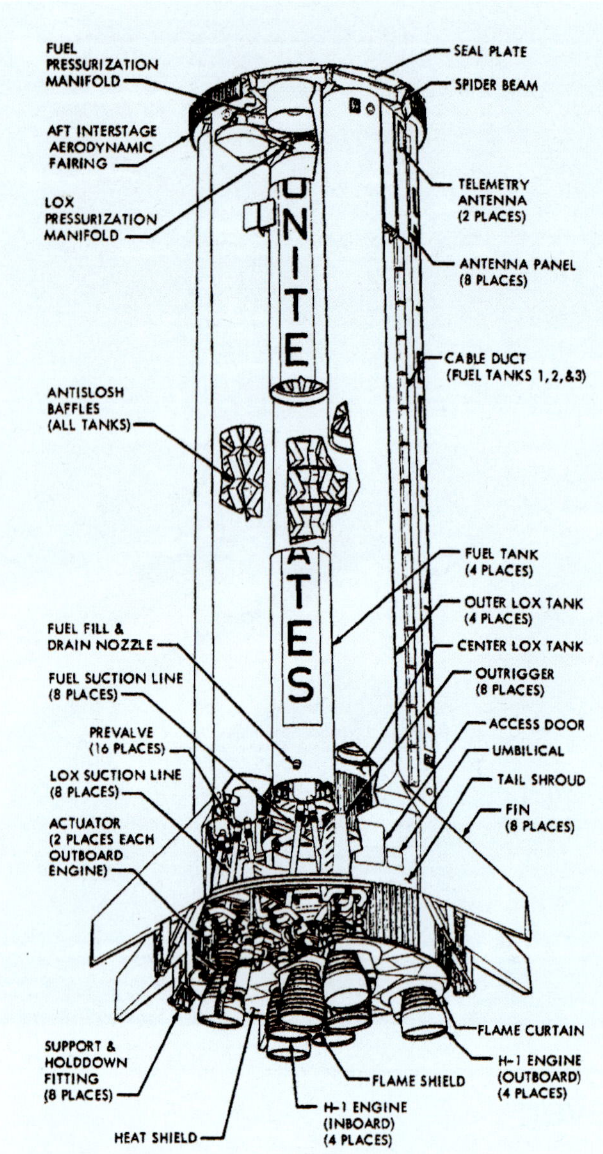

Significantly upgraded from the Saturn I programme, the S-IB is most visibly noticeable by the eight modified fins replacing the four large and four small fins on the S-I stage. *MSFC*

each H-1 engine to 911.84kN (205,000lb) and total lift-off mean thrust to 7,295kN (1.64million lb). The first five Saturn IBs had a stage thrust output of 7,117kN (1.6million lb). While the performance improvement was very little and of no productive value for planned mission requirements, NASA was keen to develop the Saturn IB into a flexible launch vehicle for multiple assignments.

Activated in 1958, Rocketdyne's Neosho plant conducted these development trials, the fourth improvement cycle to the H-1. This included modifications to the LOX pump to improve tolerances, a positive anti-rotation clamp added to the No. 8 bearing, a flat race installed in four turbopump bearings, re-trimmed LOX and RP-1 impellers for stability at higher pump speeds, a low-pressure gas generator injector for improved performance and a new high-strength metal alloy gas generator body to withstand higher temperatures and flow rates.

On 20 February 1963 Chrysler received an order to modify the S-I into the S-IB for Saturn IB. On the rescheduling of Saturn I/IB flights, in October 1963 Chrysler's contract was modified to deliver 12 Saturn IB stages in lieu of cancelled Saturn I production after SA-10. This preceded an order from MSFC on 8 November 1963 for Rocketdyne to develop the more powerful H-1 engine for the Saturn IB. Fabrication of components for the first two S-IB stages began at Chrysler in February 1964. Due to a redirection of effort and a rescheduling of mission assignment and flight objectives, on 22 April MSFC began contract negotiations with Chrysler to cut six S-I stages from the original agreement.

In terms of nomenclature, Saturn IB flights were designated in the SA-200 series, those with an Apollo payload or not were prefixed SA but the number sequence related to a mission profile and not sequence of launch; thus did AS-203 fly before AS-202. As noted by that reference (where AS-202 did not carry an Apollo payload), this convention was not adhered to everywhere, most Saturn IB references in NASA and industry manuals being prefixed with AS for all Saturn IB flights. There is a sense of hubristic possessiveness here. MSFC placed the activation of any Apollo flight opportunity as predicated wholly on the work of a Saturn rocket. Hence, they believed, all titling of such combined flights should start with S (for Saturn) followed by A (Apollo). Not unmindful of the claim from MSFC, NASA Headquarters and the newly created Manned Spacecraft Center (MSC) preferred the pre-eminence of Apollo, hence the prolific documentation denoting flights in the AS sequence.

S-IB Engineering Design

The S-IB stage structure was 2,443cm/24.4m (962in/80.2ft) long, 652.8cm/6.5m (257in/21.4ft) in diameter across the containers, 696cm/6.9m (274in/22.8 feet) in diameter across the thrust structure, and had a span of 1,239cm/12.4m (488in /40.7ft) across the fins. A tail section, nine propellant containers (five LOX and four RP-1) and a second stage adapter were structurally joined together to make up the stage. Like the S-I stage, the S-IB has a yield safety factor of 1.1 and an ultimate factor of 1.4 before collapse, which was a conservative compilation of 2-sigma and 3-sigma loads and allowances. The material used to construct the tanks for the S-IB was the 5456 aluminium alloy tempered to the H343 category. The use of this alloy, made possible by welding advancements, allowed for considerable weight reductions over the 5086 and 5052 alloys used for the Jupiter and Redstone tanks, respectively.

The most significant differences were the eight equal-size fins on the S-IB stage (the S-I Block II stage has four large and four stub fins), the elimination of the LOX-SOX disposal system and hydrogen vent lines, the moving of the retrorockets from the second stage adapter (spider beam) to the S-IVB aft interstage, redesign of the second stage adapter, and a reduced weight of approximately 38,340kg (84,525lb). The tail unit rigidly supported the aft ends of the propellant tank cluster and the vehicle on the launcher, supported the eight engines and fins and provided the thrust structure between the engine thrust pads and the propellant tanks. Other functions of the tail unit were to support the

S-IB – fabrication and assembly sequence

1	Upper and lower thrust ring
2	Engine shear web assembly
3	Barrel assembly with upper and lower thrust rings to shear webs
4	Fin support outrigger assembly with shear panels, fin and barrel connecting holes drilled
5	Thrust support outrigger assembly with shear panels to thrust beams and bulkheads installed
6	Tail section assembly incorporating thrust structure and lower shroud
7	Tail unit assembly incorporating several key systems, propellant, pneumatic and hydraulic lines
8	Central LOX container key equipment installation
9	Spider beam unit assembly with associated fittings and fixtures
10	Outer LOX containers equipment installation prior to clustering
11	Fuel containers equipment installation prior to clustering
12	Propellant tank clustering with tail unit in assembly fixture
13	Initial installation of equipment in the tail area including propellant interconnect lines
14	Installation of equipment in the tank top area
15	Conduct engine modifications and install hydraulics on outer engines
15A	Installation of inboard engines with propellant suction lines
16	Installation of additional systems in tail area including inboard turbine exhaust
17	Installation of outboard engines with associated propellant lines
18	Installation of tank top area including electrical equipment
19	Final installation of equipment in the tail area and firing test measurement systems
19A	Stage checkout, power and telemetry tests and simulated plug-drop tests
20	Cleaning, engine flameout curtains, weight determination, final test equipment installed
21	Post-static test cleaning, stage alignment checks, functional checkout
21A	Power distribution tests, instrumentation/telemetry tests, umbilical connect/disconnect tests
22	Preparation for Shipment to the launch site following final inspections/fitments

lower shroud panels, LOX and fuel bay firewalls, heat shield support beam and panel assemblies, engine flame curtains and the engine flame shield support installation. Unlike the propellant tank units, the tail unit was constructed with higher strength aluminium alloys of the 7000 series that were heat treated to the T-6 of the T-33 condition.

The 7000 series aluminium alloys used in the tail unit assembly were not recommended for welded applications because of low welding efficiencies. These alloys were used in unpressurised areas where the assembling was done with mechanical fasteners. The tail unit thrust structure configuration loaned itself to this definition; and high strength, heat-treatable 7000 series aluminium alloy forgings, extrusions, plates and sheets were fabricated into components that were joined with mechanical fasteners to construct the tail unit assembly. Because of the susceptibility of the high strength aluminium alloys to stress corrosion cracking, methods had been employed in the design and manufacture of the tail unit assembly which minimised the danger of failure due to stress corrosion cracking. These included heat treatment to the T-73 condition, heat treatment after heavy machining operations, the use of closed die forgings and the use of adequate final protective finishes.

The tail unit consisted of a barrel assembly, 266.7cm (105in) in diameter, that directly supported the centre propellant tank, enclosed the inboard engine thrust beams and acted as the hub for the four thrust support outriggers and the four fin support outriggers. The four thrust support outriggers also acted as supplementary fin support outriggers. The main fin support outriggers were similar to the thrust support outriggers but

S-IB – flight configuration, dimensions and fin details

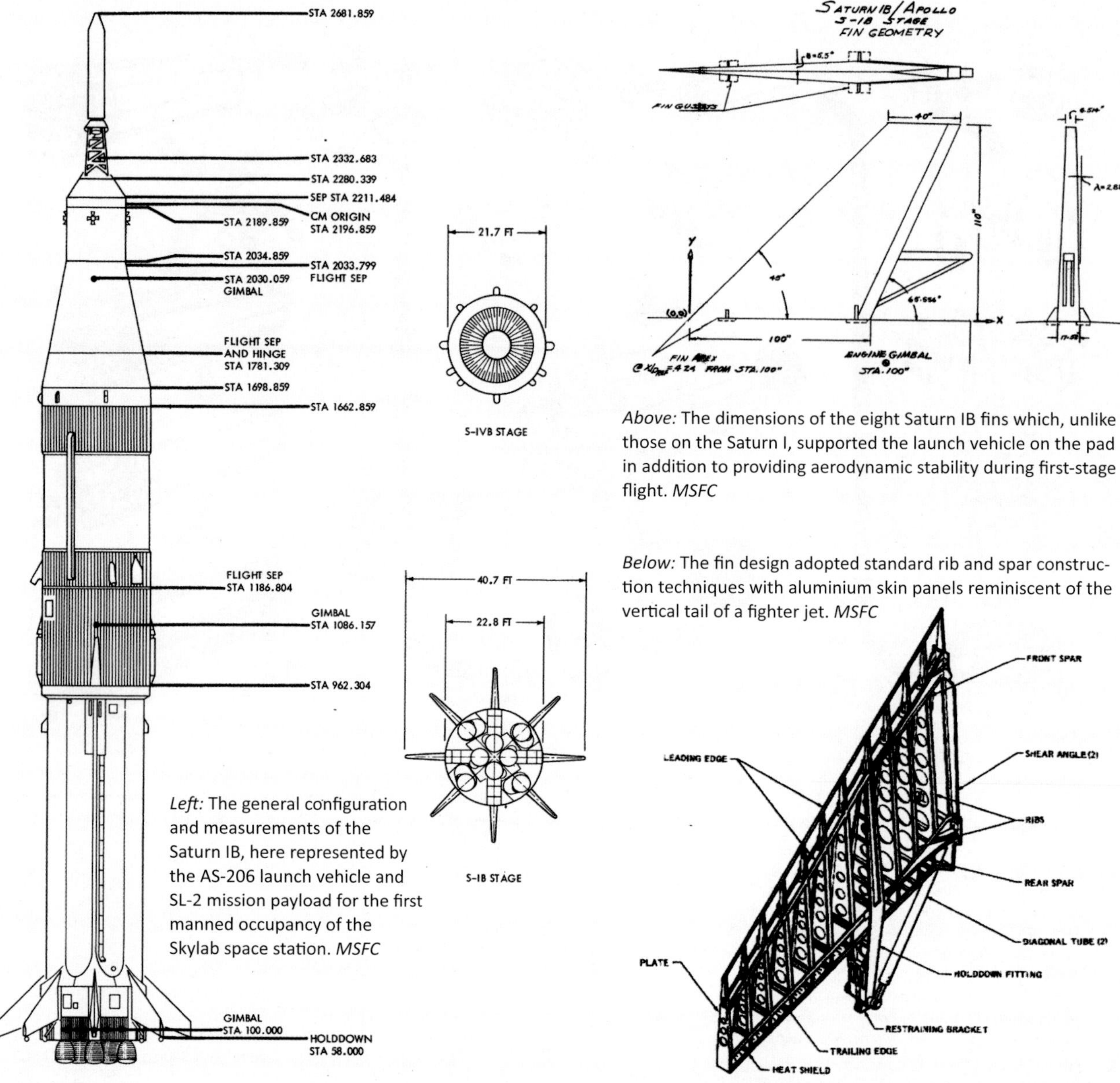

Above: The dimensions of the eight Saturn IB fins which, unlike those on the Saturn I, supported the launch vehicle on the pad in addition to providing aerodynamic stability during first-stage flight. *MSFC*

Below: The fin design adopted standard rib and spar construction techniques with aluminium skin panels reminiscent of the vertical tail of a fighter jet. *MSFC*

Left: The general configuration and measurements of the Saturn IB, here represented by the AS-206 launch vehicle and SL-2 mission payload for the first manned occupancy of the Skylab space station. *MSFC*

differed mainly in that they had no thrust support beam or actuator support beam. The outer ends of the outriggers were spanned by upper and lower ring segments and eight upper shroud panels to form the basic thrust structure. Eight smooth and eight corrugated lower shroud panels were attached to the aft end of the thrust structure to form a compartment for the eight H-1 engines. LOX and fuel bay firewall panels were installed to cover the space between the outrigger assemblies and the space over the aft end of the barrel assembly.

A reinforcing beam structure was fitted into the aft end of the lower engine shroud assembly. The heat shield panels, engine flame curtains and flame shield support installation were attached to the beam structure. Most honeycomb composites used in the vehicle utilised phenolic cores that were adhesively bonded to face sheets and limited by the upper and lower temperature constraints on the adhesive system. The heat shield honeycomb composite consisted of both corrosion-resistant steel foil cores and thin face sheets that were joined by a brazing process. The 0.63cm (0.25in) square-cell core was brazed to both the inner and outer face sheets. It had a layer thickness of 2.54cm (1.00in), and acted as the chief structural core member of the composite. The 1.27cm (0.50in) square-cell core was brazed to only the outer face sheet, had a layer thickness of 0.6cm (0:25in) and acted as the thermal insulation retaining member of the composite structure. M-31 insulation was trowelled into the retaining core cells.

Laboratory tests had generally demonstrated that, compared with adhesively bonded honeycomb composites, brazed honeycomb composites were over 100 per cent greater in tensile strength, over 75 per cent greater in core shear strength, over 20 per cent greater in edgewise compression strength and equal in flatwise compression strength. This heat shield design provided a lighter panel with increased stiffness which greatly improved the retention of the M-31 insulation material.

S-IB – tail section

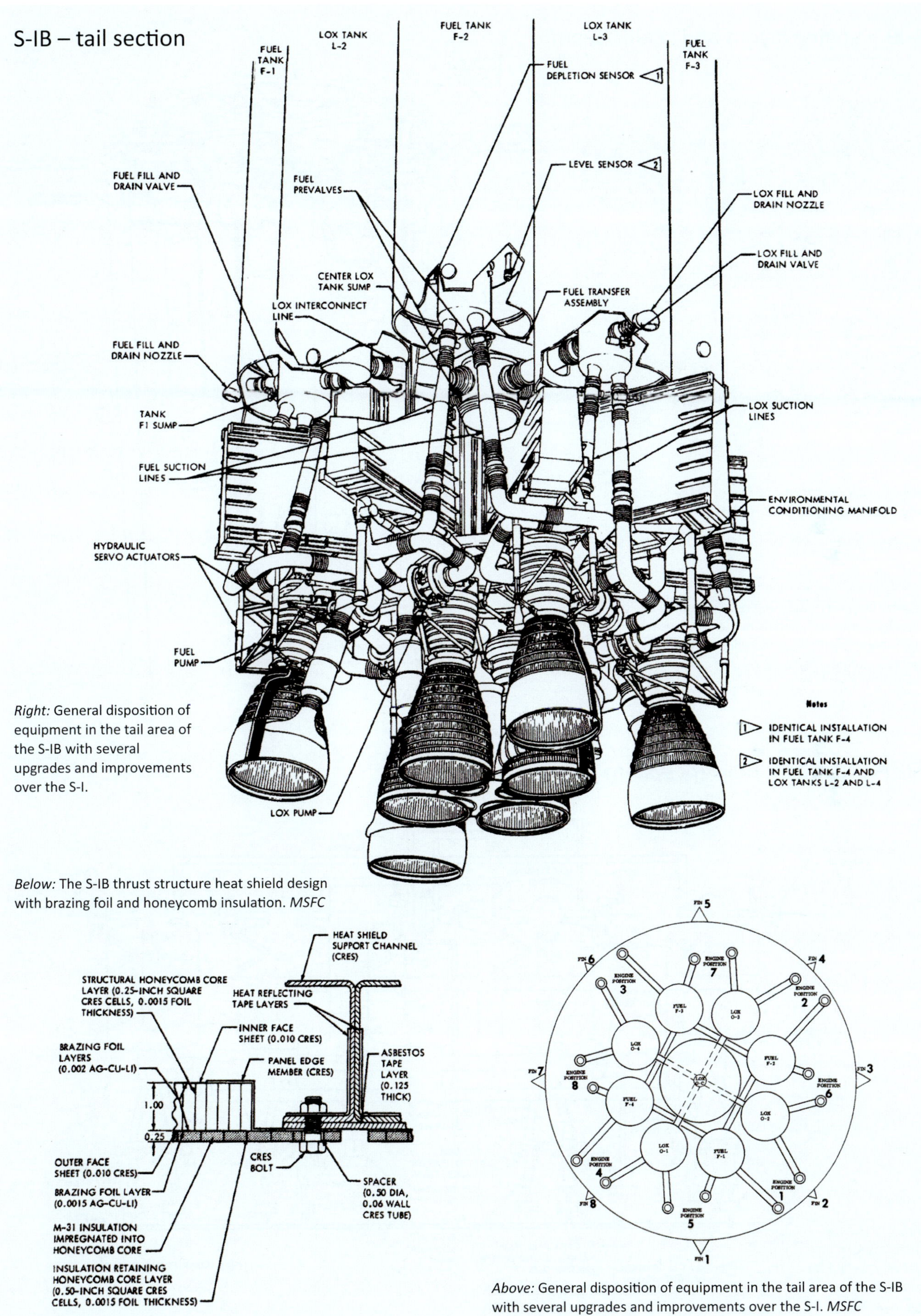

Right: General disposition of equipment in the tail area of the S-IB with several upgrades and improvements over the S-I.

Below: The S-IB thrust structure heat shield design with brazing foil and honeycomb insulation. *MSFC*

Above: General disposition of equipment in the tail area of the S-IB with several upgrades and improvements over the S-I. *MSFC*

S-IB – engine layout and spider beam

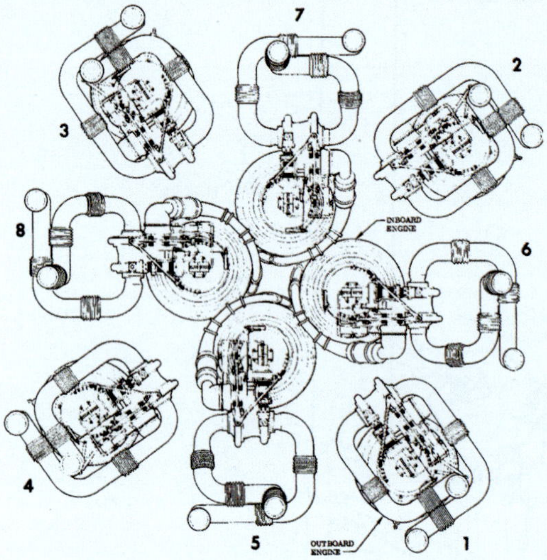

Above: The orientation and layout of the inboard and outboard engines. *MSFC*

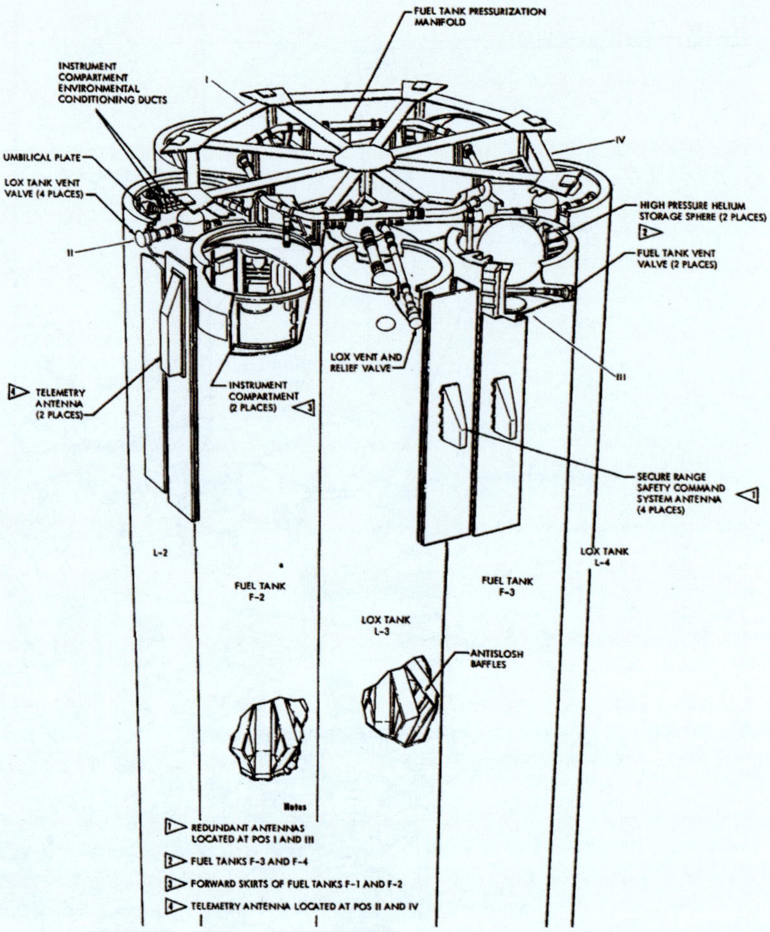

Right: The spider beam and pressurisation manifold were modified from the S-I stage. Note the relative positions of the telemetry and range safety antennas. *MSFC*

SIB – propellant delivery system

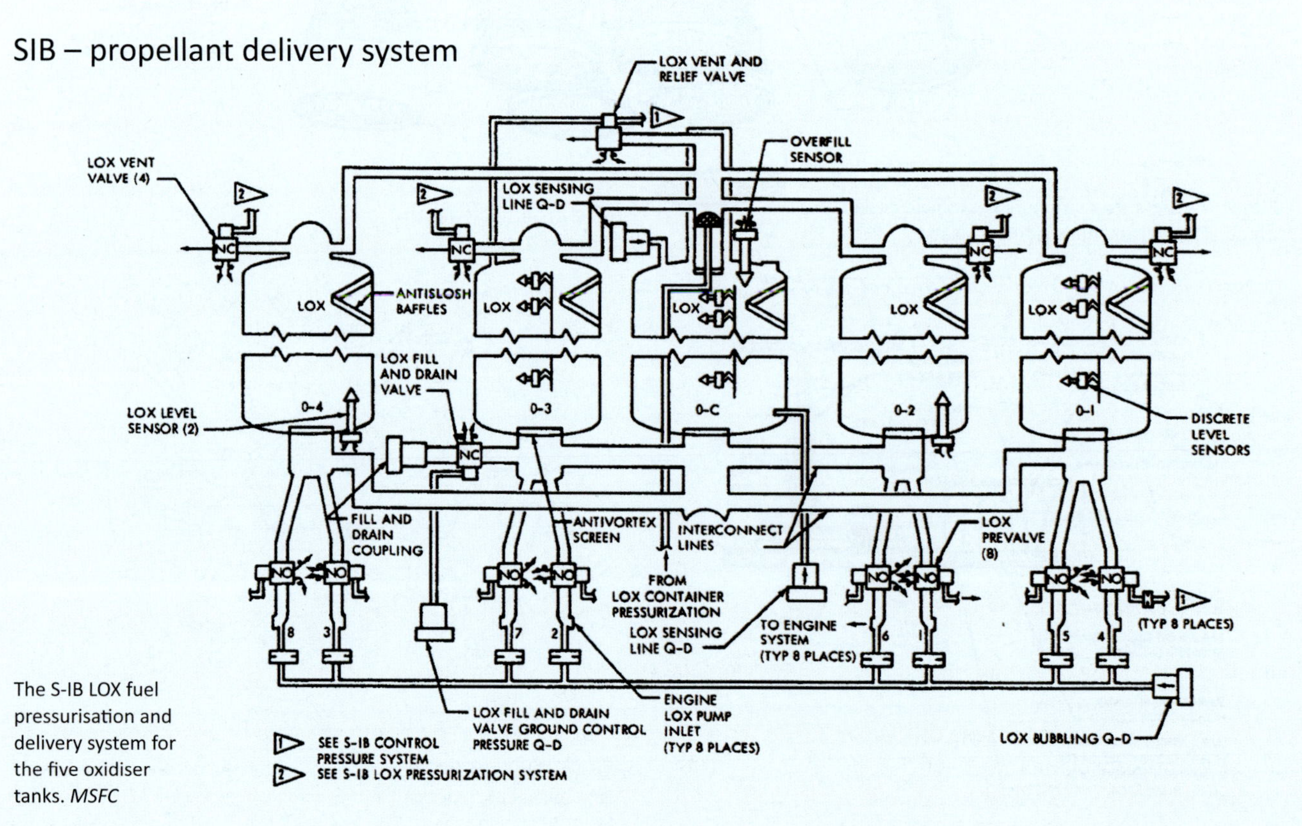

The S-IB LOX fuel pressurisation and delivery system for the five oxidiser tanks. *MSFC*

S-IB – propellant pressurization system

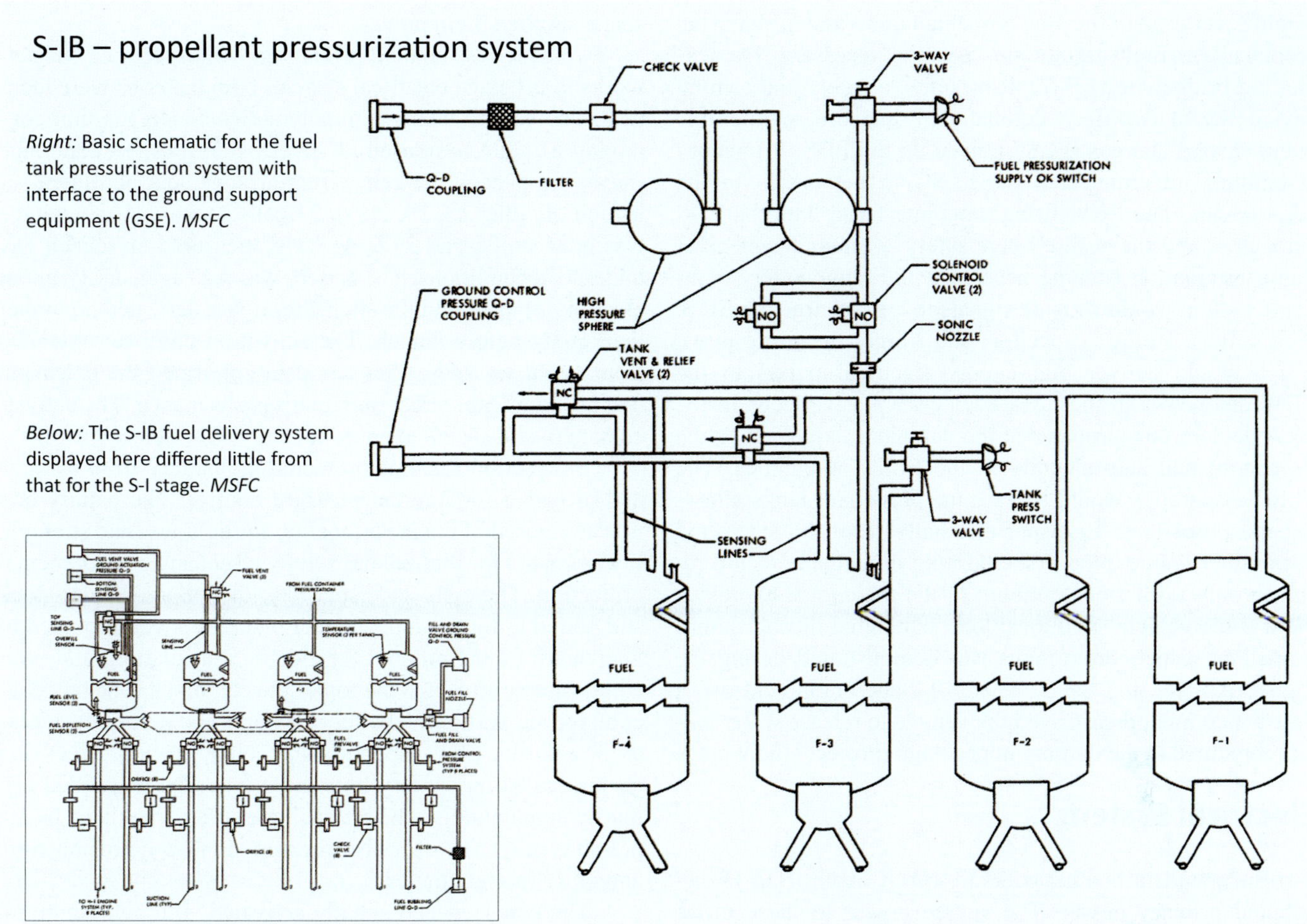

Right: Basic schematic for the fuel tank pressurisation system with interface to the ground support equipment (GSE). *MSFC*

Below: The S-IB fuel delivery system displayed here differed little from that for the S-I stage. *MSFC*

Successful results of laboratory testing and static tests of the S-I-10 and S-IB-3 to S-IB-7 stages fully qualified this heat shield panel design. In addition to the new configuration heat shield assembly, the tail unit incorporated new fin attachment fittings, gauge reduction of sheet-metal and framing and removal of the engine skirts from the lower shroud assembly.

Eight aerodynamic fins were attached to the tail section. There were only minor configuration differences between the S-IB stage for Saturn IB and the S-I stage for Saturn I. The fins were equally spaced around the periphery of the tail section. Each fin had an area of approximately 5m^2 (54ft^2). The leading and trailing edges were swept back 45deg and 25deg respectively. With the exception of the leading edge which was steel, the fins were constructed of aluminium alloy. The exterior of the fins was coated with an ablative insulation. The second stage adapter was similar to that for the S-I stage except for the deletion of the 45deg fairing and the cantilevered ends of the spider beam radial members. More specific payload and mission definition had resulted in less severe design loading conditions on the S-IB stage than on the S-I stage. The result was a lighter weight structure with the principal reductions being in the spider beam, propellant container skirts and thrust structure.

The spider beam unit held the propellant tank cluster together at the forward end and attached the S-IB stage to the S-IVB aft interstage. The five LOX tank units were rigidly attached to the spider beam while the four fuel tank shells were attached with sliding pin connections. Structurally, the spider beam consisted of a hub assembly, to which eight radial beams were joined with upper and lower splice plates by mechanical fasteners. The outer ends of the radial beams were spanned by crossbeams and joined with upper and lower splice plates by mechanical fasteners.

Like the tail unit thrust structure the spider beam was constructed of extrusions and fittings made of high strength, heat-treatable, aluminium alloys of the 7000 series, heat-treated to the T-6 condition. To form an aft closure for the S-IVB stage engine compartment, twenty-four honeycomb composite seal plate segments of approximately 1.27mm (0.05in) thickness were fastened to the forward side of the spider beam. The seal plate honeycomb composite consisted of 5052 aluminium alloy foil core material adhesively bonded to 7075-6 aluminium alloy face sheets to form thinner and lighter panels than those used on the S-I stages.

During qualification testing of the S-IB stage spider beam, a failure of the LOX tank fitting occurred. An accompanying illustration shows the radial beam reinforcing angle and bracket, crossbeam web stiffening brackets, and reinforced mounting stud flange incorporated to fix each of the eight LOX tank fittings. Other changes incorporated into the spider beam design for the S-IB stages were the reduction of beam gauges and the removal of retrorockets, 45deg fairing and radial beam tips.

The S-IB Stage Propellant Dispersion System (PDS)

would sever each of the nine propellant tanks and disperse the propellants if flight termination became necessary. Two exploding bridgewire (EBW) firing units, a safety and arming device (S&A), two EBW detonators, Primacord, and flexible linear-shaped charges (FLSC) made up the PDS. Primacord assemblies interconnected the FLSC to detonators in the S&A device. The EBW firing units interfaced with the PDS detonators and the secure range safety command system. If flight termination became necessary, the range safety command system would provide signals to arm (charge the EBW high voltage storage capacitor) and trigger the firing units, which would deliver high-energy electrical pulses to the EBW detonators in the S&A device. Explosive leads in the S&A device rotor propagated the detonator explosion to the Primacord and subsequently to the FLSC assemblies. The FLSC assemblies would rupture the propellant tanks allowing the propellants to disperse radially from the stage and burn rather than explode. The burning propellants would result in only a fractional amount of the theoretical explosive yield if the vehicle should explode. The reliability of Saturn propellant dispersion systems was demonstrated during the flights of SA-2 and SA-3. After S-I stage engine cut-off, a destruct command destroyed the vehicle to release water ballast contained in the dummy upper stage (Project Highwater).

Electrical System

Two independent bus networks (+ 1D1 1 and + 1D2 1) distributed primary power. The source power to these buses originated from ground-based electrical support equipment during prelaunch operations and transferred to stage batteries 50 seconds before lift-off. A separate battery supplied each bus network. Generally, one battery (D10) powered operational systems that drew high transient current and generated bus voltage transients. The other battery (D20) powered the measuring system and supplied parallel power for critical functions such as engine cut-off. This dual isolation feature ensured more than adequate electrical power. Primary power varied in the 27-30vdc range. Line loss was less than 2vdc from stage bus to stage load. A total of 13 cables interconnected the electrical system above and below the propellant tanks. Four cables were in the cableway on fuel tank F1; nine cables were in the cableway on fuel tank F2. The structure of the centre LOX tank provided a unipotential ground path (ID-COM). All electrical equipment incorporated design provisions in accordance with MIL-E-6051 (Electromagnetic Compatibility Requirements) to prevent electromagnetic interaction between electrical systems.

Stage electrical equipment operated independently of power sources from other stages. However, some buses extended into the stage to form a feedback loop to indicate certain events. For example, the S-IVB stage powered a bus (+ 4D1 1) that indicated physical separation of stages and consequently initiated J-2 engine start logic; the S-IU stage powered buses (+6D91, +6D92, and +6D93) that looped the H-1 engine thrust-OK pressure switches to the emergency detection system. There were other extra-stage buses, but none had drain on the S-IB stage electrical system since the requesting stage supplied them power.

Two batteries supplied primary power to separate bus networks in the stage electrical system; both batteries were identical. The battery was a manually activated storage unit containing 21 cells composed of silver oxide (positive) and zinc (negative) plates. The cell wiring arrangement permitted selection of either 18, 19, 20, or 21 cells so that battery voltage was between 28 and 29.6vdc when measured on launch day under load conditions. The battery was packaged dry (without electrolyte) to extend the shelf life; it was activated no longer than 168hr before launch. The activation time was extended from 120hr to 168hr after test data confirmed the extension had no significant effect on battery performance. The activated battery was installed approximately 57hr before lift-off.

Construction features protected the battery from damage and provided for internal pressure control. The battery box was composed of magnesium alloy for lightness and strength. It was coated for thermal emissivity effect and environmental protection. The cells were spaced with Neoprene shim-stock and potted to form a unitised cell block. A cover, moulded of Scotchcast (a 3M product for electrical insulation) and filler composition, protected the top of the cell block. The box cover added support to the battery box unit and permitted access to the cell block for inspection and cell activation. An O-ring gasket sealed the battery box and cover. The seal aided primarily in maintaining the internal pressure as regulated by the pressure relief valve; it also prevented entrance of moisture and other foreign matter.

The battery was manually activated with potassium hydroxide (KOH) electrolytic solution having a specific gravity of 1.40, +/-3 per cent at 25°C (77°F). Each cell was activated with 90cc (+/-1cc) of solution premeasured in individual containers. The battery chosen for flight use could not be older than 36 months. There were two methods for adding electrolyte to the battery: drip activation and vacuum activation. Drip activation required a rack that held 21 individual activator cases, each containing 90cm^3 (5.49in^3) of electrolyte. A drain needle, located centrally within each case, facilitated insertion into the respective cell filler hole.

The rack was positioned to allow gravity feed into the individual cells. Vacuum activation also required a rack with individual activator cases, but also included a vacuum pump as part of the assembly. Outlet filler tips, one for each cell, extended from the lower side of the activator case. Filler adapter screws threaded into each cell filler port and connected flexible tubing to transfer the electrolyte. The vacuum system consisted of a pump, gauge, regulator valve, and relief valve to evacuate and fill the battery cells.

After activation, each cell vent valve (part of the filler cap assembly) was installed. The activated battery was subjected to a 10-20amp load for 15 to 30 seconds to verify proper activation and performance; this test occurred 4hr after installing the cell vent valves. The installed vent valves allowed internal cell pressure to vent at 13-41kPa (2-6psi) but prevented leakage of electrolyte regardless of battery position. The activated battery was stored at 20-26.6°C (68-80°F) until ready for installation. Battery installation and the last case isolation test occurred no sooner than 12hr after installing the vent valves. Should the activated storage time exceed 168hr, the battery had to be replaced.

Electrical power assemblies and instrumentation equipment were located in the upper and lower skirts of fuel tanks F1, F2, F3 and F4. The upper portions of tanks F1 and F2 had extended skirts that housed most of the equipment. A bulkhead sealed this area above each tank and formed an air-tight compartment that was conditioned for increased reliability of electrical components. Compartment No.1 (F1) contained mostly electrical power equipment while compartment No. 2 (F2) contained mostly telemetry and tracking equipment. The lower portions of all tanks were open and housed propulsion and measuring equipment. Electrical cabling reached the upper and lower tank areas through two cableways, one on tank F1 and the other on tank F2.

S-IVB Stage

The S-IVB stage structure had a diameter of 660cm/6.6m (260in/21.7 feet) and a length of 1,800cm/18m (709in/59.1ft) with an average weight of 12,610kg (27,800lb). An aft interstage, an aft skirt, a thrust structure, two propellant containers and a forward skirt were structurally joined to make up the stage, all to an ultimate safety factor of 1.4. The aft and forward skirts were similar but had been modified because of lower design loads. The aft interstage was a completely different design.

Aluminium alloys (2014-0 heat treated to 2014-T6, 2014-T651, and 2014-T652) were used in fabricating the welded propellant tank. The skirts, thrust structure, and aft interstage assemblies were skin-stringer-ring frame fabrication. Aluminium alloys (7075-0 heat treated to 7075-T6 and A356) were used and joined by mechanical fasteners. The common bulkhead was a honeycomb composite structure constructed of welded spherical formed 2014-T6 aluminium alloy face assemblies, bonded to heat resistant phenolic honeycomb core material.

The loads from the first stage were transmitted to the S-IVB stage through the aft (S-IB/S-IVB) interstage. The aluminium-alloy interstage was cylinder with the same diameter as the main stage and a length of 570cm/5.7m (224.5in /18.7ft). External longitudinal hat section stringers carried the axial load and bending moment and the skin carried the shear load. The interstage skin and stringers were supported by an aft ring, seven internal intermediate rings and a forward ring. Mating surfaces for the first stage and aft skirt were provided by the aft and forward rings, respectively. The aft interstage was attached by a field splice to the first stage of the launch vehicle (at MSFC station 962).

The interstage aft ring, attached to the first stage at eight places on a 558.8cm/5.5m (220in/18.3ft) diameter bolt circle, transmitted concentrated loads to eight longerons, which sheared the load into the skin. The load was uniformly distributed to the forward ring by the stringers. Loads were transmitted to the aft skirt through the forward ring. Four retrorockets were mounted on the interstage aft of the separation plane. Attached to the aft end of the interstage was a skirt with a diameter of 660cm/6.6m (260in/21.7 feet) and a length of

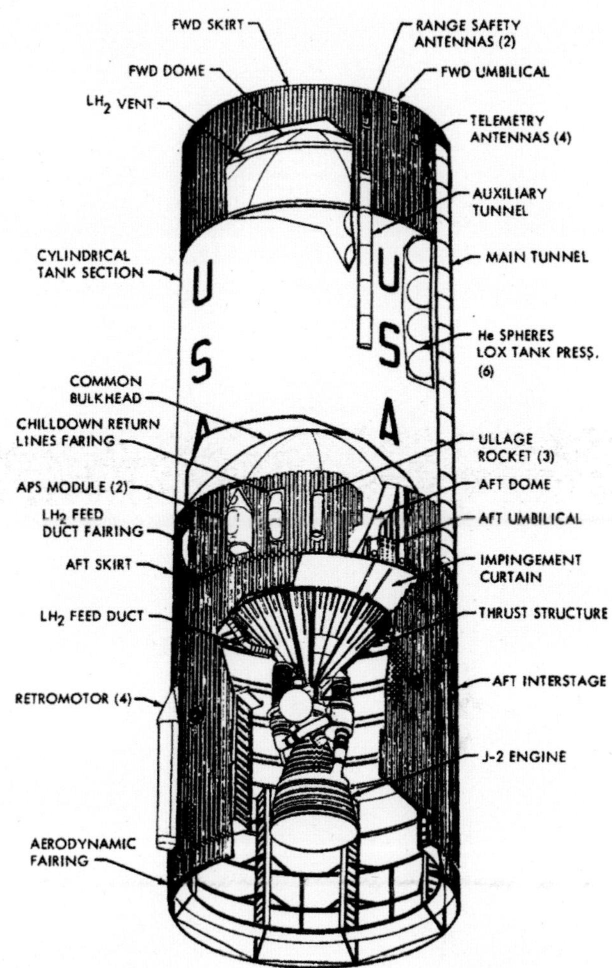

Preceded only by the S-IV and Centaur stages equipped with essentially the same engines, the S-IVB provided a test of the Rocketdyne J-2 cryogenic motor which would also equip the second and third stages of the Saturn V, albeit in modified form. MSFC

68.58cm (27in) which shrouded the S-IB stage spider beam.

The propellant tank structure consisted of a cylindrical section with a length of 680cm/6.8m (268in/22.33ft) constructed by butt-welding seven cylindrical section segments longitudinally. The segments were fabricated from 1.9cm (0.75in) plate, first mechanically milled to make a waffle pattern and then brake-formed to a 330cm (130in) radius, the pattern consisting of pockets 24cm (9.5in) on a centre, orientated +/-45deg with respect to the longitudinal axis. The pockets were surrounded by ribs 1.59cm (0.627in) high and 0.91cm (0.36in) wide with the skin at the bottom of the pocket 0.3cm (0.123in) and the weld land areas at the edges of these segments 0.64cm (0.252in) thick.

The forward dome was fabricated from nine pie-shaped gore segments welded together and to the jamb manhole at the apex, which afforded access to the LH_2 tank. The segments were fabricated from 0.38cm (0.15in) thick aluminium alloy sheet stock, formed to a 330cm (130in) radius and chemically milled to a skin thickness of 0.15cm (0.06in) with a thickness up to 0.29cm (0.118in) at the weld-land areas to maintain stress levels below the yield point of the material in

S-IVB – primary systems and constructional details

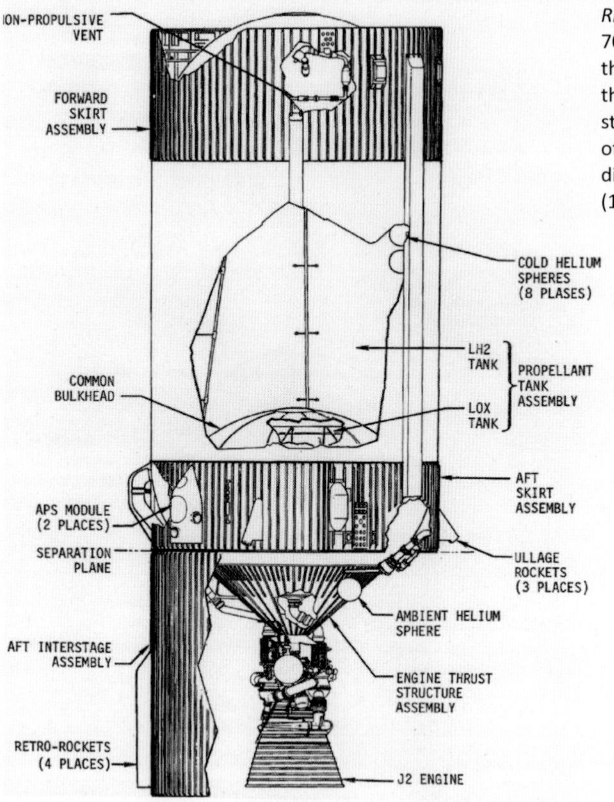

Right: Fabricated from 7075-T6 aluminium, the truncated cone of the S-IVB thrust structure had a height of 210cm (83in) and a diameter of 426.7cm (168in). MSFC

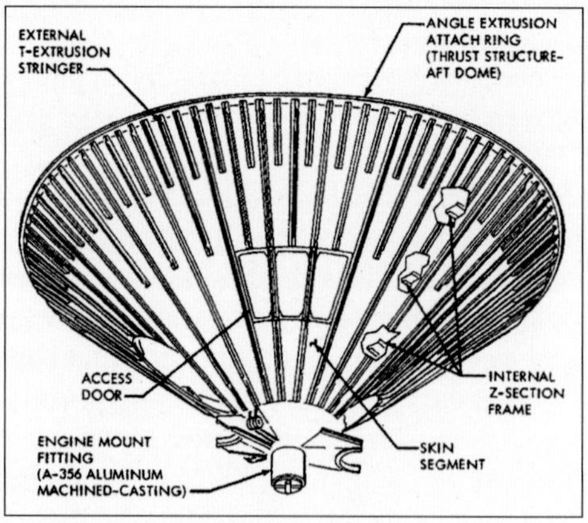

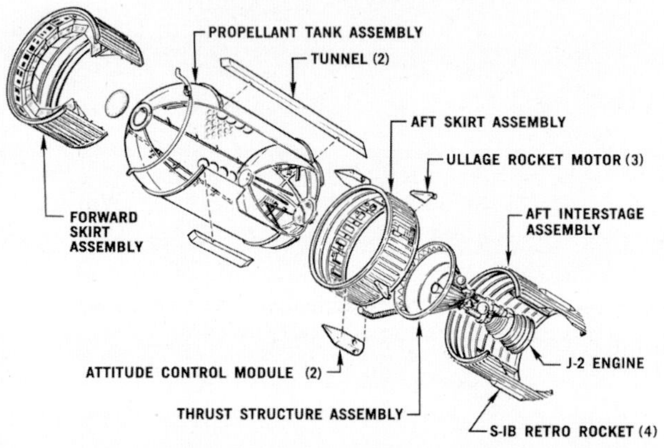

Above: Exploded view of the primary elements. *Douglas Aircraft Co.*

the welded state.

Like the forward dome, the aft dome was similarly fabricated from nine gore segments made from 0.71cm (0.28in) thick sheet stock and chemically milled after forming. The weld-land areas, including rings around the dome where the bulkhead and the common bulkhead and the thrust structure were attached, were milled to 0.485cm (0.191in). The liquid hydrogen portion of the dome was chemically milled into a waffle pattern with a web thickness of 0.21cm (0.082in) with a rib height a minimum of 0.63cm (0.245in). The 1.58cm (0.625in) wide ribs were located approximately 13.9cm (5.5in) on centre at the maximum and were orientated at 0deg and 90deg with respect to the vehicle's longitudinal axis. The web area was milled to a thickness of 0.218cm (0.086in) and 0.23cm (0.092in) for the areas forward and aft of the thrust structure joint, respectively. Access to the LOX tank was through the sump jamb in the bottom of the dome.

The forward and aft skirts and the aft interstage were, similar in configuration, cylinders constructed of 7075-T6 aluminium alloy skin, external stringers, and internal frames. Their diameter was 660cm (260in) but their lengths varied. The forward skirt had 108 stringers while the aft skirt and aft interstage each had 112 stringers. The forward skirt assembly extended forward 309.9cm (122in) from the forward skirt attaching ring of the propellant tank structure. The forward skirt provided an attachment plane for the Instrument Unit and mounting provisions for electronic equipment on thermo-conditioning panels. After vehicle assembly on the launch pad, the Instrument Unit access door provided entry to the forward skirt interior.

The aft skirt assembly extended aft 217cm (85.5in) from the aft skirt attaching ring of the propellant tank structure to the aft interstage. Ullage rockets and the Auxiliary Propulsion System (APS) were mounted on the aft skirt exterior, electrical equipment panels being mounted on the skirt interior. The aft interstage assembly was 570cm/5.7m (224.5in/18.7ft) in length and provided the interface between the S-IVB stage and the S-IB stage. The aft interstage had eight vertical internal reaction beams that translated the loads to the S-IB at 558.8cm/5.5m (220in/18.33ft) diameter structure points. Four retrorockets were mounted externally. The separation plane for the S-IVB stage was the aft end of the aft skirt.

The auxiliary propulsion system (APS) provided attitude control for the S-IVB stage/IU and payload during both powered and coast phases. During the powered phase, the APS provided only roll control; gimballing the J-2 engine provided yaw and pitch control. During the coast phase, the APS provided pitch, yaw, and roll control. It consisted of two aerodynamically shaped modules, each 203.2cm (80in) high, installed on the aft skirt 180deg apart at positions I and III on the stage. Each module contained three liquid-propellant rocket engines, a fuel and oxidiser storage and supply system, a high pressure 20,685kPa (3,000psi) helium storage and regulator system and control components for pre-flight servicing and inflight operation.

The APS used nitrogen tetroxide (N_2O_4) as oxidiser, and

monomethylhydrazine (MMH) as the fuel, a total load per module of 28.5kg (63lb) of hypergolic propellants, which did not require an ignition system. Attitude corrections were made by firing the 667N (150lb) thrust engines, individually or in any combination, in short bursts of approximately 65millisec minimum duration. Commands from the flight control computer through the attitude control relay modules actuated fuel and oxidiser solenoid-valve clusters that admitted propellant to the engine combustion chambers. Helium pressure exerted against stainless steel bellows assemblies, which contained the fuel and oxidiser, forced propellant into the engine.

The configuration of the thrust structure was similar to the skirt assemblies in that 7075-T6 aluminium alloy skin, external stringers, and internal frames were utilised. The structure measured 210.8cm (83in) high by 426.7cm/4.2m (168in/14ft) at the base. The J-2 engine was attached at the small diameter end of the thrust structure through the use of an A-356 aluminium casting. The forward end of the thrust structure was attached to the aft dome of the propellant tank assembly. Electro-mechanical and mechanical systems equipment mounts were located on the inside and outside surfaces of the thrust structure with two doors providing access to the interior of the thrust structure.

The longitudinal tunnels (main tunnel and auxiliary tunnel) housed wiring, pressurisation lines and shaped charges for the propellant dispersion system. The tunnel covers were made of 7075-T6 aluminium alloy skin stiffened by internal ribs. These structures did not transmit primary shell loads but acted only as a fairing reacting to local aerodynamic loading. The aerodynamic fairing on the aft end of the interstage was a short cylinder, 660cm/6.6m (260in/21.6ft) in diameter, made from eight 7075-T6 corrugated aluminium alloy skin panels. This structure did not transmit primary shell loads but had to maintain structural integrity when loaded by local aerodynamic pressure.

Another design requirement for the propellant tank and common bulkhead was that they provide insulation for the LH_2 tank and in the case of the common bulkhead that it prevent LOX freezing due to low temperatures during a ground hold. Generally, the insulating tile (a polyurethane foam and fibreglass composite) was bonded to the tank skin with an epoxy base adhesive system. In certain areas, balsa pads were used in conjunction with tiles to line the tank walls with insulating materials. An insulating gap filler material consisting of a glass fibre and polyurethane adhesive composite was used to fill gaps between tiles and around fasteners. Liners consisting of woven glass impregnated with polyurethane adhesive were bonded to the tile and balsa insulating material. In critical areas around brackets and joints the liner was covered by doublers of the same material.

A debris shroud surrounded the forward dome of the propellant tank, 101.9cm (40.125in) above the equatorial plane of the dome. Eight nylon cloth shroud segments 22.8cm (9in) wide by 226cm (89in) long were stitched together with Dacron threads and reinforced at their joints with eight nylon cloth splice segments, 10.1cm (4in) by 22.8cm (9in). The shroud segments were attached to the dome with Velcro tape adhesively bonded to the dome.

A retrorocket impingement curtain was used to shield, at stage separation, the area between the aft dome and aft skirt. The curtain installation spanned the area between the aft end of the aft skirt and the engine thrust structure junction with the aft dome. The basic material used in the construction of the curtain was glass cloth. Tape was used to seal openings in the curtain around boots, slots in aluminium and other openings.

S-IVB – layout and dimensions

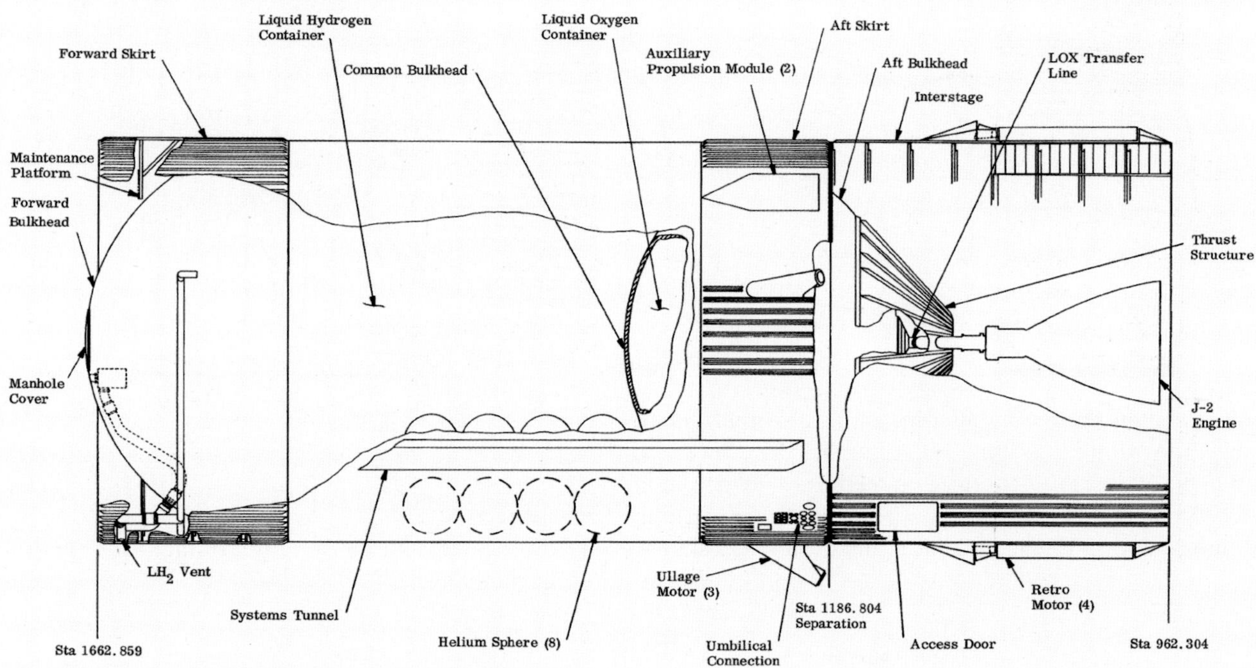

S-IVB – propellant tank

Right: Structural layout of the integrated S-IVB propellant tank with common bulkhead separating fuel and oxidiser sections. *MSFC*

Below: Crross section of the common bulkhead separating the two cryogenic compartments of the S-IVB propellant tank. *MSFC*

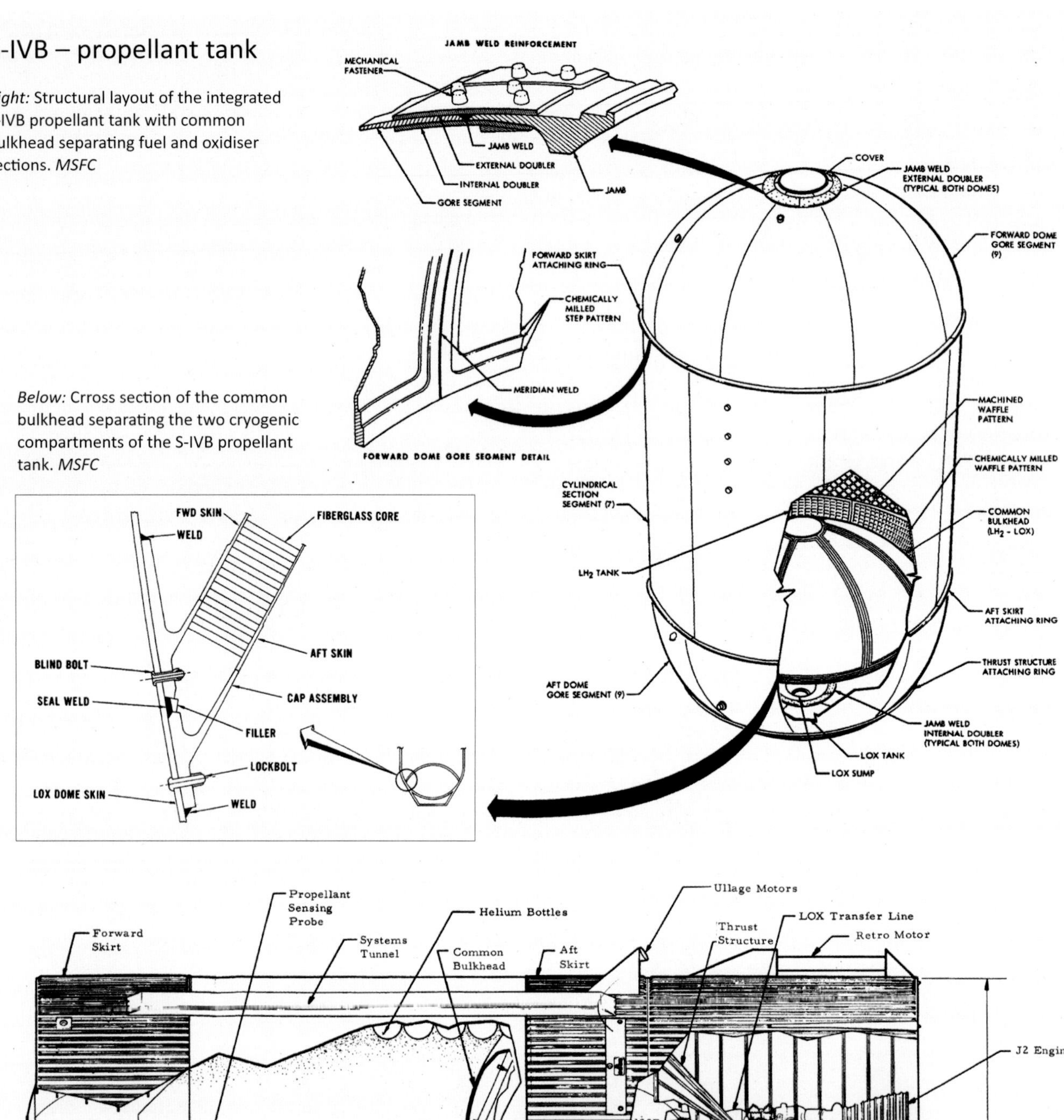

Above: Relative location of Auxiliary Propulsion System (APS), ullage rockets and retrorockets (on the Interstage). *MSFC*

S-IVB – cryogenic insulation

Above: A grainy photograph of insulation being applied to the interior of the S-IVB propellant tank. *MSFC*

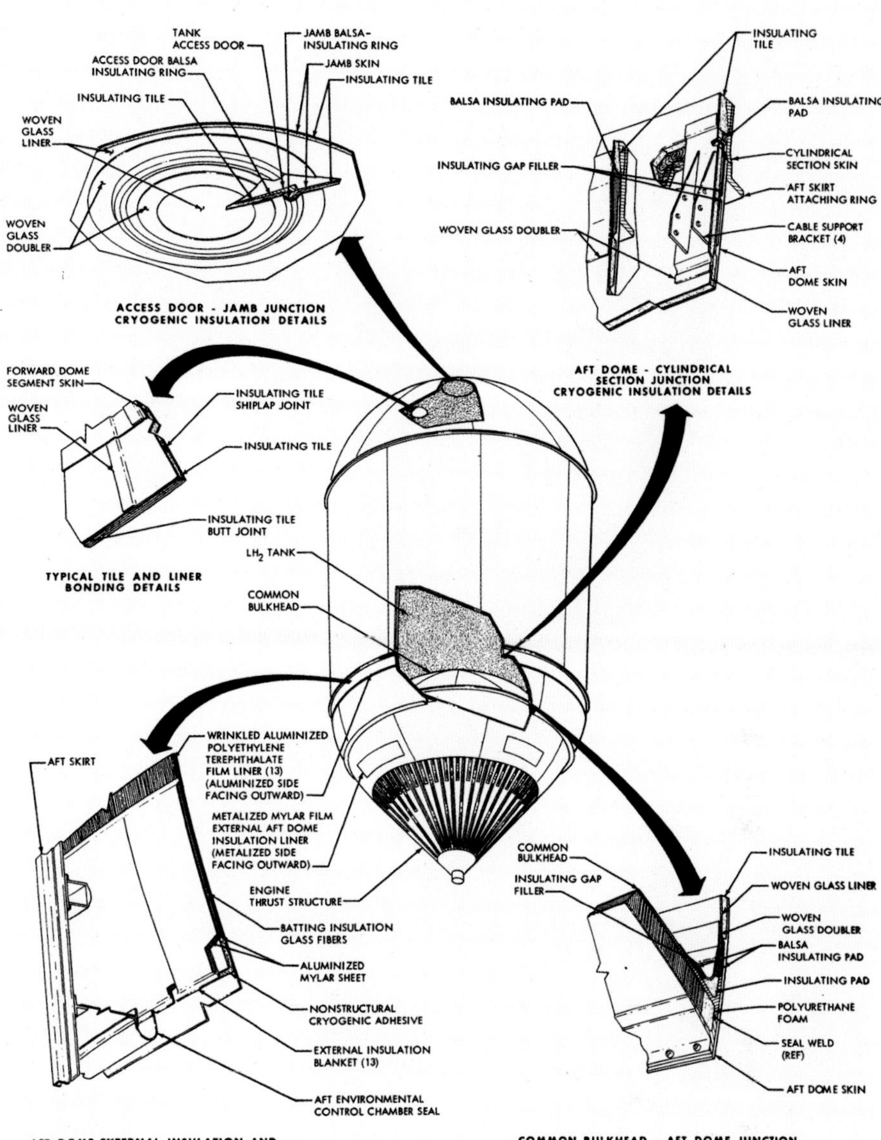

Right: Passive thermal insulation on the S-IVB with cross-sections showing various materials. Not shown are the ablative insulation panels located on the exterior to protect the stage during ascent. *MSFC*

The LH_2 was stored in an insulated tank with a maximum capacity of approximately 20,095kg (44,300lb) mass at a temperature of -252.7ºC (-423ºF). An ullage volume of approximately 1.2m (4ft) was maintained at this load level; however, for Saturn IB Earth orbit missions a full load of LH_2 was not carried. The approximate LH_2 load for Earth-orbit missions was 17,240kg (38,000lb), which included quantities required for programmed mixture ratio operation (5.5:1 for approximately 325 sec) and unusable propellants.

The LH_2 tank was pressurised to between 193kPa (28psi) and 213.7kPa (31psi). Pressure was provided by a ground supply of helium for pre-pressurisation and was maintained by a hydrogen bleed from the engine during the burn phase. Tank venting and relief was accomplished through parallel valves installed in a top-mounted vent system which exited through non-propulsive vent outlets in the forward skirt. An anti-vortex screen was installed over the engine feed duct inlet.

LOX was stored in a tank formed by the aft dome and common bulkhead with a capacity of approximately 88,000kg (194,000lb) liquid oxygen. The LOX tank was loaded to capacity for Saturn IB missions. The tank ullage pressure was maintained between 255kPa (37psi) and 281kPa (40.8psi) during boost and engine operation using gaseous helium. Two parallel valves provided venting and overpressure relief through the aft skirt. An anti-vortex screen was installed over the engine feed duct inlet.

The Saturn IB structure for the Instrument Unit was similar to that of the Saturn V, the major difference being the location of cut-outs in the sandwich panels. The Instrument Unit was attached to the S-IVB stage and payload in field splices located at stations 1663 and 1699, respectively. The first four Instrument Units were the responsibility of MSFC with IBM doing the assembly and testing at its Huntsville facility. The first IU produced entirely by IBM flew on AS-205.

Four solid propellant retromotors, mounted at 90deg intervals around the aft interstage assembly decelerated the S-IB stage and aft interstage assembly during the stage separation sequence.

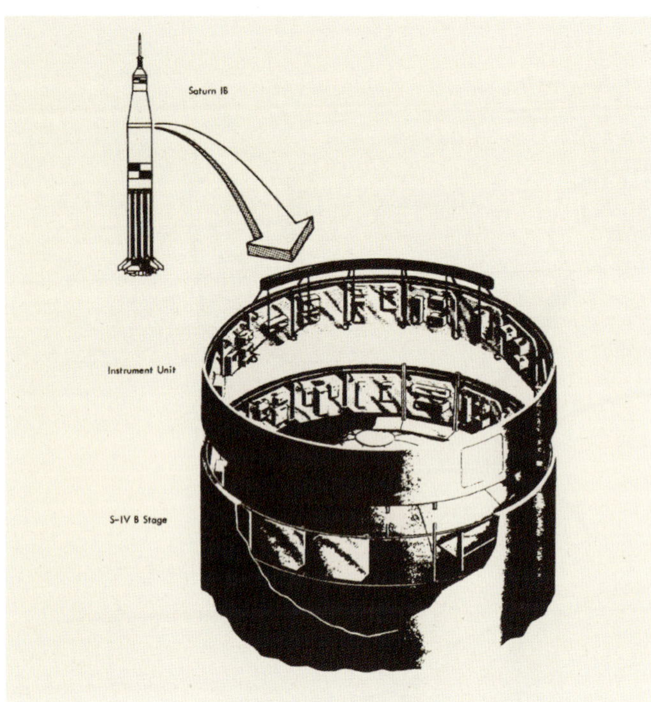

Above: The Saturn IB Instrument Unit (*above*) and (right) the location of systems and subsystems by reference to their orientation within the IU with regard to the assembled stack. *MSFC*

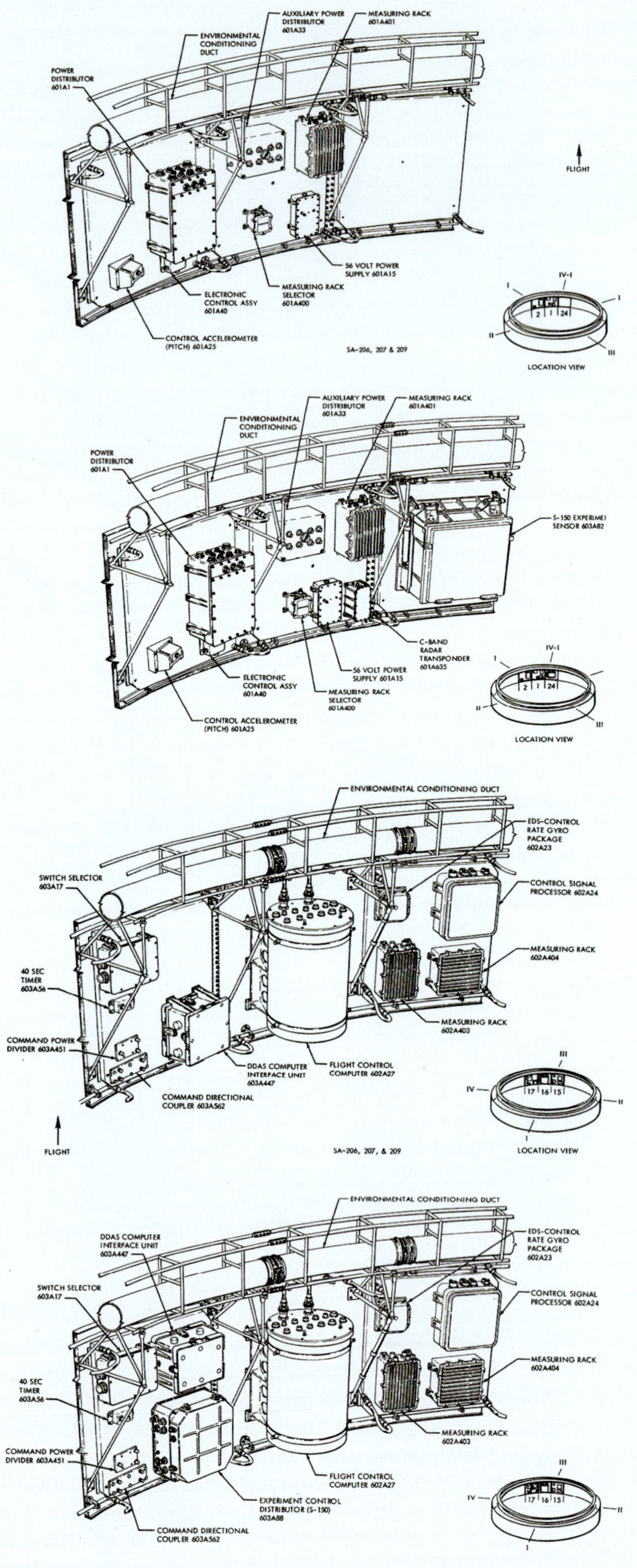

Nose, centre, and aft aerodynamic fairings enshrouded each retromotor. The aft and centre fairings were permanently installed; however, the nose fairing jettisoned when the retromotor fired to expose the motor nozzle. Exhaust gases, acting against the internal surface of the fairing, sheared a retaining pin that secured the forward end of the fairing to the aft interstage assembly. The fairing then assumed an aftward rotation, pivoting about a hinge and hook arrangement. After an approximate 70deg rotation, the fairing separated from a hook on the centre fairing and fell away from the vehicle.

Each retromotor nozzle was canted 9.5deg outboard from the motor centreline to direct the exhaust plume away from the S-IVB stage. With an average burn time of 1.52sec, each retromotor developed a thrust of 163.3kN (36,720lb). Thiokol Chemical Corporation, Elkton, Maryland, manufactured the E17029-02 Recruit motors used as the S-IB stage/aft interstage assembly retromotors. Each motor weighed 170.5kg (376lb) including 121kg (267lb) of propellant having a tapered, internal-burning, 5-point-star configuration. Independent ignition systems, consisting of two electrical bridge wire (EBW) firing units and two EBW initiators, ignited the retromotors. An igniter, which was a part of each retromotor, had two threaded receptacles for initiator installation. A pair of EBW firing units were mounted inside the aft interstage assembly at each retromotor position. An electrical cable, which was an integral part of each firing unit, connected the firing unit to its respective initiator. At separation command, each firing unit delivered a 2300vdc pulse to its initiator. The initiators detonated and fired the igniter, which directed hot particles and gases to the solid propellant surface, thereby igniting the retromotor.

Three solid-propellant ullage rockets, mounted approxi-

S-IB – Instrument Unit (IU) segments

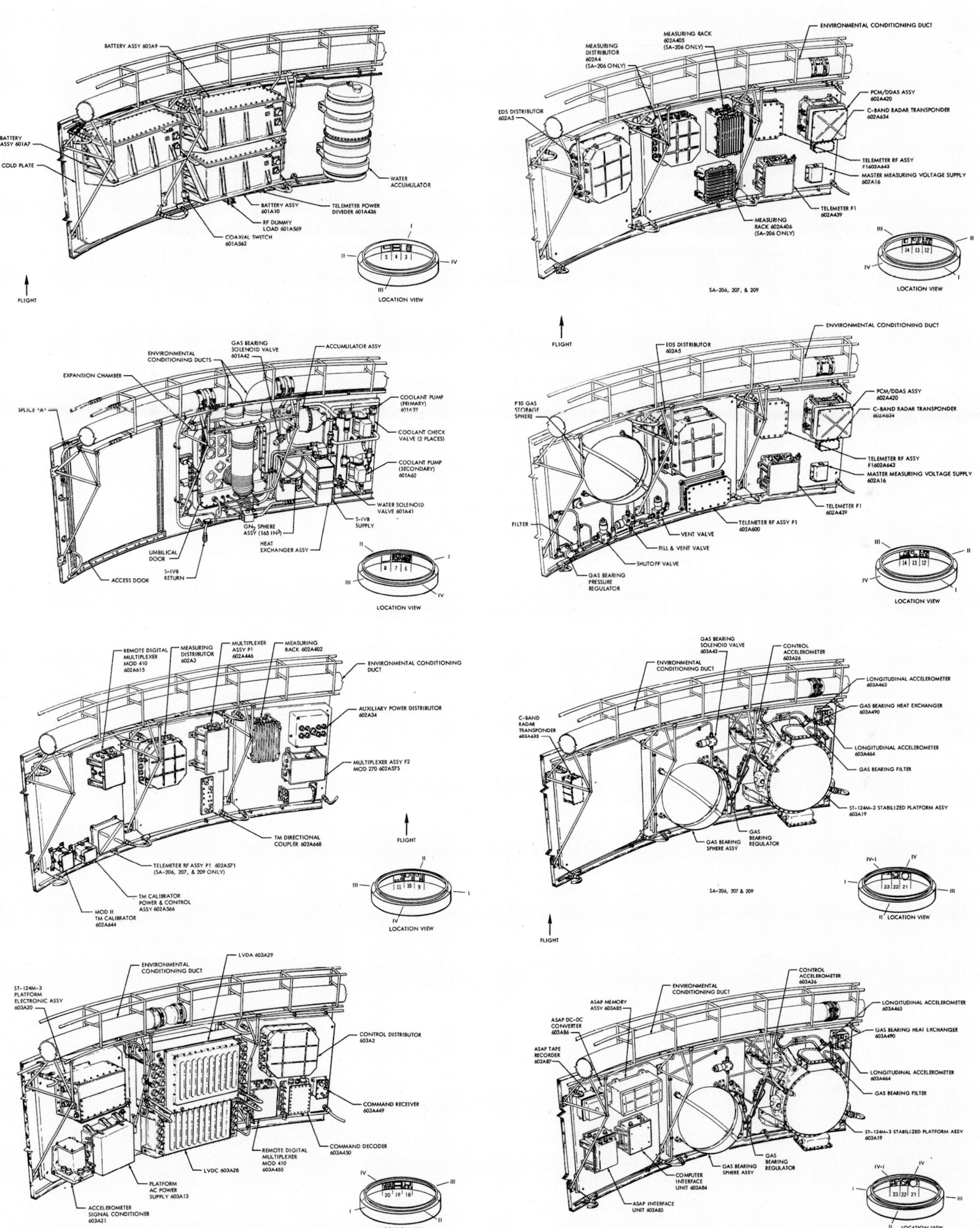

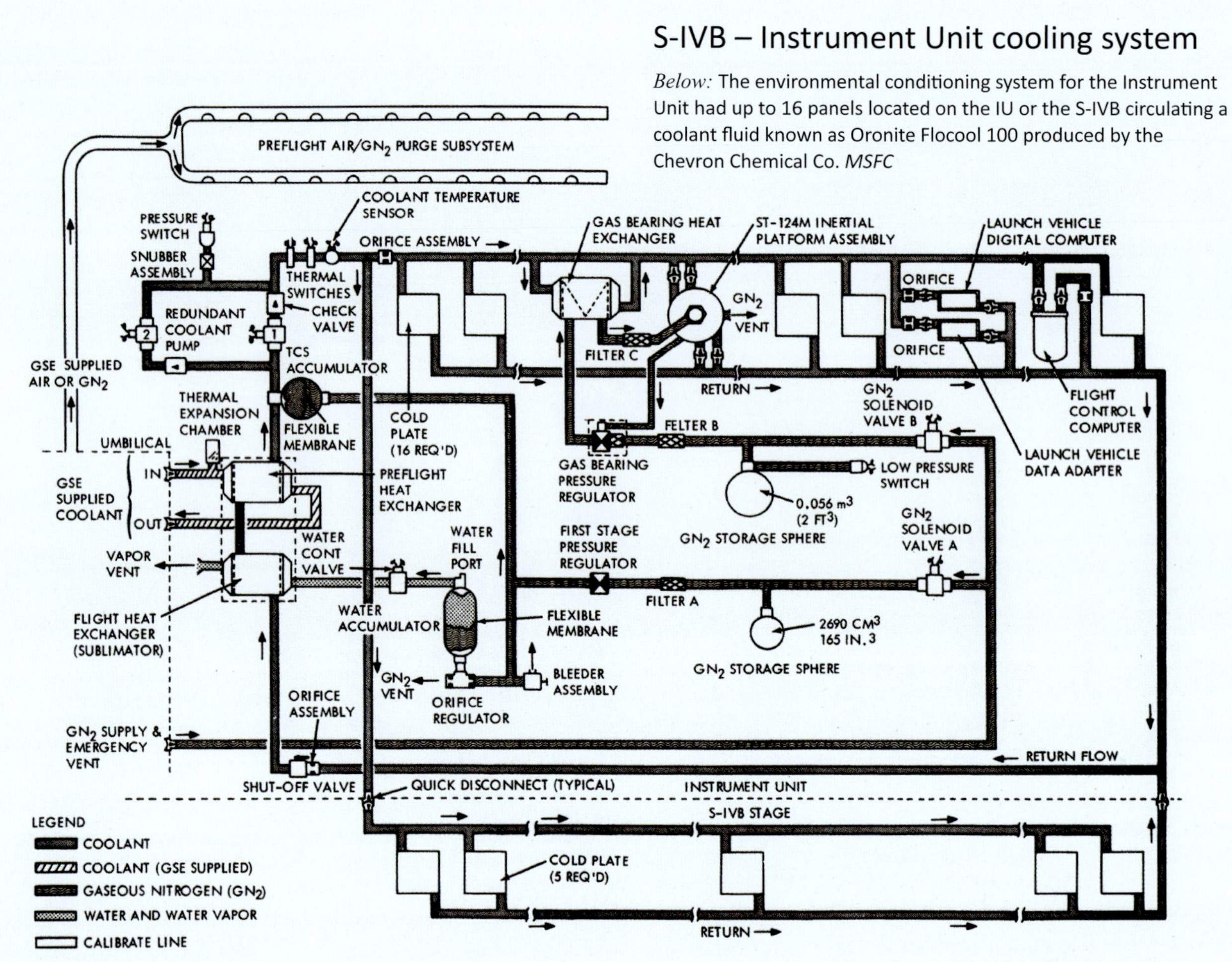

S-IVB – Instrument Unit cooling system

Below: The environmental conditioning system for the Instrument Unit had up to 16 panels located on the IU or the S-IVB circulating a coolant fluid known as Oronite Flocool 100 produced by the Chevron Chemical Co. *MSFC*

mately 120deg apart of the S-IVB stage aft skirt induced a slight forward acceleration to the S-IVB stage and payload during stage separation. The motors begin firing 0.1sec before separation and terminated shortly after J-2 engine ignition. By providing continuous acceleration to the S-IVB stage during stage separation system operations, the S-IVB stage propellant remained properly seated in the bottom of the tanks for J-2 engine start. Aerodynamic fairings housed the ullage rockets and provided for attachment to the aft skirt. The fairing assemblies canted each ullage rocket centreline 35deg outward from the vehicle longitudinal axis, thus directing the exhaust gases away from the S-IVB stage. Each fairing assembly also housed a transducer for chamber pressure measurement and two EBW firing units for ullage rocket ignition.

Thiokol Chemical Corporation manufactured the TX-280-10 rocket motors, which were 21.122cm (8.316in) in diameter and a maximum of 93.98cm (37in) in length. The internal-burning, 5-point-star-configuration solid propellant developed a thrust of 15.39kN (3,460lb) and a firing time of 3.9sec. An igniter, which was part of the rocket motor and was installed in the forward end of the motor casing, had two receptacles for EBW initiators. Two initiators, the Thiokol TX-346-1 and the Aerojet-General AGX 2008, had been approved for use with the ullage rockets. The output cables from the EBW firing units in each fairing assembly were attached to the respective ullage rocket initiators.

The S-IVB stage switch selector issued the ullage rocket ignition command 0.1sec before stage separation. The command simultaneously triggered the two EBW firing units at each ullage rocket position. Use of redundant EBW firing units and EBW initiators ensured ullage rocket ignition. A 2,300vdc pulse from the EBW firing units fired the initiators installed in the motor igniter. Thirteen ullage rocket motors were tested in motor qualification and all were successful. Eleven were fired utilising a dual ignition system and two were fired by a single ignition system. To demonstrate EBW initiator/igniter compatibility, six dual ignition tests and four single ignition tests were conducted successfully.

Approximately 15sec after stage separation, the ullage rockets and their fairings were jettisoned to reduce stage weight. To accomplish this operation, the jettison system used two EBW firing units, two EBW detonators, a detonator block, two confined detonating fuse (CDF) assemblies, six frangible nuts and three spring-loaded jettison assemblies. The S-IVB stage switch selector issued the ullage rocket jettison command, which triggered the EBW firing units. The 2,300vdc output pulses from the firing units detonated the EBW detonators and the explosion propagated from the detonators through the CDF assembly, simultaneously breaking all six frangible nuts. The spring-loaded jettison assemblies then propelled the ullage rockets away from the S-IVB stage. The EBW firing units, EBW detonators, and detonator block were mounted at

S-IVB – propellant system

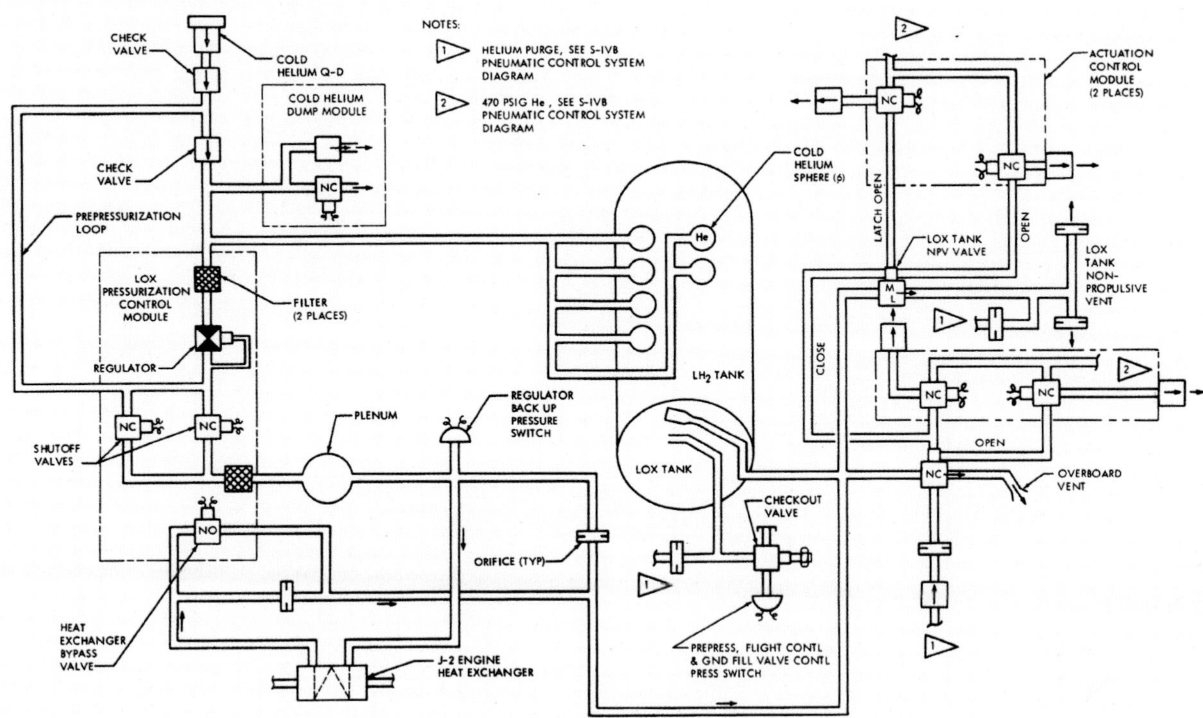

Above: Schematic of the S-IVB LOX tank pressurisation and vent system together with the interfaces for ground support equipment. *MSFC*

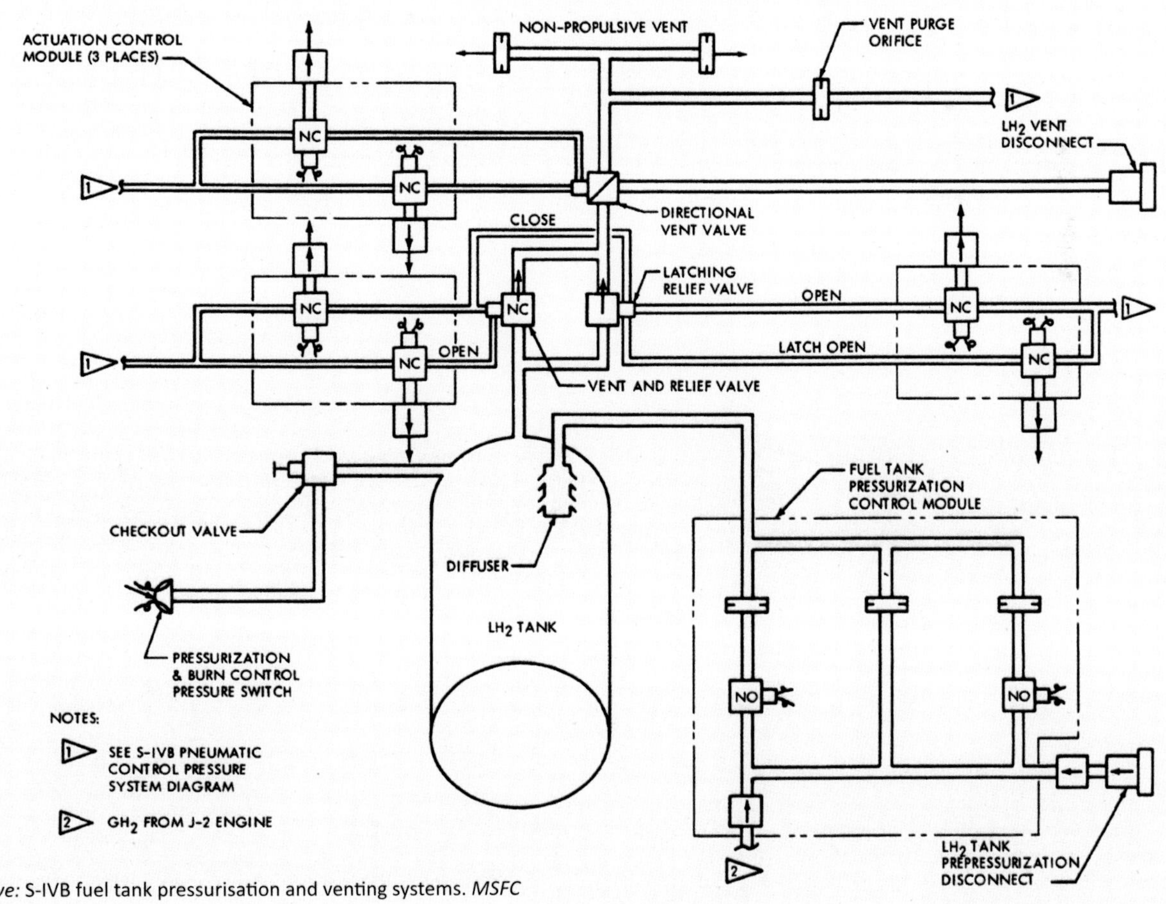

Above: S-IVB fuel tank pressurisation and venting systems. *MSFC*

S-IVB – auxiliary propulsion systems (APS)

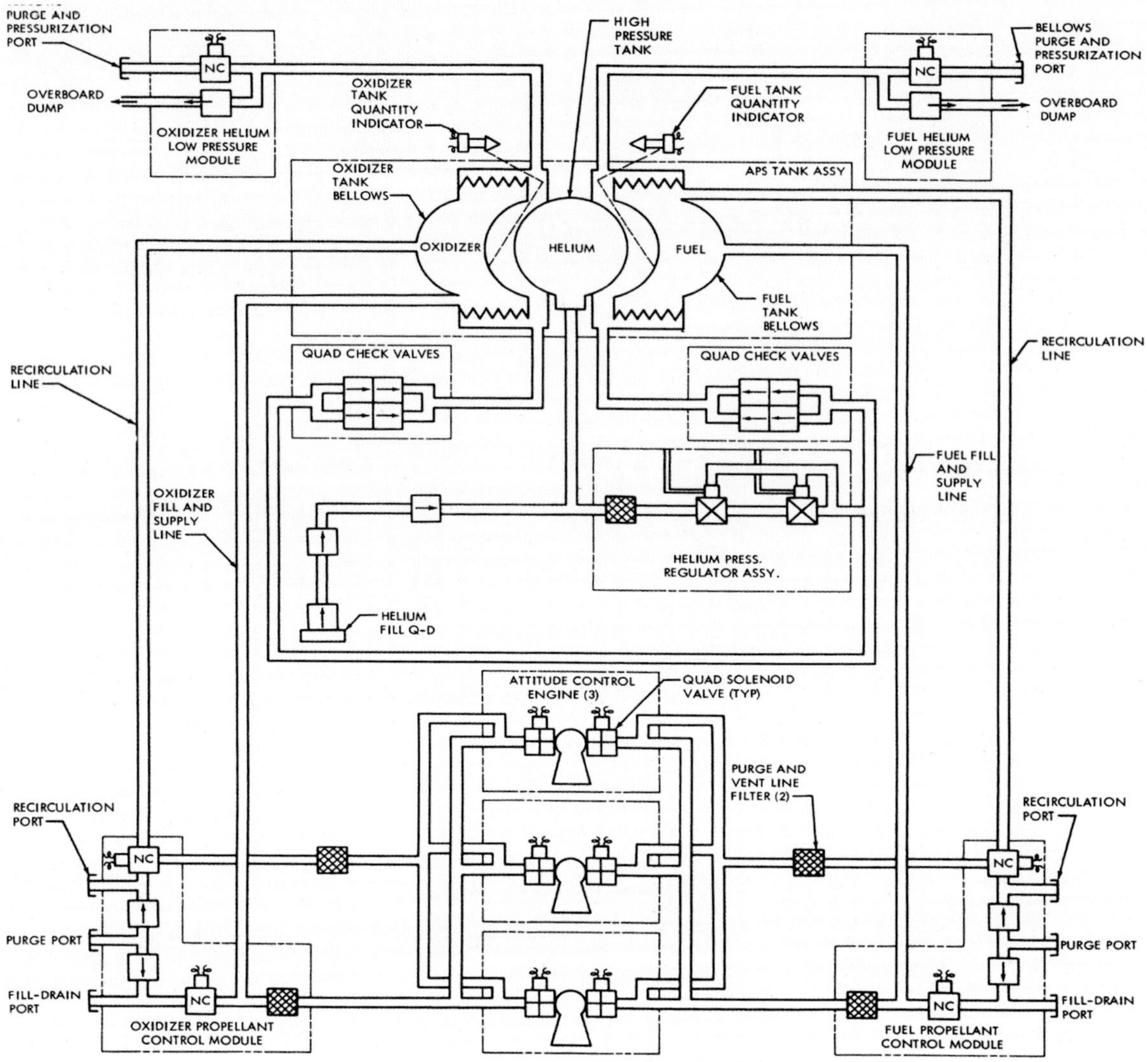

Above: The APS operated on hypergolic propellants delivered by a helium pressurisation system from a spherical tank sandwiched between the monomethyl hydrazine fuel and the nitrogen tetroxide oxidiser vessel. *MSFC*

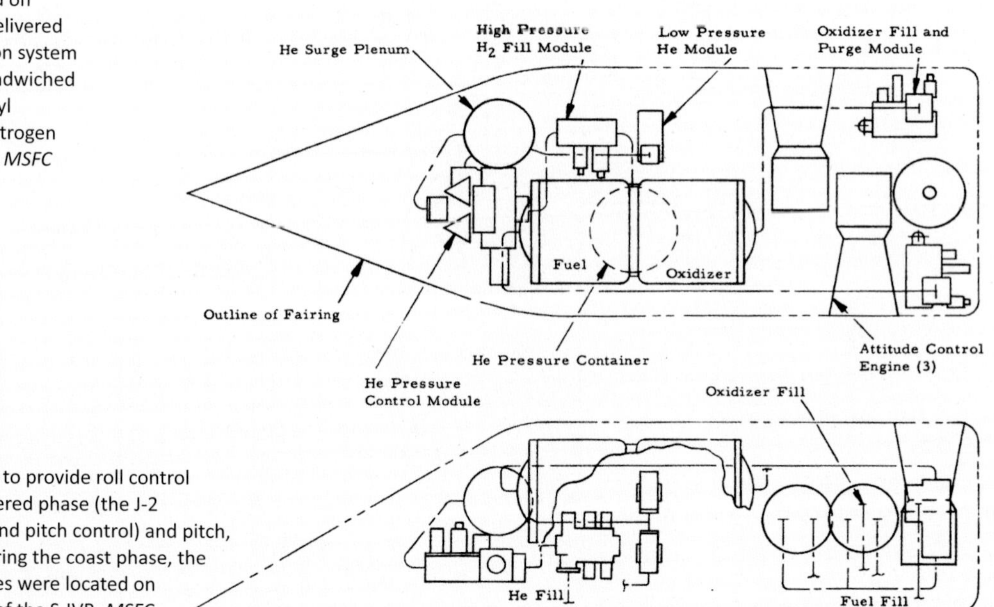

Right: Designed to provide roll control during the powered phase (the J-2 providing yaw and pitch control) and pitch, roll and yaw during the coast phase, the two APS modules were located on opposing sides of the S-IVB. *MSFC*

S-IVB – ullage and retrorockets

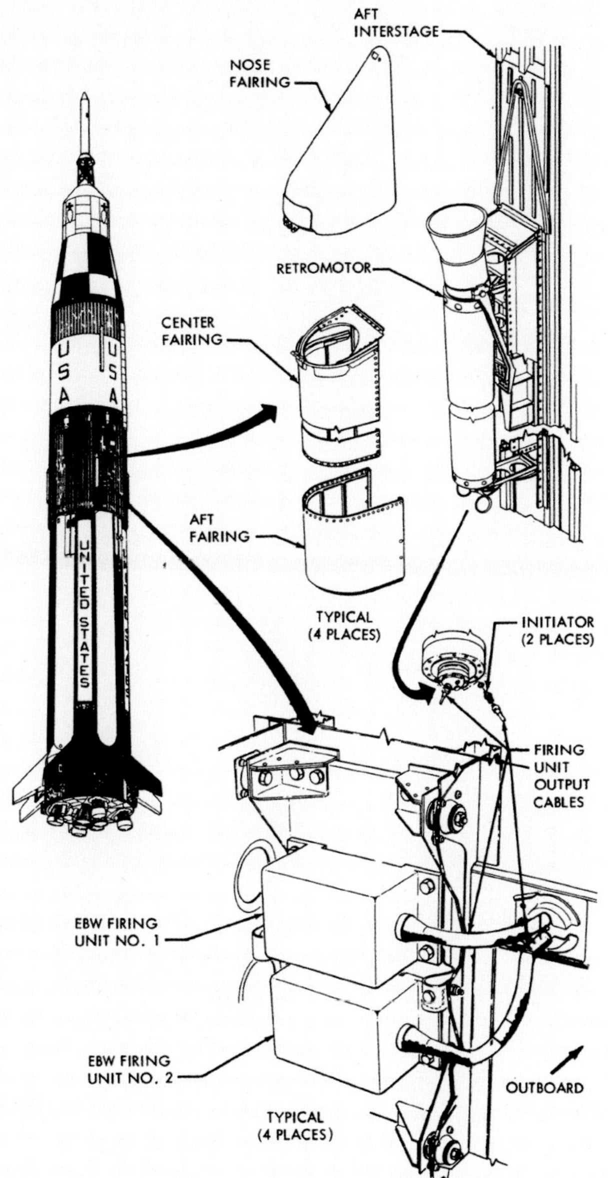

Left: The retrorocket and electrical bridge-wire (EBW) stage separation elements were attached to the Interstage and the exhaust nozzle was protected by a blow-out fairing as shown on this schematic. *MSFC*

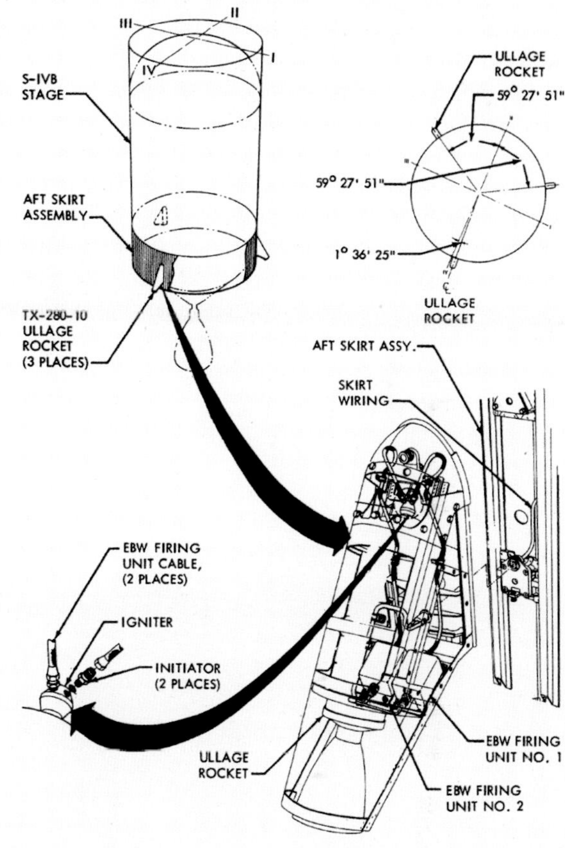

Above: Location and disposition of the three solid propellant S-IVB ullage motors attached to the aft skirt assembly. *MSFC*

panel position 18 in the aft skirt assembly. A tray with quick-release clamps, installed on the aft skirt's inner periphery just forward of the separation plane, secured the CDF assembly.

J-2 Engineering

The J-2 engine was the second major propulsion system developed using liquid hydrogen as the fuel. Much of the experience and technology gained on the RL-10 programme was directly applicable to the J-2 engine. Since the J-2 was to be used on a manned vehicle, considerable attention had to be given to achieving high reliability through extensive component and engine systems ground testing to resolve any potential problems before flight testing. In the design of the J-2, attention had been focused on potential failure modes and inherent design characteristics which could prevent these failures. Welded joints were used throughout the engine to prevent leaks. Dual seals, with intermediate bleeds, were used at all hot gas and propellant separable connections.

The engine envelope was 203cm (80in) in diameter and 294.6cm (116in) long. Nominal thrust, vacuum specific impulse and weight were 889.6kN (200,000lb), 418 seconds and 993.4kg (2,190lb), respectively for SA-201, -202 and -203. Specific impulse for SA-204 to SA-207 was 419sec. The basic design of the engine was based around the thrust chamber, which served as a mount for the injector and dome assembly, the gimbal assembly and the augmented spark igniter (ASI). The engine featured a single-tubular wall, bell-shaped thrust chamber, and independently driven, direct drive turbopumps for LOX and LH_2. Each turbopump utilised the same propellants as the main thrust chamber.

The thrust-chamber body consisted of a cylindrical section,

a narrowing throat section and an expansion section. The body was constructed of longitudinal brazed stainless steel tubes. An intake manifold routed fuel through the tubing, cooling the thrust chamber and converting the fuel to a gaseous state before injection into the combustion chamber. The fuel turbopump was an axial flow pump consisting of seven stages in addition to an inducer. It was a direct turbine drive, operating in the 25,000rpm range, self-lubricated and driven by the exhaust gases from the gas generator. The turbine shaft turned the inducer, forcing LH_2 through a series of seven stages. The oxidiser turbopump was a single stage, centrifugal pump operating at 6,000rpm, self-lubricated and self-cooled with direct turbine drive.

A gas generator supplied the hot gases that drove the turbines and consisted of a combustor, an injector, oxidiser and fuel poppets and two spark igniters. The gas generator supplied sufficient energy to operate the fuel and oxidiser turbopumps which together required 6,338kW (8,500hp). An electrically operated, motor driven, propellant utilisation valve provided for simultaneous depletion of the propellants. During engine operation, propellant level sensing devices in the propellant containers controlled the position of the valve. Oxidiser modulation was accomplished by bypassing LOX back into the pump inlet.

The concentric-orificed, porous-faced thrust chamber injector atomised and mixed propellants to produce the most efficient combustion. Oxidiser ports were machined to form part of the injector and threaded fuel nozzles were installed over the oxidiser ports to form the concentric orifices. The injector face, formed from a sintered metallic material, was welded at its outside and inside edges to the injector body. Each fuel nozzle was swaged to the injector face. An oxidiser inlet elbow, integral with the dome and injector assembly, admitted LOX from the turbopump and injected it through the oxidiser ports into the thrust chamber combustion area. Fuel entered the injector from the upper fuel manifold and flowed through orifices concentric with the oxidiser orifices. Approximately 3-4 per cent of the fuel flowed through the sintered injector face to cool the injector. Combustion zone pressure acted upon the injector face area producing the thrust force, which was transmitted through the gimbal to the vehicle structure.

Below: J-2 engine on test at the Marshall Space Flight Centre. *MSFC*

The tubular-walled, bell-shaped thrust chamber consisted of a cylindrical section where combustion occurred, a narrowing throat section, and an expansion section. The thrust chamber body was constructed of longitudinal stainless steel tubes brazed together with bands around the tubes for external stiffening. A fuel inlet manifold on the engine bell admitted fuel to 180 down tubes which carried the fuel to the return manifold at the base of the thrust chamber. From the return manifold, 360 tubes carried fuel to the thrust chamber injector. Fuel flowing through the thrust chamber jacket cooled the thrust chamber and at the same time absorbed heat converting from liquid hydrogen to gaseous hydrogen for injection into the combustion chamber.

The gimbal assembly included a universal joint consisting of a spherical, socket-type bearing with a Teflon-fibreglass composition coating to provide a dry, low-friction bearing surface. The gimbal assembly, installed on the thrust chamber dome, attached the engine to the thrust structure and transmitted engine thrust to the vehicle. A boot with a bellows configuration made of silicone-impregnated fibreglass material protected the gimbal assembly from dust, water, and other foreign matter without interfering with the gimballing action. Two hydraulic actuators, attached to the engine and thrust structure, provided the force to gimbal the engine +/-7deg for thrust vector control.

The augmented spark igniter (ASI) assembly was mounted on the thrust chamber injector. It consisted of a fuel and LOX manifold, an injector assembly, and two spark igniters and cable assemblies. The ASI assembly received the initial flow of LOX and LH_2 and ignited the propellants by discharging electrical energy through the two spark plugs. At engine start, spark exciters in the electrical control package transformed 28Vdc into 27,000V pulses that discharged across the spark plug gap at a minimum 40 sparks/per sec. Hermetically sealed transmission cabling and connections were pressurised with gaseous nitrogen to ensure operation at high altitudes.

The propellant feed system consisted of a fuel turbopump, oxidiser turbopump, main LOX valve, main fuel valve, augmented spark igniter oxidiser valve, mixture ratio control valve, propellant bleed valves and propellant feed ducts. This system transferred and controlled propellant flow from the stage tanks to the thrust chamber and the gas generator.

Rocketdyne J-2 engine

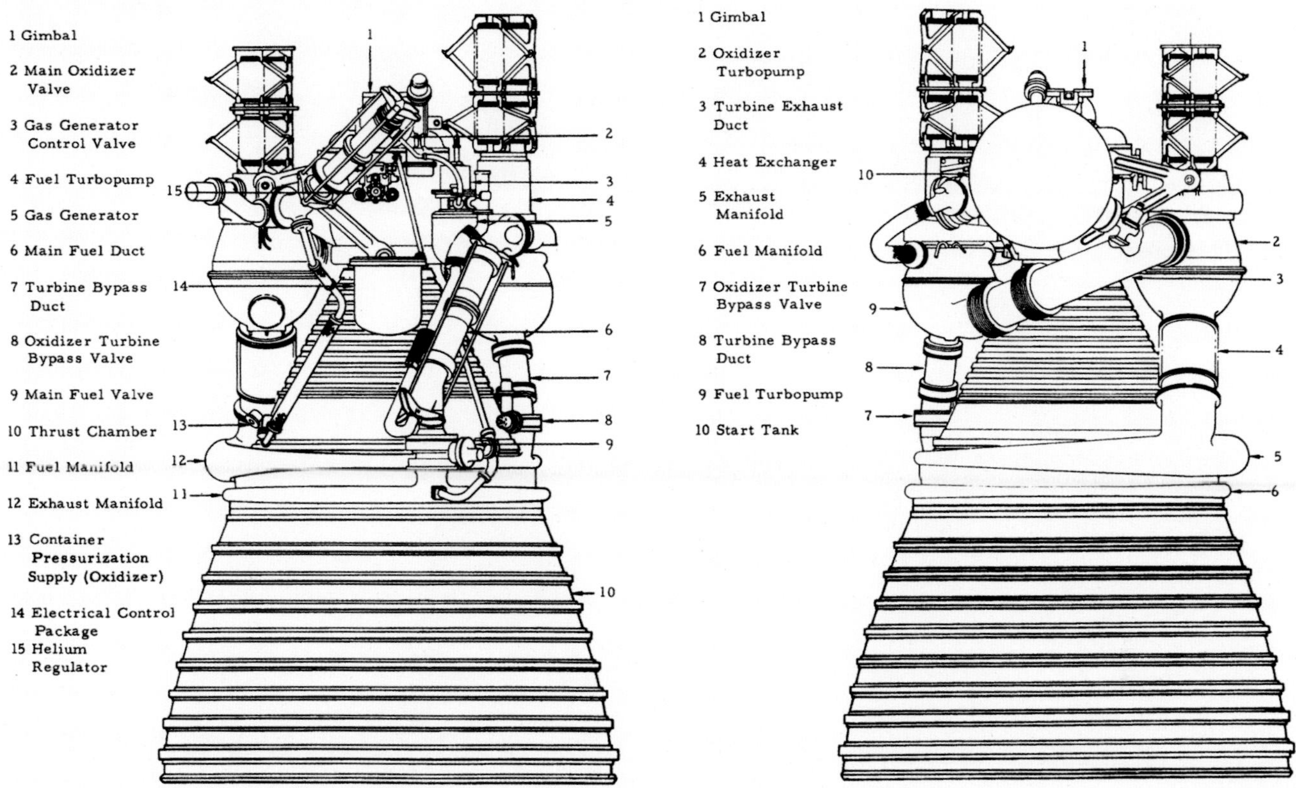

Above: Views of the J-2 engine for the Saturn IB S-IVB stage with basic elements identified, each view 180deg to the other. *MSFC*

Below: The J-2 injector and augmented spark igniter (ASI) showing fuel and oxidiser orifices. *Rocketdyne*

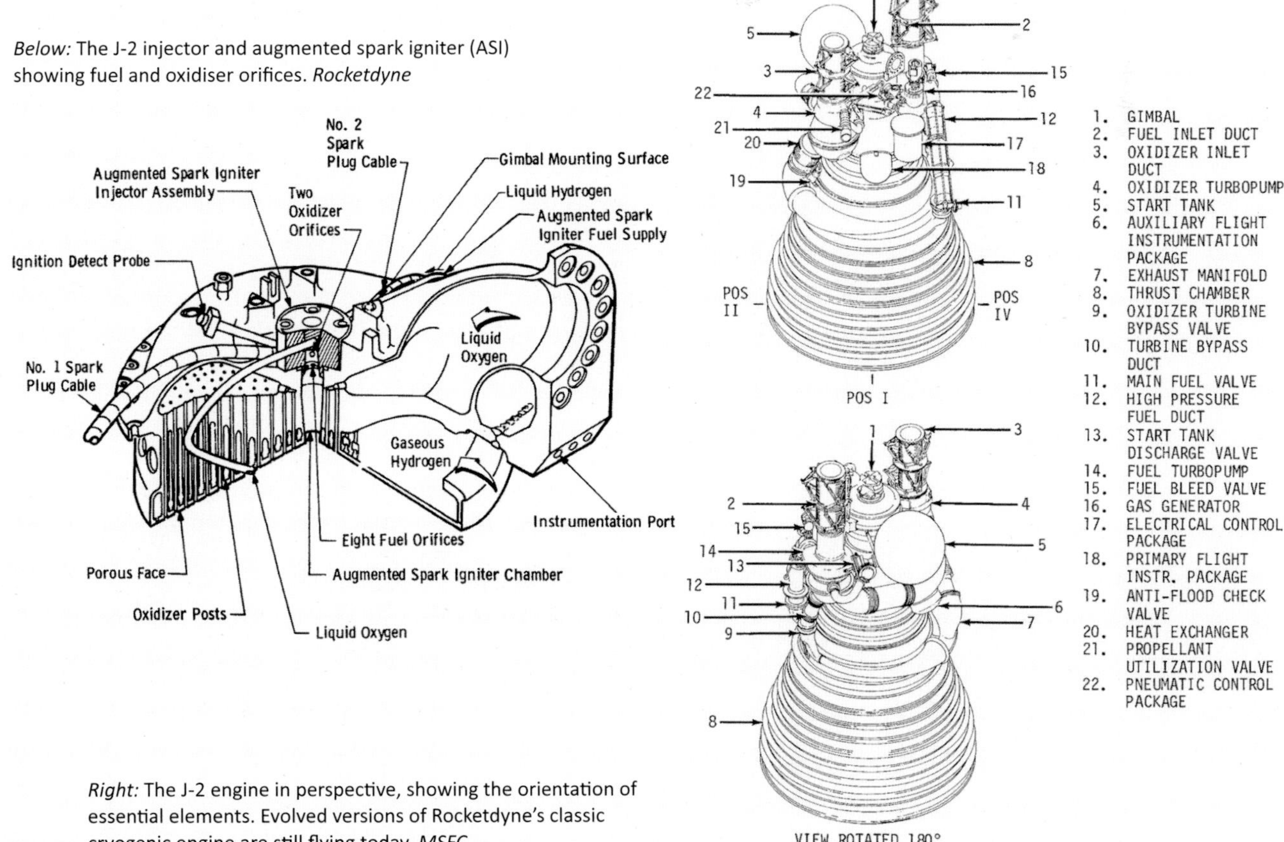

Right: The J-2 engine in perspective, showing the orientation of essential elements. Evolved versions of Rocketdyne's classic cryogenic engine are still flying today. *MSFC*

Above: Completed S-IVB stages in storage, awaiting allocation to a specific launch vehicle. *MSFC*

The fuel turbopump, a turbine driven, axial-flow pumping unit consisting of an inducer, a seven-stage rotor and a stator assembly, increased the pressure and flow rate of LH_2 entering the thrust chamber. The pump was self-lubricated and self-balanced with LH_2. The high speed, two-stage turbine, driven by hot gas from the gas generator, drove the inducer and the one-piece, seven-stage rotor assembly. Gas entered the turbine inlet manifold and passed through nozzles where it expanded and was directed at high velocity through the first-stage turbine wheel and then through stator blades that redirected the gases through the second-stage turbine wheel. Exhaust ducts directed the gases to the oxidiser turbopump turbine and also to the oxidiser turbine bypass valve. Three dynamic seals in series prevented LH_2 and turbine gases from mixing. The inducer increased the LH_2 inlet pressure at the pump.

Each stage of the seven-stage rotor contributed to the build-up of pressure which forced the fuel through diffuser vanes and an outlet volute into the high pressure duct. A self-compensating balance piston absorbed axial thrust loads developed by the rotor. Liquid hydrogen lubricated and cooled the two ball bearings on the rotor shaft. Temperature measurements of the fuel turbine inlet and the fuel pump discharge, a fuel pump discharge pressure measurement, and the pump speed measurements were telemetered to ground receiving stations for recording and post-flight evaluation.

LOX was delivered to the thrust chamber at increased pressure and flow rates by a single-stage, direct-turbine-drive centrifugal pump. The pump was self-lubricated and self-cooled with LOX. A high speed, two-stage turbine, driven by exhaust gases from the fuel turbine, provided power for LOX turbopump operation. One static and two dynamic seals in series prevented LOX and turbine gases from mixing. Turbine exhaust gases entered the oxidiser turbopump turbine manifold and flowed through nozzles and the first-stage turbine wheel. Stator blades redirected the gases, which passed through the second-stage turbine wheel and then through the turbine exhaust ducting, a heat exchanger and through the thrust chamber.

A pump shaft transmitted the turbine power to the inducer and impeller. LOX entered the turbopump through the inducer, which increased pump inlet pressure, and flowed through the impeller into the outlet volute. Passages from the outlet volute permitted LOX to flow to the two pump shaft ball bearings for lubrication. A screen filtered the LOX entering the bearing area. Seven measurements of turbopump operation were taken during flight. The turbine inlet and outlet temperatures and pressures, the pump discharge temperature and pressures, and the pump speed were telemetered to ground receiving stations and recorded for post-flight evaluation. A magnetic pick-up located behind the turbine manifold sensed pump speed as 12 equally spaced slots in the rotor assembly rotated through the magnetic field. An accessory drive adapter on the oxidiser turbopump drove the hydraulic pump for the two engine actuators.

The main oxidiser valve (MOV) was a butterfly-type device, spring-loaded to the closed position, pneumatically operated to the open position and pneumatically assisted to the closed position. The MOV, installed between the high pressure duct and the thrust chamber injector oxidiser inlet, controlled the LOX flow into the thrust chamber. Pneumatic pressure from the normally open port of the mainstage control valve plus spring pressure maintained the valve closed. During the engine start sequence, the mainstage control valve opened and applied pressure to the first and second stage opening control ports on the MOV actuator. Pressure acting against a small surface in the first stage actuator, plus helium flowing through an orifice and acting on a large surface piston which actuated the valve gate, provided a ramp opening of the valve. Pressure exhausting from the closing actuator exited through an orifice in the actuator housing and through an orifice-check valve to help accomplish the ramp opening.

The main fuel valve (MFV) was also a butterfly-type valve, spring-loaded to the closed position, pneumatically operated to the open position, and pneumatically assisted to the closed position.

The MFV, installed between the fuel turbopump high pressure discharge duct and the fuel inlet manifold, controlled LH_2 flow to the thrust chamber and to the ASI assembly. The ignition phase control valve on the pneumatic control package opened the MFV during engine start. When pneumatic pressure had opened the valve to 90 per cent, the actuator mechanically opened a sequence control valve mounted on the MFV actuator. The sequence valve permitted helium flow also from the ignition phase valve to the start tank discharge valve control valve. The MFV and sequence valve arrangement ensured that the MFV would be open before the STDV opened to deliver gaseous hydrogen to start the fuel turbopump. A position switch assembly operated by the gate shaft provided analogue position signals of the gate and discrete signals corresponding to its open or closed position.

The mixture ratio control valve allowed the engine to operate at either one of two fixed mixture ratios to achieve maximum vehicle performance. The valve changed mixture ratio by routing a portion of the oxidiser flow from the oxidiser turbopump outlet back to the pump impeller inlet. The valve had an actuator assembly and a gate assembly. The actuator was two-position, electro-pneumatic and was spring-loaded to keep it in the high engine mixture ratio position (valve closed). Pneumatic pressure was directed to the actuator piston by a three-way pneumatic control valve energised by a stage signal. The gate assembly consisted of a rotating sleeve within a stationary outer sleeve. Each sleeve had three elongated holes; by rotating the inner sleeve (valve gate) the holes were aligned or misaligned, to control the amount of oxidiser flow through the valve. The valve position indicator was mounted on the valve shaft and consisted of a rotary-motion, variable resistor and open and closed position switches.

The mixture ratio control valve had two distinct stops, to allow engine operation at engine mixture ratios of either 5.5:1 or 4.8:1 (LOX to LH_2 by weight). Pneumatic pressure was supplied to the valve from the engine pneumatic system when the engine helium control valve was energised. At a preselected time during engine operation, a control signal, supplied by the stage, energised the solenoid control valve. This allowed pneumatic pressure to enter the valve and apply force to the actuator piston, to overcome the spring tension and move the piston in the direction to rotate the gate to the low engine mixture ratio position (valve open). Opening the valve resulted in a reduced oxidiser flow to the thrust chamber. If either the pneumatic pressure or the electrical command was lost, the valve would move to the high engine mixture ratio position (valve closed). The position indicator arm rotated with the gate shaft, to remotely indicate valve position.

The gas generator, which produced the hot gases to drive the oxidiser and fuel turbines, consisted of a combustor containing two spark plugs, a control valve containing oxidiser and fuel poppets, and an injector assembly. When engine start was initiated, spark exciters in the electrical control package were energised, providing power to the spark plugs in the gas generator combustor. Propellants flowed through the open poppets of the control valve to the injector assembly and into the combustor where they were mixed and burned, resulting in hot gases that passed through the combustor outlet and were directed to the fuel turbine and then to the oxidiser turbine.

The gas generator control valve was pneumatically operated, spring-loaded to the closed position. The oxidiser and fuel poppets were mechanically linked by an actuator. The purpose of the gas generator control valve was to control the

Right: The S-IVB stage originally destined for the AS-206 lunar module test mission, but eventually flown six years later on the Skylab 2 mission of May-June 1973. *NASA*

flow of propellants through the gas generator injector. When the mainstage signal was received, pneumatic pressure was applied against the gas generator control valve actuator assembly which moved the piston and opened the fuel poppet. During the fuel poppet opening, an actuator contacted the piston that opened the oxidiser poppet. LH_2 and LOX from the bootstrap lines flowed through the control valve into the combustion chamber. Orifices in the bootstrap lines controlled the propellant flow rate to the gas generator and a line from the housing to the sequence valve vent port on the MOV provided a vent to equalise pressure in the housing to prevent premature opening of the fuel control poppet.

The gas generator injector assembly consisted of a circular metal plate containing a normally closed, spring-loaded oxidiser poppet valve and injector, centred within a fuel injector ring. The purpose of the injector assembly was to distribute propellants into the gas generator combustor. The injector was welded to the gas generator combustor, and the oxidiser poppet and injector was threaded into the gas generator injector assembly. During operation, fuel entered the injector assembly fuel inlet, filled a manifold in the top of the combustor, and flowed through drilled passages in the fuel injector ring. Oxidiser pressure displaced the oxidiser poppet valve and allowed oxidiser flow through the injector to impinge on the fuel flowing through the fuel injector ring. The gas generator combustor was a cylindrical chamber in which the propellants were mixed and burned. Two spark plugs initiated combustion. The inlet port mated with the gas generator injector assembly and the outlet port and short duct section was welded to the fuel turbine manifold. Propellants entering the combustor were ignited by the spark plugs; combustion hot gases passed through the combustor outlet into the fuel turbine manifold.

For starting the J-2, the S-IVB stage switch selector issued the engine start signal to the electrical control package. It required only dc power and start and cut-off commands for operation. The start command put power to the spark exciters in the sequence controller that provided electrical energy to spark plugs in the gas generator and the augmented spark igniter (ASI). A start tank discharge delay timer (0.640-030sec) started and the helium control valve and ignition phase control valve received power simultaneously. Helium from the

Rocketdyne J-2 – operational schematic

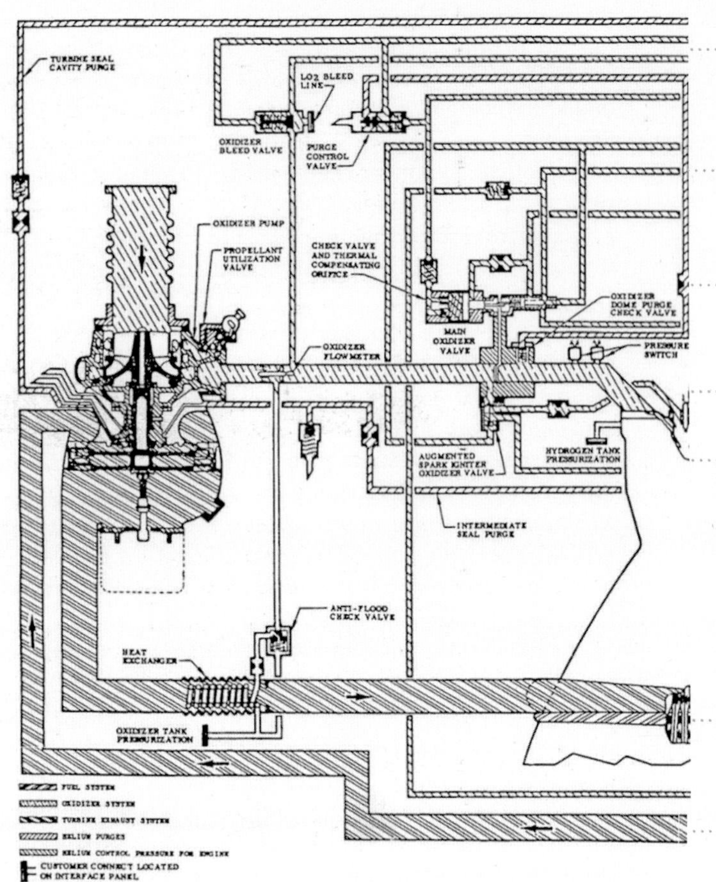

Above: The basic operating principle of the J-2 cryogenic engine showing propellant in liquid and gaseous state with the fuel largely across the bottom of the schematic and oxidiser across the top. *MSFC*

helium tank at 20,685kPa (3,000psi) flowed through the helium control valve into the pneumatic control package. Roughing, primary, and control regulators reduced the pressure to 2,758kPa (400psi), +/-172kPa (+/-25psi) for valve control. Helium from the primary regulator outlet flowed to continuously purge the LOX turbine intermediate seal during J-2 en-

Rocketdyne J-2 – thrust chamber cooling

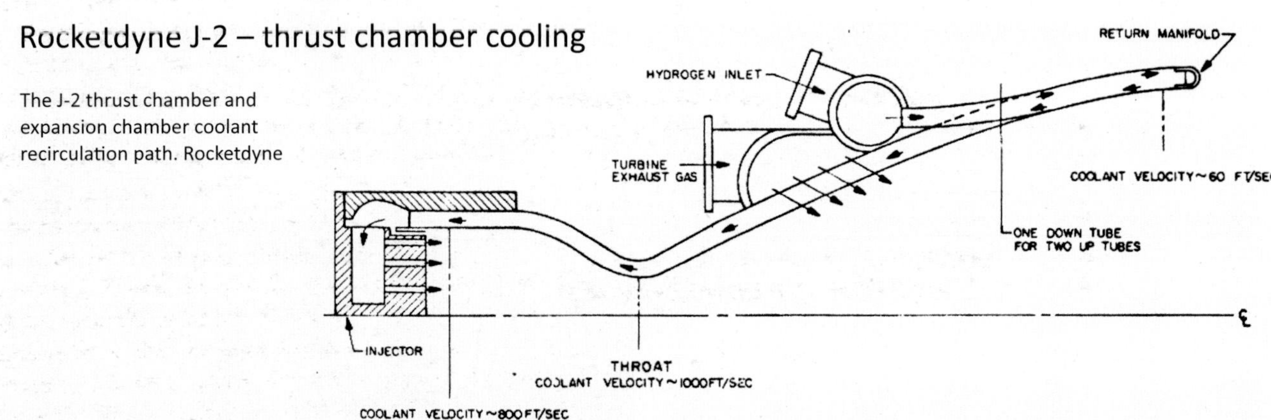

The J-2 thrust chamber and expansion chamber coolant recirculation path. Rocketdyne

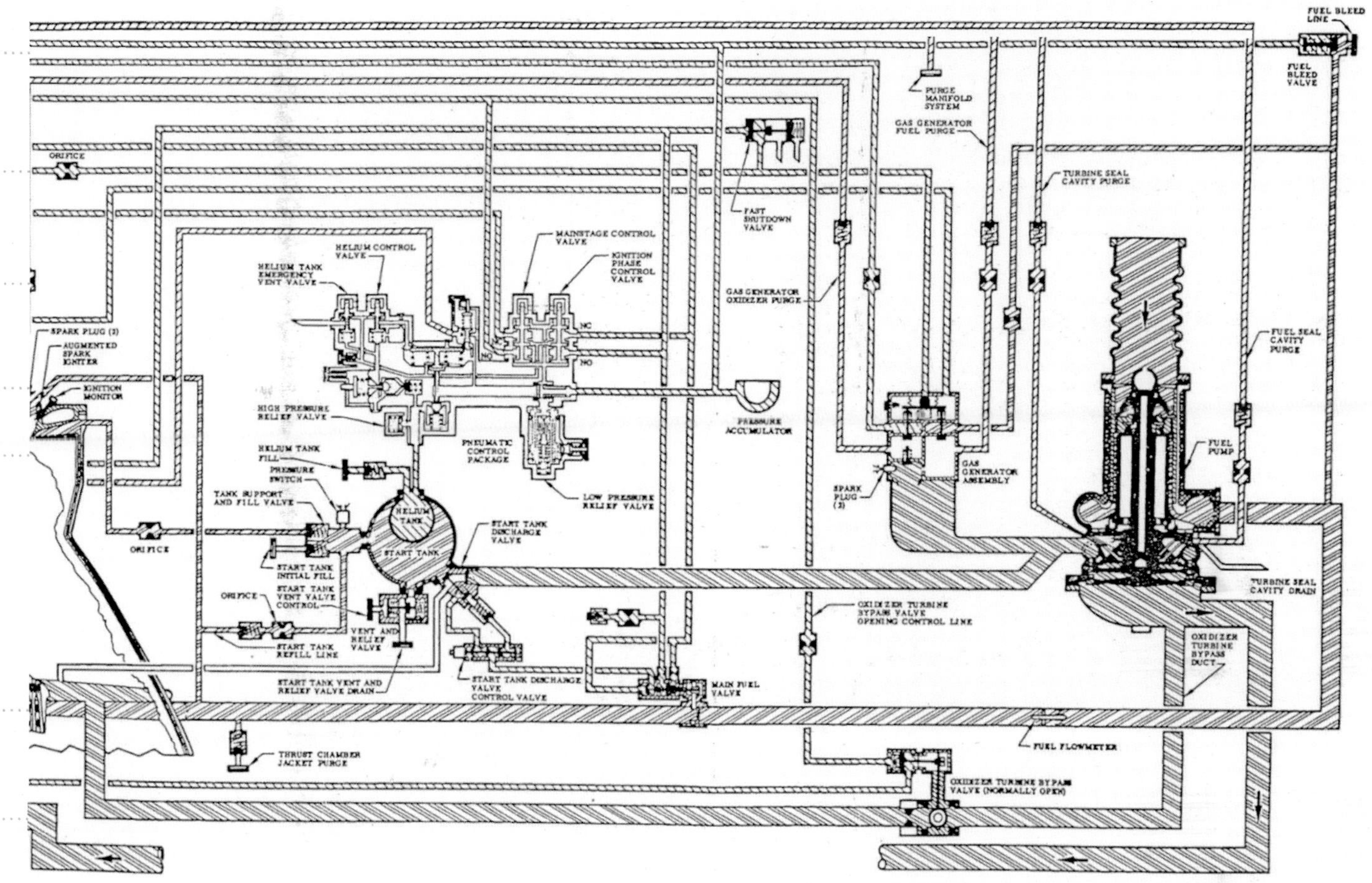

gine operations. Regulated helium from the pneumatic control package flowed into the jacket of the primary flight instrumentation package, which served as an accumulator for the control pressure system. In the event of supply pressure failure, a check valve in the regulator outlet would prevent control pressure loss and the accumulator would then supply a large enough volume of helium to effect safe shut-down of the engine. The helium also closed the fuel and LOX bleed valves, and flowed through the purge control valve to purge the LOX dome and gas generator oxidiser injector.

Helium flowed through the normally open port of the mainstage control valve to the main LOX valve closing port, to the opening ports of the purge control valve and to the LOX turbine bypass valve. Helium also flowed through the energised ignition phase control valve (a normally closed port) to open the main fuel valve and ASI oxidiser valve, and to the sequence valves on the main fuel and LOX valves. LH_2, under tank pressure, flowed through the turbopump and main fuel valve into the thrust chamber fuel manifold. Bootstrap fuel, tapped off the fuel manifold, flowed to the ASI assembly. LOX, under tank pressure, flowed through the LOX turbopump and ASI valve to the ASI assembly where it flowed through the down tubes to the return manifold and then through the up tubes to the injector where it emerged in a gaseous state.

The fuel cooled the thrust chamber as it flowed through the jacket to the combustion chamber. LOX entered the oxidiser manifold in the injector and flowed through equally distributed nozzles where it mixed with fuel and burned. The main fuel valve (MFV) sequence valve opened when the MFV opened to 90 per cent, permitting helium flow to the closed start tank discharge valve (STDV) control valve. The start tank discharge timer expired after approximately 1.0 second, and an ignition phase timer (0.450, +/-0.030 sec) started. The STDV control valve energised, admitting helium to the STDV opening port. Gaseous hydrogen from the integral start tank flowed through the series turbine drive system, accelerating the fuel and oxidiser turbopumps to the required levels to permit power build-up of the gas generator and to deliver propellant for ASI ignition. The normally open LOX turbine bypass valve routed a percentage of the GH2 from the oxidiser turbine during start and later controlled the relationship of fuel and oxidiser turbine speed by an orifice in the valve gate.

Expiration of the ignition phase timer initiated the command for the sparks timer (3.30, +/-0.20 sec), closing signal for the start tank discharge valve, and mainstage command. Energising the mainstage control valve vented pressure from the main oxidiser valve closing port, closed the purge control valve terminating the LOX dome and gas generator oxidiser injector purges. It vented the LOX turbine bypass valve opening pressure and applied pressure to the first and second stage opening port of the MOV. An orifice in the second-stage

Above: An S-IVB stage under restoration at Kennedy Space Center, Florida. *Evergreene Architectural Services*

opening line, series orifices in the actuator closing line, an orifice in the actuator, and an orificed check valve provided a controlled ramp opening of the MOV.

The MOV sequence valve opened with the MOV and supplied helium pressure through orifices to close the LOX turbine bypass valve and to open the gas generator control valve. Two spark plugs in the gas generator ignited the LOX and LH_2 admitted by the gas generator control valve. Hot-gas products of combustion passed through the fuel turbine and through the exhaust duct to the LOX turbine, accelerating the turbopump causing increased propellant flow. The turbine exhaust gases exited through a heat exchanger and into the engine exhaust manifold. The propellant entered the thrust chamber where main propellant ignition occurred. As turbopumps accelerated to operational speeds, oxidiser injection pressure increased, actuating two thrust-OK pressure switches (TOPS). Either of the two TOPS, which actuated at 3,447kPa (500, +/-30psi), would issue a main-stage-OK signal that had to be present before the sparks de-energised timer expired and automatically issued a cut-off command.

During engine operation, gaseous hydrogen tapped off the fuel injection manifold maintained LH_2 tank pressure. Turbine exhaust gases flowed through the heat exchanger and heated helium from the cold helium bottles for LOX tank inflight pressurisation. A two-position mixture ratio control valve (MRCV) bypassed LOX from the oxidiser pump outlet to the pump inlet and allowed the engine to operate at either of two fixed mixture ratios as described earlier. The MRCV was commanded to the low engine mixture ratio position prior to engine start and to the high position after 90 per cent thrust had been achieved.

Cryogenic component design in rocket engines using liquid oxygen has profited from the fact that bare metal containing liquid oxygen forms a coating of frost from the atmosphere which is a surprisingly good insulation against heat. Accordingly, with this additional knowledge, it has been common practice to design such components without added insulation. In contrast, bare metal containing liquid hydrogen is so much colder that it does not form frost. But air liquefies on the surface. The resulting streams of liquid air constitute a serious heat leak as well as being an unacceptable annoyance. It was therefore necessary to provide insulation for most engine components containing liquid hydrogen. This was conveniently accomplished for the hydrogen ducts and pump by either good, moisture-sealed insulation or by vacuum jackets. The J-2 engine employed vacuum jackets for these parts. The most difficult design problem in this regard was the liquid hydrogen inlet duct, which had to take the motion incident to 10.5deg gimballing of the engine. Compression and extension 138.4cm (54.5in) in the 53.3cm (21in) duct were required, in addition to angulation and twist. In this duct the vacuum jacket was provided by a double bellows and stabilisation was provided by scissors-type external supports.

Development of the J-2 engines was specific for the Saturn programme, adding powerful cryogenic upper stage engines for the Saturn and the Saturn V. The only difference between engines for the two launchers was in the restart capability for the Saturn V's S-IVB stage. Rocketdyne received a contract for the J-2 in 1960, with the first test at the Santa Susana test stand on 11 November. The definitive design was completed shortly thereafter and tests with the turbopump began on 9 November 1961 and with the gas generator on 24 January 1962, followed by a full duration test on 1 March. NASA awarded a production contract to Rocketdyne, initially for 55 engines, on 24 June 1964 and increased this to 157 on 24 August. Development testing was undertaken at the Santa Susana Field Laboratory (SSFL) which was located within a short distance of the company headquarters and development facility at Canoga Park.

Development problems stalked the J-2 programme but on balance no more so than a lot of other engines of the period. Several technical changes were made as the test programme evolved, with design improvements to the thrust chamber and changes to the way the fuel pump was configured when it showed a tendency to stall. A significant milestone was achieved on 11 December 1964 when the J-2 demonstrated restart capability, applicable only to the S-IVB-500 series stages for the Saturn V. Identified within the J-2000 series, the first J-2 for a flight-rated S-IVB was J-2015 fitted to S-IVB-201 for the first Saturn IB flight. It had seven pre-acceptance firings for a total of 548sec and was accepted by the customer on 16 March 1965 and fitted to the flight stage for the AS-201 mission which launched on 26 February 1966. This and successive engine numbers can be found in the missions descriptions information contained in Chapter 11. Overall, Rocketdyne produced 152 flight J-2s of which nine were on the S-IVB stages for the Saturn IB.

Chapter Eight

Growth Options

The Saturn I was developed in the late 1950s, with the potential as an early heavy-lift capability for military payloads, specifically at that time as a contender for the Advent US Air Force communications satellite which was eventually cancelled. Saturn I quickly matured into a launcher capable of supporting post-Mercury manned space vehicles in the Apollo programme and demonstrated reliability, ruggedness and flexibility. So much so that thoughts quickly turned to how it could be used as the base building block for a more capable system filling a void unoccupied by any other launch vehicle.

After President Kennedy announced the Moon goal in May 1961, it was seen as potentially the DC-3 of the rocket world, able to provide logistical support for advanced missions on the surface of the Moon as well as building, then servicing, manned

Above: The diversity of options for improving and expanding the capabilities of the Saturn IB mirrored progress with upper stage evolution as demonstrated by the Atlas-Centaur, the first of which is seen here on LC-36A prior to its initial flight attempt on 8 May 1962. *NASA*

space stations in Earth orbit. It never quite happened that way and there are many reasons mooted for that. However, in understanding the drive behind changes to the Saturn programme forced upon NASA in the second half the 1960s, pivotal to the story of all the Saturn launch vehicles, it helps to place the programme in broader context.

In its first full year of operation, in fiscal year 1960 NASA had a budget of $524million of which $462million was spent on space programmes and the rest on aeronautics. At that time the fiscal year began six months prior to the start of that calendar year, with FY1960 starting on 1st July 1959. That increased to a space-related budget of $926million the following year. As a result of the Apollo decision of May 1961, the FY1962 budget grew to $1,797million reflecting a start on funded development and as contracts awarded for Saturn, Apollo and the expensive facilities infrastructure got under way. Within three financial years the NASA space budget had peaked at $5,138million (almost $34billion in 2021 money), bringing FY1965 in as the highest in real-year terms, as well as percentage of GDP, of any in the agency's history to date.

Warnings about excessive space expenditure had been rife across both Houses of Congress during 1963 and Kennedy was talking openly of finding some way to roll back on his ambitious commitment. He held talks with NASA Administrator James Webb about cutting out all elements of the space programme not directly supporting the Moon goal and made speeches before the United Nations offering the Russians to 'go to the Moon together'. There was a recognisable level of concern at NASA regarding these overtures and it helped little that, upon seeing the assembled stack for the first orbital flight of a Block II Saturn I (SA-5) during his third visit to Cape Canaveral on 16 November 1963, his dream of overtaking the Russians in lift capacity appeared to have been satisfied. Six days later, Kennedy was assassinated.

These were pivotal days and months for NASA's manned space projects and for the Saturn programme in particular. The Apollo goal had already forged development of an interim spacecraft between Mercury and Apollo and the Gemini spacecraft would fly 10 manned missions during 1965 and 1966. But there was strong opposition in Congress to further budget increases and pressure for reductions was mounting as the war in Vietnam intensified. President Lyndon Johnson, who had been largely responsible for pushing the NASA bill through Congress and who had steered Kennedy to a Moon decision in May 1961, now had more urgent pressures, such as retaining the confidence of Congress for his civil rights bills. These factors were to prove pivotal in bringing about a shift in how the future space programme would evolve, unexpectedly moving away from a big-budget and highly ambitious reach for manned planetary exploration. Overall spending and economic ways of funding a sustainable space programme were now key to NASA's guideline policies and it was on this hinge that the fate of Saturn was sealed.

Key to deciphering the impact on planning occasioned by the annual budget process is the lead time required to assemble the budget and present it to Congress for approval. The FY1966 budget covering the period from 1 July 1965 was prepared by the government in the closing months of 1964 and presented to Congress for approval in the spring of 1965. When it was being prepared, the FY1966 budget was under significant stress and would begin the process of decline, for NASA and for several other government departments, also beneficiaries of discretionary spending. Thus, in late 1964 it began a remorseless reduction, even down to cancelling certain programmes.

The period 1964-1966 saw NASA budgets for fiscal years 1966 to 1968 fall from $5,016million to $4,430million. This was also reflected in the number of personnel working on NASA programmes. Since no money was spent in space, most of the budget went on wages and material here on Earth and that was a reflection of national capability. NASA-funded employment had increased from 46,786 in FY1960 (of which only 10,286 were on the NASA payroll), to a peak of 409,900 (including 33,200 NASA employees) in FY 1965. The total number working directly on Apollo contracts, also a peak year, was 365,000. By FY1968 total employment had dropped to 267,871 and would reach a low of 124,069 by FY 1977.

By the mid-1960s, several years prior to the first Moon landing, cost reduction management was a top priority at NASA. Where once it had been considered acceptable to fly several score Saturn I flights in development missions and in support of mission-proofing for end objectives, the financial limits of what was possible introduced significant changes

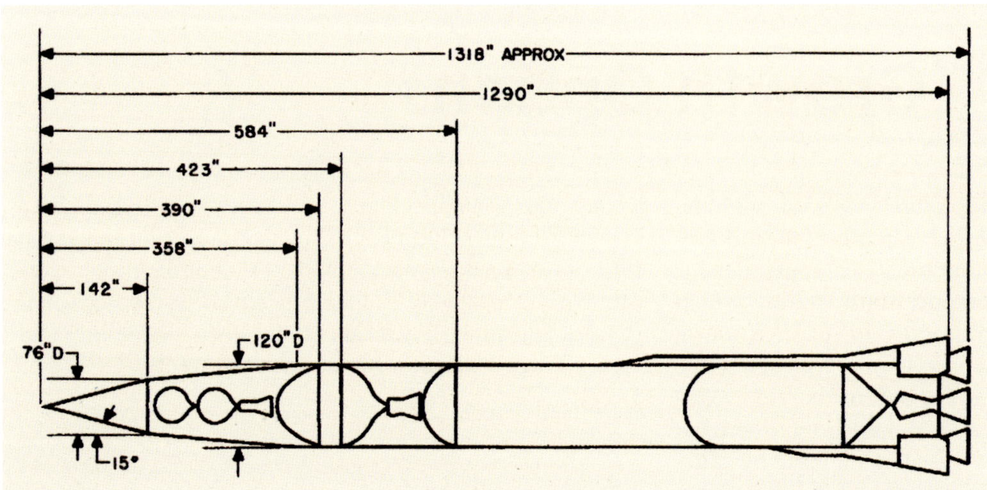

Proposed by JPL from the 6K engine for the defunct Juno IV, Vega competed with Agena for Atlas (shown here) and was cancelled when NASA became aware of the existing upper stage which had evolved through the highly classified Corona programme of spy satellites. *JPL*

Above: Vega had been proposed in parallel with Centaur as the upper stage for Atlas after a technology development for the cancelled Juno IV proposal. The mock-up seen here was as far as it got. *Author's collection*

during mid-1963 which resulted in of much of the mid-level NASA management and the complete upending of prior planning regarding testing and sequential development. Those pivotal changes have been documented in Section Two of this book. Concomitant with that seminal transformation was the concept of all-up systems testing, which influenced the Saturn flight profiling immensely, and the determined effort to reduce costs. That in turn led to a search for ways to make the Saturn IB a credible tool for long-term mission planning and cost-effectiveness.

In 1964 and over the next several years, NASA examined a possible advanced Saturn IB with a Centaur third stage and thrust augmentation, initially by two 305cm (120in) diameter solid propellant boosters. Moreover, by adding 30 per cent fluorine to the LOX for the S-IB stage, it was believed that the thrust of each H-1 engine could eventually reach 1,334kN (300,000lb), an increase of 50 per cent. But without that, in early May 1964 Rocketdyne tested an advanced H-1 engine at a thrust of 1,085kN (244,000lb) for 10sec. With the Centaur, an advanced Saturn IB could lift 10,886kg (22,400lb) to a 3,700km (2,300mls) orbit. It would also be capable of sending a 6,350kg (14,000lb) spacecraft to a circumlunar trajectory. At the time, and in the wake of a declining annual payload manifest due to cost-cutting measures, there was no immediate demand for that capacity.

These studies continued apace with the realisation that NASA would eventually have a payload gap and looked to advanced versions of the Saturn IB to fill that. So far, Centaur has not figured large in this history of the Saturn I series but it had the potential to accommodate specialised flight requirements. With its restart capability and powerful cryogenic propulsion system it had the potential value for fulfilling some deep-space missions. As such it is worthy of a synoptic description because it was seriously proposed as the launch system for a series of planetary missions. It was the Lewis Research Center, the NASA facility in charge of developing Centaur after that stage was transferred from MSFC that conducted the study based on a requirement known to exist for a planetary programme to follow the early missions to Venus and Mars flown in 1961 and 1964.

Completed and delivered on 25 February 1966, the report defined several mission profiles which were considered possible, divided into a fly-by missions, orbiter missions and orbiter-lander missions, which would leave one component to orbit the planet while another separated and went down to the surface. Under the guidance of and consultation with JPL, fast emerging as NASA's lead centre for planetary missions, launch windows for Mars in 1971, 1973 and 1975 were used as baselines, with flights to Venus in 1972 and 1973 studied for options.

At this date NASA was in a state of flux with its planetary programme and somewhat deflated by the seemingly barren and lunar-like surface features from the 21 rather blurry pictures transmitted by Mariner IV on its fly-by on 15 July 1965. It did little to pep up enthusiasm in Congress for more money! With the Apollo programme taking a significant fraction of the NASA budget, these were lean times for such esoteric side-stunts as planetary exploration. Or so thought many legislators quite prepared to fund a propaganda coup against Soviet space achievements but little else. Inside NASA, the prevailing view was that the Moon goal would soon be met and that it needed to display a broad and successful range of programmes for the longer term. These covering not only manned space spectacles but the robotic exploration of the solar system, better understanding of near-Earth space and the interaction of the Sun with the atmosphere, of space-based astronomy and the ability to study the universe in spectral bands denied to observers on Earth. The balance between what was desirable and what was affordable would define the pace and expanse of the Saturn programme and it was in an attempt to broaden the range of space activities that the Saturn IB was sought as a flexible solution to a wide range of demands. It was therefore with some degree of anticipation that the Lewis Research Center submitted its findings.

Depending on the type of trajectory chosen, a simple fly-by mission could be accomplished at Mars with spacecraft weighing up to 5,171kg (11,400lb) launched in the May-June 1971 window, reduced to 4,563kg (11,050lb) for the July-August 1973 window and increased to 4,745kg (10,460lb) for the September-October 1975 window. For orbital insertion missions utilising a Centaur and a storable insertion stage, payload delivered to Mars orbit would be 1,791kg (3,960lb), 1,574kg (3,470lb) and 1,665kg (3,670lb) for the same windows in 1971, 1973 and 1975. Assuming a 2,268kg (5,000) lander separated prior to arrival, with a Centaur and storable insertion stage, payloads for a surface lander mission would be a maximum 771kg (1,700lb), 503kg (1,110lb) and 585kg (1,290lb), respectively. The baseline missions to Venus would carry a fly-by payload of 5,130kg (11,310lb) for the March-April 1972 window and the October-November 1975 window. Orbiter missions with Centaur and insertion stage provided a lift capability of 1,859kg (4,100lb) and 2,313kg (5,100lb) for the two windows, respectively. With a 2,268kg (5,000lb) lander, respective values for maximum payloads were 762kg (1,680lb) and 1,057kg (2,330lb).

The study demonstrated that Saturn IB /Centaur had the capability to carry significant planetary payloads to Mars and Venus and that it had the capacity to expand the weight and volume of existing planetary spacecraft. When the report was submitted in February 1966 NASA had already successfully launched two fly-by missions, Mariner II to Venus in 1961 and Mariner IV to Mars in 1964, and was preparing to launch the Mariner 5 fly-by to Venus in 1967. But there was considerable debate over the next steps to Mars and little interest in pursuing further flights to Venus. The emphasis on Apollo had squeezed out money for several science missions planned prior to the Kennedy decision and this was having an effect on the choice for ambitious planetary programmes.

To this date all US planetary spacecraft flew on Atlas-Agena launchers and the next missions to Mars, Mariner's 6 and 7 in 1969, were assigned the Atlas-Centaur. They would each weigh 412kg (908lb), some 60 per cent heavier than Mariner IV yet only a fraction of the capability of the Saturn IB/Centaur, but there was insufficient money to develop a spacecraft that could have weighed eleven times as much. It was this factor that doomed the Saturn IB/Centaur but the work to develop this stage had been close to the activities at MSFC for several years, providing a stage, coupled to the RL-10 from the Lewis Center which added greater potential to the Atlas and, later, the Titan launch vehicles.

Saturn IB/Centaur

The stage had an intriguing history. It was an integral component of the universal plan to develop cryogenic upper stages which resulted in the S-IV and the more advanced S-IVB. The story began on 29 August 1958 when ARPA asked the Air Research & Development Command (ARDC) to develop an upper stage for Atlas capable of launching ten communications satellites called Advent, each of which would weigh more than could be lifted by Agena. We have already recorded how Advent also sponsored development of Saturn – perhaps with Centaur as the upper stage instead of the S-IV – but the Advent programme was cancelled while the cryogenic upper stage lived on. Designed at Convair under the inspirational guidance of Krafft Ehricke, Centaur was passed to NASA and finally to MSFC in 1959.

Ehricke incorporated a thin-wall monocoque tank structure that, like Atlas, would have no bracing or stiffening frames and would be pressurised to maintain structural rigidity. Because of the need to maintain cryogenic temperatures, the two propellant tanks would be separated by a common bulkhead comprising two thin metal sheets 1.27cm (0.5in) apart. The challenge was great, due to the juxtaposition of two super-cold structures with liquids at a temperature difference of 125° (70°C), where the much colder hydrogen would readily start to gasify the oxygen in the adjacent tank and boil off. Had Ehricke chosen to design the stage with two separate tanks it would have been necessary to increase the overall length by 1.2m (4ft).

It was a design feature with highly technical challenges. Filled with a Styrofoam insulation, the space between the two sheets was pumped clear of air and filled with nitrogen.

Above: The cryogenic Centaur had a chequered start with significant delays before it came good but Huntsville sought it as a supplement to existing S-I and S-IB stages. *General Dynamics*

When fuelled with the super-cold propellants, the hydrogen would conduct its low temperature across the thin sheet, freezing out the nitrogen and creating an almost perfect vacuum. The oxygen tank was above the hydrogen tank, the lower section of the oxygen tank in convex cross-section to lie conformal to the cylindrical hydrogen tank below with hemispherical end domes.

Centaur was planned for flights starting in 1961 and was to be powered by two Pratt & Whitney RL-115 engines, each producing a thrust of 66.72kN (15,000lb) with a specific impulse of 420sec. The combustion chamber would operate at a pressure of 2,068kPa (300lb/in^2) and with a thrust chamber ratio of 1:40. A regenerative turbopump started on tank pressure, and a high-energy electrical system would provide multiple restarts in space. After passing through the two-stage centrifugal pump and chamber cooling jacket, the hydrogen fuel flowed through the turbine, where it expanded before being introduced to the main combustion chamber. The engine itself was 173cm (68in) tall with a skirt diameter of 99cm (39in) and a dry weight of 135kg (298lb), operating at a flow rate of 16kg/sec (35lb/sec) and a demonstrated specific impulse of 433sec. The complete engine assembly would be gimballed for flight control.

Because of the low critical temperature and pressure, high specific heat and low initial temperature of the fuel, the heat rejected by the fuel in the cooling jacket provided in itself the kinetic energy to drive the turbopump, removing the need for a gas generator. Coincident with the start of turbopump rotation, propellant that flowed to the injector was electrically ignited and the engine automatically accelerated to the required level for sustained operation until the flow valves were shut to terminate the burn. The oxygen pump was gear-driven from the turbine shaft and its output passed through a throttling valve that was set before launch for the required mixture ratio.

The main engine propellant feed system consisted of the oxidiser and fuel centrifugal boost pumps with a hydrogen peroxide supply system and the bifurcated oxidiser and fuel propellant ducts which connected the individual boost pumps with the two main engines. No anti-vortex baffles were used, although boost pump inlet vanes were included. Each boost pump unit was submerged in the propellant tank outlet sump of each tank. The pumps raised the propellant pressures from the low values in the tanks (LOX at 206.85kPa [30 psia]; LH_2 at 137.9kPa [20psia]) to the higher pressures required at the main engine turbopump inlets. LOX boost pump speed was approximately 3,600rpm and was driven by the turbine through a 9.1:1 gear reduction. LH_2 boost pump speed was approximately 7,600 rpm, driven through a 5.97:1 gear reduction. Nominal steady state head rise for these two operating conditions was approximately 227.5kPa (33psi) 110.3kPa (16psi) for the LOX and LH_2 boost pumps, respectively.

The hydrogen peroxide propellant supply system used regulated helium pressure at 2,068kPa (300psia) to collapse the flexible bladder and thus force the mono-propellant out through a perforated standpipe. The peroxide was expended through a boost pump system and the use of the auxiliary propulsion reaction control system. Total nominal expended peroxide was 200.9kg (443lb). Allowing 20.4kg (45lb) for performance margin, a total capacity of 221.3kg (488lb), or 159,229cm^3 (9,715in^3) was required. A second storage sphere introduced with the AC-12 configuration – 79,819cm^3 (4,870in^3) – was added, bringing the total usable volume to 159,639cm^3 (9,740in^3) which was adequate for the needs outlined. Temperature conditioning was provided for the storage spheres during terminal ground operations by a manually controlled 28vdc heater blanket. Supply line temperatures were within a range of 4.4-48.9degC (40-120degF) by constant power line heaters. Expulsion pressure was controlled by a regulated helium supply solenoid.

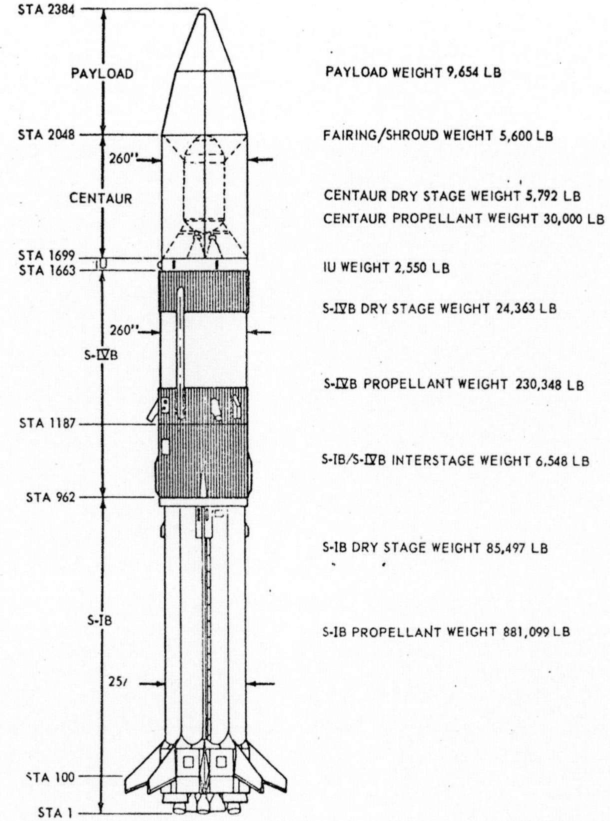

Above: A Saturn IB/Centaur concept was examined for deep space missions, and where a terminal stage was required for advanced missions, but the cancellation of Saturn IB production stalled that possibility. NASA

No modifications to the basic engine system were required for Saturn IB/Centaur although engine pump chilldown was required prior to initiation of the main pump flow to prevent pump stall or combustion instability during the start transient. The system considered for Saturn IB/Centaur employed the essential features of the Atlas/Centaur system. Some new design was required in this system as a result of the new shroud and S-IVB/Centaur interstage structure.

Chilldown occurs in two operations, ground and flight related. Ground chilldown was accomplished by flowing cryogenic helium at flowrates up to 4.5kg/min (10 lb/min) and temperatures below -234degC (-390degF) through the LH^2 pumps. Control parameters included the hydrogen pump surface temperature which had to be held below -218degC (-360degF) and the helium supply temperature held below -234degC (-390°F) both for a 10 minute period prior to launch.

Inflight chilldown occurred either by prestart LOX flow through the pump and vented overboard through the engine injector or prestart LH_2 flow which circulated through both pump stages and would be dumped overboard through a vent system. The inflight venting operation required close control and sequencing of events relative to times when vented gas ignition could occur. Dumped hydrogen vapours were ducted beyond the S-IVB stage envelope. Overall propulsion chilldown sequence included boost pump chilldown and propellant feed system chilldown. The boost pumps were started before the initiation of the main engine chilldown and propellants were circulated through the feed lines and tanks in a closed loop. Other methods could have been used to accomplish engine chilldown that required a lesser penalty associated with the dumped vehicle propellants. These may have included propellant circulation concepts, usage of pump flow thermal conductivity internal coatings, etc. The method outlined here was selected, however, because of its use in the existing Centaur programme.

A major design step was taken with the elimination of flexible pipe and conduit sections. In the LR-115 design the use of rubber and plastic seals was replaced by metallic seals. The high-energy electrical start system for ignition took many parts and components from the J57 turbojet engine, which had itself been subject to tests as to whether it could be made to run on hydrogen fuel. A single signal source for the ignition of the paired engines in the Centaur stage ensured identical start times, unachievable in reality when separate starters are used in parallel motors.

By 1960 the Air Force was thinking of using Centaur on Titan and NASA was planning to incorporate it into cryogenic upper stages for the Saturn I, with P&W commissioned to produce a more powerful version, the RL-119 with a thrust of 77.84kN (17,500lb). For application with Atlas, Centaur was quickly assigned to a series of lunar and planetary missions to follow the Ranger series with Agena B. By the end of 1960 and with NASA driving Centaur, the LR-115 (which bore a P&W/Air Force designation) had been re-designated RL-10. But when technical and management problems hit, the programme slowed and ran into numerous postponements.

The first Atlas-Centaur (AC-1) was launched on 8 May 1962, long after the originally expected launch date, by which time the cryogenic RL-10 engines had been test-fired 700 times, accumulating 60,000sec of running time. To alleviate loads, the propellant tanks were only 40 per cent full and a Christmas tree of sensors was installed to measure the movement of the liquids during flight. A 100kW strobe light illuminated the interior with images recorded on tape every two seconds and assembled into a running movie.

Within 57 seconds of launch the flight was over when external weather shields tore loose causing the hydrogen tank to overheat and explode under increasing pressure, destroying the complete stack. The weather shields had been thought necessary to prevent the liquid hydrogen from boiling off while on the pad and during the early phase of the ascent. But to prevent the panels from cold-welding to the skin of the Centaur stage a helium purge system had been fitted, the 612kg (1,350lb) panels having to be jettisoned as soon as practicable to offset the increased weight.

Three months later, in August 1962 NASA transferred management of Centaur from MSFC to its Lewis Research Center. Paradoxically, von Braun, only a reluctant convert to the tricky vicissitudes of cryogenic fuels, had campaigned strenuously for the cancellation of Centaur, while accepting the necessity of using a version of the RL-10 in his Saturn rocket upper stages. While the stage management went to Lewis, the management of the RL-10 stayed with Marshall and a major development programme for the stage itself was implemented, the first several flights being with A, B and D models of the Centaur.

The second flight of an Atlas-Centaur, AC-2 on 27 November 1963, went well and was followed by a further test, AC-3 on 30th June 1964. An improved engine, the RL-10-A-3, had been introduced on AC-2, with a slightly reduced thrust of 65.6kN (14,700lb) but a specific impulse of 444sec. Due to skirt modifications which raised the expansion ratio from 40:1 to 57:1, the engine was 249cm (98in) in length. It was this time that MSFC got seriously interested in adding Centaur, as either a second or third stage to the Saturn IB. Launch of the AC-6 flight on 11 August 1965 heralded the introduction of the Centaur D that would remain the definitive stage platform from which further modest variations were applied in successive years.

Success with the Centaur came slowly and only after sustained improvements and modifications to the stage and the propulsion system over several years. An early beneficiary of the improved lift capacity of the Atlas was the Surveyor lunar soft-landing programme, with demonstration missions beginning with the second launch as noted above. Several flights later, AC-6 was followed by AC-8 on 7 April 1966 but this failed prior to the next launch, AC-10 on 30 May 1966, which successfully sent the Surveyor 1 spacecraft to a soft-landing on the lunar surface. Thereafter, the remaining five Surveyor spacecraft were successfully sent on their way by Atlas-Centaur, the last on 7 January 1968 but two failed long after separating from the launcher.

Solids and Hybrids

During the second half of the 1960s reimbursable launches were beginning to emerge, opportunities for friendly foreign powers to buy the services of NASA for launching satellites or space probes on a 'cost only' basis. But only for non-US payloads judged by the US State Department to offer no competitive commercial challenge to US interests. Chrysler put together a sales package for attracting customers to the Saturn IB but also to promote the capabilities of this launcher to other divisions within NASA and external US government agencies. Internal studies at MSFC and Chrysler itself offered two, three and four-stage versions. It offered a range of auxiliary payload options and boasted man-rated reliability for a group of derivative options capable of satisfying a wide range of customers.

The philosophy adopted was to advertise space for supplementary payloads, small packages of instruments or experiments located in places on the launcher where they would not interfere with the primary NASA mission but which afforded volume where they could be carried. These included locations in the Apollo spacecraft, the Lunar Module, the Instrument

Right: Development of the Titan II ICBM as a space launch vehicle evolved through the use of solid propellant boosters, as evidenced by this Titan IIIC, the first to carry solids but here represented by the type used to carry the Manned Orbiting Laboratory boilerplate and Gemini spacecraft on its only test launch. *USAF*

Unit or the S-IVB and the Centaur if that should be carried for a specific mission. Optional shroud designs were available for these payloads, or piggy-back modules of relatively small size and low weight, even including small satellites for deployment in an appropriate orbit commensurate with the primary mission of the launcher. This was an early expression of space commercialisation and opened the market to strap-on options for small and proportionate fees.

Within NASA, a more determined effort was made in 1966 to examine optional alternatives on the basis that any future space programme after the initial Apollo Moon landings would come to rely on an intermediate launch vehicle. They were for payloads between the Thor and Delta class and the Atlas and Titan space vehicle derivatives for small and intermediate-size payloads and the colossal Saturn V for large or massive payloads destined for low Earth orbit or deep space. At this date, post-Apollo prospects were still strong, albeit reduced in scale from early expectations of a consistently sustainable manned space programme foreseen for the 1970s and 1980s. The potential for the Saturn IB was great in the perceived market for launcher requirements. Because some of the anticipated Saturn IB missions for the Apollo Applications Program would not take up all the space or mass available on any single launch, great potential existed to market that up to the capacity of the vehicle and this was specifically defined in the marketing material supplied by Chrysler and NASA.

As related earlier, added potential was made available with the prospect of a three-stage Saturn IB, the Centaur added on top of the S-IVB to provide a capability of sending 5,580kg (12,300lb) to the vicinity of the Moon or 4,355kg (9,600lb) to the nearest planets. There were much wider opportunities when solid propellant strap-on boosters were offered utilising existing motors developed for the Titan programme and space-launchers developed out of the Titan II ICBM. Identified by the prefix MLV (Medium Launch Vehicle), several classes of launcher were studied and approved as potential upgrades to the Saturn IB.

Designated MLV-11.5, this upgraded Saturn IB launcher had four United Alliance 304.8cm (120in) solid rocket motors (UA-1205) added to the first stage with a zero-stage ascent profile in which the eight H-1 engines were ignited at altitude, much as had been planned for Titan III programme from whence came the solids. With these, the launcher had a capability of placing 35,380kg (78,000lb) in low Earth orbit, 5,216kg (12,000lb) to an escape trajectory or 9,980kg (22,000lb) with an added Centaur stage above the S-IVB. In basic configuration without the Centaur, the MLV-11 would have had a total mass of 1,549,791kg (3,416,648lb) with an optimised payload of 33,566kg (74,000lb) but an enhanced S-IB stage thrust of 7,978kN (1,793,672lb) due to the higher altitude at ignition.

The configuration which would have been used with the Saturn IB had a lift-off thrust of 20,460.8kN (4,600,000lb) from the four solid rocket motors. Flight control would be effected by the thrust-vector-control (TVC) system on the solids with control switching to the Saturn IB's guidance control system prior to ignition of the H-1 engines, 1min 48sec after lift-off and before immediate separation of the solids 7sec later. The advantage in starting the H-1 engines before jettisoning the solids ensured consistent axial orientation during the thrust

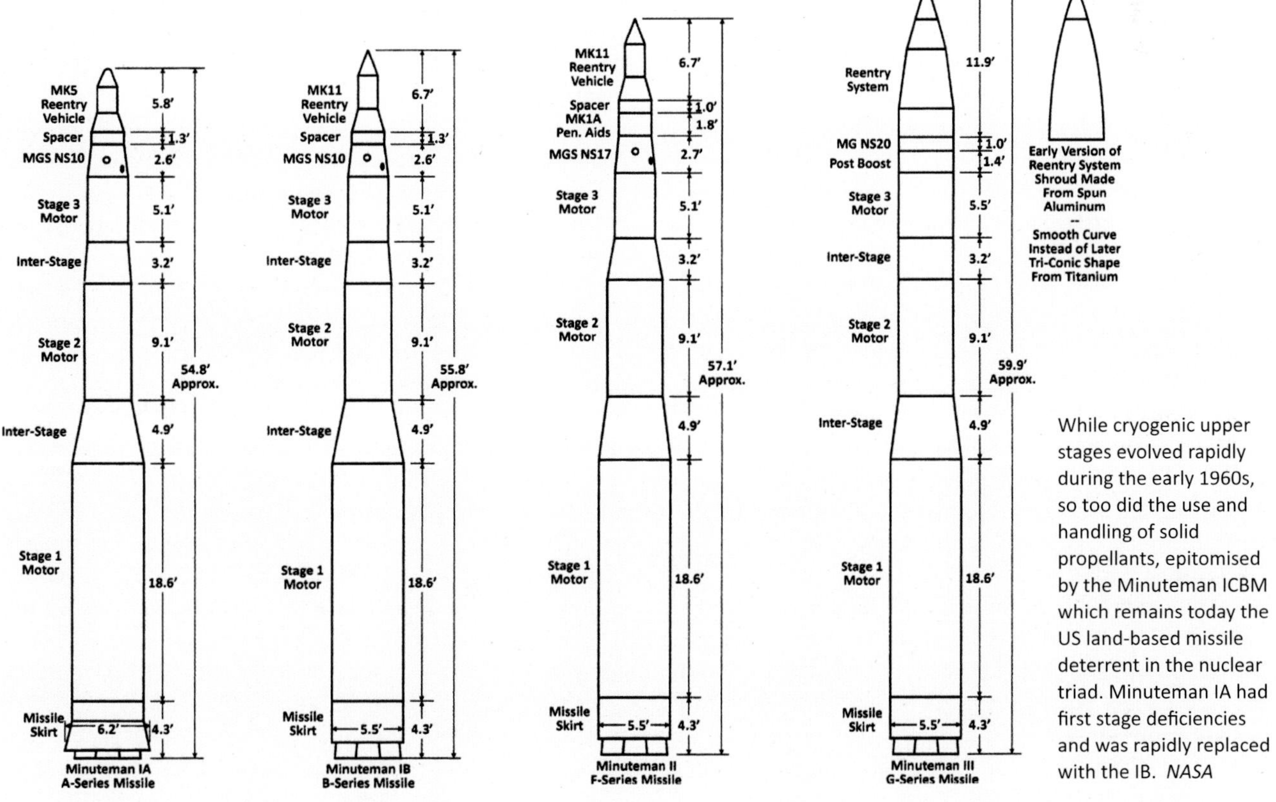

While cryogenic upper stages evolved rapidly during the early 1960s, so too did the use and handling of solid propellants, epitomised by the Minuteman ICBM which remains today the US land-based missile deterrent in the nuclear triad. Minuteman IA had first stage deficiencies and was rapidly replaced with the IB. *NASA*

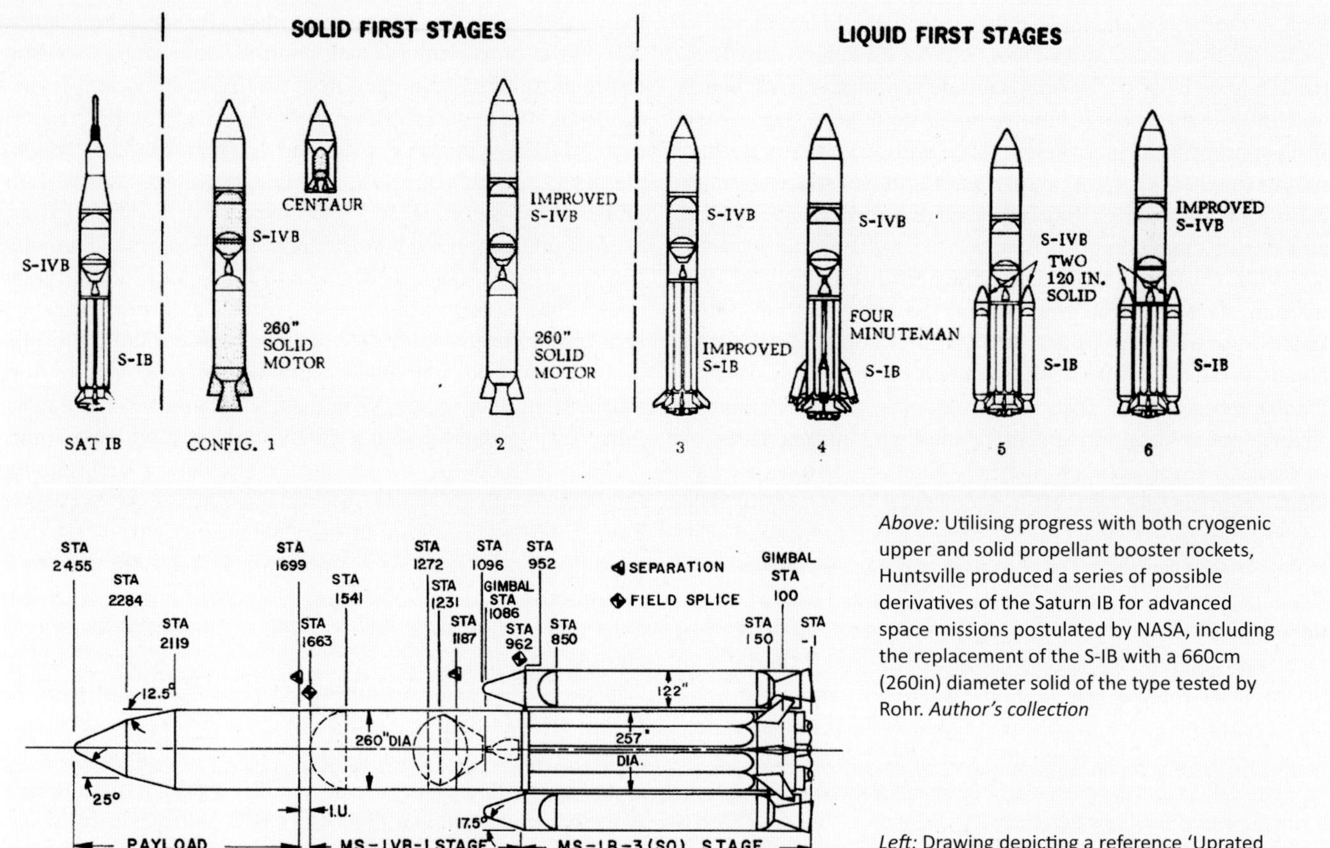

Above: Utilising progress with both cryogenic upper and solid propellant booster rockets, Huntsville produced a series of possible derivatives of the Saturn IB for advanced space missions postulated by NASA, including the replacement of the S-IB with a 660cm (260in) diameter solid of the type tested by Rohr. *Author's collection*

Left: Drawing depicting a reference 'Uprated Saturn IB' as derivatives were known. It incorporates the same design of solid rocket boosters as those employed for some Titan space launchers. *Author's collection*

tail-off and optimum momentum at S-IB ignition and gimbal-control of attitude. A derivative variation on this configuration, the MLV-11.5A would have used the same 5-segment solids to extend the burn duration on the S-IB by increasing its length by 6.1m (20ft) to raise the payload capability to 39,917kg (88,000lb) to low Earth orbit.

In another option, the MLV-12 had only four H-1 engines and those and the four solids were ignited on the pad. The base fins were deleted on both MLV-11.5 derivatives. The MLV-13 arrangement had only two solid rocket motors. MLV-14 and MLV-15 were boosted by four and eight Minuteman SRMs respectively, each delivering a thrust of 889.6kN (200,000lb). S-IB stage fins were retained for these, each Minuteman first stage being sufficiently small to allow space at the base of the Saturn stage for the solids to be attached. All these configurations were made to the definitive S-IB stage modifications including the 911.8kN (205,000lb) thrust H-1 engines, although some propellant mixture ratio shift was utilised for the S-IVB stage.

By 1967 a thorough examination of all these possible configurations had been run through exhaustive analyses scanning each on credibility for payloads, maximum dynamic pressure, solid motor impact points after jettisoning, vehicle controllability and lift-off motion. Parametric analyses examined prime variables for S-IB propellant tank extension and the number of segments (five or seven), profile of the ignition sequence and the optimum jettison time during ascent. A very wide range of optional variations were possible but only five made it through the filters and these included MLV-11.5 with UA-1205 (5-segment) solids, MLV-11.5A, MLV-11.7A with UA-1207 (7-segment) solids, MLV-13.7 and MLV-14. All are shown on these pages. The most powerful was the MLV-11.7A which boasted S-IB tank extension, seven-segment solids and a lift-off thrust of 24,908.8kN (5,600,000lb) with a maximum payload capability of 48,082kg (106,000lb).

The many and varied mission options to which these various launch vehicle concepts were attached is too lengthy and detailed to discuss here and will be left to a separate volume. Suffice to say that these variants were judged highly applicable to Earth orbit operations for supply and logistical deliveries, to support for space stations such as the Manned Orbiting Telescope (MOT), the National Research Observatory (NRO) and a Manned Orbital Research Laboratory (MORL); Over time and after several decades, the MT would appear as the Hubble Space Telescope and the MORL would go through several generations of design iterations to become the International Space Station (ISS). It is interesting to speculate, as we will do in a companion volume to this, that the MLV-11.7A could have been used to launch a reusable shuttlecraft while less powerful variants could have lifted the separate elements of the station for very much less that the coast of the Space Shuttle and associated assembly and resupply flights flown by the US and its international partners.

To put the Saturn IB derivatives into context we must get into launch costs here and that is a murky pool in which to swim. Expectations of Space Shuttle costs were increasingly, and somewhat outrageously optimistic, as the programme evolved between 1968 and 1972, when the final commitment to fund it was made. Adjusting all figures to 2021 values, as assessed historically, the average price of a Shuttle mission across the 30 years of flight operations was $1.125billion, whereas today the cost of a Saturn IB MLV-11.7A would be $500million. However, the Shuttle's payload limit to low Earth orbit of 27,200kg (60,000lb) was only 57 per cent of the lift capacity of the Saturn 11.7A with its four solid boosters. If non-recurring development costs are included that figure reduces to 45 per cent discounted for infrastructure costs for the Saturn launcher as well.

Saturn IB MLV-11.7A would have had 177 per cent the payload capacity of the Shuttle for a price per kilogramme of $10,399/kg ($4,717/lb). The Shuttle had an equivalent price ticket of $45,955/kg ($20,833/lb) when expressed in 2021 money. None of these cost-equivalent matrices were valid at the time because the econometric analysis of the Shuttle was still several years ahead and the decision to abandon the Saturn I/IB had already been made. NASA wanted to eliminate unnecessary infrastructure and removing two launch sites which had been dedicated to those launchers helped contain outlays. These issues were pivotal to whether launch vehicles came or went and occurred before NASA began preparations for a radical transformation in the direction of future manned space flight operations.

Despite the political and policy shifts impending, cost analyses showed that a significant payload increase could be had by the adoption of four or eight strap-on Minuteman solids, as in the MLV-14 and MLV-15 designs. At the other end of the performance enhancement, the MLV-11.7A would lift-off with the four solids and the four outer H-1 engines firing, with ignition of the four inner H-1s at 1min 55sec just prior to thrust tail-off of the solids and their jettison 7sec later. The separation system and the dynamics of how that would be achieved closely followed the process utilised in the Titan IIIC profile from whence the solid themselves came. Inner H-1 cut-off would occur at 4min 11sec followed by the four outers 6sec after that and 6sec prior to S-IVB ignition. Orbital injection would occur at 11min 45sec. Cost savings would have been great had these solids been used to supplement the Saturn IB if due only to their large actuarial spread of recurring and non-recurring development and manufacturing costs.

The Saturn IB also offered outstanding flexibility not displayed by any other vehicle or arrangement of configurations. Merely by engineering an S-IB tank extension for extended burn duration, and making available Titan or Minuteman solids, the size and cost of a specific mission could be closely tailored to match the actual needs of the user, with little or no wasted volume or lift capacity. Throughout these exhaustive studies, MSFC and Chrysler demonstrated a capability for mission flexibility unmatched in launcher history – to this day, except perhaps for the Thor and Delta derivatives. That they were not developed or utilised was down entirely to a decision to cancel further development, and use, of both the Saturn IB and the Saturn V at a time when NASA was moving inexorably toward the development of an entirely new way of operating the national space programme – a reusable vehicle, the Space Shuttle.

The potential for varied payload requirements reflected the expectations for future missions as understood in the mid-1960s. Most of the AAP planning structure divided the establishment of a permanent presence on the lunar surface into three discrete phases. Initially, surface exploration would consist of small bases with temporary occupancy over several places on the lunar surface. It was expected that sufficient time would be allowed between sequential lunar landings to make adjustments and to modify specific equipment based on results from the previous flight. The second phase would shift to support for a base crew of approximately six people remaining at the surface for six months to two years. From there, major scientific investiga-

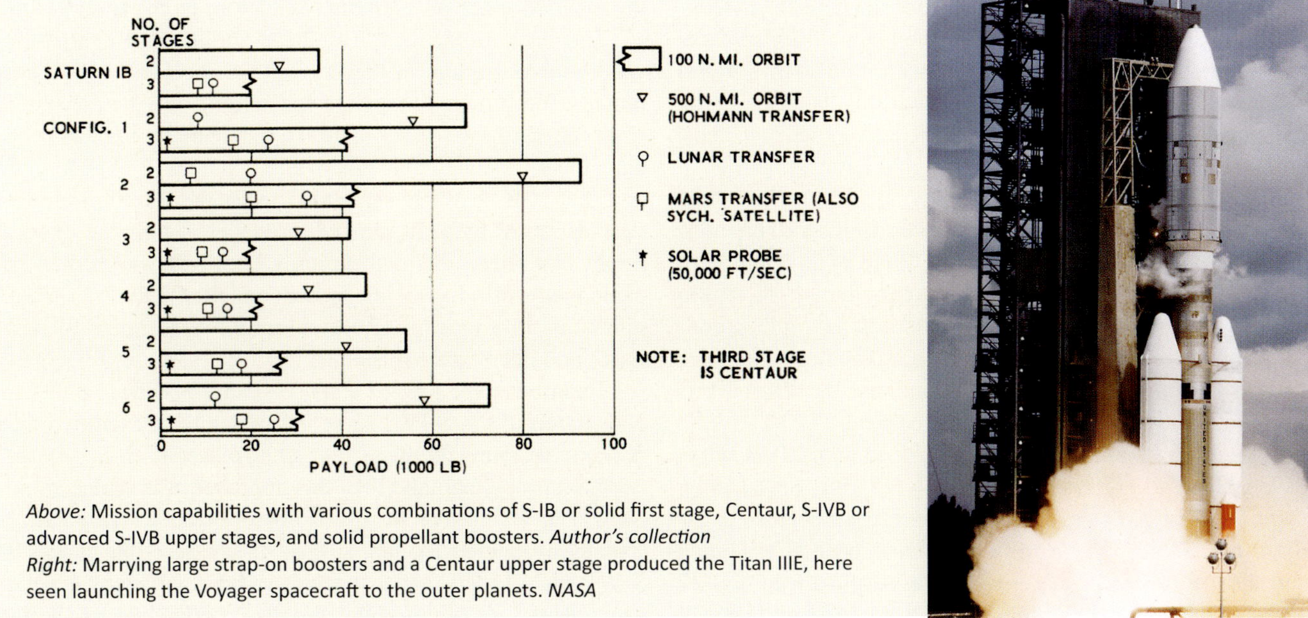

Above: Mission capabilities with various combinations of S-IB or solid first stage, Centaur, S-IVB or advanced S-IVB upper stages, and solid propellant boosters. *Author's collection*
Right: Marrying large strap-on boosters and a Centaur upper stage produced the Titan IIIE, here seen launching the Voyager spacecraft to the outer planets. *NASA*

Saturn IB – Solid rocket motor supports

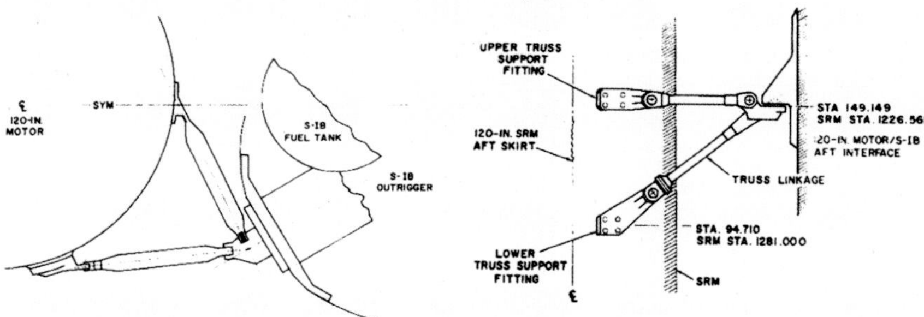

Above: Support for the zero-stage solid rocket motors employing truss members attached to the aft skirt support, with perspectives looking forward (*left*) and from the side (*right*). *Author's collection*

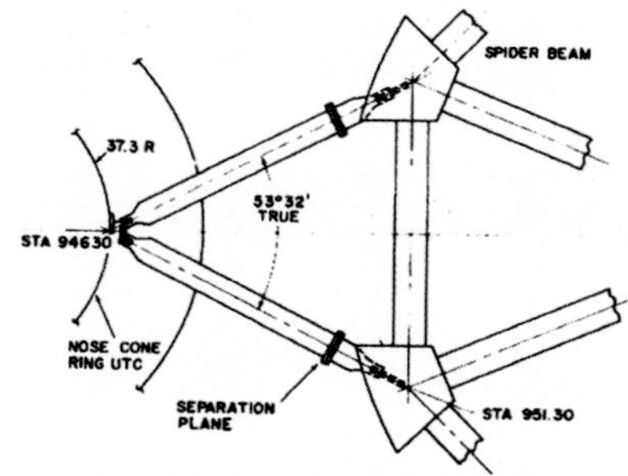

Above: The solid rocket motor forward support structure with location of the attachment point at station 946 at the spider beam position on the S-IB. *Author's collection*

tions and periods of geologic exploration would build a highly detailed examination of a wide area around the base. Sustained and reliable logistics and resupply flights would be essential for maintaining the facility. Phase three would build a base enduring for between five and ten years, supporting up to 12 people for major geoscientific examination.

During this period, other visits to bases established in the second phase would increase the number of lunar flights to an average of one every two months. If that was to be realised there were numerous payload requirements for supplying these bases and other orbiting facilities in the overall deployment of AAP objectives. For AAP and for other heavy-lift requirements, the only competitor in the mid-1960s was the Titan IIIC, a developed evolution of the Titan II with two solid rocket boosters and a payload capability of 1100kg (28,900lb) to low Earth orbit or 1,200kg (2,650lb) to the vicinity of Mars, much less than the basic Saturn IB. There was no other launch vehicle to compare with the flexibility and the performance of the growth variants described here. But several companies were excited by the inherent possibilities and conducted studies, most on behalf of NASA looking at a broader range of possibilities.

One intriguing proposal merged from a detailed study conducted by Bellcom which identified the potential for replacement of the conventional S-IB stage with a 660.4cm (260in) diameter monolithic solid propellant stage, that launcher being designated Saturn IB-5A. This had first been mooted in a report from Douglas Missiles & Space Systems Division earlier in 1966 and followed a Saturn IB Improvement Study commissioned by MSFC in early 1965 which was itself based on internal study reports from MSFC's Future Projects Office in September 1964. The solid stage had a total length of 41.8m (137.4ft) and came directly out of a proposed Saturn IB-5 concept. The developed variant was approximately one-third longer with a tailored grain design to optimise the thrust levels and more closely match the performance need.

Including a hypothetical Apollo payload and S-IVB stage, this launcher had a total height of 85.9m (281.9ft) and a launch mass of 1,872,000kg (4,127,000lb) providing a payload capability of 42,945kg (94,675lb) to low Earth orbit. The launch environment was such that a water-basin spur of the LC-37 site at Cape Canaveral was required to dampen acoustic shock waves. Some may question the logic of replacing a trusted first stage with a much bigger and heavier launcher capable of less payload lift than the Saturn IB MLV-11.7A. The logic was buried within the cost savings and reduced pre-flight preparation time with the all-solid first stage which, while much less efficient and therefore less payload-capable, was more than offset by cost reduction. Hence, as cost savings crept ever higher up the priorities list at NASA, consideration of that as a cheaper option achieved leverage. But without the S-IB stage, was it a Saturn IB at all?

In summary, much of the motivation for expanded capabilities came on the back of expectations that NASA would move forward with a sustained period of lunar exploration, while simultaneously supporting manned space stations and large bases in Earth orbit, in geosynchronous orbit and in lunar orbit. These would require continued production of the Saturn IB and the Saturn V – only economical if supporting a manifest populated with high annual launch rates for a wide range of missions. From the mid-1960s studies informed plans brought before Congress but since these were made at a time when the NASA budget was still rising inexorably, and funds were sought for the war in Vietnam, support drained away and the manned space programme came close to collapse.

The background to all of these studies and evaluations was the expectation, in the first half of the 1960s, for a sustained commitment to manned space flight and the expanding exploration of the solar system. To the decision-makers of the day, it was inconceivable that events would unfold as they did and that NASA would reset the clocks and regroup along a very different path. It allowed, however, a glimpse of what might have been, defined through Apollo-X which then became the Apollo Extension System and finally the Apollo Applications Program (AAP), involving a wide range of near-Earth and deep-space missions.

Tentative steps toward a space station produced plans for conversion of the S-IVB stage into a habitable facility, initial concepts adapting the stage after it had placed itself in orbit and expelled propellant, and then this approach was replaced by a decision to fit it out on the ground before launch on a Saturn V. Eventually named Skylab, suffice to say here that it was an S-IVB initially assigned to the Saturn IB SA-212 vehicle which was adapted into the Orbital Workshop launched by the last

Saturn IB – potential variations

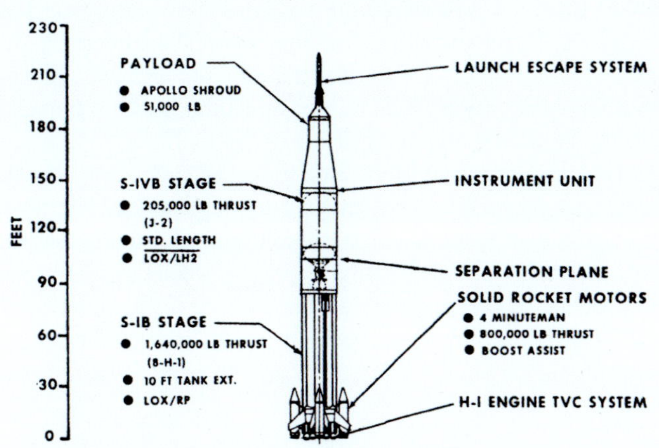

Extending the length of the first stage and adding four Minuteman solid boosters increased payload by 25 per cent.

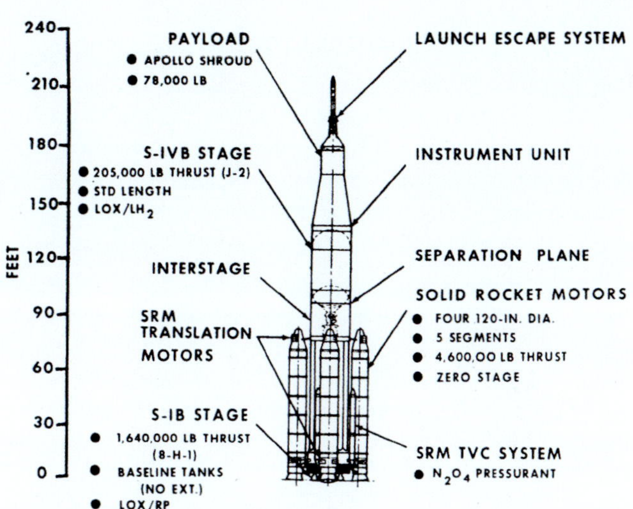

With four solid propellant five-segment boosters from the Titan III programme, payload capacity grows by 95 per cent.

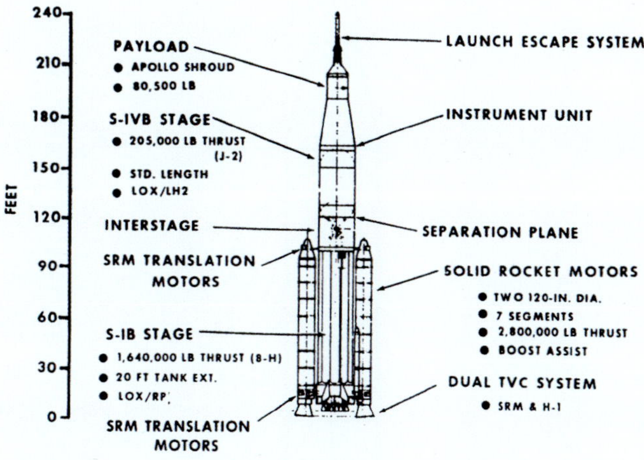

Incorporating a 6.1m (20ft) first stage tank expansion and two seven-segment Titan boosters, payload increases by 100 per cent.

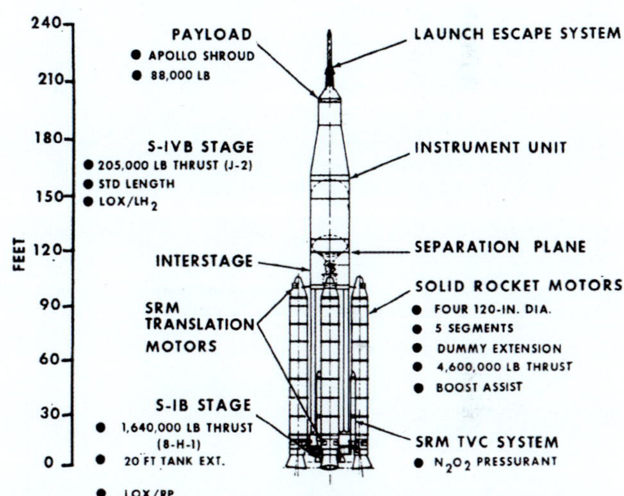

With and S-IB tank extension and four five-segment boosters, payload capacity increases by 120 per cent.

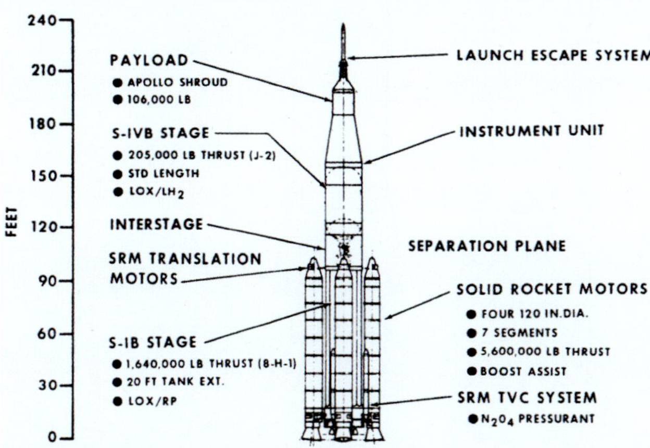

With the tank extension plus four seven-segment solid boosters, lift capacity is increased by 165 per cent – a capability midway between the standard Saturn IB and the Saturn V.

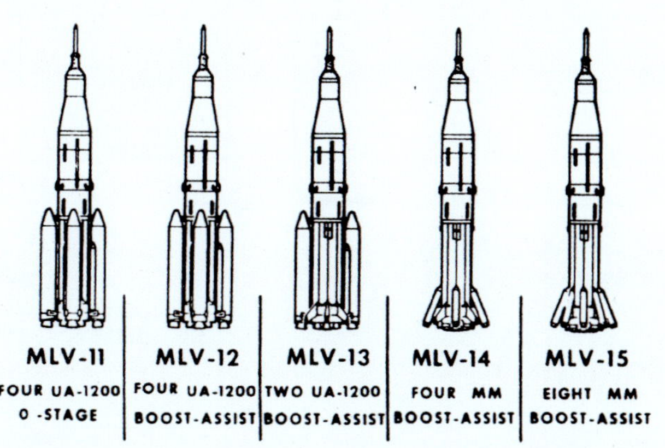

Five configurations were adopted for further possible incorporation into the roster of launch vehicles for deep-space or heavy low Earth orbit missions. MLV-11 would have provided air-start for the S-IB stage.

Saturn V (AS-513) in 1973.

As emphasis focused on being able to make a landing in 1969, plans for future Moon missions began to look uncertain. Prospects for further production of both the Saturn IB and the Saturn V looked bleak, all future manned flights being pegged to the remaining hardware in storage or in production. Crafting a fiscal budget in line with a lacklustre response in Congress, NASA Administrator Webb commented that 'The future is not bright'. The original plan had been to push on with Saturn IB rockets beyond the 12 fully funded vehicles but as part of preparations for submitting the Fiscal Year 1970 budget, on 1st August 1968 NASA stopped work on long lead procurement for SA-215 and SA-216. Long lead contracts for Saturn Vs SA-516 and SA-517 were also brought to a halt.

Five complete Saturn IBs for SA-209, SA-211, SA-212, SA-213 and SA-214 were built but never flown and two S-IB stages were built and stored for what would have been SA-215 and SA-216. No S-IVB stages or Instrument Units were assembled for those stages. SA-209 served as the back-up launcher for Skylab SL-IV and the ASTP mission.

Of interest to note, some small components completed for some of those rockets were secreted away for special memories of a sad retreat from deep space exploration for many decades to come. A few are to be found in private collections and some not insubstantial items are in the back lots of former employees within the Saturn family, which is where they should be.

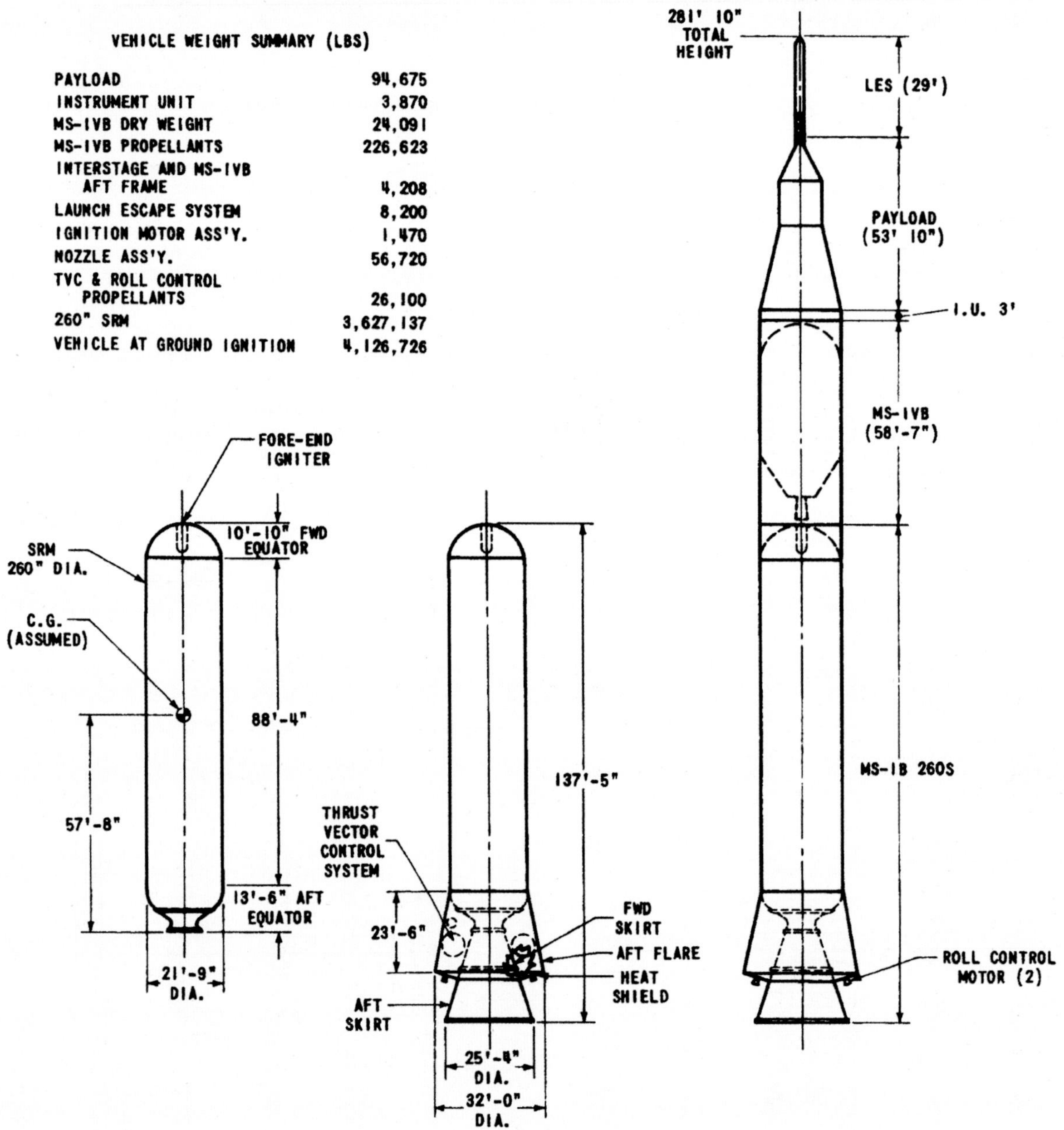

A 1966 study by Douglas defined a successor to the standard Saturn IB replacing the S-IB stage with a 660cm (260in) monolithic stage and fixed nozzle plus yaw motors and a standard cryogenic S-IVB stage. It was believed that considerable cost savings could be made based on a solid first stage. *Author's collection.*

Section 4

Saturn Flight Results

Chapter Nine

Launch Sites

When it came down to it, all the years of hard work would only be fulfilled when the mighty Saturn rocket took to the skies. But there was a lot of development work in those 19 demonstration and operational flights flown by the Saturn I and the Saturn IB. Preceding that, as described in Section One, development of the A-4/V2 began at White Sands Proving Ground and continued there with evolution to the Redstone and then the Jupiter missiles before launch operations switched to Cape Canaveral. It was from Cape Canaveral Air Force Station that all the Saturn I flights, and five of the nine Saturn IB flights began with the last four launching from the Kennedy Space Center.

Launch site development and evolution was a crucial part in getting the rockets operating successfully and much had to be engineered into their design which came through from previous experience and, in the case of Saturn, from intuitive awareness of the wide range of parameters that had to be satisfied. The challenges were immediate with the use of LOX but when cryogenic LH_2 was introduced the engineering requirements entered a new and untested area. It was because of the complexity of these varied propulsion demands that the story of the respective launch sites is at least as intriguing as that of the rockets and launch vehicles themselves.

When the German engineers from the A-4/V2 programme arrived in the United States they brought a unique asset – the experience of having built and launched several thousand missiles over a 10-year period. This gave the Americans a head start in the operational testing of large rockets and quickly defined the way in which launch pads and test sites would be built and integrated into a comprehensive test and development structure. Equally, the challenges in management and opera-

Above: Launch Complex LC-34, Cape Canaveral, Florida – site of the Saturn I's first launch. The LC-34 complex supported a single launch pedestal specifically designed for the Block I Saturn I with associated blockhouse (domed building in foreground), launch gantry and rail-mounted service structure, seen here looking east. *NASA*

tional techniques were developed as the rockets evolved and were every bit as evolutionary as the design of the hardware and the introduction of new technology.

Told in greater detail in the Saturn V story, a companion volume to this, the shift from assembly and checkout on the launch pad to stacking and checkout in a weather-proof building required the development of a system for moving the assembled launch vehicle to the pad. Initially established for the Saturn V, the Saturn IB was integrated with that process for its last four missions. This was required so that closeout of LC-37 could proceed, future flights originating from the LC-39 complex in support of the Skylab space station and the ASTP mission.

White Sands Proving Ground (WSPG), New Mexico

The establishment of a proving ground at White Sands was itself a part of the determination to learn as much as possible about the defining technology behind ballistic missiles. Its construction, however preceded the arrival of the Germans in late 1945, although the major construction works were based on the requirements of the A-4/V2 rocket. The location was one of the chosen sites for an aircrew training ground for the Royal Air Force. In April 1941, Major General 'Hap' Arnold had met with Air Marshal Sir Guy Garrod, Air Member for Training, to set up the British Overseas Training Programme, under which large swathes of the American West would be assigned to training grounds and bombing ranges for aircrew driven from the UK by a German invasion which never came.

On 13 April 1941, Alamogordo Army Air Field (AAAF) was established and by October the government ordered local ranchers to begin selling off their livestock and moving out for the whole area to be adapted into a bombing range. When Japan attacked Pearl Harbor on 7 December, and America declared war on Japan, Germany and Italy, ranchers in 55 townships in four New Mexico counties were notified that grazing licences had been revoked for the new Alamogordo Bombing and Gunnery Range. New construction began on 6 February 1942 and was completed on 1 June, a massive influx of military personnel inflating the region with an extra 45,000 trainee aircrew.

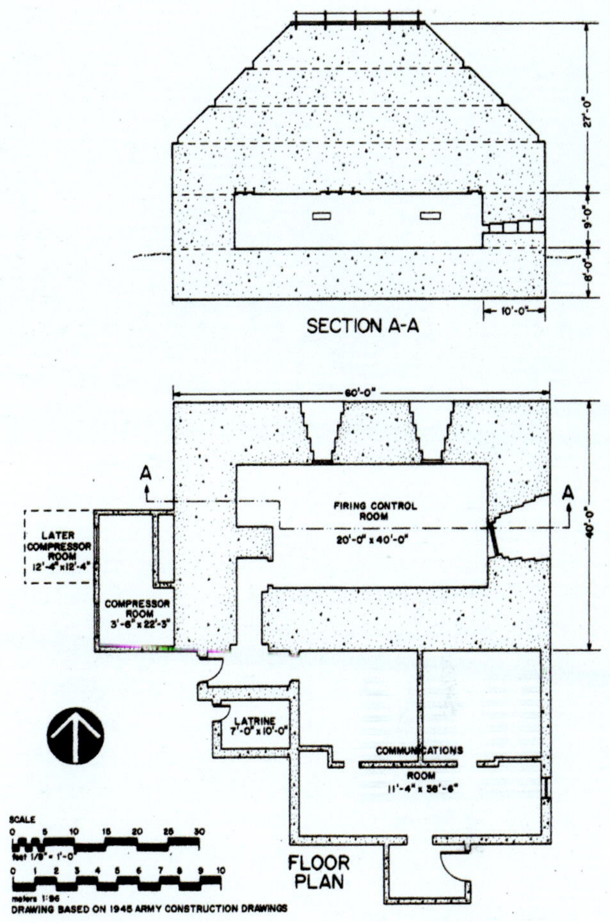

Above: The LC-33 at White Sands blockhouse was built specifically to support flights with the A-4/V2 rocket, construction starting in July 1945 in anticipation of the arrival of German rockets and their personnel. *US Army*

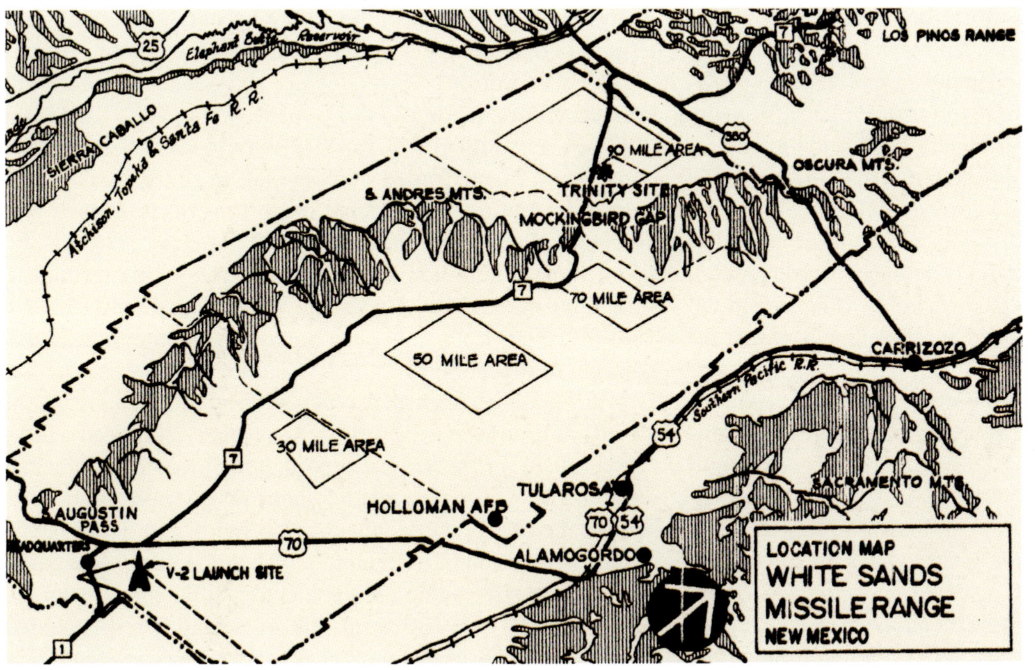

Left: Designated as a weapons test site in September 1943, the post-war rocket facilities at White Sands, just across the Texas border in New Mexico, were equipped with static test rigs and pads in a configuration based on the experience of the German scientists from Peenemünde. *US Army*

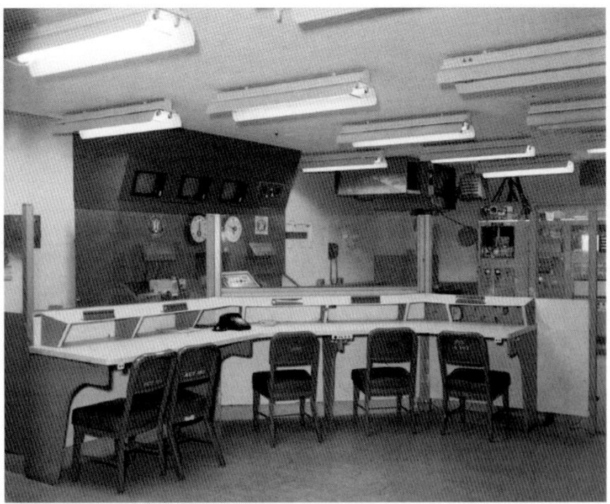

Above left: The LC-33 blockhouse: the first built in the United States for testing ballistic missiles and the place where the Army learned all about operations with the A-4/V2. *US Army*

Above right: Bereft of any technical equipment, the firing control section followed a layout modified by German scientists at Peenemünde and later served as a model for the LC-34 and LC-37 launch facilities built at Cape Canaveral. *US Army*

In June 1942 the development of an atomic bomb undertaken by the Manhattan, New York, Engineer District (Project Manhattan), was moved to a secret site at Los Alamos, New Mexico, some 388km (241mls) away by air and almost directly north from White Sands. On the establishment of the Army Ordnance Corps in September 1943 the site was considered to be the home for testing small guided projectiles then in development. Prior to its selection the US Army had searched several possible sites where development firings could take place.

Eliminating several candidates, the Army settled on a large area around the White Sands National Monument where the isolation, dry air and shielded topography was ideal for firing rockets. In February 1945 the Army ordered the existing military ranges, parks and private land within the boundary to be contained as a military test site. At this date the war in Europe was rapidly coming to an end but there was concern about the length of time it would take to defeat Imperial Japan and on 25 June the Corps of Engineers began preparing the site for rocket tests, just nine days after the world's first atomic bomb had been successfully detonated at the nearby Trinity site, close to the town of Alamogordo.

Headquarters for the site was located south of the White Sands-El Paso highway and an initial job for the Corps of Engineers was to drill a half-dozen shallow wells south and east to provide drinking water until more complete deep wells could be sunk on site. Water for mixing concrete came from the array of flooded mines in the surrounding hills and mountains. Huts made from plywood were erected, each housing three men, and a number of buildings were acquired from Sandia Air Base near Albuquerque. Old cavalry buildings from Fort Bliss provided timber for other structures. Framing a parade ground, several large barracks were erected where the officers and the 40 Germans permanently stationed at White Sands would reside; the other Germans were at Fort Bliss.

In February 1946 a second building was under construction and this became the 'Mill Building', a three-bay, rectangular steel structure with a length of 55m (180ft) and a width of 36.5m (120ft) where the German rockets would be assembled with the help of a crane running the entire length of the structure on rails laid in the floor. Lean-to bays for sub assembles and offices were provided against opposing walls. Flow tests with the rocket motors were conducted in an outdoor structure known as the Propulsion Unit Calibration Stand using water pumped through the motor at 1,834kPa (266lb/in^2). To protect personnel from the volume being discharged at such a high pressure, a small concrete hut was built close by and dubbed 'von Braun's blockhouse'!

The initial launch structure consisted of a single concrete slab from which rockets could be fired, with operators housed in a large concrete blockhouse. With the imminent availability of the V2, however, a second pad was built – Launch Complex-33 (LC-33) – set up 10.5km (6.5mls) from the headquarters area. This consisted of a rectangular area 457m (1,500ft) by 312m (1,025ft) with the long axis in the same north-south alignment as the proving ground itself. A 7.6m (25ft) wide service road surrounded the inside of the perimeter in an approximate square 167m (550ft) by 183m (600ft). Construction of the blockhouse started on 10 July 1945 and was specifically designed to support the A-4/V2 rocket which had such terminal impact on striking the ground that a direct hit would dig out a 9.1m (30ft) crater. The blockhouse walls were 3m (10ft) thick containing a work space 6.1m (20ft) by 12.2m (40ft) with a height of 2.7m (9ft). Pyramidal in form, it was constructed of reinforced concrete, 8.2m (27ft) thick at the apex with the long side facing north toward the launch pad and incorporating two observation slits, each 91cm (36in) by 15cm (6in). The entrance was on the south side through a steel door buffered by a concrete wall and a 4.5m (15ft) by 15m (50ft) concrete slab was poured outside the blockhouse adjacent to the entrance as a blast shield.

The second launch pad was set up approximately 298m (980ft) north of the LC-33 blockhouse and comprised a 21m (70ft) square for the WAC Corporal which would be integrated with the A-4/V2 as second stage for Bumper flights. To service the rocket when launched on its own, a 31m (108ft) steel tower was built, triangular in plan and with vertical guide rails for it to slide along until achieving sufficient velocity for stable flight. There had been talk of testing a rail-launched A-4/V2 but that was never taken up and there is no record of a serious plan to do so. A separate rail track was laid between the blockhouse and the Corporal launch rail and this was accompanied for about 122m (400ft) by a second track. The first rocket launched from the WSPG was a modified Tiny Tim, a booster for the WAC Corporal, on 26 September 1945, followed by a WAC Corporal A on 11 October. The 1st Guided Missile Battalion was formed the same day and stationed at what, since 9 July 1945, had been designated the White Sands Proving Ground (WSPG).

Only a few weeks after construction began, the arrival of 300 rail cars packed with German A-4/V2 structures and components and rocket motors took the Corps of Engineers by surprise and forced a rethink on plans for the complex. Nobody had told them of the magnitude of the operation. The entire operation was still a secret to prevent news leaking out and alerting the Russians. Completed in September 1945, plans provided for an initial A-4/V2 launch platform located 9m (30ft) west of the Corporal platform and consisting of a simple concrete slab 16.4m (54ft) by 17.7m (58.5ft). It was without a tower since the rocket's guidance system was effective from lift-off. Both the WAC Corporal and A-4/V2 pads were connected to the blockhouse by underground cables.

In December 1945 plans were completed for a combined test and launch facility comprising a platform and subsurface blast pit which could be used for 88.96kN (20,000lb) thrust rocket motors together with a gantry crane for moving hardware on and off the platform. Work started early in 1946 and incorporated an excavated blast pit 9.1m (30ft) deep, a V-shaped flared incline rising back up to the level of the desert floor 29m (98ft) away. Through this would be channelled the exhaust efflux from rocket motors. The walls were lined with prefabricated concrete and armoured with overlapping steel plates, the platform over the blast pit having an opening 2.1m (7ft) square. Later, a sprinkler system was added for quenching the blast and reducing the effect of the efflux on the steel facing.

An integral part of the test pad was a gantry moved along two parallel rails, 8.2m (27ft) apart extending 167m (550ft) north. It comprised two open steel towers, 18m (60ft) tall mounted on two pairs of rail wheels. The towers were connected at the top by a steel truss and integrated platform surmounted by a 13.6tonne (15ton) chain hoist for lifting equipment or even small upper stages. Work platforms were provided for access to various levels, hinged to rotate down and surround the rocket and accessed via steps and ladders. The sizing of the gantry was based on an A-4/V2, clearing a space 4.8m (16ft) and 16.4m (54ft) high. The work platforms were a great improvement over the system employed at Peenemünde and afforded ease of access and work stations adjacent to critical components on the vertical rocket, a system which would serve as a template for the more sophisticated launch complexes built at Cape Canaveral for bigger rockets including the Saturn series.

Two observation towers were drawn up on the original plans, Tower No.1 located 6.1m (20ft) behind the blockhouse with a roofed and partially enclosed platform on a square steel structure 12.2m (40ft) high from where visual observation could be maintained and where cameras and other recording devices could be installed. Tower No.2 was located 9.1m (30ft) west of the launch pad, although this one was only partially assembled before it had to be removed for enlarging the A-4/

Below: The LC-33 rail-mounted gantry displays the familiar design carried forward to the Redstone pads at Cape Canaveral. *US Army*

V2 pad to connect it with the Corporal pad to the east and extend it 17.6m (58ft) to the west. As part of this rework, the service road was widened and concreted and a spur was added to the rail track for the gantry to reach the A-4/V2 pad. The static tests of the A-4/V2 took place at the 100K (or 100,000lb) static test track located several kilometres away.

The original blueprints for the complex and its then radical design and layout were completed by the Albuquerque District of the Corps of Engineers but von Braun and Kurt Debus were behind the designs and provided the knowledge and experience to lay out the site. The US Army had no experience with test or launch facilities of rockets anywhere approaching the size and thrust output of the A-4/V2 and the German engineers had learned through much trial and error the way these facilities could be made effective and efficient. Great emphasis was placed on the ability to access the rocket on the launch pad and this set in train a major design trend which would lead to the gantry designs at Cape Canaveral Air Force Station (CCAFS), Florida and culminate in the Mobile Service Structure (MSS) for the Saturn V at LC-39, work on which began 15 years later.

This combined test and launch facility at White Sands, originally designated US Army Launch Area No.1, was never used for an A-4/V2 launch although it was used for other rockets, the first major launch from there being the Navy's Viking rocket on 3 May 1949. It was tested bolted down and unbolted for launch following a successful static test. The concept of static testing was itself something insisted upon by the von Braun team and this would become an entrenched part of launch preparation and pre-flight procedures. The first A-4/V2 pad was essentially a concrete slab since the rocket was largely independent of any support or guide-rail structure.

In January 1947 further plans were registered for modifications, setting up a second launch pad and adding facilities to the blockhouse for communications equipment and an air compressor. Communications equipment was to be contained in a room 11.9m (39ft) by 9.1m (30ft) and attached to the south side. It had three equipment rooms and was entered through doors in the south wall and in the west wall. Tower No.1 was taken down to make room for this addition and for the air compressor in a separate room 9.1m (30ft) by 4.1m (13.4ft) attached to the west side of the blockhouse. A second viewing slot, 1.5m (5ft) by 33cm (13in), was cut into the east wall but at a slight angle for better observation of the additional pad. Associated with this upgrade was removal of the two tracks out to LC-33.

Above: Upper levels of the LC-33 gantry with moveable work platforms, a key feature of all launch support structures throughout the Redstone, Jupiter and Saturn programmes. *US Army*

Below: Work huts at White Sands where Army personnel developed the flight test programme alongside the German scientists and rocket engineers. *US Army*

Arguably the most important initial capability at White Sands was a rig where the A-4/V2 engines could be tested. As described earlier, of the approximately 2,000 rockets left behind in Germany, in various states of assembly and condition, not a single complete rocket was brought to the United States. They all had to be put together on site, which was reason enough for the several score German engineers essential to this task; without them not a single rocket would have been fired. Although the A-4/V2 motor developed a thrust of 251.7kN (56,600lb) in vacuum, consideration for growth and for bigger engines called for a 100K (in lb thrust) stand supporting engines with almost twice that capability. Although it was possible to test functional elements of the engine with water flow methods, there was no place where the engine could be fired.

Again, this was a facility designed and drawn up by the Germans themselves. It was completed in early 1946 together with a test stand, a small machine shop, water storage tanks (for flow tests and quenching) and blockhouses for control, observation and monitoring. It was laid out in consideration of tests involving the completely assembled rocket, which was to be erected by *Meillerwagen*, a transporter/erector, moved to the site and elevated. A test would involve full monitoring and a record made with high speed still and motion cameras. The place selected was isolated from the launch site and situated in the foothills of the Organ Mountains on a slope facing northeast and approximately 2.4km (1.5mls) from the headquarters building. The base was embedded in the surrounding earth and with the fourth side facing northeast out across the desert. The test stand had to accommodate great forces, with the exhaust efflux reaching a speed of 7,240kph (4,500mph) and a plume temperature of 1,982ºC (3,600ºF).

To retain the A-4/V2 during static, full duration test firing, a special carriage consisting of two steel pylons 15.2m (50ft) in height supported restraining girdles completely encircling and gripping the rocket and accepting the thrust loads transmitted through the body of the missile. The design of the rocket prevented any boattail restraint as would become standard, designed in to later rockets. The carriage itself supported three hinged work platforms which could be lowered for access to critical elements. A ladder on the rear-facing side of the carriage provided access to those work areas. The test blockhouse consisted of a 4.5m (15ft) by 12.2m (40ft) reinforced concrete structure with walls 45.7cm (18in) thick incorporating a control and communications area with two slit windows, each 61cm (24in) by 10cm (4in) in size.

The adjacent instrumentation room would monitor pressure, temperature, vibration, acceleration and event sequences. Test equipment including Baldwin load cells for measuring thrust, an Esterline Angus 20-channel sequence recorder, oscillographs, Wianecko carrier systems and an eight-channel Consolidated System D amplifier. These names and this equipment represented the adaptation of the direct method of testing installed at Peenemünde to a combined all-analogue and digital approach. There were 86 separate data channels which transmitted information between the instrumentation room and the test stand through 12 underground conduits. The machine shop was set up in a light metal building 6.1m (20ft) by 7.3m (24ft) with a series of work benches and machine tools for conducting on-site maintenance and repair. Separately, a 378,500lit (100,000USgall) water tank was positioned southwest and about 61m (200ft) up the hill from the test stand capable of providing up to 3,406lit (900USgall) of water for flame quenching in the exhaust diverter at the test stand.

While all this was going on, during 1946 the German rocketeers at Fort Bliss were confidently predicting the possibility of building an engine with a thrust of 1,334kN (300,000lb), more than five times the thrust of the A-4/V2 motor. This, they said, would be the minimum required to propel a nuclear-tipped missile to its target. There was reluctance on the part of the Army hierarchy to accept this as a serious possibility – no effective operational experience with any rockets had at this date been either demonstrated or catalogued. However, the successive series of A-4/V2 tests conducted from April 1946 showed sufficient promise for approval to build a 500K Static Test Stand, capable of supporting engine tests of up to 2,224kN (500,000lb) of thrust, approval being given in 1948 when construction got underway. Unfortunately, plans for the rocket motor itself, first mooted by the Germans, failed to gain approval. The test stand was not completely designed and then only in modified form before 1950, but it was not used until the fol-

Below: A perspective view of the LC-33 servicing gantry. *US Army*

Left: The Mill Building at White Sands with the A-4 assembly plant at extreme right. *US Army*

Right: The 20K engine test facility with servicing gantry. *US Army*

lowing year and only then for long duration tests of graphite control vanes exposed to hot exhaust plumes.

The 500K stand was built several kilometres from the headquarters building, a huge structure set into unfaulted granite strata at the base of the Organ Mountains with a steel carriage inclined at 60deg with integral propellant tanks and pumps. Two separate pump houses were placed behind and above the motor carriage, each 9.1m (30ft) by 12.2m (40ft) separately dedicated for moving the fuel and the oxidiser, respectively. Two 56,775lit (15,000USgall) tanks were placed in barricade structures to divert any overspill to concrete flow channels. Separate propellant lines, 86cm (34in) thick transferred propellant from the dewars to the test stand where they would be directed to the engine. Situated uphill, a 378,500lit (100,000USgall) water tank would cool the metal-plated, concrete flame diverter at the foot of the motor carriage.

Below: The 500K engine test facility with control and support equipment looking out over the desert floor. *US Army*

The control room was a 27.87m² (300ft²) building set into the side of the mountain and for safety reasons had no direct view of the test stand but reflected views through moveable mirrors which provided visual sight. Most of the interior of the control room was a copy of the 100K control station and a round pillbox on the opposite side of the room provided access via an underground tunnel. Beginning in 1951, modifications to the facility prepared it for tests of the Redstone rocket motor with the 60deg inclined carriage replaced by an 18.3m (60ft) vertical tower incorporating work platforms, an access stair and a lift. The original pumps were replaced with smaller ones for the Redstone engine and one of the pump houses was converted to a machine shop and the other to a storage area. After Redstone, the 500K test stand was used for bigger motors, none of which were associated with the Saturn programmes.

In January 1948 the AAAF was re-designated Holloman Air Force Base and missile test activity continued to expand. In the five-year inclusive period 1946-1950, White Sands hosted 564 missile launches of many different types, of which 329 were conducted by the Air Force and the remainder by the Army and the Navy. On 6 February 1948, General Electric launched the first electronically controlled missile – A-4/V2 No. 36 – and on 11 June, No. 37 carried the first rhesus monkey (Albert I) to a height of 62.75km (39mls) but it died during the flight. On 13 July 1948, the Air Force Project MX-774 launched the first of three Hi-Roc test rockets following a contract to Convair for a missile capable of delivering a warhead over long distances. Powered by a four-barrel Rocketdyne motor with a total thrust of 35,584N (8,000lb), it had integral tanks and gimbals for directional control without guide vanes in the exhaust efflux – the first to incorporate what would be a definitive solution for attitude control. The other two launches occurred on 27 September and 2 December 1948, after cancellation of the project. It would be resurrected as the Atlas ICBM several years later.

Supporting tests with the A-4/V2 rocket in the Bumper programme which began on 13 May 1948, planning for the High Speed Test Track (HSTT) began in October that year with Northrop and Hughes at Holloman, which had by this time been reorganised under Air Materiel Command. As noted previously, the first Viking sounding rocket was launched from LC-33 on 3 May 1949 prior to further land use extension under government sequestration. The von Braun team departed WSPG in 1950 and most tests related to the Redstone and Jupiter missiles moved to Cape Canaveral.

Michoud Assembly Facility (MAF)

Responding to the major expansion programme authorised by the decision to send astronauts to the Moon, on 7 September 1961 NASA announced that it would be taking over a government manufacturing plant at Michoud, near New Orleans, Louisiana, for fabrication of the Saturn launch vehicles.

The plant originated in World War Two for building Liberty ships, transport aircraft and engines for tanks, with a production building 10.67m (35ft) high and a manufacturing floor space of 12.5ha (31acres). MAF had only come into widespread use during the Korean War when Chrysler manufactured tank engine cylinder heads there but it never achieved its total production potential. It was the first of two outstations of the Marshall Space Flight Center, Mississippi Test Facility (MTF) built to support the Saturn V programme and details can be found in the companion volume on that launch vehicle. Located on a water route to the Gulf of Mexico, MAF was perfect for transporting large Saturn rocket stages to Cape Canaveral. It was also close to sparsely populated land which could be used for test facilities.

The MAF was originally purposed to support manufacture of the first two stages of the Saturn I and Saturn IB rockets and the huge S-IC, the first stage for Saturn V. The complex incorporated 360ha (890acres) which accommodated 19.51m² (210ft²) of engineering and office space, 181.99m² (1,959ft²) of manufacturing space and 38,553m² (415,000ft²) for storage, handling and maintenance. Initially, Chrysler settled in to manufacture the Saturn I stages and Boeing used the facility for the S-IC. Mason-Rust did all the support work and kept the place furnished with services and plant.

The facility was originally designated Michoud Operations on 18 December 1961 and assembly of the first Chrysler-built Saturn I first stage (S-I-8) began on 4 October 1962, with acceptance by NASA of the first two stages built by industry on 13 December 1963. On 1 July 1965 the plant was renamed Michoud Assembly Facility.

Michoud quickly filled up with personnel, rising to a peak of just over 12,000 civil service and contractor personnel by mid-1965, more than half being the Boeing personnel working the Saturn V's first stage. Slidell had a computer responsibility there with a total of just over 200 people working the tele-computing support services. They provided the housing and facilities for the digital and analogue computers together with a dedicated computer development and programming centre. Slidell were located about 32km (20mls) from Michoud and about 24km (15mls) from the Mississippi Test Facility, now known as the Stennis Space Center. Overall, staffing levels at the MAF would slowly decline as production of the Saturn vehicles ran down, dropping to around 10,000 by mid-1967.

Sea and Air Transportation

Transporting S-I and S-IB rocket stages between MAF and MSFC and to Cape Canaveral was primarily the role of the barge *Palaemon*, built in Houston, Texas, and delivered to MSFC on 22 November 1960. It was used for moving test and flight stages as well as dummy S-V stages. Attention was made to the forces and stresses imposed on Saturn stages by the forces involved in moving them by water on this barge. It was modified in 1964 for conveying the S-IB stages and the opportunity was taken to add a pilot house and a wing bridge for better control of the vessel and observation of proximal obstructions either on the waterway or adjacent banks.

In addition, in 1961 NASA acquired a Navy barge which it named *Promise* and this was modified for conveying Saturn stages and brought into use the following year. This was used

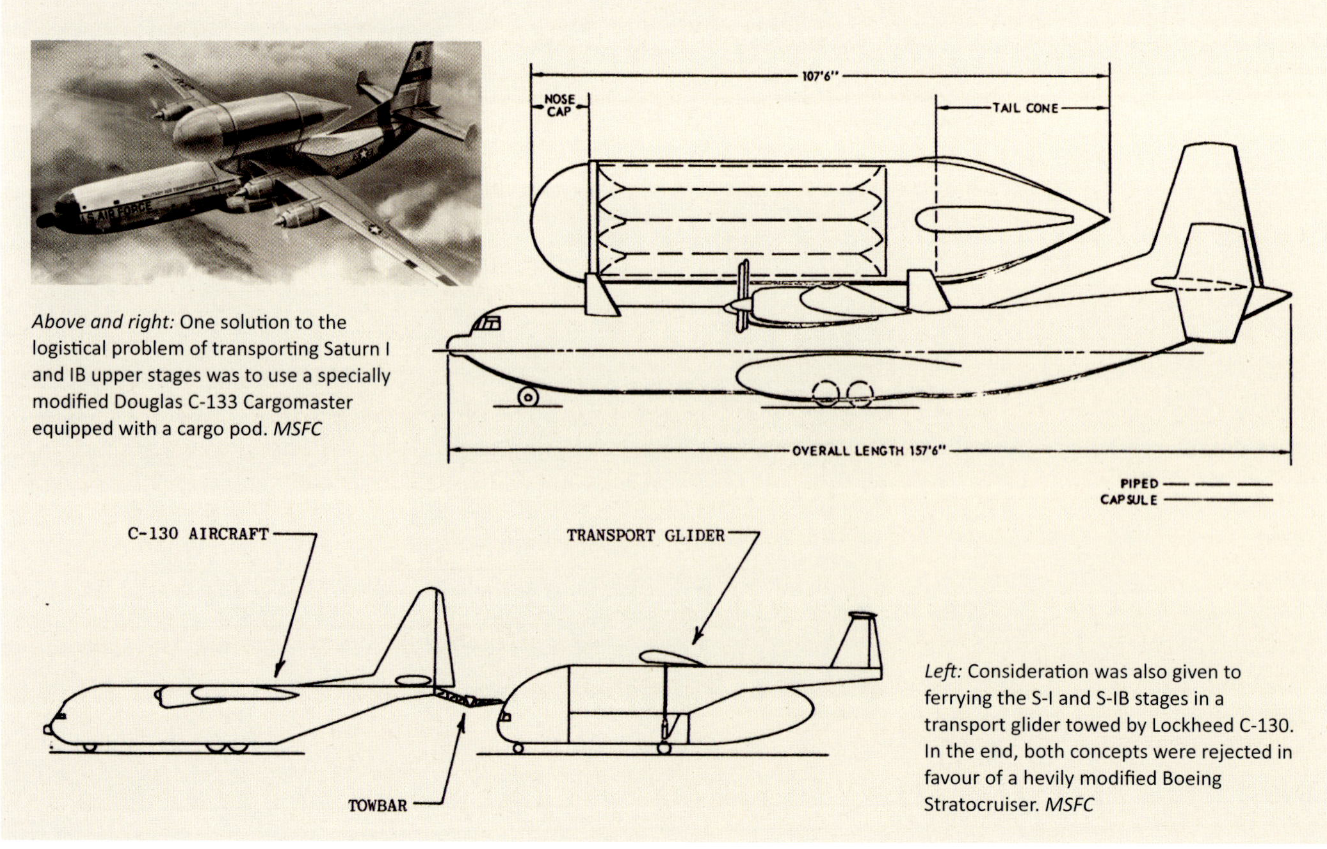

Above and right: One solution to the logistical problem of transporting Saturn I and IB upper stages was to use a specially modified Douglas C-133 Cargomaster equipped with a cargo pod. *MSFC*

Left: Consideration was also given to ferrying the S-I and S-IB stages in a transport glider towed by Lockheed C-130. In the end, both concepts were rejected in favour of a hevily modified Boeing Stratocruiser. *MSFC*

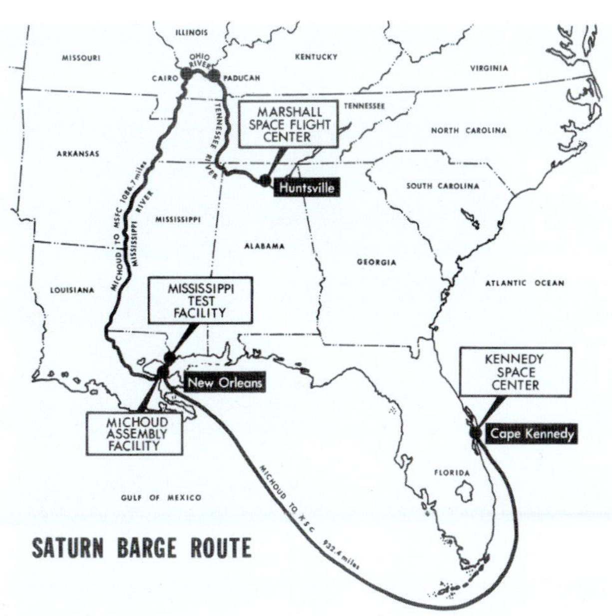

Above: The barge *Palaemon* carrying a Saturn IB stage. Barges conveyed Saturn stages between MSFC and MAF and then on to Cape Canaveral, the latter a distance of 1,500km (932mls) along the Gulf of Mexico and up through the Atlantic Ocean. *Author's collection*
Right: Logistical challenges posed by the separation of the Huntsville facility from Michoud and Cape Canaveral are evident in this graphic which shows the barge route taken by the S-I and S-IB stages. *MSFC*

for conveying stages until 1967. For bigger journeys and for moving stages from California to Cape Canaveral, NASA obtained the AKD *Point Barrow* from the Military Sea Transport Service and conducted sea trials during 1963 to assess it for use in that role. Modifications to the *Point Barrow* were completed during 1964 and it was primarily used at first for conveying S-IVB stages between locations on the coast of California and thence from there to Florida. As a sea-going covered ship it served well as a protected structure for environmentally protecting the stages it carried, retired off in the 1970s to serve the Navy supporting deep submersible research and test activity.

Moving stages by air was only possible for the S-IV, S-IVB and S-VD (dummy SV) stages and that was handled by the Pregnant Guppy and Super Guppy aircraft modified from Boeing 377 Stratocruiser passenger airliners. This was one of the most bizarre and unusual conversions and provided quick and efficient movement for upper elements of the Saturn I and Saturn IB. Designated 377-PG, the Pregnant Guppy grew out of ideas for several air-lift conversions from the basic Stratocruiser, itself a development of the B-29 Superfortress of the type which conducted large-scale bombing operations against Japan in 1944-45 and which dropped the two atomic bombs. This type was developed into the KC-97 aerial tanker from which came the double-deck airliner pioneered by PAA with an order in November 1945.

Below: When it was decided that SI and S-IB stages would travel to the Cape by barge, a Boeing 377 Stratocruiser, converted by Aero Spacelines and known as the Pregnant Guppy, was used to fly S-IV stages around the country from California to Cape Canaveral. *NASA*

The aircraft selected for conversion was the second Pan American Airlines aircraft (N1024V), Boeing 377-10-26, taken charge of by Aero Spacelines Incorporated for a major modification by On-Mark Engineering of Van Nuys, California. Wind tunnel tests validated the concept in which the upper fuselage would be opened up to allow a large, bulbous upper section providing a cavernous volume into which the S-IV stage, with a diameter of 5.49m (18ft), could be carried. Various aircraft were considered for the job but the cost of the used Stratocruiser was low at the time and offered the best option available at the price. Modifications included lengthening the fuselage by 5.08m (16.66ft), a straight section spliced in just aft of the wing from another Stratocruiser (G-AKGJ *Cambria*) previously owned by the UK airline BOAC but sold back to Boeing on 26 January 1959.

To evaluate the aerodynamics a superstructure was built up above the existing fuselage to provide an external contour exactly the same as a completely reworked aircraft. This was flown to verify its acceptability before cutting the top out and building up a permanently re-formed cargo area. Equally radical, after modification the aircraft was cut in half and connecting locks fitted to the circumference so that the aft section could be removed and set aside, vacating the aft entry for payloads elevated to height by a special lifting device on the ground. The first flight took place on 19 September 1962 and a series of test and certification flights followed during which it was rated to carry a payload of 15,422kg (34,000lb). It was

Left: Aero Spacelines modified another Boeing 377, gave it a lengthened fuselage and a front-loading capability, which NASA used to lift the S-IVB stage between locations. *NASA*

first used in support of the Saturn programme when it carried S-IV-5 from Huntington Beach to Cape Canaveral on 20 September 1963.

To obtain a similar capability for the bigger and wider S-IVB with a diameter of 6.6m (21.6ft), YC-97J (52-2693) was acquired, a turboprop-powered former aerial tanker acquired after being sold back to Boeing. At first known as Very Pregnant Guppy before it emerged as the Super Guppy, it had a substantial rebuild retaining little of the original airframe. When completed as civilian-registered N1038V, it had an internal length of 28.8m (94.5ft) and a payload capacity of 24,000kg (54,000lb). A major difference was in the hinged forward section behind the nose leg allowing the entire front end to rotate sideways for loading S-IVB stages. First flown on 31 August 1965, it began a series of training flights in January 1966 and flew a dummy S-IVB before carrying flight hardware in support of the Saturn IB programme.

Cape Canaveral Air Force Station (CCAFS)

Formed in 1953, the Missile Firing Laboratory (MFL) at the Redstone Arsenal reported directly to von Braun's Development Operations Division which was responsible to NASA HQ after the shift of ownership in 1960. It had Kurt Debus as its Director with Hans Gruene as his deputy and R. Heiser as technical assistant. The MFL originally managed the ballistic missile, its responsibilities extending to field troop training for Redstone, Jupiter and space launchers. It also maintained facilities and equipment at the Atlantic Missile Range.

NASA's Saturn operations at Cape Canaveral would be embedded in military management of the site, which had been established as the Joint Long Range Proving Ground (LRPG) on 1 October 1949 after formal approval from President Truman dated 11 May. As defined in its title, it was a joint undertaking of the Army, Navy and Air Force, with the latter service retaining executive control through the Chief of Staff of the Air Force. From 16 May 1950, the LRPG was placed under exclusive control of the Air Force. That August, the LRPG Air Force Base was renamed Patrick Air Force Base in honour of General Mason M. Patrick, who led the US Army Air Service during World War 1 and in 1926 became the first chief of the Army Air Corps.

When the military arrived at Artesia's Cape Canaveral there had been nothing with which to construct a launch site. Local materials scavenged from old beach houses, huts and a dressing room for swimmers sufficed to set up a makeshift pad for sending V2 rockets into the stratosphere. The Air Force had been there since 1 September 1948 when it took over the facilities of the Banana River Naval Air Station, an area on the east coast of Florida midway between Miami and Jacksonville covering 60km² (23mls²), which was chosen as the site for a missile test range. Prior to that, the area was a backwater of scattered fishing communities, untouched beaches and scattered houses – and a lighthouse which for several years into the space age remained a fixture reminding visitors of a long-gone past. Operating since July 1894, the current lighthouse is further inland than its predecessor completed in 1873.

The local Coast Guard opened up its small area at the Cape to missile use with construction work commencing on 9 May 1950, work undertaken by the Duval Engineering Company being completed on 20 June. These first two sites were designated Launch Complexes 3 and 4 (LC-3 and LC-4), the former becoming operational first. Scaffolding provided by painters sufficed to access the vertical V2 rocket with plywood platforms laid across the shaky structure. The swimmers' hut became the firing room, protected by sandbags piled across the

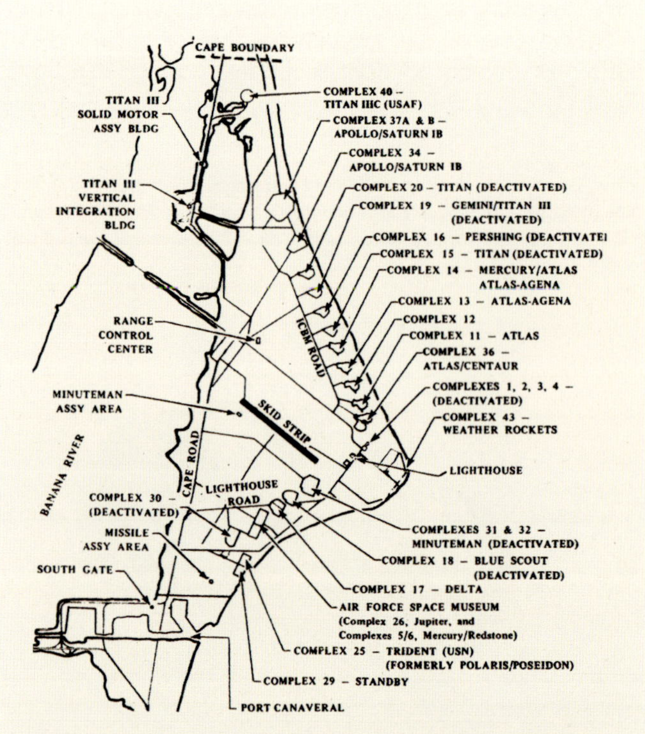

The growth in launch facilities at Cape Canaveral was exponential throughout the 1950s and early 1960s, with pads extending north up to the Saturn I and IB facilities and to the Air Force Titan complex further up the coast. *NASA*

Above: From the outset, launch operations were frequently hampered by encroaching thunderstorms, demonstrated here by close lightning strikes on LC-5 and LC-6 from where Redstone flights took place. *NASA*

Complex 56 - Pad 5
Emergency Vehicle, Gate and Roadblock Locations

Frangible Fence
Pad 6
Mobile Tower
T-55 to T-0
Service Structure
T-55 to T-0
Liftoff Position
Transfer Van
T-160 to T-75
Mobile Tower
T-390 to T-55
Fire Truck 2
T-390 to T-270
and
T-220 to T-30
Gate
Gate
Frangible Fence
Lighthouse Road
Gate
M-113
(30. 3, UHF, MOP 1. 9)
T-145 to T-45
600 ft
Pad 5
Complex 30
M-113
(30. 3, UHF, MOP 1. 9)
T-45 to T-0
Roadblock 8
Fire Truck 2
T-30 to T-0
Roadblock 8W
Fire Truck 2, T-270 to T-220
Transfer Van, T-75 to T-0
Water Tank Fire Truck, T-30 to T-0
Fire Truck 1
(30. 3)
T-150 to T-0

Right: The layout for LC-5, alongside Lighthouse Road at the southern end of 'missile row', as it became known. *NASA*

Left: Adjacent to LC-5/LC-6, LC-26 was used for Jupiter-C and Mercury-Redstone launches, as seen here. *NASA*

front just 122m (400ft) from the rocket, and a line of trucks and trailers provided communications and links to tracking sites. The pad itself was a 30.5m x 30.5m (100ft x 100ft) square, reinforced concrete slab 20cm (8in) thick. The firing room was mostly underground, a 6.1m x 6.1m (20ft x 20ft) construction with periscopes for observation of the rocket.

Initially, Launch Complexes 1, 2, 3 and 4 were set up to support early flights and it was from these that the Bomarc, Matador (a winged weapon resembling a large fighter propelled into the air by a solid propellant motor and sustained to its target by jet engine), Mace (an advanced version of Matador), UGM-27 Polaris, Redstone and some X-17 launches took place, although they were not all activated at the same time.

Of the area first mapped out for use as a proving ground, 4,746ha (11,728acres) was acquired from the US Coast Guard by the military and construction of a blockhouse and launch pad was completed on 20 June 1950. It was from there that the first rocket was fired, Bumper No 8 on 24 July equipped with a WAC Corporal upper stage and followed by Bumper No. 7 five days later. This marked an end to the age of V2 test firings and presaged a significant shift toward a broader range of military missiles and rockets which would find use much later when adapted for space research.

However, test firings with weapons continued beginning 25 October 1950 with the first launch from this site of the Lark surface-to-air missile (SAM), which had been developed toward the end of the Second World War as a defence against Japan's Kamikaze bombers. The missile was quite small, with a length of 1.91m (6.25ft) and a range of 55km (34mls) but it was never much use. Test firings stopped on 8 July 1953 after 40 shots and Lark would be replaced by the Terrier SAM.

Activity really began to pick up pace with test flights of America's first pilotless cruise missile, the TM-61 Matador which had a range of 1,125km (700mls). Classified as a tactical missile, Matador was 12m (39.5ft) in length with an 8.7m (28.6ft) wingspan. It was powered by a 20kN (4,600lb) thrust turbojet engine supplemented by a solid fuel rocket motor producing a thrust of 244.6kN (55,000lb) for two seconds to get the missile airborne. The first flight of the Matador had taken place at White Sands on 20 January 1949 but tests shifted to Cape Canaveral from 20 June 1951 where 191 shots had taken place by 30 November 1956. Test shots with the Mace cruise missile began in 1956 and moved to CCAFS from 29 October 1959.

On 30 June 1951, the LRPG was renamed the Air Force Missile Test Center (AFMTC) to more properly signify the central role that service played in the development of a national ballistic missile force and associated test programmes. A significant step was taken on 21 July when the US government signed an agreement with Britain giving access to a 1,600km (1,000mls) corridor utilising UK tracking networks, extending the range from Point Jupiter in Florida to Grand Bahamas Bank

Below: From the right looking left, test stands for Redstone, Jupiter and Saturn S-I stages at Redstone Arsenal, later the Marshall Space Flight Center. *MSFC*

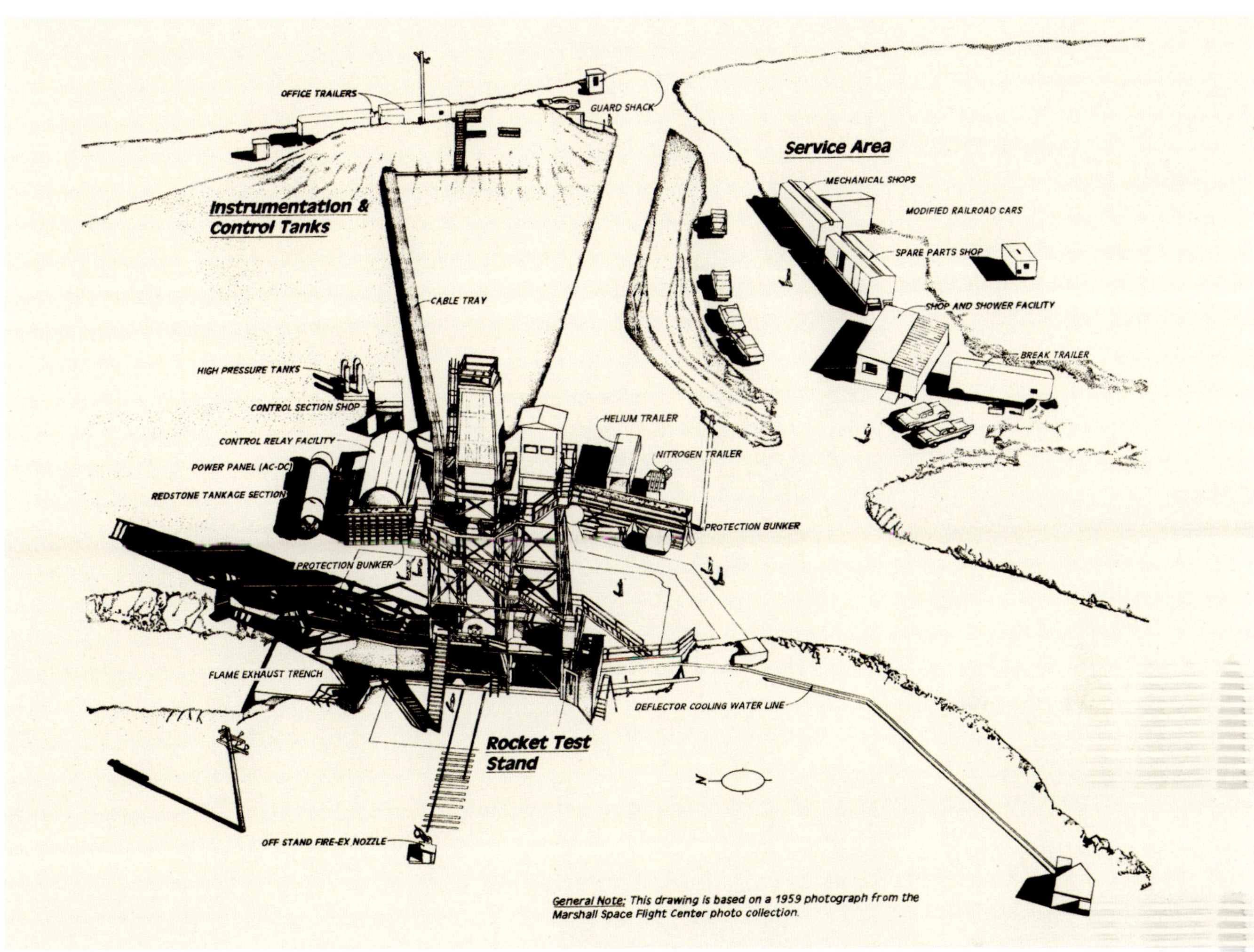

Above: Based largely on requirements for testing the Redstone missile, Redstone Arsenal built a test stand and supporting infrastructure which was cutting edge at the time with facilities for workers and test personnel where they could take a break, have lunch or confer in buildings situated south of the test site. *MSFC*

and Grand Turk Island. Eventually, this agreement would be extended, increasing the range to 8,000km (4,971mls) by taking in the stations on Ascension Island.

On 29 August 1952 test flights had begun with the Snark cruise missile. Matador and Mace carried only fission warheads with a yield of 40KT but Snark was designed to deliver a 3.8MT warhead across a range of up to 10,185km (6,330mls). It failed to live up to its promises and was cancelled in 1961, the last of the subsonic long range cruise weapons, replaced by the much more efficient ICBMs support for which would occupy the Cape for many years. Hard on the heels of tests with Snark, on 10 September 1952 the Air Force began test firings of the Bomarc surface-to-air missile, an integrated air defence system of which 1,140 were built and deployed between 1959 and 1972. But as the total number of test firings of subsonic cruise missiles and rockets from Cape Canaveral began to build up, with 5 in 1950, 18 in 1951 and 41 in 1952, the first of the new generation of ballistic missiles began to emerge for trials which would increase in number over the next several years.

Redstone and Jupiter Pads

In terms of first-use, LC-3 was to have been followed by LC-4, work starting on 5 September 1951 ready for the first launch of a Bomarc on 10 September 1952. After this single launch of a Bomarc, the engineers moved in on LC-4 during 1953 to adapt it for the Redstone. It hosted the first Redstone launch on 20 August 1953 with the flight of RS-1, five more following from this pad with the last, RS-8 on 9 February 1955. LC-4 would continue to be used for Bomarc launches until 15 April 1960 when it was retired. It was similar to LC-3 but with one side of the concrete hard stand 61m (200ft) long with another area, 55.5m x 91.5m (182ft x 300ft) surrounding the pad paved area.

But these pads were not specifically built for the ABMA to test Redstone and later Jupiter rockets, the first of those inheriting the mobile launch concepts drawn up by Kurt Debus, with LC-5 and 6 for Redstone and LC-26A and LC-26B primarily intended for Jupiter missiles. With an orientation to permit flights directly down the Atlantic Missile Range, from the south up were pads 5, 6, 26A and 26B in a section of the CCAFS 853m (2,800ft) in length and 244m (800ft) on a north-east/south-west axis. Each of the four pads was 61m x 61m (200ft x 200ft) in surface area, with centre-points 152m (500ft) apart and 122m (400ft) from the perimeter fence line. The northernmost LC-6 was 305m (1,000ft) apart. Equidistant

from LC-5 and LC-6 and offset to one side, the common blockhouse was 91m (300ft) in a direct line from the centre of each pad. A common rail track was laid in 1957 facilitating access to all four pads, with bypass spurs added in 1958. An original A-frame gantry was replaced in 1959 with a larger H-style gantry which was initially used at LC-6 for Jupiter flights from that site.

Built in 1954-55 to support Redstone and early Jupiter A flights, LC-6 became active with the first launch on 20 April 1955. The complex was the first to use a mobile launcher concept developed by Dr Kurt Debus at Redstone Arsenal, engineers erecting a Missile Service Structure (MSS) to support Redstone launches. The MSS had a height of 41m (135ft), a width of 7.9m (26ft) and a length of 18.6m (61ft). Set on rail tracks so that it could be moved up to the missile on its pad, the MSS weighed 139,700kg (308,000lb) and supported a 13.6tonne (15ton) crane and four movable work platforms with an air conditioned work room, elevators and a standby power plant. This amazing improvement over the facilities at LC-3 was well received by the workforce in this humid, mosquito-ridden swampland. LC-6 also supported Explorer and Pioneer flights as well as the Jupiter test shots. There were 43 launches from LC-6 between 20 April 1955 and 27 June 1961.

Left: Adaptation of the ABMA test stand in 1959 for Saturn I and IB where first-stage rounds were fired. *MSFC*

Below: The general layout of the test facilities for Saturn first stages and, later, for the F-1 rocket motors built for Saturn V. *MSFC*

LC-5 was the second in a paired pad shared with LC-6 but was first used more than a year after the first shot from that adjacent location when the Army launched one of its Jupiter A rockets to a height of 90km (55mls) in the early hours of 19 July 1956. The range of 264km (163mls) was a little long but this goes into history as the first Jupiter fabricated and assembled by Chrysler, all previous missiles being turned out from the Redstone Arsenal. The second launch, on 20 September, was a Jupiter-C re-entry vehicle test which performed as planned in almost all regards, except for an early shut-down due to an error in providing sufficient propellant in the tanks.

Several military flights followed from LC-5 before the launch of NASA's Explorer III on 26 March 1958. After an Army Redstone launch on 17 May, another Jupiter C put Explorer IV in orbit, followed by Explorer V on 24 August. Launched on 23 October 1958, the Beacon satellite was lost when the first stage failed. Some interesting flights followed, including Pioneer 3 sent up on a Juno II launcher on 6 December as an Army mission managed by NASA, one of the Pioneer Moon probes. Following an Army Jupiter IRBM shot, Pioneer 4 was sent off on 3 March 1959, the last of the four military Moon probes subsequently managed by NASA. Explorer VII followed a Juno II launch on 13 October before the first Mercury-Redstone shot (MR-1) took place on 21 November 1960.

On 31 January 1961, three years to the day after America's first satellite, MR-2 lifted off from LC-5 and sent the chimpanzee Ham on the ballistic trajectory of the type astronauts

would fly reaching a height of 251km (155mls) before falling back to a splashdown in the Atlantic Ocean. Mercury-BD was successfully flown on 24 March, 18 days before the flight of Yuri Gagarin. Astronaut Alan Shepard flew MR-3 on 5 May and returned to a hero's welcome. The second manned Mercury shot on 21 July carried Gus Grissom on a repeat flight, the last of the 23 launches from LC-5, by which date LC-6 had already seen its last shot.

LC-26A was the first of two on a dual pad site with a common blockhouse and is arguably second only to LC-39 in fame as it was the location from which the US Army launched America's first Earth-circling satellite. Construction of this dual complex had begun in 1956, with the first flight, a Jupiter IRBM, on 28 August 1957, notable for the first demonstration of separation from its thrust unit. It achieved historic recognition on 31 January 1958 (EST) by the launch of Explorer I, followed on 5 March 1958 by the failure of another Jupiter-C carrying Explorer II. A succession of Redstone and Jupiter IRBM test shots followed, not all successful, some of which were NATO combat training launches to qualify Italian and Turkish crews for Jupiter deployment in Italy and Turkey. The 14th and last launch from LC-26A occurred on 22 January 1963.

LC-26B became operational on 23 October 1957 when a Jupiter IRBM was sent on a 2,037km (1,266mls) flight down the AMR. The first flight of a Juno II from LC-26B was launched on 14 August 1959, a failed attempt to place a Beacon satellite in orbit. The last of 22 launches from this pad occurred on 24 May 1961 when a Juno II attempted the failed flight of a Beacon satellite (Explorer S-45). Both LC-26 pads were decommissioned in the mid-1960s and the area is now home to the Air Force Space and Missile Museum.

By mid-1959, the first 10 years of activity at Cape Canaveral had supported the launch of 903 rockets and missiles from launch pads owned and operated by the US Air Force. Of this total, little more than 3 per cent had anything to do with the exploration of space. More than a quarter were development and test flights by the MGM-1 Matador cruise missile. Other cruise missiles boosted to 421 the total number of this class launched for the Army and the Air Force. Only 29 launches were for satellites or space probes, the remainder being for development of ballistic missiles such as Redstone, Jupiter, Thor, Atlas and Titan, and Polaris for the Navy. These figures do not account for rockets that blew up on the pad or momentarily lifted off and then exploded, of which there were several!

The Saturn I/IB would fly from three launch complexes: LC-34, LC-37 and LC-39. The blockhouse concept evolved rapidly in the early space programme and as launchers got more complex and the data-driven test flights more demanding on engineers and launch controls teams, the number of personnel involved grew, some 250 being housed in the blockhouse control centres for LC-34 and LC-37, with more than 400 for operations from LC-39. LC-34 is historic because it was the first complex built by NASA for a dedicated space launch vehicle and it supported the first four ballistic test shots of the S-I stage.

LC-37A and B were completed in August 1963 to support

Below: The completed test stand incorporating both Jupiter and Saturn programmes. *MSFC*

Left: A Redstone rocket sits inside the test stand, a historic tribute to the first major test facility at Redstone Arsenal. *MSFC*

flights with cryogenic upper stages on the Saturn I and IB. Each pad had a common blockhouse and a servicing tower running between launch stands and moved between each as required, with fixed umbilical towers at each pad. LC-37A was never used but LC-37B supported SA-5 to SA-10 plus the AS-203 flight in 1966 and AS-204 in 1968, only eight flights in all. By the end of 1968 the irreversible decision had been made to decommission LC-37B, focusing all future manned launches on LC-39.

Arguably the most famous launch complex in the world after the pad from which Sputnik 1 (1957) and then Yuri Gagarin (1961) were launched, LC-39A and B were designed and built to support Saturn V launches which took place there between 1967 and 1973, carrying men to the Moon between 1968 and 1972. LC-39A was ready by October 1965 followed by LC-39B in November 1966. There had been plans for a third pad and speculative concepts for a further two more but these were never built.

After the decommissioning of LC-34 and LC-37B, four Saturn IB flights took place from LC-39B between 1973 and 1975, closing the curtain on Saturn I/IB flight operations. For these, the Saturn IB was placed on top of a pedestal to raise the height of the S-IVB to the relative position it would occupy had it been installed in a Saturn V configuration. It was easier to adapt the complex to support the hydrocarbon S-IB stage than the cryogenic S-IVB, which would make use of the existing propellant feed lines and arms configured for, and at the height of, the S-IVB stage on the Saturn V.

Dedicated Launch Sites

Planning for the Juno V/Saturn I launch complex took on new life when Kurt Debus flew to Cape Canaveral on 26 September 1958 to discuss what the ABMA wanted from the Air Force. Debus briefed Maj. Gen. Donald N. Yates, the AFMTC commander on the magnitude of the Juno V/Saturn I system and proposed a new facility close to LC-26, the site for the launch of Explorer I and embracing two pads, A and B. But the Air Force considered this too close to other sites and that safety concerns argued against a location at that position. Yates offered areas adjacent to LC-20 and LC-11 and suggested consideration of a new area to the north, already assigned to future and much larger launch vehicles.

For the first 12 years of its tenure in the United States, the von Braun team (as it was usually referred to) operated as a

Below: Looking north up the coastal strip (left), LC-34's two yellow-painted flame deflectors can be seen at lower right. Looking south down 'missile row' (right) toward the place where the first launch pads were erected beginning in the 1950s. *NASA*

cohesive unit, each specialist within the organisation subsumed to some extent by the overarching charisma of its leaders. But there were many equally capable managers within the group who were only recognised for their unique skills when out from under the von Braun umbrella. Debus was singularly one of those. Bringing great experience from his work at Peenemünde, Debus was the architect of launch complex design for the Redstone and Jupiter rockets and now for the more formidable challenge posed by these giant rockets.

The requirements were an order of magnitude greater than those for Redstone and Jupiter rockets. The definitive Saturn I would have both launch mass and lift-off thrust 10 times that of the Jupiter IRBM. It would spend four times the on-pad duration of Jupiter and require a facility capable of handling almost 12 times the quantity of propellant per rocket. Control systems were a generation ahead of those installed on the pad for Jupiter, with a requirement to handle almost five times the amount of telemetry data through more than three times the number of RF links. Moreover, due to the clustering adopted for Saturn I, the vehicle would have to be held down until engine measurements demonstrated success.

For the next two weeks, the Debus team and the Air Force evaluated five potential locations at Cape Canaveral under the guiding experience of James Deese, who had been selecting places in which to expand the launch base since 1950. One of the most potent concerns was in the blast effect of a booster of this size exploding on the launch pad, an energy estimated to be 50 per cent of the total mass in terms of equivalent TNT. The full Juno V configuration would require a safety radius of 1,650m (5,413ft) and placement of the new launch complex had to take account of the launch azimuth range of 44deg to 110deg and the hazard of overflying existing pads should the rocket tumble within a few seconds of lift-off.

Full authority for MFL to design, develop and manage the launch activity for Juno V/Saturn I was granted by von Braun's Development Operations Division on 20 October 1958. Clearly not as dramatic or spectacular as the rocket itself, the launch facilities were, nevertheless, key to a successful operational capability. These were times of very steep learning-curves and considerable effort was expended on looking at all appropriate design criteria for facilities which had to be reliable and allow for growth and expansion as the programme matured. Safety was a crucial priority but so was operational fluidity, maintaining an efficient countdown and risk mitigation strategy would be fundamental to consistency and reliability.

Unique challenges were raised by the magnitude of the vehicle itself. The high cost of each rocket argued against a large series of test launches, such as had been the luxury with Redstone and Jupiter missiles. The height of the rocket would be up to three times that of the earlier rockets and the supporting infrastructure would have to accommodate a very great number of telemetry channels. The sheer quantity of propellant required was an order of magnitude beyond anything required so far and the time spent on the pad would almost treble that catered for in Redstone and Jupiter launches. And in deciding upon the physical support system, the rocket would have to be held down on the pad until verification of engine performance cleared the stack for lift-off.

In November 1958, MFL set up a Special Project Staff, a Scientific and Technical Staff, a Military Support Office and a Program Coordination, Engineering Services and Administration office. First order of business was to consult the experts and in that month the MFL sat down with personnel from the District Corps of Engineers and with the architect and engineering firm of Maurice Connell and Associates based in Miami, Florida. A set of site requirements was drawn up and development criteria laid down.

Two months later, ARPA got involved and visited the AFMTC to gain first-hand knowledge of the requirement and to receive a briefing from Deese and his team, which recommended a pad 300m (985ft) north of LC-20 with common use of that pad's blockhouse. Built for the Titan ICBM, LC-20 was to be the last of four pads constructed specifically for test shots with this ICBM and construction was already under way and would be completed on 10 September 1959. Not before 25 February 1960 would the first launch attempt be made from LC-20. The Air Force objected strongly to co-location, pointing out safety issues in the event of an explosion anywhere closer than 600m (1,970ft) but also the degradation to monitoring equipment in a voltage drop along that distance. It was finally decided to draw up plans for a launch complex some 1,463m (4,800ft) from LC-20 which the Air Force had preferred from the outset. It was designated LC-34.

Launch Complex 34 (LC-34)

A construction contract for LC-34 was awarded on 3 June 1959 and the pad and blockhouse were completed within a year, ready for flights SA-1 to SA-4 between 1961 and 1963. LC-34 was modified to support operations with the cryogenic S-IVB and used for the first and third flights of the Saturn IB in 1966. LC-34 was retired after the fateful Apollo fire of 1967 and the AS-205 manned Apollo flight in 1968, having supported only seven launches. The pedestal remains intact with the mournful words 'ABANDON IN PLACE' stencilled on the side of the concrete hardstand, an enduring reflection on the inherent dangers of space flight. There had been plans for a second pad (B) at LC-34 but that was never implemented.

The entire area encompassed by LC-34 occupied 18.2ha (45acres). The pad, or hardstand, stood upon a central area of reinforced concrete, 133.5m (438ft) in diameter and 20cm (8in) thick. The foundation contained 3,364m^3 (4,400yds^3) of 20.3cm (8in) thick reinforced concrete and 526tonnes (580tons) of steel which was placed to a depth of 2.4m (8ft) at the centre and 1.2m (4ft) at the circumference. The orientation of the axial alignment of the pedestal and the umbilical tower was decided by the requirement for an unobstructed line of sight from the launch vehicle to the radar antennae and the on-site telemetry stations located at the industrial area some 3km (1.8mls) to the southwest.

It had at its centre a pedestal consisting of a flat platform on four legs, 12.8m (42ft) square and 8.2m (27ft) in height, at the top on which was positioned a circular hole to allow exhaust to flow from the base of the rocket down to a deflector below the pedestal. It was open on all four sides to allow a flame deflector to channel the exhaust in either one of two opposing directions.

Launch complex 34 (LC-34)

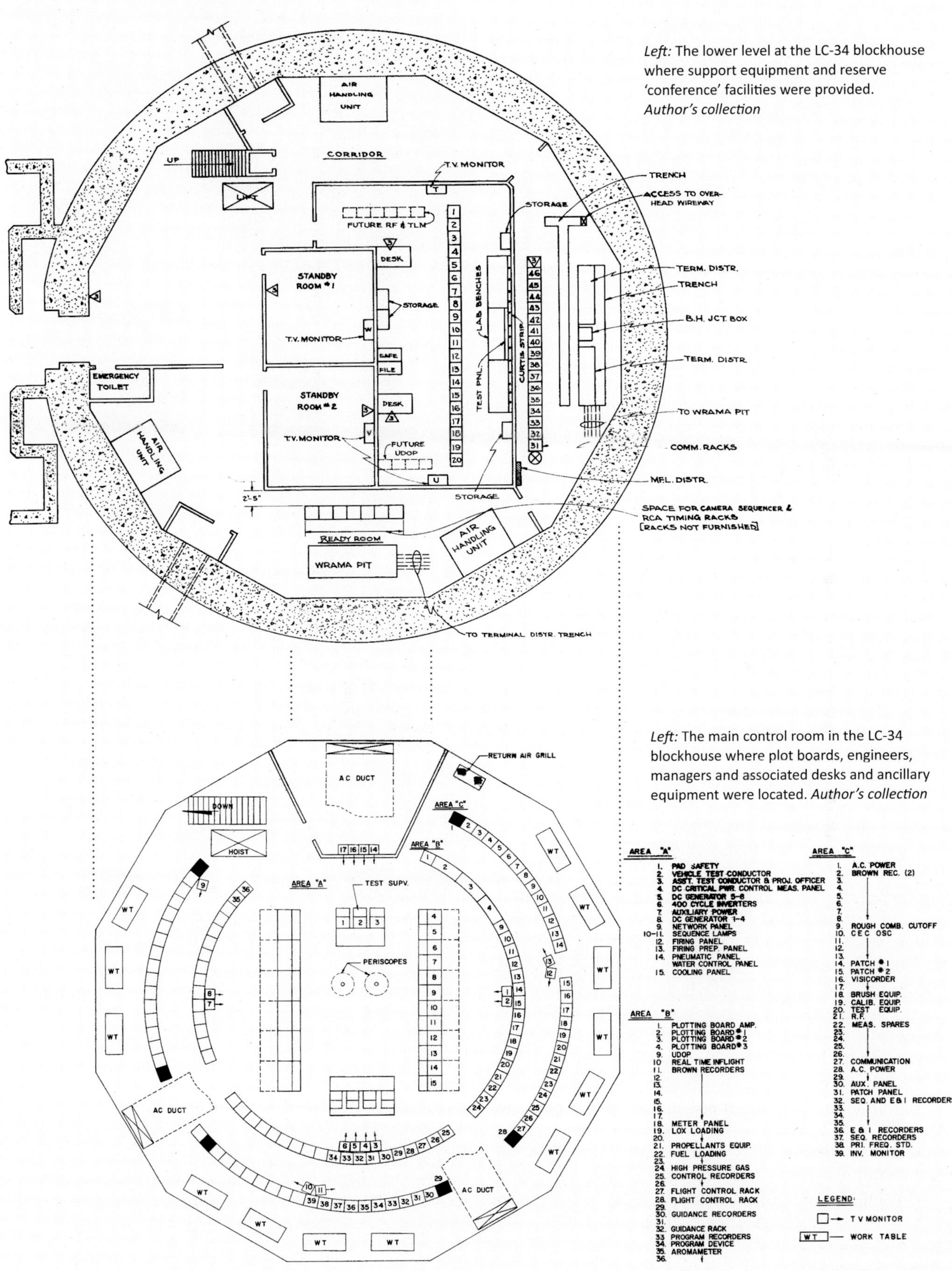

Left: The lower level at the LC-34 blockhouse where support equipment and reserve 'conference' facilities were provided. *Author's collection*

Left: The main control room in the LC-34 blockhouse where plot boards, engineers, managers and associated desks and ancillary equipment were located. *Author's collection*

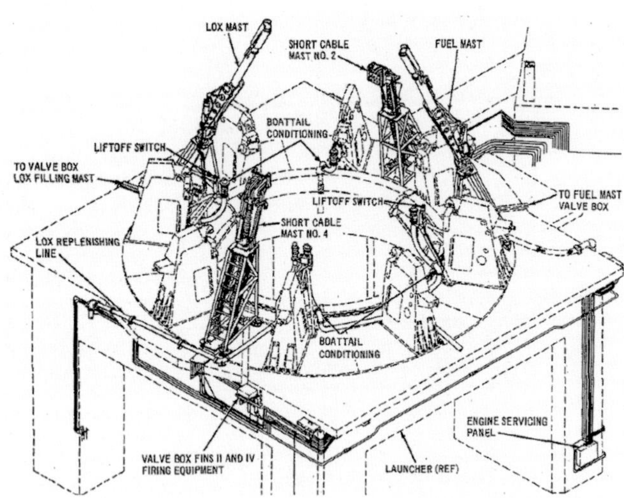

The umbilical connections and tie-down positions on the LC-34 launch pedestal for the Block I Saturn I launches. *Author's collection*

The top of the pedestal supported an outer ring to which was attached eight arms for restraining the rocket until released for flight, of which four were for supporting it prior to lift-off. There was some concern about the design of the restraining arms, which held back the final configuration for a few weeks.

Recognising the high failure rate with rockets at the time, the need to verify satisfactory combustion of all eight engines was always going to precede the release of the hold-down arms to avoid an immediate disaster. The four support arms would retract horizontally at ignition as the mass of the vehicle began to offload during the build-up of thrust. This was necessary so that the rocket's tail shrouds would clear the restraints as it began to rise. But if a single engine shut itself down within three seconds after ignition, just before lift-off, the entire stage would shut down and settle back. To ensure the retracted arms were back in place to accept the uncompensated weight of the stack, the four restraint arms would have to move back in again. A special nitrogen pneumatic system would close the four arms back in within 0.16sec of the given command but the design of that was a problem for some time.

An Umbilical Tower with a height of 8.2m (27ft) tall was utilised for the first two launches. A much taller tower with a height of 73m (240ft) and a base area 7.3m (24ft) square was installed beginning with the third Saturn I launch in preparation for its operational function with the Block II vehicles. The tower at LC-37 had a height of 79.2m (260ft). The towers were located 12m (40ft) from the pedestal and incorporated a 2,268kg (5,000lb) boom hoist which was equipped with a trolley extending the hook 8.2m (27ft) from the pivot point with a full 360deg rotation capability for manoeuvring relatively light loads. Swing arms were attached to the Umbilical Tower connecting the rocket to the ground facilities, each hinged at the joint to the tower with a rotary hydraulic actuator and lock pin, automatically uncoupling and moving aside at lift-off. These service points were important for keeping the vehicle electrically and hydraulically connected to the ground and for delivering propellant to the stack when required, filling the tanks from ground supplies.

An Apollo access arm was situated at the 67m (220ft) level which consisted of a swing-arm assembly, an extension and an environmental chamber, or 'white room', forming an integral assembly and allowing access to the interior of the Apollo spacecraft. The launch facilities envisaged manned Apollo flights on the Saturn I and provision for this formed an integral part of the initial designs. While the arm would move away some time prior to ignition, it could be moved back to envelop the Apollo hatch within 30 seconds of activation. The Umbilical Tower incorporated an elevator which could rapidly move the crew down to safety and was capable of carrying 1,360kg (3,000lb) at a descent rate of 137m/min (450ft/min).

Above: The LC-34 rail-mounted servicing gantry with its multiple work levels for access to key locations on the launch vehicle. *MSFC*

Beneath the pedestal upon which the rocket stood, a rail-mounted, A-shaped flame deflector would channel the rocket's exhaust to opposing sides of the pad and protect the surface from the 2,760°C (5,000°F) flame temperature. The deflector had a width of 9.7m (32ft) and a length of 13m (43ft) weighing 136tonnes (150tons), directing the exhausted gases along a northwest/southeast axis.

The design of this deflector brought some challenges and initial uncertainty as to whether it should have two or four sides. The V2 and Bumper launches employed a small flame deflector placed beneath the single rocket engine in the form of a four-sided cone, channelling the exhaust flame away from direct impingement on the ground and preventing the rocket digging a hole for itself! This type of deflector would prevent a clear view of the base of the rocket for post-launch analysis and could endanger equipment which had to be close by, so a two-sided exhaust flow diverting gases in opposing directions was chosen. The definitive dry-flame deflector was constructed with trusses formed at an 80deg angle covered with 2.54cm (1in) steel skin supporting a 10.2cm (4in) thick refractory coating, mounted on rails and parked 185m (607ft) to the northeast.

Attending to the Saturn I during its erection on the pad and to the countdown required a Service Structure capable of erecting the various stages and payload elements of the vehicle, providing weather protection and conducting checkout operations. The design inherited much from the design of the launch site at White Sands and consisted of two vertical towers with a connecting truss structure across the top, which had a total height of 94.5m (310ft) and weighed 2,540tonnes (2,800tons). At 17m (56ft) the space between the two legs was sufficiently wide to straddle the concrete pad itself. With a base dimension of 39.6m (130ft) by 21.3m (70ft), it supported a travelling bridge crane with a main 54.4tonne (60tons) main hoist and a 36tonne (40tons) secondary hoist with hook heights of 74.7m (245ft) forward and reaches of 8.5m (28ft) forward and 6.1m (20ft) laterally. Two travelling hoists were installed at the 92.5m (303.5ft) level with a traverse capability of 28.3m (93ft) for attaching the Apollo Launch Escape System (LES) to the spacecraft.

The Service Structure had seven fixed work platforms within the legs and an additional eight platforms retractable platforms for enclosing the vehicle at required stations. Two additional retractable platforms were located at the level of the LES and hurricane protection was provide by four hinged doors, each 13.4m (44ft) tall to encase the first stage. The upper stage was protected by silo sections which also encased the Instrument Unit and the Apollo spacecraft. Two elevators were located in each leg of the service structure, available for personnel and the movement of light equipment up or down or between work platforms. Pneumatic and hydraulic equipment was situated in two 7.6m (25ft) high sections at the base, together with four traction motors allowing the structure to move by itself on a dual-track rail to and from the pedestal and where it could be parked 182.9m (600ft) away, controlled from a driving position 8.2m (27ft) above the surface level of the pad. The entire structure was mounted on four 12-wheel trucks, with anchor points for stability when stationary, with hydraulic jacks extending to raise the structure on the wheels. At the time it was the largest moving structure on Earth and made its first motion during a test on 26 March 1961.

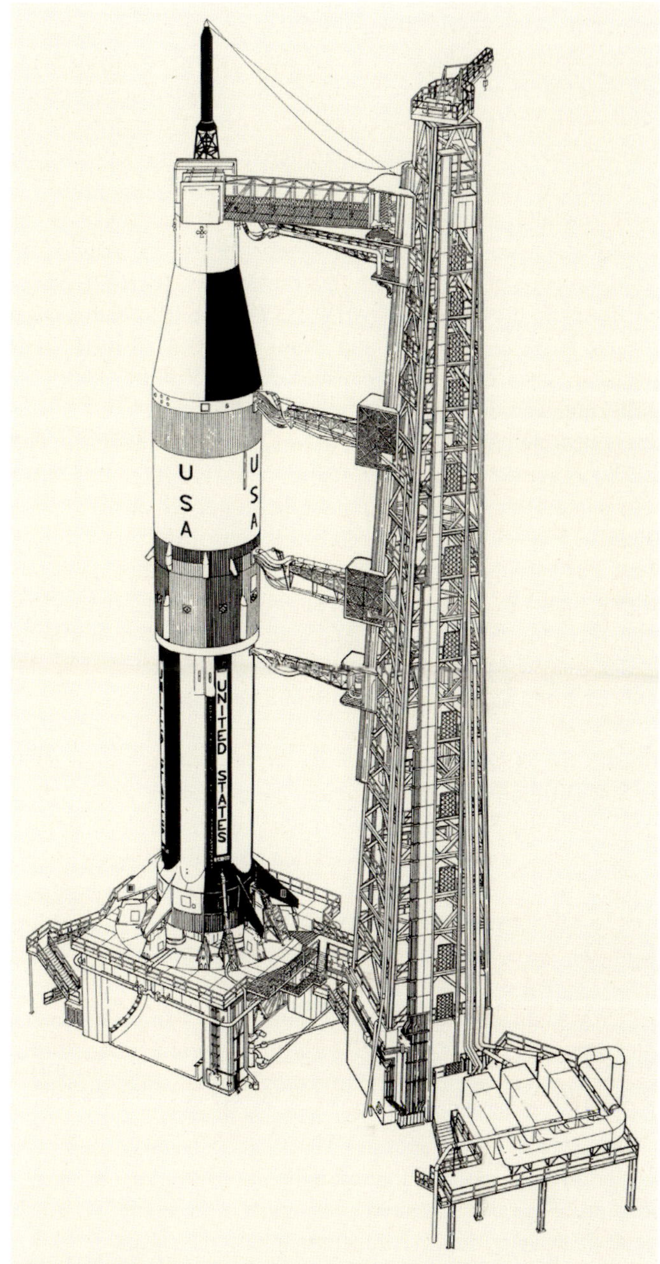

Above: The addition of an Apollo Command Module access arm and 'White Room' as fitted to LC-34 for the Apollo 1 and Apollo 7 mission launches. *Author's collection*

All operations at LC-34 would be monitored and controlled from the blockhouse, a protective structure which drew its design from LC-20, a reinforced design situated 320m (1,050ft) from the pedestal. The domed roof would comprise an inner layer of reinforced concrete 1.5m (5ft) thick supporting a layer of earth between 2.1m (7ft) and 4.2m (14ft) in depth surmounting which was a 25cm (10in) layer of shotcrete. It was designed to withstand a blast pressure of 15,086kPa (2,188lb/in^2). The blockhouse had 929m^2 (10,000ft^2) of protected floor space and an additional 232.2m^2 (2,500ft^2) of unprotected space which was not occupied during launches. The interior diameter was 24.4m (80ft) with a height of 7.9m (26ft). The ground floor was occupied by contractor personnel involved in communications with the vehicle and telemetry data collation. The next floor up was for launch personnel, with firing consoles, test consoles and conductor consoles and a selection or display and recording panels. A separate room allowed observation through a glass window and activity on this level could be observed from a small balcony above. Construction had begun in June 1959 and the blockhouse was ready for use by July the following year.

Propellant and pressurisation fluids were contained in spherical tanks around the periphery of LC-34. Along the southern edge of LC-34 were four tanks – two for RP-1 situated closest to the ocean and two for LOX. The RP-1 tanks were located close to the beach and 290m (950ft) from the launch pedestal. They each had a capacity of 113,550lit (30,000USgalls) and were filled by trucks. Fuel was pumped to the rocket via lines delivering 7,570lit/min (2,000USgalls/min). The LOX storage area was further to the southwest of the fuel tanks and some 198m (650ft) from the pedestal. It consisted of a main 473,125lit (125,000USgall) capacity insulated primary tank and a 49,205lit (13,000USgall) vacuum-jacketed replenishing tank. Two 15.2cm (6in) supply lines delivered a flow rate of 9,462litres/min (2,500USgalls/min) to the umbilical tower and the LOX filling mast. Each propellant would be delivered to the Saturn rocket in 40 minutes.

Development of this massive complex came at a price, estimates for LC-34 increasing from $8.7million in March 1959 to $38.1million a mere 17 months later. Most of the inflation occurred with the ground support equipment, the physical pad, pedestal, umbilical tower and service structure being predictable and based on experience by the Air Force and its contractors in other pads across CCAFS. But the shift in ownership from the Army to NASA allowed greater expenditure, sanctioned by the exclusivity of priority for launcher development which was the enabling key to an ambitious plan for space exploration. Under the Army, Saturn I had a problematic future, carrying no real priority for Army programmes and not a single funded mission beyond the series of test flights.

Under the direct control of MSFC, the arsenal concept was imported to LC-34 and that brought labour troubles when unions protested that Marshall's personnel were doing too much construction and supervision compared to their usual work for the Air Force. Under the management of the LOD, the level of intervention and control over the contractors was unique in the experience of the companies familiar with carrying out work elsewhere at Cape Canaveral and that brought a series of strikes, primarily from electricians, ironworkers and carpenters with 800 man-days lost between August and November 1962. It all flared to a head when LOD people began installing cables, of which several were for Saturn I (SA-1), through conduits and fitting out the control centres with consoles.

In all, LC-34 supported seven flights, four between October 1961 and March 1963 and three with the Saturn IB (AS-201, AS-202 and AS-205) between February 1966 and October 1968. Operative for a mere seven years and after the first flight of a manned Apollo spacecraft, NASA issued a directive on 29th November 1968 to close down subsequent operations, applicable to LC-37 also. When designed, it had been expected that LC-34 would support Saturn I/IB manned flights with Apollo and conduct several Earth orbit operations prior to the deep-space lunar missions. None of that occurred and operations switched to the Saturn V until NASA once again reverted to Earth orbit flights, sending astronauts to the Skylab space station and on a docking flight to the Russian Soyuz spacecraft in 1975, those flights from LC-39B.

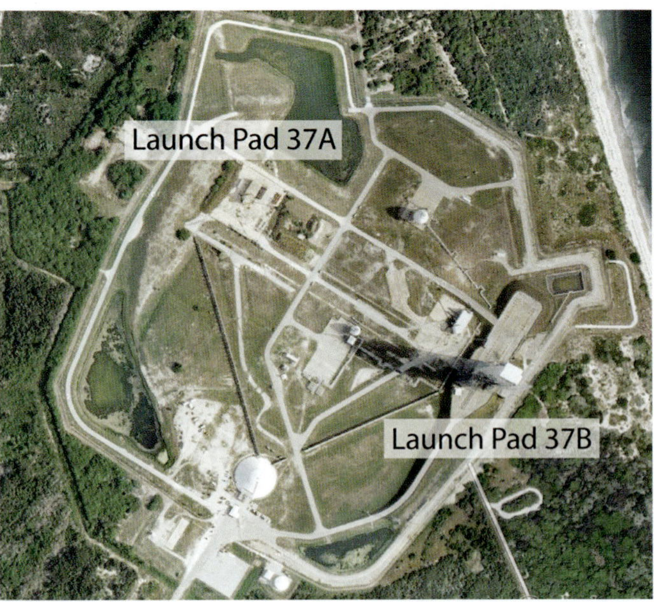

Above: An aerial view of LC-37 showing the relative location of the two pads and the close proximity of the Atlantic coastline. *NASA*

Launch Complex 37 (LC-37)

During 1959 and 1960 the magnitude of NASA future mission planning became clear and it was obvious that a second launch site would be required for the Saturn I and for ground systems to handle the migration on to cryogenic upper stages. Silverstein was working schedules and programmes for NASA with a degree of interchange to speed preparations for flight operations expected to begin in 1961. Calculations by Debus' people indicated that if a pad explosion occurred, LC-34 would be out of commission for a year. But at a separation distance from LC-20 of 730m (2,342ft), a second pad (LC-34B) would push too close to the Air Force complex for safety. The only conclusion was to construct a completely separate pad, LC-37. Initially, plans were to fly the first three Saturn Is as ballistic shots with dummy upper stages and to have LC-37 ready as a back-up to SA-2. The schedule was later changed, adding a fourth ballistic shot (SA-4), and at this date the Saturn C-2 was also a candidate and expected as a three-stage launch vehicle.

There was a great deal of uncertainty about the number of Saturn flights required, some estimates by future mission planners at MSFC estimating up to 100 utilising evolved variants of the Saturn I and none of this would get settled until the Kennedy Moon decision in May 1961 forced NASA to focus on the immediate goal. At this time too, several different Saturn-series rockets were in contention, Saturn I considered only a test rocket for greater capabilities developed when programme plans stabilised. The story of the Saturn C-2, C-3, C-4 and C-5, together with proposals for a Nova vehicle are told in full within the companion volume to the book on the Saturn V. Yet, although construction costs for LC-34 were escalating, there was

Right: The two LC-37 pads with fixed umbilical towers and the mobile service gantry capable of supporting only one launch at a time. *NASA*

little doubt that a second complex was necessary and on 29 January 1960, Debus asked Rees for permission to build it. It was to follow as closely as possible the basic configuration of LC-34 but with two pads, LC-37A and LC-37B.

The cost was estimated at \$20million and the preferred location was 1,220m (4,000ft) north of LC-34 and 425m (1,400ft) inland from the sea. Located to the north, this brought excessive demands on basic utilities as well as challenges for laying out all the required ground infrastructure. But plans to have the complex ready by January 1962 were confounded when a decision was made to change the service structure from the design used at LC-34 to one which was deemed necessary for supporting launches of the C-2 configuration.

Throughout March-June 1960, a Service Structure Review Committee met several times to consider a wide range of configurations deemed suitable for the C-2, including a launch pedestal below ground, thus reducing the height of the Service Structure. That was deemed unacceptable due to the cost and the high water table at Cape Canaveral. There was deeper concern over the clearance required to compensate for drift which the launcher might experience due to side winds as it lifted away from the pedestal. These concerns and the conclusions of a separate team examining other factors tipped the decision to go with a mobile service structure versus a fixed design which had been favoured to date. As for the umbilical tower, after looking at various alternatives, it was decided to go with a free-standing unit towed to the service structurc for support during high winds.

In August 1960 von Braun approved a request for a second pad at LC-37 but the distance between the two was questioned due to uncertainties about the explosive yield of a cryogenic stage. Air Force analysis on the Centaur stage prompted an increase in the separation space from 183m (600ft) to 365m (1,200ft). Having two pads, however, meant that a second Saturn launch could be conducted without having to wait four weeks for the anticipated pad clean-up work. Construction bids were invited in March 1961 with site preparations getting under way the following month, construction completed on LC-37B on 7 August 1963 followed by LC-37A ten days later.

Each pad was 27.87m² (300ft²) in area with the pedestals 16.7m (55ft) square and 10.7m (35ft) high incorporating eight identical hold-down arms. The flame deflectors were 13.1m

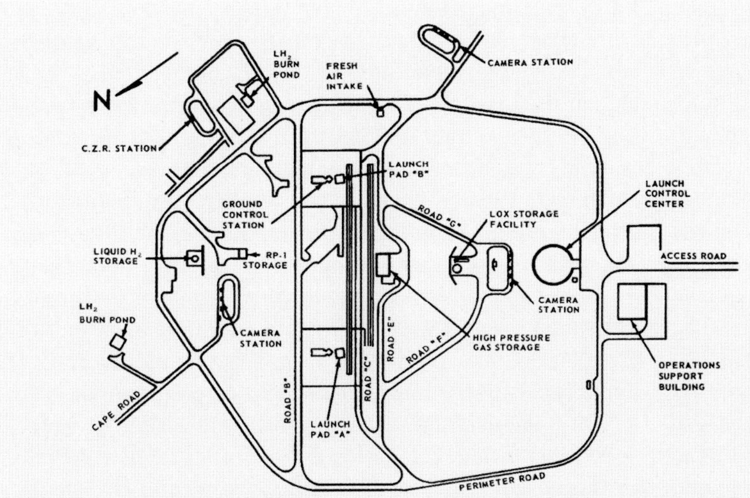

Above: Built with two launch pads, LC-37 featured a single service gantry capable of connecting pads A and B. Only the latter was ever used for Saturn I Block II and Saturn IB flights. The complex was also built to accommodate the Saturn C-2, which was never built. *Right:* The gantry for LC-37 had weather protection for the work platform levels. *NASA*

Above: A comprehensive view of the LC-37B site as the first Block II Saturn I (SA-5) prepares for launch. *NASA*

(43ft) in length, 9.8m (32.3ft) wide and 6.47m (21ft) high, weighing 136tonnes and were fabricated from ASTM A36 steel of dry, roof-truss structure. Each of the two pads was distinctive with its own launch stand and umbilical mast but with a common servicing tower which moved between pads along a rail track. When built it was the largest moving structure in the world and jacks were available at each pad to take the weight off the wheels when in position. The umbilical towers had a height of 81.68m (268ft) tall with a base 9.75m (32ft) square and 10.67m (35ft) tall, tapering to 4.26m (14ft) square at the top. The two fixed towers for respective pads were 274m (900ft) apart and carried one swing arm for the first stage, two for the S-IV stage and one for the spacecraft.

The single Service Structure eventually built had a height of 91.4m (300ft) excluding the top trapezoidal derrick, with a base 36.57m (120ft) square with a total weight of 4,717tonnes (5,200tons), a structure moving between pads as required along the 365.76m (1,200ft) rail. Mounted on four trucks, the self-propelled structure had 64 flanged wheels on the drive trains down one side powered by four electric motors. The other, unpowered side, had 24 flanged wheels. Overall, the Service Structure was designed for taller rockets, especially for the C-2, with a stiff-leg derrick on top for lifting stages and equipment onto the pedestal. With a 23m (75ft) radius of movement, the derrick mounted a single 54.4tonne (60tons) hook, a single 9tonne (10tons) jib hook and a 36.3tonne (40 tons) auxiliary hook. The working machinery for these, and three 1,361kg (3,000lb) capacity working elevators, were housed in the base.

Protective curtains were provided around the pedestal between the 10.7m (35ft) and 19.8m (65ft) levels, with six split silo enclosures from there to the 75.6m (248ft) level. Anchor points were co-located at the concrete hardstand and at the base of the service structure to fix it to the ground when positioned at the pad, and the entire structure was designed to withstand winds of 56kph (35mph) during movement between pads. Now this Service Structure outclassed that for LC-34 and succeeded it as the largest moving structure on Earth! But the cost of LC-37 had been high, coming out at $65million, three times the estimate.

RP-1 fuel was stored in an earth-revetted horizontal tank with a length of 20.4m (67ft) and a diameter of 3.65m (12ft) with a capacity of 164,647lit (43,500USgall). Liquid oxygen was stored in two spherical double-walled tanks with an outside diameter of 12.87m (42.5ft) and a capacity of 473,125litres (125,000USgall). Liquid hydrogen for the S-IV and S-IVB stages was contained in a double-walled spherical tank with a diameter of 38.8ft (11.8m) and a capacity of 473,125lit (125,000USgall).

The launch control centre consisted of a two-storey domed building located 365.76m (1,200ft) from the pad capable of surviving a blast pressure of 15,086kPa (2,188lb/in^2). The inner shell was of 1.52m (5ft) thick reinforced concrete, earth-revetted 2.1m (7ft) at the top and 12.5m (41ft) at the base. An internal layer of 5cm (2in) sound proofing material helped control the acoustic effect of launch. Close by, an operations support building had a length of 31m (101.9ft) and a width of 12.19m (40ft) and a separate spare parts storage facility had a length of 49.37m (162ft) and a width of 12.8m (42ft).

LC-37A was never used but LC-37B supported a total of eight flights including six Saturn I (AS-5 through AS-10) and two Saturn IB (AS-203 and AS-204) launches between January 1964 and January 1968. There had certainly been expectations of greater things and a grander vision. Even as late in the operational phase for this launcher as 1966, plans existed for 26 flights with the Saturn IB as part of the Apollo Applications Program supporting six orbital workshops in low Earth orbit and geosynchronous orbit.

The entire complex was retired when the decision was made to move remaining Saturn IB flights to LC-39 and launch the last four Saturn IB missions from there between May 1973 and July 1975. Effective for just four years, the complex was deactivated following an announcement on 20 November 1968 on the basis that future requirements for Saturn IB could be supported from a modification of one launch pad at LC-39, originally designed for Saturn V operations.

There had never been the large-scale use of this complex envisaged when it was designed, a victim of over-ambitious expectations at a time when deep-space exploration was an implicit part of the future manned space programme.

On 8 January 1998, Boeing received a right of entry to begin converting LC-37 for the Delta IV launch vehicle, a new generation of very powerful rockets, and re-designated SLC-37. Completely reconfigured, with a giant building for preparing the Delta rockets in a horizontal configuration before roll-out and erection to a vertical position on the pad, the first flight from SLC-6 occurred on 20 November 2002 and it has been in continuous use ever since.

Chapter Ten

Saturn I Flight Results

The first flight of the Saturn S-I stage was conducted little more than three years after full authorisation from ARPA, several months before NASA became active, and little more than four years after studies of a clustered booster rocket began. The first four flights would carry dummy S-IV and S-V upper stages and were to be followed by six orbital flights, the last three of which would carry a payload attached to the S-IV stage for gathering information about micrometeoroid population levels in low Earth orbit. Consisting of the Centaur stage, the speculative S-V was cancelled in May 1961 although the dummy stage was retained to preserve the aerodynamic configuration and mass balance as tested in wind tunnels. This plan was part of the phased approach to a graduated qualification of first, second and third stages.

Expectations of success were relatively low; big questions remained regarding the ability to control all eight H-1 engines in flight. In the late 1950s and early 1960s rocket failures were common, and frequently with less challenging designs and less demanding performance requirements than those posed by Saturn. The introduction of cryogenic propellants for the S-IV stage had many problems which brought their own development challenges. Thus it was that almost all Saturn programme engineers from the top down felt that the effort would be a success if 50 per cent of the ten flights achieved their objectives.

As described in an earlier chapter, the initial, highly conservative approach to flight testing changed significantly in late 1963 and early 1964 when the planned series of six manned Saturn I Apollo flights was cancelled, along with further flights with this vehicle after SA-10. But to understand the true value of this launch system during the period 1961-1965 the relevance to activities with other launch systems defines the contributions this launcher made not only to the

Above: Protected by weather shrouds, Saturn I flight SA-3 sits on Launch Complex LC-34 with the flame deflector in place. *MSFC*

Apollo programme but to the larger space effort.

When the first Saturn I flew, the Centaur stage with its two RL-10 engines was proving deeply troublesome and as the upper stage to Atlas, in its initial configuration as AC-1, the Atlas-Centaur was not to fly successfully before 29 November 1963. It would not achieve another successful flight before 11 August 1965, by which time the Saturn I programme was over and six successful flights of the S-IV placing payloads in orbit, had been completed. The Saturn I, with its six LR115 (RL-10) engines, demonstrated the overall success of the cryogenic propulsion system. However, back in mid-1961, when the first Saturn I was making ready to fly, NASA was looking with not a little concern at Centaur and possible options for upper stage applications.

Alternatives being considered included a Titan II with a Rocketdyne Nomad cryogenic upper stage, with or without large solid rocket boosters to increase payload capacity, an Atlas topped with an S-IVH stage utilising a Rocketdyne Atlas sustainer engine converted to burn liquid hydrogen, or a Thor-Diana using the basic Thor booster with two solid rocket boosters supplementing the first stage, powered by a single LR-115 engine as second stage. While Thor was eventually developed into the Delta launch vehicle and Atlas carried a wide range of upper stages, none of these initial proposals were deemed necessary. The Atlas-Centaur came good and began to do the heavy-haul on lunar and planetary spacecraft,

Above: The S-I-1 stage with a good view of the partially fitted thermal shield, the areas around the engine to be installed after checks. Note the seven personnel working inside the thrust structure, a potential diorama subject for modellers! *NASA-MSFC*

in addition to launching heavy satellites to Earth orbit.

Overall the Saturn I programme can be described in three evolutions: the initial Block I Saturn I in which the S-I and early development H-1 engines were tested, Block II in which the S-IV demonstration and evaluation phase successfully achieved orbital flight and the third phase in which the basic vehicle was developed into the Saturn IB, with more powerful S-IB stage engines and the much more powerful cryogenic S-IVB stage.

Throughout the Saturn I programme, the launcher moved from being a development path for intermediate and heavy payloads to a support element for the Apollo Moon-landing programme. The Block II vehicles became development tools for the S-IVB and Saturn V programmes and the Saturn I lost its original purpose as a medium-class booster. The missions for which it could have been relevant, and which were originally proposed, simply withered away, in the wake of Moon-fever which focused attention on getting to the lunar surface. With evaporating public and political interest, and the declining budgets which ensued, Saturn became a victim of success in other directions.

In the following diary of launch vehicle events, quoted engine numbers in the initial table identify the engines in place for launch. The narrative history beneath each vehicle identifies those engines which were installed initially but were for some reason changed out prior to flight.

Saturn I flight SA-1

Launch date:	27 October 1961
Lift-off time:	10:06:03.89 EST (15:06:03.89 UTC)
Launch pad:	LC-34
Launch azimuth:	100deg
Length:	49.60m (162.745ft)
S/L thrust:	5,959kN (1,339,800lb)
Dry mass:	140,450kg (309,635lb)
Propellant mass:	288,185kg (635,328lb)
H-1 engines:	H-1011; H-1012; H-1013; H-1015; H-1016; H-1017; H-1019; H-1021
Ignition mass:	429,55kg (946,551lb)
Lift-off mass:	421,269kg (928,725lb)

On 26 May 1960, MSFC began assembly of the S-I-1 stage in Building 4705 and it was transferred for functional checkout on 16 January 1961. The dummy S-IV stage was manufactured at MSFC and the dummy S-V stage, fabricated by Convair, arrived at Huntsville on 8 February. Three static firings were conducted at Static Test Tower East on 29 April for a programmed duration of 30sec, on 5 May for just over 44sec and on 11 May for 111sec. After being removed from the MSFC test stand, S-I-1 was delivered up for final checkout and acceptance testing, a process beginning on 12 June 1961.

The flight hardware for SA-1 arrived at Cape Canaveral on 15 August 1961 on the newly adapted barge *Compromise*. Booster erection on LC-34 took place five days later followed by the upper stages on 23 August. A series of tests and checks were made over the next several weeks with some changes necessitated by the inaugural use of LC-34 with flight hardware. An unexpected test of the stability of the structure on the pad occurred after the Service Structure had been rolled back and a squall blew through with 102kph (63mph) winds without any ill effect.

The inert S-IV stage weighed around 11,340kg (25,000lb) and was ballasted with 40,608kg (89,525lb) of water while the S-V mock-up had a dry weight of almost 1,360kg (3,000lb) and carried 46,267kg (102,000lb) of water ballast. The extreme nose was a dummy payload section comprising the nose section of a Jupiter missile. SA-1 carried 610 channels of telemetry, of which 510 measured flight parameters, with the balance monitoring systems, subsystems, parameters in the dummy upper stages, propulsion systems, propellant subsystems and electrical systems. Because of the unprecedented power of the Saturn I, 50 acoustic measurements were made from the pad itself to a radius of 16km (10mls). Sound is destructive and can shatter solid objects; it was vital to understand the acoustic effects on structures and buildings around.

The primary objective for this flight was to prove out the launch facilities including all the servicing, propellant loading and launch events, to demonstrate satisfactory performance of the eight H-1 engines and to verify the structural integrity of the stack. Secondary objectives included verification of the aerodynamic qualities of the vehicle, demonstration of the modified ST-90 guidance platform, and a range of engineering requirements. Pre-flight preparations were intense and busy, the first full-scale demonstration of electrical, pneumatic, hydraulic and propellant connections, transfer cables, conduits and pipes being integrated with flight hardware for the first time.

Range radar and telemetry checks took place during the second half of September and the service structure was moved back on 15 September. In the first week of October the launch date of October 18 was set after the dummy S-IV and S-V stages had been filled with water ballast. Originally targeting 12 October for launch, by the beginning of the month that had slipped to 20 October. The first major test of the LC-34 systems had occurred with the LOX loading test on 4 October, hooked to a real live rocket for the first time, this being followed by checks on the outer engine gimbal motion and a simulated flight test but a further delay on the planned launch date was driven by some concerns from MSFC that there was a need for more sensors at the base of the rocket to monitor responses which the vehicle would incur as it passed through a critical bending mode between 10sec and 35sec after lift-off.

Targeting for a lift-off at 09.00hr EST on 27 October, the countdown began at 23.00hr the preceding day (T-600min) and 30 minutes later electrical circuits were activated on the

Left: A throwback to the days of testing as S-I-T fires up the two H-1 engines installed in the mock-up at Static Tower East on 28 March 1960 for the first time. The first test with all eight engines installed occurred on 26 May 1960. *NASA-MSFC*

UNITED STATES
DORSET

Left: The S-I-1 stage arrived at Cape Canaveral on 15 August 1961 and was erected on LC-34 five days later, hoisted into position and straddled by the service gantry. *NASA-KSC*

S-I stage. Fuel had been put into the four outboard tanks on 25th October. LOX fill began at 03.10hr with 18,144kg (40,000lb), or 10 per cent, going into the centre tank and four of the outboard tanks. This enabled leak checks to be carried out and lines and valves to get a chilldown. On the day of launch a continental polar air mass hung over the entire Eastern Seaboard bringing a cold front moving south across Cape Canaveral and clouds and high winds aloft could cause delays. Generally, weather conditions were regarded as good, however. The platform in the guidance system was turned on at 04.30hr EST (T-270min) followed 10min later by the stack going on internal power for a check on the flight batteries. At T-120min the massive Service Structure was rolled back some 183m (600ft) from the stack. The first hold was called at 07.00hr for 34 minutes but a further hold, for 32 minutes was called at 09.14hr.

The LOX tanks were filled to the level required for this flight between T-60min and T-30min and the automatic sequences went in at T-6min 4sec followed by a nitrogen purge of the engine compartment at T-5min, the vent valves on the five LOX tanks closing two minutes before ignition. The Saturn switched to full internal power at T-35sec and 10sec later the 21m (70ft) cable boom between the S-I and dummy S-IV stages was ejected. The single solid propellant charge ignited the gas generator and began to spin up the turbopump on each H-1 engine. Hydraulic valves opened and propellants flowed into the combustion chamber.

The eight H-1 engines were started in pairs at 30-100millisecond intervals – outboard engines 5 and 7 first (30millisec after the ignition command) followed by outboards 6 and 8 (110millisec later), then inboards 2 and 4 (100millisec after that), lastly inboard engines 1 and 3 (100millisec later). It took just 1.4 seconds to build to full thrust, during which the thrust frame flexed 2.5cm (1in). The four restraint arms held the rocket for 3.92sec while the automatic checkout equipment watched for combustion instability and engines, pumps, tanks and hydraulic pressures were checked. At ignition the weather systems for several hundred metres (yards) around were affected by the blast, the wind changing direction and increasing velocity for about 30 seconds. As the vehicle rose slowly from LC-34 it shook the surrounding area with a power more than three times that of any previous rocket sent aloft from Cape Canaveral. It was a chill morning but the sky was clear and to those watching it seemed like a miracle of science, engineering and technology – all rolled into one giant candle lighting up the morning sky with a pronounced yellow flame and white billowing exhaust.

Immediately after lift-off the vehicle yawed to the right at 0.2°, an angle maintained until 32.5sec and probably caused by wind. At 17sec, SA-1 began to pitch over until, 83sec later, it held pitch at 43° to the vertical and continued on its trajectory at that inclination. To shape the trajectory, the pitch programme operated in five time segments: 17.80-20.89sec (2deg); 32.39-42.89sec (7deg); 46.79-85.79sec (26deg); 86.99-95.99sec (6deg); and 97.19-100.19sec (2deg). The stack ran through the area of maximum aerodynamic pressure (a point known as Max-Q) at 61sec, some 11km (7mls) downrange – the point launch operations director Kurt Debus believed which, if passed successfully, would make this mission a 100 per cent success.

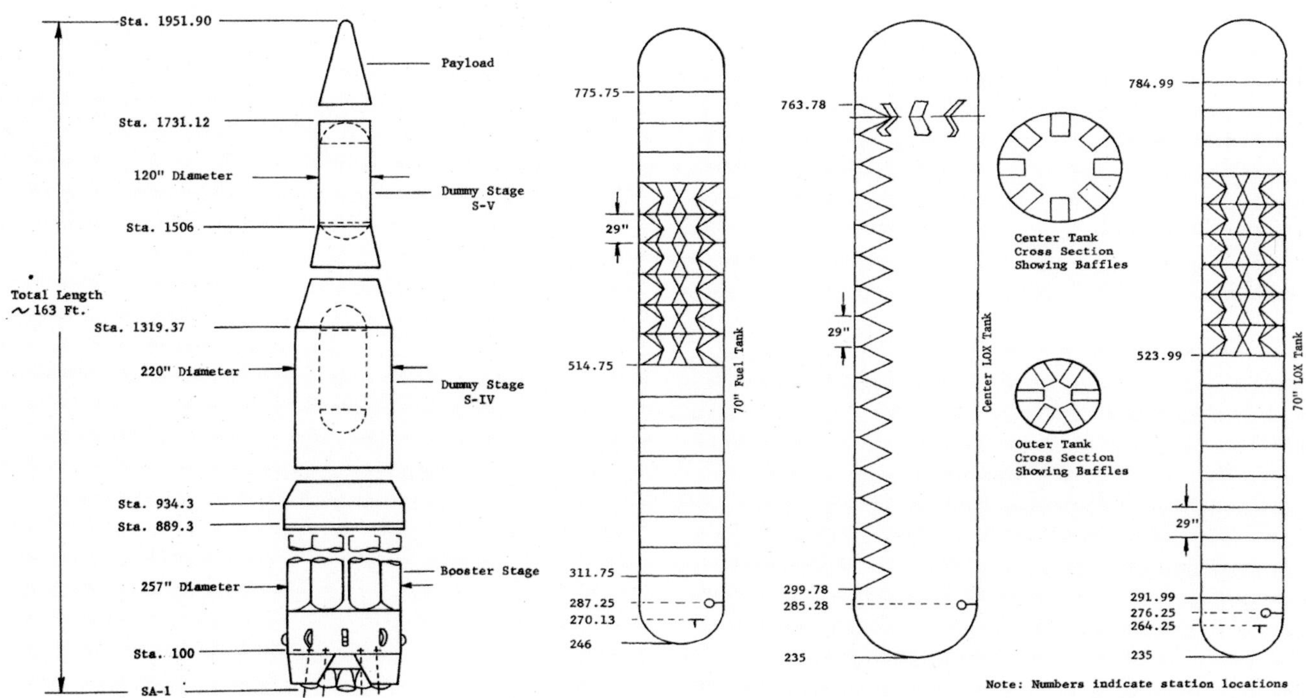

Below: Measurement stations for the SA-1 rocket (*left*) in a compressed view of the live first stage and mock upper S-IV and S-V stages. Based on projections regarding the movement of liquid propellant, tank baffles for the SA-1 rocket (*right*) would be modified on later stages. *MSFC*

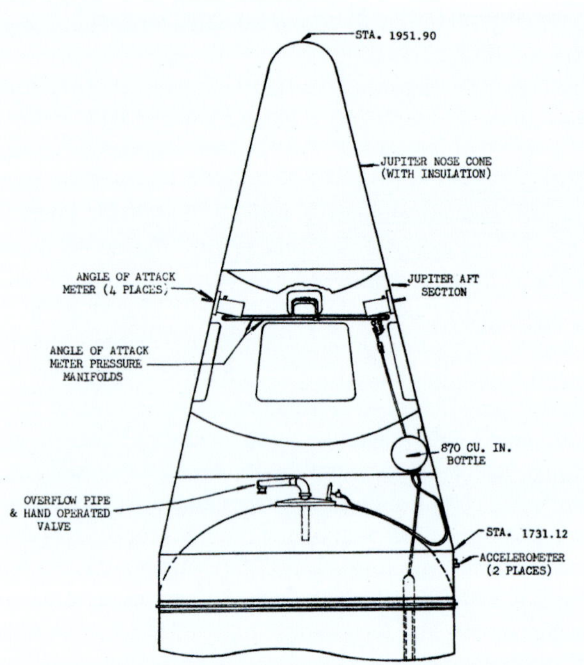

The nose cone for the Block I Saturn I flights was fabricated from that for the Jupiter with an additional section forward of the transition section to the dummy S-V. *MSFC*

At that instant wild pandemonium overwhelmed the tension in the pad blockhouse as shouts and screams broke out. All knew what passing that point successfully meant; with stresses reducing on the giant rocket, it was an uphill but safer ride thereafter. Uncertainty surrounded just how well this initial flight would go, many asserting that it had a 25 per cent chance of never getting off the pad and that if it did only a 30 per cent chance of completing its mission. Outside the NASA enclaves and the fraternity of engineers and technicians who had been working this concept for almost four years, local Cape residents as far afield as Cocoa Beach feared that a failure would cause widespread damage. NASA's own internal assessment remained firm – if it did explode on the pad, LC-34 would be out of commission for a year.

With Max-Q survived, there were about 48sec in which to watch and to monitor events as the giant rocket climbed inexorably toward shut-down. Emotions certainly ran high and were expressed in the excited voices monitoring events in the blockhouse. With events relayed live to Huntsville, personnel at MSFC unable to attend the launch but for whom the achievement had been a focused commitment, verification that the Saturn I was well on its way was a palpable relief. Nobody had expected such a flawless performance, breaking the tendency for maiden flights to go badly. The consequences of a failure with SA-1 would have had serious consequences for NASA's long-term programmes.

As planned, engine cut-off came in two separate events, the inboard engines shutting down at 1min 49.37sec followed by the outboards at 1min 55.15sec, events staggered to prevent pendulous oscillations causing an abrupt and uncontrolled shut-down. Peak altitude of 136.455km (84.8mls) was achieved at 4min 9.24sec with loss of telemetry at 6min 49.3sec and an altitude of 19.59km (12.17mls) on the way to

Below 1-4: The first Saturn I lifted off from LC-34 on the morning of 27 October 1961, opening a new era in US launch capability with the first of four ballistic flights to qualify the S-I stage for carrying the cryogenic S-IV on orbital missions. *NASA-KFC*

splashdown some 346km (215mls) from the Cape. The total flight time was 8min 3.6sec, maximum velocity was 5,803kph (3,607mph) and impact occurred within 11km (7mls) of the target area, about 110km (70mls) north east of Little Abaco island. Up aloft, chase aircraft had included a Lockheed C-130 at 7,620m (25,000ft), a Douglas A3D at 12,190m (40,000ft) and a Lockheed U-2 – its precise altitude classified due to the nature of this exotic spy-plane.

To those on the ground, most watching from a distance of 3.2km (2mls), the sound was less than had been anticipated and some said they felt it was not as loud as the Atlas or Titan, rockets fired at frequent intervals across the Cape. And those 50 acoustic microphones agreed, measured levels being less than expected. Temperature sensors on the rocket's aluminium glass-fibre heat shield also showed that the design ably protected the lower section of the stage from the fiery heat of all eight H-1 engines, verifying that the curtain had been well designed, insulating adjacent areas from the searing 2,480°C (4,500°F) temperature of the exhaust plume.

In analysis of the trajectory, and in the post-flight review, it was noted that had one inboard engine failed it would have reduced performance by a mere 3 per cent, allowing the other seven to consume the propellant not used by the lost engine. There was good news on the area cleared for launch, which had stood at 3km (1.9mls). Pad and blast assessment reduced that range for future launches to 1.53km (0.95mls) compared with 0.96km (0.6mls) for an Atlas or a Titan.

Setting this event in context, in October 1961 NASA was still struggling with a wide range of possible launch systems for the very ambitious plans already laid before Congress. It was vital that test operations continued on a generally successful course – with many challenges ahead.

Saturn I flight SA-2

Launch date:	25 April 1962
Lift-off time:	09:00:34.41 EST (14:00:34.41 UTC)
Launch pad:	LC-34
Launch azimuth:	100deg
Length:	49.60m (162.745ft)
S/L thrust:	5,959kN (1,339,800lb)
Dry mass:	140,450kg (309,635lb)
Propellant mass:	287,113kg (632,966lb))
H-1 engines:	H-1028; H-1029; H-1032; H-1033; H-1034; H-1035; H-1036; H-1038
Ignition mass:	427,991kg (943,543lb)
Lift-off mass:	420,380kg (926,673lb)

The second flight of the Saturn I was similar to that of SA-1 but with the additional experiment of deliberate destruction shortly after the S-I stage was shut down, releasing water ballast into the ionosphere to determine the effect of that dispersed in the upper atmosphere. In addition, due to post-flight analysis of SA-1, an extra 136kg (300lb) of baffles was placed in the lower sections of the eight outer propellant tanks to suppress sloshing observed on the first flight and potentially disruptive for attitude control and the flight path. SA-2 was equipped with 622 channels of telemetry information of which 527 were flight instrumentation. The ST-90 stable platform was carried as well as an ST-124 platform of the type planned for the Pershing IRBM.

Project High Water, the release of water ballast in the ionosphere, was sponsored by NASA's Office of Space Science and as a precursor calibration to the experiment, on 2 March

3

4

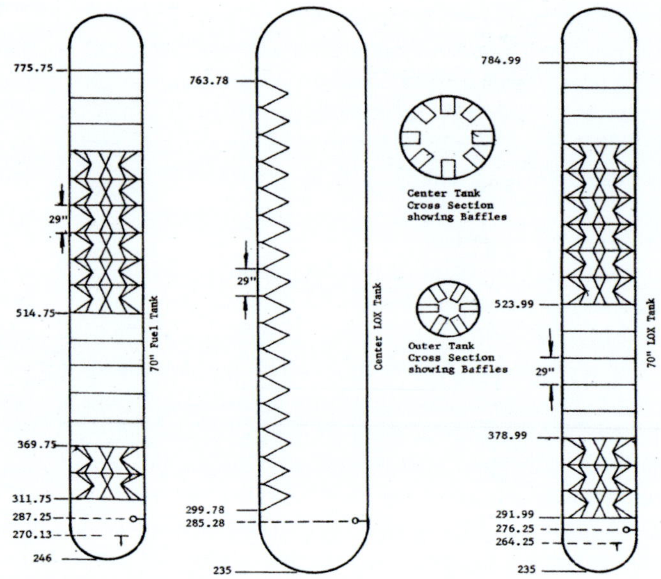

Above: Significant sloshing and attitude alignment due to the moving mass of unspent liquid propellant with SA-1 forced a redesign of the slosh baffles in the tanks of SA-2 and subsequent boosters. *MSFC*

Left: Tests underway on SA-2 with protective shrouds over the S-I tanks. *NASA-KSC*

1962 a Nike Cajun sounding rocket fired from Wallops Station, Virginia released 18kg (40lb) of water at an altitude of 104km (65mls), forming an ice cloud visible as far north as Washington, DC.

Assembly of the S-I-2 stage got under way on 27 December 1960, approximately two months behind the planned schedule due to delays in component manufacture, and it was transferred to the checkout building on 1 August 1961, the dummy S-V having been delivered by Convair in June. The S-I-2 stage was at the Static Test Tower East from early October for about four weeks during which two firings took place, one of 32sec on 10 October and one for 119sec on 24 October. Stage delivery to CCAFS began on 17 February 1962 with arrival at CCAFS ten days later, all elements aboard the barge *Promise* (the renamed *Compromise*) with the S-I stage taken to LC-34 the same day. The stage was erected on 1 March followed by erection and mating of the dummy S-IV and S-V stages next day. This was followed over the next week by fuel loading simulations. A wide range of tests, checks, simulated cable ejection procedures and electrical switching occurred throughout the rest of March and much of April. Throughout the pre-flight preparation phase, a raft of minor problems were encountered but minimal on their impact as both processes and procedures matured.

Fuel was loaded two days before launch and while the tanks would only be partially full at ignition, loading was initially to 15.7 per cent at 757lit/min (200USgall/min) to allow a leak check, following which the tanks were filled to 98 per cent at 7,570lit/min (2,000USgall/min) whereupon the level settled to 92.7 per cent due to the uneven filling of the tanks. Following this, a slow fill topped it to 100 per cent at the rate of 757lit/min (200USgall/min). LOX loading took place on launch day when the tanks were filled to 14 per cent for a leak check, with the system stabilising at these conditions for approximately 4hr 50min. The tanks were then filled to 100 per cent at T-60min at a rate of 14,383lit/min (3,800USgall/min). Final computer checks brought the levels down to the required values for flight, 59,000kg (130,000lb), almost 17 per cent, less than full.

Weather conditions were excellent, with only a haze restricting visibility to 13km (8mls) and the temperature at a comfortable 24°C (76°F) with a light breeze flowing up the coast from the south. The official countdown began at 22.30hr EST on 24 April for a scheduled launch at 08.30hr

Right: SA-2 lifts off on the morning of 25 April 1962, completing a second successful ballistic flight in the Block I development series. *NASA-KSC*

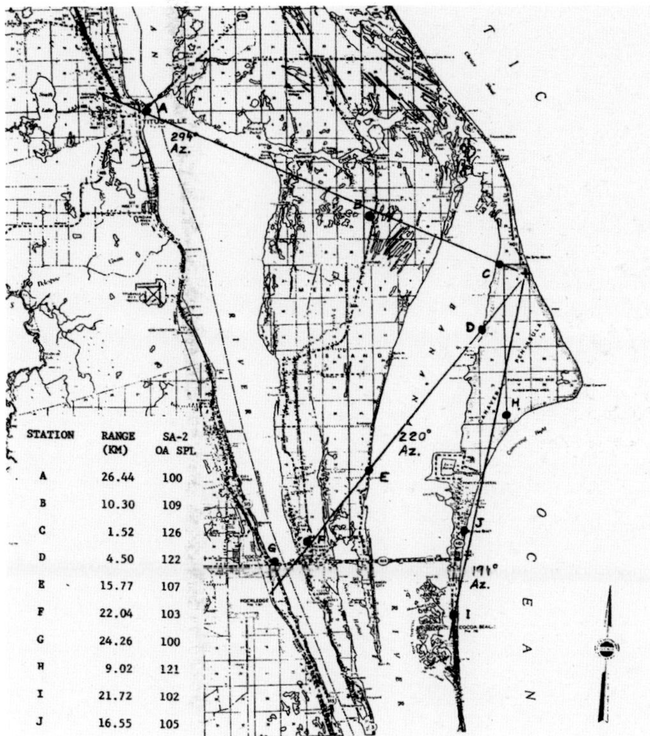

STATION	RANGE (KM)	SA-2 OA SPL
A	26.44	100
B	10.30	109
C	1.52	126
D	4.50	122
E	15.77	107
F	22.04	103
G	24.26	100
H	9.02	121
I	21.72	102
J	16.55	105

Left: Critical assessments were made regarding the acoustic impact of sound waves at ignition and during launch and ascent. This map displays where those recordings were made for future consideration when planning support infrastructure and bigger launch vehicles. *MSFC*

the following morning but a 30 minute hold was necessary at T-10min to wait for the removal of a ship from the danger area, resuming at 08.50hr with the automatic countdown sequence at T-6min 14sec.

Ignition occurred at T-3.61sec with launch commit at 0.13sec. The eight H-1 engines were started in pairs at 40-110millisecond intervals – outboard engines 5 and 7 first (40millisec after ignition command) followed by outboard 6 and 8 (90millisec later), then inboards 2 and 4 (110millisec after that), lastly inboard engines 1 and 3 (90millisec later). The tilt manoeuvre commenced at 10.41sec with the vehicle going supersonic at 46.88sec and Max Q at 59sec followed by tilt arrest at 99.5sec at an angle of 43deg to the vertical.

Inboard engines shut down at 1min 51.29sec and an altitude of 50.5km (31.38mls) and a surface range of 25.1km (15.6mls) when the vehicle mass had dropped to 157,470kg (347,158lb). It was followed by the outboards at 1min 59.84sec, an altitude of 57.8km (35.9mls), a velocity of 6,022kph (3,742mph) and a vehicle mass of 150,065kg (330,832lb). The destruct signal for Project High Water was delivered at 2min 42.56sec at an altitude of 105.3km (65.4mls) dispersing 72,161lit (19,065USgall) into the atmosphere. The debris fell to the Atlantic Ocean 362km (225mls) downrange of Cape Canaveral.

The release of the water was accomplished through the vehicle command and destruct system, which employed strings of primacord to cut each fuel and oxidiser tank longitudinally. Additional primacord was spliced in to the system to cut the two water tanks. In each case, the primacord ran through a conduit within the tank, but adjacent to the tank wall. The tank comprising the dummy third stage could be very effectively ruptured by the explosives. The dummy second stage presented a greater difficulty since its walls were 0.63cm (0.25in) thick steel. Here it was practical to produce four 142cm (56in) by 203.2cm (80in) ports in the outer wall. The inner tank containing the water would be effectively ruptured by its charge. The entire pyrotechnic train was initiated by a radio link on command from the ground.

A wide variety of optical instrumentation was available for recording phenomena subsequent to the release of the water. Cameras were located as far south as Vero Beach and Melbourne Beach with one located down at Grand Bahama. The primary object was the release of water but in addition some 3,400kg (7,500lb) of RP-1 and 4,600kg (10,140lb) of LOX was released in the destruction of the vehicle and the rupturing of the propellant tanks. The cloud was visible for about 12 seconds, racing faster across the sky than the spent debris of the Saturn I, producing an electrical discharge similar to that produced by a thunderhead. Project High Water proved to be a success, considerable data being obtained about the physics of near-Earth space and outer extremities of Earth's atmosphere.

Around the time of the SA-2 launch, NASA got enthusiastic about flying at least 10 manned Apollo missions on the Saturn I. The successful preparation, countdown, launch and performance of the Saturn I on its second mission, although a ballistic shot, augured well for the future of the programme and MSFC settled in to planning a third and fourth flight in the same configuration before utilising uprated engines and added stability fins on SA-5 which would also carry the first live S-IV stage for the Block II configuration.

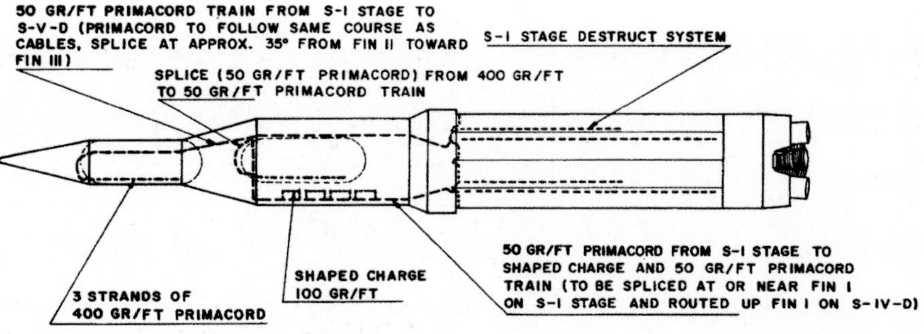

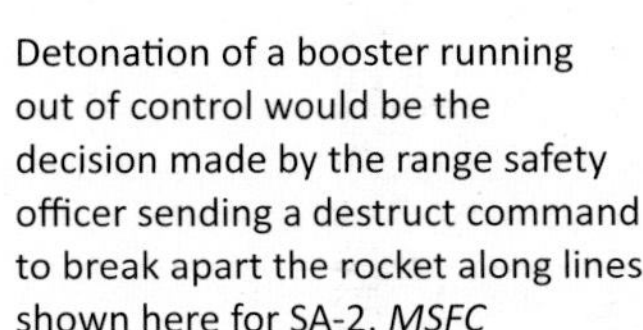
Detonation of a booster running out of control would be the decision made by the range safety officer sending a destruct command to break apart the rocket along lines shown here for SA-2. *MSFC*

Saturn I flight SA-3

Launch date:	16 November 1962
Lift-off time:	12:45:02 EST (17:45:02 UTC)
Launch pad:	LC-34
Launch azimuth:	100deg
Length:	49.60m (162.745ft)
S/L thrust:	5,959kN (1,339,800lb)
Dry mass:	143,598kg (316,580lb)
Propellant mass:	355,601kg (783,966lb)
H-1 engines:	H-1040; H-1041; H-1042; H-1043; H-1045; H-1047; H-1048; H-1049
Ignition mass:	499,683kg (1,101,614lb)
Lift-off mass:	492,182kg (1,085,076lb)

In preparation for the first Block II series, the third flight of a Saturn I was planned as a stiffer test of the eight-engine cluster and served as a test bed for systems in development for SA-5. With full propellant tanks, the eight H-1 engines would burn longer and fly a higher and longer trajectory. Because of the heavier weight of SA-3, greater demands were placed on the flight control and propulsion systems, with an engineering model of the ST-124 stable platform flown in an open-loop configuration, together with equipment to monitor its performance. Engine cut-off was to be controlled by a propellant utilisation system with quantity level switches to shut down the four inboard engines after 95 per cent of the propellant had been consumed. The four outer engines would run to propellant depletion, around seven seconds longer.

Also tried out with this flight, the new 73.1m (240ft) tall umbilical mast on LC-34 would be used for testing Block II vehicles and would carry power instrumentation and air conditioning for flights from SA-5 onward. SA-3 carried 716 telemetry channels, of which 612 were in-flight channels reporting on engine turbine temperatures and rpm, temperature on engine bearings, heat exchanger outlets, tail skirt and turbine exhausts, pressure in the combustion chambers, propellant tanks, dummy upper stages, maximum dynamic forces on the vehicle, performance of the ST-124 platform, engine gimballing, propellant levels correlated with engine cut-off,

Above: Framed by the service gantry and with the new umbilical tower for Block II flights in the background, SA-3 presents an aesthetic view as preparations for launch get under way. *NASA-KSC*

MANNED SPACE FLIGHT SCHEDULE
SATURN
CURRENT FLIGHT SCHEDULE

MILESTONES
CY 63 CY 64 CY 65 CY 66 CY 67 CY 68 CY 69 CY 70
1 2 3 4 (each year)
1
2
3 SATURN I 4 5 6 7 9 8 10 111 112 113 114 115 116
4
5
6 SATURN IB 201 202 204 206 207 208 209 210 211 212
7 203 205
8
9 SATURN V 501 502 504 506 508 510 511 513 515
10 503 505 507 509 512 514
11

LEGEND
○ UNMANNED FLIGHTS
● MANNED FLIGHT

By the end of 1962 a schedule had been set calling for six manned flights of the Saturn I and eight of the Saturn IB, followed by six unmanned test flights and nine manned missions of the Saturn V by the end of 1968. In reality, by the end of the programme in 1975, there would be no manned flights of the Saturn I, just four of the Saturn IB and 10 of the Saturn V. *NASA*

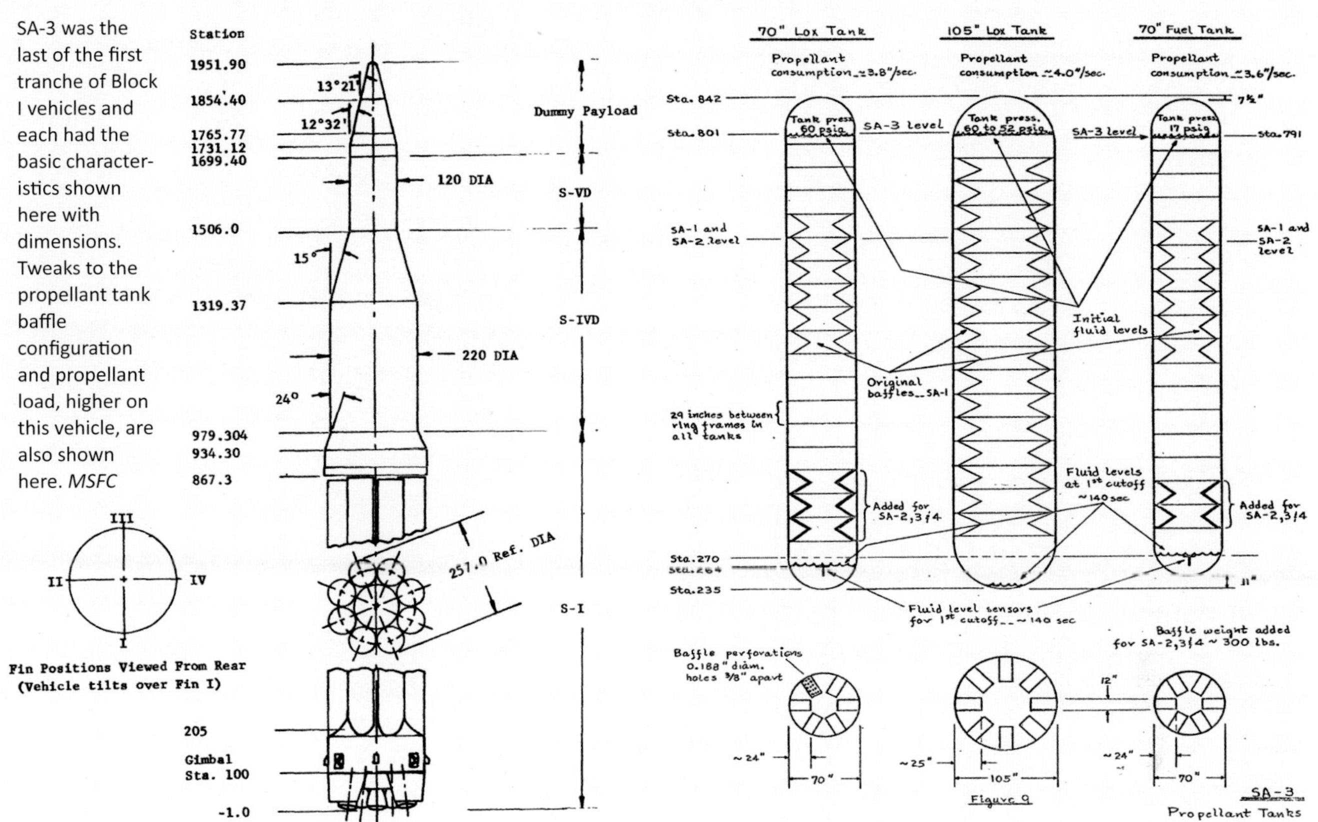

SA-3 was the last of the first tranche of Block I vehicles and each had the basic characteristics shown here with dimensions. Tweaks to the propellant tank baffle configuration and propellant load, higher on this vehicle, are also shown here. *MSFC*

battery voltages and current and inverter frequency. It also carried a pulse-code-modulated (PCM) transmitter in the S-I stage and a UHF transmitter in the payload.

Also new for this flight, SA-3 carried four solid propellant retrorockets which would fire two seconds after shut-down of the four inboard engines but before the outboard cut-off. As indicated, SA-3 carried a very different engine shut-down sequence whereby the tank level sensors would activate a sequencer on the inboard engines, 350millisec after detecting receding propellants. Previously, on the first two Saturn I launches a timer shut down the four outboards six seconds after the inboards. On SA-3, the timer would command the tie-in to a single circuit of the thrust-OK switches on the four inboards. These sensors would measure a drop in pressure in the combustion chamber and a fall of more than 10 per cent would normally initiate shut-down. Under this new operating procedure, the four outboards would continue to suck in propellant until cavitation, inducing a pressure drop of more than 10 per cent which would initiate shut-down.

This would also be the second Saturn I to fly the Project High Water experiment, including 87,784kg (193,528lb) of water ballast, with the destruct profile at higher altitude and further downrange due to the longer burn time of the S-I stage.

Started on 31 July 1961, S-I-3 stage assembly was completed on 26 December and taken to the Static Test Tower East after pressurisation tests. Three firings took place, the first for 31sec on 10 April 1962 but unusually high vibration in a turbopump was discovered, prompting a repeat firing on 17 May. A third test lasting 119.4sec took place one week later, after which the stage was removed from the test stand and departed MSFC on 9 September. Hardware began arriving at Cape Canaveral on 19 September 1962, with the S-I on the barge *Promise*, the stage being erected on the pad two days later after a delay caused by incessant rain and high winds as a tropical storm moved through the Cape area. Even as the task of installing S-I-3 on LC-34 went ahead, technicians were working high up on the Service Structure in winds of 37kph (23mph).

Dummy S-IV and S-V stages were erected on the S-I on 24 September followed by the usual pre-launch procedures including rollback of the service structure on 19 October. The weather abated and the retrorockets were installed on 9 November, with a simulated flight test four days later followed by RP-1 fuelling on the 14th. The countdown began at 02.00hr EST on 16 November but a single hold of 45 minutes was necessary to replace part of a ground power generator at T-75min. In the area, weather was now good with no clouds, no rain and visibility of at least 16km (10mls). Light winds were from the south-west and the morning was balmy to say the least. It was an idyllic morning with a tropical feel to the air.

The automatic countdown sequence started when the firing commend went in at T-6min 3.45sec, later than with SA-2 due to a time difference for LOX tank pressurisation. The eight H-1 engines were started in pairs at 100-110millisecond intervals – outboard engines 5 and 7 first (10millisec after ignition command) followed by outboard 6 and 8 (110millisec later), then inboards 2 and 4 (100millisec after that), lastly inboard engines 1 and 3 (100millisec later). Mach 1 was passed at 1min 8.1sec at an altitude of 19.3km

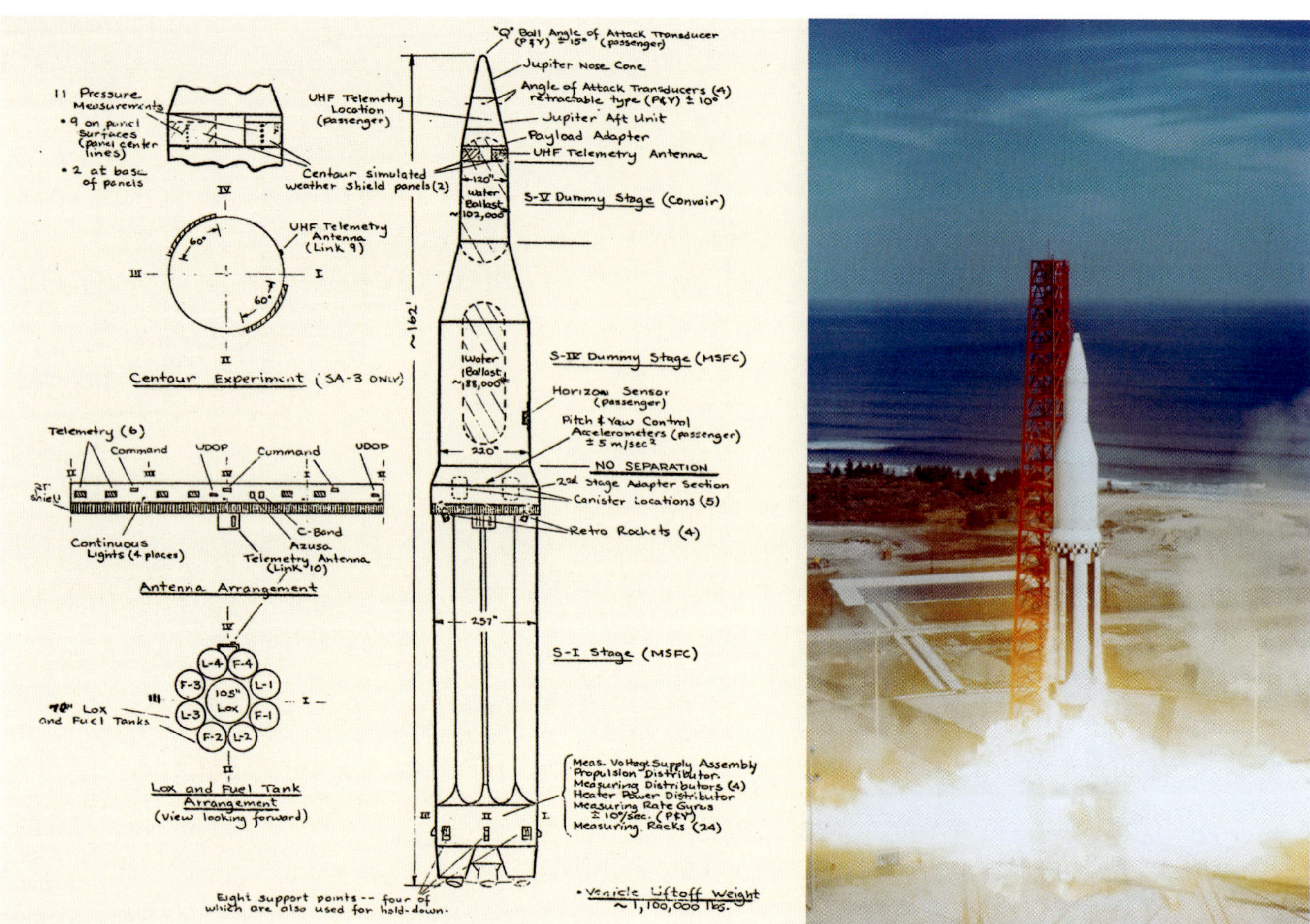

Above: General configuration and specific characteristics of SA-3.
Above right: With a heavier propellant load, SA-3 gets under way. *MSFC*

(12mls) followed by Max-Q at 2min 21.8sec, achieving maximum velocity of 6,521kph (4,053mph) at 2min 29.5sec.

Inboard engine cut-off was detected first in tank No.4, at 2min 21.66sec into flight and an altitude of 61.4km (38.15mls) but 1.32sec later than predicted. Outboard No.3 first detected the drop in pressure and with all thrust-OK switches on the same circuit, initiated shut-down of all four outboard engines at 2min 29.1sec and an altitude of 71.1km (44.2mls). Approximately 348,225kg (767,692lb) of propellant was consumed during the burn period, leaving 6,037kg (13,309lb) remaining in the tanks. Note that the shortfall between these totals and the figures at the head of this section is accounted for by the burn time between ignition and lift-off. The retrorocket ignition command was sent at 2min 23.66sec for a burn duration of 2.15sec, delivering a total average thrust of 648.2kN (145,750lb).

The destruct command releasing the water was given at 4min 52sec with the vehicle at an altitude of 167.2km (103.9mls), 211.4km (131.4mls) downrange across the Atlantic Ocean.

A mission strategy evolved from the successful demonstration of the propellant depletion technique tested on SA-3, a decision was made that it would be repeated on SA-4 but not adopted for SA-5 – which would be the first Block II and the first to carry a live S-IV stage for an orbital demonstration. The special demands of ignition procedures for the six cryogenic RL-10 engines required a precise pre-ignition chill-down sequence which called for an exact time for S-I shut-down. However, MSFC held open the possibility that if SA-5 was a success, the depletion-trigger method demonstrated by SA-3 could be re-introduced on SA-6.

An unintended advantage of SA-3's performance was an unexpected, but perfect, roll without dispersions in pitch or yaw, giving tracking stations the opportunity to accurately measure the propagation patterns for the 27 telemetry antennas at various positions around the body of the vehicle. It was also noted that five telemetry links continued to transmit after the destruction of the vehicle for Project High Water, providing valuable information about the motion of the fragments. But two new telemetry links, a pulse-code modulated (PCM) and a UHF transmitter, operated well, albeit noisily. But FM-FM telemetry antennas in the 230-260MHZ band dropped 30db during the firing of the retrorockets.

After launch, pad inspection revealed that very little additional damage had been caused by the slower rise due to the heavier lift-off mass. Film coverage showed that it took 3.2 sec longer to reach a point 93m (305ft) above the complex, keeping the vehicle in close proximity to the pad and greater duration of exposure to the flame plume, for 30 per cent longer than with the lighter SA-1 and SA-2 vehicles. Nevertheless, pad damage was about the same as that recorded for the previous shots and the only real change was damage to the pedestal water deluge system.

Saturn I flight SA-4

Launch date:	28 March 1963
Lift-off time:	15:11:00 EST (20:11:00 UTC)
Launch pad:	LC-34
Launch azimuth:	100deg
Length:	50.43m (165.45ft)
S/L thrust:	5,959kN (1,339,800lb)
Dry mass:	88,452kg (195,500lb)
Propellant mass:	340,200kg (750,000lb)
H-1 engines:	H-1051; H-1052; H-1053; H-1054; H-1055; H-1056; H-1057; H-1058
Ignition mass:	429,355kg (946,551lb)
Lift-off mass:	421,269kg (928,725lb)

The fourth flight of the Saturn I and the last of the Block I, was to be a full-scale demonstration of the much acclaimed 'engine-out' capability, whereby the vehicle could continue on its way should one engine stop operating. To test this, it was deliberately programmed into the flight profile with the No 5 inboard engine being shut down at a pre-determined 1min 40sec. The plan here was for the other seven engines to draw upon propellant no longer being consumed by the inactive engine. This was unique in the history of rocketry and has never been actively tried as a deliberate test since. But the reason for this is that so few launch vehicles have clusters of more than five engines, and taking 20 per cent of the thrust away at any time during ascent has serious impact on guidance navigation requirements. Much less so with eight engines. In the modern era, such capabilities can be calculated digitally for a virtual simulation.

The plan was for 38.556kg (85,000lb) of propellant to remain in the nine tanks with tank No 1 supplying outboard engine 1 and inboard engine 5. When the latter was shut down the RP-1 was expected to transfer through the cross-manifolds interconnecting the base of the four fuel tanks to the other seven H-1 engine pumps. The same process would ensure that LOX was transferred and the entire system was expected to stabilise within 10sec. There was some concern that the inactive engine may disintegrate due to a lack of cooling and the heat generated by the other engines. To measure this, temperature sensors were attached to the rim of the nozzle on the No 5 engine. Added ground tests allayed fears that a backflow of heat would destroy that engine causing damage, or destroying, adjacent engines.

Other changes included the addition of four dummy ullage rockets, chill-down hydrogen ducts and a cable trunk tunnel on the inert S-IV stage and the interstage. These would be an active feature on the next flight. As with SA-3, four retrorockets were installed on the S-I stage and operated on the same cycle, despite there being no stage separation to pull away from. Because there had been a temporary blackout in telemetry when these rockets fired on SA-3, a small tape recorder was added to collect data on 10 channels through this period and transmit back about a minute after engine shut-down. As previously attached, an ST-124 stabilised guidance platform was flown in an open-loop configuration, active guidance being obtained from the ST-90 platform. It was decided not to fly the ST-124 closed-loop until SA-7.

Also, a radar altimeter planned for the Saturn V was flown on SA-4 with the capability of precisely determining the flight trajectory up to a distance of about 402km (250mls). And a MISTRAM (Missile Trajectory Measurement) transponder was carried as a test during this device's development programme for the General Electric tracking system. An electronic measuring system that would determine the position and velocity of a vehicle by means of interferometry and triangulation techniques, the MISTRAM system was being developed for the Atlantic Missile Range. And in development of flight control systems, a Q-ball transducer sensing angle of attack was attached in the tip of the modified nose cone, similar to a device carried on the X-15 hypersonic research aircraft. About 45.7cm (18in) long, an adapter connected it to the nose cone which extended the length of the overall stack. SA-4 carried 719 measurements of which 617 were through flight channels and 102 from the blockhouse during the countdown.

On this flight there would be no Project High Water, as there had been on the previous two vehicles, the trajectory being a little more depressed than earlier flights. There had been suggestions that it would be a stiffer test of the guidance and navigation system should one of the four steerable outboard engines be subject to an intentional shut-down. That was abandoned due to the use of two flight-controlling accelerometers rather than angle-of-attack sensors on previous vehicles sensing in pitch and yaw planes which would send corrections to deviations through to the guidance computer. It was necessary to get shut-down profiles from a more conservative and representative flight profile rather than pushing the operability of the flight control system. Another reason was the need to preserve the overall flight in its entirety rather than risking total loss of the vehicle.

Assembly of the last Block I S-I stage began on 2 January 1962 and was completed on 26 May with transfer to the test stand where it arrived on 1 August. In preparation for a firing test some foreign objects were discovered and in clearing that out

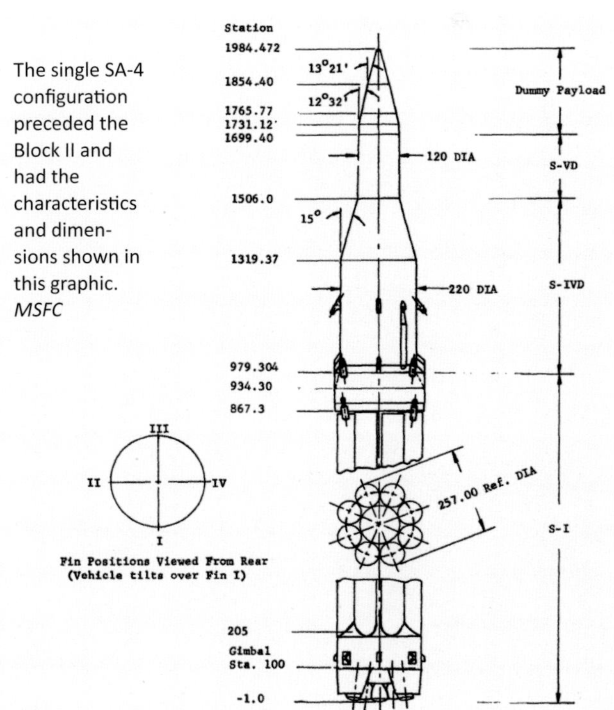

The single SA-4 configuration preceded the Block II and had the characteristics and dimensions shown in this graphic. *MSFC*

some areas were found to be contaminated with oil, delaying the schedule by several weeks. The first test was run on 11 September for 31.5sec, witnessed by President Kennedy and Vice President Johnson during a visit to Marshall. The second firing took place on 26 September, the inboards cutting off at 121.5sec followed by the four outboard engines six seconds later.

Removed from the stand on 1 October, S-I-4 was delivered up for quality control inspections three weeks later and left MSFC for the Cape aboard the barge *Promise* on 20 January 1963. With S-I-4, dummy upper S-IV and S-V stages and the water ballast tank, *Promise* had the hardware at CCAFS on 2 February. Two days later the stack was assembled on LC-34 and final pre-launch work got underway.

Flight preparations began at 03.30hr EST on 28 March, proceeding toward a launch scheduled for 13.30hr EST without major concerns but with three holds totalling 1hr 41min, the first lasting 19min caused by checks on the ST-90 guidance platform, the second for 40min to replace a faulty theodolite and the third for 42min to attend to a faulty micro-switch which could have threatened the countdown. The tilt programme began 10.58sec after lift-off followed by Mach 1 at 50.15sec at an altitude of 6.55km (4mls) and Max-Q at 1min 2.1sec and a height of 10.7km (6.6mls). The cut-off signal for engine No 5 occurred at 1min 40.62sec at a height of 35km (21.7mls) with thrust decaying out within 1.8sec. The tilt programme ended at 1min 45.58sec. The four inboards cut off at 1min 53.49sec, at an altitude of 47.95km (29.8mls), with the outer engines at 2min 1.1sec, at an altitude of 56.62km (35.2mls). Retrorocket ignition occurred at 2min 5.46sec and the apex of the trajectory, 129.66km (80.5mls), was reached at 2min 47.8sec, two seconds earlier than predicted, imparting a velocity decrement of 7.8m/sec (25.6ft/sec). At outer engine cut-off the altitude was 1.55km (0.96mls) higher, with velocity 45.7kph (28.4mph) greater than nominal. Loss of telemetry occurred at 6min 38.05sec and an altitude of 26.95km (16.75mls). Impact occurred approximately 370km (230mls) downrange from Cape Canaveral.

With completion of test launches for the overall configuration and qualification of the basic S-I stage, it was now time to move on with the Block II configuration introduced on the next vehicle with a live S-IV stage and dummy S-V before the fully operational capability was reached with SA-8.

Below: SA-4 lifts off, 28 March 1963. *MSFC*

Saturn I flight SA-5

Launch date:	29 January 1964
Lift-off time:	11:25:01.41 EST (16:25:01.41 UTC)
Launch pad:	LC-37B
Launch azimuth:	100deg
Length:	49.86m (163.6ft)
S/L thrust:	6,689kN (1,504,000lb)
Dry mass:	85,370kg (188,240lb)
Propellant mass:	431,510kg (951,300lb)
H-1 engines:	H-2001; H-2002; H-2003; H-2004; H-5002; H-5003; H-5004; H-500
RL-10 engines:	Unknown but in the P6418 series 5
Ignition mass:	516,880kg (1,139,720lb)
Lift-off mass:	508,794kg (1,121,680lb)

Much preparation had gone into the development of the Block II vehicle with its significant number of new systems and engineering developments, outlined previously. For SA-5 these included the new S-IV stage, spec-rated H-1 engines for a stretched-tanks S-I stage, longer burn time for the first stage and added tail fins for greater aerodynamic stability with the high lift-off mass. In addition, an Instrument Unit would be carried for the first time. This would be the first launch into orbit for a Saturn I and the first time the mission would not end with the S-I stage breaking up on its way back through the atmosphere.

In this configuration, the useful payload for orbital insertion was around 9,072kg (20,000lb) but for this flight the total orbited mass would be 17,240kg (38,000lb) counting the mass of the inert S-IV which would place itself in orbit. The propellant loads for this mission were far greater than for any previous vehicle – of any type – with the S-I carrying 385,560kg (850,000lb) of LOX/RP-1 and the S-IV carrying 45,950kg (101,300lb) of LOX/LH_2.

As the programme moved toward the Block II phase, where boilerplate Apollo spacecraft would be carried on the last three flights, the Radio Frequency Systems Section of the Instrumentation Development Branch of MSFC's Astrionics Division planned to introduce a TV camera for observing the separation of the S-I and the S-IV stages. An evolution of the camera carried on Mercury Redstone 2, which had launched on 31 January 1961, it would take 30 pictures/second on a composition of 525 lines for evaluation as an optical tool and with the view that future flights might carry two or more TV cameras. SA-5 also carried eight motion picture cameras mounted in pairs and attached to four beam ends of the eight-legged spider beam assembly in the top of the S-I stage. Four cameras would look down via fibre-optic bundles, one scanning an outer LOX tank and a second into the centre LOX tank. Two more would look into the interstage area observing chill-down of the RL-10 engines and stage separation. Each camera weighed 27.2kg (60lb) and would be ejected after S-I shut-down to be recovered downrange by individual paraballoons.

SA-5 would support 1,183 measurements, of which the S-I would account for 616, the S-IV 362, the Instrument Unit 189 and the payload 16. In addition, 201 blockhouse measurements were received during the countdown and terminated at lift-off. These were largely for the use of launch personnel on a selective basis. The vehicle had 13 flight telemetry systems, six on the S-I, three on the S-IV and four on the IU. In addition, 50 measurements were made at LC-37B and at other locations across Cape Canaveral up to a maximum distance of 24km (15mls) from the pad.

The overall length of the S-I was reduced from 25.1m (81.6ft) to 24.4m (80.2ft) despite an elongation of the propellant tanks by about 1.8m (6ft), thinning of the walls compensating for the additional weight. The area above the S-I tanks where the instrumentation was carried had been eliminated and the S-IV sat directly on top of the spider beam and instruments previously located in this area were moved to the IU or relocated to the two small instrument compartments above two fuel tanks. The four additional fins, described in the previous technical section, provided attachment points for the hold-down arms. Other changes included a disposal system preventing potential detonation of LOX or solid oxygen (SOX) during the chill-down phase prior to RL-10 ignition. Gaseous nitrogen would flow from storage tanks through manifold rings to the six combustion chambers to keep the oxygen from freezing, breaking away and falling from the S-IV engines.

Also, a hydrogen vent system was applied to remove chill-down fuel which would start to flow through the plumbing 40sec prior to stage ignition. The hydrogen would exit through three 30.5cm (12in) ducts leading down the sides of the S-I/S-IV interstage at 90deg locations around the S-I in line with stub fins 2, 3 and 4, dumping it into the jetstream. Other changes included the replacement of the 48 gaseous nitrogen propellant pressurisation system tanks with two 0.566m³ (20ft³) stainless steel tanks.

In previous S-I stages, LOX had been piped from the centre tank through the outer tanks and thence to the engines. From SA-5, a sump tank had been welded to the bottom of each tank with flanges for interconnect lines welded to the sump. A Y-arrangement would carry propellant from both the centre engine and the other tanks direct to the engines for better propellant utilisation and improved performance.

A few structural changes were introduced, including new beam and panel outrigger assemblies rather than pure beam outriggers, providing attachments points for the fin supports. Also, the S-IV stage adapter was now covered with a seal plate to keep hydrogen and oxygen used for S-IV chill-down from accumulating in the space between the separate S-I tanks and thus eliminating a potential explosive hazard from escaping propellants. As said above, SA-5 was the first Saturn I to carry an Instrument Unit, previous stages having instrument canisters. This, and SA-6 and -7 had a pressurised compartment that provided a conditioned environment for navigation, guidance and control equipment.

Where previous Saturn I test launches had been flown at 5-7 month intervals, preparation of the first Block II vehicle raised a gap of 10 months. This was the first use of LC-37B, much larger and more heavily instrumented than LC-34 used on the first four Saturn I missions and described earlier. With the introduction of the cryogenic S-IV upper stage, integration with the new launch site was another step on the learning curve. It was, in several respects a trial run for the expanded challenges with the Series-200 S-IVB stage for the Saturn IB (and the Series-500 S-IVB for Saturn V) and, in terms of propellant handling and transfer operations over at LC-39, operations with the much bigger cryogenic S-II stage.

Assembly of the S-I-5 stage began on 7 July 1962, three months after the H-1 engines arrived at Marshall, a start on the stage work being delayed after propellant tanks fabricated by Chance Vought were found defective and in need of replacement. After work to uprate the H-1 engines giving the stage its

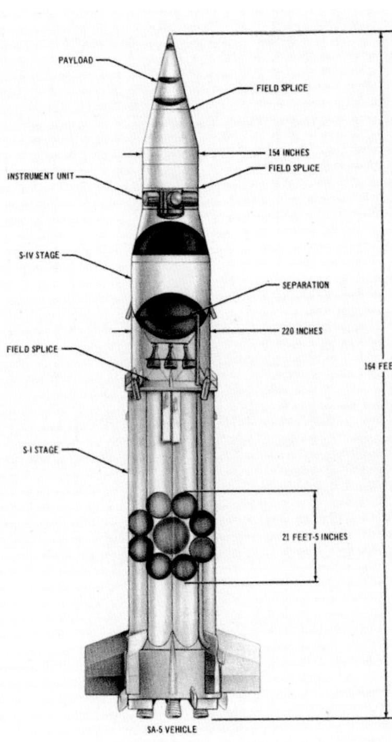

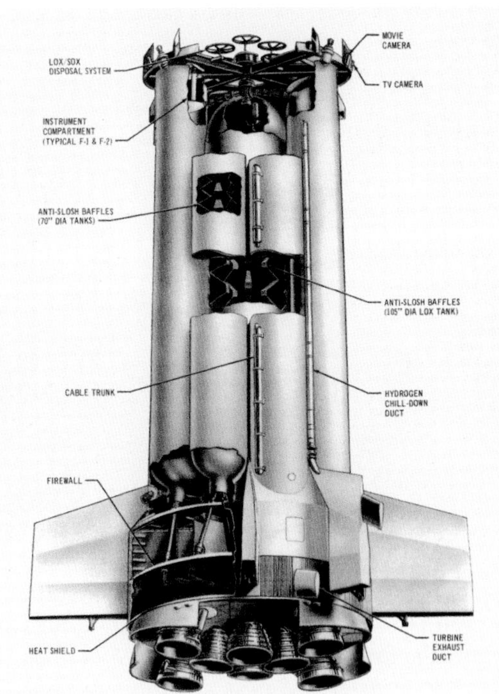

General configuration and dimensions for the SA-5 stack incorporating the S-IV (above) for the first time. The Block II features extended S-I tanks for more propellant, the additional weight partially offset by lighter materials and fins for stability. *MSFC*

design thrust level for the first time, the final outer engine was installed on 11 October. Placed in Static Tower East on 28 January 1963, the first firing took place for 31.9sec on 27 February. This had been delayed due to a suspected problem with the original engine H-5006, which was replaced by H-5005. The second firing test on 13 March lasted 144.4sec but revealed some issues prompting a second long-duration for 143.4sec, 14 days later. However, after S-I-5 was removed from the stand on 2 April further work was required to correct other issues. The stage was electrically mated to the new Instrument Unit with

Left: President John F. Kennedy visited the new LC-37B pad on 16 November 1963 and this was as close as he was supposed to get to SA-5. With him in the back of the car was NASA Administrator James Webb (hand raised) and Wernher von Braun.

no major issues and was placed in temporary storage awaiting availability of the S-IV stage.

Development of the S-IV was paced by several factors, not least because of the new RL-10A-3 motors which, although installed with the Centaur stage had not yet flown into space. At Douglas, stage build-up got under way in late 1961, delayed when some damage was caused to a segment of the common bulkhead. The completed tank was ready for insulation in October 1962, the same month the RL-10 engines arrived at Santa Monica and were then taken to Sacramento for further updates, held up for two months by a problem with spares. Douglas declared all six engines installed on 1 March 1963 with gimbal tests completed almost four weeks later. The next step was to transfer the completed S-IVB from the manufacturing facility at Santa Monica to the SACTO site, where two test areas – Alpha and Beta – were available. S-IV-5 left on 15 April 1963 and was transferred to SACTO first by barge from Los Angeles to Courtland Dock near Sacramento, arriving on 21 April by road.

A succession of delays, caused by checkout, simulated flight test, electrical integration and compatibility and by some technical problems, preceded a successful steam blow-down test on 20 July. Further cold-flow tests on the RL-10 turbines were held before the first static firing at Alpha TS-2B on 5 August, aborted after 63.6sec when overly conservative thermal sensors registering the temperature of hot steam indicated a fire. A second test seven days later ran the full 476.4sec. Removed from the stand, S-IV-5 was flown out by Pregnant Guppy aircraft on 20 September 1963, arriving at CCAFS the following day, one month after the rest of SA-5.

Hardware had begun to arrive at the Cape on 21 August 1963 with the S-I and the IU having been in transit from Marshall for the preceding 10 days. These were erected on LC-37B beginning 23 August and lasting three days during which the new fins for added stability were attached and a makeshift S-IV dimensional simulator placed on top of S-I-5 in lieu of S-IV-5, which had yet to arrive. When it did, the dummy spacer was removed and replaced with the flight-ready S-IV-5 which was mated to S-I-5 on 11 October followed by the Instrument Unit three days later.

President Kennedy visited Cape Canaveral on 16 November 1963. He was shown LC-37 and the SA-5 launch vehicle during a tour of the facilities, pausing to look back at the rocket several times before moving out for a helicopter ride to the ship *Observation Island* from where he witnessed the flight of a Polaris missile launched from the submerged submarine USS *Andrew Jackson*. To all appearances, Kennedy was personally fascinated, if not enthralled with the sight of the SA-5, which would soon exceed the (then) lift capacity of Soviet space launchers. For Kennedy, this represented fulfilment of an urgent desire to put the Soviet Union on the back foot and demonstrate the ability of the United States to achieve great things, albeit on the driven challenge from outstanding scientists and engineers in Russia. His bold initiative was already stimulating a major expansion of NASA activity and in April 1964, from LC-19 the first unmanned Gemini spacecraft would launch on an unmanned test flight.

At Cape Canaveral, Kennedy was shown a mock-up of the two-man Gemini and from the pristine LC-37 blockhouse he was given a detailed briefing on Saturn and Apollo with large models of Saturn I, Saturn IB and Saturn V. At 11.00am local time, von Braun briefed Kennedy alongside the SA-5 on LC-37B vehicle and presented him with a model of that launcher, which now resides in the JFK Presidential Library and Museum at Columbia Point, Boston, Massachusetts. Secret Service people had wanted to keep Kennedy away from the giant rocket but he walked right underneath and stood immediately below the eight engines, discussing details with officials, including NASA Administrator James Webb.

Planning to return to the Cape the following year to watch the launch live, Kennedy invited von Braun and his wife to a dinner at the White House scheduled for 25 November. By that date, as a consequence of the tragic events of 22 November, Lyndon Johnson had assumed the mantle of office. On 14 March 1964, Jacqueline Kennedy wrote to NASA's Robert Seamans thanking him for the model and expressing her deep desire for her son, John Fitzgerald Kennedy Jr, to understand what had motivated his father to play with aeroplane models and toy powder rockets when he himself had been growing up.

A potential delay occurred on 24 January when the All Systems Vehicle (ASV) S-IV test stage on Test Stand 1 at SACTO exploded prior to the start of the firing sequence. The ASV carried an abundance of sensors and test instrumentation for validating the flight-rated stages. Because the ASV was identical to the S-IV-5 stage there was concern that this failure could suspend SA-5 flight preparations until the cause had been determined. Fears were quickly allayed when it became clear that the explosion, which demolished the stand, had been caused by ground equipment and by some inappropriate procedures by a few test personnel. It was decided not to build another ASV and to shift all tests to the flight stages for the standard test cycle. Originally targeting a launch on 27 January, another failure, on the part of workmen to remove a diaphragm from a ground

Right: To the alarm of his security escort, an excited Kennedy got out the car and walked up to the first stage as far as the yellow flame deflector. Associate Administrator Robert C. Seamans is at extreme left. *John F Kennedy Presidential Library*

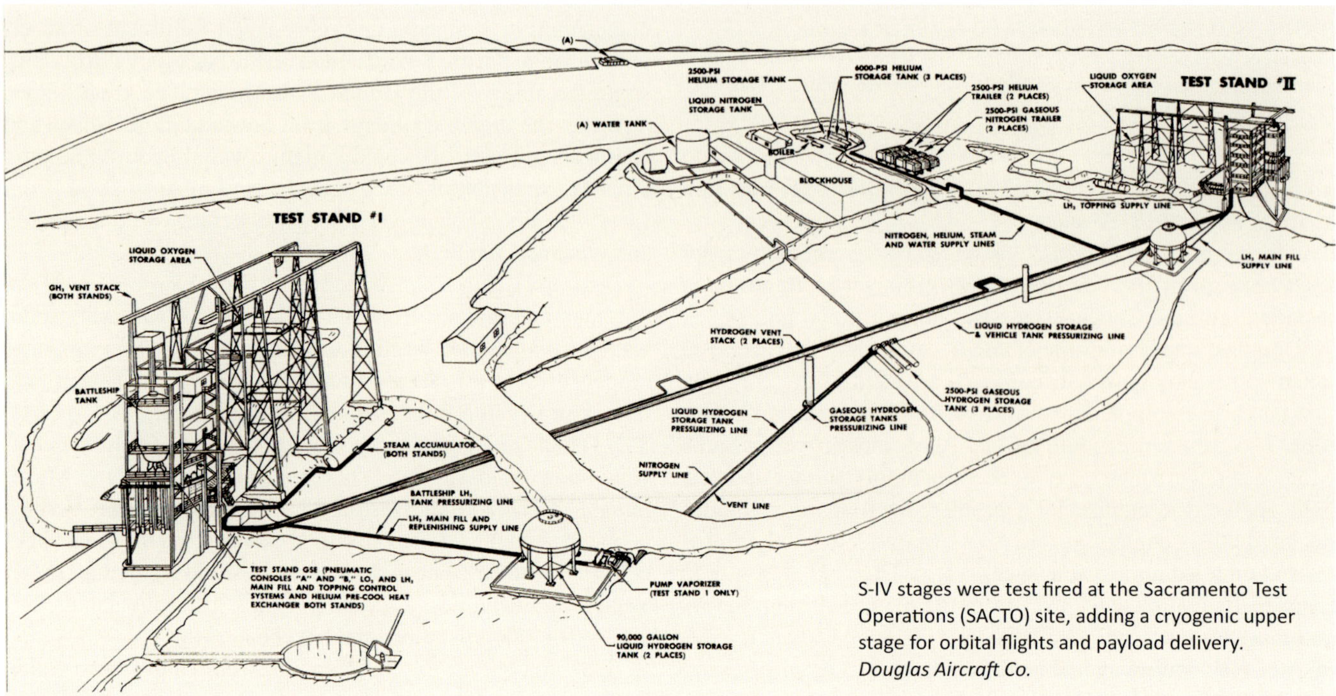

S-IV stages were test fired at the Sacramento Test Operations (SACTO) site, adding a cryogenic upper stage for orbital flights and payload delivery. *Douglas Aircraft Co.*

LOX replenishment line with the AS-5 vehicle at LC-37B, prevented the final 18,144kg (40,000lb) top-up being loaded, and caused a delay of two more days. On launch day, a total of 90min accumulated holds were logged, correcting interference from the C-band tracking radar and solving a problem with the command destruct receiver.

Lift-off occurred into clear skies with few clouds, a light breeze and a cool 17.8ºC (64ºC), the roll manoeuvre starting at 8.8sec and ending at 13.05sec. It was followed by the tilt manoeuvre starting at 15.71sec. The vehicle passed through Mach 1 at 54.7sec and a height of 7km (4.3mls) and Max-Q at 1min 9.2sec at 12km (7.4mls). The complexity of a two-stage ascent sequence began at an elapsed time of 1min 47.7sec, with preparations for LH_2 chill-down and switching on the six movie cameras. The S-IV LOX pre-start began at 2min 19.8sec followed at 2min 20.75sec with inner engine cut-off at an altitude of 67.5km (41.9mls). The four outer engines shut down at 2min 26.73sec at an altitude of 75.18km (46.7mls), with ullage motor ignition almost precisely one second later.

Post-flight analysis indicated a residual 5,806kg (12,801lb) of LOX/RP-1 propellant remaining. Stage separation and S-I retrorocket ignition occurred at 2min 27.14sec, with S-IV ignition 2min 28.84sec after lift-off. The ullage rockets were jettisoned at 2min 47.1sec and the apex of the S-I trajectory (175.5km/109mls) was reached at 5min 3sec. In clear skies, shut-down of both sets of H-1 rocket motors had been seen by a long range telescope at the USAF-Boston University connected to a TV vidicon tube.

Cut-off of the six RL-10 engines came at 10min 29.97sec with the vehicle at an altitude of 262.4km (163mls). Orbit insertion was defined precisely 10sec later after thrust tail-off, placing the vehicle in an orbit of 771.4km (478.3mls) by 261.2km (162.6mls) providing a predicted lifetime of 451 days versus the 286 days pre-flight. Tracking of the S-I stage was lost at 6min 7sec with telemetry going down at 7min 56.7sec and impact with the ocean at 10min 26sec, 913km (567miles) from the pad. The orbital parameters were 129.22km (80.3mls), higher than predicted at apogee, and 4.12km (2.56mls) high at perigee. Moreover, at orbit insertion the vehicle was 22.3km (13.85mls) to the left and 6.5km (4mls) higher than expected, with an inclination of 31.4deg and a period of 94.8min.

The mass placed in orbit comprised the 6,486kg (14,300lb) of the spent S-IV stage, the 2,359kg (5,200lb) IU and guidance module attached to the top of the S-IV, the 1,814kg (2,500lb) Jupiter IRBM nose cone and 5,262kg (11,600lb) of sand ballast together with miscellaneous equipment, tracking beacon and temperature probes weighing 45kg (100lb). Unused propellant made up the difference to the orbited mass of 17,240kg (38,000lb). The stage decayed out of orbit on 30 April 1966 after 791 days.

Right: The first Block II Saturn (SA-5) heads for orbit on 29 January 1964, a launch the late President Kennedy had pledged to attend. *NASA*

Saturn I flight SA-6

Launch date:	28 May 1964
Lift-off time:	12:07:00.4 EST (17:07:00.4 UTC)
Launch pad:	LC-37B
Launch azimuth:	105deg
Length:	58.06m (190.5ft)
S/L thrust:	6,689kN (1,504,000lb)
Dry mass:	69,401kg (153,000lb)
Propellant mass:	431,827kg (952,000lb)
H-1 engines:	H-2006; H-2007; H-2008; H-5007; H-15008; H-5009; H-501
RL-10 engines:	Unknown but in the P6418 series 50
Ignition mass:	518,155kg (1,142,336lb)
Lift-off mass:	511,931kg (1,128,615lb)

This was the first mission dedicated to flight development of Apollo hardware, carrying a boilerplate Command/Service Module (BP-13) and a prototype Launch Escape System (LES), all equipped with instrumentation for engineering analysis and evaluation. As such, it was also known as Apollo mission A-101. This was the first of two Apollo boilerplate flights designed to verify the integrity between launch vehicle and spacecraft and to obtain data from instrumentation recording flight information, stress levels, aerodynamic properties and thermal characteristics of the ascent and orbital phases. No attempt would be made to recover the boilerplate but the LES was a fully flight-rated unit which would have performed the same function it was designed to in the event if a threat to the integrity of the payload, up to and including planned jettisoning of the tower during ascent.

At launch, the total mass of the usable payload was 10,679kg (23,543lb) but in orbit that was reduced to 7,722kg (17,023lb) after jettisoning the Launch Escape System, with a mass of 2,957kg (6,520lb), during ascent. This did not account for the inert mass of the S-IV and all other associated elements for an additional 9,373kg (20,664lb). For this mission, SA-6 carried 1,310 telemetry measurements of which 633 were on the S-I-6 stage, 355 on the S-IV, 210 on the Instrument Unit and 116 on the BP-13 spacecraft, the data being transmitted over 13 telemetry links.

The vehicle had eight Milliken motion picture camera packs and two TV cameras in a repeat of the SA-5 configuration, with one monitoring inboard engine No 2 and the other mounted on the forward end of the S-I booster for observing stage separation and S-IV ignition. Two of the movie cameras would observe flow phenomena in the interiors of two LOX tanks to better define slosh characteristics, four filmed S-I retrorocket firings and the S-IV ullage motors, flutter on a blow-out panel in the interstage structure and separation, coast and ignition of the S-IV. The seventh camera was focused on one of the RL-10 engines and one filmed operation of the LOX-SOX disposal system. The eight 16-mm film cameras were in containers built by Cook Electric. The Saturn I programme achieved full maturity with the responsibility for carrying this first Apollo payload to orbit and the busy countdown and pre-launch preparation time involving several hardware sets reflected that.

Above: Aiming to carry the first active Apollo boilerplate to orbit (BP-13), SA-6 awaits launch on LC-37B. *MSFC*

In a sequence outpacing the flight schedule, assembly of the S-I-6 stage began on 25 September 1962 but slowed when subcontractors were late delivering components, with checkout only completed on 3 April 1963. The first test firing occurred on 15 May but at shut-down, following a 33.75sec run, the main LOX valve on engine H-2009 failed to close, causing a fire. The engine was replaced with H-2007 with a second stage firing on 6 June for 142.37sec, preceded at 136sec by a planned shut-down of the four inner engines. Removed from the test stand eleven days later, analysis showed that engines H-2005 and H-2007 had 3.6 per cent higher thrust levels than expected. Further work on the stage included replacing some broken slosh baffles in the centre LOX tank. Strain gauges were to be incorporated in those baffles on S-I-7. Hardware for the SA-6 flight began to arrive at the Cape on 18 February 1964 when the barge *Promise* brought the S-I-6 stage and S-IU-6, erected on LC-37B two days later.

S-IV-6 manufacture had begun in 1962, followed by final assembly early the following year. Saturn historian Alan Lawrie has chronicled the extensive difficulties experienced by Douglas in fabricating and assembling the various elements of the S-IV stage and notes that difficulties with S-IV-6 caused by shortage of parts and delayed tests put the stage behind schedule. Not inconsequentially, he notes that several items had to be replaced after damage during tests which prevented it from being delivered to SACTO, where it eventually arrived on 27 September 1963. Installed on TS-2B three days later, the acceptance firing was conducted on 22 November and lasted 461sec. A tirade of troubles afflicted the schedule and included de-bonded insulation on the fuel side of the common bulkhead, a leaking hydraulic valve, another leak on the LOX side and several other minor problems prevented final checkout until 17 February 1964.

The journey to Cape Canaveral began when the S-IV-6 departed aboard the Pregnant Guppy on 21 February, arriving at

SA- 5 – Apollo boilerplate details

Right: Delayed by two days, SA-6 finally gets off the pad. *NASA-KSC*

Below: BP-13 had active sensors and equipment for measuring the space environment and the structural integrity of the Command Module and Service Module on Apollo mission A-101. *MSFC*
Below right: The physical dimensions of BP-13 together with the Launch Escape System. *MSFC*

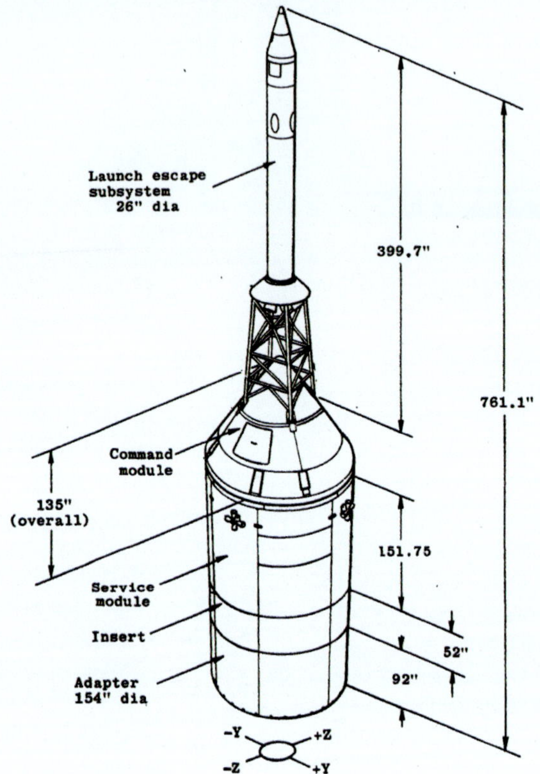

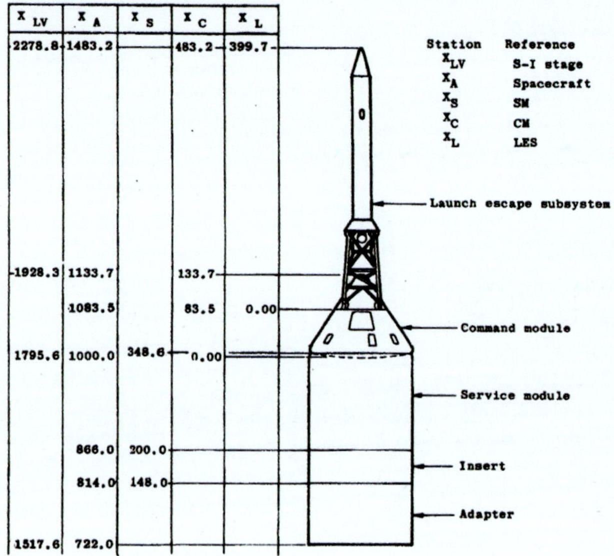

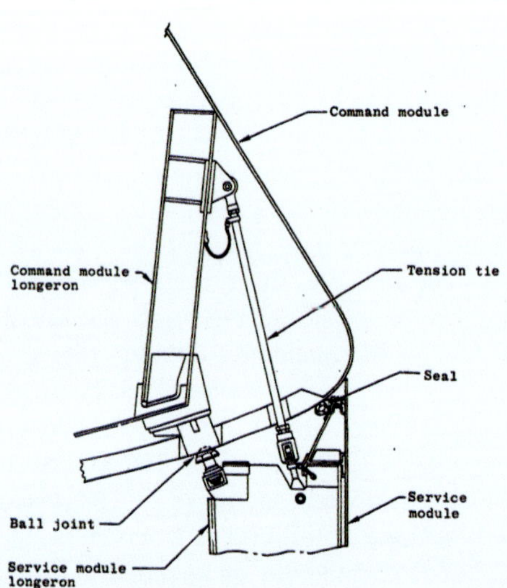

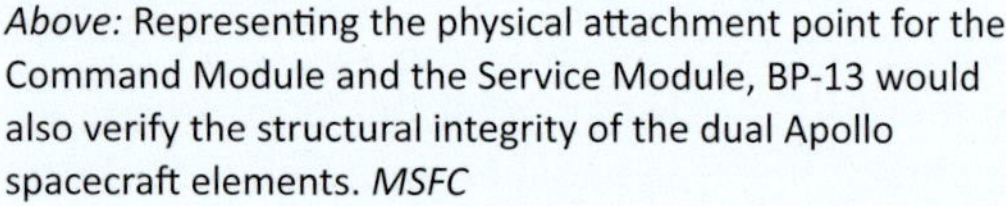

Above: Representing the physical attachment point for the Command Module and the Service Module, BP-13 would also verify the structural integrity of the dual Apollo spacecraft elements. *MSFC*

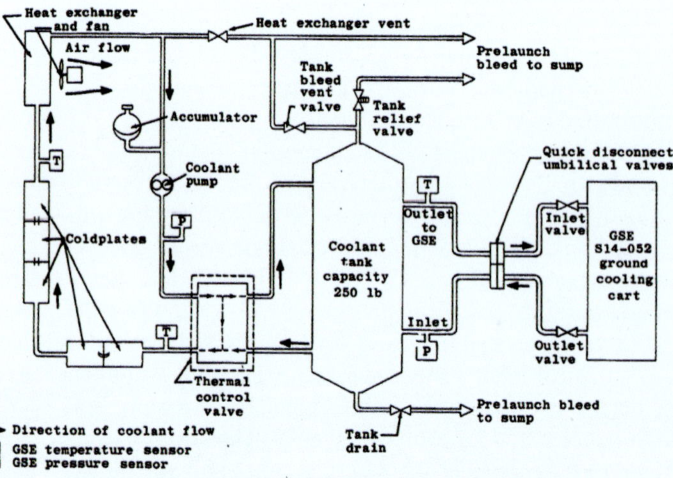

Above: Boilerplate spacecraft often carried a wide range of technical equipment. The environmental control equipment for BP-13 was designed to provide information on the temperature gradients during ascent and orbital operations providing vital engineering data for the Block I CSM. *MSFC*

its destination the following day. Held over in Hangar AF, further repairs were necessary when transverse cracks were noted on the LOX side of the common bulkhead. The stage was stacked on S-I-6 on 19 March followed by the IU on top four days later. Fabricated at North American, the BP-13 boilerplate Apollo was mated to the adapter and the LES installed on the SA-6 launcher on 2 April. A test of the RP-1 fuel loading procedure was conducted four days later. Across the Cape, momentum was picking up. On 8 April the first of two unmanned test flights of the Gemini spacecraft roared into orbit from LC-19. Back at LC-37B, during propellant loading tests on 11 May, the LOX sump screen mesh in the S-IVB stage failed and had to be replaced. First stage RP-1 was loaded on 23 May.

Aiming for launch on 26 May, the count began as planned at T-17.25hr (3.50am EST) on the day of launch, with a hold beginning at T-13hr 15min. The scheduled countdown required 17hr 15min with a rest period of 12hr 15min for launch teams, effectively dividing the activity into two portions consisting of 8hr 10min on F-1 day and 9hr 5min on F day. The final count for launch day began 20 minutes late on 26 May due to a power failure at the blockhouse. Further technical delays were encountered and the launch was re-scheduled, the final count starting at 11.55pm EST on 27 May, this time for a launch at 10.37am EST the following morning but a hold was called at T-41sec. After 1hr 14min, the countdown resumed at 11.52am for lift-off at 12.07pm.

The pitch profile began at 15.2sec and terminated at 2min14.2sec, after which a fixed attitude was maintained until active guidance began 18.6sec after separation of the stages. At 1min 57.28sec, the No. 8 engine shut down 23sec prematurely from the programmed cut-off for all inboard engines at 2min 20.4sec. A slight underperformance of the S-I stage resulted in a velocity 33.2m/sec (108.93ft/sec) less than it should have been at that point. The remaining inner engines shut down at 2min 23.2sec, 3sec later than planned, followed by the outer engines at 2min 29.23sec with the ullage ignition signal 0.27sec later and stage separation at 2min 29.68sec. Staging had occurred 2.98sec later than scheduled with the velocity 362kph (225mph) slower, and altitude 338m (1,110ft) lower, than planned.

Retrorocket ignition occurred milliseconds after separation, followed by S-IV ignition at 2min 31.3sec. The ullage motor jettison signal went in at 2min 41.6sec, all fired but one got hung up and failed to release, but the Launch Escape System was successfully jettisoned at the same time. At 2min 43.6sec the guidance commands switched from the ST-90S to the ST-124 platform and active ST-124 commands went in from 2min 48.2sec. S-IV cut-off came at 10min 24.8sec followed by notional orbit insertion 10 seconds after that. No separation of the Apollo boilerplate was planned and the orbited payload remained as a single unit.

A higher S-IVB thrust level than expected reduced S-IV burn time by 4 seconds. Residual S-I propellants remaining at staging were 2,749kg (6,062lb) of LOX and 351kg (775lb) of RP-1 versus the 90kg (200lb) of LOX and 850kg (1,874lb) of fuel expected, a difference explained by a lower LOX density than calculated. The No 8 engine failure was due to a gear problem, the Mk III turbine flying on this vehicle for the first time, with subsequent vehicles getting the improved Mk III H. The higher thrust level on the S-IV was caused because the thrust controller on the No. 4 RL-10 failed in the fully closed position resulting in higher chamber pressure, thrust and acceleration.

The S-I stage impacted the Atlantic Ocean at 9min 37.2sec, 767.5km (476.9mls) from Cape Canaveral. The total mass at end of S-IV thrust was 17,095kg (27,687lb). At insertion, the payload had an orbit of 227km (141mls) by 181.8km (115mls) with an inclination of 31.8deg and a period of 88.6min. The S-IV re-entered the atmosphere and impacted the Pacific Ocean at 12.31am UTC on 1 June at 21.2deg N by 167.7deg E. All eight film camera pods were recovered by the US Air Force Air Rescue Service and no attempt was made to recover the TV cameras.

Overall, the mission had been an outstanding success, for all the reasons identified above. For the second time, an ascending Saturn I had managed an engine shut-down and this was the first unintentional exercise of that kind. But the flight path and parameters were far from the planned profile, altitude at orbit insertion was 2.4km (1.49mls) lower and 41.4km (25.7mls) shorter than the desired values. The orbit was different by some margin, giving the payload a life of 3.3 days, 1.5 days shorter than planned. Deviations were worked through a still evolving use of the Instrument Unit and the guidance platforms but these were correctable given development time. Another reflection of the value in flying Saturn I test flights with elements of hardware utilised in the Saturn IB and the Saturn V, a great advantage with the commonality concept.

Below: Wernher von Braun consults the launch countdown handbook during launch preparations for SA-6. *MSFC*

Saturn I flight SA-7/A-102

Launch date:	18 September 1964
Lift-off time:	11:22:43.26 EST (16:22:43.26 UTC)
Launch pad:	LC-37B
Launch azimuth:	105deg
Length:	58.06m (190.5ft)
S/L thrust:	6,689kN (1,504,000lb)
Dry mass:	87.545kg (193,200lb)
Propellant mass:	431,827kg (952,000lb)
H-1 engines:	H-2010; H-2012; H-2019; H-2021; H-5013; H-5015; H-5016; H-5027
RL-10 engines:	Unknown but in the P6418 series 50
Ignition mass:	519,566kg (1,145,448lb)
Lift-off mass:	513,327kg (1,131,693lb)

Also designated A-102, this mission would be similar to that of SA-6, carrying a live S-IV and an Apollo boilerplate, BP-15. The usable payload of the Apollo Command/Service Module and the adaptor was 7,816kg (17,231lb), a little heavier than BP-13 with SA-6. Total payload at launch with the LES was 10,813kg (23,838lb). Before the hardware left the Downey plant of North American Aviation the mass had been reduced by removing 725kg (1,600lb) of ballast but still leaving 1,108kg (2,442lb) in place. For this flight the ST-124 platform would be in sole control, determining the steering commands for the flight trajectory right up to orbital insertion, rather than for S-IV thrusting alone as configured for the SA-6 mission.

There was also a change to the S-I engine cut-off system which left less propellant in the tanks. Level sensors placed closer to the bottom of the four outer LOX and two fuel tanks would initiate a signal to shut down the four inboard engines. When the pressure dropped in the four outboard tanks another signal would shut all four down simultaneously, expected about six seconds after the inboards. On SA-6 the interval between inboard and outboard cut-off was fixed by a timer but on SA-7 the new sequence, known as LOX starvation, was expected to leave only 3,855kg (8,500lb), rather than the 6,350kg (14,000lb) left in the tanks on S-I-6.

Left: SA-7 during a radio frequency interference test. *MSFC*

Several subtle changes were applied to the S-IV stage, including non-propulsive vents for the LOX and hydrogen vent lines, located 180deg apart for each propellant. And, as noted for SA-6, the Mk III gearbox was installed on all eight H-1 engines. Previously, S-IV-5 and -6 carried extra helium as a back-up to the helium heater system providing gas for LX tank pressurisation but as it had worked satisfactorily on all previous flights, this was removed, saving 450kg (1,000lb).

SA-7 carried 1,378 measurement points with 661 on the S-I stage, 397 on the S-IV, 187 in the Instrument Unit and 133 on BP-15, making this the most heavily instrumented launch vehicle yet flown. There were also more than 200 blockhouse measurements for the use of launch personnel. As with SA-6, it had 13 flight telemetry systems. Eight film cameras and one TV camera were installed, viewing the interior of two LOX tanks, S-IV separation, retrorocket firing and S-IV ullage motor operation.

The colour film cameras were in pods on the top of the S-I, three externally mounted operating from 40 seconds before staging and up to 20 seconds after. A fourth camera would start at the same time and continue operating through pod ejection and until the pod struck the water, estimated 853km (530mls) downrange. The two cameras viewing the LOX tanks would start operating 25sec before separation; one, running at 7.31m/sec (24ft/sec), would run out of film before separation, a second at 3.65m/sec (12ft/sec) until separation, and two at 19.5m/sec (64ft/sec) for 25sec until separation. The TV system would operate from lift-off until S-I impact with video recorded at the ground stations.

The SA-7 stage carried a new, non-propulsive propellant venting system which was designed to prevent the stage tumbling after its work was done. This was specially developed by Douglas for the last three Saturn I missions which had unique post-boost requirements in that each would carry a Pegasus micrometeoroid detection payload. The solution to stage tumbling was two opposing vents for each of the two propellant tanks, boil-off being routed via 7.62cm (3in) diameter lines perpendicular to each other. Evaluated on SA-7, the S-IV stage held to within 1-2.5deg/sec, considerably better than the specified requirement for 6deg/sec.

Assembly began at MSFC in late 1962 and on 7 January 1963 tank clustering began and S-I-7 was delivered for checkout on 25 May. Checks on the LOX tank baffles were conducted to ensure their integrity following the experience with SA-6. The Static Test Tower East hosted a 33.78sec firing on 2 October but engine H-5014 was replaced with H-5027 due to a minor discrepancy. A second test lasting 138.93sec took place on 22 October, following which a fuel tube leak necessitated changing engine H-2015 with H-2021. Post-firing inspections, and the problem with S-I-5, prompted replacement of critical tube assemblies with checks, completed on 12 May 1964 followed by transfer of S-I-7 and S-IU-7 to the barge *Promise* and departure

SA- 5 – Apollo boilerplate details (contd.)

BP-15 was slightly heavier than BP-13 on the preceding Saturn flight but there were a few changes to the configuration of the instrumentation for this mission.

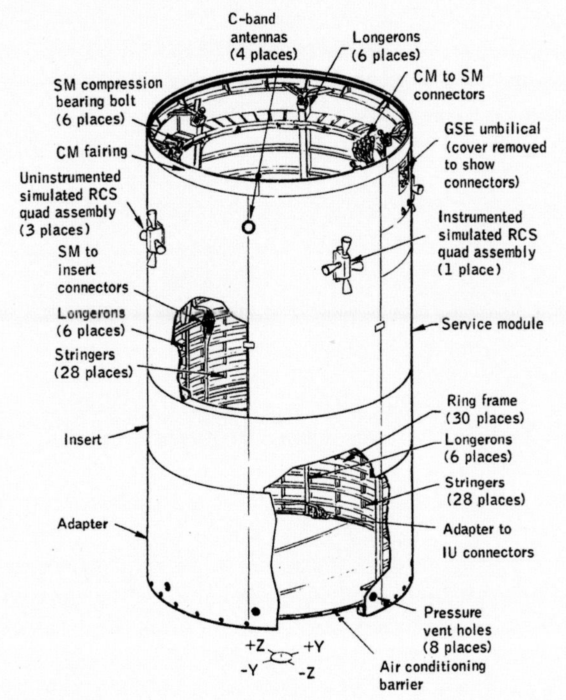

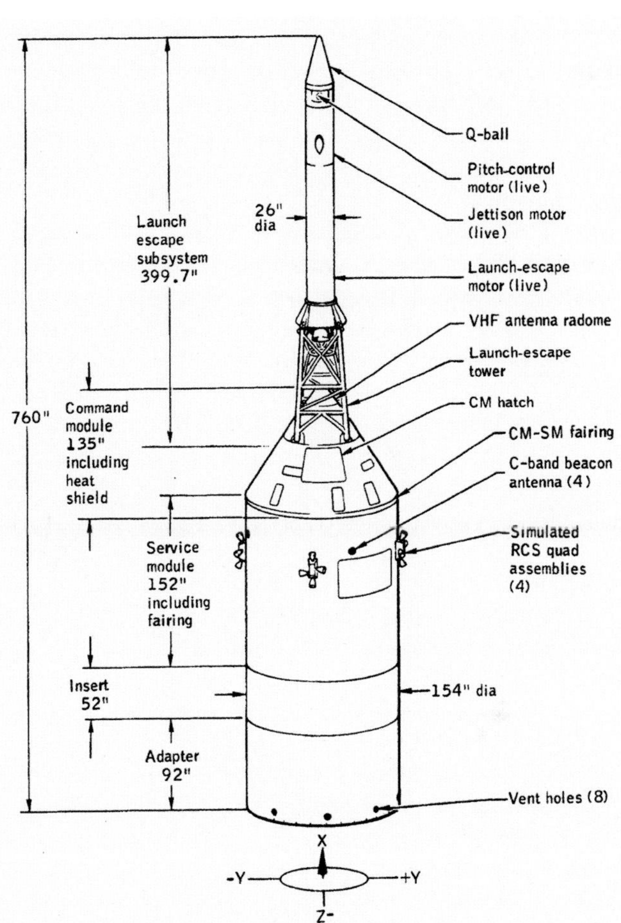

from MSFC sixteen days later. *Promise* arrived at the Cape on 7 June with S-I stage erection on the pad two days after that.

Fabrication of S-IV-7 began in early 1962 and, while plagued with the problems set out previously for S-IV-6, not until October 1963 were the six engines installed. With check-out completed by January 1964, S-IV-7 was installed in TS-2B at SACTO on 22 February, delayed a week while repairs to a damaged crane were carried out. The static firing was further delayed due to a missing wire, not shown on the engineering drawings, and then due to a faulty umbilical disconnect fitting. The test was finally conducted on 29 April with a 484sec burn of the six engines and the stage was loaded aboard the Pregnant Guppy, planned for departure on 9 June. Weather prompted a 24hr postponement in departure and a sequence of eventful episodes including blown tyres at intermediate refuelling stops delayed arrival at Cape Canaveral.

The S-IV-7 stage finally arrived on 12 June and was mated to the first stage for SA-7 seven days later. BP-15 was delivered to the Cape on 15 June and placed in position on 26th. However, during inspection, a 7.6cm (3in) crack was discovered in the LOX dome of engine H-2011, resulting in its replacement by H-2019 and the replacement of LOX domes on all remaining first stage engines. They were returned to Rocketdyne between 4 July and 20 July where the material used for LOX domes was changed from 7079T-6 aluminium to 7075-T73. The engines were checked out and given full pressure checks on 4 August.

On 27 August, Hurricane Cleo moved through the Cape area and LC-37B was secured to ride out the storm. A second tie-down was called for when Hurricane Dora passed by on 9 September. A countdown demonstration test (CDDT) was held 14-15 September followed by the first of the two-phase launch countdowns on 17 September, a sequence broadly similar to that of SA-6. The second part of the count began at 11.25pm EST, 17 September at T-9hr 5min and continued until T-4hr 5min when a 1hr 9min hold was called after a premature release of cooling water dowsed one of the S-IV umbilical connectors, which had to be dried out.

Activities proceeded as planned to the built-in 21min hold at T-30min but in this interval a problem emerged with the S-IV LOX pressure regulator and a 4min extension was given. At T-12min a malfunctioning S-I hydraulic pump caused another hold but this was resolved after 20 minutes and the count resumed. At T-5min a range safety hold was ordered due to a problem with the radar at Grand Turk Island but added technical problems with battery power appeared and in the intervening period further correctable problems emerged before the count resumed after 49 minutes, having been recycled back to T-13min before continuing to lift-off.

Generally, the trajectory followed a similar profile to that flown by SA-6, with Mach 1 at 55.2sec, Max-Q at 1min 13sec, inboard engines shutting down at 2min 41.5sec and an altitude of 65.6km (40.7mls). The outboards shut down at 2min 27.6sec at an altitude of 72.9km (45.3mls), followed by ullage motor ignition on the S-IV, 0.7sec later. S-IV stage ignition occurred at 2min 30.1sec with the ullage motors and the Launch Escape

System departing the stack 10.3sec later, stage cut-off occurring at 10min 21.4sec. The initial orbital parameters were 234.1km (145.5mls) by 180.2km (111.9mls) at an inclination of 31.7deg with a period of 88.64min. The total mass in orbit was 17,754kg (39,141lb) including the BP-15 boilerplate Apollo and the S-IV stage together with its supplementary structures.

The S-I stage reached its ballistic apex of 159.4km (99mls) at 4min 53sec and impacted the ocean at 6min 56.8sec, 883.6km (549mls) downrange. The eight motion picture and two TV cameras carried by SA-7 were not recovered at the time due to heavy sea conditions caused by hurricane Gladys but two were discovered 50 days later. The S-IVB stage re-entered the atmosphere over the Indian Ocean at 7.59am EDT on 22 September on the 59th orbit, 17 orbits later than expected. Post-flight analysis revealed some deviations at various points in the trajectory, with the orbit insertion parameters showing -0.99km (0.61mls) error in altitude, +6.18km (3.84mls) in apogee and -0.74km (0.46mls) in perigee compared with the pre-planned trajectory. While the nominal thrust at lift-off is given at the top, the reconstructed thrust profile showed an actual sea-level thrust of 6,730.64kN (1,513,108lb), an increase of 1.24 per cent over the anticipated value. Large pressure disturbances in engine No 3 (H-5015) were noted during ignition but without measurable effect on stage performance at lift-off or on the trajectory.

As noted at the top, a new thrust termination sequence devised to reduce wasted propellant was expected to lessen the amount left in the tanks and measurements indicated that the residual was actually less than expected: 3,462kg (7,632lb), little more than half that remaining in the SA-6 tanks. The value in this was added data to allow a modest increase in payload on later flights due to less propellant required to achieve a given goal. But some additional data-gathering on subsequent flights was essential to improve the veracity of this measurement.

As with SA-6, this flight tested the enhanced guidance and navigation system to control the later and more advanced S-IVB/Apollo on its way into orbit, a requirement driven by the demands which would be placed on the S-IVB for the Saturn IB and Saturn V flights to come.

Below: SA- 7 lofts Apollo boilerplate BP-15 into orbit. *MSFC*

Saturn I flight SA-9/A-103/Pegasus 1

Launch date:	18 September 1964
Lift-off time:	11:22:43.26 EST (16:22:43.26 UTC)
Launch pad:	LC-37B
Launch azimuth:	105deg
Length:	58.06m (190.5ft)
S/L thrust:	6,689kN (1,504,000lb)
Dry mass:	87.545kg (193,200lb)
Propellant mass:	431,827kg (952,000lb)
H-1 engines:	H-2010; H-2012; H-2019; H-2021; H-5013; H-5015; H-5016; H-5027
RL-10 engines:	Unknown but in the P6418 series 50
Ignition mass:	519,566kg (1,145,448lb)
Lift-off mass:	513,327kg (1,131,693lb)

Considered to be the first operational flight of the Saturn I, confusingly SA-9 was launched before SA-8. Last of the Saturn Is built at MSFC, SA-9 was defined by its hardware set and the numerical sequence of launches was irrelevant. As the first of the later Block II vehicles it had an unpressurised Instrument Unit more closely configured to the design which would be incorporated on the top of the S-IVB stages flown on Saturn IB and Saturn V. It would evaluate the accuracy of the closed-loop guidance system and was the first flight to utilise the iterative guidance scheme.

Also designated A-103, SA-9 was carrying Apollo boilerplate BP-16 and a modified Service Module into which was integrated a Meteoroid Technology Satellite (MTS) developed by NASA's Office of Advanced Research and Technology. In truth, it was not an independent satellite because it would remain permanently attached to the S-IV but this was also to be the first demonstration of the separation of the BP-16 Command Module in orbit. The primary objective for this mission was to deploy the MTS payload and evaluate its operability in space and to evaluate a closed-loop guidance system with the first flight of an iterative guidance package.

The functional purpose of the MTS programme was to assemble a detailed record of the frequency and distribution of micrometeoroid particles within an orbital swath between 499km (310mls) and 748km (465mls). It would do this by deploying in orbit a pair of 'wings' which would record the impact of tiny particles. NASA was concerned about the danger these hyper-velocity particles could pose, potentially puncturing a spacecraft or destroying a satellite. The first investigation into this threat had been made by Explorer 13 launched on 25 August 1961 but it remained in orbit only three days. Launched on 16 December 1962, Explorer 16 operated for eight months and Explorer 23, launched on 6 November 1964, operated for one year and a day before finally decaying into the atmosphere on 29 June 1983.

NASA began developing plans for a much larger experiment in 1962, proposed by MSFC's Ernst Stuhlinger, leading to a request for industry proposals that December for two large satellites with extendible 'wings' covering a surface area of 185m² (1,991.3ft²). On 5 February 1963 the Fairchild Stratos Corporation (Fairchild Hiller from 1964) was chosen for a de-

Left: Launched out of numerical sequence, SA-9 stands on LC-37B with an active micrometeoroid detection payload called Pegasus installed in the much modified BP-16 Service Module. *NASA-KSC*

velopment and manufacturing contract which was awarded on 4 March. By the end of the year, MSFC examined the possibility of using SA-10 for the launch of a third satellite and in April 1964 it was approved by NASA HQ leading to a contract to The Martin Company from NASA's Langley facility for an advanced array potentially launched by Saturn IB/Centaur on a flight to the vicinity of the Moon for evaluating micrometeoroid impact levels likely to be encountered by Apollo astronauts. The Saturn IB option was never taken up and Fairchild built a third Pegasus which was assigned to SA-10.

The satellites were officially named Pegasus, after the mythical flying horse, on 21 July 1964. As built they each consisted of two solar panel arrays folded compactly to form a box 5.27m (17.3ft) tall, 2.13m (7ft) wide and 2.4m (7.9ft) deep. This was attached to the Service Module adaptor on top of the S-IV stage and provided for a mounting upon which to attach the wings. The centre section of the structure consisted of an open frame of square, drawn aluminium tubing and in addition to providing the supporting structure for the wing frame members it also supported the deployable solar cell array for providing electrical power. It was attached to the Service Module adapter at the separation end between it and the Service Module by a six-point mounting pad support system.

Each wing consisted of seven hinged frames, spring-loaded to open like an accordion with the action controlled by a scissors link attached to a motor and torque tube. Each panel measured 51cm (20in) by 102cm (40in) with the six frames containing 16 panels and one providing eight panels for a total of 208 in all. When deployed, the wings spanned 29.3m (96ft) with a width of 4.1m (13.45ft) and a total surface area, on both sides, of 200m² (2,152.8ft²). The outer surface of the panels consisted of thin sheets of aluminium with the thickness of the panels (detectors) varying from 0.04mm (0/015in) to 0.4mm (0.157in).

The way it worked was through a 40-volt electrical charge across the surface of each side of each panel, established between the outer aluminium skin and the inner copper coating. When a panel was penetrated by a tiny particle the material would vaporise, forming a conducting gas discharging the capacitor. As the gas dissipated the capacitor would recharge with the detailed data providing information on the number of hits per area, hits per panel thickness, direction of trajectory and the orientation of the MTD with respect to the Sun, plus the precise time it was recorded. Three would be launched on the last three Saturn I launch vehicles as Pegasus A, B and C (SA-9, SA-8 and SA-10, respectively).

There was a plan at one point for a two-man Gemini spacecraft to rendezvous with the last Pegasus to fly, conducting a spacewalk to remove the subpanels for inspection on the

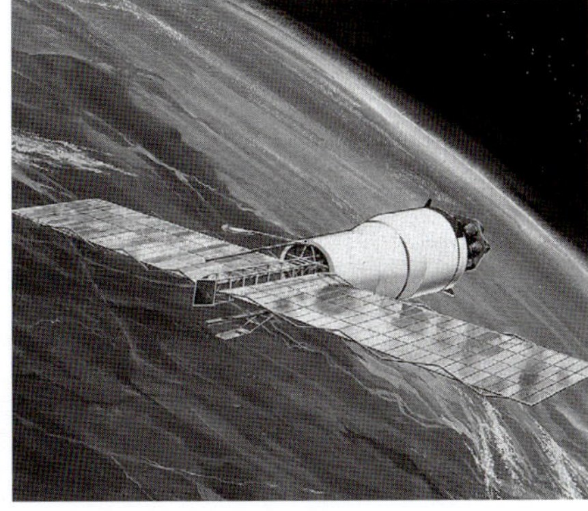

Left: Revealing its enormous size when opened from its launch container, here the Pegasus payload (*above*) is deployed on test at the Fairchild-Hiller facility, Hagerstown, Maryland. The A-shaped 'roof' over the centre structure contains solar cells for electrical power. *MSFC*

ground. This idea was rejected in late 1965 as being impractical as it would require an excessive amount of manoeuvring propellant for station-keeping with Pegasus and imposing costs for additional retrorockets to bring Gemini back down from the higher operating altitude of the S-IV stage and its payload.

SA-9 carried no film camera but retained a single TV camera mounted on the exterior of the Service Module adaptor to provide live pictures of Pegasus deployment. The BP-16 Service Module, which served as a cylindrical shroud over the Pegasus payload, was attached to the adaptor by six explosive nut assemblies mounted on two guide rails, 4.47m (12.6ft) long and 180deg apart, with three roller-sleeve assemblies per rail. An additional explosive nut was located at the forward end of the Pegasus payload. The ejection mechanism consisted of four negator springs, each exerting a constant force of 178N (40lb) through a distance of 3.96m (13ft) and 12 compression springs, each with a spring constant of 6,553N/cm (580lb/in) and a stroke of 4.3cm (1.7in).

Hardware build-up for SA-9's first stage began in late 1962 and on 4 June 1963 the tanks began to come together in

SA- 9 – the Pegasus flights

The general configuration (*right*) of the last three Saturn flights, with dimensions and (*below*) the Instrument Unit for SA-9, representative of SA-8 and SA-10 which followed. *MSFC*

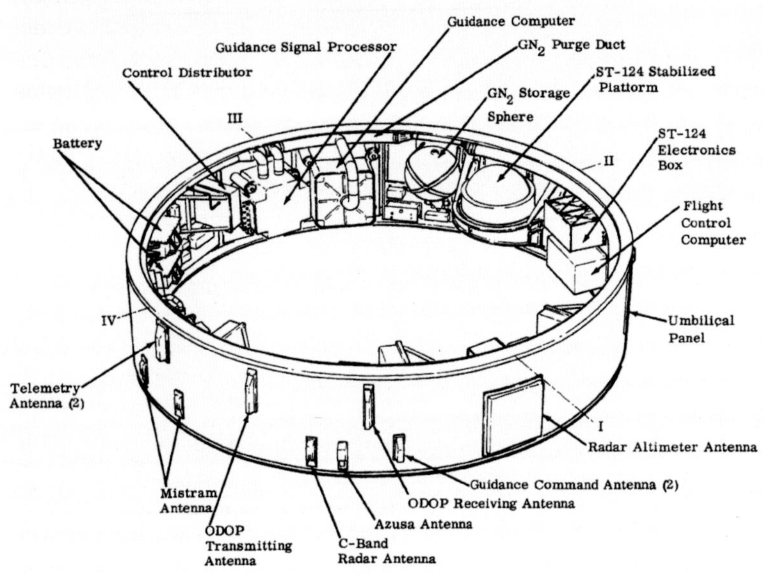

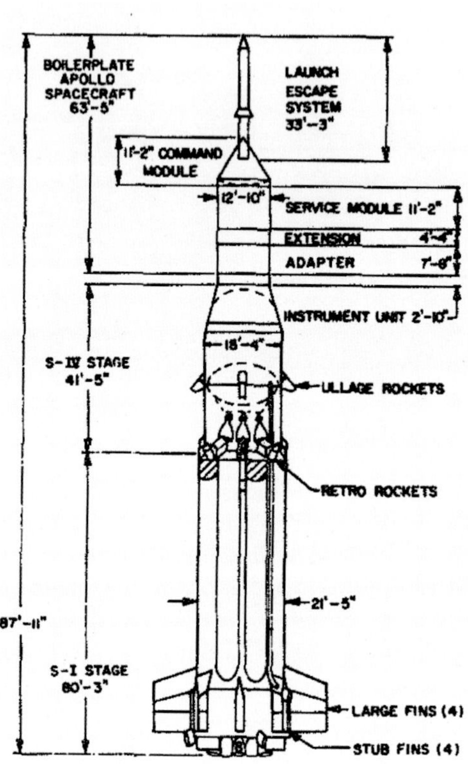

Right: The base launch pad at LC-37B with propellant filling masts as indicated. *MSFC*

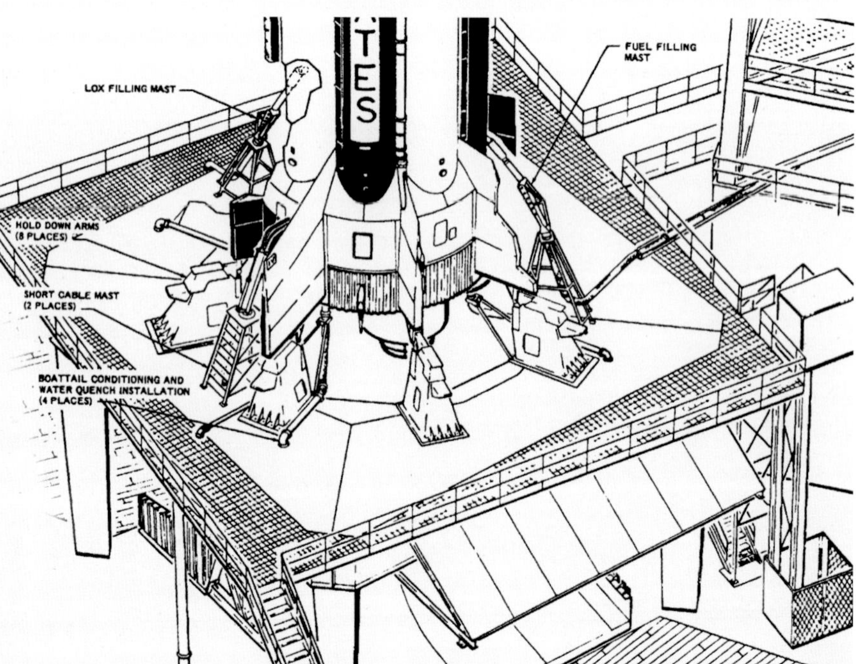

Below: Technicians in the control room prepare for the launch of the first Pegusus flight aboard SA-9. *MSFC*

a cluster, a process which was completed 15 days later. MSFC completed fitting all eight engines on 28 August but quality checkout was interrupted by the need to replace critical tubing assemblies resulting from the SA-5 discovery, that being completed on 27 January 1964. The initial firing test took place on 12 March but a technical problem caused a pre-ignition abort, the planned 35.22sec firing being completed the following day. Thrust was low on almost all engines, only one being within tolerances and further re-orificing was necessary prior to the second, full duration test of 142.21sec on 24 March. On that day, NASA Administrator James E. Webb hosted Lady Bird Johnson, the President's wife to the firing at the Static Test Tower East.

During stage checkout after removal on 8 April, cracks were discovered in the LOX domes of all eight engines and they were removed and returned to Rocketdyne for dome replacement and new fuel shaft seals, arriving back at MSFC in August. There, the engines were re-installed and some additional upgrade work on the stage was completed prior to shipment due to a delay of one month in the scheduled launch. S-I-9 and its separate fins together with the Instrument Unit left Huntsville on 19 October on the barge *Promise* and docked at Canaveral on 30 of October. The stage was erected on LC-37B four days later.

The S-IV-9 stage was slightly different for this flight, in that it had a blow-down vent in the forward section of the interstage to supplement the existing non-propulsive vent already present. In other respects it was identical to S-IV-7. Production of S-IV stages ensured that manufacturing was well under way by mid-1962, with assembly of the LOX tank by early 1963 followed by final engine installation in January 1964. Some delay had been experienced due to parts shortage and additional work required on the Pratt & Whitney engines. A further delay was incurred due to protracted testing with S-IV-7, preventing delivery to SACTO before 8 May. During pre-test checks, some 23 cracks were noted in the common bulkhead welds necessitating a repair procedure. This, and a succession of problems involving test equipment and the stage proper, delayed the acceptance firing until 6 August 1964 when a full duration run of 398.9sec was carried out. Post-test checks revealed further cracks in the fuel tank insulation liner, delaying removal from the stand exactly three weeks after the acceptance firing. Finally, S-IV-9 departed for Cape Canaveral aboard the Pregnant Guppy on 21 October, arriving the following day, seven days before S-I-9 and IU-9 docked aboard the barge *Promise*.

The S-I-9 stage was erected on LC-37B on 3 November followed by the S-IV and the IU on the 19th. The Pegasus payload arrived on 29 December and after an extensive checkout it was erected on 13 January, followed by the Command Module the following day. The CDDT was completed on 12 February and the terminal countdown began three days later. Countdown schedules ran close to those for SA-7 with the terminal activities starting at 21.55hr EST on that day. A hold was called at 6.40am EST the following morning to fix a suspect battery charging circuit in Pegasus and the count picked up again at 7.10am EST until 7.55am EST when the flight safety computer experienced a power failure, the count resuming at 9.11am EST after which events cycled down to lift-off.

Flight parameters were close to those of SA-7, with the pitch and roll commands going in at 8.9sec, Mach 1 at 53.2sec (7.1km/4.4mls), Max-Q at 1min 6sec (11.6km/7.2mls), inner engine cut-off at 2min 20.2sec (80.9km/50.3mls) with the outer engines shut down at 2min 25.6sec (89km/55.3mls). Ullage rocket ignition occurred at 2min 26.4sec followed immediately by stage separation and retrorocket ignition with the S-IV start command at 2min 28.1sec. The stage guidance cut-off occurred at 10min 21.7sec. The ballistic S-I stage impacted the ocean at

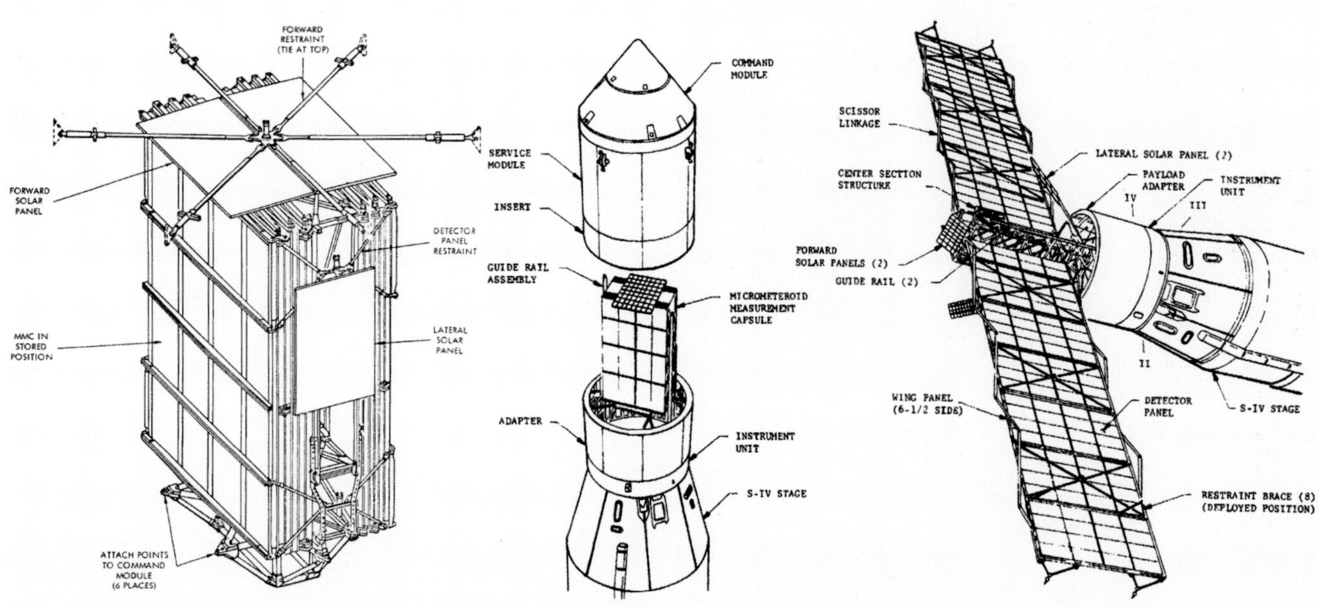

The container for Pegasus was integral to the BP-16 Service Module and contained the automated deployment mechanism. The container attached to the adapter sectionand was exposed in space by release of the BP-16 elements to allow deployment of the detector wings. *MSFC*

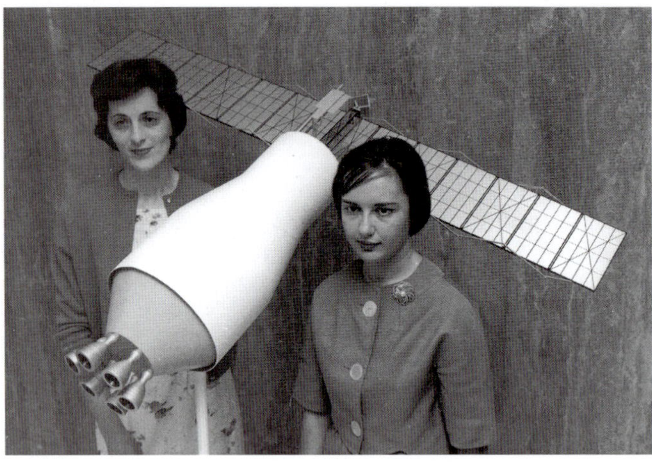

11min 58.9sec, 961.3km (597.3mls) from the pad.

At insertion, the payload was in an orbit of 745km (463mls) by 496.5km (308.5mls) at an inclination of 31.76deg with a period of 97.1min. The orbited mass was 15,370kg (33,895lb), or 10,300kg (22,700lb) after Apollo separation. Significant improvement was made in achieving low residuals remaining in the S-IV tanks, best calculations indicating 391kg (863lb) of LOX and 115kg (253lb) of LH_2. Apollo shroud separation was commanded at 13min 23sec followed by initiation of wing deployment on Pegasus, which began at 14min 23.4sec and was completed in 39sec. Following a highly successful period in orbit gathering data on micrometeoroid strikes, the S-IV/Pegasus vehicle was deactivated on 29 August 1968 and decayed back through the atmosphere on 17 September 1978.

Above: Ann McNair and Mary Jo Smith hold a model of Pegasus and the S-IV stage as it would appear deployed in orbit. McNair had been with the Huntsville engineers since 1958. *MSFC*
Below: SA-9 gets under way on the morning of 16 February 1965. *NASA-KSC*

SA-8/A-104/Pegasus 2

Launch date:	25 May 1965
Lift-off time:	02:35:01 EST (07:35:01 UTC)
Launch pad:	LC-37B
Launch azimuth:	105deg
Length:	57.28m (187.92ft)
S/L thrust:	6,689kN (1,504,000lb)
Dry mass:	66,680kg (147,000lb)
Propellant mass:	451,721 (996,044lb)
H-1 engines:	H-2016; H-2018; H-2031; H-2032; H-5019; H-5020; H-5021; H-5032
RL-10 engines:	P641855; P641856; P641860; P641861; P641862; P641863
Ignition mass:	518,824kg (1,143,812lb)
Lift-off mass:	512,640kg (1,130,178lb)

The second of three operational Saturn I flights and the first to launch in darkness, SA-8 would carry Pegasus B (renamed Pegasus 2 after orbital insertion). It was the first of two Saturn I first stages built by Chrysler and designated A-104 as it was carrying Apollo boilerplate BP-26, almost identical to Pegasus A aboard SA-9 (which see for full description). SA-8 launched after SA-9 due to delays caused by difficulties with the Pegasus B power supply and data systems. This stage together with SA-9 and SA-10 had the unpressurised prototype Instrument Unit of the type fitted to Saturn IB stages. New equipment packaging techniques permitted each assembly to be pressurised individually as necessary and eliminated the need for pressurising the entire instrument unit. SA-8 carried 1,394 measurements of which 561 were on the S-I-8 stage, 412 on the S-IV-8 and 242 on the IU with 179 on the payloads. Only one TV camera was carried to shoot views of the interior of the Service Module adaptor.

S-I-8 was the first Saturn I stage fabricated by Chrysler, with whom the von Braun team had a long association. Fabrication of components began at the company's Space Division on 4 October 1962 while the uprated H-1 engines were delivered to MSFC, there to await delivery of the commercially fabricated S-I-8 stage. Chrysler completed clustering the propellant tanks on 10 May 1963 and, after modifications specifically for the SA-8 mission, the engines were transported from Huntsville to Michoud where all eight were installed in September. Fully assembled, S-I-8 was moved to checkout on 27 October and on 13 December 1963 it was handed over to von Braun by Lynn A. Townsend, the vibrant and talented President of Chrysler Corporation in post since 1961, a mere four years after joining the company. Von Braun expressed a constrained nervousness over the manufacturing transfer to Chrysler

Tubing assemblies on S-I-8 were replaced by February, as planned, and on 17 April it was shipped to MSFC by the barge *Promise*. S-I-8 was back at Huntsville on 25 April and taken to Static Test Stand East two days later where the usual procedures got under way. The propellant loading test took place on 12 May, LOX loading on 21 May – repeated four days later due to a leak in the centre tank lower manhole cover – and the

Left: Pictured during assembly in Building 4705, the Instrument Unit for Saturn SA-8. *MSFC*

first firing of 18.94sec on 26 May. Six engines had to be re-orificed and the full-duration firing took place on 11 June lasting 145.61sec, the inboards being cut off at 139.92sec. Anomalies were seen in the thrust level with engine H-2029 twice during the test so it was replaced with H-2031. Moreover, a further concern arose with engine H-2017 which had been re-orificed but also displayed fluctuating levels; it was replaced with H-2032. To qualify the replacements for flight, H-2031 was fired up for 60sec on 18 June 1964 and H-2032 was run for 60sec too, on 27 June, and for a second time at 60sec three days later.

Meanwhile, after LOX loading tests on 17 June the first stage was returned to Michoud, arriving on the barge *Promise* on 29 June, where Chrysler put it in modification and repair following the thrust tests. In this process attachment bolts for the four LOX tank antennas were found to have sheared, as had been experienced with similar bolts on S-I-9. In persisting with change-out of the H-1 engine domes, all eight were returned to Rocketdyne to be fitted with new ones. They were back and installed by the end of September.

Some damage was sustained by the thrust chamber on H-5020 and some brazing was required to correct dents. Then a careless workman dropped a pair of pliers which collided with H-5022, fracturing two thrust chamber tubes. It was removed from the stage and replaced with H-5032. Further carelessness with handling equipment on 11 February 1965 caused a dent in a fuel tank which was smoothed out but a succession of dents and scratches were discovered further on a GOX line, which was replaced together with suction line bellows. Finally, the S-I-8 stage left Michoud for the Cape on 22 February 1965, arriving six days later. But a further set of repairs were carried out and contaminated H-1 gas generators had to be returned to Michoud for cleaning before reinstallation back at Cape Canaveral.

Manufacture of the S-IV-8 stage began in early 1963 and proceeded largely without any major issues through fabrication and assembly. Moved to SACTO on 7 August 1964, the stage followed S-IV-9 into the test stand on 28th after that stage had been removed the previous day, the acceptance firing taking place on 20 November and lasting 475.8sec. Departing SACTO on 23 February 1965, S-IV-8 arrived at the Cape the following day, two days before S-I-8 which was installed on LC-37B on 2 March. The Instrument Unit arrived six days later and both that and S-IV-8 were erected on the 17th. BP-26 and the Apollo adapter arrived on 10 April and were followed by Pegasus B (Pegasus 2 after orbital insertion) on 15 April, those elements being mated to SA-8 on the 28th.

The pace of delivery and spacecraft checkout was gathering momentum across the Cape, as NASA flew the first two-man Gemini mission out of LC-19 to the south and far out in space, Mariner IV was closing in on Mars, the world's first spacecraft to reach the Red Planet. Down at LC-34, preparations for arrival of the first Saturn IB hardware were underway and to the north, up along the coast, the gargantuan shape of the Vehicle Assembly Building was nearing completion. From there, Saturn V rockets would begin flight tests with the Apollo spacecraft in little more than two years.

The countdown for launch began at 15.55hr EST (T-10hr 5min) on 24 May 1965, the only hold being a planned, built-in 35min pause which was not needed. This was the first countdown that proceeded without any technical holds and sustained only a few engineering issues which were quickly resolved. The winds on launch morning were relatively calm and there were only a few cirrus clouds with some cumulus. A very pleasant night and a comfortable temperature of 23°C (73°F) but with 93 per cent humidity and light winds coming across the sea from the south-east. With it, a faint hint of

Right: SA-8 gets ready for launch from LC-37B with the second Pegasus satellite as a payload fixed to the adapter on top of the S-IV stage. *MSFC*

Above: Streaking spaceward, the penultimate Saturn I paints a curve of light, silhouetting the three Saturn V Mobile Launch Platforms and the Vehicle Assembly Building (VAB) to the north. *MSFC*

cloud drifting leisurely at a height of 760m (2,500ft) with visibility seemingly going to the horizon but measured at 16km (10mls). Throughout the early hours of darkness, lights burned bright around this penultimate Saturn I. From the air other lights, lining intersecting roads and at active and inactive launch pads, illuminated the night.

The flash of light was very bright as SA-8 shook the area around LC-37 with its eight H-1 engines. Following lift-off, the slow ascent appearing magisterial in its pace, the pitch and roll command went in at 8.6sec followed by Mach 1 at 54sec, Max-Q at 1min 7sec, inboard engine cut-off at 2min 22sec (79.2km/49.2mls) and the four outboard engines six seconds later (89.2km/55.4mls). Separation occurred at 2min 28.9sec with S-IV ignition at 2min 30.6sec for a burn terminated at 10min 24.1sec. Orbital insertion occurred 10sec later with an apogee of 748.5km (465.1mls), a perigee of 506.5km (314.7mls) at 31.78deg inclination and an orbital period of 97.3min. Estimated life was 1,220 days, or 10 days less than the pre-flight prediction. The S-I stage impacted the Atlantic Ocean at 12min 0.9sec some 980km (608.9mls) downrange. The orbital mass prior to shroud jettison was 15,473kg (34,113lb). The Apollo shroud was jettisoned at 13min 25.9sec and full array deployment on Pegasus was completed 1min 40sec later. The orbited mass accounted for 10,478kg (23,100lb). It decayed back through the atmosphere on 3 November 1979.

SA-10/A-105/Pegasus 3

Launch date:	30 July 1965
Lift-off time:	08:00:00 EST (13:00:00 UTC)
Launch pad:	LC-37B
Launch azimuth:	107deg
Length:	57.28m (187.92ft)
S/L thrust:	6,689kN (1,504,000lb)
Dry mass:	66,747kg (147,150lb)
Propellant mass:	450,575kg (993,349lb)
H-1 engines:	H-2026; H-2027; H-2030; H-2034; H-5028; H-5029; H-5030; H-5031
JRL-10 engines:	P641817; P641864; P641865; P641869; P641884; P168486
Ignition mass:	517,348kg (1,140,556lb)
Lift-off mass:	511,159kg (1,126,913lb)

The final flight of a Saturn I looked forward. Not only would it carry Apollo Boilerplate 9 and be designated A-105, it would also have a trajectory selected for an orbital altitude chosen to allow Gemini XII astronauts to rendezvous with the Fairchild Hiller Pegasus 3 satellite and possibly retrieve some of the 48 specially adapted panels, returning samples for analysis. As mentioned previously (SA-9) this was an option not adopted by the Gemini Program Office for aforementioned reasons but the trajectory designed for SA-10 preceded that decision.

In all, Pegasus 3 carried 352 specimens of 45 different thermal-control coatings, including zinc oxide-silicone, silicon dioxide and gold on aluminium. It was felt that by retrieving some of the panels struck by micrometeorites, much could be learned about the effects on materials, as space missions became longer and thoughts began to turn toward the assembly of a permanently manned space station and other, long-duration structures which could be vulnerable to unexpected strikes.

The flight of SA-10 came after the successful launch of the first two manned Gemini spacecraft, on the second of which the first American spacewalk had been conducted. Ahead were a further eight missions during which astronauts would practice rendezvous and docking, experience flights of up to 14 days in duration and conduct more spacewalking, or EVA (Extravehicular Activity). It made sense to integrate an existing mission with the possible value in achieving both a rendezvous and docking and a spacewalk for sample retrieval from Pegasus. It turned out that the risk to the crew in moving across to an unstable target, plus the digression from what was already a highly self-contained set of hardware for Gemini experiments, argued against that option. At the time of flight it was still an open possibility.

Chrysler began fabrication of the S-I-10 stage on 4 October 1962 with tank clustering underway by 25 July 1963. After Rocketdyne delivered the H-1 engines, Chrysler continued with stage make-up and delivered it to checkout on 4 May 1964. After a small fire from a drill, checks were completed on 13 July and it was moved from Michoud to MSFC aboard *Promise* on 24 July. After arriving at MSFC on 31 July, S-I-10 went in to Static Test Tower East but shortly thereafter the eight H-1 engines were removed and returned for new LOX domes.

Returned to MSFC, all were back in S-I-10 by 29 August. The first, short-duration test firing was targeted for 16 September but on that day an attempt to get a 35sec run was aborted by a leak at the main LOX tank rear cover flange. And when it was finally run, on 22 September, the planned 35sec was cut short by 3sec due to an automatic cut-off when the signal indicating that all engines were operating failed to get power. An offending valve was replaced and a second run two days later went for 35.08sec. The mandatory full duration test was run on 6 October and endured for 154.48sec, the inboard engines for 4.5sec less as planned despite H-2026 showing pressure oscillations which caused excessive heat. S-I-10 was removed from the stand on 29 October.

Loaded on the barge *Palaemon*, the stage departed MSFC on 2 November and arrived at Michoud five days later. But a tool bag was accidentally dropped on to engine H-5030, damaging the tubing. It was sent back to MSFC where it was repaired and re-certified with a 60sec firing in the facility's Power Plant Test Stand on 10 December. Back at Michoud two days later, it was re-installed and sent for final checkout during which some attention was needed to correct some technical issues before entering storage on 9 March 1965, awaiting the call to leave. That came on 26 May when the stage left for the Cape on the barge *Promise*, arriving at its destination on 31 May, where it met up with S-IV-10 which had been there for three weeks.

In preparing the last of its type, Douglas Aircraft Company completed fabrication of the basic S-IV-10 structure during

Above: The Apollo boilerplate and Pegasus-3 spacecraft are mated to SA-10 at Launch Complex LC-37B. *NASA*

Below: S-I first stages for the SA-8 and SA-10 missions undergo final assembly at the Michoud Assembly Facility. SA-8 was the first flight to entrust construction of the S-1 stage to an outside contractor. *Chrysler*

September 1963, at which date it was moved to the vertical assembly tower at Santa Monica where welding and substructures were emplaced on this, the last of six cryogenic stages for the Saturn I. The RL-10 engines were installed in July 1964, with two replaced later, with final assembly the following month. S-IV-10 was installed in Alpha Test Stand 2B at SACTO on 5 December. The first test took place on 21 January 1965 after a succession of relatively minor tanking interruptions. It lasted 479.5sec and the stage was removed on 23 February, being flown out of Mather Air Force Base to CCAFS, arriving on 10 May.

The Instrument Unit was delivered on 1 June and erection of S-I-10 on the pad was completed the following day with the mating of the S-IV on 8 June. The Apollo Service Module and adaptor flew in on 21 June, the day before Pegasus C arrived, with the Apollo boilerplate Command Module (BP-9) and the Launch Escape System on 29 June. The payload was erected on the stack on 6 July and the LES was installed two days later. Tanking tests and flight readiness tests were held prior to RP-1 loading which was completed on 23 July. The CDDT began on 26-27 July and the countdown for launch began at 21.25hr EDT on 29 July. There were no problems in preparing the vehicle and it continued on down as the second Saturn I to proceed without a technical hold. But it came close. A single S-I stage hydraulic temperature measurement went off scale and MSFC had to provide a waiver for this one issue!

SA-10 carried 1,018 telemetered measurements of which 3 failed prior to launch and 12 failed in flight, with 376 on the S-I stage, 404 on the S-IV and 241 on the Instrument Unit. Overall, telemetry was good and the success rate with the sensors was very high for a space launch vehicle. The number of inflight measurements had been reduced by 145 over those on SA-8. The weather for this last Saturn I flight was marginal on wind but, at 16km (10mls) it shared good visibility and breezes from central Florida pushed cirrus clouds above a warm summer morning at a balmy 26°C (78°F) and relatively dry air, unusual for the Florida coastline at this time of year.

Launch occurred on time and the vehicle passed through Mach 1 at 54.8sec (7.22km/4.48mls) with Max-Q at 1min 8.7sec (12.2km/7.58mls). The inner engine shut-down occurred at 2min 22.2sec at an altitude of 79.3km (49.2mls) with the outboards at 2min 28.3sec and an altitude of 89.5km (55.6mls). The ullage motors ignited at 2min 29.3sec followed immediately by staging and ignition of the S-IV at 2min 30.8sec. S-IV cut-off occurred at 10min 30.2sec with orbital insertion 10sec later, apogee of 531.9km (330.5mls) and perigee of 528.8km (328.5mls).

The total mass in orbit was 15,621kg (34,438lb) and S-IV residual propellants remaining were 454kg (1,001lb) of LOX and 87kg (191lb) of LH_2. The S-I stage impacted the ocean at 12min 5.8sec, 985km (612mls) from LC-37B. The Apollo shroud separated at 13min 32sec and the Pegasus panels deployed within 40sec. Orbited mass remaining was 10,324kg (22,761lb), of which approximately 6,577kg (4,500lb) was the spent S-IV stage and 1,423kg (3,138lb) was the Pegasus structure, the remainder being the Instrument Unit and associated adaptors and equipment. The mission ended on 4 August 1969 when the payload decayed back into the Earth's atmosphere and was destroyed. The Saturn I programme was over.

At the Cape there was a sense of achievement but great excitement over the approaching launch of the first Saturn IB across at LC-34. Hardware was expected for that within the next few months but the unique nature of the 10 Saturn I flights had introduced a reliable and effective operational advance with the six S-IV stages. In the bars and motels across Brevard County, and certainly at Cocoa Beach, billboards and signs acknowledged the pioneering efforts of the Saturn I teams and in that strange juncture between reflection and anticipation, there was a great feeling of achievement and that even better things were to come. In diners and bars and at condominiums across the Cape area, Saturn-burgers were all the rage – with a special hot sauce for those with the taste buds to survive that ordeal! The excitement was electric, filled with anticipation but a controlled sense that the job was not yet done and that greater challenges lay ahead.

Left: The last Saturn I, SA-10 is launched from LC-37B with the third Pegasus payload on 30 July 1965. *MSFC*

Chapter Eleven

Saturn IB Flight Results

The original purpose of the Saturn IB was to support a unique and dedicated set of missions which had evolved from the beginning of the concept out of the early C-series launchers (C-1 through C-4). The adoption of the Douglas S-IVB mated with an uprated S-IB stage by MSFC and Chrysler, was for it to fill a niche as a medium-lift launcher between the existing inventory of missile-based vehicles (Thor, Atlas and Titan) with their different upper stages, and the C-4/C-5 class capable of sending 40,825kg (90,000lb) to the Moon. Numerous hypothetical mission profiles were set for Saturn IB, constrained initially only by the useful payload of around 13,610kg (30,000lb), about 40 per cent greater than the S-I/S-IV combination.

Much in the schedules and mission manifests would change significantly around late 1963 and early 1964. Just before this upheaval, which saw many flights cancelled, all-up testing introduced and a more robust approach to engineering development, the Saturn I was scheduled to complete its 10 research and development flights in December 1964 (SA-10) and from March 1965 fly the first of six manned flights (SA-111 to SA-116), the last in June 1966, interleaved with the first few Saturn IB launches. Note that operational manned flights of the Saturn I would be numbered in the 100-series for mission identification. Saturn IB flights would be ranked in the 200-series. The Saturn Vs had the reserved 500-series allocation.

By mid-1963, before those major NASA changes, the fifth manned Saturn I flight (SA-115) was assigned a launch in March 1966. These manned flights were to have developed operating techniques for Apollo Command/Service Module ac-

Above: The AS-204 payload apex cover is moved into position for mating with the SLA. *NASA-KSC*

tivities, also evaluating the procedures for new guidance systems and for recovery operations. In the same month, the fourth Saturn IB flight (AS-204) would have been a potential candidate for a manned launch but the formal end to Saturn IB development activities would be with AS-205 scheduled for May 1966, one month before SA-112 in June. And so it would proceed, with the last manned Saturn I flight (SA-116) coming after AS-205 which, depending on the success with previous Saturn IB flights, may have been the first one carrying a manned Apollo spacecraft. On this schedule, seven manned Saturn IB flights would flow along at three-month intervals, the last (AS-212) being in February 1968. Note that at this time, Saturn IB flights were identified by a SA prefix, shown here as AS for consistency.

At this stage (mid-1963), the Saturn V was expected to make its first flight in March 1966 but not assigned to carry a manned Apollo vehicle before at least five test shots had been successfully completed, with the first tentative expectation of that for AS-506 in April 1967, but more likely AS-507 two months later. In this remarkably ambitious manifest, the Apollo programme would have supported up to 25 manned Apollo missions on those three launch vehicles between March 1965 and October 1968, when the last of the 15 Saturn Vs (AS-515) would have been launched. And these on top of six Mercury flights and 10 manned Gemini missions to raise the grand total to 41. Partly for this reason, NASA recruited a further 23 astronauts in 1962 and 1963.

This ambitious schedule reflected production expectations before the cancellation of Saturn I vehicles SA-111 to SA-116. Readers may note the novelty of retaining the SA prefix, indicating the prime place in mission objectives taken by the rocket itself. This would soon change, to give the Saturn IB 200-series an AS prefix, as NASA shifted the alphabetical pre-eminence from the Saturn rocket to the Apollo programme which it was now serving. As the Saturn I programme came to an end, in 1965 NASA planned eight Saturn IB flights in support of the Apollo programme, the remainder being notionally allocated to the Apollo Applications Program (AAP).

As planned, AS-201 and AS-202 would fly unmanned ballistic test shots of the Apollo spacecraft for increasingly severe trials with the Command Module to simulate the energy dissipated from lunar-return velocities in qualification of the heat shield. AS-203 would be a test of the S-IVB stage to examine

Below: NASA's Lewis Research Center hosts a Saturn IB test model for its 3.05m x 3.05m (10ft x 10ft) wind tunnel. *NASA*

the behaviour of liquid hydrogen in microgravity conditions and would not carry an Apollo spacecraft. AS-204 and AS-205 would fly manned Apollo spacecraft on Earth orbit shakedown tests before AS-206 would carry an unmanned but fully fuelled Lunar Module for automated checkout. AS-207 and AS-208 would each carry manned Apollo and Lunar Module spacecraft, respectively, but with only a fraction of their propellant on board due to limited lift capacity.

At this date (mid-1963), the first two missions were to fly from LC-34, the remainder from LC-37B. Thereafter, manned Apollo flights in support of the Moon landing would be conducted by Saturn V. By mid-1966, the test schedule had been tweaked. The all-up tests of the Apollo CSM/LM on AS-207 and AS-208 were changed to dedicated payloads: AS-207 carrying a loaded Apollo CSM with AS-208 carrying an unmanned Lunar Module which would dock together in orbit. The manned Apollo CSM would dock with the LM in orbit much like the Gemini had docked with the Agena target vehicle. For these flights, having supported AS-201 and AS-202, Complex 34 was to support the AS-206 and AS-208 Lunar Module flights while the manned missions (AS-204, AS-205 and AS-207) would fly from LC-37B. Modifications to LC-37B for the Apollo CSM combination were cheaper than converting LC-34 to take the Lunar module.

By mid-1966 the first missions in the Apollo Applications Program were being planned. Based on Saturn IB flights, they carried the prefix SAA (Saturn Apollo Applications). SAA-209 (formerly AS-209) would carry a crew of three to conduct an extensive reconfiguration of the S-IVB stage launched by SAA-210. The crew would adapt it for habitation by means of an airlock in which they could extend the 14-day flight of Gemini VII in December 1965 to a 28-day stay on what would become NASA's first space station. SAA-211 would follow for a 56-day stay at the station followed by the launch of an Apollo Telescope Mount (ATM), a battery of solar instruments, on SAA-212, to be docked to the station. Those plans would be reworked after the Apollo fire of January 1967 and the AAP missions would eventually become the Skylab programme, centred on an S-IVB fitted out not in space but on the ground prior to launch on a Saturn V (AS-513) in May 1973.

By November 1966 delays to the first manned Apollo flight (AS-204) drove postponement from a target date of December 1966 to no earlier than 16 January 1967, with that date highly optimistic. The following month, NASA shifted the AS-204 mission to no earlier than 21 February due to persistent problems, with what was considered by some engineers to be a flawed design for the environmental control system, and delays to mission preparation. And while subsequent flights had been expected to fly at intervals of three or four months that had now stretched to six months.

The general sequence after AS-204 had called for AS-206 to carry an unmanned LM on an Earth-orbit shakedown test followed by a dual flight of a manned Block II CSM on AS-205 for rendezvous and docking with a LM on AS-208. A further change redirected the AS-205 flight to a repeat of the AS-204 Earth-orbit mission but with a more complex suite of experiments. After that all Apollo support operations for the Moon objective would cease and remaining Saturn IB launchers given over to the Apollo Applications Program, initially assigned flights to a space station from a converted S-IVB stage. But there were concerns about the clutter of missions threatening to stretch out the timeline for getting to the Moon and about delays to the Lunar Module which was threatening to allow an early checkout in orbit with a crew launched separately.

Above: Here represented as a scale model in wind tunnel tests, the Saturn IB had an infinitely smarter first stage with more powerful engines and a higher capability together with a more powerful and flexible upper stage, supported by an advanced electronics and avionics suite. *NASA*

On 15 November 1966, the AS-205 mission was cancelled. The crew for that flight (Schirra, Eisele and Cunningham) were to become the back-up to the AS-204 mission and the original back-up for AS-204 (McDivitt, Scott and Schweickart) would fly the first dual flight of both CSM and LM but launched on a Saturn V. Events would unfold to cause further delays and the fire on LC-34 that took the lives of astronauts Grissom, White and Chaffee. In the aftermath, attention focused on the Moon goal and with declining budgets the previous bipartisan support for AAP in Congress withered away, further eroding the potential for flights with the Saturn IB.

Optimism was high that the tragic events of 1967 could be left behind and, while the agony and suffering of family and close friends of the deceased crew would endure for some time, the Apollo programme made a significant step forward with the first flight of a Saturn V (AS-501) on 9 November 1967. The first all-up test of a completely new launch vehicle, it was an outstanding success and raised morale to a new level. The only remaining major element left untested was the Lunar Module, which was assigned to the AS-204 launch vehicle that would

U
S
A
UNITED STATES

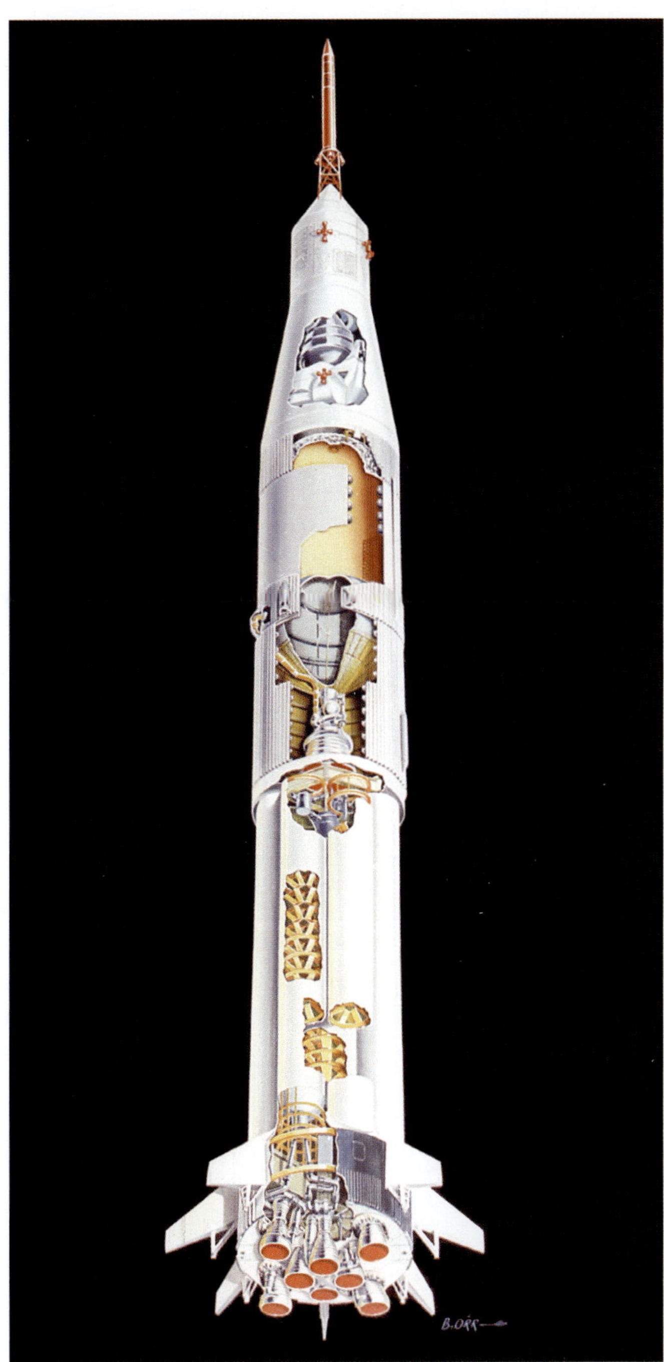

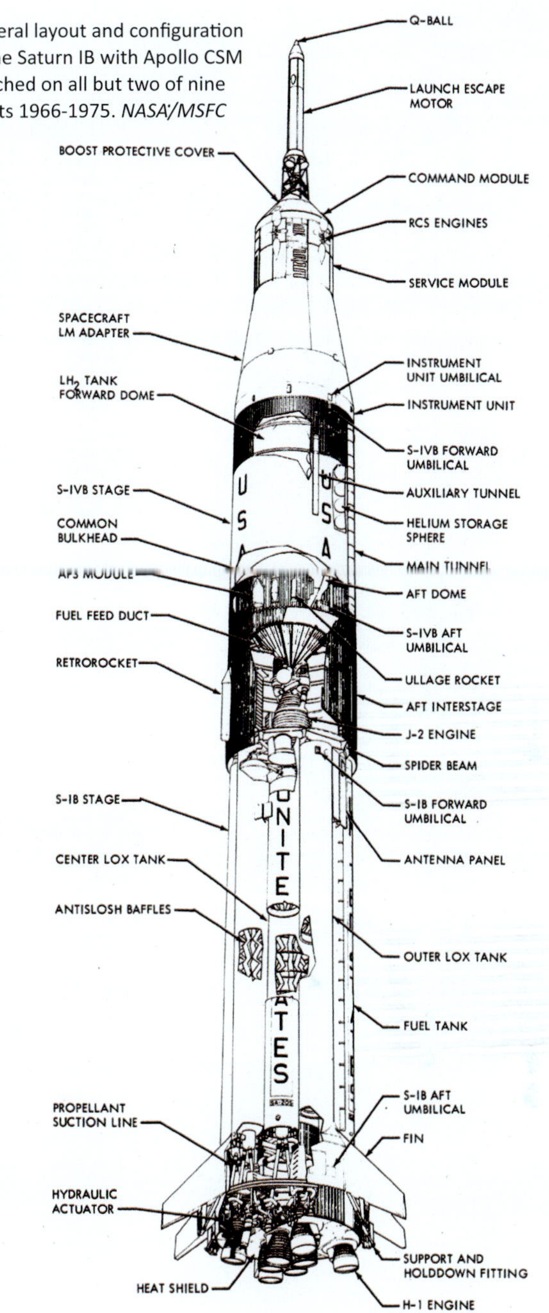

General layout and configuration of the Saturn IB with Apollo CSM attached on all but two of nine flights 1966-1975. *NASA/MSFC*

have carried the crew of the mission which had previously been designated Apollo 1.

After the fire, and not wishing to use that identifier again, a new series of mission designations was mandated by manned space flight boss George Mueller on 24 April 1967: the AS-501 mission became Apollo 4, with AS-204 designated as Apollo 5, AS-502 as Apollo 6 and the first manned flight of the Apollo spacecraft, AS-205, as Apollo 7. This was the last Saturn IB flight in support of the Apollo Moon programme, Saturn V taking over for those remaining missions. It was on 18 July, two days after the launch of Apollo 11, that NASA officially announced that the AAP programme (soon to be renamed Skylab)

Left The launch of the first Saturn IB marked a return to duty for LC-34 – the first time since SA-4 in March 1963. *NASA/KSC*

would use the hull of an existing S-IVB stage, converted on the ground into a habitable structure, to comprise the Skylab space station. The Saturn IB would only be used to carry astronauts to and from the orbiting facility.

Saturn IBs AS-206, -207 and -208 would support three expeditions to Skylab with AS-209 on standby for a rescue on the last Skylab visit and for the ASTP mission. During the Skylab programme, the next flight in line served as a potential rescue should the Apollo spacecraft at Skylab be rendered inoperable. AS-210 was the last shot Saturn IB flight and would fly the US component of the Apollo-Soyuz Test Project (ASTP). On 26 November 1971, NASA modified the contract with Chrysler to refurbish the four S-IB stages which would be utilised in the Skylab programme, taking them out of long-term storage and delivering them to KSC as well as supporting launch opera-

S-IB stage – location unit labelling

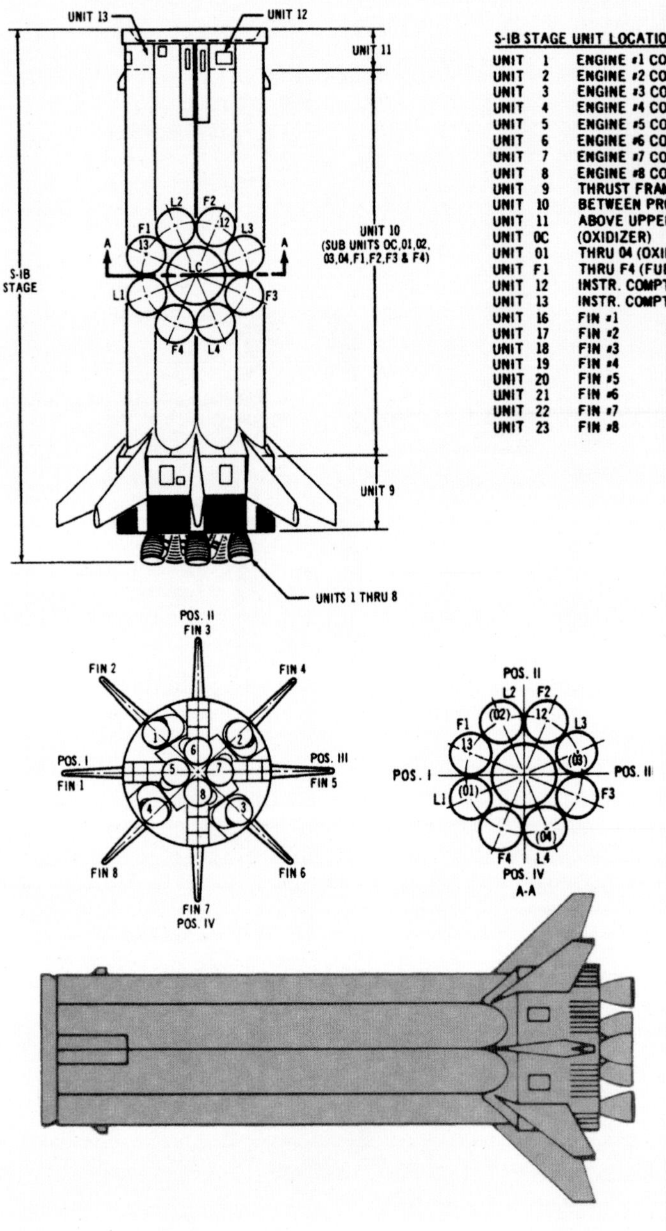

S-IB STAGE UNIT LOCATIONS

UNIT 1	ENGINE #1 COMPARTMENT
UNIT 2	ENGINE #2 COMPARTMENT
UNIT 3	ENGINE #3 COMPARTMENT
UNIT 4	ENGINE #4 COMPARTMENT
UNIT 5	ENGINE #5 COMPARTMENT
UNIT 6	ENGINE #6 COMPARTMENT
UNIT 7	ENGINE #7 COMPARTMENT
UNIT 8	ENGINE #8 COMPARTMENT
UNIT 9	THRUST FRAME AREA
UNIT 10	BETWEEN PROPELLANT TANKS
UNIT 11	ABOVE UPPER TANK BULKHEA
UNIT 0C	(OXIDIZER)
UNIT 01	THRU 04 (OXIDIZER)
UNIT F1	THRU F4 (FUEL)
UNIT 12	INSTR. COMPT. F2
UNIT 13	INSTR. COMPT. F1
UNIT 16	FIN #1
UNIT 17	FIN #2
UNIT 18	FIN #3
UNIT 19	FIN #4
UNIT 20	FIN #5
UNIT 21	FIN #6
UNIT 22	FIN #7
UNIT 23	FIN #8

Left: S-IB stage design, manufacturing and location 'Units', plus engine and fin positions. Units were used to identify associated components on the assembly line. *MSFC*

Below: The Unit-400 series aided identification of relevant areas around the upper stage. Originally the 500-series would have gone to the S-V stage. The 600-series relates to the Instrument Unit. *MSFC*

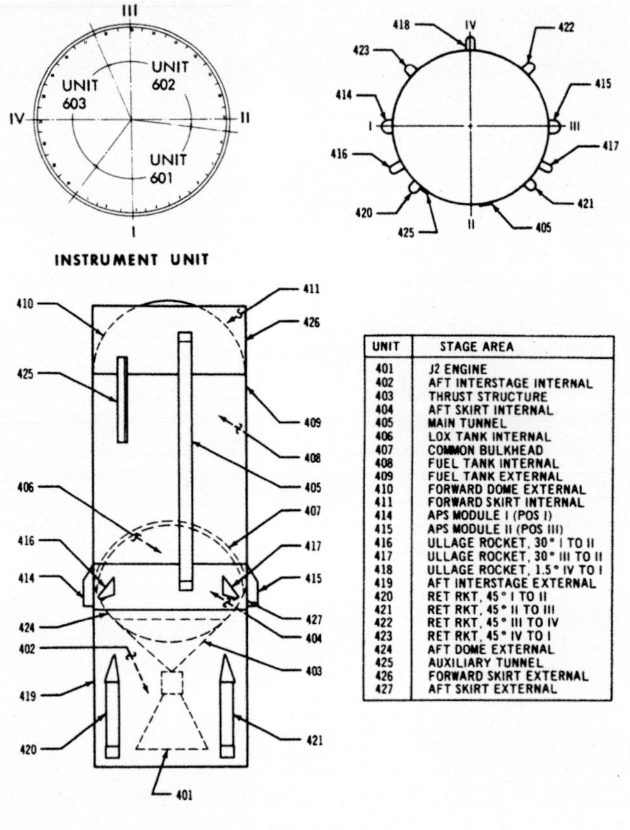

INSTRUMENT UNIT

UNIT	STAGE AREA
401	J2 ENGINE
402	AFT INTERSTAGE INTERNAL
403	THRUST STRUCTURE
404	AFT SKIRT INTERNAL
405	MAIN TUNNEL
406	LOX TANK INTERNAL
407	COMMON BULKHEAD
408	FUEL TANK INTERNAL
409	FUEL TANK EXTERNAL
410	FORWARD DOME EXTERNAL
411	FORWARD SKIRT INTERNAL
414	APS MODULE I (POS I)
415	APS MODULE II (POS III)
416	ULLAGE ROCKET, 30° I TO II
417	ULLAGE ROCKET, 30° III TO II
418	ULLAGE ROCKET, 1.5° IV TO I
419	AFT INTERSTAGE EXTERNAL
420	RET RKT, 45° I TO II
421	RET RKT, 45° II TO III
422	RET RKT, 45° III TO IV
423	RET RKT, 45° IV TO I
424	AFT DOME EXTERNAL
425	AUXILIARY TUNNEL
426	FORWARD SKIRT EXTERNAL
427	AFT SKIRT EXTERNAL

tions. Three Saturn IB launchers, AS-209, -211 and -212, were never flown but the S-IB stages of AS-213 and -214 were built before being scrapped. Propellant tanks for AS-215 were fabricated and stored before they were disposed of.

The operational life of the Saturn IB had been brought to an abrupt end when the Apollo Applications Program was cancelled for everything bar the Skylab space station. The S-IVB built for AS-212 had been converted to Skylab but there had been notions of a second station, Skylab B. A direct copy of the first station, it was under serious consideration and had been converted from another S-IVB. On 16 August 1973 NASA announced that there would be no Skylab B, which was largely inevitable due to the shrinking budget and the approval to build the Space Shuttle which had been granted approval by President Nixon in January 1972. There would be no more orbiting research facilities until the International Space Station (ISS) as a cooperative venture with Russia, the European Space Agency, Japan and Canada, assembly of which would not begin before 1998.

Nevertheless, Skylab B was built as a conversion from the S-IVB for AS-515, the last Saturn V built, and that now resides in the National Air & Space Museum in Washington, DC. At one point, AS-209 was assigned to a Skylab visit to re-boost the station and extend the orbital life prior to a Shuttle visit but that too was cancelled.

In the following diary of launch vehicle events, quoted engine numbers in the initial table identify the engines in place for launch. The narrative history beneath each vehicle identifies those engines which were installed initially but were for some reason changed out prior to flight.

AS-201

Launch date:	26 February 1966
Lift-off time:	011:12:01 EST (16:12:01 UTC)
Launch pad:	LC-34
Launch azimuth:	105deg
Length:	68.1m (223.33ft)
S/L thrust:	7,116.8kN (1,600,000lb)
Dry mass:	66,747kg (147,150lb)
Propellant mass:	519,610kg (1,145,544lb)
H-1 engines:	H-4044; H-4045; H-4047; H-4052; H-7046; H-7047; H-7048; H-7049
J-2 engines	J-2015
Ignition mass:	599,179kg (1,320,965lb)
Lift-off mass:	592,684kg (1,306,646lb)

The primary test objectives for this first flight of a Saturn IB focused on the launch vehicle and the S-IVB stage as well as the first of the Block I Apollo Command and Service Modules (CSM-009) destined for high-velocity, re-entry tests on a suborbital trajectory. Only one Block I CSM had flown before, CM-002 on a Little Joe II abort test on 20 January 1966. AS-201 was the first of three research and development flights qualifying the new uprated H-1 engines, the new and much larger cryogenic upper stage and several refinements in supporting elements such as the Instrument Unit.

Spacecraft objectives included verifying the operation of hardware separation of the Command and Service Modules from the Spacecraft Lunar Module Adaptor (SLA), attached to the top of the Instrument Unit, and the separation of the Command Module from the Service Module. In addition, the primary objective for the spacecraft was to evaluate the heat shield during a high-velocity flight path through the atmosphere, assisted by operation of the Service Propulsion System (SPS) in the Service Module to accelerate the speed of the descending spacecraft. This was a major event in evaluating the performance of the Apollo spacecraft in that it was the first test of the SPS engine in space and this had been tagged as a key mission objective second only to verification of the Command Module heat shield.

The SLA was common to all Saturn IBs carrying an Apollo payload and to all Saturn V launchers. It was a truncated cone 8.53m (28ft) long with a forward diameter of 3.91m (12ft 10in), the diameter of the Apollo CSM, and an aft diameter of 6.6m (21ft 8in) for alignment with the diameter of the Instrument Unit. It was divided into four separate panels with the aft set fixed and the forward set deployable. Each panel consisted of a 4.3cm (1.7in) thick aluminium alloy honeycomb structure bonded to aluminium face sheets.

The most recent launch from LC-34, only the fourth since its construction, had been the last of the ballistic Saturn I SA-4 in March 1963. Since then it had been under modification to accommodate the cryogenic S-IVB of Saturn IB, necessitating new swing arms on the launch tower and to receive 'man-rating', which involved an escape system should crews of later missions need to escape a launch vehicle believed to be in imminent danger of exploding.

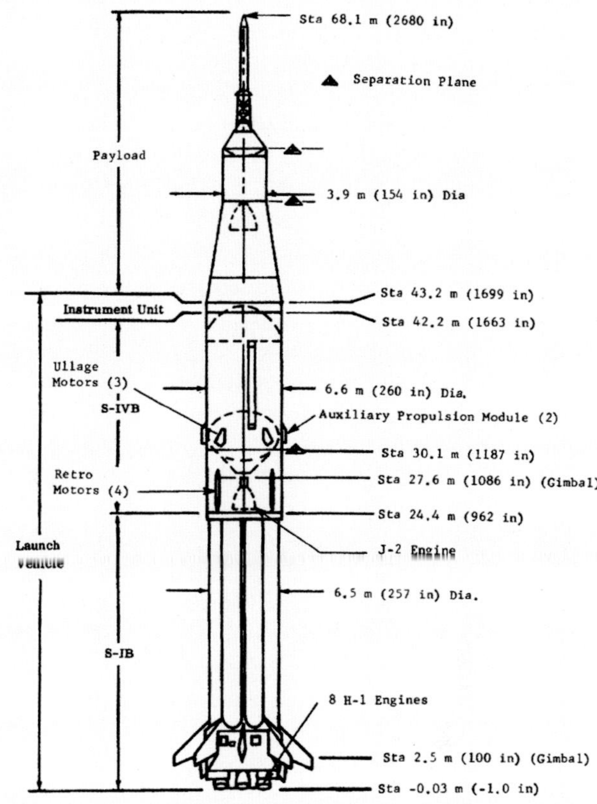

The precise characteristic dimensions of the AS-201 launcher. *MSFC*

AS-201 supported 1,922 telemetry measurements of which 545 were on the S-IB stage, 470 on the S-IVB stage, 300 on the Instrument Unit and 607 on CSM-009. Two wide-angle 16-mm Milliken recoverable cameras were installed to film separation of the two stages and ignition of the S-IVB. Each had a 160deg lens and an operating range of 130deg and would record events at 128 feet per sec. With one black and white and one colour, they would start three seconds before staging and eject 25 seconds after separation about 483km (300miles) downrange. They would deploy paraballoons for flotation with radio beacons and dye markers for recovery. Because of the extreme light conditions they were likely to encounter, the cameras used a special film with triple-layer emulsion and the capacity for wide exposure latitude. Each would be processed three times.

The pre-launch preparation and countdown would benefit from a new automated checkout system using RCA 110-A computers. This was known as the Acceptance Checkout Equipment for Spacecraft (ACE/SC) which integrated computers, consoles, recording equipment and liaison with the NASA Manned Spacecraft Center in Houston, Texas, where continuous monitoring and control of CSM-009 would take place. This automated equipment was part of a series of sequential development steps leading toward more extensive operations for the Saturn V.

Fabrication of components for the S-IB-1 stage began in July 1963, before NASA made a significant shift in forward planning and cancelled the Chrysler contract for six Saturn I launchers after S-I-10 on 30 October 1963. This was part of the strategic shift and reorganisation of how to rationalise the flight manifest for pushing forward with the Apollo Moon landing goal, stripping manned flights from the Saturn I and minimising reliance on the Saturn IB before shifting to the Saturn V. The immediate effect was to convert the contract to one supporting the first Saturn IB vehicle and to incorporate those components in S-IB-1. Had the eleventh Saturn I flown it would have carried a manned Apollo Block I and the first stage would have been identified as SA-111.

Integrating components originally intended for SA-111, structural assembly of S-IB-1 began at Michoud in March 1964, with tank clustering underway by 18 June and completed on 24 July. The newly uprated engines arrived by the end of the following month and were installed, only to be removed and returned to Rocketdyne for new LOX domes and changes to the injector assembly. Sans engines, assembly of the S-IB-1 stage was completed on 20 November and checkout activity was signed off on 2 February 1965. With the engines back at Michoud, the outboards were installed by 10 February with all the outboards five days later. The barge *Promise* departed MAF with S-IB-1 on 6 March and arrived at MSFC eight days later, when it was taken to Static Test Tower East. Numerous checks and some technical issues later, the first firing occurred on 1 April for 35.29sec, the second on 13 April lasting 145sec, the inboards cutting off 6.8sec earlier. Extensive post-firing checks forced removal of engine H-4046, repurposed for S-IB-2, and replaced with H-4052.

During its removal from the test tower on 19 April, some ripples were noted in a fuel tank which had to be smoothed out, after which the stage was returned to Michoud by the barge *Palaemon*, arriving there on 24 April. Some change-outs were necessitated on inspection and the stage left for CCAFS on 9 August aboard the barge *Promise*, arriving five days later. The S-IB-1 was moved directly off the dock to Hangar AF, a facility located on the Hangar Road in the CCAFS Industrial Complex and providing 6,147.2m^2 (66,170ft^2) of work area with a reinforced concrete foundation and a concrete block and aluminium exterior.

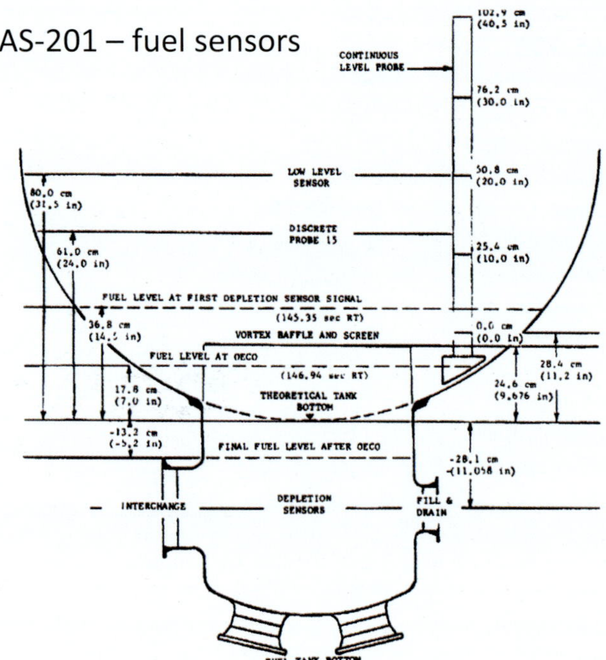

Fuel utilisation for S-IB-201 was 98.41 per cent and various levels were calibrated for cut-off. Fuel depletion sensors were installed in tanks F2 and F4 to prevent an oxygen-rich engine at fuel starvation. *NASA-KSC*

The stage was taken to LC-34 on 18 August. It was there as a height spacer for the S-IVB-F (Facilities) stage to verify the coupling of LOX and LH_2 loading systems but on 10 September fuel tank No 1 in the S-IB stage was reversed during over-pressurisation and it was replaced on 29 September, a day after de-stacking of the S-IVB-F stage from where it was eventually transferred for facilities checkout in support of the Saturn V at LC-39A.

Fabrication of elements of the Douglas S-IVB-201 stage started with manufacturing in December 1963 at Santa Monica and by January 1965 it was ready for installation of an interim J-2 engine (J-2012) to continue checkout until the flight engine became available, the replacement swap being completed on 14 April. Two weeks later, on the last day of the month, the stage was shipped by the barge Orion to SACTO by way of Courtland where it arrived on 6 May for installation in the Beta III test stand. There, preparations were made for a static test after checkout and a simulated firing sequence. A test on 31 July was aborted due to problems with ground equipment but a full duration firing occurred on 8 August lasting 452sec, a milestone in that this was a fully automated

Left: Framed by the servicing gantry, the first of only three Saturn IB rockets launched from this site. *NASA*

Right: Framed by the umbilical tower and the mobile service gantry, AS-201 introduced significant improvements over the Block I and Block II Saturn I vehicles. *NASA-KSC*

process from initial checkout to propellant loading and firing.

The formal handover to NASA occurred in a low-key ceremony at Beta III test site on 31 August prior to transport to Courtland Dock three days later from where it was moved to the Naval Shipyard in San Francisco Bay. There, it was loaded aboard the freighter *Steel Executive* and shipped to the Cape, arriving on 19 September for transfer to Hangar AF prior to being moved to LC-34 and stacking with the S-IB-1 stage on 1 October. The Instrument Unit docked at the Cape on 20 October and was erected five days later, three days after initial power had been applied to the S-IB-1 stage using an interim operating system programme from the RCA-110A computer. CM-009 arrived at the Cape on 25 October and was followed by its Service Module two days later, the day after power had first been applied to the S-IVB since stacking.

During November, the full electrical mating procedure was completed, the primary tape for the RCA-110A operating system arrived and checkout began on the guidance and control system. CSM-009 was erected on 26 December, the SLA mating being completed the following day. Numerous tests and checkout occupied pad teams during January, with the mating of the LES on the 24 followed by a full stack electrical mating sequence four days later. A plugs-out test was completed on 2 February with the dry Count-Down Demonstration Test (CDDT) on 8 February and a wet CDDT the following day. In this procedural shift, the Saturn IB and Saturn V flights would follow similar lines of pre-countdown preparations, the 'dry' CDDT being a full rehearsal for flight but without propellants, the 'wet' equivalent including full propellant loading. The Flight Readiness Test was completed on 12 February.

Due to the added power of the S-IVB stage, AS-201 flight operations were slightly different compared to Saturn I Block II timelines. Much was hanging on the demonstration of that stage in flight, procedures crucial to the entire Apollo Moon landing programme. It had to demonstrate a capacity for restart, as would be required when it was utilised as the third stage of a Saturn V with responsibility for propelling the Apollo/LM stack to a cislunar trajectory. The S-IVB had experienced several technical problems in development but the added advantage of cryogenic rocket motor technology was the enabler for the Saturn V mission, without which – even with a LOX/RP-1 upper stage – would have been impossible on that vehicle. This alone rendered AS-201 one of the most important missions to date in the evolution of lunar landing capability.

Originally planned for launch in late December, various problems pushed it first to January and then to February. The planned launch on 23 February was put back three days by poor weather. First stage propellant loading began at 14:20hr EST on 19 February and replenished at 06:45hr on launch day. The launch countdown had begun at 17:15hr EST on 25 February and proceeded without major problems but accumulated 3hr 27min of unscheduled holds due to a gaseous nitrogen control system achieving insufficient flow from the ground supply through a 0.16cm (0.4in) line at T-35sec; plans were made to install a back-up for future flights. Some dispute broke out when Mission Director Brig. Gen. Carroll H. Bolender in Mission Control-Houston ordered a scrub for that day, a decision contested by Dr. Kurt Debus in the LC-34 blockhouse. Lighting requirements for recovery of the Command Module demanded a launch window cut-off at 11:20hr EST. Bolender did not believe the work could be carried out in time. He was wrong.

The weather was subject to a high pressure ridge moving into the area, with winds from the north-west and the temperature distinctly cool. Winds were higher than had been experienced for any other Saturn launch vehicle to this date, measuring 70m/sec (230fps) at an altitude of 13.75km (8.5mls). This was considerably higher than the previous record of 51.8m/sec (170fps) at an altitude of 13km (8mls) during the launch of SA-4.

First stage start intervals lit up the H-1 engines in pairs of Nos 5 and 7, 6 and 8, 2 and 4 and 1 and 3 at 14msec, 114msec, 214msec and 314msec from the initial ignition command, respectively. Apparent in the data and through observation, the temperatures and vibration were substantially higher than had been experienced with the Saturn I and greater levels of damage were experienced. The drama of launching the first Saturn

Above: AS-201 departs LC-34 on the morning of 26 February 1966 for a suborbital flight to test the CSM-009 Block I Apollo Command and Service Modules. *NASA*

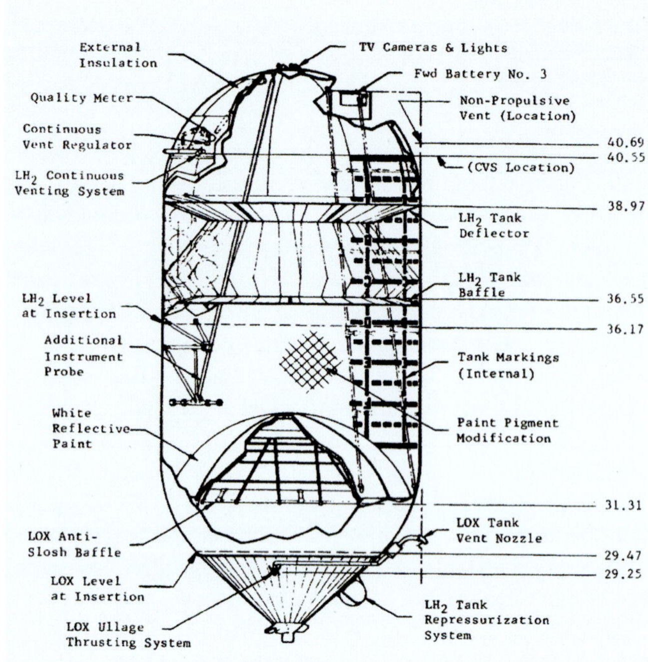

The essential features of the S-IVB-201 stage as prepared for the first Saturn IB mission. *NASA*

IB and fully equipped Block I Apollo spacecraft was heightened when a universal power failure blacked out LC-34, LC-37, the Converter Compressor Facility and several stations along the Eastern Test Range. Caused by vibrations shaking loose a high voltage fuse, the power outage affected the water quenching system, adding to pad damage from plume impingement and allowing small fires to break out.

Unimpeded by the drama unfolding on the ground, AS-201 continued on its way. Pitch and roll commands went in at 11.2sec followed by roll disable at 20.5sec and tilt arrest at 2min 14.4sec. The stack encountered Mach 1 at 1min 3.5sec with Max-Q at 1min 16sec, each event slightly later due to the increased mass and slightly increased fraction of ascent mass to thrust ratio. The four inner engines shut down at 2min 21.4sec at an altitude of 52.8km (32.8mls) followed by the four outers at 2min 26.9sec, at 58km (36mls). Ullage ignition and stage separation occurred within one second of S-IB shut-down and retrorocket ignition, thus ensuring adequate separation for ignition of the single J-2 engine on its inaugural outing.

The S-IVB ignition command came at 2min 29.3sec at a chamber pressure of 4,357kPa (632psi) and a LOX/LH_2 mixture ratio of 5:1 but this increased to 5,516kPa (800psi) at a ratio of 5.5:1 for the first 2min, producing a vacuum thrust of 1,000kN (225,000lb). For the remainder of the 7min 33sec burn duration the J-2 operated at chamber pressure back at the ignition levels. Shortly after S-IVB ignition the Launch Escape System again performed as scheduled, jettisoning at 2min 52.6sec. Due to a 1.63 per cent under-thrust versus predicted values, the stage ran for 10.3sec longer than planned with shut-down at 10min 3sec. S-IB stage impact had taken place at 8min 43sec, 465km (289mls) downrange. Including the S-IVB stage, the total orbited mass was 33,154kg (73,780lb) of which 15,333.7kg (33,805lb) was the CSM-009 payload. S-IVB residual propellant was an estimated 3,452kg (7,611lb), 3.3 per cent of a total tanked quantity of 104,256kg (229,845lb) at launch. Immediately after orbital insertion the assembly pitched down 180deg with the apex of the Command Module pointing generally toward the Earth, using the Auxiliary Propulsion Modules to achieve orientation. Spacecraft separation occurred at 14min 4.9sec, 10.5sec

AS-201 – key flight details

A
SM to SLA umbilical disconnect
B
Detonator and shield
A
B
C
-Y
+Z
-Z
+Y
C
Panel thruster

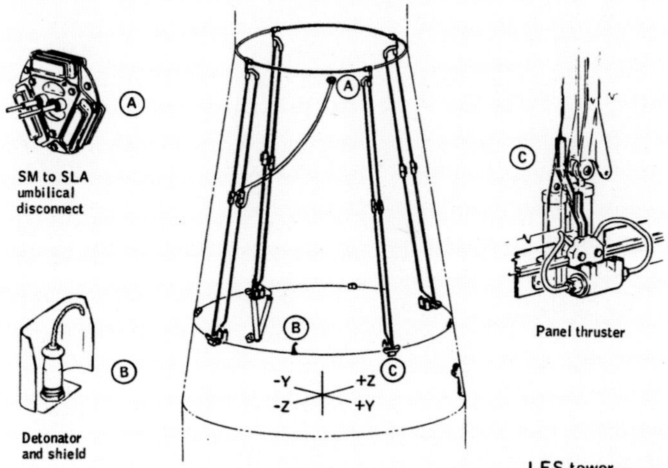

Left: Dynamic operation of the Spacecraft Lunar Module Adapter (SLA) panels incorporated pyrotechnic severance detonators to explosively detach the CSM-009 and to deploy the four segments. *MSFC*

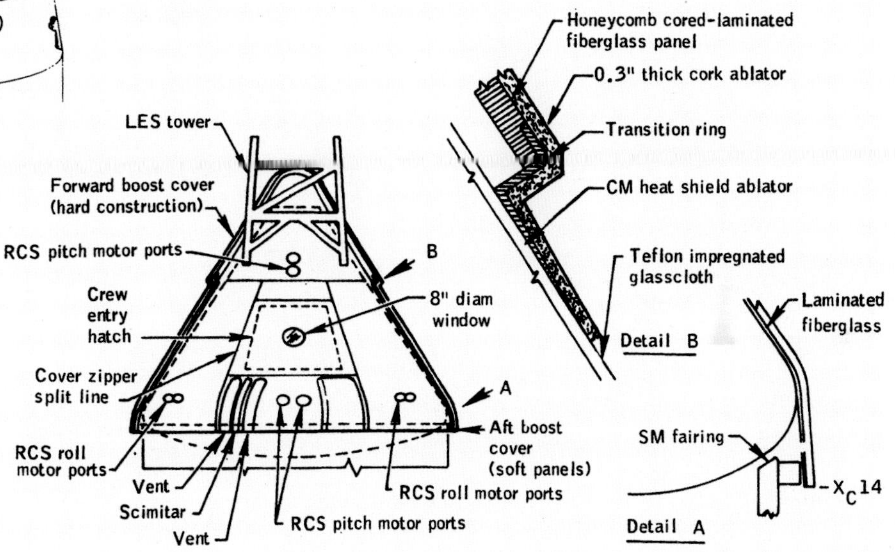

LES tower
Forward boost cover (hard construction)
RCS pitch motor ports
Crew entry hatch
Cover zipper split line
RCS roll motor ports
Vent
Scimitar
Vent
B
8" diam window
A
Aft boost cover (soft panels)
RCS roll motor ports
RCS pitch motor ports
Honeycomb cored-laminated fiberglass panel
0.3" thick cork ablator
Transition ring
CM heat shield ablator
Teflon impregnated glasscloth
Detail B
Laminated fiberglass
SM fairing
$-X_C 14$
Detail A

Right: Evaluation of the Block I Boost Protective Cover (BPC) which protected the Command Module during ascent and was later jettisoned, was a crucial part of the secondary objectives. Note the step below the LES tower and the overhang at the interface with the Service Module. *MSFC*

later than expected due to the extended S-IVB burn duration. The CSM-009 Reaction Control System (RCS) thrusters took control of attitude orientation and the Service Propulsion System started its burn at 20min 11.2sec which continued for 3min 4sec.

The manoeuvre was expected to add 1,432m/sec (4,700fps) to the velocity but as chamber pressure decayed the thrust level dropped and the increment was only 1,234m/sec (4,050fps). A second SPS burn ran for 10sec from 23min 30.7sec but failed to make up the difference, resulting in an entry velocity of 7,925m/sec (26,600fps) instead of the planned 8,306m/sec (27,250fps). The Command Module separated from the Service Module at 24min 15sec and re-entered the atmosphere, effecting a full lift/drag ratio of 0.334, deploying the drogue parachutes at 30min 54sec followed by the main parachutes 53sec later for a splashdown at 37min 19.7sec.

The flight of AS-201 boosted confidence in NASA's ability to handle cryogenic propellants in a major upper stage and verified the operability of its J-2 engine, the S-IVB stage, the APS attitude system and the basic functionality of the Apollo spacecraft and the Service Propulsion System. But lessons were learned, and verification of the automated flight sequence when all ground power was lost at lift-off profoundly vindicated the insistence from von Braun that rocket flights should not be subject to 'piloted' control. Either from the ground or within the payload by a crewmember on manned missions. The Air Force had been insistent that human control at every step of the flight was imperative, a legacy of the approach taken with manned aircraft and the role of the pilot in powered flight. Events showed that as the complexity and sophistication of automated systems grew, human intervention within the sequential operation of highly integrated systems was impossible.

SEQUENCE OF EVENTS
1 S-IVB CUTOFF
2 S-IVB/CSM SEPARATION
3 APOGEE
4 ULLAGE IGNITION
5 ULLAGE CUTOFF/SPS IGNITION
6 FIRST SPS CUTOFF
7 ULLAGE IGNITION
8 ULLAGE CUTOFF/SECOND SPS IGNITION
9 SECOND SPS CUTOFF/CSM SEPARATION
10 400,000FT
11 PARACHUTE DEPLOYMENT
LATITUDE (DEGREES)
WEST LONGITUDE, (DEGREES)
30
20
10
0
-10
-20
-80
-70
-60
-50
-40
-30
-20
-10

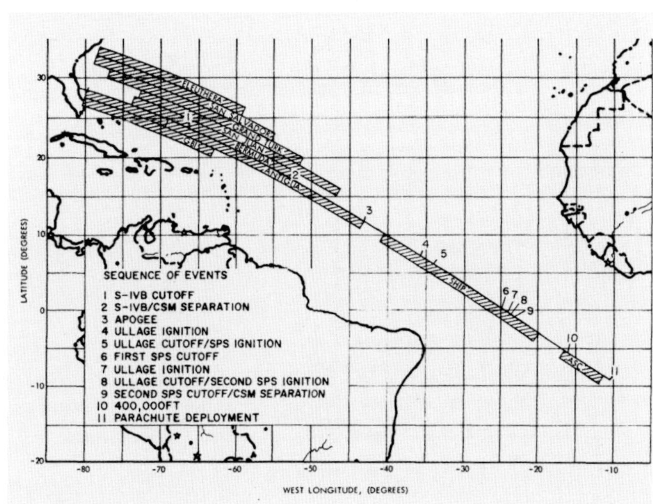

Right: The trajectory for AS-201 showing primary events on the ballistic flight down the Atlantic Missile Range with coverage from tracking stations indicated in the cross-hatching. *MSFC*

AS-203

Launch date:	5 July 1966
Lift-off time:	09:53:17.86 EST (14:53:17.86 UTC)
Launch pad:	LC-37B
Launch azimuth:	72deg
Length:	52.7m (173ft)
S/L thrust:	7,116.8kN (1,600,000lb)
Dry mass:	58,423kg (128,800lb)
Propellant mass:	485,765kg (1,070,906lb)
H-1 engines:	H-4053; H-4054; H-4056; H-4057; H-7056; H-7058; H-7059; H-7060
J-2 engines	J-2019
Ignition mass:	544,213kg (1,199,765lb)
Lift-off mass:	537,998kg (1,186,062lb)

We take this flight in chronological sequence where it had originally been assigned the third launch slot but a delay in delivering the payload for AS-202 put AS-203, with its hardware ready, first of the two to make it into space. The primary objectives were focused around the performance of the launch vehicle, particularly the S-IVB stage and the behaviour of cryogenic propellants, especially liquid hydrogen. The rationale for flying a dedicated cryogenic liquid hydrogen test flight began to take form in late 1964 after the successful flight of the S-IV stages on Saturn I.

As preparations for AS-203 picked up pace for launch, to the north of LC-37B, the AS-500F facilities checkout vehicle had been moved from the giant Vehicle Assembly Building to LC-39A where, from 25 May 1966, it would remain until 14 October while fit-checks and handling assessments were made between the giant replica-rocket and the ground servicing and support equipment. Personnel working at LC-37B would take trips up to watch the rollout and to see for themselves the magnitude of the effort underway across the length and breadth of Cape Canaveral, NASA's own particular part of the site where the Saturn V and its facilities could be found, being named the Kennedy Space Center by President Lyndon Johnson in an Executive Order on 29 November 1963 in commemoration of President John F. Kennedy who had been assassinated seven days before.

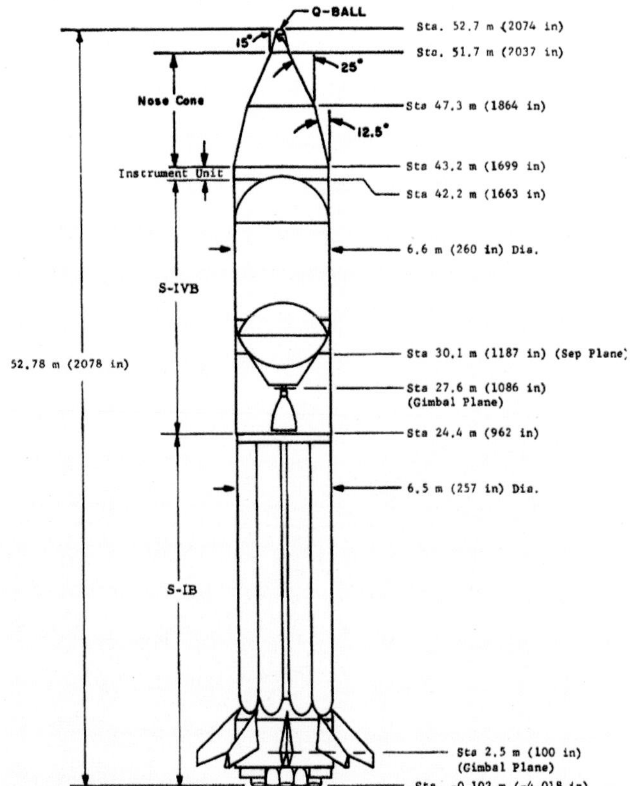

The general configuration and dimensions of AS-203. *MSFC*

The S-IVB stage for Saturn V missions to the Moon would require a dormant period in Earth orbit before firing up the stage a second time for Trans Lunar Injection (TLI). To maintain a positive acceleration for settling the otherwise weightless propellants, the residual LOX would be vented to simulate the effect on a lunar mission of hydrogen boil-off vented through two aft-facing nozzles in a propulsive discharge. Rather than add ullage rockets to the AS-203 stage, the small amount of LOX vented would simulate that calculated to be sufficient from LH_2 venting to maintain propellant at the bottom of the fuel tank, albeit at an infinitesimally minute 1/50,000th of 1g. In this way, engineers could verify that the heat transfer in orbit would be enough to induce boil-off at a level sufficient to provide this positive effect.

The test would require a full load of LH_2 at launch but only 65 per cent of the usual load of LOX so that at thrust termination there would be a residue of 8,165-8,618kg (18,000-19,000lb) of hydrogen but only a very small quantity of LOX. This would prevent having to instrument the LOX venting activity as well as that from the LH_2. Due to the unique nature of the mission's roles and responsibilities, the weights and timeline were very different to those of a normal Saturn IB flight delivering a payload to orbit; the S-IVB stage itself being the purpose of the objective. Nevertheless, this was to be the heaviest US object placed in orbit thus far. Clearly, there would be no requirement for an Apollo spacecraft, boilerplate or otherwise, but there were two launch configurations under consideration. Configuration 1 consisted of an Apollo payload shell but without a Launch Escape System, in essence a steel panel assemblage conforming to the precise dimensions of an Apollo spacecraft. Configuration 2 comprised a specially designed, tapering aerodynamic shell, unique to this mission. In the case of Configuration 1, the available liquid hydrogen in orbit would be at least 680kg (1,500lb) less than with Configuration 2, due to a 1,270kg (2,800lb) increase in the jettisoned mass and different aerodynamic drag losses. It was decided to adopt the aerodynamic shell which would have a weight of 1,587kg (3,500lb), a diameter of 6.6m (21.7ft) and a height of 8.5m (28ft).

Several modifications were made to both stages, in the case of the S-IB most notable being the re-routing of the exhaust ducts along the thrust chamber instead of through the stub fins, which allowed their removal, together with redesign of the honeycomb heat shield panels and the inboard engine flame curtains. Some weight reduction measures were carried out too on the skirts, bulkheads and the milled skin area of the LOX

AS-203 – S-IVB stage details

Above: The S-IVB-203 stage had several unique features for this test mission as identified on this graphic. *MSFC*

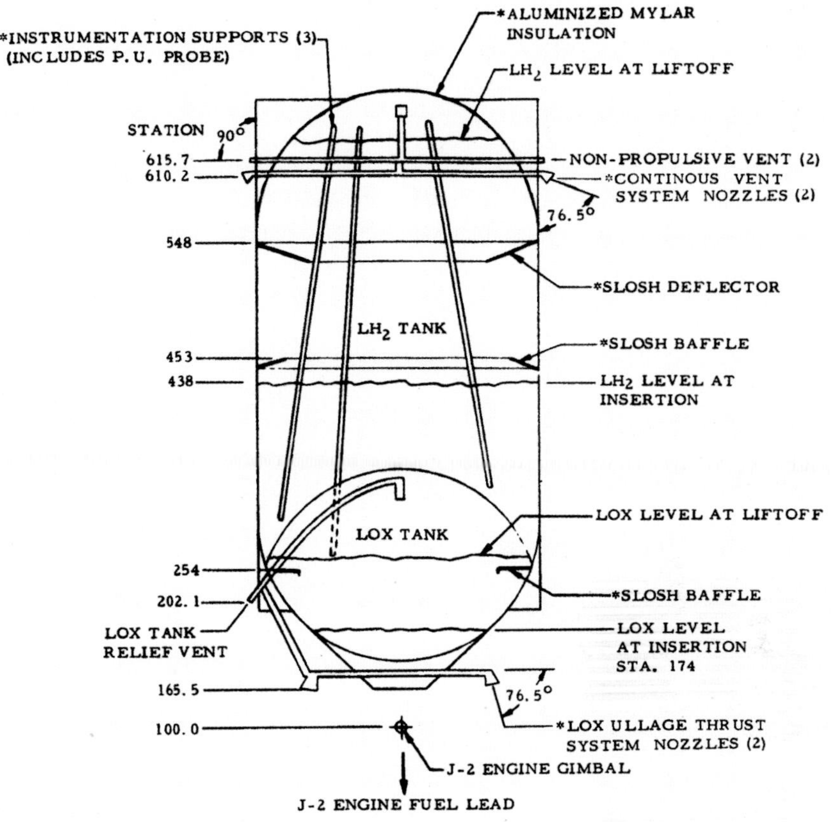

Above: Comparison with the S-IVB-201 stage shows that S-IVB-203 had several modifications and special equipment for the cryogenic tests that underpinned the mission. *MSFC*

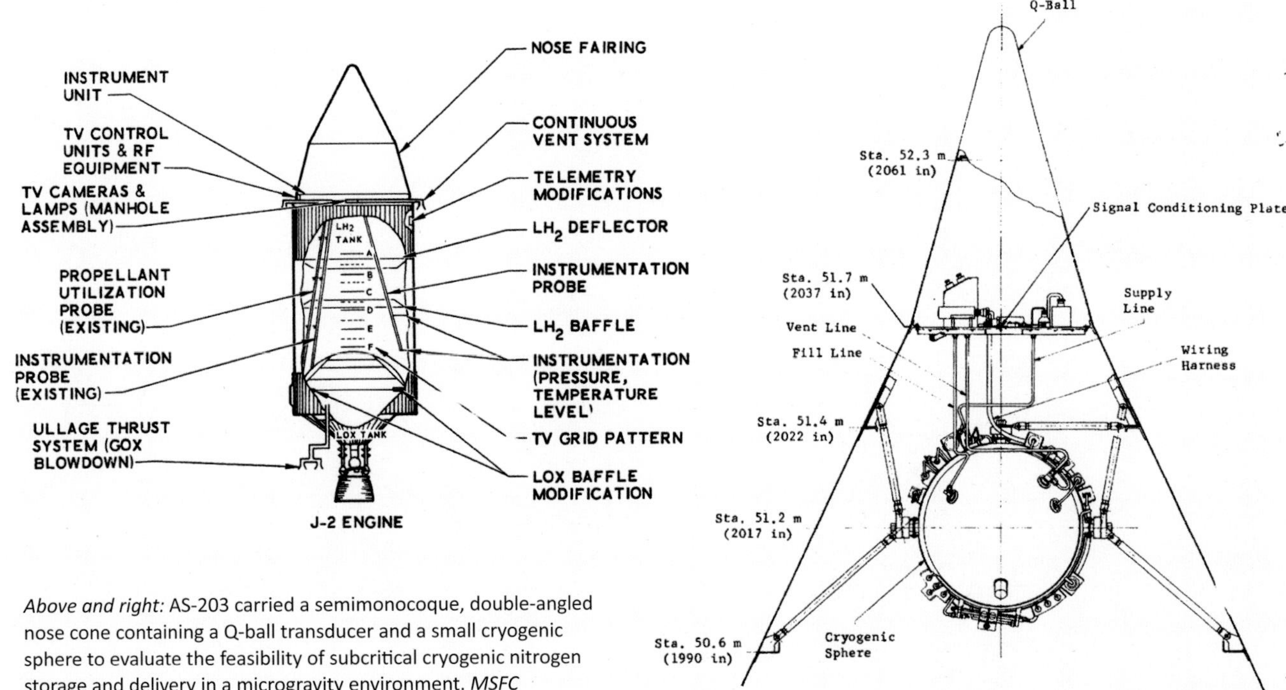

Above and right: AS-203 carried a semimonocoque, double-angled nose cone containing a Q-ball transducer and a small cryogenic sphere to evaluate the feasibility of subcritical cryogenic nitrogen storage and delivery in a microgravity environment. *MSFC*

AS-203 – S-IVB stage details (contd.)

Right: A detailed graphic showing the internal equipment including diagnostic sensors and cameras for recording the behaviour of the cryogenic liquid hydrogen. *MSFC*

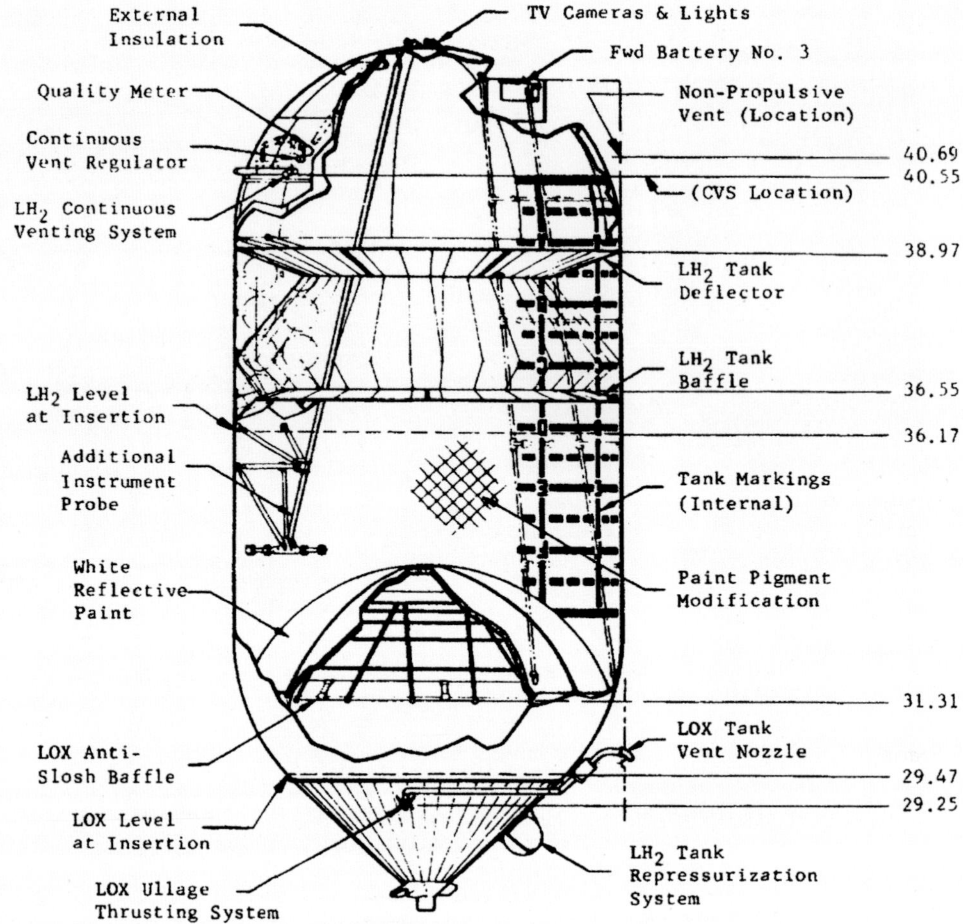

Below: A lot of engineering tests were crammed into the AS-203 mission, including the vital provision of an S-IVB chilldown sequence to prepare the J-2 for an orbital restart. *MSFC*

Below right: A continuous vent system was installed on AS-203 for providing a positive acceleration of 2 x 10-5 g. *MSFC*

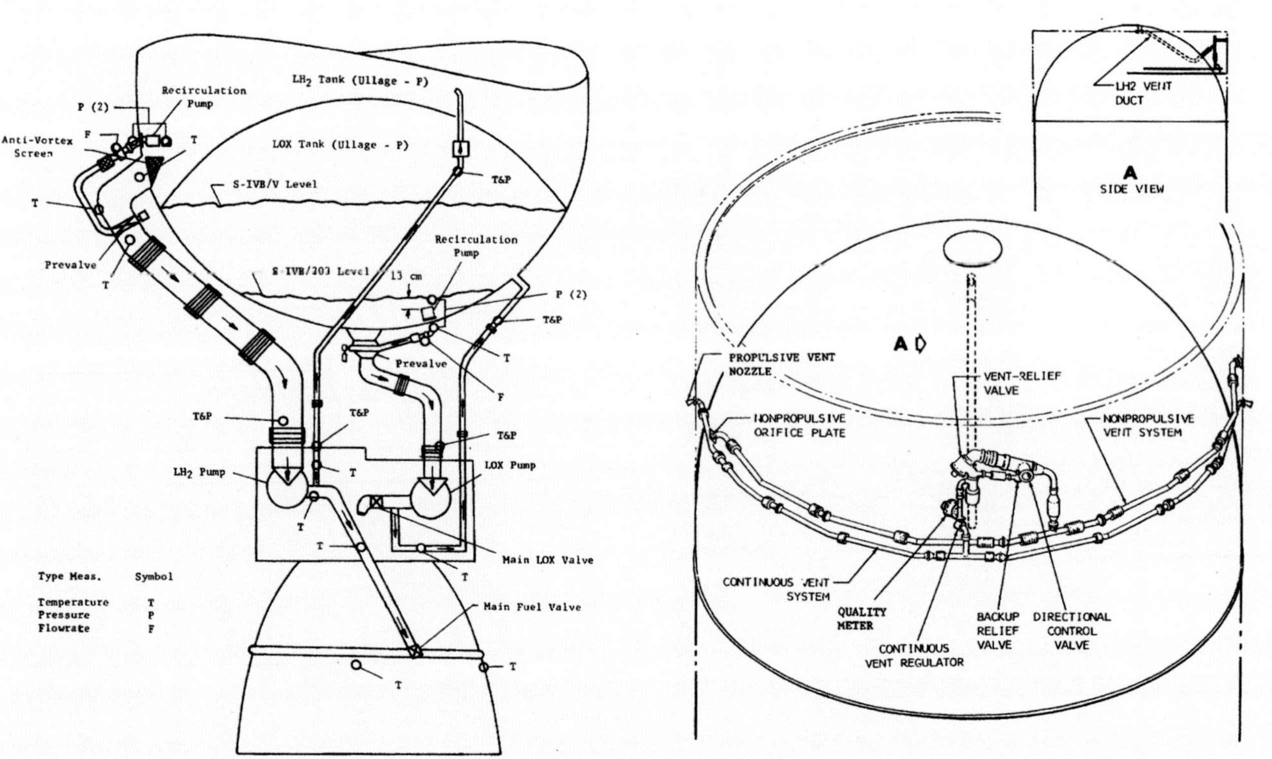

tanks. Changes to the S-IVB stage brought it more into line with the configuration the Saturn V S-IVBs would require so that evaluation of procedures and operating processes essential for a second engine start could be completed. More on these below in the S-IVB reporting paragraphs.

Taking advantage of the Saturn IB as a test vehicle for common Saturn V systems, the Instrument Unit was basically the same as that which would control the flight of the first Saturn V. The nose cone configuration comprised a semi-monocoque, double-angle nose with an overall length of 953cm (375in). The nose cone was constructed with ring frames and skin stringers with a Q-ball transducer at its apex and a single conditioner plate together with a small cryogenic sphere containing nitrogen for an experiment in the feasibility of storing subcritical cryogens in a low-g environment.

AS-203 carried 1,491 measurement points, 550 on the S-IB, 581 on the S-IVB, of which 83 were dedicated to studying the behaviour of the liquid hydrogen, and 360 on the Instrument Unit. Two Kin Tel, 0.39cm, 700-line TV cameras were carried in the forward dome of the LH_2 tank to observe the effects of flight and subsequent weightlessness on the fuel, providing exceptionally sharp views of the propellant as globules separated and drifted around the interior of the void.

Development of the definitive S-IB stage from the Saturn I evolved across three sets of hardware, with S-IB-3 finally carrying all the minor changes introduced with the first two vehicles, bearing in mind that this was the third to be assembled and the first not to incorporate components initially fabricated for a Saturn I stage. Changes effected with this stage compared to S-IB-1 included reduced metal thickness in the sheet metal and the machined elements of the tail section, a redesign of the brazed honeycomb in the heat shield and re-routed exhaust ducts along the thrust chambers instead of through the stub fins. This allowed removal of the four turbine exhaust fittings in the form of stub fins from the exterior of the lower shroud, reorientation of the heat shield support structure, redesign of the inboard honeycomb heat shield panels and the inboard engine flame curtains and redesign of the flame shield and its support structure.

Some sheet metal elements of skirts, bulkheads and milled skin areas of all the LOX tanks were reduced in thickness and the sump cover in the 2.67m (105in) central LOX tank was inverted and the gasket redesigned. Also, the GOX line and diffuser for that tank was re-routed to allow additional topping during propellant fill. Some alterations were made to the LOX tank upper manifolds and tube apertures in the upper skirts, necessitating some design changes to the LOX vent. For this stage the retrorockets were moved 16.5cm (6.5in) farther outboard and made parallel to the vehicle skin so that the loss of a single rocket would not cause a collision during separation due to unbalanced forces causing the stage to skew.

Changes to the H-1 engines included a redesigned inboard engine turbine exhaust system to re-route the exhaust gases into the area enclosed by the four inboard engines instead of through the stub fins, as noted above. Some other minor changes were made to improve operability and open the options for safer operation, including a reduction in the diameter of the heat exchanger orifice, redesigned electrical harness potting to incorporate connectors with a metal sleeve and replacement of the two individual unitised check valves with a single valve to reduce complexity and potential failure. A third thrust-OK switch was added for redundancy.

Innovatively, a reduction of ullage volume in the LOX tanks from 1.7 per cent to 1.5 per cent allowed an additional 680kg (1,500lb) of propellant to be loaded and a dome vent was added to each of the four outboard tanks. A new relief valve was added to the centre LOX tank to vent and relieve overpressure. Changes to the turbine exhaust system on the H-1 engines also pushed redesign of the GOX interconnects from the inboard engines. Changes affecting the electrical system altered the logic for outer engine cut-off so that two outer engines would have to fail, rather than one, to trigger an automatic shut-down of the remaining six. Data from all previous eleven flights showed that despite their steering role, 50 per cent of the outer engines could fail without resulting in a catastrophic failure.

Build-up of S-IB-3 began at Michoud in July 1964 with tank clustering getting underway on 21 January 1965 and installation of all eight H-1 engines completed on 21 April. The stage was moved out for checks on 17 June but problems with engine H-7058 forced a need for a new turbopump shaft seal before that activity was completed on 14 August.

Stowed aboard the barge *Palaemon*, the stage left Michoud on 9 September but Hurricane Betsy forced a sheltered stop near Baton Rouge that night, *Palaemon* finally arriving at MSFC on 16 September. A day later, S-IB-3 was in Static Test Tower East with simulated flight tests and propellant loading tests underway. The first of two test firings took place on 12 October and lasted 35.3sec, followed by the second, full duration test on 26 October lasting 146.2sec with no problems encountered during a smooth and uneventful sequence. The stage was transferred to Michoud, arriving there on 9 November for post-test modifications. It departed for the Cape on 7 April 1966 aboard the barge *Promise* and arrived five days later.

Because S-IVB-203 was required to conduct several unique procedures and to monitor the behaviour of liquids in microgravity conditions, it had several unique changes incorporated during manufacture. Most aimed to simulate Saturn V operations as closely as possible, although several major stage systems would be different for that application. The main differences between this stage and S-IVB-201 included a modified aft skirt to accommodate heavier loads, a deflector and ring baffle integrated into the LH_2 tank at two critical locations so as to control the propellant sloshing at J-2 cut-off, and the anti-slosh baffle on the LOX tank extended 61cm (24in) to increase the damping effect by 3 per cent.

Related to the J-2, changes included two recharging lines to insure greater charge probability for the start tank, a strengthened and more flexible fuel tank pressurisation line and better insulation on the thrust chamber to reduce the warm-up rate during boost. Modifications to the fuel tank pressurisation system included the non-propulsive vent carried by AS-201 to a continuous vent system of the type which would be fitted to the S-IVB stages used with Saturn V (a more comprehensive description of the S-IVB for that launch vehicle is carried in the companion volume on Saturn V). The new system included two 2.54cm (1in) lines routed around the inside of the forward skirt to two nozzles located 180deg apart in the longitudinal axial direction. This would allow data to be gathered on the behaviour of the liquid hydrogen and the effect on the action of that liquid when the engine shut down and restarted.

A LOX ullage thrusting system was added so as to vent residual GOX through two nozzles located 180deg apart and maintain ullage control during the 50sec after J-2 cut-off, assisting in settling the propellant. They were to simulate the ullage effect of the 311N (70lb) thrust Gemini manoeuvring engine which would be utilised on the Saturn V Auxiliary Propulsion System (APS) modules. The propellant utilisation system was also configured to operate in the open loop mode instead of in a closed loop configuration and this would help maximise the propellant residuals. In another evaluation of a Saturn V system, the LH_2 auxiliary chill-down pump was redesigned to provide a differential pressure measurement, with application to the Saturn V's S-II stage as well. Changes were also incorporated into the flight control system, the electrical system and the environmental control system.

The major structural assemblies began to come together during July 1964 and the engine arrived on 1 May 1965 but during checkout a few cracks were noted around the manhole cover in the LH_2 tank. After repairs, the stage passed to a complete inspection cycle which concluded on 14 October. It was loaded on the barge Orion on 29 October and delivered to Courtland Docks on 1st November from where it was trucked to SACTO for firing tests. A full duration firing of 284.9sec took place on 26 February and it was removed on 19 March for a final, post-firing checkout. It was delivered to the Cape by Super Guppy on 6 April.

S-IB-3 arrived at the Cape on 12 April 1966 and went straight to Hangar AF before transfer to LC-37 and erection of the first stage on the pad six days later. Three horizontal fins were attached on 15 April and installation of all fins was completed the day after the stage stood erect. The S-IVB stage was mated to the S-IB on 21 April together with the Instrument Unit and the nose cone assembly prior to mechanical and propulsion checks the following day and an electrical mate of the entire vehicle on the 25. Various propellant dispersal, vehicle pressure and LOX simulated malfunction tests occupied May, LOX and LH_2 loading being tested on 7 June. Preparation of the AS-203 vehicle was arduous and demanding, given the poor quality of some of the electronic equipment; more than 8,000 printed circuit boards had to be replaced prior to clearance for launch. It had been a problem inherited from the previous vehicle and would persist throughout several more flights. The RCA-110A computers were becoming a pacing item on the flight readiness agenda.

In this month of June there were three Saturn rockets on their launch pads: in addition to AS-203 on LC-37B, AS-202 was waiting hardware on LC-34 and the AS-500F Saturn V facilities checkout vehicle was on LC-39A to the north up the shoreline. The two countdown demonstration tests for AS-203 were conducted on 29 June and 1 July, prior to the launch countdown starting at 20:30hr EST on 4 July at the T-11hr 30min mark, heading toward a lift-off time of 08:00hr EST on 5 July. A total of 1hr 53min 17sec was lost on unscheduled holds due to a wide variety of technical and range telemetry issues for this unique launch.

Following a successful lift-off 0.86sec after range zero, Mach 1 was passed at 51.5sec with Max-Q at 1min 10sec. The four inner engines cut off at 2min 19.2sec and an altitude of 62.4km (38.7mls), with the outboard engines shutting down at 2min 22.6sec at 66km (41mls). Separation of the S-IB stage came at 2min 23.4 and an altitude of 66.8km (41.5mls) followed by S-IB impact with the ocean at 9min 44.3sec, 809km (502.7mls) downrange. The single J-2 on the S-IVB fired up at 2min 24.9sec and shut down at 7min 13.3sec. At cut-off the orbit was 187.3km (116.3mls) by 185km (114mls) with a period of 88.21min, an inclination of 31.9deg and a mass of 26,723kg (58,913lb). The S-IVB had 11,079kg (24,504lb) of propellant remaining, equivalent to that calculated to be in the tanks during a Saturn V flight. Which was of course the objective, as the behaviour of the LH_2 inside the stage on orbit during a Moon mission was the functional purpose of AS-203.

After orbit insertion the LOX tank vent remained open for 77sec. Before closing the LOX vent, the fuel tank continuous vent system was opened to produce a continuous longitudinal acceleration of approximately 2×10^{-5} g. In a simulated restart of the J-2, fuel re-pressurisation began late during the first orbit, 1hr 25min 7.55sec after J-2 cut-off, and ended 70sec later with an initial pressure rise to 156.5kPa (22.7lb/in^2). It continued to rise and reached 174kPa (25.38lb/in^2) 3min 17sec later, higher than anticipated, when the continuous vent system was activated. S-IVB-203 carried a single re-pressurisation bottle whereas the Saturn V S-IVBs would carry seven bottles, which was thought sufficient to obviate any problems.

Primary interest focused on the behaviour of the liquids when the acceleration was suddenly reduced. Toward the close of the first orbit, a chill of the engines feed lines and engine components was conducted on both propellants and this lasted for about five minutes. In this period the hydrogen venting was terminated and the tank partially re-pressurised to obtain a net position suction for the recirculation pump, the LOX thruster system used to provide acceleration. On the second orbit the process was repeated except that only the hydrogen feed system chill was conducted but without re-pressurising the tank. With this, the tests related to Saturn V operations were concluded.

The next two orbits were committed to experiments for obtaining data on alternate venting systems and on the design criteria necessary for future cryogenic stages. They were a detailed observation of the behaviour of liquid hydrogen without any longitudinal acceleration and tests to determine the rate of pressure rise in a closed cryogenic container.

Early in the fourth orbit, at an elapsed time of 4hr 44min 52sec, 20sec after the LOX ullage system had been opened to provide positive acceleration, the fuel tank continuous vent system was closed for the remainder of the flight so as to provide data on the rise rate of the LH_2 tank pressure and thereby derive the differential required to collapse the common bulkhead separating the two propellant tanks. About 7,258kg (16,000lb) of fuel remained in that tank when the vents were closed. Heating of the fluid caused the fuel tank pressure to rise but it was to have been maintained below 289,6kPa (42lb/in^2) by the pressure relief system. By the beginning of the fifth revolution, as the vehicle passed over Cape Canaveral, the hydro-

AS-203 – S-IVB tank cameras

Right: A continuous vent system was installed on AS-203 for providing a positive acceleration of 2 x 10^{-5} g. *MSFC*

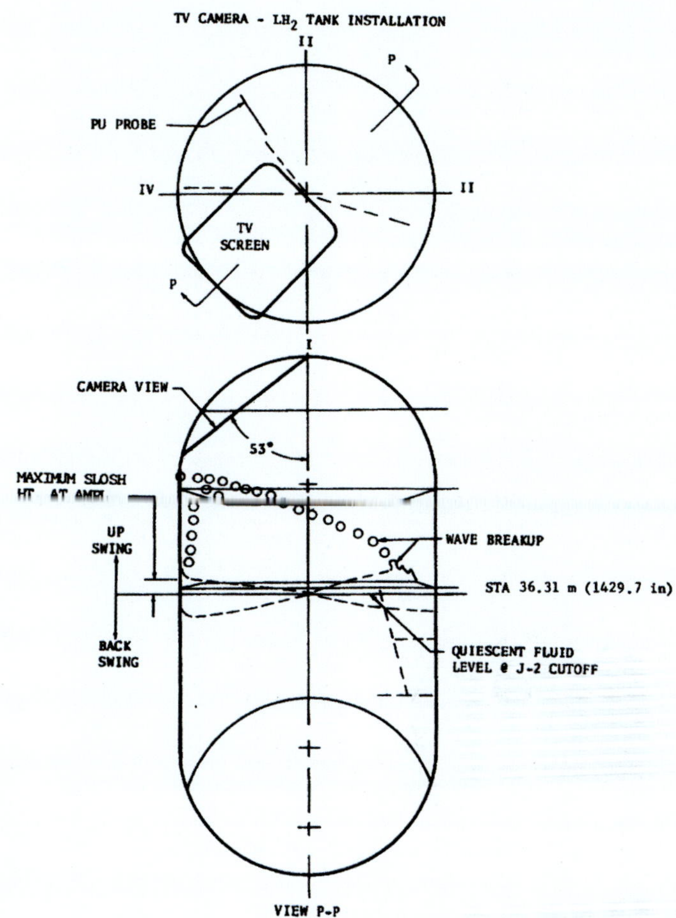

Above and below: A system of cameras and lights was installed in the forward section of the S-IVB stage (pictured here separating from the S-IB stage to record the behaviour of the liquid hydrogen under such conditions). *NASA-MSFC*

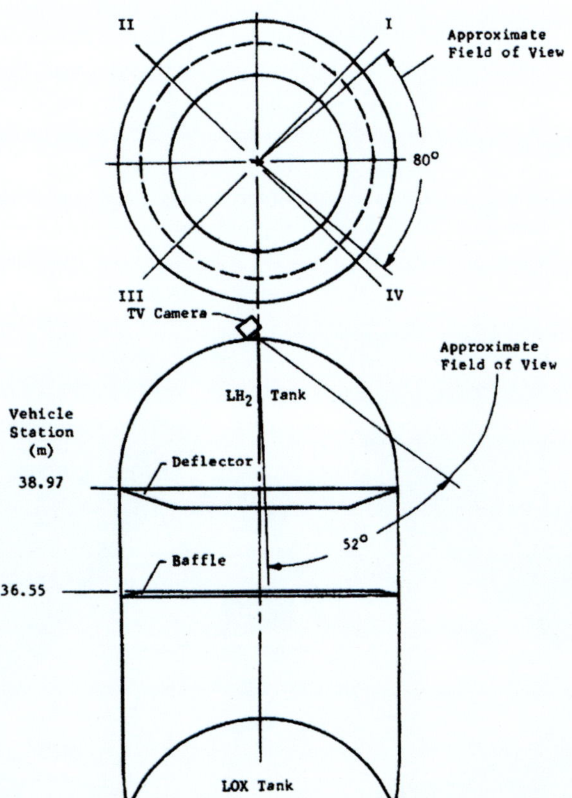

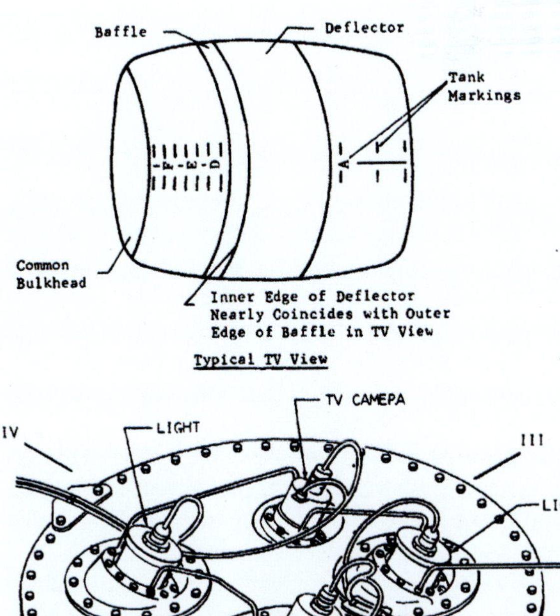

Above: AS-203 takes to the skies to test the behaviour of liquid hydrogen under a variety of accelerations and microgravity conditions. *NASA*

gen tank measured 275kPa (39.9lb/in²) and the LOX tank showed 34.47kPa (5lb/in²). Two minutes after losing contact with the Cape, when the Trinidad site picked up a radar fix, what remained of the stage was in a large number of pieces. Pressure differential had collapsed the common bulkhead and it was surmised that a spark had ignited the contacts, destroying the vehicle.

This flight was all about evaluating the performance of liquid hydrogen in a large stage and one specially configured for supporting cislunar escape trajectories. A wide range of tests were conducted from launch and throughout the mission to measure acceleration levels, thermal conditioning and cryogenic propellant behaviour when in a microgravity condition. Engine operations were conducted to demonstrate proper chill-down procedures which would normally be applied prior to second ignition of the J-2. Some work-around was necessary to simulate Saturn V operations with the stage, and the payload capability of the launch vehicle was insufficient to replicate weights and forces which would be imposed on a Moon mission, limitations which also prevented a restart of the J-2.

The tests had been an outstanding success and provided both verification of calculated responses under a variety of conditions and confirmation that conditions could be met for the role of the S-IVB on the Saturn V. With the flight of the first Saturn V little more than sixteen months away, this dedicated S-IVB test mission was one of the most important in the series of Apollo-related flights, yet it is given little note in most history books. What it lacked in excitement and drama, it more than made up for in a slice of solid engineering achievement.

AS-202

Launch date:	25 August 1966
Lift-off time:	12:15:32 EST (17:15:32 UTC)
Launch pad:	LC-34
Launch azimuth:	105deg
Length:	68.1m (223.33ft)
S/L thrust:	7,116.8kN (1,600,000lb)
Dry mass:	83,462kg (184,000lb)
Propellant mass:	514,739kg (1,134,786lb)
H-1 engines:	H-4046; H-4048; H-4049; H-4050; H-7050; H-7051; H-7052; H-7054
J-2 engines	J-2016
Ignition mass:	598,593kg (1,319,671lb)
Lift-off mass:	592,215kg (1,305,610lb)

The second flight of a Block I Apollo CSM on a Saturn IB had been delayed considerably beyond the ready-date for the launch vehicle because of late delivery of the payload (CSM-011) from North American Aviation Space & Information Systems Division at Downey, California. There had been deep problems within the company regarding schedules, accountability and quality control of hardware on both the manned spacecraft and the S-II second stage for the Saturn V. CSM-011 was to have shipped out to KSC on 28 February but by that date it was not expected before early April, after which it would have to go through about 19 weeks of checkout before it would be ready for launch.

The purpose of this unmanned test flight was to test the Block I Command Module (CSM-011), specifically its heat shield, in a 'roller-coaster' suborbital re-entry path simulating a return flight from the Moon. It was also a test of the recovery operations involving the US Navy, as it had been during recovery of CM-009 launched by AS-201. While AS-201 had demonstrated high heat rate on the heat shield, AS-202 would evaluate high heat loads such as those anticipated on lunar flights.

A small TV camera was mounted on the cross beam, on top of the Instrument Unit to view the movement of the four petal

Below: The much-delayed CSM-011 is inspected at KSC. *NASA-JSC*

AS-202 – launch configuration

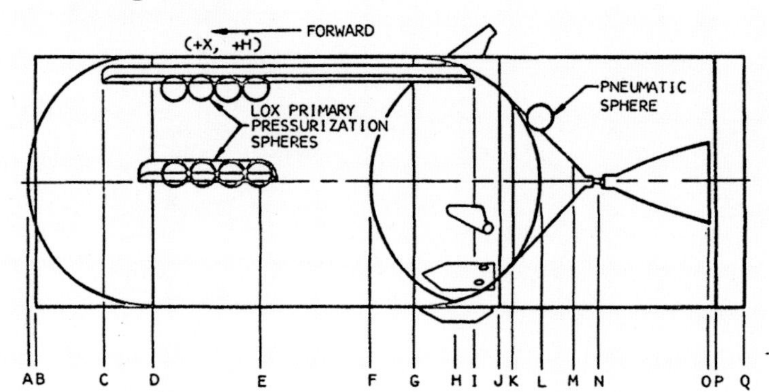

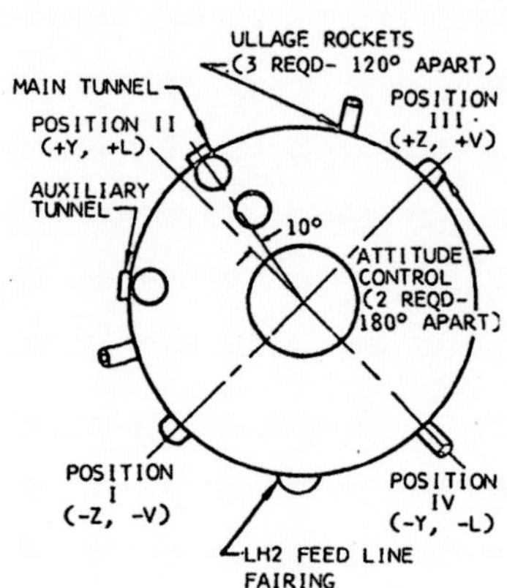

Right: Primary layout of the S-IVB-202 stage with principle systems and subsystems as appropriate. *MSFC*

	INSTALLATION	STATION		INSTALLATION	STATION
A.	FWD BULKHEAD (FWD END)	684.6		AUXILIARY PROPULSION MODULE (AFT END)	205.1
B.	FWD SKIRT (FWD END)	676.7		ULLAGE ROCKET FAIRING (AFT END)	204.6
C.	MAIN TUNNEL (FWD END)	607.7		LH2 FEED LINE FAIRING (AFT END)	202.7
D.	FWD SKIRT (AFT END)	554.7	J.	AFT SKIRT (AFT END)	200.6
	FWD BULKHEAD (AFT END)	553.0		INTERSTAGE (FWD END)	200.6
E.	COLD HELIUM SPHERES (8 REQD - 26.9" ℄ TO ℄)	454.0	K.	THRUST STRUCTURE/AFT BULKHEAD TANGENT POINT	186.9
F.	COMMON BULKHEAD (FWD END)	335.2		AMBIENT HELIUM SPHERES	158.6
	AFT BULKHEAD (FWD END)	287.8	L.	AFT BULKHEAD (AFT END)	156.3
G.	AFT SKIRT (FWD END)	286.1	M.	THRUST STRUCTURE SKIN (AFT END)	121.3
	LH2 FEED LINE FAIRING (FWD END)	286.1	N.	GIMBAL STATION	100.0
	AUXILIARY PROPULSION MODULE (FWD END)	285.5	O.	ENGINE NOZZLE (AFT END)	-16.0
	ULLAGE ROCKET (FWD END)	245.3	P.	INTERSTAGE (AFT END)	-23.9
H.	COMMON BULKHEAD (AFT END)	244.4		AERODYNAMIC FAIRING (FWD END)	-23.5
I.	MAIN TUNNEL (AFT END)	224.1	Q.	AERODYNAMIC FAIRING (AFT END)	-50.9

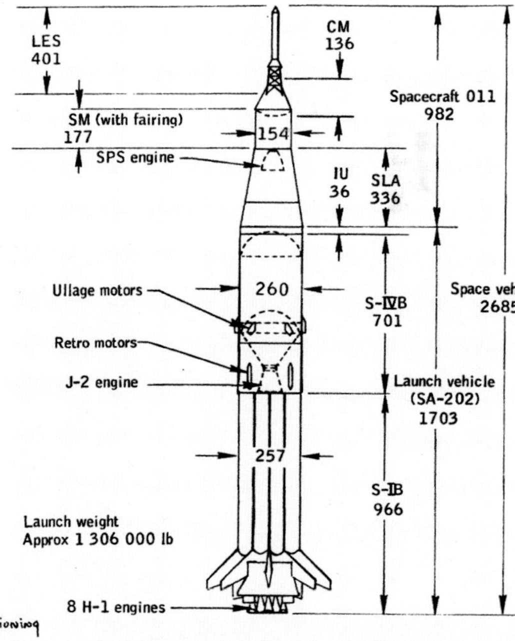

Above: The configuration of AS-202 was similar to AS-201 in all essential aspects. *MSFC*

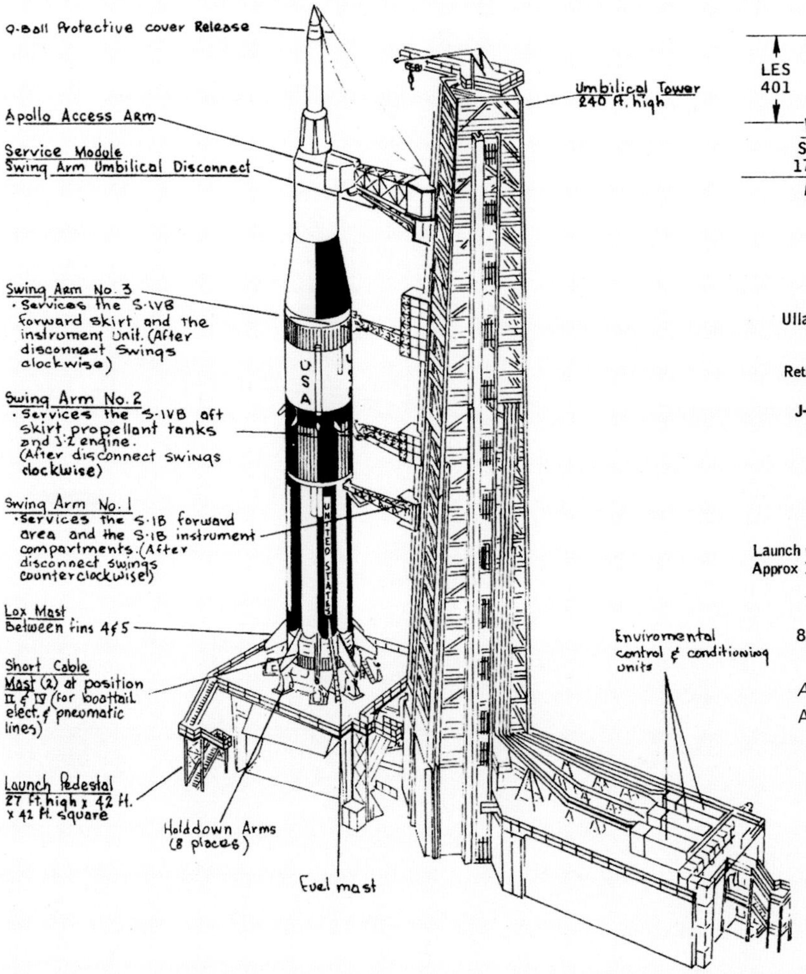

Left: Generally representative of the LC-34 support infrastructure, AS-202 is depicted with its three swing-arm assemblies and Apollo access arm. *MSFC*

AS-202 – launch details

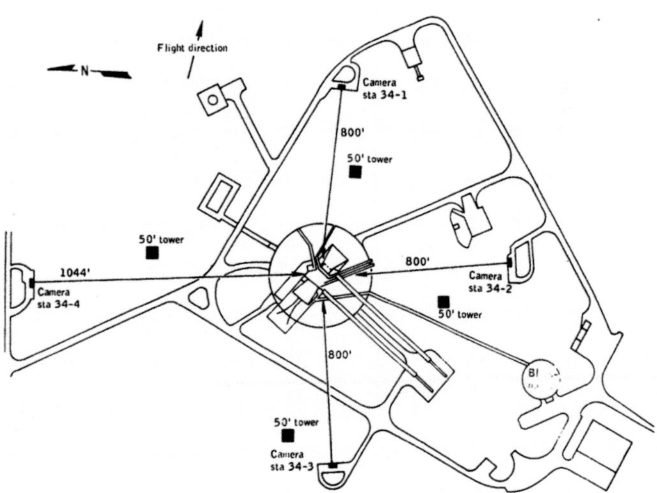

Above: With every aspect of a Saturn IB flight recorded from several positions, the site geometry for cameras recording the lift-off and ascent of AS-202 at LC-34 is shown here.

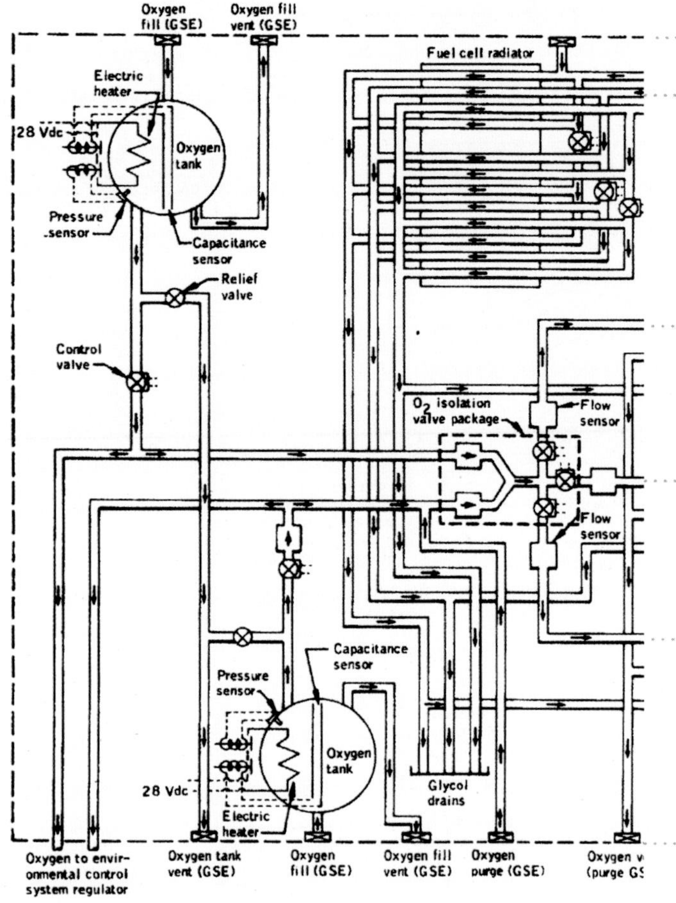

Right: A vital test before astronauts could fly the Block I Apollo spacecraft was demonstration of operation with critical life-support functions including the provision of electrical power through the fuel cells, as displayed here for the CSM-011 spacecraft.

Below: The ascent ground track for AS-202 with associated tracking stations. Event times are those predicted before flight. *MSFC*

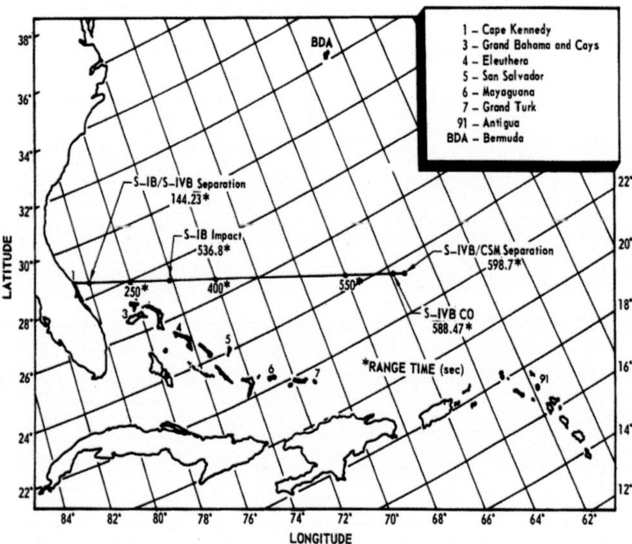

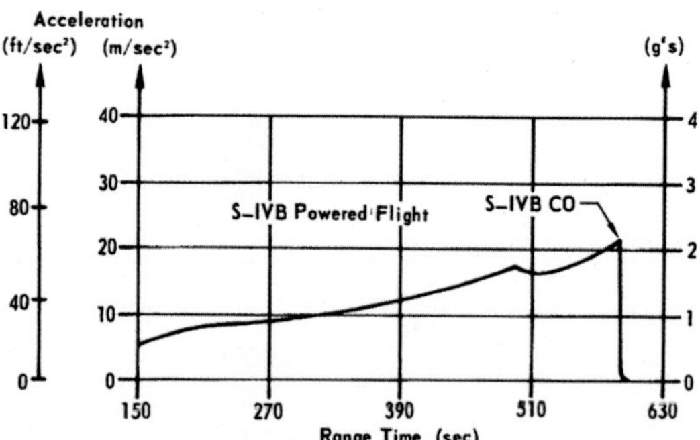

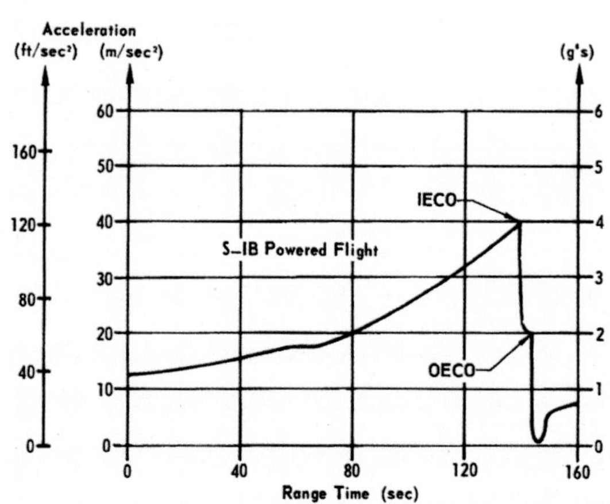

Right: In a typical ascent phase, AS-202 displays acceleration and g-force levels over time for the S-IB stage (bottom) and S-IVB stage. Note the maximum 4g acceleration on the first stage and 2g on the second stage. *Far right:* Earth-fixed velocity and elevation angle for a typical Saturn IB flight as with the trajectory flown by AS-202. *MSFC*

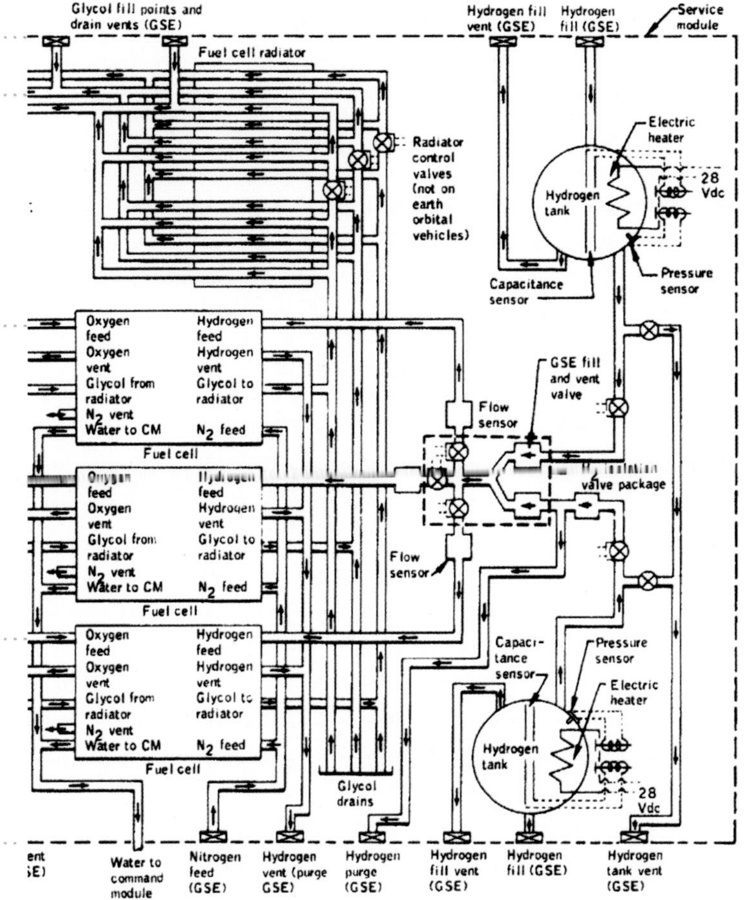

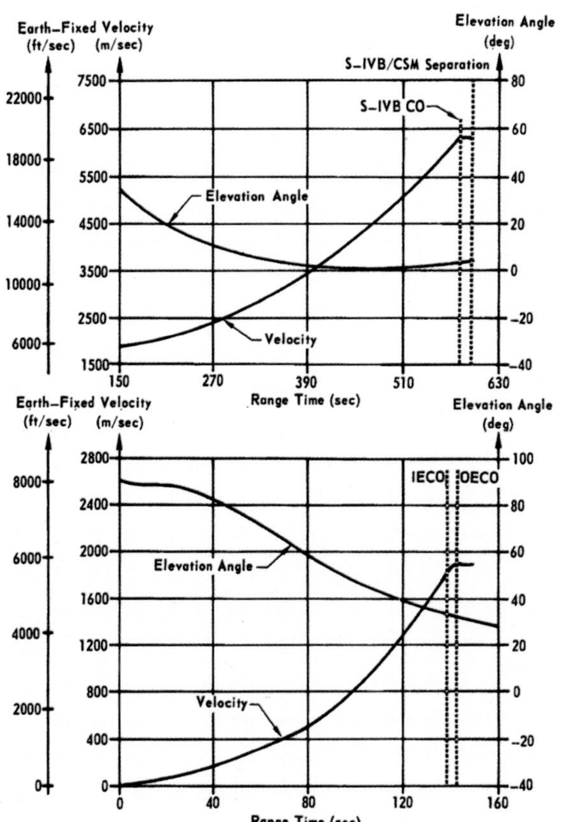

doors of the Spacecraft Lunar Module Adaptor (SLA) connecting the Apollo spacecraft to the top of the rocket and which were hinged to fold outward 35deg to expose the area where the LM would be on an operational flight. Eight mirrors would allow the camera to view all four panels, with four lights on each, to illuminate the area. Operating from lift-off, the camera would scan at commercial rates of 525-lines and 30 feet per second. Two 16-mm Milliken film cameras were mounted on the forward end of the S-IB stage to record staging and J-2 ignition, starting three seconds before separation. Mounted in waterproof containers, they would be ejected 25sec after staging and descend on paraballoons for recovery. Four Milliken cameras were also installed in the Command Module for filming key events, including re-entry. A total of 1,242 measurement points were assigned, 506 on the S-IB stage, 435 on the S-IVB and 301 on the Instrument Unit.

Although not a part of the launch vehicle and its operation, notable advances with the spacecraft flown on this mission included the new guidance and navigation system and the Service Module carried the first fuel cell electrical production system in an Apollo spacecraft. Only two of the three fuel cells were operative for the flight, due to an incorrectly wired on/off switch in the in-line heater. By this date, the fuel cell had proven to be a credible way to produce electrical energy at the required power levels and against the weight constraints that prevented the use of batteries. Moreover, tapping into an existing requirement for oxygen to maintain habitability and taking advantage of the water produced as a by-product was revolutionary.

Incorporating components already built for the cancelled Saturn I SA-112 mission, Chrysler began assembly of S-IB-2 in March 1964 followed by tank clustering in September, finalising build-up on 26 February 1965. Inspection revealed damaged thrust chambers on the inboard H-1 engines and they were returned to Rocketdyne for attention. Reinstalled by 27 April, preparations began for shipping it out of Michoud and across to Huntsville. Those plans changed when inspectors noted a large dent in the aspirator of engine H-7053 caused by collision with a vertical platform lift and this was replaced with H-7050, installed on 1 June. Finally, the stage was prepared for transport and shipped out of Michoud on 12 June, arriving at MSFC seven days later.

On 19 June, S-IB-2 was placed in Static Test Tower East on 21 with an initial test firing on 8 July, aborted after 3sec of a planned 35sec when the thrust-OK switch on engine H-7054 failed. The second attempt resulted in a successful run on 9 July lasting 35.3sec followed by a full duration run lasting 144.3sec on the 20th. No re-orificing was necessary, the engines having performed well and within specifications. However, several ripples were noted in tank F-3 and severe scorching was reported on the aspirator lips of engines H-7050 and H-7054 but no further action was required. Removed from the test stand on 29 July and transported back to Michoud on the barge *Palaemon*, S-IB-2 entered post-firing checkout where technicians discovered that gear case oil on engine H-4051 had leaked through the LOX pump cavity check valve. That engine was replaced with H-4046 but the work was not done until 65 tube assemblies had been removed, cleaned and reinstalled. The stage left for the Cape on 1 February 1966.

The cryogenic S-IVB-202 stage for this important qualification and test flight entered fabrication at Douglas' Santa Monica facility in February 1964 with installation of the J-2 engine on 28 April 1965. Checkout of the stage at Huntington Beach got under way two days later and absorbed some late hardware changes which had originally been slated for the post-checkout phase. In August the stage was conveyed to the SACTO facility where it was set up in the Beta III stand ready for checkout and firing. An initial attempt was foiled by incompatible batteries and the first firing took place on 2 November. After 0.41sec a problem with the combustion stability monitoring system caused premature shut-down and a second attempt was made on 9 November lasting 307sec, only partially successful due to a failing mass sensor for the liquid hydrogen. Repeated on 1 December, the full run lasted 463.8sec as planned and the stage was removed and taken for cleaning and checkout. Shipped out on 15 December, S-IVB-202 arrived at the Cape on 31 January 1966, the first of the major flight elements to check in.

The S-IB stage arrived on 7 February 1966, offloaded from the barge and taken to Hangar AF for inspection and the installation of three fins. The stage was erected on LC-34 on 4 March where the remaining fins were attached and the hardware joined by the S-IVB six days later after initial propulsion checks on the S-IB. The Instrument Unit was erected and mated to the S-IVB on 11 March. For the next six weeks power and propulsion checks were completed, with a partial LOX loading test on 30 April. Not before 2 July was CSM-011 and the SLA erected on the top of the vehicle, with RP-1 loading on 18 August and S-IVB APS propellant loading three days later. Propellants were loaded aboard the Service Propulsion System on 20 August.

After the usual pre-flight preparations, the launch countdown began at 22:30hr EST on 24 August, with lift-off scheduled for 11:30hrs on 25 August incorporating two scheduled holds of 30 min each. A series of short, unscheduled holds accumulated total unscheduled delays of 45min threatened a launch day wave-off due to the window closing at 13:00hr because of hurricane preparations at Antigua. But the weather was good at the Cape for countdown and launch, light winds from the southwest and good visibility with high, broken cumulus clouds, and some rain.

Lift-off was followed by the roll and pitch programmes going in at 10.3 and 10.7 sec respectively, followed by the end of roll at 16.3sec and pitch arrest at 2min 18sec, shortly before inner engine cut-off at 2min 19.6sec at an altitude of 54.4km (33.8mls) and the four outers at 2min 23.4sec, and 58.2km (36.2mls). Stage separation occurred at 2min 24.2sec with associated ullage/retrorocket firing and S-IVB ignition at 2min 25.6sec. The 3,906kg (8,611lb) Launch Escape System was jettisoned at 2min 48.9sec with J-2 cut-off at 9min 48.4sec and a terminal velocity of 6,800m/sec (22,311ft/sec, or 15,212mph) at an altitude of 217.23km (135mls) and a downrange distance of 1,557km 967.5mls). Residual propellants in the S-IVB totalled 2,551kg (5,624lb).

Left: AS-202 gets away from LC-34 on the afternoon of Thursday 25 August 1966 as the second test of a Command Module heat shield. *NASA-KSC*

Above: Successfully down but a little out of place, CM-011 bobs about in the Pacific Ocean after a skip re-entry to demonstrate a lunar return profile.

The total mass of the injected payload was 21,866kg (48,205lb) of which 22,482lb was tanked propellant including 494kg (1,090lb) reaction control system propellant for the four quads in the Service Module and the two loops in the Command Module. The SPS propellant in the Service Module was less than half that which would be carried aboard a fully loaded spacecraft due to the limited payload capacity of the Saturn IB. The total mass exceeded the orbital payload capability of AS-202 because it did not need to achieve orbital velocity in order for it to carry out the flight objective of this, a fully equipped Block I CSM short of crew-specific equipment in the Command Module which was regarded as the final qualification for manned flight on the next mission.

The Apollo spacecraft separated from the S-IVB at 9min 58.7sec. The first SPS burn began at 10min 9.7sec and lasted 3min 35.9sec, boosting it on its way to high altitude, following which the spacecraft was orientated apex-down for a little more than 33 minutes, a coast period in which an apogee of 1,142.6km (710mls) was achieved at 41min 14sec elapsed time. The spacecraft was re-orientated for the second SPS burn which began at 1hr 5min 56.1sec and lasted 1min 28sec. Two additional burns of 3sec each began at 1hr 7min 34.5sec and 1hr 7min 47.5sec, adding a further 1,890m/sec (6,200fps) to the velocity. A further orientation pointed the spacecraft for Command Module separation which occurred at 1hr 11min 4sec after which the CM turned to re-entry attitude and encountered the atmosphere at 1hr 12min 28sec at a velocity of 8,690m/sec (28,512fps) with a communications blackout period lasting 11min from 1hr 13min 36sec.

Flying a skip trajectory to terminal descent, all the programmed events occurred without failures, CM-011 achieving splashdown at 16.1deg N by 168.9deg E in the Pacific Ocean northeast of Wake Island, 1hr 33min 2.2sec after lift-off some 370km (230miles) from the recovery ship USS *Hornet*. Recovery occurred at a mission time of 10hr 2min. The shortfall in the range of the descent trajectory was in part due to marginal errors in the guidance system and in a slightly different L/D, 0.32 to 0.36 versus the design goal of 0.5.

AS-204/Apollo 5

Launch date:	22 January 1968
Lift-off time:	17:48:08 EST (22:48:09 UTC)
Launch pad:	LC-37B
Launch azimuth:	72deg
Length:	55.16m (181ft)
S/L thrust:	7,116.8kN (1,600,000lb)
Dry mass:	71,122kg (159,000lb)
Propellant mass:	518,630kg (1,143,384lb)
H-1 engines:	H-4058; H-4060; H-4061; H-4062; H-7062; H-7063; H-7064; H-7065
J-2 engines	J-2025
Ignition mass:	591,065kg (1,303,076lb)
Lift-off mass:	584,394kg (1,288,368lb)

Shortly after analysis of the AS-202 mission had verified the critical system for clearing the Apollo CSM design for manned flight, in October 1966 the AS-204 mission was set for launch to take place on 5 December. It was to have been a 14-day flight in which the crew of astronauts Virgil I. 'Gus' Grissom, Edward H. 'Ed' White II and Roger B. Chaffee were to conduct the first manned Apollo mission in Earth orbit and test spacecraft systems including the critical Service Propulsion System. Delays to Apollo CSM hardware pushed that back first to 16 January 1967 and then to 21 February, for a launch window opening at 11:00hr and closing at 15:30hr EST. Planned for an open-ended duration of up to 14 days, involving eight tests with the Service Propulsion System, the flight of AS-204 was to have launched from LC-34 and qualify all systems for more advanced rendezvous and docking operations with a Lunar Module comprising dual flight AS-205/AS-208. The plan was then to move on to Block II flights on Saturn V.

Fabrication of S-IB-4 started in October 1964 with all the changes and modifications imported to S-IB-3 applied to this stage too. Uniquely incorporated to improve safety for ground personnel, a drag-in cable door was introduced to the A-IB/S-IVB interstage, which prevented ground cables having to be routed around the existing access door. One of the two fibreglass gaseous nitrogen spheres was taken out and this decreased the volume, an opportunity provided by removal of a mass spectrometer and sensor. The flight control system's potentiometer in the servo actuator printed circuit was replaced with a standard insulated wire, which would prevent signs of corrosion noted on earlier vehicles, and the actuators were painted to reduce the possibility of stress corrosion. The changes to the position and orientation of the retrorockets noted for S-IB-3 were also applied to this stage and in addition insulation was added and inhibitor coating applied after the final trim to the nozzle end with sealant used to fill the space between the adaptor and the grain.

Changes to the H-1 engines included a shortening of the aspirator shells by decreasing the extension past the chamber into the exhaust stream by 5.52cm (2.25in) in an attempt to prevent buckling in the lower part. The change made in the LOX seal on S-IB-3 was also applied to this stage. Changes were also made to the Solid Propellant Gas Generator (SPGG) in the ignition sensing system, which allowed for engine shutdown in the event of a premature ignition or if a valve had fired prior to the ignition command going in. Deleted were the recoverable cameras and their associated equipment and some attendant changes were made to increase the reliability of the engine cut-off circuits. Again to increase reliability, redundant electrical circuits were added to the separation systems and to the power switch selectors.

In Michoud, Chrysler finished clustering the propellant tanks on 9 August 1965 with build-up completed on 6 October, all engines installed by 3 August and checkout completed on 9 November. S-IB-4 was shipped out on the barge *Palaemon* on 7 December, arriving at MSFC six days later where it was unloaded the next day and transported to Static Test Tower East for the standard firing tests. Preceded by simulated flight pressure checks and propellant loading tests, the first firing took place on 17 January 1966 for 35.2sec, followed on 21 January with a second firing of 147.1sec, the inboard engines shutting down 3.4sec earlier as planned. Performance was good, preventing the need for any re-orificing and the stage was taken down on 28 January and loaded aboard the barge *Palaemon* on which it left for Michoud the following day where post-test checks would be conducted. More than six months later it left MAF aboard *Promise*, arriving at the Cape on 14 August.

Meanwhile, Douglas had begun sub-assembles of S-IVB-204 during the second half of 1964 and, after the usual build-up sequence, the J-2 engine was installed on 30 September 1965. But there had been problems with the stage build-up, due in part to incorrect sizing for critical subassemblies and welding deficiencies. Vertical inspection of the completed stage was conducted on 20 December with post-manufacturing checkout the following day. Delivered to SACTO two weeks later, the stage went in to the Beta III stand on 15 January 1966 where it performed a full duration firing test of 451.2sec on 18 March before it was removed from the stand eight days later. Checkouts were completed before a simulated flight test after which the stage was placed in storage at SACTO. Held back due to postponements in the anticipated launch date for AS-204, S-IVB-204 left for the Cape aboard the Super Guppy on 6 August, arriving next day.

Below: The Apollo 15 payload assembly is lifted to the top of the AS-204 Instrument Unit at LC-37B. *NASA-KSC*

Delays to spacecraft manufacturing, test and verification had plagued the Apollo programme throughout 1965 and 1966. After the major structural and administrative improvements at NASA in the second half of 1963, management processes had forced change at a time when hardware flow was in progress. But the contractors for both the Apollo spacecraft and the Lunar Module had some responsibility for not fully accommodating challenging shifts in requirement from the customer – NASA. Much of the delay was attributable to changes to specifications and engineering items added as the space agency climbed the learning curve. Much too was still being learned about how to build and operate highly complex machines in space, all of which had to have the highest levels of safety and reliability. And a lot of the problem was in the prototyping nature of the hardware, where improvements and modifications were rife on the production line.

Considerable differences existed between S-IVB-204 and previous stages of this type, embracing the structure, the J-2 engine system, propellant system, tank vent/pressurisation system, pneumatic control system, hydraulic system, the APS, electrical system, emergency detection system and the range safety system. In summary, these included the same external insulation as that applied to S-IVB-203 but with an increase in the area of the forward skirt vent and stiffeners added to the auxiliary tunnel cover as applied to S-IVB-203. In addition, stringers and joints were redesigned on the aft skirt to increase their load-carrying capacity by 20 per cent. The skin thickness of the fuel tank was also reduced due to a lower ullage pressure limit reduced from 289.6 to 268.9kPa (42 to 39lb/in^2). One change introduced after analysis of the performance of the J-2 on the AS-504 (Apollo 4) Saturn V flight on 9 November 1967 was an overboard drain line from the engine LOX pump seal cavity to prevent venting into the closed area of the interstage.

A wide range of minor changes were made to the engine and supplementary and ancillary systems, significant being the relocation and redesign of the thrust chamber struts, stiffening of the gas generator housing, stronger helium tank cover bolts to prevent cover plate deflection under maximum relief valve pressure and a re-routed Augmented Spark Igniter line to prevent interference with the restraining arm. The reference mixture for propellant utilisation was adjusted to a baseline of 4.7:1 and the hydrogen feed ducts were reworked with chromate primer to inhibit corrosion. Equally significant was a rework of the propellants tanks relief valves for better performance, operability and reliability. The pneumatic control system got a bigger ambient helium sphere providing 0.127m^3 (4.5ft^3) instead of the 0.0149m^3 (0.525ft^3) of the previous tanks. Numerous changes to the Instrument Unit for AS-204 were made to the structure, the guidance system, the flight control system, the thermal conditioning system and the gas bearing system.

AS-205 – launch details

Right: When planned as the first manned Apollo flight, typical of all Saturn flights AS-204 had a series of abort landing locations as shown here for the mission which was never flown but equally applicable to AS-205 (Apollo 7). *MSFC*

Below: The dimensions and configuration of CSM-012 configured for AS-204 and quoted to the second decimal place. *MSFC*
Below right: After the fire which destroyed CSM-012, all Block I CSM flights were dropped in favour of Block II (which would fly on AS-205). Instead of carrying what would have been numbered Apollo 1, AS-204 was moved to LC-37B and given the task of flying the first Lunar Module (LM-1) on an unmanned test flight in the configuration shown here as Apollo 5 (Apollos 4 and 6 being the two unmanned Saturn V flights. *MSFC*

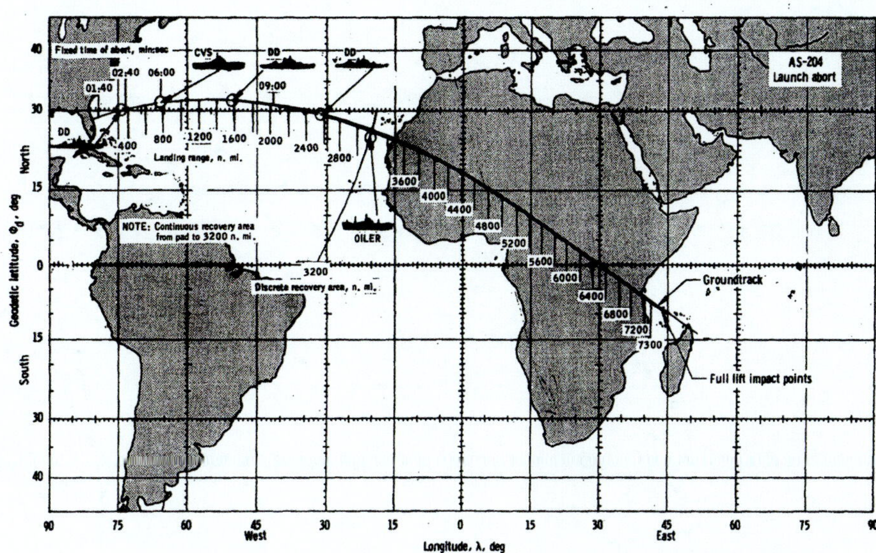

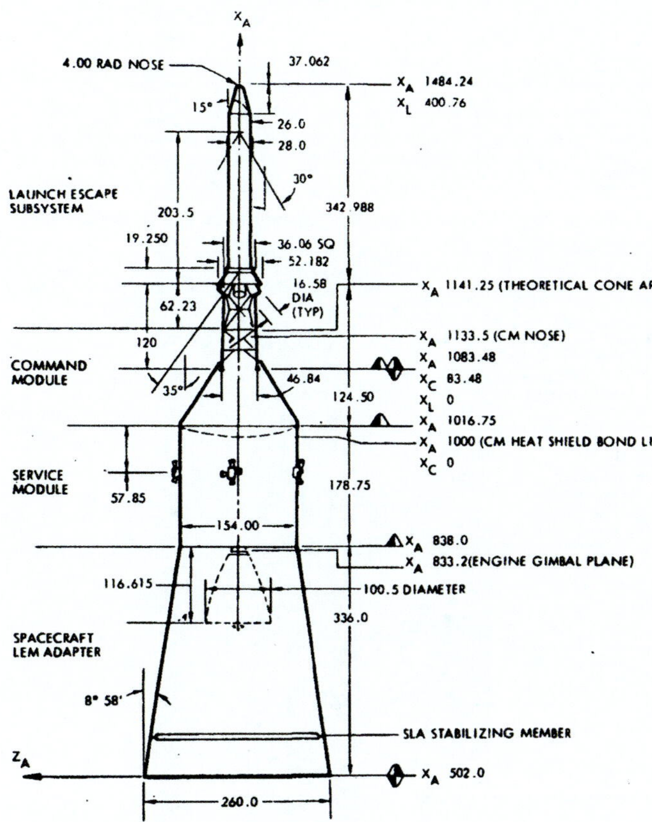

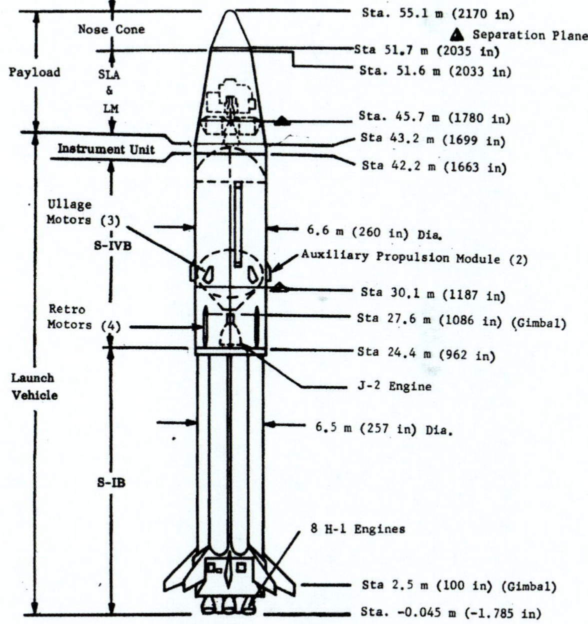

Notes: Separation Plane

Field Splice

X_A Apollo Spacecraft (CSM) Coordinate System

X_C CM Coordinate System

X_L LES Coordinate System

All linear dimensions are in inches.

Left: LM-1 and the SLA are brought together at the Kennedy Space Center before moving the payload to LC-37B. *MSFC*

AS-205 – launch details (contd.)

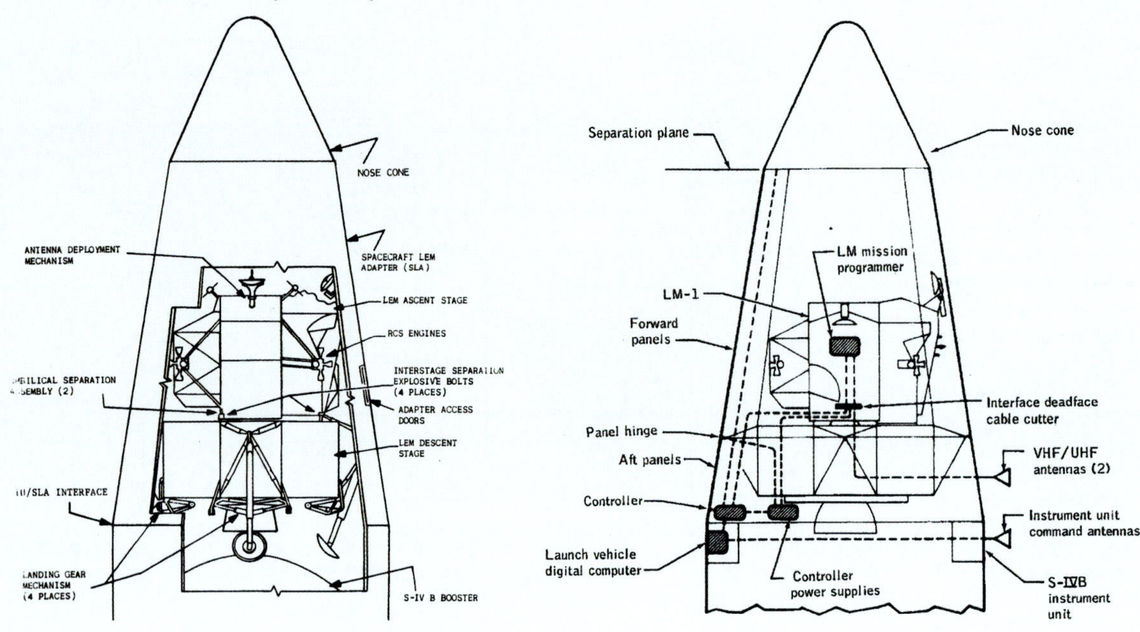

Above: Initially, it had been assumed that LM-1 would fly with its four landing legs folded for launch as shown here. However, due to problems involving LM-1 and weight considerations involving centre of mass, the landing legs were removed for the Apollo 5 mission as shown in its final configuration. *MSFC*

Below and left: The LC-37B pad support infrastructure for the unmanned AS-204 mission with umbilical swing-arm assemblies. *MSFC/NASA*

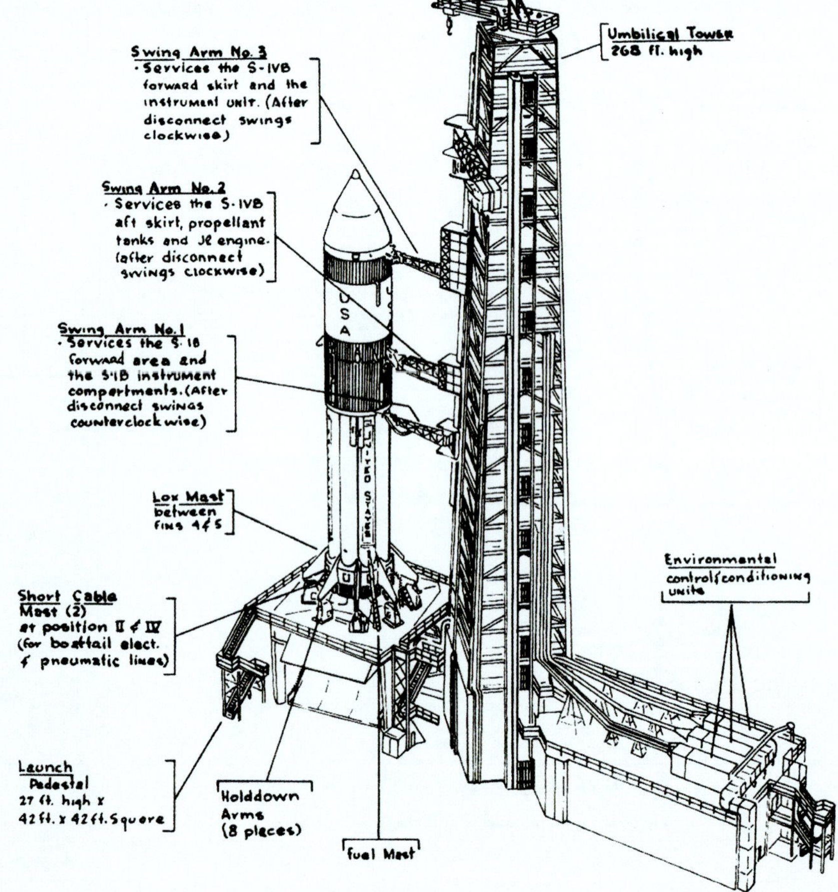

The guidance, navigation and control system for AS-204 was the same as that flown on previous Saturn IB flights except for minor components but it was the first space vehicle to carry the Apollo Standard Coordinate System. The stable platform was the ST-124M-3 iteration. It was also the first to have the capability for live navigation updates, although elements had been test flown on previous Saturn launch vehicles. AS-204 carried a total of 1,226 measurements with 396 on the S-IB, 515 on the S-IVB and 315 on the Instrument Unit.

As noted above, S-IVB-204 was the first major piece of hardware for this mission delivered to the Cape. It arrived by air on 7 August 1966 followed by the S-IB-4 stage on the 14 and the Instrument Unit on the 16 August. The Apollo CSM-012 began fabrication at NAA's Downey, California, plant in August 1964, with the basic structure completed thirteen months later. It was delivered to the Cape on 26 August 1966 and an intensive series of tests and checks ensued. It was mated to the AS-204 launch vehicle at LC-34 on 6 January 1967 in preparation for a launch no earlier than 21 February. After delivery, numerous technical problems had pushed the planned launch date further out than expected and these persisted during the several checks conducted with the stack on the launch pad, to the annoyance of the crew and the frustration of ground personnel.

During an intensive series of tests in a highly pressurised, pure oxygen environment on the evening of 27 January during a plugs-out test in which electrical power on the stack was derived from battery supplies simulating inactive fuel cells, a fire broke out in the Command Module. Within 14sec of the fire starting the pressure in the Command Module reached 200kPa (29lb/in^2), splitting the base of the heat shield and causing flames to rush out on to the platform where technicians were attending the test. All three crewmembers succumbed to smoke inhalation. Immediate fears that the flames could ignite solid propellants in the LES were allayed when the fire was rapidly extinguished by the dense cloud of carbon dioxide produced by the conflagration. But the launch vehicle was unaffected by the fire, the Instrument Unit above the S-IVB stage being 13m (42.75ft) below the line where the Command Module was attached to the Service Module.

After a re-analysis of mission plans and the flight sequence after manned launches resumed, following an exhaustive internal review stipulating a rigorous series of modifications to material in, and procedures with, the Apollo spacecraft, it was decided to fly the first Lunar Module (LM-1) on the undamaged AS-204 launch vehicle. This mission and its flight objectives had originally been planned for AS-206. Because LC-34 could not support flights of the unmanned Lunar Module, the two stages of the Saturn IB were de-erected on 6 April 1967 and transported to LC-37B where the S-IB stage was installed the following day and joined by the S-IVB on 10 April. There was a long wait of more than nine months for the arrival of the payload, LM-1. In 1966 it had been expected that LM-1 would be delivered to the Cape on 16 November of that year but it did not arrive before 27 June 1967, the delay caused by development problems with the spacecraft and fallout from changes, checks and tests imposed by the review board looking into the fire.

When it became apparent that AS-204 would be sitting on LC-37B for some time, MSFC worked with Cape engineers to study the effect of extended periods in a stacked configuration. A periodic set of inspections were carried out to monitor the effects of the salty atmosphere. No items were replaced and there was no corrosion of critical elements, the launch vehicle itself being continuously purged with nitrogen to protect the engines and other vulnerable items. Not so lucky were the engineers working to get LM-1 ready, where after mating and de-mating the ascent and descent stages of the spacecraft several times, and returning several items to the contractor Grumman, they had to repair leaks in the ascent engine and ship several items back for further work. Repeatedly taken apart and reassembled, the two stages were eventually cleared of technical problems.

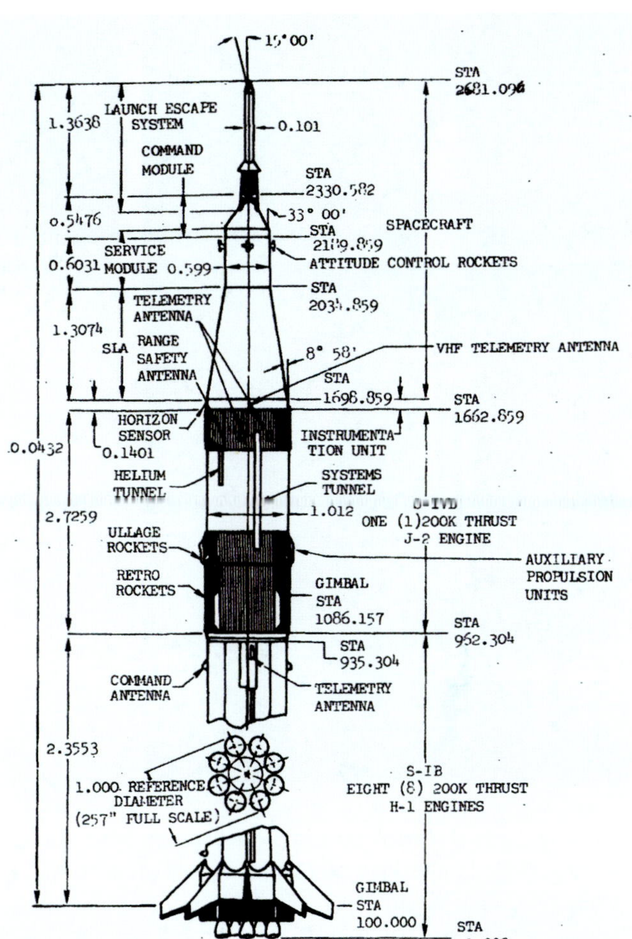

Dimensions for AS-204 – set up to carry CSM-012 and the first manned Block I Apollo flight in February 1967 before the fire of 27 January 1967 that took the lives of Grissom, White and Chaffee. *MSFC*

In the new configuration, the LM would be aerodynamically protected by an enclosure consisting of the four SLA panels and a bespoke nose cone consisting of a 25° semi-monocoque cone-shaped structure. This had an overall length of 343cm (135in) and a base diameter of 391cm (154in), constructed with ring frames and skin stringers. Separation would be effected by 16 springs placed symmetrically around the base and the forward SLA panels. A mild detonating fuse would be ignited by two detonators 180deg apart.

LM-1 was mated to the launch vehicle on 19 November

Above left: AS-204 stands on LC-37B, to which it had been moved to facilitate servicing of LM-1 after its mission changed from carrying the first manned Apollo flight. *NASA-KSC*

Above right: Eleven months after it was due to carry the first manned Apollo crew into orbit, AS-204 lifts off from LC-37B. *NASA-KSC*

1967 and the CDDT commenced on 16 January 1968, successfully completed three days later with a formal move to start the countdown. That began at 10:00hr EST on 21 January with a planned 6-hr hold starting at 04:30hr the following morning. After the count picked up, a problem with the cooling system caused a hold which required access to the pad and that halted cryogenic loading before further technical problems introduced another hold, the count resuming at 15:18hr, total time delays being 3hr 48min.

Following a successful lift-off, event times were close to those for the preceding flight, with Mach 1 at 59.5sec, Max-Q at 1min 11.5sec and inner engine cut-off at 2min 18.9sec followed by the four outboard engines at 2min 22.2sec. S-IB-4 stage thrust was 1.3 per cent higher than predicted with engine cut-off times very close to the planned values. S-IVB separation occurred 1.3sec later at an altitude of 64.1km (39.8mls) with ignition of the J-2 at 2min 25sec. The S-IVB stage had a 1.38 per cent higher thrust level than calculated. S-IB impact came at 9min 42.7sec, 497km (309mls) downrange. S-IVB cut-off occurred at 9min 53.3sec with a total orbited mass of 31,769kg (70,039lb), of which 14,302kg (31,530lb) was LM-1 and the S-IVB stage 13,103kg (28,887lb), the balance being the Instrument Unit, SLA and the 484kg (1,067lb) nose cone. At this point the orbit was 221.5km (137.6mls) by 157.6km (97.9mls). The S-IVB had residual propellants of 2,052kg (4,523lb) on orbit.

Separation of LM-1 was made by the LM RCS engines and this occurred at 53min 55.2sec. The physical disturbances produced little attitude instability either in the S-IVB or the LM. On the third orbit the LM Descent Engine was fired for a planned duration of 48sec but the guidance shut it down after 4sec due to slow thrust build-up. This triggered an alternate mission which began at 6hr 10min 41.7sec with a Descent Engine burn lasting 33sec, a pause for 32sec and a further burn of 28sec. The Ascent Propulsion System was then fired simultaneously with separation of the Ascent Stage from the Descent Stage at 6hr 12min 14.7sec, simulating an abort while descending to the lunar surface on a future Moon mission. In this, the Descent Engine shut down and staging occurred as planned and the Ascent Engine fired for 60sec.

The second Ascent Engine burn started at 7hr 44min 12.7sec and lasted 5min 50sec as the propellant was burned to depletion, a manoeuvre which began to send the Ascent Stage back down through the atmosphere where contact was lost and impact occurred 644km (400mls) west of the coast of Central America on 24 January. The Descent Stage came down on 12 February falling into the Pacific Ocean southwest of Guam.

As noted earlier, and at the top of this chapter, mission objectives and designations were in a great state of flux made more disruptive by the Apollo fire. By the time AS-204 flew, the first manned flight of Apollo (AS-205) had been designated to fly after the three unmanned Saturn V flights, which was to be followed by the second unmanned test of the Lunar Module (LM-2). But in consideration of the delays in the Lunar Module programme and the encroaching target date for getting on the lunar surface, an intensive analysis of the Apollo 5 mission persuaded managers to forgo the flight of LM-2. They would go straight to LM-3 flown on the third Saturn V flight with an Apollo spacecraft and the McDivitt crew, thus eliminating AS-206.

AS-205/Apollo 7

Launch date: 11 October 1968
Lift-off time: 11:02:45 EDT (15:02:45UTC)
Launch pad: LC-34
Launch azimuth: 72deg
Length: 68.1m (223.33ft)
S/L thrust: 7,116.8kN (1,600,000lb)
Dry mass: 75,070kg (165,500lb)
Propellant mass: 509,594kg (1,123,655lb)
H-1 engines: H-4063; H-4064; H-4065; H-4066; H-7066; H-7067; H-7068; H-7069
J-2 engines J-2033
Ignition mass: 592,670kg (1,306,678lb)
Lift-off mass: 586,392kg (1,292,994lb)

The flight objective for this launch vehicle was to place in orbit the first manned flight of an Apollo spacecraft. Almost all the objectives focused on the payload, CSM-101, much modified and in a far better space-worthy condition than the Block I type in which the three astronauts died more than 20 months earlier. After Mercury-Redstone, Mercury Atlas and Gemini-Titan, Saturn IB was the latest space launch vehicle to demonstrate man-rating certification. Its crew of Schirra, Eisele and Cunningham would determine, through their shakedown of the Block II CSM, whether NASA should proceed to a lunar orbit mission two months later on the first manned Saturn V.

For this flight, however, the weight of the CSM had to be kept within the payload capability of the Saturn IB. A loaded Apollo CSM carrying full tanks for the Service Propulsion System would weigh around 28,800kg (63,500lb), more than 50 per cent greater than the lift capability of a Saturn IB. For the Apollo 7 Earth-orbit flight, CSM-101 would carry 4,416kg (9,737lb) of SPS propellant, a little less than 25 per cent of that required for a Moon mission, bringing the total mass of the CSM to 16,519kg (36,419lb) at insertion.

Several vehicle changes were made too on this, the second operational flight of the Saturn IB. The amount of telemetry and measurement instrumentation and sensor packages were reduced considerably, saving weight and appreciably increasing payload capability. AS-205 would carry 720 vehicle measurements, with 260 on the first stage, 260 on the second stage and 200 on the Instrument Unit, a total which was more than 500 fewer than on the preceding Saturn IB, in effect providing only those measurements required of an operational launch vehicle. No tape recorders were carried and telemetry systems were reduced by two in the first stage, four in the second stage and two in the Instrument Unit.

The single J-2 engine on the second stage was modified to reduce the possibility of a premature cut-off, which had happened during the second flight of Saturn V (Apollo 6) on 4 April 1968 when two J-2 engines in the S-II stage shut down. On that flight, technically considered a failure because it was unable to match some primary objectives, an Augmented Spark Igniter (ASI) line had ruptured due to vibration, causing a second J-2 engine to shut down due to a faulty crossed-wire installation. The ASI on the AS-205 S-IVB J-2 had new, modified propellant lines. An addition for this mission was the propellant dump scheduled for around 1hr 34min to increase the safety of the stage for a possible rendezvous by the Apollo spacecraft.

Configuration changes to the AS-205 vehicle as a result of previous flights included a new load distribution structure in the smaller LOX tanks, arising directly from the AS-203 flight. A new LOX seal cavity drain line was added to the H-1 engines to allow fitting of a 3-element temperature probe. A second source auxiliary pump was also added to pressurise the hydraulic system for pre-flight test and readiness, sufficient for ground servicing only and inactive during flight. Removed from the S-IB-5 stage were continuous liquid level probes together with adapters which were deleted from future stages also. On 21 February 1968 during a stage test on S-IB-11, engine H-4095 exploded, caused it was determined by, a leaking LOX seal. S-IB-5 incorporated a temperature interlock on the drain seal to initiate shut-down on first indication of a sensitive temperature rise should a similar event occur on a flight stage. S-IB-11 never flew as there was no mission assigned.

Changes to S-IVB-205 incorporated all those listed previously for S-IVB-204 but with some subtle amendments to incorporate lessons learned from that flight. Results from the S-IVB-502 stage flown on the second Saturn V mission (AS-502) on 4 April 1968 had a modified Augmented Spark Igniter

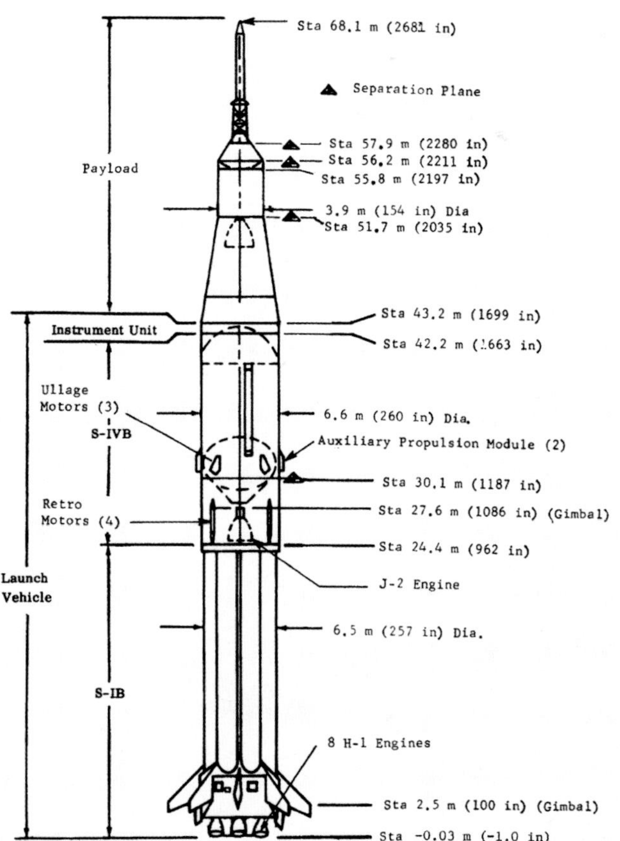

Right: The fifteenth flight of a Saturn rocket and the first manned Apollo mission, AS-205 was visually similar to AS-201 and AS-202 but had subtle differences brought on by the unique mission requirements and the high level of risk mitigation throughout the programme. *MSFC*

AS-205 – launch details

Below: Communication with the spacecraft during ascent and in flight would go through VHF and S-band antennas on the Command Module and the Service Module.

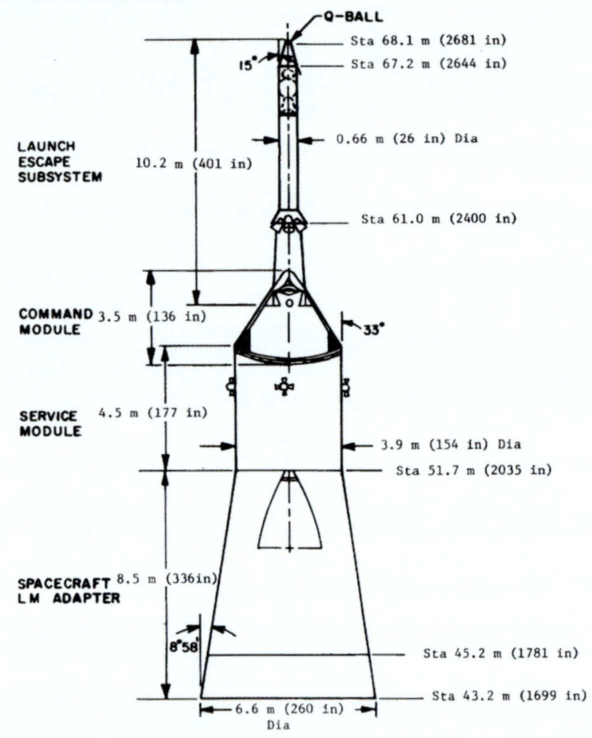

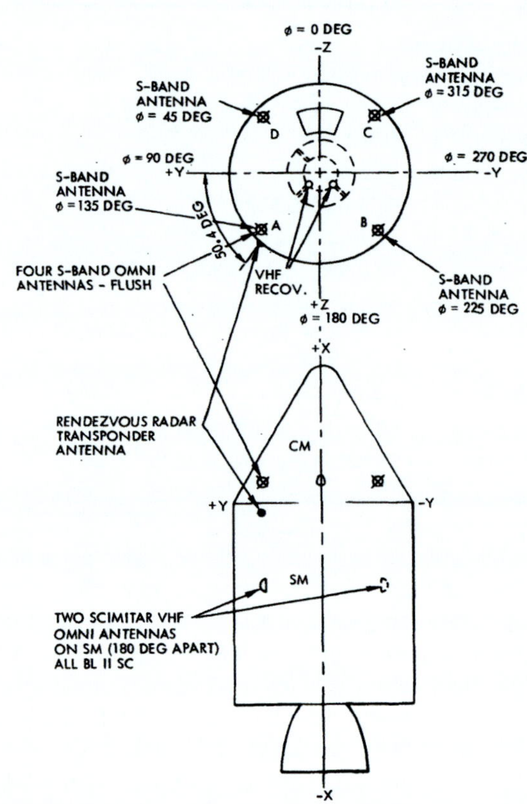

Above: The CSM-101 payload for AS-205 was the first Block II spacecraft flown, essentially capable of a full lunar mission profile but without the specialist equipment. *MSFC*

Left: A comparative view of CSM-101 in the Manned Spacecraft Operations Building at NASA Kennedy Space Center. *NASA-KSC*

feed system installed to prevent anomalies by replacing the flexible lines with rigid lines, single welds with double welds, reducing the total number of welds (and thereby reducing weak points) and relocating the orifice in the engine start tank liquid refill line downstream of the check valve. The refill orifice was replaced with a blank orifice to prevent recharging while the engine was operating and a redundant emergency dump valve was added for improved safety. Several changes and adjustments were made in the propellant system, in the LOX pump, the orbital safing kit and the propellant utilisation system.

Changes to the Instrument Unit were incremental on the results of preceding flights and modifications were made to the ST-124M platform, with the most notable for this flight being modifications to the flight control system so that astronauts could take control of the vehicle during S-IVB thrust and coast modes. When the crew opted to take over manual operations, limiters were switched allowing the astronaut to programme a maximum attitude error of 2.5deg in pitch and yaw and 3.5deg in roll. The DC amplifier gain change would correct for different scale factors between the launch vehicle and the spacecraft.

Assembly of S-IB-5 began in January 1965 with all engines installed by 27 October before it was shipped on the barge *Palaemon* from Michoud to Huntsville, where it arrived on 27 February 1966. Taken to Static Test Tower East,

Above: S-IB-5 hoisted into position (left) at LC-34, the last time this pad would be used for a Saturn launch.

Above right: Locating CSM-101 and the SLA onto the top of the AS-205 stack. *NASA-KSC*

the first firing occurred on 23 March for 35sec and conducted a full duration run eight days later, lasting 144.6sec. Due to the slow pace of flight schedules, the stage was placed in storage at Michoud on 10 August 1966 and covered with light protecting plastic. A complete inspection and refurbishment began on 31 August 1967 which involved removing all the engines and checking on any turbopump lube oil seal leakage, which was found to be absent from all engines. There was some concern too over the correct use of the appropriate oil but inspection revealed that nothing was amiss. The stage left Michoud for the Cape on 25 March.

Douglas began assembly of S-IVB-205 in March 1965 and the engine arrived on 16 August prior to inspection and checkout with installation on 11 January 1966. Arriving at SACTO on 14 April, it was installed in the Beta III stand the next day for pre-test checkout and a 437.5sec firing was completed on 2 June. Taken down from the stand on 5th July, the stage was placed in storage for the same reasons that S-IB-5 had been held at Michoud, before departing by air for the Cape on 6 April 1968.

The first launch vehicle hardware to arrive at KSC was the S-IB stage which was delivered by barge on 28 March 1968, erected on LC-34 on 16 April and followed on the 18 by the S-IVB and a boilerplate CSM, BP-30. This incorporated the boilerplate Command Module previously identified as BP-18, which had been delivered to the Cape on 6 January 1968. The purpose of the boilerplate was to rehearse in advance several procedures involving the flight spacecraft. Also, procedures such as propellant loading for the propulsion systems on the flight Service Module that could be run through on the pad to save time. CSM-101 would not arrive before May so there was much that could be done. Unfortunately, one event would add further delay to the first manned flight.

On 21 April a test was run utilising BP-30 and its Service Module to check hypergolic flow of the oxidiser, nitrogen tetroxide ($N_2 0_4$), and the fuel, a blended hydrazine. Hooking the boilerplate mock-up to the ground supply system, and following an established safety procedure involving the use of protective suits, technicians began to apply a vacuum to the simulated Command Module tanks. These required a vacuum to remove any void between the flexible bellows and the rigid tank wall. After the Service Module tanks had been filled, the Command Module RCS tanks received the next attention but under the mistaken impression that one of the oxidiser tanks had been emptied from a previous test, a technician began to apply a vacuum which pulled $N_2 0_4$ into the pump's oil sump which then released the toxic liquid in a fountain spray.

Standard procedures were followed but this was an unusual event and in dowsing the oxidiser the technicians discharged some 1,514lit (400USgall) of water through a hose which in mixing with the oxidiser produced diluted nitric acid running down the inside of the void occupied by the SLA, across and into the electronics of the Instrument Unit and down to the top of the S-IVB and around the hemispherical end dome of the liquid hydrogen tank. Although this was a procedure associated with the spacecraft and not the launch vehicle, it significantly affected the Saturn IB, causing extensive damage which set back the launch date several weeks.

The boilerplate and Instrument Unit had to be removed and a lengthy and extensive clean-up on the various modules in the IU which were contaminated. Wiring inside the IU had to be removed and cleaned, several cables and external antennas had to be replaced and five electrical connectors had to be rebuilt on site. Engineers removed the forward skirt of the S-IVB to allow internal access for cleaning up the stage and one of the S-IB fins had to be replaced as a thorough inspection of the entire stack looked for potential signs of corrosion from the diluted nitric acid. The supply of fuel and oxidiser had been from two trailers 61m (200ft) at the base of the launcher and manually controlled via control panels connected nearby to electrical power cables and related equipment. Procedures were fully re-evaluated and lessons learned from the accident, which had threatened to seriously delay the launch of AS-205.

After installation of the IU, the SLA and CSM-101 on 9 August, integrated systems under way six days later and the CDDT was held 12-16 September. Before all this, NASA set a launch date of 20 September for AS-205/Apollo 7 but that began to drift into the following month. The Flight Readiness Test began on 25 September and went through to T-0 two days later with RP-1 tanking under way on 4 October. The countdown for Apollo 7 embraced a wide swathe of new activity associated with the complexity of the first manned spacecraft for a Saturn IB and brought more media attention than any previous launcher in the Saturn I/IB series for the obvious reason that it was the first crewed Apollo mission and, following on from the fire, attracted considerable public attention.

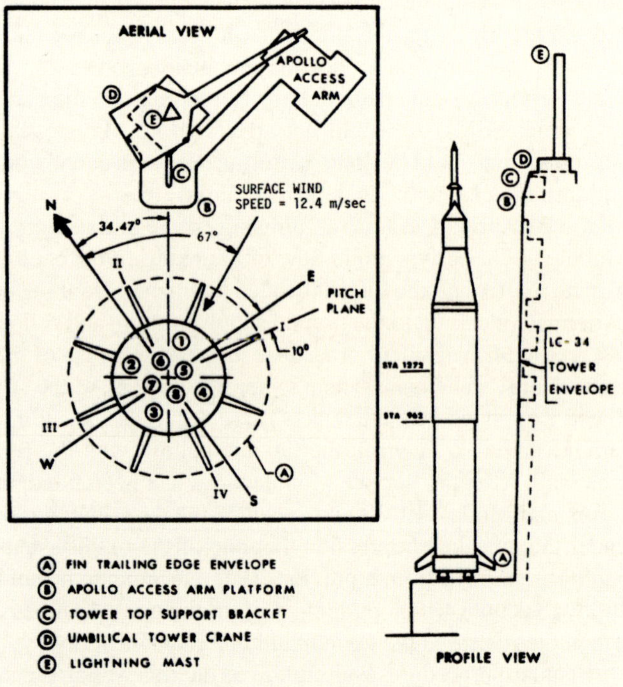

A record number of news people checked in at Cape Canaveral for this flight and a surge in demand for interviews accompanied the electronic media as they followed events unfolding toward this first manned Apollo mission. Greater interest still arrived with the rollout of the third Saturn V (AS-503) to LC-39A for Apollo 8, two days before the launch of AS-205 and Apollo 7. It had already been decided that if Apollo 7 went well the next mission, the first manned flight of a Saturn V, would carry Bormann, Lovell and Anders to the Moon.

Four abort modes were available to the Apollo 7 crew for use in contingencies where there was a high probability of catastrophe due to a single or uncontained set of failures. Mode I was effective from the moment of engine ignition on the first stage to shortly after ignition of the second stage, covering the first 2min 43sec of flight where the LES would lift the Command Module away from the stack and carry the capsule and its crew to a splashdown up to 741km (460mls) away. Under normal flight, the LES would be jettisoned and Mode II would take over up to 9min 33sec shortly before orbital insertion. In this, the entire Apollo spacecraft would separate from the SLA followed by the Command Module Reaction Control System (RCS) thrusters firing for 20sec to propel it away from the second stage and IU. After which the CM would separate and conduct a conventional re-entry and recovery operation. Mode III would be identical to the Mode II but involve a retrograde burn by the SPS followed by conventional recovery, in both modal cases splashdown occurring 5,930km (3,680mls) downrange. Mode IV would begin at 9min 27sec, lift off from the stack and use the SPS to propel the spacecraft into orbit where further judgement could be made about continuing with the planned mission or conducting an early return to Earth.

The countdown began at T-4 days and 5hr (T-101hr), starting at 15:00 EDT on 6 October and with three planned holds totalling 15hr. The countdown went well and only a single unscheduled hold was called, at T-6min 15sec lasting 2min 45sec, as it continued down to launch. With a temperature of 28.3° C (82° F) and clear visibility beyond 16km (10mls), the weather was not untypical of the month. But winds were gusting from the east at 49kph (30mph), which was higher than that experienced on any previous Saturn launch and very close to allowable limits for launch, as a high pressure system sat over Nova Scotia. Wind speed increased to 150kph (93mph) at 56.5km (35mls) although such a velocity was not unusual and the effect was minimal due to the lower pressure at that altitude. Nevertheless, Schirra protested over the risk and set in motion a conflict of decision as to the overall command responsibility which spilled over into the mission itself.

Ignition and lift-off proceeded as expected, with the vehicle going through Mach 1 at 1min 2.1sec and an altitude of 7.6km (4.7mls), encountering Max-Q at 1min 15.5sec and 12.1km (7.5mls). This was the first time astronauts had experienced a Saturn flight and the initial impression was one of restrained alarm as, unexpectedly the entire stack shook vio-

The orientation and relative position of AS-205 on LC-34, the last Saturn rocket to fly from this facility. *MSFC*

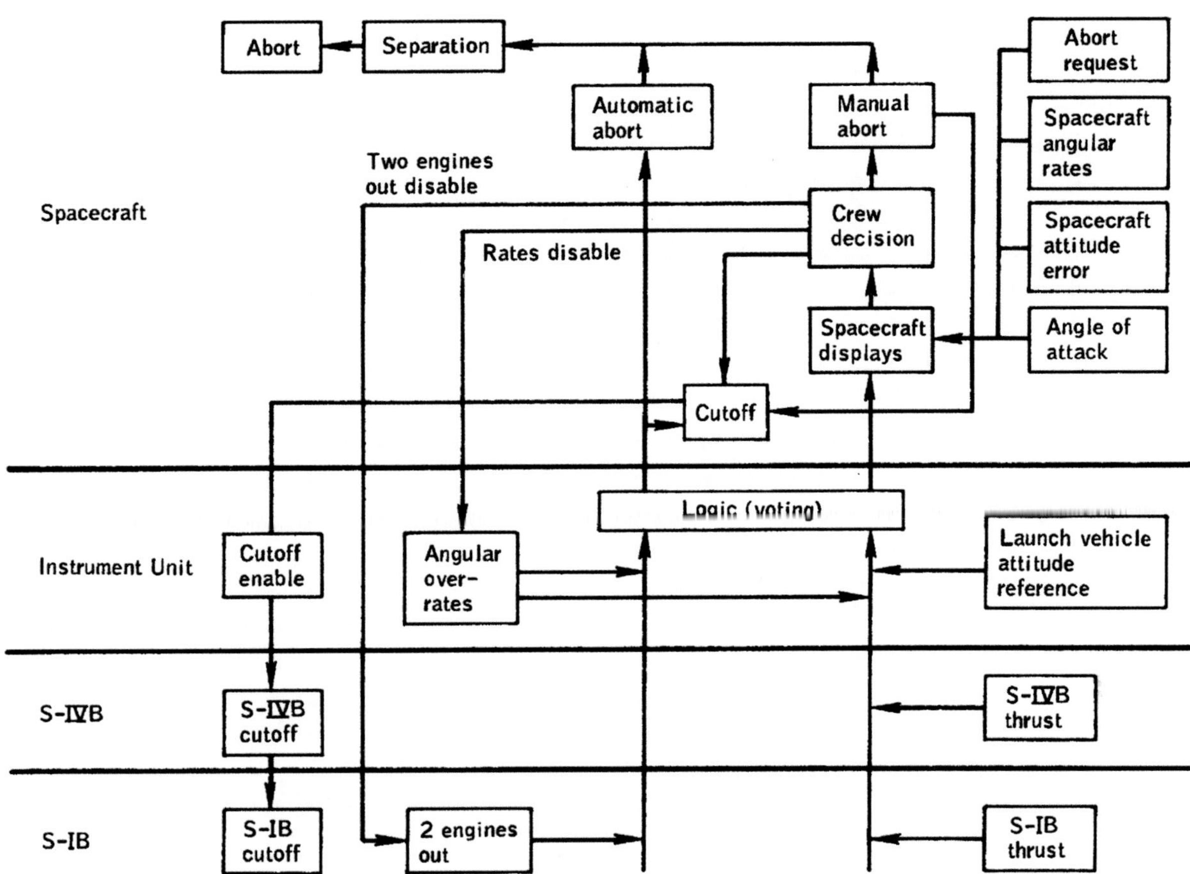

Abort tree showing automatic and discretionary flow through the two stages, the IU and CSM-101 should a malfunction occur during ascent. *MSFC*

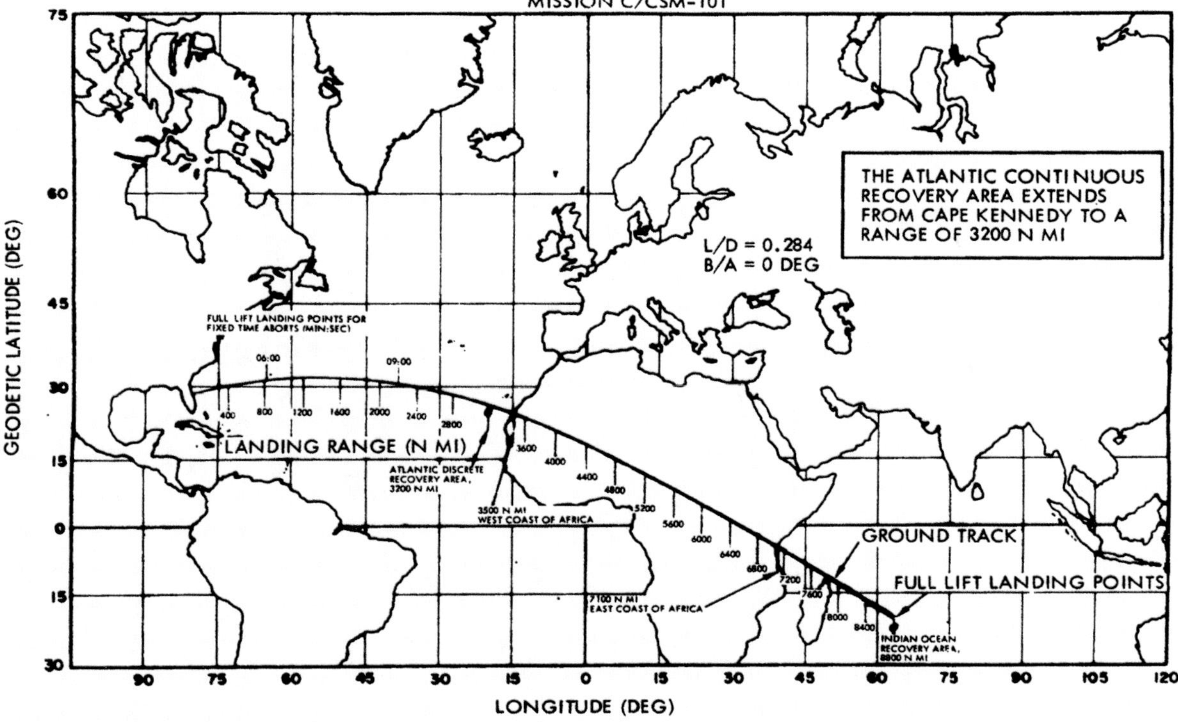

Ascent abort trajectories for AS-205 showing a full-lift re-entry for the Command Module avoiding the continent of Africa. *MSFC*

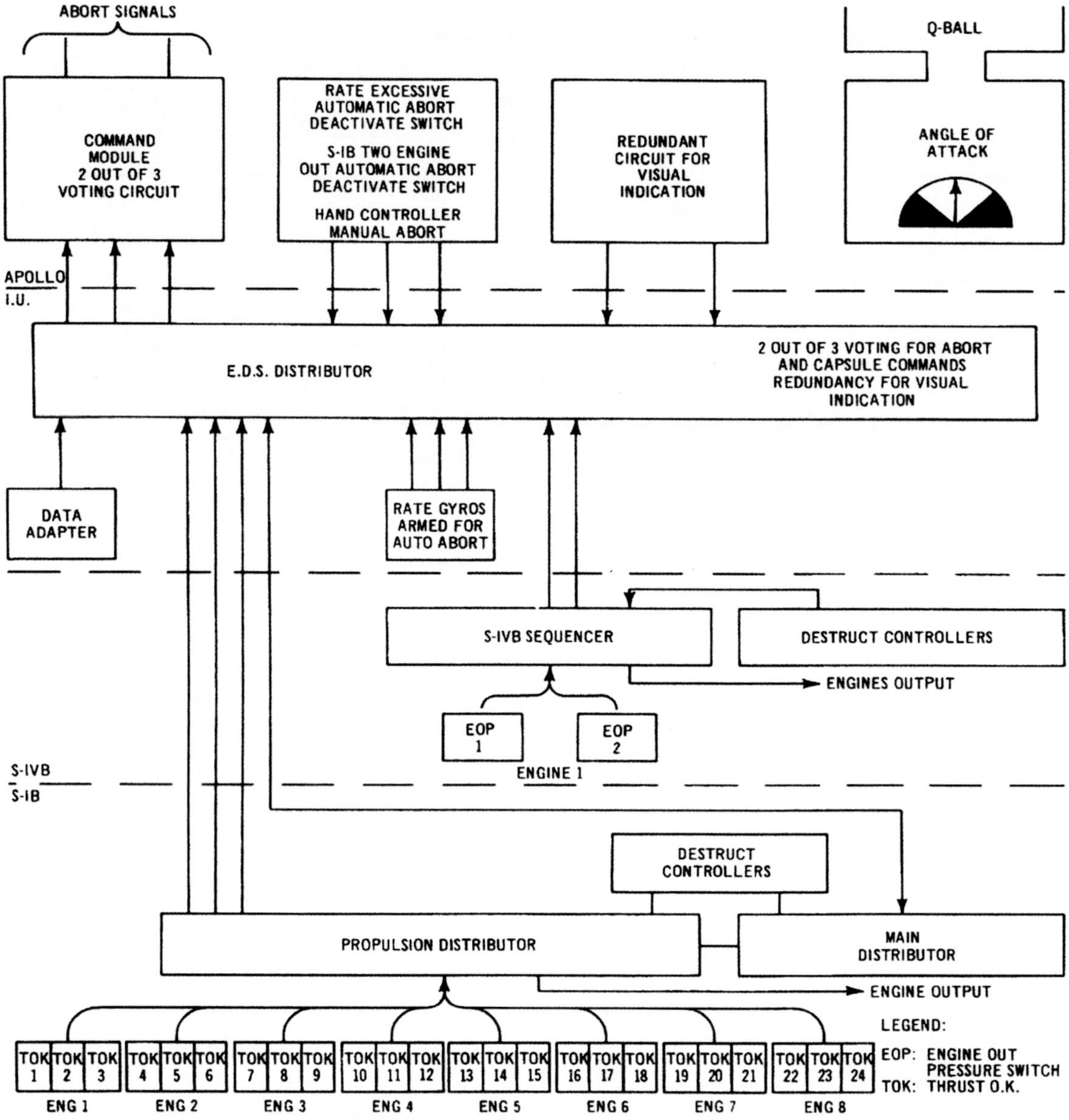

ABORT SIGNALS
COMMAND MODULE 2 OUT OF 3 VOTING CIRCUIT
RATE EXCESSIVE AUTOMATIC ABORT DEACTIVATE SWITCH
S-IB TWO ENGINE OUT AUTOMATIC ABORT DEACTIVATE SWITCH
HAND CONTROLLER MANUAL ABORT
REDUNDANT CIRCUIT FOR VISUAL INDICATION
Q-BALL
ANGLE OF ATTACK
APOLLO
I.U.
E.D.S. DISTRIBUTOR
2 OUT OF 3 VOTING FOR ABORT AND CAPSULE COMMANDS REDUNDANCY FOR VISUAL INDICATION
DATA ADAPTER
RATE GYROS ARMED FOR AUTO ABORT
S-IVB SEQUENCER
DESTRUCT CONTROLLERS
ENGINES OUTPUT
EOP 1
EOP 2
ENGINE 1
S-IVB
S-IB
DESTRUCT CONTROLLERS
PROPULSION DISTRIBUTOR
MAIN DISTRIBUTOR
ENGINE OUTPUT
TOK 1 TOK 2 TOK 3 TOK 4 TOK 5 TOK 6 TOK 7 TOK 8 TOK 9 TOK 10 TOK 11 TOK 12 TOK 13 TOK 14 TOK 15 TOK 16 TOK 17 TOK 18 TOK 19 TOK 20 TOK 21 TOK 22 TOK 23 TOK 24
ENG 1 ENG 2 ENG 3 ENG 4 ENG 5 ENG 6 ENG 7 ENG 8
LEGEND:
EOP: ENGINE OUT PRESSURE SWITCH
TOK: THRUST O.K.

Above: AS-205 would carry the definitive Emergency Detection System applied to what was already a conservative engineering design with wide margins and high risk mitigation solutions. This EDS tree shows the double 'engine-out pressure switch' sensors on the S-IVB's J-2. *MSFC*

1 – Cape Kennedy
3 – Grand Bahama and Cays
4 – Eleuthera
5 – San Salvador
6 – Mayaguana
7 – Grand Turk
91 – Antigua
BDA – Bermuda
BDA
S–IB/S–IVB Separation 144.23*
S–IB Impact 536.8*
S–IVB/CSM Separation 598.7*
S–IVB CO 588.47*
250*
400*
550*
*RANGE TIME (sec)
LATITUDE
LONGITUDE
38° 36° 34° 32° 30° 28° 26° 24° 22°
22° 20° 18° 16° 14° 12°
84° 82° 80° 78° 76° 74° 72° 70° 68° 66° 64° 62°

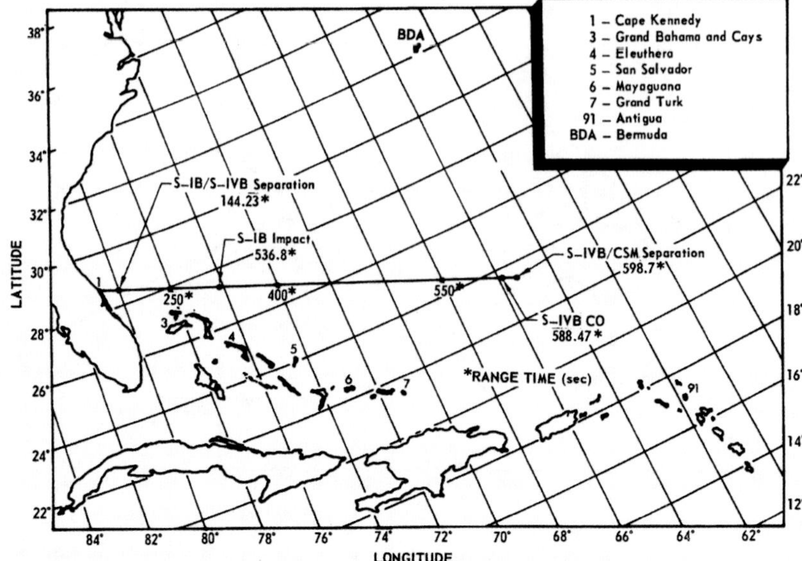

Right: Ground track trace for AS-205 during the ascent phase to orbit. *MSFC*

Right: The first manned flight of a Saturn rocket. AS-205 lifts off from LC-34 on 11 October 1968, the start of a 10-day mission which successfully qualified the rocket and the Apollo spacecraft for future flights, clearing the way for the next mission, the first Saturn V (see companion volume) to send three men into lunar orbit just over two months later. *NASA-KSC*

lently at ignition and for the first few seconds of flight. The inner engines shut down at 2min 20.6sec and a height of 56.7km (35.2mls), outer engines cutting off at 2min 24.3sec and an altitude of 60.5km (37.6mls). The separation command went in at 2min 25.6sec at the time of retrorocket ignition and staging was complete at 2min 26.6sec, followed by S-IVB ignition at 2min 27sec for a burn duration of 7min 50sec.

Several times during the ascent phase the crew reported on how smooth the ride was, veiling initial concerns which came as a great surprise to Schirra and to the two 'rookies' on board, the acceleration gradually building to a peak of 4g at S-IB cut off, falling back at staging before rising slowly again to 2.5g by S-IVB cut-off. Only Schirra had prior experience with rocket rides, his being on the Atlas aboard the third, and penultimate, Mercury (MA-8) flight in October 1962 and the Titan rocket for Gemini VI-A in October 1965. Maintaining a consistent flight separation of three years between rides, Schirra was again on his way into space!

S-IVB cut-off came at 10min 16.7sec, 1.95sec later than predicted with the vehicle at an altitude of 228km (141.6mls). The mass at orbital insertion for the S-IVB and payload was 30,694kg (67,669lb) including a residual propellant mass of 1,866kg (4,113lb). At lift-off, the weight of the payload was calibrated at 20,545kg (45,302lb) including the Launch Escape System. At orbital insertion, the mass of the Apollo CSM was 16,519kg (36,419lb), 6.2 per cent of the lift-off mass. The S-IB had impacted at 9min 20.2sec at 75.72deg W by 29.76deg N, into the Atlantic Ocean 490.8km (305mls) downrange. The initial orbit imparted by the S-IVB was 289.2km (179.7mls) by 222.6km (138.3mls). The S-IVB propellant dump began at 1hr 34min 28.9sec and ended 14min 1sec later and this modified the orbit to an apogee of 309.2km (192.1mls) and a perigee of 223.1km (138.6mls).

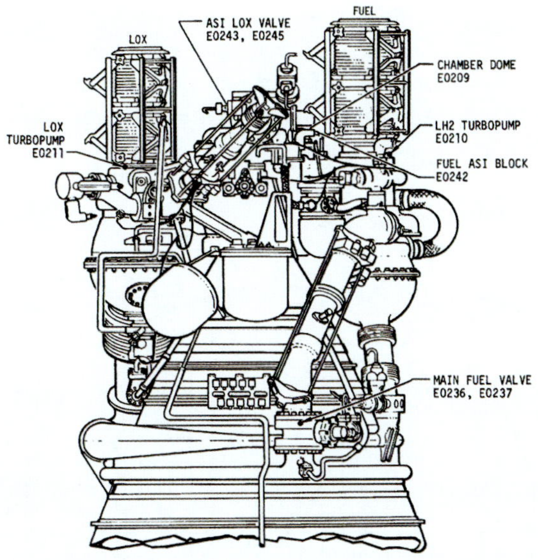

Eight vibration measurements were made on the J-2 engine attached to S-IVB-205, of which five were new on this flight: two on the main fuel valve, one on the fuel ASI block and two on the ASI LOX valve. Also flown on previous launches were sensors on the combustion chamber dome, the fuel turbopump and the LOX turbopump. *MSFC*

With the S-IVB effectively safed, the Apollo spacecraft separated from the S-IVB/IU/SLA at 2hr 55min 2.4sec on a manual signal delivered by the crew over Hawaii on the second orbit. Before separation, the crew manually controlled the attitude of the S-IVB/CSM combination and verified the operability of the flight control systems as well as their ability to determine the pointing angles of the docked vehicles. After separation, the crew performed a simulated transposition and docking manoeuvre as crews would be required to perform when extracting a Lunar Module from its stowed position within the SLA.

The last major event involving the S-IVB began when CSM-101 conducted a phasing manoeuvre at 3hr 20min 9.9sec using the Reaction Control System thrusters in a 17sec burn to effect a separation of 139km (86.3mls) at 26hr 25min. At that point the crew began a 6-hr rendezvous manoeuvre back to the S-IVB but the orbital decay of the stage had been greater than expected and a second phasing manoeuvre with the RC thrusters began at 15hr 52min 00sec, a burn lasting 18sec. The first SPS burn was a corrective combination manoeuvre which began at 26hr 24min 55sec and lasted 10sec. The second was a coelliptic manoeuvre which began with SPS ignition at 28hr 00min 56sec and lasted 9sec, bringing the spacecraft up for a Terminal Phase Initiation (TPI) burn start-

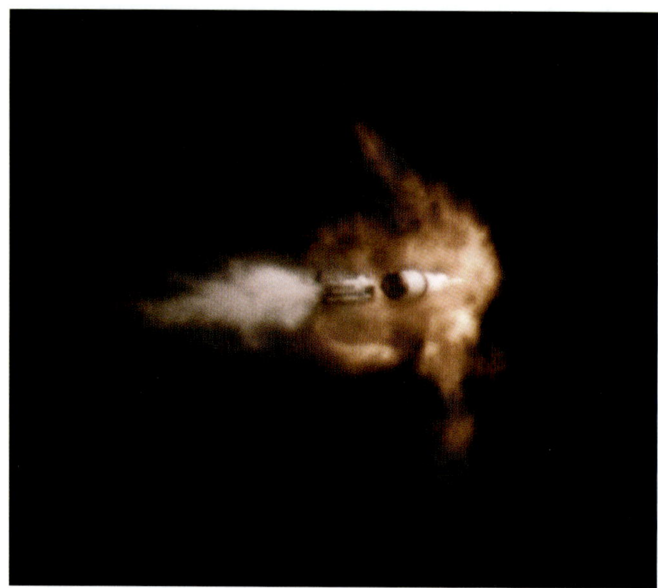

Left: Staging as S-IB-5 falls away from the upper stack and ignition for the S-IVB-205 stage. *KSC*

ing at 29hr 16min 33sec. The spacecraft began a series of braking burns at 29hr 43min 55sec on results from sextant tracking of the S-IVB with station-keeping just 21m (70ft) away from the inert stage about 30 hours after lift-off. These manoeuvres relative to the S-IVB/IU demonstrated the guidance and control computer, the RCS thrusters, the SPS engine and the ability of the crew to manoeuvre around, and maintain station-keeping with, the inert stage.

The S-IVB stage remained in orbit for a total of 6.77 days, returning through the atmosphere and impacting the surface of the Earth at 81.6deg E by 8.9deg S during the morning of 18 October, 162hr 27min 15sec after launch. As for the crew of Apollo 7, they would conduct a further six SPS burns, send TV transmissions from orbit to entertain the global public and perform a wide range of tests and crew evaluations to wring out every potential problem that they, or future crewmembers, might encounter. Most significant were the propulsion systems, the cryogenic fuel cells and the guidance and navigation equipment, including the sextant and the telescope. Prior to re-entry at the end of the mission, after consumption of SPS propellant during the flight, the CSM weighed 10,638kg (23,453lb). The Apollo 7 mission ended at an elapsed time of 260hr 9min 3sec with splashdown of the Command module, weighing 5,175kg (11,409lb), at 27.55deg N by 64.66deg W. The crew were retrieved by the carrier USS *Essex*.

Much has been written about the crew's attitude when tasked with additional activities, over and above the detailed content of the flight plan and about the acerbic spats which frequently broke out between the spacecraft and Mission Control. Readers will find incisive and opinionated accounts of that verbal fracas but the crew performed what was universally accepted as making Apollo 7 a '101 percent success' and that was the defining conclusion of engineers working on the programme. It had demonstrated that the Saturn IB, the Apollo spacecraft and the Lunar Module could perform the tasks for which they had been designed.

It is here that the story of the Saturn I/IB as a development vehicle for the Moon landing programme departs from the narrative followed thus far as the programme transitioned to the sole remaining element of the Apollo Applications Program – Skylab. With the end of support flights for the Apollo Moon landing programme, LC-34 and LC-37B were closed and decommissioned – LC-37A never had been activated. All Apollo Moon operations would switch to the Saturn V launched from LC-39A and LC-39B. It would be four years and seven months before another Saturn IB was launched, the next one on LC-39B at the Kennedy Space Center and in support of America's first space station. The remaining chapters in the Saturn saga as it relates to getting astronauts on the Moon are taken up in our companion volume on the Saturn V.

Below left: The pre-flight plan for separation and rendezvous from the S-IVB to demonstrate the CSM propulsion system and the on-board crew calculations for returning to an object in space. *NASA*

Below right: The post-flight trace of the rendezvous sequence showing the 'actual' trajectory prediction based on orbital parameters achieved and the flown flight path between the Corrective Combination burn and Terminal Phase Initiation (TPI) executing the rendezvous manoeuvres. *NASA-JSC*

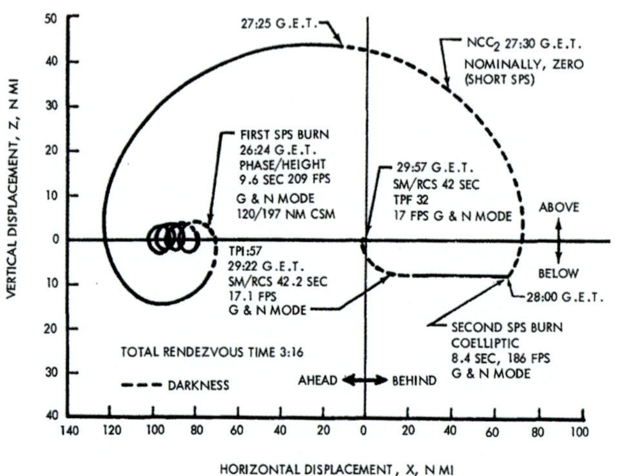

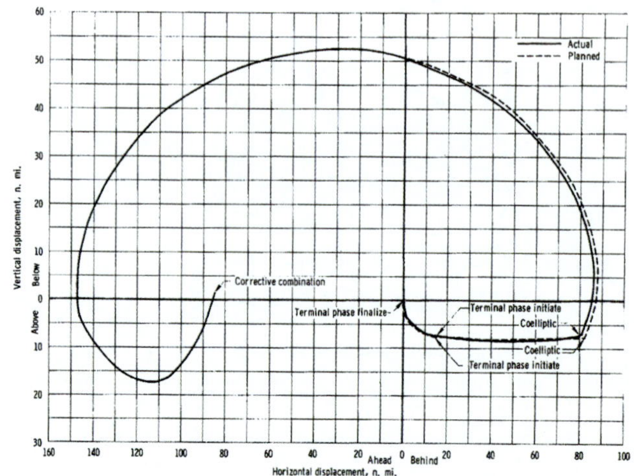

Top: The S-ICB-205 stage as viewed by the crew of CSM-101 with the open upper SLA panels deployed following separation. *NASA*
Above left: An aesthetic shot of S-IVB-205 over Cape Canaveral. *NASA*
Above right: A good shot of the rendezvous radar antenna providing a return for cooperative manoeuvres executed by the crew. *NASA*

AS-206/SL-2

Launch date:	25 May 1973
Lift-off time:	09:00:00 EDT (13:00:00 UTC)
Launch pad:	LC-39B
Launch azimuth:	47.5deg
Length:	68.1m (223.33ft)
S/L thrust:	7,294.7kN (1,640,000lb)
Dry mass:	73,256kg (161,500lb)
Propellant mass:	519,120kg (1,144,446lb)
H-1 engines:	H-4068; H-4069; H-4070; H-4072; H-7071; H-7072; H-7073; H-7075
J-2 engines	J-2046
Ignition mass:	592,709kg (1,306,679lb)
Lift-off mass:	586,390kg (1,292,378lb)

In its nine flights the Saturn IB would support only five manned missions but the lapse between the first and the second would be a long four years and seven months, bringing challenges for hardware already produced to a schedule driven by the need to fulfil a national objective. With responsibility for supporting the lunar programme lifted from the Saturn IB and picked up by Saturn V, it would be called upon to support a very different requirement. But the change in manifesting the different launchers created a backlog from hardware prepared far ahead of planned missions which introduced its own unique problems. For Saturn IB, that meant a significant shift in requirements.

From the truncated Apollo Applications Program emerged Skylab, a converted S-IVB stage fitted out before launch for habitation in space, occupied by three teams of three astronauts, each team launched sequentially at intervals on three Saturn IB launch vehicles. The first stage was now powered by the uprated H-1 engines each producing a thrust of 910kN (205,000lb). But the launch concept troubled some engineers and managers who saw high risk in the way the Saturn IB would send crews to Skylab.

As mentioned earlier, decommissioning of the original LC-34 and LC-37B for Saturn I and Saturn IB rockets made it necessary to bootstrap flights from LC-39B for the Skylab and ASTP missions. The haste with which the former pads were shut down reflected the urgency of redirecting the future manned space programme and the certainty that NASA could not continue to operate these expendable launch vehicles

Below: Erected on the 'milk stool' on ML-1, AS-206 is moved to LC-39B. *NASA-KSC*

indefinitely. By 1968 it was clear that some type of reusable shuttlecraft flying routinely back and forth to Low Earth Orbit would be central to the future of manned space flight in the United States.

The only way to launch a Saturn IB/Apollo combination from the Saturn V pad was to lift it so that the upper stage and spacecraft support swing-arms, umbilical connections and emergency evacuation equipment was contiguous with those on the Launch Umbilical Tower (LUT) for the equivalent elements of a Saturn V. Only the S-IB stage would require adaptations for position and location of electrical, hydraulic, pneumatic and propellant lines and conduits. LC-39A would support the last Saturn V, lifting the Skylab space station into orbit in 1973, following which extensive modifications would be made for supporting the Space Shuttle programme, formally announced by President Richard Nixon in January 1972.

While on paper the design and certification of a pedestal – dubbed the 'milk stool' – was a simple engineering task, there were concerns about the degree of stability the assembly would have in high wind and the age of the launch vehicle itself factored in with several unknowns. Fatigue and some corrosion was anticipated and a special programme of inspections was set up to discover any areas of concern.

In every essential respect, the configuration of AS-206 was the same as that for AS-205 but with subtle changes in that the Apollo spacecraft was modified for its role in Skylab. As an operational vehicle, AS-206 carried only a minimal 735 instrumented measurements. There were several changes to the S-IB stage worthy of note. Most distinctive was the F-8 fuel tank from S-IB-1 damaged during over-pressurisation with AS-201 and noted as such on 29 September 1965. It had been changed out for a tank from S-IB-6, the repaired F-8 now becoming F-1 on the stage for AS-206. Most of the other configuration differences involved re-routing cables or conduits, in light of previous experience, and removal of some redundant and unnecessary equipment or line items.

AS-206/SL-2 – launch complex changes

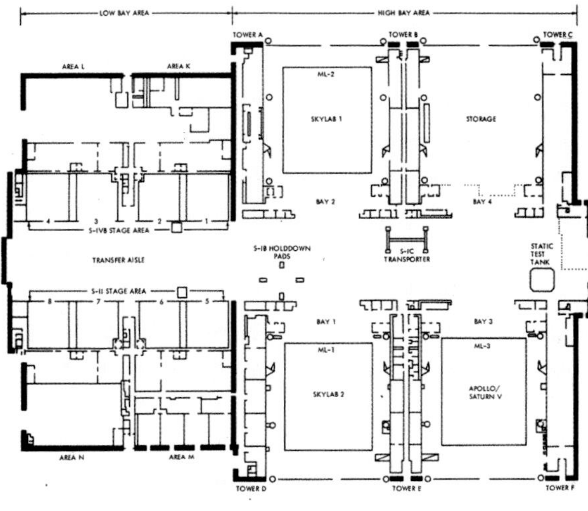

Above: A plan of the Vehicle Assembly Building at NASA's Kennedy Space Center from where all remaining four flown Saturn IB missions would be prepared before rollout to LC-39B. The AS-206 launcher is in High Bay 1 on Mobile Launcher 1 (ML-1) while the Skylab space station is on AS-513 at High Bay 2 on ML-2. Note that in this time frame, AS-512 was in High Bay 3 on ML-3 for the Apollo 17 lunar landing mission. *Author's collection*

Below: The pad configuration of ML-1 and the Mobile Service Structure (MSS) during preparation for the launch of AS-206 for SL-2, the first manned flight to the Skylab space station. Note the selected application of various platforms on ML-1 for this and the two subsequent Skylab flights. *Author's collection*

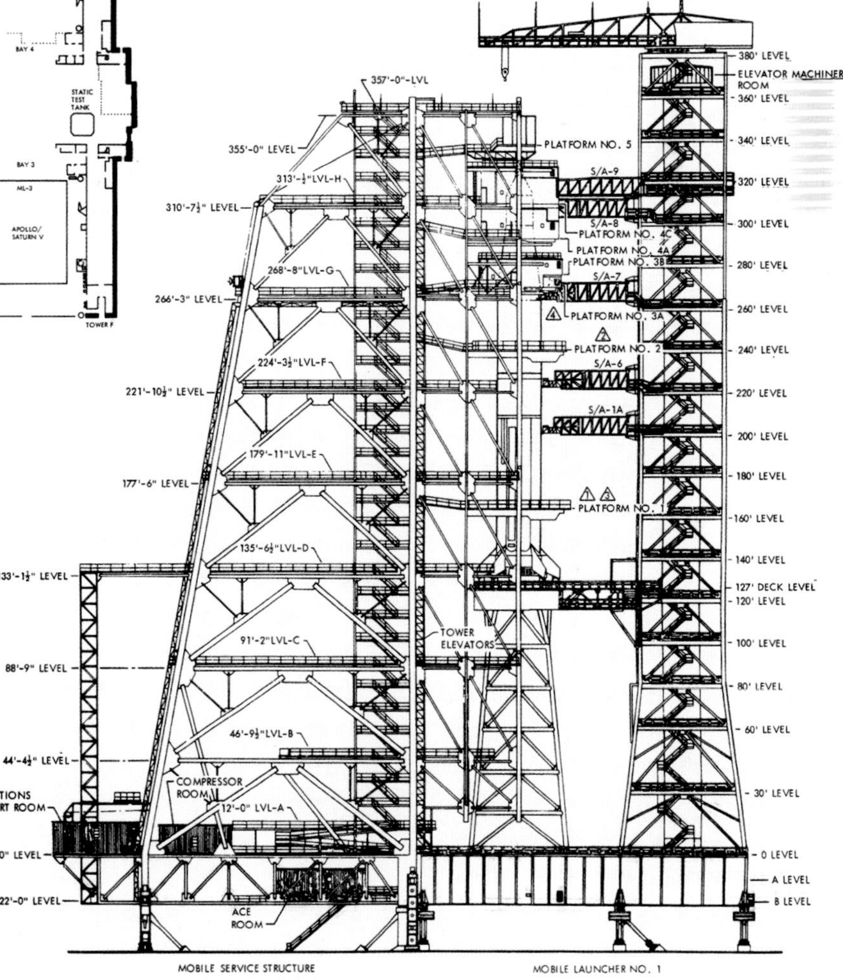

⚠1 PLATFORM NO. 1 WILL NOT BE USED FOR C/O OF SL-2 BUT AFTER MODIFICATION WILL BE USED FOR C/O OF SL-3 & 4.

⚠2 PLATFORM NO. 2, BY USE OF (4) NEW PORTABLE ACCESS PLATFORMS, WILL BE USED FOR C/O OF SL-2.

⚠3 THE (4) NEW PORTABLE ACCESS PLATFORMS DESIGNED FOR PLATFORM NO. 2 WILL BE MOVED TO PLATFORM NO. 1 FOR CHECKOUT OF SL-3 & 4.

⚠4 (2) ACCESS PLATFORMS WILL BE ADDED TO PLATFORM NO. 3A FOR ACCESS TO THE IU COMMAND ANTENNAS.

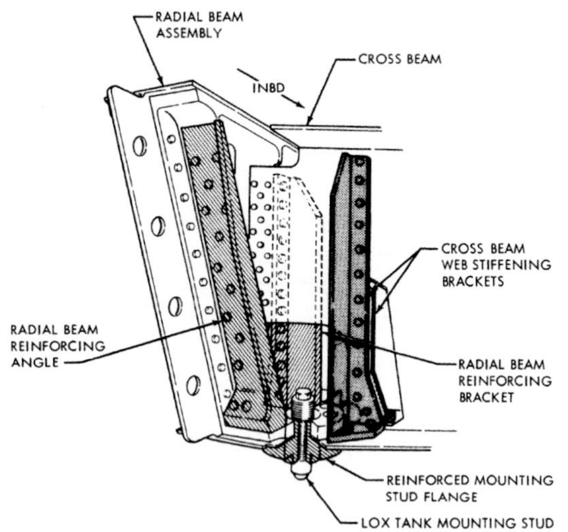

Above: During tests a LOX tank fitting on the spider beam failed so reinforcing angle and brackets and crossbeam web stiffening brackets were applied to all fittings. *MSFC*

Stage assembly got under way in early 1965 with tank clustering on 25 October, all of the upgraded engines installed by 28 January 1966. After a stage checkout following assembly, the stage was moved out of Michoud on 19 May, arriving at MSFC nine days later. Installed in Static Test Tower East on 31 May, the first firing took place on 23 June lasting 35.58sec, good performance shown by all engines and no re-orificing was necessary. The second, full duration firing of 141.2sec – the outboards cutting of 2.8sec earlier as planned – occurred on 29 June. Although generally trouble-free, engine H-4071 revealed a split thrust chamber tube and it was replaced by H-4072, brought in from Michoud and installed on 1 July. This was the only stage in which one of the engines was never test fired before flight and although engines H-4069, H-7071 and H-7072 were seen to have some tubes partially separated they were deemed acceptable.

S-IB-6 was removed from the test stand on 8 July and was transported back to Michoud the next day, arriving on 13 July. Five months later to the day, post-static checks had been completed and the stage left for the Cape on the barge *Palaemon*, arriving on 18 December 1966. S-IB-6 was erected on LC-37B during the third week in January 1967 and received the S-IVB-206 stage on the 23rd with the anticipation of receiving LM-1, which was of course significantly delayed. However, after the Apollo fire with AS-204 on 27 January, the AS-206 stack was de-erected in the first week of March and loaded aboard the barge *Promise* on 31 March for Michoud, arriving there on 10 April where it went into storage.

On 16 October it was removed for some modification and minor attention before being placed in a checkout phase between 21 November and 2 January 1968. The decision had been made to have this stage on standby for launching LM-2 in support of a support flight with a CSM on another Saturn IB but after the successful flight of LM-1 and the relegation of AS-205 to the first manned Apollo flight, on 18 December 1968 it went into environmental storage, not to emerge before 18 October 1971 for a period of modification to support the SL-2 Skylab flight. These activities were completed on 12 February 1972 and a special programme of post-storage rework kept technicians busy until 6 June. It departed Michoud on 17 August 1972 bound for the Kennedy Space Center on the barge *Orion*, that journey lasting five days.

Stage assembly for S-IVB-206 began in 1965 and imported the common bulkhead from S-IVB-207 because of welding problems with the planned element. The J-2 engine arrived with Douglas on 20 December 1965 and installed in the stage on 31 March 1966. One important upgrade was to ex-

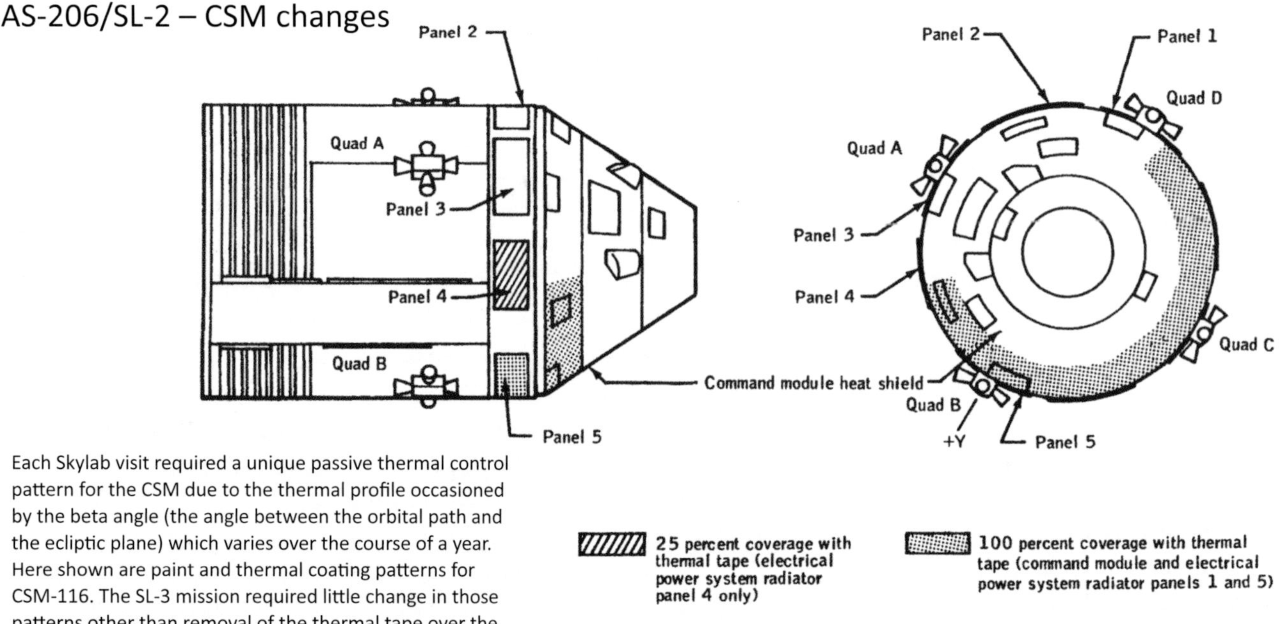

Each Skylab visit required a unique passive thermal control pattern for the CSM due to the thermal profile occasioned by the beta angle (the angle between the orbital path and the ecliptic plane) which varies over the course of a year. Here shown are paint and thermal coating patterns for CSM-116. The SL-3 mission required little change in those patterns other than removal of the thermal tape over the radiators for the electrical power system. *NASA-MSFC*

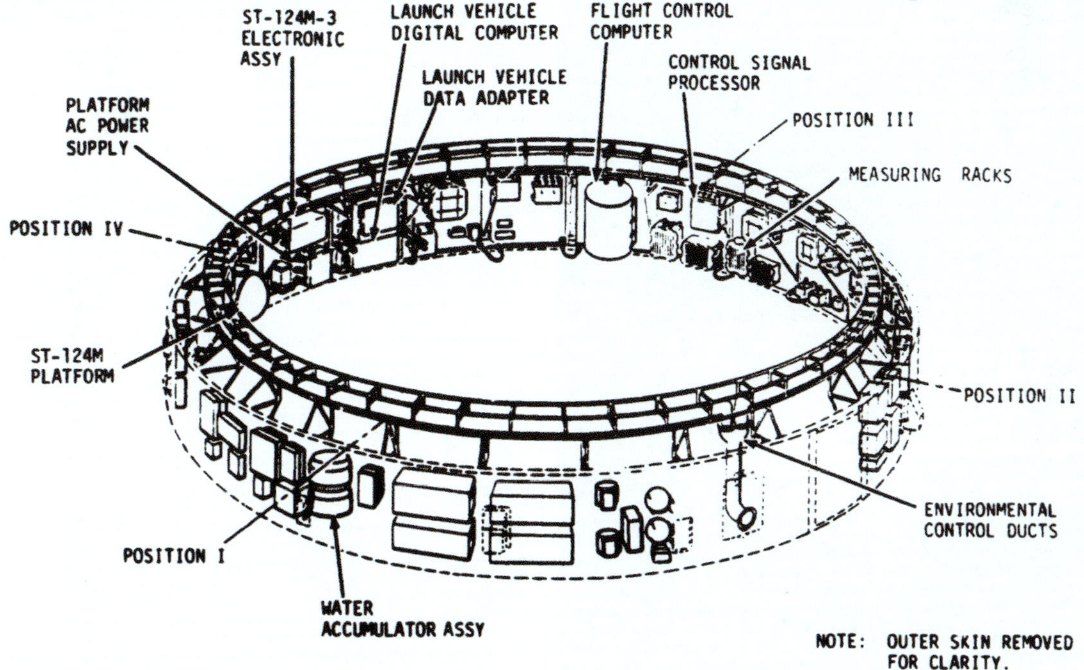

Above: The Instrument Unit for the AS-206 launch vehicle during assembly. The IU-206 carried some minor modifications due to improvements developed from the AS-205 configuration. *MSFC*

tend the coast capability from 4.5hrs to 7.5hrs through a series of changes and modifications which would be applied to the next stages in line also. Checkout was completed by the end of the third week in May 1966 and was sent across to SACTO for tests where it arrived on 1 July for installation in the Beta III test stand six days later. The simulated firing tests were completed on 12 August with a live firing of 433.7sec on 19 August but a large quantity of metallic particles were found in the turbopump, which was replaced. A re-qualification firing of 66.6sec occurred on 14th September and the stage was removed from the stand on 4 October for an all-systems test, completed one calendar month later. The stage arrived at KSC by air on 14 December and was trucked to LC-37B and stacked to the S-IB stage on 23 January 1967.

Left: The launch of AS-206 on 25 May 1973 as SL-2. This first Skylab visit was delayed due to problems with the orbiting station and preparation of additional equipment carried aboard CSM-116. *NASA*

After the AS-204 fire the S-IVB-206 stage was de-erected and flown out on 13 April to long-term storage at SACTO pending a decision on its use. On 22 November it was moved to the newly modified Beta III stand for post-storage checks which were completed on 29 December prior to its removal on 11 January 1968 when it was placed in storage once more. It was moved to the McDonnell Douglas Huntington Beach facility on the 3rd of August 1970 and from there to KSC, arriving on 24 June 1971 where it was placed in long-term storage from that October to 17 April 1972. A check on the propulsion system was conducted by 21 August.

CSM-116, the Apollo spacecraft payload for AS-206, arrived at the Cape on 19 August 1972, followed by the S-IB stage and the Instrument Unit three days later. As the first Saturn IB to be launched from LC-39B, the pre-flight preparations were different to those of preceding flights with this launcher. Stacking of the stages included Boilerplate BP-30, used as a spacer for fit checks, which took place in the Vehicle Assembly Building beginning on 31 August 1972 atop Mobile Launcher 1 (ML-1). The stages were placed on a trestle structure to raise the entire launch vehicle and mate the cryogenic S-IVB swing arms on the Launch Umbilical Tower with the upper stage.

The two stages were rolled out to the pad on 9 January 1973, where fit checks and alignments were made, the stack returning to the VAB on 5 February where BP-30 was removed and CSM-116 was erected in its place. The completed assembly was moved to LC-39B for the final time on 26 February. In preparation for launch, RP-1 was loaded on the S-IB stage on 23 April and the CDDT was conducted over two days (3-4 May), anticipating a launch one day after the Skylab space station had been delivered to orbit from LC-39A.

Designated Skylab-1 (SL-1), the AS-513 Saturn V launched at 13:30:00 EDT (17:30:00 UTC) on 14 May but a failure in the Skylab workshop micrometeoroid shield during ascent resulted in the thermal shield being torn off and, on reaching orbit, one of the two solar array wings being ripped away by thrust from the S-II retrorockets impinging on the dislodged beam. Uninhabitable without an effective shield to reduce temperatures on the exposed hull of the converted S-IVB, the Skylab space station could not receive the crew of SL-2 until a repair procedure had been put in place. The launch planned for 15 May was scrubbed and plans made for the SL-2 crew to erect a sunshade and free the remaining solar array boom which had been fouled by debris from the torn micrometeoroid shield. With repair equipment and a makeshift sunshade on board the launch was rescheduled for 25 May.

The terminal countdown began with a clock time of T-59hr, two days before planned lift-off. Scheduled holds to-

Right: As seen in orbit docked to Skylab, CSM-116 displays the unique colour patterns due to the different thermal conditions experienced on this extended Earth orbit mission. *NASA*

talling 1hr 15min were programmed but no further holds were necessary. Carrying the SL-2 crew of Conrad, Kerwin and Weitz, the stack passed through Mach 1 at 60.5sec and an altitude of 7.66km (4.75mls) with Max-Q at 1min 15.5sec and 12.8km (7.9mls). The inner engines shut down at 2min 18.7sec at 54.7km (34mls) with the outer engines cutting off at 2min 22.26sec and a height of 55.3km (34.3mls). Staging occurred at 2min 23.7sec and an altitude of 59.34km (36.8mls). The single J-2 engine on the second stage began its job at 2min 25.9sec and the S-IVB shut down at 9min 46.2sec. Insertion occurred 10sec later with the vehicle in an orbit of 352km (218.7mls) by 149.99km (93.2mls) with a period of 89.53min and an orbital inclination of 50deg.

With a mass of 16,333kg (36,007lb), Apollo CSM-116 separated from the SLA at 16min 0.28sec and a height of 172.3km (107mls). The S-IB stage had impacted the ocean at 8min 55.3sec, 493km (306mls) downrange. Residual propellant in the second stage was calculated to be 2,311kg (5,096lb). The S-IVB was de-orbited by a 7min 40sec LOX dump beginning at 5hr 24min 20sec and a fuel dump of 2min 5sec starting at 5hr 32min 31sec with stage impact logged at 6hr 00min 7sec. And here the launch vehicle events ended.

The SL-2 mission proved highly successful, in that the crew were able to free the snagged solar array boom and erect a sunshade 'umbrella' through a small scientific airlock in the hull of the Skylab workshop. Priority tasks for the early part of crew activity focused on improving the long-term habitability of the station and performing scientific experiments. SL-2 ended with splashdown of the Command Module on 22 June 1973 after 28days 49min 49sec. The work conducted by Conrad, Kerwin and Weitz had established the station as an operational facility which would have been impossible without the attentions of the crew. The mission also validated the purpose of the second and third visits, building on the achievements of the first crew and raising the endurance record to four weeks, doubling the previous record of Gemini VII in December 1965.

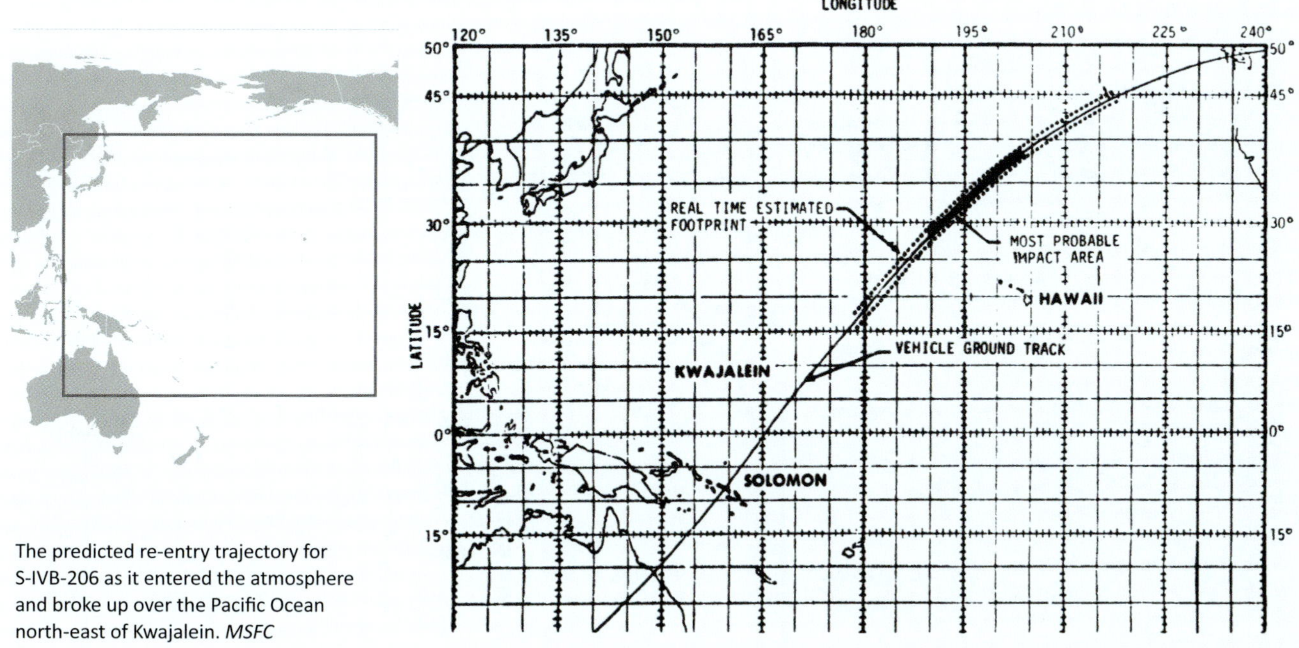

The predicted re-entry trajectory for S-IVB-206 as it entered the atmosphere and broke up over the Pacific Ocean north-east of Kwajalein. *MSFC*

AS-207/SL-3

Launch date:	28 July 1973
Lift-off time:	07:10:50 EDT (11:10:50 UTC)
Launch pad:	LC-39B
Launch azimuth:	45deg
Length:	68.1m (223.33ft)
S/L thrust:	7,294.7kN (1,640,000lb)
Dry mass:	73,710kg (162,500lb)
Propellant mass:	519,605kg (1,145,515)
H-1 engines:	H-4074; H-4075; H-4076; H-4078; H-7074; H-7076; H-7078; H-7085
J-2 engines	J-2056
Ignition mass:	594,299kg (1,310,207lb)
Lift-off mass:	587,775kg (1,295,802lb)

The mission for AS-207 was to carry the second Skylab crew to the space station in CSM-117 for them to erect a more permanent sunshade within a planned two-month long expedition conducting numerous experiments and expanding upon the work completed on the first visit. The launch was scheduled for a few weeks after the return of the SL-2 crew, carrying Alan Bean, Owen Garriott and Jack Lousma on a visit to Skylab which was expected to last 56 days. The flight profile for SL-3 was very close to that for SL-2, which would be completed when the S-IVB stage had been de-orbited into the Pacific Ocean by propellant dumping for a retrograde impulse to bring the stage back down into the Earth's atmosphere. Only those instrumented systems contributing toward the monitoring of the vehicle and its operational requirements were emplaced, AS-207 having 821 for flight, with 266 on the S-IB, 239 on the S-IVB and 316 in the Instrument Unit. The total payload weight of the spacecraft was 20,124kg (44,365lb) including the Launch Escape System and associated ancillary items. At orbital insertion the mass of CSM-117 would be 16,362kg (36,073lb).

Very few changes were logged for AS-207. The only one of note on the S-IB being modification of the premature ignition detection circuit with the auto-igniter links on each engine's gas generator wired in parallel so that both would have to open on parallel to actuate the circuit. This would avoid a false premature ignition signal. Changes to the S-IVB-207 stage were limited to those incorporated from S-IVB-206 together with an increase in the size of the purge control valve to eliminate potential valve instability during an emergency shut-down of the accumulator. Numerous improvements, upgrades and modifications were logged for the Instrument Unit, which had some unique adaptations for supporting Earth-orbiting space station support missions.

S-IB-7 stage build-up began during 1965 with all original engines installed by 11 April 1966. During post-assembly checkout engine H-7079 had to be removed on 13 July and replaced with H-7076 on 23 July when it exhibited a thrust chamber leak at the aspirator band. The stage departed Michoud aboard the barge *Poseidon* on 4 August and arrived at MSFC seven days later where it was taken to Static Test Tower East and installed there for the first of two test firings on 1 September lasting 35sec. It was followed on 13 September by a full duration firing of 140sec. Removed from the

Right: AS-207 on the launch pad with visible venting. *NASA-KSC*

During LOX replenishment in the countdown sequence, gases were observed exiting the four outboard oxidiser tanks between 5.5hrs and 2min 43sec prior to launch. This graphic shows the LOX levels in the centre tank (right) and a windward outer tank, which displayed a level 6.8cm (2.7in) above the level in the centre tank. *MSFC*

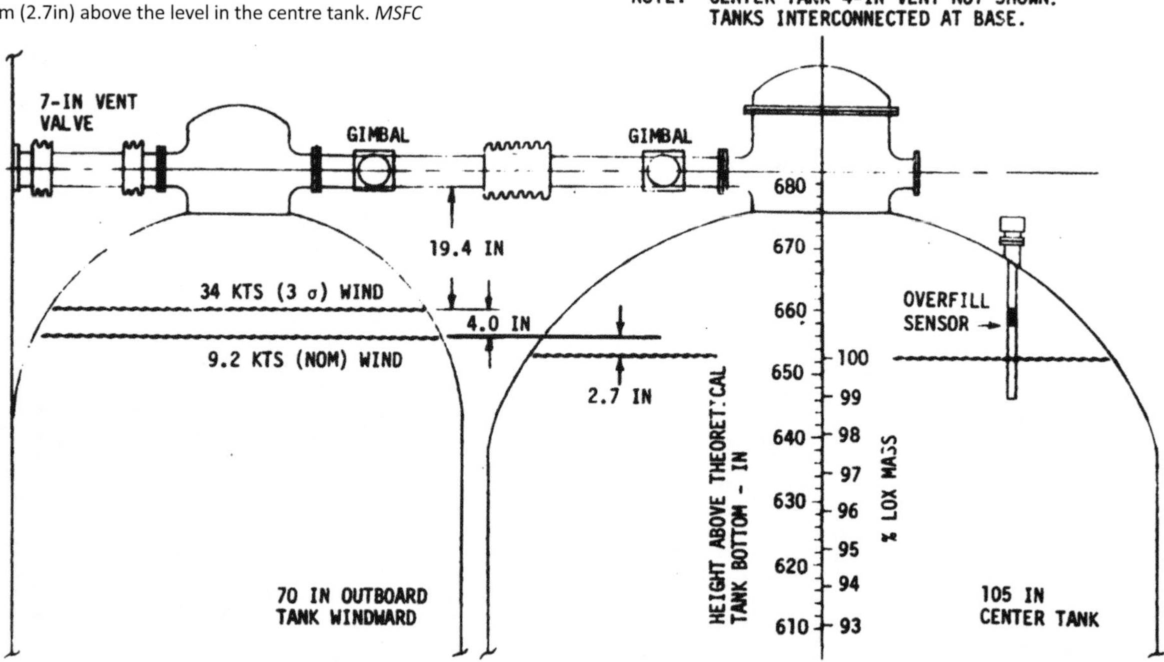

tower, S-IB-7 was placed aboard the barge *Palaemon* on 20 September on which it was delivered back to Michoud on the 25th. Post-static firing checks began on 1 November but a succession of problems caused several engine changes.

When several pieces of Teflon were found lodged behind the injector plate on engine H-7080 it was removed and replaced with H-7074 on 16 November. It was also found that the performance of the engine had changed between engine-level tests and stage tests and this was attributed to the contamination. Suspecting that engine H-4073 had also been contaminated, it too was taken out and replaced with H-4078 but only after a six month hiatus. The contaminated engine (H-4073) had already been removed from S-IB-8, on 21 November, on suspicion of defective turbine blades but was then found to have five blades manufactured from the wrong material. Following recertification firings on 2 February 1967, H-4078 was delivered back to Michoud and installed in the stage on 10 June. But this was not the end of the story.

Between removal of H-4073 and its replacement with H-4078, engine H-7078 was found to have three turbine blades fabricated from incorrect material and its turbine wheel replaced by Rocketdyne before a test firing on 13 January 1967. Returned to Michoud fifteen days later, it was back in S-IB-7 on 16 February. Coming little more than two weeks after the Apollo fire, the stage got caught in the suspension of flight activity pending a detailed review of flight manifests and the clearance of operations to resume manned flights. On 18 July S-IB-7 was placed in long-term storage and subject to inspections, on one of which a socket head was discovered in engine H-7077, whereupon it was removed and replaced with H-7085, an engine which had already been removed from S-IB-8 on 15 December 1966. Re-evaluation firings were conducted on this engine and it was back at Michoud on 14 August 1967 and installed on the stage on 16 October 1968.

With long-term decisions regarding the balance between Saturn IB and Saturn V flights in support of the Apollo programme now defined, on 14 January 1969, S-IB-7 was placed in environmental storage, a step up from the long-term storage which it had entered on 18 July 1967. With decisions over the scheduling of Skylab operations now in hand, S-IB-7 was removed from this condition on 31 January 1972 with detailed post-storage inspection and checkout which lasted from 24 May to 28 August. The stage left Michoud for the Kennedy Space Center on 24 March 1973.

It was now nearly eight years since assembly of the stage had begun and much had changed within the space programme over that time. All the Moon landings had come and gone and, in its original guise, the Apollo programme was over, only the remnants of available hardware re-crafted into what had begun as the Apollo Applications Program, the sole example being Skylab.

Meanwhile, stage build-up on S-IVB-207 began during 1965, the same year as S-IB-7, and reached a milestone with installation of the J-2 cryogenic engine on 11 May 1966. Checks were completed on the stage across the usual cycle of post-manufacturing inspections and the stage was shipped out from Huntington Beach on 30 August, arriving at SACTO next day. Installed in the Beta I test stand, it was subjected to integrated systems checks before simulated firing procedures and a live test on 19 October 1966 for 445.6sec. After post-test cleaning, checks and further inspection, it was placed in long-term storage at SACTO on 19 January 1967, its future uncertain within the Apollo programme. More than two years later, on 1 May 1970 it was flown to Seal Beach and the following day it entered temporary storage. On 11 December, S-IVB-207 was finally moved to temporary storage back at Huntington Beach. From there it was moved again, to KSC on 26 August where it was placed in long-term storage on 12 November 1971. With a defined mission ahead supplying the second crew to Skylab, it was finally removed from storage on 28 November 1972.

Apollo CSM-117 arrived at KSC on 1 December 1972, followed by the S-IB stage on 30 March 1973 which was erected on the ML-1 'milk stool' stand in the Vehicle Assembly Building on 4 April. The Instrument Unit (S-IU-208) arrived at KSC on the 8 May, S-IVB-208 was erected on 28 May followed by the IU the next day. Conducted over the

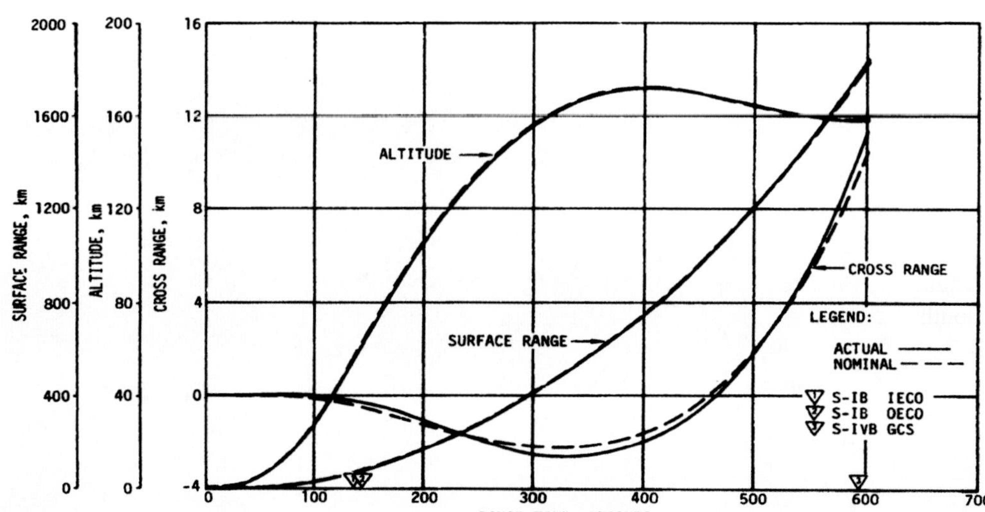

Ascent parameters showing surface range, altitude and cross range for AS-207 with predicted versus actual values determined by range tracking, accelerometers and data from the Instrument Unit. *MSFC*

next several days, electrical mating and associated tests were completed on the 2nd of June. This cleared the stack for rollout to the pad nine days later followed by electrical mating between the launcher and ground systems the day after. Further checks and tests were conducted before the flight readiness test was completed on 29 June.

The first stage was fuelled on 11 July and the CDDT was conducted on the 20th. The countdown began at the T-59hr point on 25 July with two scheduled holds at T-3hr 30min and T-15min but no unscheduled interruptions to the flow. Intermittent measurements from the resistance thermometer to tank F-1 prompted a change to the way the temperature was measured, the two lower sensors being connected in parallel provided a summed output which was used instead. At about 8hrs 30min before launch the fuel levels in the four tanks were raised from 1,524cm (600in) to 1,618.5cm (637.2in), the point measured by the overfill sensor. This provided a measure indicating that the level was 314lit (83USgall) above that originally set, which was equivalent to an error of 252kg (550lb), a bias which was integrated with the launch vehicle computer.

For several hours before the vents closed at T-2min 43sec, LOX was observed venting around the S-IB stage from undefined tanks but inspection of video located it to tanks O-1 and O-2 but with emanations via tanks O-3 and O-4 with no indication of any boil-off from the centre tank O-5. Data showed no loss from any of these tanks and the repressurisation time of 73sec was consistent with normal tanked quantities. Various atmospheric changes on even a subtle basis could account for this effect but there was nothing in the meteorological data to show that this could be a reason. Similar occurrences sought from video of CDDT and countdowns for AS-206 and AS-207 were undetected. Procedures for the countdown with AS-208, however, would include a reduction in the LOX level prior to the astronauts boarding should similar venting occur.

The day before launch saw numerous thunderstorms break over Florida and in the vicinity of the Cape with total cloud cover overnight and throughout launch day. Forecasters were optimistic but conditions needed monitoring; the lightning strike on AS-507 (Apollo 12) after its launch on 14 November 1969 had caused weather rules to be rewritten to prevent a recurrence of that incident. At the time of launch, broken 10/10th altocumulus cloud lay across the sky at a base of 4,600m (15,000ft) with cirrus clouds at 6,400m (21,000ft). A grey fog had covered the launch site, some areas covered in 180m (600ft) blankets, as dawn rose with visibility down to 4.8km (3mls). But there was no lightning, no threat of imminent thunderstorms and the temperature was a relatively modest 23° C (74° F) with 93 per cent humidity, a dampness overhanging the Cape. But winds were light, drifting from west-southwest, the air barely moving at ground level.

Following launch, events proceeded close to those of AS-206, with Mach 1 passed at 59sec, Max-Q at 1min 15sec and inner engine cut-off at 2min 17.4sec followed by the outer outboard engines at 2min 20.7sec and an altitude of 58.7km (36.5mls). Stage separation occurred at 2min 22sec at an altitude of 59.97km (37.2mls), the S-IB impacting the ocean at 8min 59.9sec, 496.6km (308.6mls) downrange. Ignition of the

Above: The second visit to Skylab gets under way with the launch of AS-207 on 28 July 1973 carrying CSM-117 and three astronauts. *NASA-KSC*

J-2 came at 2min 26sec with S-IVB engine cut-off at 9min 53.2sec. The orbit at insertion was 226.3km (140.6mls) by 149.8km (93.1mls) with a period of 88.26min. The CSM separated at 18min 0.4sec, about 2min 4sec later than nominally scheduled. The S-IVB de-orbit propellant dump was modified real-time so that tracking stations on the island of Kwajalein in the Pacific Ocean could use it as a target. LOX dumping began at about 5hr 20min 57sec with impact at 6hr 00min 50sec of the fragmented remains at 23.75deg N by 171.6deg W.

SL-3 rendezvoused with the Skylab space station and docked within nine hours of launch and the crew entered the orbital workshop. Prior to docking the crew were aware of a leaking RCS thruster on their Apollo spacecraft but were able to contain it. On day 6, another leak was detected and the crew was alerted by the CSM caution and warning system. On the morning of 2 August, Skylab Program Director William (Bill) Schneider called the Kennedy Space Center and requested accelerated activation of the AS-208 vehicle and CSM-118 for standby in case it should be needed for a rescue (see next entry). If the RCS in the Service Module deteriorated or was potentially unable to support the return of the crew nearly two months later, the crew would jettison the CSM and stand by in Skylab for a rescue with the next Saturn IB /CSM in line.

The danger of astronauts being stranded in space had been discussed over many years. North American Aviation had conducted several studies and analyses of how a crew, stranded in lunar orbit, might be rescued by an Apollo spacecraft launched in response. It was found that two lightweight couches could be installed beneath the three standard couches. With a two-man crew on the standard couches, it was

Right: CSM-118 is stacked on AS-208 in preparation for a possible rescue mission to recover the Skylab crew after a suspected leak in the reaction control system of CSM-117 docked to the space station but a move eventually judged unnecessary. *NASA-KSC*

theoretically possible to mount a rescue to a three-man crew stranded in Earth orbit where time from launch to station-keeping would be several hours at most. Such a plan was not possible for a CSM stranded in lunar orbit as the flow for the next launch plus the three-day transit time would not have supported a rescue flight before consumables aboard the stranded Apollo spacecraft ran out. But it was quite possible for an Earth-orbit mission to be supported by a rescue plan where the crew had a safe haven in the form of the space station itself. Accordingly, a special requirement for such a procedure was issued on 17 May 1972 so that the sequential flow of hardware could be synchronised making the next vehicle in line available to serve in that function as a rescue flight.

Bean and Lousma had been on standby to serve as a rescue crew for SL-2, were that to be needed. Now, astronauts Brand and Lind (part of the back-up crew for SL-4) were on call to provide the same service for SL-3 using AS-208 and CSM-118. Because the flow was designed to provide a rescue service, the launch vehicle and spacecraft for SL-4 were erected on ML-1 on 31 July and 1 August. When the call came to activate the SL-R option the stack was moved out to LC-39B on 14 August but the emergency in space dissipated over the next several weeks and the SL-R option was abandoned; working in simulators, Brand and Lind had adequately demonstrated that a safe de-orbit and re-entry could be made even in the event of a disabled Service Module as the Command Module RCS thrusters could have performed a retro-burn if required. Had NASA adopted the SL-R plan and launched AS-208, it would have required AS-210 to be prepared as a potential rescue for SL-4, the last Skylab visit.

The SL-3 mission progressed, building a learning curve for both astronauts and flight controllers during which some contested flight plan schedules taxed the patience and the tolerance of both parties, productively as it turned out. A great number of science and technology experiments were carried out as well as successful operations of the remote sensing equipment in the Multiple Docking Adaptor and the instruments in the Apollo Telescope Mount, where observations were made of the Sun. The crew returned to Earth on 25 September 1973 after a flight lasting 59days 11hr 9min 4sec.

Below: CM-117 returns to Earth and a splashdown in the Pacific Ocean on 25 September 1973. *NASA*

AS-208/SL-4

Launch date:	16 November 1973
Lift-off time:	09:01:23 EST (14:01:23 UTC)
Launch pad:	LC-39B
Launch azimuth:	53.781deg
Length:	68.1m (223.33ft)
S/L thrust:	7,294.7kN (1,640,000lb)
Dry mass:	74,163kg (163,500lb)
Propellant mass:	519,914kg (1,146,196lb)
H-1 engines:	H-471; H-4077; H-4079; H-4080; H-7079; H-7081; H-7082; H-7096
J-2 engines	J-2062
Ignition mass:	594,686kg (1,311,036lb)
Lift-off mass:	587,968kg (1,296,226lb)

The original Skylab schedule had the final crew launch to the space station for a stay lasting approximately 56 days but that changed significantly after the initial problems with the launch on 14 May 1973. Initially, there was concern that the programme would be able to fulfil its original goal of three visits, the first lasting 28 days, the second and the third each remaining for 56 days. It was to these cycles that the hardware had been scheduled for launch but during the recovery process conducted by the SL-2 crew, and the expansion of capability effected by the SL-3 astronauts, plans for the third and final visit were considered flexible and open-ended, contingent on the consumables available and the condition of the station.

Fabrication of elements for the S-IB-8 stage began in 1966 with all engines installed by 22 June 1966 but during checkout engine H-7084 was seen to have a defect and was replaced with H-7081. Post-manufacturing inspection and checkout was completed on 26 September and it was shipped across to MSFC on 17 October where it arrived eight days later. Installation into Static Test Tower East was completed on that day followed by a sequence of simulated tests including propellant loading. Before a live firing, one of the Rocketdyne welders noticed a dent on a tube on engine H-4078 and further concerns were raised when the rotating turbine gave out a distinct clicking sound. The first firing, of 35.4sec, was conducted on 16 November but H-4078 displayed a low thrust level of 818.4kN (184,000lb) instead of the specified 911.84kN (205,000lb).

The engine was removed five days later with a plan to replace it with H-4071. This engine had already experienced the two standard test firings when originally installed in S-IB-6 and the day after the S-IB-8 test it was delivered to Marshall and installed on that stage during 23 November. The other engines performed as expected and no re-orificing was necessary. The second, a full duration, test firing took place on 29 November for 145.3sec, again and as normal practice, seeing the inboards shut down 2.5sec earlier. Some problem had been experienced with H-4079 in that the turbine inlet pressure was below normal but this was found to be fuel and carbon clogging the inlet screen. With these tasks accomplished the stage was shipped to Michoud, arriving there on 14 December.

As the stage entered post-static test checks and inspections, word came from Rocketdyne that they had found incorrect material used in the troublesome H-4078 engine. Instead of Haynes Stellite 21, the turbine blades had been manufactured from 316 stainless steel. This triggered an engine-wide search for other misapplication of materials, finding that engines H-7083, H-7081, H-7085 and H-4077 had a total of 51 blades fabricated from the stainless steel. The engines were returned to Rocketdyne for replacement blades and each engine static fired again before returning them to Michoud where they would all be re-installed.

This work was completed on 26 January 1967 with three of the four engines back in their original locations. When engine H-7085 had been removed from the stage it was decided to replace the turbine wheel and after it had been test fired at Rocketdyne it was transferred to the S-IB-7 stage, its place taken on S-IB-8 by H-7096, which had been removed from S-IB-11 before that stage was static fired. It was delivered to Michoud on 19 June 1967 and installed on 10 July. Post-static checks had been signed off on 17 April 1967 but H-7083 was removed from S-IB-8 on 20 July due to a contaminated LOX seal cavity, its replacement being H-7079, which had previously been removed from S-IB-7 before the static firing of that stage. It was installed on S-IB-8 on 1 August 1967. Finally, with only three of the originally assigned engines retained, this was the engine configuration for launch more than six years later.

Rocket engineer and historian Alan Lawrie has pointed

Right: AS-208 is rolled out to the pad for the third and final visit to Skylab. NASA-KSC. *MSFC*

AS-208/SL-4 – remedial changes

Below: During inspection a crack was discovered in the lower web of the upper E-beam to the outrigger fin 4 position. A coupon of cracked material was removed and a spacer and splice plate fitted (*left*). After the CDDT all eight fin assemblies were found to have stress corrosion cracks and were replaced, with reinforcing blocks installed about the mounting bolts to provide an alternate load path (*right*). *MSFC*

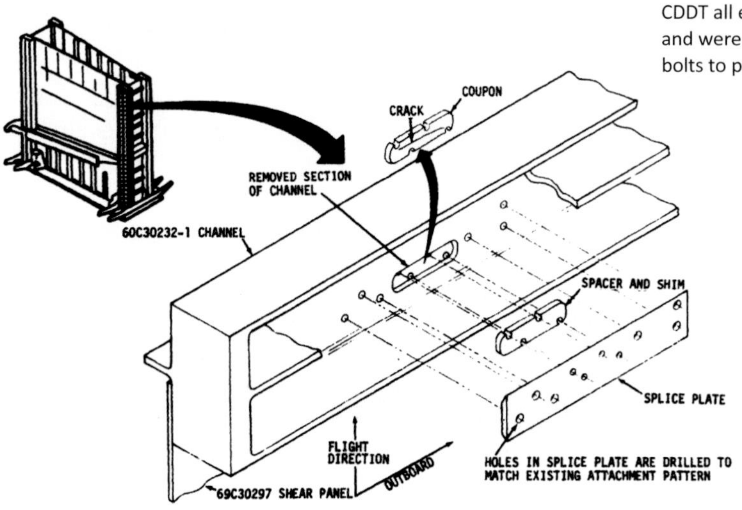

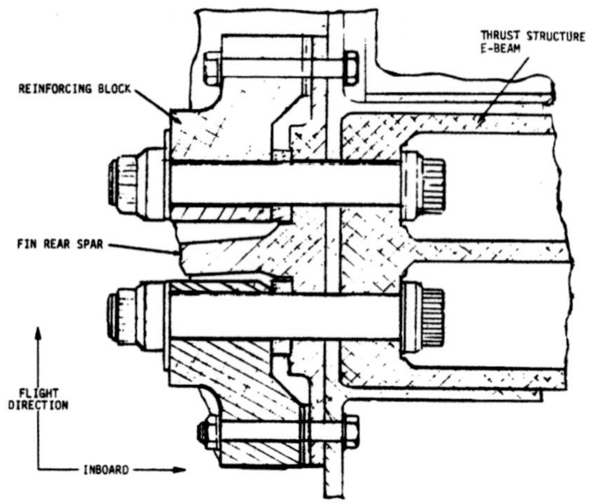

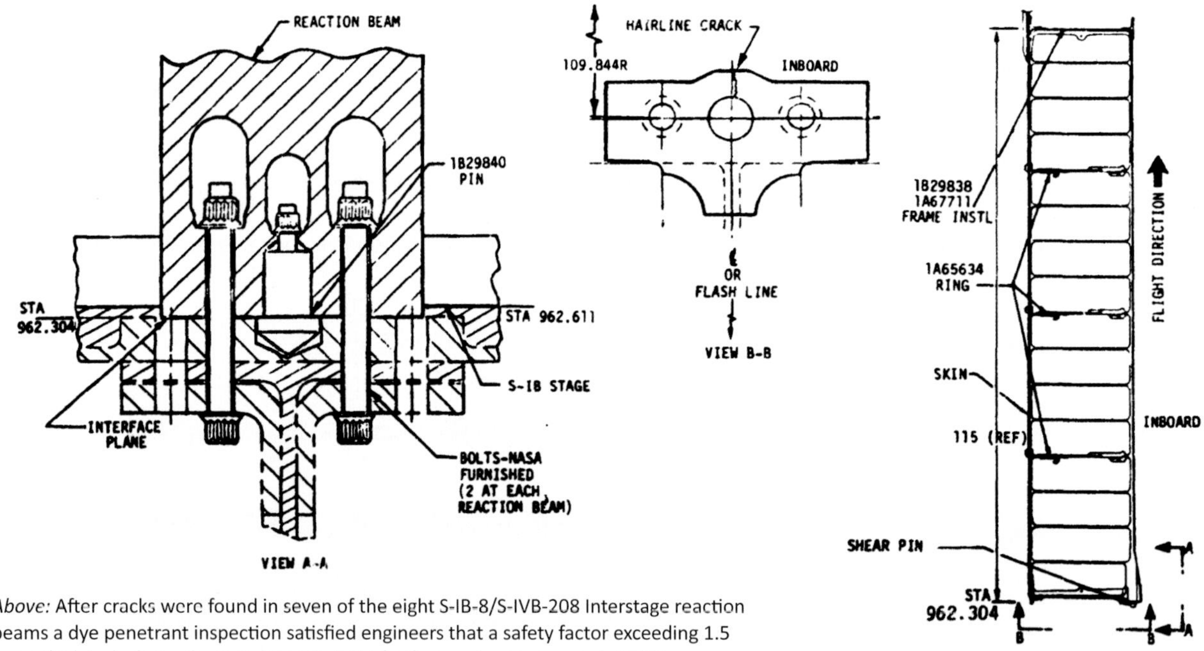

Above: After cracks were found in seven of the eight S-IB-8/S-IVB-208 Interstage reaction beams a dye penetrant inspection satisfied engineers that a safety factor exceeding 1.5 was calculated, above the mandatory 1.4. No further work was required. *MSFC*

out that the problems of incorrect material being used in engine fabrication also affected engines for the Atlas and Thor rockets, the latter accounting for 10 of the 28 turbine wheels inspected and found to have affected turbine blades. Ten H-1 engines were so identified along with six research and development engines and two identified during production and scrapped. By the time the engine changes had worked through the system, the fate of the AS-208 mission hung in the balance and S-IB-8 was placed in deep environmental storage at Michoud on 29 August 1968. With its assignment to Skylab confirmed, the stage was removed on 22 March 1972, given extensive checks and inspections and restorative attention before it was cleared on 16 January 1973 and sent to KSC on 15 June for a five-day journey.

Assembly of the S-IVB-208 stage began in 1966 with installation of the engine on 10 August followed by checkouts and inspections prior to shipping it to the SACTO facility where it arrived on 2 December. Installed in the Beta I stand three days later, the acceptance firing occurred on 11 January 1967 and lasted 426.6sec. Nine days later, the S-IVB-503 stage which had been assigned to fly the first manned Saturn V mission (Apollo 8) exploded on the Beta III stand. Located a mere 567m (1,860ft) from the Beta I stand, allaying fears that it might have sustained damage, the S-IVB-208 stage and

AS-208/SL-4 – CSM changes

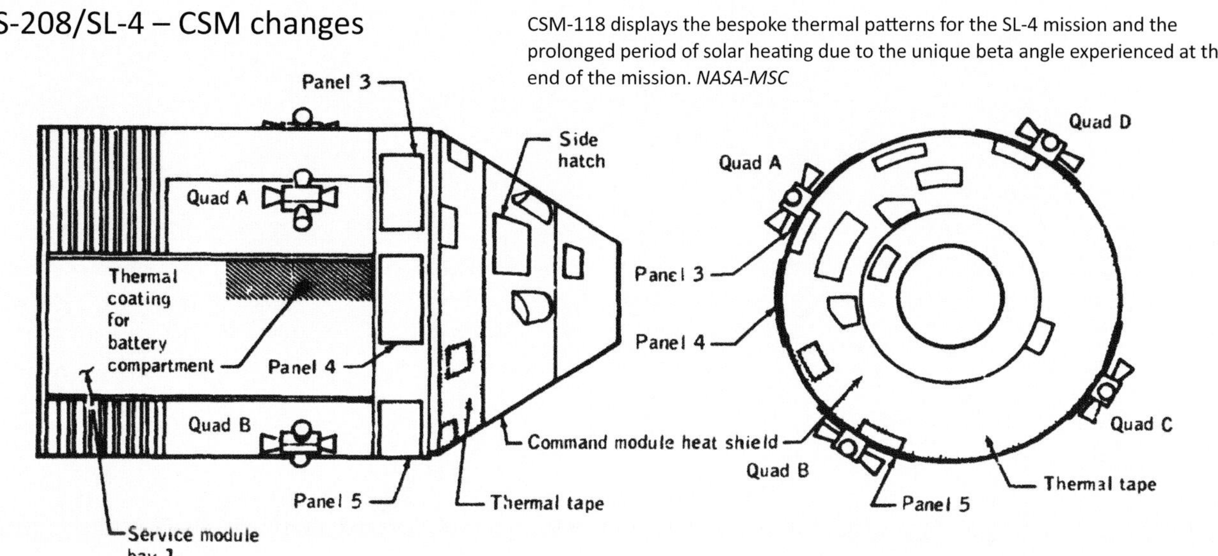

CSM-118 displays the bespoke thermal patterns for the SL-4 mission and the prolonged period of solar heating due to the unique beta angle experienced at the end of the mission. *NASA-MSC*

engine were found to be unaffected by debris or the acoustic wave from the explosion.

It was processed through the usual post-test procedures, completed on 22 March, by which time the effects of the Apollo LC-34 pad fire determined that the stage go into storage at SACTO. On 13 October 1970 it was transferred to the Douglas facility at Huntington Beach and from there to long-term storage at KSC where it arrived on 4 November 1971, its encapsulation beginning the following day. By this date its assignment had been purposed by the Skylab missions plan and it was removed for the AS-208/SL-4 mission on 28 November 1972. The notional launch date was set for 10 November 1973 and the hardware schedule was laid down for that event.

With the S-IVB-208 stage already at KSC, that element was followed by the Instrument Unit on 12 June 1973 and the S-IB-8 stage on 20 June, with stacking completed by 1 August and transfer to LC-39B on 14 August. As noted above (AS-207/SL-3), the launch vehicle for this mission was brought to a state of readiness for a possible rescue flight to SL-3 but as that was not required there had been debate as to whether to launch SL-4 on 24 September while the existing crew were still in space. This was considered a potential saving on the time caused by one crew de-activating Skylab to put it into a quiescent state before the next crew arrived to go through the reactivation process. Several systems aboard the station were showing stress and some managers felt it prudent from an engineering viewpoint to maintain them at a consistent operational state.

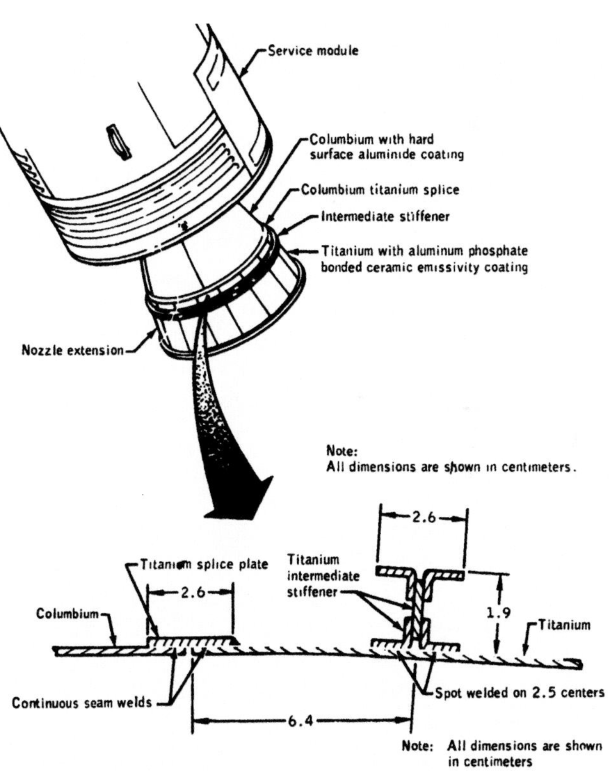

Modifications to the Skylab CSM included a nozzle extension in the area of the columbium/titanium interface and the intermediate stiffening ring which exhibited some distortion in on-orbit pictures due to the thermal shock of the extremely short duration firings of the SPS engine. *NASA-MSC*

To achieve that overlap, CSM-118 would have been launched before the existing crew departed, dock to a radial port and join the SL-3 crew in Skylab for a formal handover. The SL-3 crew would return to Earth in CSM-117, already docked to the axial port, and the SL-4 crew would then get back in CSM-118 and relocate their spacecraft to the port vacated by the outgoing crew. This plan had never been seriously considered during the development stage of Skylab and was not adopted, forestalled in any event by engineering problems with the S-IB-8 stage which prevented launch before the SL-3 crew departed.

With the AS-208 stack already on LC-39B, on 16 August flight plans focused on a launch on 9 November, so that the crew of Gerald Carr, Edward Gibson and William Pogue could observe Comet Kohoutek from space and make scientific measurements as it reached its closest pass around the

Above: Carrying CSM-118, AS-208 gets under way from LC-39B on the morning of 16 November 1973. Designated SL-4, it was the last mission to the Skylab space station, the penultimate flight of a Saturn IB. *NASA-KSC*

Sun on 28 December. By late September the launch date had moved to 11 November due to orbital perturbations of the Skylab station. Further tracking refinement moved that back up to 10 November with a launch at 11:41 EST (16:41 UTC). A Flight Readiness Test was held on 11 October with RP-1 loading twelve days later. During a normal gravity drain on the same day the upper bulkheads of two fuel tanks experienced a reversal of curvature because vent covers had not been removed. Sucked in by the partial vacuum, the tanks were reversed to a normal position by the application of a positive pressure in the ullage volume two days later.

On 2 November, a CDDT passed all tests. The formal launch countdown was scheduled to begin at 07:30 EST (12:30 UTC) on 8 November. Two days before that event, cracks were discovered in the aft attachment fittings on fins at the base of the first stage. Three cracks had also been observed on the forged aluminium fins on S-IB-9 assigned to AS-209 (which see) and this had led to the detailed inspection of equivalent areas on S-IB-8. The fin hold-down fitting passed compressional loads through the fourth rib position from the structural interface, with fin attachment provided by front and rear bolt blocks at respective spar positions where they butted with the thrust structure E-beam assembly.

Cracks were found on all eight fins with two on seven fins and one on an eighth, the worst being 3.8cm (1.5in) in length. Several weeks before, a crack had been discovered in the lower web of the upper E-beam of the outrigger assembly at one fin position. The E-beam was forged from 7178-T6 aluminium alloy and a 2.54cm x 7.62cm x 1.90cm (1in x 3in x 0.75in) coupon of cracked material was removed from the lower web and a spacer and splice plate installed to restore integrity.

The decision was made to replace all eight fins and by early on 7 November the launch had been put back to 9.37am EST (14:37 UTC) on 15 November. By that evening all fuel had been drained to lighten the stack and work on removing the first fin began at 14:33 EST (19:33 UTC) on 8 November and was completed 35 hours later. The countdown was now scheduled to begin on 02:30 EST (07:30 UTC) on 13 November. But at 11:30 EST (16:30 UTC) more cracks were discovered, this time in seven of eight S-IB/S-IVB interstage reaction beams but because they were in compression and not in tension there seemed little necessity to perform work on them. Although up to 0.63cm (0.25in) deep and 6.35cm (2.5in) long, stress analysis on those cracks indicated a safety factor of 1.5, just inside the 1.4 requirement. Nevertheless, this again put back the launch, to 09:01 EST (14:01 UTC) on 16 November.

As fin replacement continued, the mating surfaces on the thrust structure showed pitting and roughness necessitating a clean-up before installation of the replacement. But these too showed signs of long-term storage, with roughness and globules developing in the paint. Cleaned with nitric acid, buffed and treated with a zinc chromate primer, the interface was ready for mating with the stage proper, after shimming had taken out the roughness from the thrust structure. By 07:04 EST (12:04 UTC) on 13 November, all eight fins had been replaced. With reinforcing blocks installed about the mounting bolts at each fitting, this would ensure an adequate load path if cracks developed in the fittings after final inspection.

Never had such comprehensive work been conducted on a Saturn rocket at the launch pad and the conditions were challenging, technicians performing arduous and precise work in an unfamiliar environment, much of which took place in the open air and on top of the 'milk stool'. But it certainly got the attention of the media, who found something exciting to carp about and raise fearful predictions, musing on whether the rocket would fall over! Most were keen to point out to their readers that the rocket that was about to launch three astronauts to Skylab had been in post-manufacturing and checkout for more than seven years. For its part, NASA was at equal pains to urge upon the public that there was no danger and only launch director Walter J. Kapryan brought a tone of levity when he remarked, however, that: 'The metallurgists tell me…that age does make a difference!'

Preparation for SL-4 involved two launch vehicles, the second being AS-209 which was made available for a possible rescue flight. The two stages for that rocket had been fabricated in 1966 with stage testing for the S-IB later that year and in early 1967 for the S-IVB. The S-IB-9 stage arrived at KSC on 20 August 1973 and was stacked with the S-IVB-9 stage which had been in storage at the Cape since 12 January 1972. The alert triggered by the discovery of fin cracks on S-IB-8 prompted inspection and replacement of all eight fins on the S-IB-9 stage before stacking in the Vehicle Assembly Building at KSC. The AS-209 vehicle and CSM-119 were moved to LC-39B on 3 December 1973, three weeks after the launch of AS-208/SL-4. Upon completion of this last Skylab mission, the stack was rolled back to the VAB and disassembled for storage in April 1974.

Meanwhile, the terminal countdown for AS-208 resumed on 14 November at T-42.5hr, with 1hr 2min of scheduled holds and no unexpected delays or waivers. The total payload weight including the LES was 20,878kg (46,028lb). The launch took place on a bright and sunny day but with high winds aloft, AS-208 passing through Mach 1 at 59.5sec and Max-Q at 1min 9.5sec. Inner engine cut-off occurred at 2min 17.8sec with the outers shutting down at 2min 21.2sec and an altitude of 57.35km (35.6mls). Staging occurred at 2min 22.5sec with the stack at 58.54km (36.37mls) followed by the S-IVB ignition command going in at 2min 24sec. The S-IB stage impacted the ocean at 8min 54.2sec, 492.5km (306mls) downrange. S-IVB stage shut-down came at 9min 37.1sec, with the payload in an orbit of 227km (141mls) by 150km (93.3mls) and a period of 88.26min. Separation of CSM-118 occurred at 18min 00sec at an altitude of 169.6km (105.4mls). At separation the CSM-118 weighed 15,554kg (34,291lb). The propellant dump for de-orbit began at 5hr 10min 37.7sec and involved a LOX bleed lasting 7min 55sec and an LH_2 dump of 86sec. Impact occurred at 6hr 3min 56sec at 24.5deg N by 172.3deg W.

The last visit to Skylab was a great success. A considerable amount of experience was gained which added measurably to the scheduling of mission tasks, timelines and general routines on visits to the International Space Station around 25 years later. The SL-4 mission was extended beyond the notional 56 days originally planned and supported a record number of experiments and scientific activities. The crew returned to Earth in CSM-118 on 8 February 1974, having logged 84days 1hr 15min 31sec from lift-off to splashdown.

Thus ended the operational flight activities in support of the Apollo programme, which since its conception in 1960 had always envisaged flights to and around the Moon, and in support of an Earth-orbiting space station. Which, it turned out, was the world's first orbiting laboratory to be revisited for habitation on successive occasions.

AS-210/ASTP

Launch date:	15 July 1975
Lift-off time:	15:50:00 EDT (19:50:00 UTC)
Launch pad:	LC-39B
Launch azimuth:	45.2deg
Length:	68.1m (223.33ft)
S/L thrust:	7,294.7kN (1,640,000lb)
Dry mass:	66,747kg (147,150lb)
Propellant mass:	450,575kg (993,349lb)
H-1 engines:	H-4087; H-4089; H-4090; H-4104; H-7091; H-7093; H-7094; H-7099
J-2 engines	J-2087
Ignition mass:	517,348kg (1,140,556lb)
Lift-off mass:	511,159kg (1,126,913lb)

The very existence of this flight was due to the development of a cooperative venture with the USSR whereby an Apollo spacecraft would dock to a Soyuz spacecraft. It would be the last flight of an Apollo CSM, the last flight of the Saturn I/IB series and the last time a US crew would launch on a US-built ballistic spacecraft for almost 45 years; between 1981 and 2011, NASA would launch 135 reusable Shuttle vehicles returning to a runway landing on solid ground. Only after that would ballistic capsules resume the transport of astronauts from US soil, carrying crew to and from the International Space Station starting in 2020.

The groundwork for a possible joint venture had been laid by President Kennedy in 1962 when he invited the Soviet premier Nikita Khrushchev to talks which would be continued by leading scientists and officials on each side long after his death. Initially limited to an exchange of scientific data on Earth observations and weather, by the early 1970s there was a thaw in East-West tensions leading to a period of détente.

Below: Preparation of the 'milk stool' for AS-210 as activity begins to wind down on the Saturn IB programme. *NASA-KSC*

U S A
UNITED STATES

Right: The S-IVB-210 stage hovers above S-IB-10 shortly before mating, as the AS-210 launcher is prepared for the final flight in the series, fourteen years after the first Saturn I was prepared for a flight on LC-34. *NASA-KSC*

The mood was sympathetic and both President Richard Nixon and foreign policy adviser Henry Kissinger positioned NASA to conduct discussions on behalf of the White House and an agreement for a joint flight was signed in April 1972. With an impending end to operations with the Apollo spacecraft, it was a unique opportunity to use available hardware. It was also an opportunity for desk-bound astronaut Donald K. 'Deke' Slayton to finally reach orbit, accompanied by Vance D. Brand and mission commander Thomas P. Stafford. Selected as one of the 'Mercury Seven' astronauts in April 1959, an erratic heart beat discovered during training kept him away from crew assignment until a personal health plan alleviated the problem and from 13 March 1972 he was a contender for space flight.

The two spacecraft were very different and bringing them together for mutual occupation would bring significant, but not unsurmountable, engineering challenges. Soyuz used a two-gas, oxygen/nitrogen, atmosphere for a near sea-level environment, while Apollo used a pure oxygen atmosphere at one-third sea level pressure. To allow movement between the two vehicles, the pressure inside Soyuz would have to be reduced by one-third and transfers made via a Docking Module (DM) which was manufactured by North American Rockwell, the company that built the Apollo spacecraft. The DM was carried inside the SLA where a Lunar Module would be on Moon flights, attached to the lower section of the SLA and supported on a truss assembly at four points where on Saturn V flights the LM would be attached. It would be extracted by the CSM after separation and docked to the forward docking collar on the Apollo spacecraft. The opposite end of the DM would be used to dock with Soyuz.

The spacecraft selected was one of the old H-series which would have followed Apollo 14 into space had the last of the unmodified Block II types, prior to the extended-duration J-series missions (Apollo 15, 16 and 17), not been cancelled.

Left: Unique for the three manned Skylab missions and the one ASTP flight, the elevated launch position of the Saturn IB was distinctive and counter-intuitive to see, in a paradoxical way a reflection of just how big was the Saturn V for which LC-39 had been built. *NASA-KSC*

Below: The flight objective was to place CSM-111 in orbit and to eventually rendezvous with, and dock to, the Soyuz spacecraft previously sent into space using a Docking Module carried inside the SLA. *NASA-MSC*

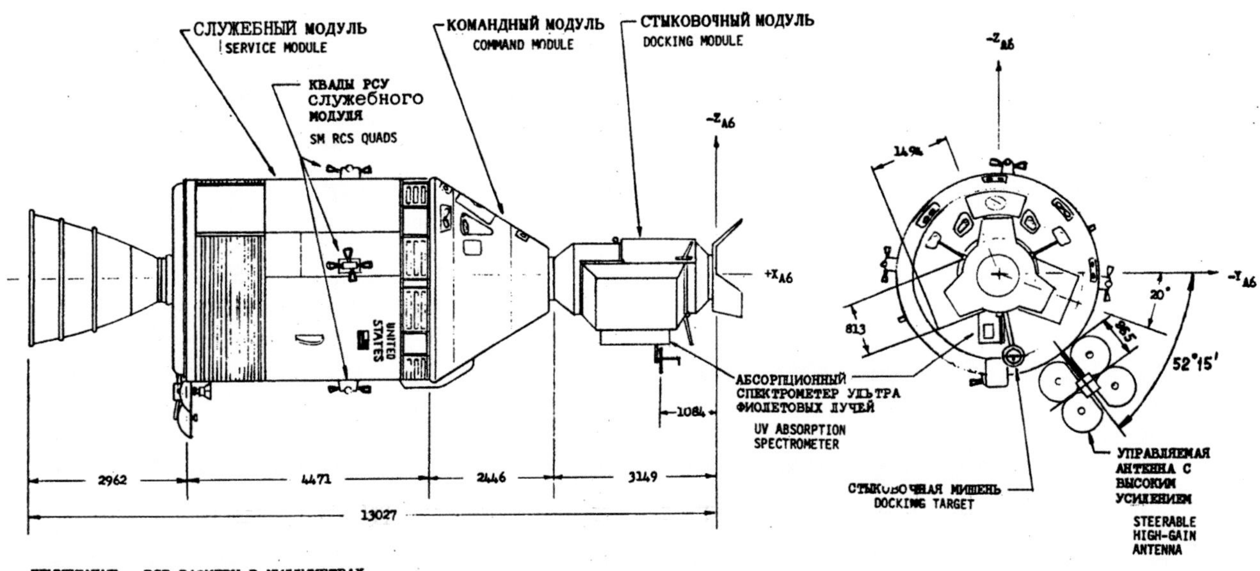

AS-210/ASTP – Key modifications

Right: The dimensions and specific characteristics of SLA-22 employed for the AS-210 flight with the same load positions as previously utilised for a Lunar Module. *MSFC*

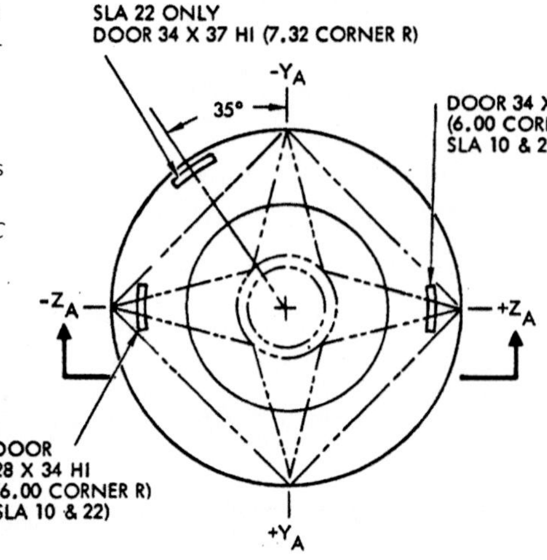

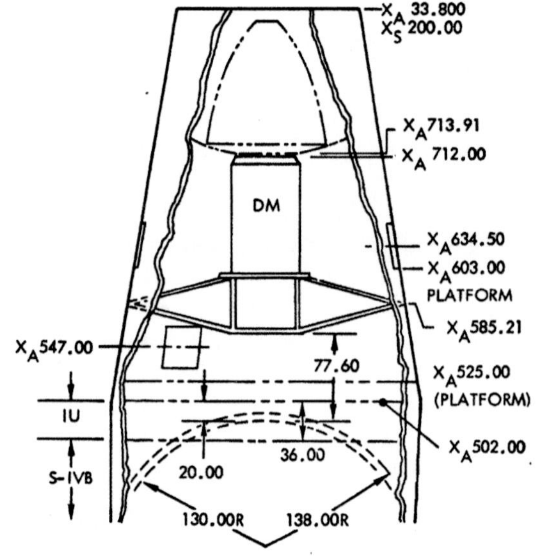

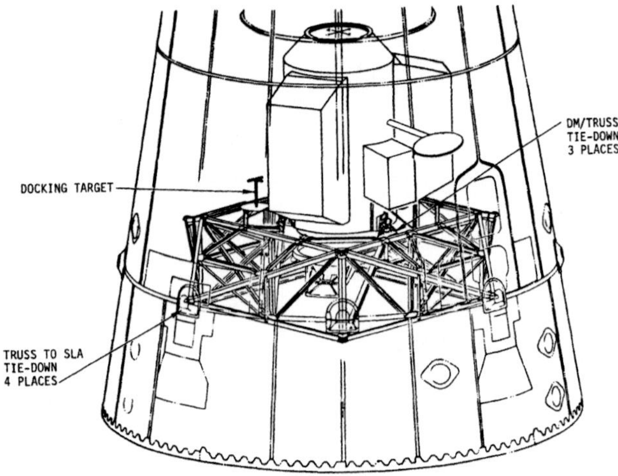

Above: The Docking Module attachment truss and tie-down fittings for the SLA panels which were jettisoned at deployment. *MSFC*

Below: Detail on the SLA panel fixture, deployment and jettison assembly with release of the panels occurring at a 45deg tangent to the longitudinal axis of the S-IVB stage. *MSFC*

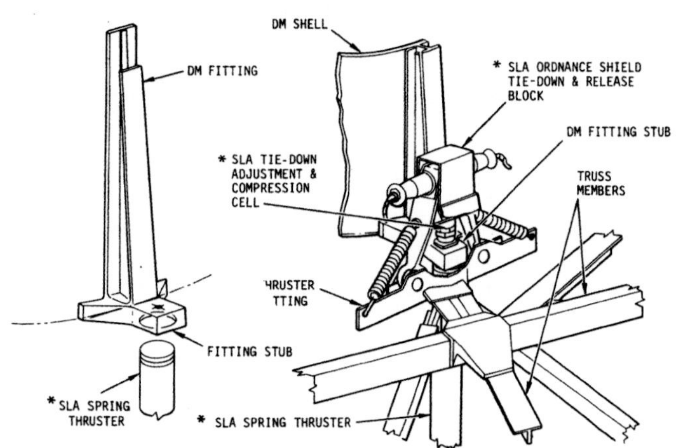

CSM-111 manufacturing began in 1967, the spacecraft entering storage until 1972 when it was brought out for modifications fitting it for ASTP. It was delivered to KSC on 7 September 1974 for launch on AS-210.

The last flight of a Saturn IB involved a combination of rocket stages which had been manufactured nine years previously, fabrication of the S-IB-10 stage starting on 6 September 1966. The eight engine were installed by 5 December that year. When incorrect materials were discovered to have been used on engine turbine blades for the S-IB-8 stage, engine H-7092 in the S-IB-10 stage was determined to have two blades similarly affected and the engine was removed on 13 December and replaced with H-7099, delivered on 10 October and installed on 20 December. Stage checkout was completed on 7 March 1967 and it was sent to MSFC on the barge *Palaemon* on 31 March to arrive seven days later.

The stage was placed in the Static Test Tower East on 10 April to conduct its first, 35.4sec, firing on 9 May. Noting that H-4090 had a low thrust of 932.5kN (209,645lb), the orifice was replaced. Politics got in the way of the second, full duration firing, which had to wait for five days to conduct the test during Vice President Humphrey's visit on 22 May. That test lasted 145.7sec, the inboards cutting off 3sec earlier as planned. Taken from the stand on 7 June, S-IB-10 was on its way to Michoud the following day on the barge *Palaemon* where it arrived on the 13th. It was placed in long-term storage on 30 August 1967, where it remained until removed on 30 October 1972, when called upon to fly the ASTP mission in 1975. However, during post-incubation modification and inspection engine H-4088 was removed and replaced with H-4104. That engine had last been tested on 8 May 1967 and had been in storage at Michoud since 3 July that year. After a lengthy post-storage checkout phase, S-IB-10 was signed off and left Michoud for KSC on 19 April 1974 and arrived on the 24th. Placed in environmental storage five days later, it was removed on 3 December 1974 for pre-flight preparations and eventually for stacking in the VAB.

Assembly of the S-IVB-210 stage began in February 1966

with the engine being delivered to Huntington Beach on 11 November and installed on 20 January 1967. Some work was required during the checkout between 3 February and 22 March 1967 to replace a deformed helium sphere and to change out another over concerns about its condition. With systems tests and leak checks completed, without a specified mission the stage was put into long-term storage on 23 April 1967 in Huntington Beach and not subject to any engine firing until flown in support of ASTP, the only S-IVB stage to launch without a static test. Periodically checked and inspected, it was finally removed from storage on 27 January 1971. But it took a year to complete post-manufacturing checkout, conducted a second time due to the length of time encapsulated, before release on 11 January 1972. Departing from Huntington Beach on 7 November on the Super Guppy, it arrived at KSC the following day. It was held there in storage between 14 December 1972 and 11 February 1974 at which point it was checked again before a further spell between 29 May and 3 September on which latter date it was prepared for joining S-IB-10.

Because fin corrosion had been discovered late in the launch preparation phase for AS-208, special attention was given to the possibility of stress corrosion on the first stage for AS-210. Compressive stress has been placed on the structure by a method known as rod-peening, whereby small bundles of rods were pounded against the material by a pneumatic device. Also, engineers had rounded out each bolt hole in an attempt to remove potential stress points.

The first stage was erected on the Mobile Launcher pedestal on 13 January and the S-IVB was mounted the following day, followed by the Instrument Unit on the 16th. Timely inspection of the eight S-IB-8 fins revealed a hairline crack in two of the 16 hold-down fittings on 19 February and before the end of the month similar cracks had been discovered on the other six fins. All eight were replaced by new fins tested at MSFC before delivery to KSC. A boilerplate CSM was added on 5 March before Apollo CSM-111 was erected in place fourteen days later and the complete stack rolled to the LC-39B on 24 March. Integral to the assembly had been the Docking Module secured within the SLA and to which the Apollo spacecraft would conduct transposition and docking prior to rendezvous with the Soyuz spacecraft. The CDDT was successfully completed on 3 July.

One notable addition to the MLP was the fitting of a 2.4m (8ft) insulated fibreglass mast with a 1.27cm (0.5in) diameter cable running from the top to positions 305m (1,000ft) either side of the mobile launcher. The mast itself was attached to the top of the Launch Umbilical Tower. This new installation would carry lightning strikes clear of the stack and reduce the risk of electrical interference with equipment aboard the launch vehicle or the spacecraft. Sudden electrical storms are common in Florida during July. Additional safeguards were in place, however. Several years previously, NASA had asked the National Oceanic and Atmospheric Administration (NOAA) to examine the potential danger from storms in the vicinity of Cape Canaveral.

Each launch had introduced sophistication to the task of monitoring for electrical activity but the final Saturn IB launch reached a new level of preparedness. If electrical force built up to a danger point, a piston-engine Convair T-29 aircraft would seed the clouds with chaff and discharge the current before it could threaten an ascending rocket. The chaff consisted of several million fibreglass, aluminium-coated, needles 10cm (4in) long each capable of producing a 1-microamp corona in a charged cloud. Altogether, eight aircraft were stacked up between 1,524m (5,000ft) and 12,500m (41,000ft) to monitor electrical activity, moving into position one hour before launch and clearing the area just five minutes before lift-off.

The AS-209 stages, IU, SLA and CSM-119 which had been prepared as a possible rescue vehicle for AS-208 were in the VAB at the time the ASTP mission flew but there was little that this vehicle could do to effect a launch in time to save a stranded crew. What had been possible for Skylab, where the station itself provided safe haven for an extended period, would not work for this mission. There were insufficient consumables on CSM-111 to support life long enough for that

Below: AS-210 rolls out the VAB as the stack heads for LC-39B (left). Special lightning protection was provided for the AS-210 flight, with a mast attached by crane from the top of the Vehicle Assembly Building after the stack had cleared the door (right). *NASA-KSC*

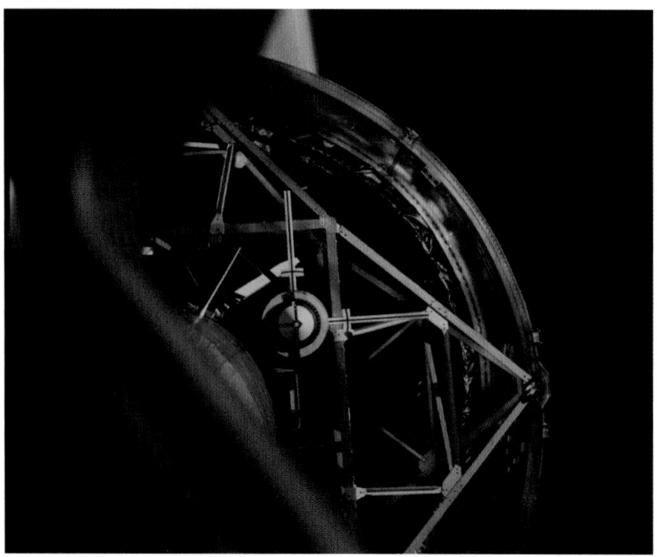

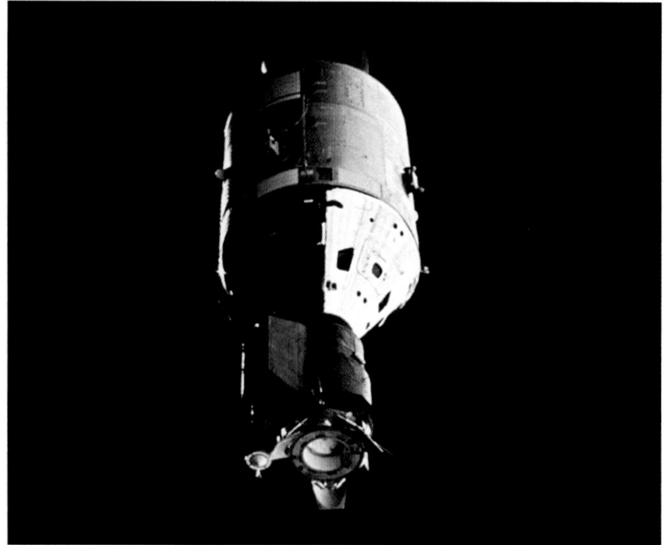

Above: A view of the Docking Module (*left*) as CSM-111 moves in to engage the module and extract it from SLA-22 and (*right*) CSM-111 as viewed from the Soviet-era Soyuz spacecraft. *NASA-KSC*

launch vehicle to go through preparations and make it into space in time. But the concept of a rescue vehicle became enshrined in future missions planning and would become an integral part of operational capabilities with the International Space Station, where a second spacecraft was always docked should it be needed for a rapid return to Earth.

The manoeuvring potential of the Apollo spacecraft was much greater than that of Soyuz and CSM-111 would follow the Russian vehicles into space about 7.5hr after the Soyuz launch. This would allow the Russian spacecraft to stabilise in an orbit which would be pursued by Apollo for a rendezvous and docking. The launch window for Apollo was at most 10min each day with an optimum window of 5min. The limited life of Soyuz constrained the longevity of the launch windows. Any delay in launching Apollo on the first day (15 July) would be followed on the second and third days by similar windows. Because of the limited time Soyuz could remain in orbit, the maximum time docked would be reduced for every additional day, declining from a nominal maximum of 43.8hr for the first three launch windows to 21.6hr on day four and 7.5hr on day five (19 July). Any delays in getting the Saturn IB off the pad after that would have no purpose; having launched on 15 July, due to limited consumables Soyuz would have to return to Earth on 21 July.

It was for these reasons that some concern was attached to a problem during the countdown. About one hour prior to lift-off, technicians reported that there was a potential problem with a hydraulic system employed to move the swing-arms back from the Launch Umbilical Tower. Each swing-arm has an accumulator connected to pumps in the main hydraulic reservoir for top-up. This seemingly arcane bit of technology could abort the launch if it failed to function a tantalising three seconds prior to lift-off. It appeared that two pumps in the reservoir had failed. If they had, and the individual swing-arm accumulators ran below specified levels, they would not work when required to clear the launch vehicle for lift-off. But the pumps were needed only once every 7-10 hours and it was judged highly unlikely that they would need to draw on hydraulic fluid pumped up from the reservoir. Nevertheless, it was decided to manually command the pumps on at T-3min 20sec ensuring they would operate if required by the individual swing-arms.

There was another concern, one which the launch crews could do nothing about, and one which lived up to Florida's reputation as the 'Thunderstorm State'. Ever since Apollo 12 had drawn energy out of charged storm skies back in November 1969, knocking out spacecraft systems on the ascent, weather criteria demanded no electrical activity within a wide distance of the Cape. Each day, thunderstorms had rolled across the area with an electrical potential which could have caused a hold for launch. On this day, however, while a storm cell began to develop offshore it dissipated as it crossed the coast moving inland. A veritable fleet of aircraft patrolled the skies with lightning discharge sensors mapping the potential for a strike.

Around the world more than 10,000km (6,200mls) from Cape Canaveral, the Soyuz spacecraft was launched at 08:20 EDT (12:20 UTC) on 15 July and went into an initial orbit from where some adjustments were made to place it in an appropriate path as a target for the Apollo spacecraft. At Cape Canaveral, the final stages of the AS-210 countdown went without a hitch and ascent to orbit was trouble-free and standard for a typical Saturn IB launch. The inner engines cut off at 2min 17sec followed by the outer engines at 2min 20sec and an altitude of 58km (36mls) followed by separation. The S-IB stage impacted the ocean 486km (302mls) downrange, some 482km (300mls) off the coast of Savannah, Georgia. Ignition of the second stage came on time and the S-IVB shut down at 9min 42sec with the remaining stack at an altitude of 158km (98mls). Orbital insertion was a little off from the nominal, the initial orbit being 170.5km (106mls) by 153km (95mls), an error of 3km (2mls) high in both apogee and perigee, but with an inclination of 51.6deg.

Right: The last Saturn of any type leaves the Kennedy Space Center during the early afternoon of 15 July 1975 as AS-210 lifts off, the last Apollo spacecraft and the passing of an era. *NASA-KSC*

USA

CSM-111 separated from the SLA at 1hr 14min and after extraction of the Docking Module, the assembled spacecraft weighed 14,743kg (32,508lb) of which 2,012kg (4,436lb) was the Docking Module itself. A considerable quantity of loose hardware used in CSM-111 had been scavenged from other spacecraft and used equipment flown on previous flights. One of which was the docking probe from Apollo 14 that caused so much trouble on the third lunar landing flight to reach the Moon's surface. Refurbished by the manufacturer, it was fed back in to the Apollo inventory chain and ended up on CSM-111 where it again caused problems when one of the three capture latches failed five times before working. Earlier, the crew had wrestled with a technical problem on the probe before discovering one component in the probe head had been incorrectly wired.

The first two days of the mission proved troublesome, with a crush of flight plan activity spurring the crew on chasing a wide range of tasks and activities. A busy schedule of science activities, inserted so as to placate politicians critical of the need for this mission at a time of budget cuts, laboured the schedule. Some legislators said the $210million it cost to mount this mission could have been better spent on a second Skylab station rather than conducting a costly public-relations exercise which had no legacy for US operations.

For this mission, CSM-111 carried only 1,233kg (2,727lb) SPS propellant for major manoeuvres and orbital changes, less than 7 per cent the total other Apollo spacecraft had carried for a lunar mission. A series of rendezvous manoeuvres preceded docking with the Soyuz spacecraft, which came at 12:09:09 EDT (16:09:09 UTC) on 17 July, followed by a hard dock with closed latches 3min 21sec later. There followed the heavily orchestrated sequence of prepared exchanges and apolitical sentiments as the crew shared (pre-planned) 'thoughts' with their global audience. Undocking occurred at 08:12 EDT (12:12 UTC) on 19 July with a second docking less than 30 minutes later. Final undocking occurred at 11:26 EDT (15:26 UTC) that day and Soyuz returned to Earth at 06:51 EDT (10:51 UTC) on 21 July, ending a mission which, for Soyuz, lasted 5days 22hr 31min.

During their own descent three days later, the CSM-111 crew had to manually deploy the drogue parachutes but because they had not been activated automatically the thrusters continued to fire, causing violent oscillation as the Command Module descended on parachutes. Some of the toxic vapour from the thrusters entered the spacecraft causing some crew discomfort for the remainder of the time they were in the spacecraft and for some period thereafter. The last Apollo mission ended at an elapsed time of 9days 1hr 28min.

Below: Activated as a result of the Russian challenge with Sputnik in 1957 and the first manned orbital flight in 1961, the pace of the Saturn I/IB programme had cleared the way for Moon landings and the world's first re-visited space station. In its final role, the last Saturn IB brought a measure of cooperation with the USSR as both the United States and Russia sent their spacemen to shake hands in space. *NASA*

Chapter Twelve

Stages and Missions Never Flown

It seems extraordinary, perhaps outrageous, that of all the stages and rocket motors built for the Saturn IB programme, only nine were actually flown. It can only be explained by the policy shift that required cancellation and redundancy for all the equipment development and manufactured over the previous decade. It signalled the greatest strategic shift in the history of the manned space flight programme experienced by NASA in its substantial history and nothing like it was to occur again. Even the gargantuan Space Launch System developed to underpin America's desire to put astronauts back on the Moon, leans heavily on what has gone before as a product of this monumental shift in strategic thinking played out during the first half of the 1970s.

The transition from a long-term use of Apollo hardware to a new era of reusable winged shuttlecraft brought an end to the Saturn IB and operational missions and the shift in future planning outpaced the use of contracted stages. Here, we look at the Saturn IB hardware built but never flown, all of it redundant to the post-Apollo programme. Note that the Apollo-Saturn mission numbers in the Saturn IB 200-series are adopted in the following stage information so as to group them into the pairings which were allocated when fabrication and assembly began at Chrysler and Douglas, respectively. It is not meant to imply that these were formal mission numbers with specific manifests.

AS-209

H-1 engines:	H-4087; H-4089; H-4090; H-4104; H-7091; H-7093; H-7094; H-7099
J-engine	J-2087

Although this launcher never flew it was assigned a mission number due to the role it played as a standby rescue vehicle for the third visit to Skylab. At first, however, it was to have played a role in developing flight operations for Apollo Moon landings, although that became redundant after the redirection of the manifest after the Apollo fire of January 1967. It is probable that this launcher would have launched the first crew sent to Skylab B, had that station been developed. There was a firm plan for a second station, and it was built, but never flown due to the tight financial constraints under which NASA had to operate while finding money for the emerging Shuttle programme, which entered the costly procurement phase starting in 1972. That hardware now resides, on display and with public access to the interior, at the National Air &

Above: Retired from service after only four Saturn I and three Saturn IB launches, LC-34 today stands as a memorial to the crew of Apollo 1 who perished at this place on 27 January 1967. *Author's collection*

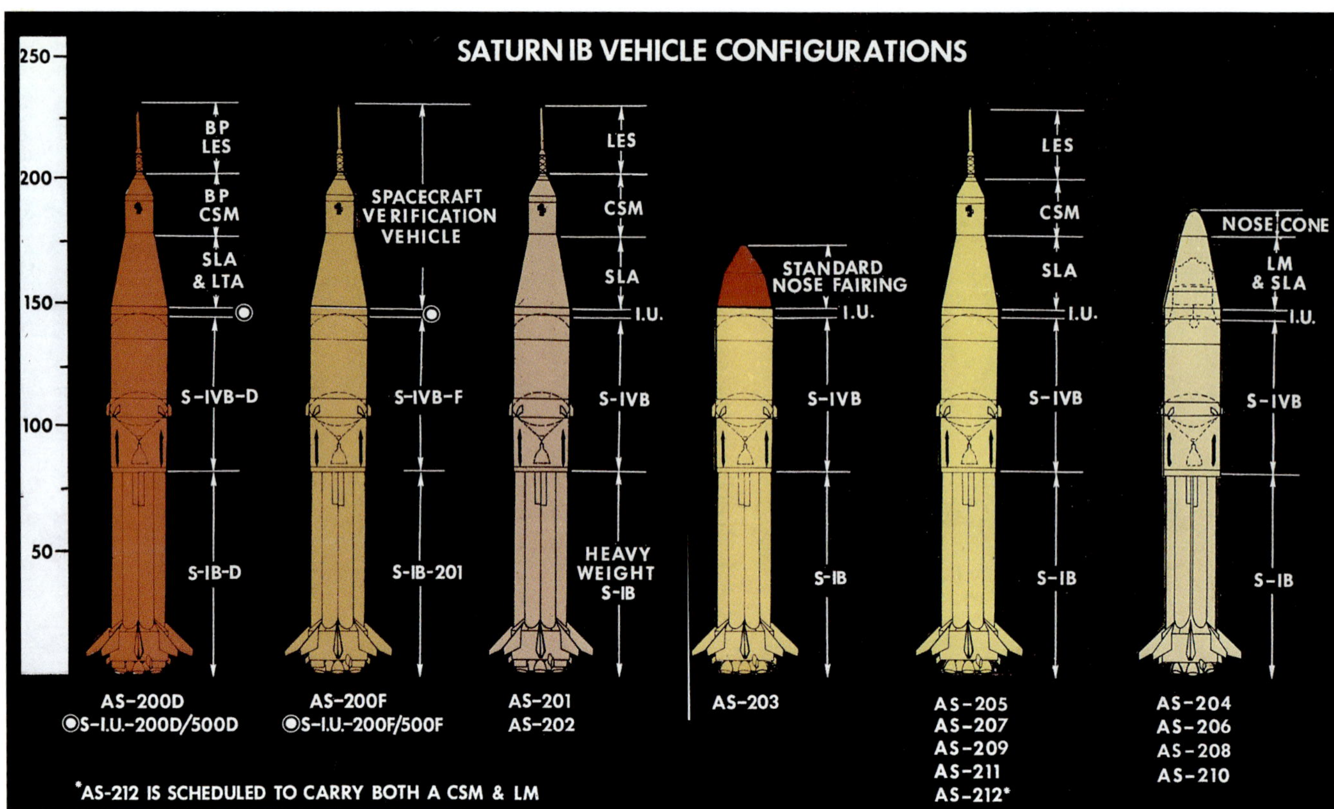

Space Museum in Washington, DC.

Assembly and manufacture of S-IB-9 began in early 1966 and the initial set of engines were installed by 12 September with the post-manufacturing phase completed on 20 December. Checks were made following the discovery of incorrect material having been used in the manufacture of turbine blades for some engines on the S-IB-8 stage and H-7086 and H-4081 had to be replaced on S-IB-9. Replacement engines H-7090 and H-4086 had been installed by 30 December. On 19 January 1967, S-IB-9 departed Michoud on the barge *Palaemon* and was at MSFC seven days later where it was installed in Static Test Tower East on 27 January 1967 – the day of the Apollo fire which, indirectly, redirected its fate.

Nevertheless, simulated flight tests were held early February and early March during which a number of components had to be replaced and some limited contamination was found in the gas generator on engine H-7090 but it was cleaned and reinstalled. During the first firing trial on 24 February the planned 35sec test was aborted after 13.5sec by a faulty digital data acquisition system but a short-duration run of 25.4sec was satisfactorily completed three days later. The third firing, planned as a full-duration run, took place on 7 March but that too was aborted at 3sec for a similar reason. The final attempt was conducted later the same day and ran for 145.488sec on the outboards, just over 3sec less for the inboard engines. Engine H-7088 exhibited a higher thrust level at 927.1kN (208,430lb) and inspection revealed burned aspirator lips on

Above: Absenting the AS-200D development and AS-200F facilities test vehicles, four different Saturn IB flight configurations were envisaged in this 1968 graphic displaying planned flights, of which AS-209, AS-211 and AS-212 never flew. *NASA*

Left: CSM-119 is removed from AS-209 when SL-4, the third and last Skylab mission, proved not to need a rescue vehicle. It could have been used to fly a short-duration SL-5 mission to Skylab to raise the station's orbit for later reactivation by Shuttle. That function was relegated to an early Shuttle mission, expected to be operational by 1978. In any event, AS-209/CSM-119 was reserved as ASTP back-up. *NASA*

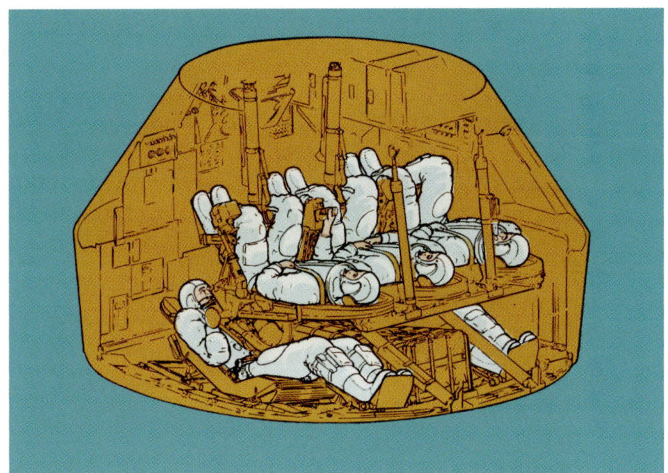

Above: Skylab operations called for a rescue function where the Command Module would be configured to fly with a crew of two, recovering the stranded crew of three with additional seats below the standard couches for return to Earth. *NASA*

engines H-7088 and H-7089, but repairs obviated the need for changing these engines.

On 14 March, offloaded from the test tower, the stage was placed aboard the *Palaemon* and it began its journey off back to Michoud the next day, arriving on the 20th. After inspection and checks, it was placed in storage on 3 August and from there to the more comprehensive environmental storage on 12 February 1968, where it remained until taken out on 6 July 1972 for modifications completed by 22 November. It was in this period that engine H-7090 was replaced with H-7110, which was in place by 27 October 1972. Removed for post storage and post modification checks, completed on 28 March 1973, S-IB-9 left Michoud for KSC on 14 August and docked there six days later.

The S-IVB-209 stage was in assembly from early 1966 and the engine was installed on 15 November. After post-manufacturing checks, the stage departed Huntington Beach on 9 March 1967 arriving at SACTO the following day. Installed in the Beta I stand it was subject to the usual cycle of further checks prior to simulated phases of a test profile culminating in a hot firing on 14 June. That was aborted due to a faulty ignition detection probe and then a second time due to a minor delay. The full duration test was successfully run on 20 June for 455.9sec, the last Saturn IB S-IVB stage so fired. A further range of post-test clean-up and checks preceded removal of the stage on 7 July. But there was no formal examination and clearance for transfer to KSC, any mission for this stage having evaporated.

It remained in storage at SACTO from 19 July 1967 to 22 July 1970 when it was flown by Super Guppy to Huntington Beach, departing on a trans-continental flight for long-term storage at KSC on 11 January 1972, where it arrived the following day. Having been assigned to the AS-209 Skylab rescue vehicle, it was removed from storage on 2 July 1973 and reprocessed for flight and stacked on the S-IB-9 stage in the VAB. For its adopted role as a potential rescue vehicle following the launch of the third Skylab visit, planned to launch

Below: The last spacecraft to stand in as a potential rescue vehicle, CSM-119 would have flown on AS-209 and is now on display at the Apollo Saturn Center, Kennedy Space Center, Florida. *Author's collection*

on 10 November 1973, preparations were accelerated for placing AS-209 in a position to retrieve a stranded crew.

But inspections during early November revealed cracks on all eight stabilising fins on S-IB-9 and that in turn led to checks on S-IB-8 which was already on LC-39A awaiting launch. While work got underway to replace affected fins on S-IB-8, the fins on S-IB-9 were replaced, with the stage still in the Vehicle Assembly Building. Readers are referred to coverage above (AS-208) showing the integrated story of these two launchers

After the delayed but successful launch of AS-208 on 16 November, the AS-209 vehicle was moved out to LC-39B on 3 December and a flight readiness test was conducted two weeks later. It remained in place on that pad until the SL-4 missions came home safely, its function no longer required. Rolled back to the VAB, the stack was de-mated from the top down. The S-IB-9 stage was placed in environmental storage at KSC beginning on 19 November 1974. There was no further use of the AS-209 stages after the end of all Apollo-related flights with ASTP. They were retired out of use and after a brief period at the 1978 Space Expo in Tokyo, Japan, moved to the KSC Rocket Garden where, in 1994 the eight H-1 engines of the first stage were removed and replaced with red engine covers. Reclining in a horizontal position for visitors to view, AS-209 stood as testimony to what might have been. Recently, it has been subject to an extensive restoration process and remains at KSC, brought back to its pristine condition.

Below: Several more flights from the LC-39B 'milk stool' may have taken place and there were plans for a Skylab B before NASA cancelled all Saturn missions as it turned toward the reusable Shuttle. *Author's collection*

AS-209

H-1 engines for static firing:	H-4091; H-4092; H-4093; H-4094; H-7086; H-7092; H-7095; H-7102
J-engine for static firing	J-2095

There was no specific mission assigned to this Saturn IB flight number but because both stages were available the sequence has been retained for convenience. However, had programme plans retained their original sequence, this launcher would have supported crew supply flights in the Apollo Applications Programme and, conceivably, supported a second Skylab space station, called Skylab B (see AS-209). If Skylab B had flown as it could have been, AS-211 would likely have been assigned to the final crew delivery flight for that facility. For this reason it is an important step in the long history of the Saturn IB.

S-IB-11 stage assembly began during 1966 with all eight engines installed by 20 February 1967, but post-installation checkout found contamination in all four outboard engines; H-7095 was replaced with H-7092 (previously on S-IB-10), H-7096 with H-7095, H-7097 with H-7086 (removed from S-IB-9 in December 1966) and H-7098 with H-7102, in that sequence between 11 May and 8 June 1967. After an extended series of inspections and checks to clear for any further contamination, S-IB-11 left Michoud aboard the barge *Palaemon* on 20 October.

At Marshall seven days later, S-IB-12 was placed in Static Test Tower East on 1 November and after the now mandatory checks had been made to ensure the appropriate materials had been used for the engine turbine blades, preparations went ahead to conduct the first test firing. But these were conducted with a different purpose in mind to previous tests on earlier stages. At this stage in the development of the H-1 engine, the higher stage thrust of 910kN (205,000lb) introduced with AS-206 was believed to carry some risk of destabilisation when all eight engines were firing together. For this reason, five tests were scheduled to measure the effect of charges placed inside the combustion chambers of two H-1 engines in the cluster, detonated during firing, and to monitor how well they reacted to this unstable effect.

The first test was of short duration, lasting 35.3sec on 19 December, but three engines displayed higher thrust levels of 979.9kN (208,517lb). It was decided to retain them and press on for the combustion instability tests but engines H-7092 and H-4094 were replaced with development engines H-T6-B and H-4067. The thrust structure above H-4067 was strengthened and a pressure bomb placed in its combustion chamber which was detonated 2.4sec into a 15.6sec firing on 25 January 1968. Instability was recorded as 26msec but high thrust output was noted from engine H-4095. Both of the test engines had bombs installed for the second test and the thrust structure above H-T6-B was similarly strengthened also. Conducted on 6 February, the second instability test lasted 15.5sec with the bombs detonated at 2.7 and 2.8sec and a

Left: Of the final Saturn IB stages only S-IB-11 and S-IB-12 were tested in Static Test Tower East at MSFC. S-IB-13 and S-IB-14 were never tested. *MSFC*

measured recovery time of 8-10msec.

The third planned instability test occurred on 14 February, which lasted 15.3sec, bombs in the test engines detonated at 2.66sec after ignition, nulled out within 9msec. The fourth test held seven days later ran for 3.4sec of a planned 15sec when H-4095 exploded and caught fire. But with bombs again detonated in the test engines, at 2.4 and 2.6sec after ignition, a solid return to stability within 11msec and 3millisec achieved the test goals. But the engine did not fail as a result of the bombs in the two test engines, rather because of a cascade of events which began with a failure in the LOX seal propagating to an explosion in the turbopump. It was replaced with H-4091. In the period while additional checks were made on LOX seal configurations while engines H-7095, H-7086, H-4092 and H-4093 were modified with an old-style vented lip seal and the other four, H-T6B, H-7102, H-4067 and H-4091, were fitted with the new type of bellows seals.

The outflow from these changes was to cancel the fifth combustion instability test and to move immediately to the qualification tests conducted on all stages. Because of their display of higher thrust than anticipated, engines H-7102 and H-4093 were re-orificed and the two test engines were removed and the original engines (H-7092 and H-4094) put back. The usual process of pre-test flight simulation was conducted and the first, short duration, burn took place for 35.3sec on 9 April 1968. Following this, bellows seals were fitted to the remaining four engines previously equipped with vented lip seals and prepared for the final seventh stage test – the greatest number performed by any Saturn stage of any type. That took place on 23 April 1968 and ran for a duration of 145.3sec, the inboard engines being cut off 3sec earlier.

Out of the test tower on 3 May, S-IB-11 left MSFC aboard the barge *Palaemon* eight days later and arrived at Michoud on the 16th where it went straight into storage, bypassing the usual post-test inspection and checks. As NASA neared the first flight of a manned Apollo spacecraft with AS-205 and a rapid transition of flights on to Saturn V, there was now virtually no hope of this stage ever being flown. It formally entered environmental storage on 20 June 1969, exactly one month before the first manned landing on the Moon. But it was still on the inventory as available for flight and on 26 March 1973 it was removed from deep storage for checks and returned to that condition on 10 January 1974.

On 17 May it was struck from the NASA inventory with

Below: An outboard propellant tank is moved across to start the clustering process around the central LOX tank as S-IB-13 is assembled at the Michoud Assembly Facility.The completed stage was later shipped to MSFC (right) where it underwent the customary firing test, the last taking place on 25 July 1968, the last of all Saturn rockets. *MAF/MSFC*

ownership transferred to Boeing Services International on 1 January 1975. From there it was moved to several locations, including MSFC, and to the Alabama border with Tennessee where it was erected in a vertical position in July 1979 as a display at the Alabama Welcome Center, a rest location on the I-95.

The second stage for what would have been AS-211, S-IVB-211 entered the manufacturing cycle in 1966 and was considered as the Orbital Workshop (OWS) for Skylab, for which it was fitted with insulation and clip-bonding. Although incorporating material which would have allowed it to have been the second Skylab, it was placed in flow for conventional duty as the second stage to S-IB-11. The J-2 engine arrived at Douglas' Huntington Beach plant on 3 January 1967 and it was installed on 11 April. The completed stage went into post-manufacturing checkout eight days later and checked out of the paint shop on 6 July. At this juncture it would have been in flow for SACTO and test firing of the engine but the inevitable redundancy of this stage cancelled that plan, to save money and time, and it was placed in long-term storage on 3 August 1967.

After a period in post-manufacturing checkout, the stage left Huntington Beach for SACTO on 17 October 1968 where it arrived next day. But there was no intention of placing it in the test tower and S-IVB-211 entered long-term storage there on 18 December. Flown back to Huntington Beach on 15 September 1970, on 5 October it was moved across to North American Aviation at Seal Beach for long-term storage. On 2 September 1971 it was trucked back to Huntington Beach where McDonnell Douglas conducted an all-systems test on the stage, completed on 28 April 1972. From there it was flown to KSC, arriving on 27 June for long-term storage.

At some point during the late 1970s the stage joined S-IB-11 and was displayed horizontally at MSFC before being moved to the Alabama Space and Rocket Center, now the US Space & Rocket Center, as an outdoor exhibit. Ingloriously perhaps, when NASA set its sights on returning to the Moon under the now cancelled Constellation programme, the stage bequeathed components and some elements in tests supporting development of the Ares launch vehicle, now superseded by the Space Launch System for the Artemis programme. At one point the J-2 engine was displayed at the Hermann Oberth Raumfahrt Museum in Feucht, Germany.

Below: A nearly complete S-IB-13 stage replete with white coating (right) as assembly gets under way on S-IB-14 in early 1969, the last Saturn IB rocket assembled. *MSFC*

AS-212

H-1 engines for static firing:	H-4073; H-4097; H-4098; H-4099; H-7098; H-7100; H-7101; H-7103
J-engine for static firing	J-2103

There was no mission assignment for AS-212 because that flight only existed on paper manifests which had been prepared long before the number of missions ran out prior to any assignment. However, it has relevance for the speculative application that this launcher had as explained below. For that reason it is appropriate to assume that the first stage for this would have been S-IB-12. There was no S-IVB-12 stage as that had been used as the Orbital Workshop for Skylab. But there was a proposition to use S-IB-12 which would have paired it with the S-IVB stage removed from AS-513 for the two-stage launch of the Skylab station.

Assembly of S-IB-12 stage began during 1966 with installation of all eight H-1 engines completed by 28 September 1967. In the original arrangement of engines, H-7102 was replaced with H-7098, the engine going to S-IB-11. Also, H-4096 showed problems with some contamination and was removed for H-4073 to be installed on 28 August, an engine which had already be removed from S-IB-7 in December 1966. After post manufacturing checkout, the stage left Michoud for Marshall on 23 April 1968 aboard the barge *Palaemon*. After arriving on 4 May, the stage went to static Test Tower East where engineers replaced all the LOX pump seals for bellows style seals, prior to propellant loading tests, a 35.4sec firing on 10 July, and a full duration run of 145.4sec on the 25th. This was the last S-IB static test firing and the last of all at Static Test Tower East.

Taken down on 7 August, S-IB-12 was taken straight to Michoud, arriving there on 12 August where it went straight to storage without the usual post-test checkout. In environmental storage from 30 June 1969 to 20 May 1970 it was transferred to Marshall on 17 July and a further session in storage until 20 September 1971 when it was returned to Michoud, arriving on 7 October. It remained there in environmental storage until 12 June 1973. Two months later all eight engines were removed for components to begin a new life as the RS-27 for Delta launches. All had been removed by 13 August at which point it was prepared for its final move to KSC on 8 May 1974. It was there that the stage was disassembled and scrapped. But there is a postscript to this.

During the second half of 1970, several months after the decision to utilise a Saturn V for launching a 'dry' workshop and other elements of Skylab, the suggestion was raised that an unmanned Saturn IB should be flight tested from the LC-39B pedestal, an adaptation of the original pad to allow the smaller launch vehicle to align the S-IVB, IU and payload as it would for a Saturn V launch. There was some concern that the new configuration should be proofed through a flight designed to verify the integrity of the pedestal so as not to compromise sending the first crew to the station, scheduled for a day after the launch of Skylab from LC-39A. With limited capacity for operating untended, the station would need to receive its crew in timely fashion.

The prospect was raised of using an unmanned Saturn IB, potentially AS-206, to verify the new pad configuration and launch procedures before committing to a manned flight. As the possibility of that was studied in greater depth it became apparent that such a precautionary launch would not be necessary, that it would add additional cost and could delay the Skylab programme itself through additional elements placed in the critical path of getting the laboratory into space. It is important to remember here that 1970-71 were critical years for NASA, as it wrestled to find an affordable configuration for the Space Shuttle against strong opposition in Congress. It needed to get on with the job of transitioning to a reusable manned launch system and a permanently manned station, which it would support with crew and logistical supplies, as quickly as it could.

But the idea sparked interest elsewhere and the Saturn IB was sought for launching a unique payload which NASA's Office of Space Science & Applications had proposed; the Smithsonian Earth Physics Satellite (SEPS). Comprising instruments for precise ranging data down to +/-2cm (0.8in), SEPS would provide a key component in NASA's Earth Physics Program which aimed to determine the motions of continental plates and measure temporal changes on the Earth's surface. Weighing 3,628kg (8,000lb), SEPS would be placed in a circular orbit of 3,720km (2,312mls) at an inclination of 55deg, for which a bespoke Saturn IB would be ideal.

Selecting SA-212 (the designation now indicating a unique category) as the SEPS launcher, a preliminary study by MSFC showed that, following cancellation of production, this would be the first Saturn IB for which there was no S-IVB stage, S-IVB-212 being withdrawn from the flight programme

Right: The last of the line, the thrust structure on S-IB-14 complete with plumbing but no engines. Those were installed in March 1969 to be removed in 1973 prior to both this, and S-IB-13 being broken up. *MSFC*

for rework into the Skylab Orbital Work Shop (OWS), scheduled for launch by the first two stages of Saturn V AS-513. At the time of the study, six complete Saturn IB/S-IVB stages were available, leaving SA-212 as the first without a second stage. It was proposed that the AS-513 S-IVB, removed from that launch vehicle so it could send up Skylab, be adopted as the second stage for SA-212 and the SEPS flight. A Saturn V S-IVB was required for a second burn capability to get the SEPS into the correct orbit. As none of the Saturn IB S-IVBs had that capability it was a perfect choice. It was also proposed that to save money and development time, the same adapter and payload shroud as that employed for AS-204 could be used, the Lunar Module attachment points being used to support the SEPS satellite itself.

Overall, the weight breakdown projected a total lift-off mass of 577,054kg (1,272,194lb) with a mass eventually placed in circular orbit at the specified altitude being 20,177kg (44,485lb), of which 16,541kg (36,467lb) was the S-IVB and associated hardware. To meet the mission requirement, a restart of the J-2 engine was necessary within 60min of launch but that contravened Rocketdyne rules that the stage have a period of 80min between firings. However, to back up the restart time, it was recommended that by lowering the vent valve relief setting and retiming the Main Oxidiser Valve, the Start Tank could reach an acceptable state for re-ignition with a 5sec fuel lead rather than the 8sec lead required for normal Saturn V operations. S-IVB-513 mixture ratios were 5.5:1 and 4.8:1. As this was the only mission seriously studied for SA-212, the timeline for getting the SEPS in orbit is of more than academic interest and defines the possibilities that could have allowed these launch vehicles to conduct many missions had they not been retired.

The initial boost phase would have placed the stack on a flight azimuth of 37.94deg east of north for a planar boost to a 55deg orbital inclination. Tilt initiation would have occurred at T+10sec, followed by tilt arrest at 2min 11.5sec followed by the inboards out at 2min18.3sec and the outer engines exactly 3sec later. With separation at 2min 22.7sec, ignition of the S-IVB-513 would occur at 2min 24sec and ullage case jettison at 2min 34.6sec with a mixture ratio shift to 4.8:1 at 7min 14sec. S-IVB cut-off would occur at 9min 37.7sec and insertion 10sec after that in an orbit of 150km by 3,720km (93.3mls by 2,312mls).

The second burn would be accomplished at first apogee and circularise the orbit at the specified altitude, preceded by the IU commanding a local horizontal attitude hold, maintained throughout the burn. J-2 restart would occur at 1hr 12min 45sec with a sustained mixture ratio of 4.8:1 and shutdown 21.2sec later followed by separation of the payload at 1hr 57min 49sec. Two burns with the Auxiliary Propulsion System would lower the orbit of the combined S-IVB and IU, the first occurring at the first descending node after separation (about 1hr 24min elapsed time) to provide a perigee below that of the satellite. The second burn at the first ascending node after separation (about 2hr 48min elapsed time) would reduce apogee to prevent a collision with the SEPS.

In attempting to satisfy the requirements defined by NASA's OSSA, the Smithsonian Astrophysical Laboratory had a defined payload which would have been well accommodated by the match of vehicle hardware proposed for the only mission of SA-212 which had credible veracity. In this report, which went direct to Dr. Eberhard Rees, then the Director of the Marshall Space Flight Center, a start on the project was speculated for January 1972 leading to a launch in April 1974, after the last Skylab visit by AS-208 and before the flight of AS-210 on the ASTP mission.

Above: The scene of five Saturn I and two Saturn IB launches, in 1972 the demolition crew moved in and began taking down the familiar LC-37B site only for it to be recommissioned as the Delta IV launch complex. *NASA*

Of course this mission never happened, one of several which were considered but shelved as overarching policy prevented further use of the hardware.

S-IB-13

H-1 engines: H-4109; H-4110; H-4111; H-4112; H-7113; H-7114; H-7115; H-7116

The stage was built but never static-fired and there was no S-IVB stage. Construction and build-up began at Michoud in 1968 and the engines were installed by 25 March 1969. With no mission assigned and no prospect of it ever being used, rather as an asset retained on the inventory, S-IB-13 was placed in storage at Michoud from 22 July. Less than a year later it was removed and on 17 July 1970, together with S-IB-14, it was sent off to MSFC where it arrived on the 28th. It remained in Building 4708 in environmental storage until 4 February 1972 whereupon it was moved back to Michoud where it arrived on 27 March. It remained there, in environmental storage until 19 January 1973. While there, all eight H-1 engines were quickly removed by 8 February and sent for recycling into RS-27 motors for the Delta launch vehicle. Following scavenging for useful components, the stage was disassembled at Michoud and elements used elsewhere or scrapped.

S-IB-14

H-1 engines: H-4109; H-4110; H-4111; H-4112; H-7113; H-7114; H-7115; H-7116

As applied to its numerical predecessor, no mission, second stage or test firing awaited the S-IB-14 when manufacturing began in 1968, engines installed by 24 April 1969. The same sad cycle set by S-IB-13 dogged this stage, which went straight to environmental storage on 29 July until 1 June 1970 when it was sent to Marshall, departing Michoud on 17 July accompanied by S-IB-13. Both stages entered environmental storage from 28 August 1970 and they remained there until 4 February 1972. Both stages left MSFC for Michoud on 8 February and were placed in storage. S-IB-14 was removed on 23 April 1973 and all eight engines taken out by 9 May. Parted out and essential or recyclable material removed, the H-1 engines were re-purposed into the RS-27 series for Delta launch vehicles.

AS-215 (S-IB-15)/AS-216 (S-IB-16)

Chrysler had been contracted to produce launch vehicles supporting missions up to AS-216 and had begun accepting propellant tanks for these two vehicles in 1969 but, with cancellation of further development, these were placed in storage at Michoud in April that year. Several long-lead items were also accepted by Chrysler and these too were placed in storage until removed over time and discarded or re-purposed. No engines were built for these stages and long-lead items for the manufacture of those, as well as components and systems built, were decommissioned or scrapped.

AS-217 REFERENCE LAUNCH VEHICLE

No hardware was built for the AS-217 configuration and there was never any prospect of its core application being realised. Nevertheless, it was a mission number.

In January 1968 the Space Vehicle Systems Branch of the Advanced Studies office at the Marshall Space Flight Center submitted a technical memorandum (X-53686) which defined the summed weight reduction and perfor-

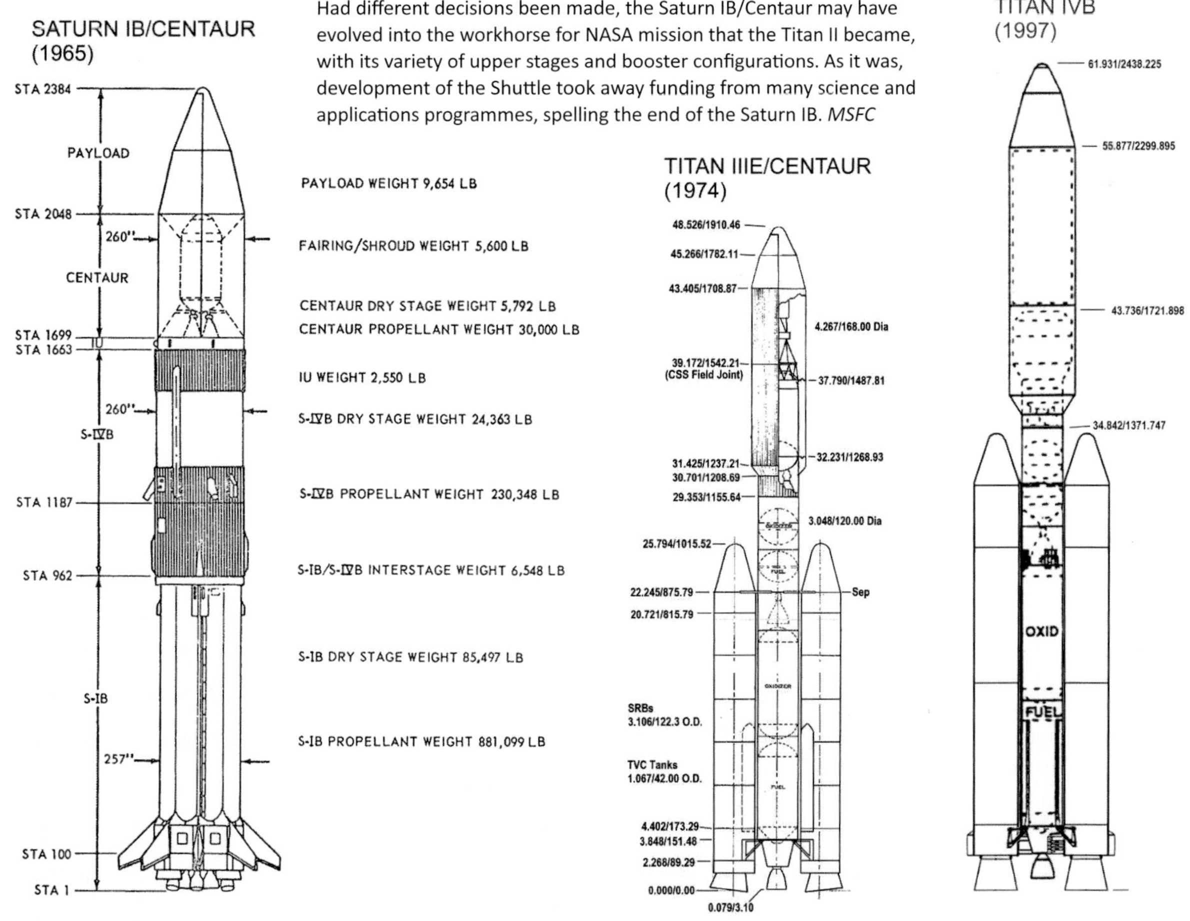

Had different decisions been made, the Saturn IB/Centaur may have evolved into the workhorse for NASA mission that the Titan II became, with its variety of upper stages and booster configurations. As it was, development of the Shuttle took away funding from many science and applications programmes, spelling the end of the Saturn IB. *MSFC*

mance enhancements that had accrued in 19 previous launches of the Saturn I series. Recognising the impending demise of the Saturn programme, MSFC wanted to project a still better and more capable launch vehicle, production of which had already ended.

With the latest improvements, the AS-217 vehicle took as baseline the AS-212 configuration and extrapolated the next launch vehicle mission number beyond the existing contract for rocket stages which ran out with AS-216, the sixteenth Saturn IB. The uprated versions of the basic launcher, utilising strap-on solid-propellant boosters and various upper stages, were explored at the end of Section 3 but the specification for the SA-217 vehicle described how the payload capability could be raised to 20,396kg (44,965lb) purely through a cumulative weight-saving programme for all elements of the stack ,including the stages and various structural elements.

The process of design refinement and performance improvement in the memorandum included the use of the J-2S engine. Beginning life as the J-2X, a designation later applied to a different evolution of this engine, the J-2S was a development programme which began in 1964 and provided replacement of the gas generator concept to a tap-off design which used a small proportion of gas from the combustion chamber to power the turbine, saving weight and improving efficiency. The engine had a modest increase in nominal thrust to 1,072kN (241,000lb) or 1,182kN (266,000lb) in vacuum but overall flexibility for operational use was obtained by providing a throttle capability by way of changing mixture ratio and a new low-thrust mode for idling the engine so that it could be used to provide a very small acceleration to the stage to settle propellants and dispense with the ullage motors. Rocketdyne ran tests with a few prototype engines and accumulated almost 31,000sec in burn time demonstrating its capabilities.

To achieve an increase in weight-lifting capacity, the memorandum asserted that 1,174kg (2,589lb) could be shaved from the S-IB stage, primarily through the redesign of slosh baffles, the use of titanium on thrust structure fasteners and the redesign of gimbal joints on the outboard engines. Weight savings of 666.8kg (1,470lb) could be achieved, it said, on the S-IVB stage alone. Here, there was a wider range of candidate savings, including installation of the J-2S engine, low-density encapsulating material, removal of the three ullage rockets and reduction in insulation on the forward dome. Residual savings could achieved through eliminating the need for LOX depletion cut-off. Further savings could be made with the interstage and the Instrument Unit.

The purpose in proposing the refined and greatly improved Saturn IB was to present other departments at NASA with a possible alternative to other payload carriers and in doing so recruit support for reopening the production line, which of course never occurred.

Right: The GOES-N meteorological satellite heads for geostationary orbit on a Delta IV launch vehicle, keeping alive the legacy of a historic location re-purposed for the 21st century. *NASA*

Chapter Thirteen

Postscript

In all, 295 H-1 engines (of which 152 were flown) and 37 J-2 engines (of which 9 flew) were built for the S-I, S-IB and S-IVB stages of this launcher, following 36 RL-10 engines flown on the S-IV stage of the Block II Saturn I. All achieved their objective in supporting 19 successful missions out of 19 attempts between 1961 and 1975. They drew upon a legacy of excellence in preceding engines and rocket stages and they served to proof-test major elements of the Saturn V vehicle, which itself flew thirteen times. In taking stock, the true value of the Saturn I and IB programmes resided in the sophisticated management practices which recruited, encouraged and supported aspirational engineers, technicians and scientists who designed, built, tested and launched what were the most powerful and sophisticated launch vehicles of their time. It would not be easy saying 'goodbye' but there was an enduring attribute at the end.

The last Saturn IB flight (AS-210) was always going to be a fitting demise to the Saturn programme – the first of the great Space Age rockets and the last to get launched before the retirement of a unique series, thundering into space to send humans to link up across the ideological divide between the Soviet Union and the United States and shake hands in space. Two nations who together forged a race that never really did have a winner, each nation excelling in its own unique capabilities. Both now starting down a long road which would eventually lead to a permanently manned space station built by Americans, Russians, Europeans, Japanese and Canadians almost 25 years later.

Spanning nearly 14 years of flight operations and 19 launches, of which there had been no failures, the Saturn I/IB programme had developed the management and operating techniques which raised the skill level of an already experienced group and produced the world's largest operational rocket – Saturn V. But without Saturn I and Saturn IB, that behemoth among giants that took men to the Moon nine times between 1968 and 1972 would not have been possible.

Was it a mistake to abandon the Saturn IB? The author thinks so. It was retired, along with the Saturn V, when the Shuttle was considered to be the sole means by which, in future, people could be sent into space and that in itself was a big mistake. When the Saturn IB made its last flight the Shuttle had been in development for more than three years, the NASA budget was in decline and support for this reusable spaceplane was recruited on the promise that it could reduce launch costs

Above: Instrumental in accelerating the development of the Saturn I and IB, President Kennedy visits the first Block II orbital version with S-IV upper stage. *John F. Kennedy Presidential Library*

by replacing all expendable rockets; the more Shuttles flown, the cheaper each flight, or so it appeared in 1975.

The user community, together with the broader space industry, would come to accept that mistake, and to amend the error when, after the *Challenger* disaster of 1986, expendable launch vehicles were brought back into production as a back-up to further failures with the reusable system.

There was an obvious logic in retiring the Saturn V; with its colossal lift capacity there would be little need when the Shuttle, and its flaunted lift capacity of 29,500kg (65,000lb), which was never realised in practice, could place segments of a station in orbit consecutively – adding more launches and, according to the perverse logic of economics, lowering costs. But more launches meant more costs and while the price-per-given mass would be less, the overall price would be higher. But Saturn IB could place a usable payload of more than 13,600kg (30,000lb) in orbit and that rested neatly between existing heavy-lifters and Shuttle.

Had the Saturn IB been retained, production Apollo spacecraft could have been used to ferry crews back and forth to a manned habitat – whatever that became. As it turned out, for nearly 10 years after the Shuttle retired in 2011, NASA spent several billions of dollars flying astronauts to the ISS in Soyuz spacecraft rented from the Russians until the commercial Crew Dragon from SpaceX came along in 2020 to do that job.

Hindsight has perfect vision but the hubris tagged on to the Shuttle denied it a survivable niche, adding pressure which triggered failures, the loss of 14 astronauts and its retirement. Today, commercial contractors, subsidised by the same amount of money that could have kept Saturn IB alive, have produced a Saturn IB-size launcher to lift an Apollo-size spacecraft and pick up the job as a taxi last conducted by AS-208 when delivering three astronauts to Skylab in 1973. Perspective is a point of view and a viewing point in itself.

Following the launch of AS-210, operations director Walter Kapryan paid tribute to the people who had created, managed and flown the Saturn I/IB launch vehicles and spoke eloquently of the group which was now to be dissolved: 'Speaking as a private individual and not a NASA official, I'm very sorry to see that happen to us. We're losing a team that I don't think we'll ever have together again.'

The then NASA Administrator James B. Fletcher maintained that the Saturn team had made 'a significant contribution to our country'. Adding: 'And now as you go your separate ways over these next few weeks, I'd like to thank all of you and wish you the best in your future ventures.'

They were indeed going separate ways. Of the 12,000 contractor employees involved with Saturn launch operations at Cape Canaveral, within a month more than 2,000 would have been laid off with 10,000 more moved to different assignments and quite a lot deciding to leave the rocket business altogether. The Saturn facilities would be kept in a position where they could be restored to status and the hardware would be mothballed against the day that it all may be called back to duty. But it was not to be.

That night in July when CSM-111 disappeared into the sky to join Soyuz already in space, there were quite a few parties, divided equally in number between celebration, reflection of past glories and a wake of sorts. Ten years earlier there had been another type of celebration, with proud anticipation that although the Saturn I was over, the impending first flight of a more capable Saturn IB would carry the name forward and open even grander opportunities. Now that dream too had gone and would never be realised.

In Huntsville the mood was muted, a great challenge posed by the Shuttle and its cryogenic External Tank, bequeathed from the technology of cryogenic Saturn stages, lay ahead presenting new opportunities for the next generation, young engineers learning innate skills and old tricks from men and women who had been the true pioneers of what had emerged as America's largest civilian rocket programme in history.

At Chrysler, the work was gone and there were new fields to plough while at Douglas there were many more expendable rockets to build and fly. And at Rocketdyne and at Pratt & Whitney, the lessons of the last 20 years placed on each the burden of responsibility they had to power the launch vehicles and spacecraft of the future. For them, deeply involved in both the Shuttle and the expendable launcher programmes of the day, the business opportunities were just beginning.

But in Florida, around Cape Canaveral, Cocoa Beach and Titusville, bars and diners stayed open late that night, for whatever the national space programme had still to accomplish in the black void of space, for Saturn people down on the ground at KSC the job was over. Many would leave colleagues, work associates and friendships forged on the back of a big silver and white rocket called Saturn. It was truly a night to remember and wandering between drinking parlours and diners, the air was full of pride, nostalgia and exuberance with hours spent sharing stories. It was a strange night, nobody wanting to be the first to end the day on which they launched the last Saturn. And a few, more than would ever admit, quietly shed a tear or two.

Right: Had it not been for the decision to commit to the reusable Space Shuttle, seen here on its first flight on 12 April 1981, Saturn launch vehicles could have continued in use until the end of the 20th Century. *NASA*

Appendix 1

Wernher von Braun

Dreams and Deliberations

Central to the story of the Redstone, Jupiter and Saturn I, Wernher von Braun stands as a catalyst for invention and innovation in the field of rocketry and space exploration, bringing to a post-war public the breath-taking image of a space-based future few outside the realm of science fiction had ever contemplated, providing Americans with an ambition to build on their pioneering spirit and sail across what John F Kennedy defined as 'this new ocean of space'. In the following paragraphs we take an overview of von Braun's impact and his pursuit of a programme of public education, defining through simple words for a lay audience the complexities of rocketry and space travel. In doing so his contribution is seminal. But it was not a message immediately accepted outside the confines of the military audience to which he initially answered.

When the US Army secured the services of the German rocket team in 1945 they knew better than most about the opportunities that lay ahead. The financial resources of the United States, aligned with its burgeoning industrial might, made a lot of things possible that were denied to the other Allied powers, reeling both societally and economically from the traumatic events brought about through five years of remorseless conflict. With different levels of concern regarding the future, the belligerent powers left standing in 1945 had scant resources to capitalise on the veritable treasure-trove of technical wizardry they found, beyond the imagination of even the most futuristic of scientists and engineers outside the boundaries of the Third Reich. Most of whom had been surprised beyond articulate response to the discoveries made across the blasted landscape of industrial Germany in all forms of military technology.

Quite quickly after the war, the freshly independent US Air Force aspired to seize control of space. But the wherewithal was missing. For its part, the Army thought it could do something very different and took advantage of its prime asset – Wernher von Braun, whose public proselytising of potential opportunities for a US space programme, including

Above: Von Braun in his office. Behind him, a fine collection of scale model rockets and missiles traces the productive designs of the ABMA with a couple developed by the Air Force. The enormous Saturn V could only be contained by leaving off the payload section! *Author's collection*

flights to the Moon and Mars, had already made him a national lobbyist for human space flight. While firmly in the employ of the US government and subject to all its restrictions about public speaking and personal propositions for national policy, the Army allowed von Braun a loose rein. They saw in this an advantageous publicity campaign for securing the only opportunity remaining for funded studies on big rockets – a massive booster for leading America forward in a space programme which they fervently believed they could own, much to the chagrin of the Air Force.

Von Braun had a strong and personal ambition in pushing the Army toward a leading role in a very different application of the rocket to that which he had worked on for the German army. For long, he had wanted to mastermind a programme which would realise his dream of missions to Mars, even though he recognised that getting to the Moon would be the first objective for deep-space explorers. To do that, he opined, first it would be necessary to build space stations in Earth orbit from where massive interplanetary ships could be assembled through a series of launches with big and powerful rockets much larger than the Atlas and Titan ICBMs in development by the Air Force since the mid-1950s. With these, the exploration of the Moon and then Mars could begin. This became known as the 'von Braun paradigm' – space stations, the Moon, then on to Mars.

At the core of requirements for a massive rocket was the ingrained belief that humans, not unmanned robotic space vehicles, would pioneer a path to the planets. Writing in the October 1945 issue of *Wireless World*, British science fiction writer and futurist Arthur C Clarke had done much to publicise the idea of using satellites in geostationary orbit to relay communications around the world. Placed at such an altitude, 36,000 km (22,370mls), so that the period of their orbit would equal the rotation rate of the Earth, geostationary satellites at this distance from the Earth would appear to be stationary to an observer on the surface. Launched by giant rockets, with three such satellites placed at 120º intervals in an orbit in the plane of the equator it would be possible to transmit radio signals around the world uninterrupted, linking them together by radio signal.

But Clarke imagined that these satellites would have human operators and require pressurised areas where technicians could live and work in rotation, periodically replaced by different workers launched off the ground. In the 1940s it was inconceivable that electronics and radio communication would achieve autonomy and that unmanned stations would be possible. And so it was with other aspects of potential space applications. The idea of the communication satellite had been born but it had yet to take the form in which it would emerge in the early 1960s.

But in the mid-1950s, with no major drive toward a major space programme of any kind, von Braun convinced the ABMA that it could be the organisation not only to launch and service these orbiting communications satellites but to send Americans to the Moon or the planets, about which little was known at the time. In these pre-Sputnik years, there was simply no other programme around to achieve such ambitious goals. So it was that in November 1956 the Army's strategy

Above: Von Braun's identity badge when working with the Army Ballistic Missile Agency. *Author's collection*

was to make a quantum leap forward and lay out a plan for a large booster development programme, serving the military interests of its own but also providing the United States with a world-leader in space engineering and rocket technology.

By the late 1950s, across the period that saw the emergence of the Space Age driven by Russia's Sputnik flights, a detailed analysis was conducted as to how that could happen and much debate ensued. Von Braun was aware of some very large thrust rocket motors being talked about, both by the Air Force and by Rocketdyne, the rocket motor division of North American Aviation Inc. But there were alternatives for achieving a faster pace and a reduced development cycle. In those pivotal years, between the era in which it was believed that humans were required for every operation considered practical from space and serious planning for major tasks from Earth-orbiting satellites, the development of automation and electronics had retired the notion that humans were essential in all circumstances.

While the technical efficiency of the personnel who coalesced around von Braun was crucial to the success of missile programmes, the personality of von Braun himself was vital for charging the group with excitement about the future. The expansion of the military missiles programme was one of the more significant hallmarks of US preparedness for defence and national security, as tensions with the Soviet Union showed no sign of abating. The Korean War was in suspension, an armistice signed in 1953 which would endure for at least the next 70 years, but there was a large amount of intelligence information which indicated a surge in progress by Russian rocket engineers. It followed on the heels of the infamous 'bomber gap' myth of the early to mid-1950s in which it was universally believed by the general public that the Russians were building an unstoppable offensive air force. Many citizens were convinced that the US could face another Pearl Harbor, in which an aggressor could strike with impunity, flying massed bomber formations against vulnerable targets.

The surge in political support for an expansion of US offensive power was induced by this and by the Air Force's adherence to the Stanley Baldwin dictat that 'the bomber will always get through', tipping US policy toward offensive rather than defensive strategies. It had served the British well during the war with Germany, enabling them to deploy a passive defence net (the Chain Home radar stations) while placing great emphasis on the development of RAF Bomber Command to take the war to the enemy's heartland. Strongly supported by President Eisenhower, the need for more intelligence information pushed along the Lockheed U-2 spy plane and the development of a Top Secret spy satellite programme, which would eventually result in the Corona missions, veiled in public sight by the name 'Discoverer', purporting to be a science research programme – as many early satellite programmes were!

There is perhaps a contradiction in von Braun's drive for a peaceful space programme, tinged with a level of pragmatism that takes precedence over preference; von Braun's push to influence the military in deciding on the future use of rockets is not, however, contradictory to that objective. He chose pragmatism in persuading the Army to indulge him in his grand ideas, visionary to most but sufficiently robust in its scientific and technical excellence as to be entirely credible. But again, there is a display here of the lengths he was intended to go to in setting an inevitability of destiny before his military masters. Had it been so in Nazi Germany where he become trapped by his own persuasive arguments? Motivation has many masters within the human soul and is owned by a conflicting array of sources to mobilise the drive to fulfil an inner passion.

A Matter of Personal Choice

On 27 August 1949, Len Carter, the Executive Secretary of the British Interplanetary Society (BIS) had written to von Braun inviting his acceptance of an Honorary Fellowship of that respected organisation. As related earlier, the BIS had been formed in 1933 as a think-tank and advocacy group supporting rocketry and the exploration of space, one of three such organisations with others formed in Germany and the United States during the inter-war years. He received a reply on 29 September accepting the honour and on 2 November that year the noted British rocket engineer Val Cleaver wrote to von Braun with his support for that award.

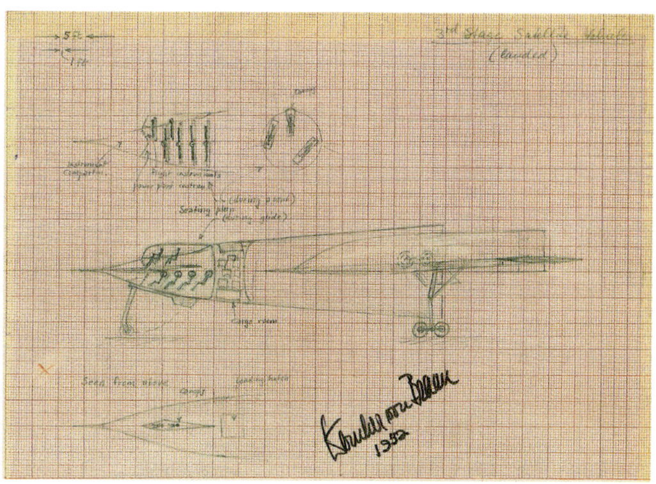

Above: Von Braun's sketch of his shuttlecraft dated 1952. *Author's collection*

While being a significant rocket engineer and a talented manager of men, von Braun was a creative thinker and expressed his talent through writing. Prolific since his days at Peenemünde, in 1948-49 he wrote his first book, titled Mars Project, and accompanied it with a technical appendix of 120 pages including contributions from former German rocket engineers. The grand plan he envisaged came straight out of deliberations from the 1930s and owes much to the great thinkers and futurists of that decade, including the great Hermann Oberth. In his imagination, von Braun envisaged ten giant spaceships and 70 crewmembers, with large winged entry gliders to carry astronauts down to the surface of the Red Planet.

Laid out in the form of a novel, it had sound engineering calculations but lacked realism only in the vast scale of the expedition he envisaged. In a mood of anticipation, during late February 1950 von Braun wrote to Kenneth W. Gatland of the British Interplanetary Society and explained that he had sent draft copies to several publishers and that he was convinced that a publication contract was 'just around the corner'. A contract never arrived, commissioning editors finding it totally unsuitable as a matter of fiction but in 1952 von Braun published the technical appendix as *The Mars Project.* That book was eagerly snapped up by those for whom rocketry and the Space Age were just around the corner.

For von Braun, the stimulus in this drive to a space-based future was inherent within his lifelong passion. It had begun in Germany when he was a boy of tender years and it endured in his new country. After failing to get a novel published in 1950, von Braun made a better success in attracting public acclaim through his inspiring writings in *Collier's* magazine, by contributing to a series of TV programmes and appearing with Walt Disney to promote the concept of a space-based future. It was *Collier's* that first gave public voice to von Braun's aspirations for space travel through a series of eight

Below: Von Braun's natural charisma and deeply-felt conviction for the pursuit of space exploration underpinned his ability to communicate complex ideas in a simple and relevant context. *Author's collection*

Above: In 1954 Heinz Haber (left), von Braun and Willy Ley discuss the concept of a 'bottlesuit', enabling an astronaut to move around outside his vehicle. *NASA*

visionary articles published between 22 March 1952 and 30 April 1954. They were written by a group of experts in astronomy and space, including in addition to von Braun, Fred L. Whipple, Joseph Kaplan, Heinz Haber and Willy Ley.

Fred Whipple was an astronomer who would make major contributions to the science and who remained with Harvard College Observatory from 1931 until late in the century, making a name for himself in several related fields, including the discovery that small strips of aluminium foil ('chaff') when dropped from aircraft would confuse enemy radar. Joseph Kaplan was a Hungarian-born physicist who chaired the US National Committee for the International Geophysical Year and served as adviser to President Eisenhower. Heinz Haber was a German physicist and writer involved with several of von Braun's projects and helped gather material for publicising the possibilities of space travel. Venerated during his lifetime, Willy Ley was largely self-educated but quite early in his life became a proponent of space travel and joined the *Verein für Raumschiffarhrt* (German Society for Space Travel) and wrote extensively on the history of rocketry before fleeing Nazi Germany to eventually end up on the United States.

Through the imaginary art of Chesley Bonestell, together with Fred Freeman and Rolf Klep, the *Collier's* articles came alive, sending a new awareness of the coming Space Age to a populace weaned on Buck Rogers and fictional tales of interplanetary travel. Realised now by one of the world's leading experts on rocketry and the theoretical possibilities of space flight, it became apparent that these were events brought straight from the pages of comic books into the workaday lives of ordinary individuals. Following von Braun's Mars book, *Collier's* provided a bigger platform and a wider audience. It spoke the language of the citizen and gave voice to ideas once thought to be the idle dreams of science fiction writers. It was, if nothing else, a legitimisation of a new technocracy and a society whose anticipation and excitement was raised to new levels. Later, these articles were expanded into three books: *Across the Space Frontier* (1952); *Conquest of the Moon* (1953) and *The Exploration of Mars* (1956).

If *Collier's* gave validity to the integration of previously purposed space fiction with the realities of space fact, Walt Disney was to launch von Braun on a national campaign extolling the magnificence of the coming age – as he saw it. To von Braun there was no equivocation, space exploration was the quintessential essence of a future in which humankind would become an interplanetary society and Disney would give it colour and movement through his projection of Disney World as a signpost to the future. Located in California, the theme park would embrace four separate sections: Fantasyland, Frontierland, Adventureland and Tomorrowland. It was the latter for which von Braun was recruited and whose charisma made it real.

Ward Kimball, Disney's senior producer, seized on the German rocket engineer to fill the last of these sections who, together with Willy Ley and Heinz Haber, the latter studying the new field of aerospace medicine, defined the shape of a future in which humans went to the Moon and walked the dusty surface of Mars. As ever an arch-exponent of the moving image, Disney and Kimball gave von Braun a platform across three TV series. Where *Collier's* reached a readership of 4 million, TV had a potential audience several times that number through the 15 million sets already in American homes by this time.

Hosted by ABC and introduced by Disney himself, the first programme, *Man In Space*, aired on 9 March 1955 and exploited to the full Disney's now famous animation techniques, giving motion to images of a space-world future. From an animated bust of Isaac Newton, basic principles of theoretical principles were brought to life and given stereotypical images of scientific possibilities. The theme moved from animation to reality as von Braun faced the cameras and spoke in simple and explanatory tones, given emphasis by his German accent: 'If we were to start today on an organised, well supported space programme, I believe a practical passenger rocket could be built and tested within ten years.'

The second programme, broadcast on 28 December, was titled *Man and the Moon* and quickly moved from animations and cartoons of how people had perceived our nearest celestial neighbour from ancient times to the present, to von Braun with slide-rule in pocket narrating his visualisation of a trip to the Moon. It would he said, require a space station in Earth orbit with crew and cargo supplied by reusable shuttlecraft from where an expedition to the Moon would depart. Imagining a wheel-shaped station rotating for artificial gravity, von Braun introduced rigid pods into which a single occupant could fit and perform tasks outside in space using manipulators as arms.

Aired on 4 December 1957 – two months to the day after the launch of Sputnik I – the final programme, Mars and Beyond, extended the von Braun roadmap to the exploration of Mars with animated depictions of ancient lore concerning the Red Planet and legends associated with its colour. From there flowed a dramatic depiction of fleets of Mars ships departing Earth on expeditions to survey and settle the planet, enlivened by the dramatic colour animations from the Walt Disney studios. By this time the TV audience had swelled to 42 million people, their attention triggered by the Soviet achievement in sending a dog into orbit four weeks earlier.

It is difficult to over-emphasise the importance of these productions, which sealed von Braun as the portent of future space exploration where new worlds would be visited and colonised. It was a 20th century paradigm, retelling for a new age and a different people the drive West in the 19th century, but for Americans who could reconnect with traditional values of pushing back the boundaries and striving for great possibilities. Tinged with the creeping fear in America that the Soviet Union was about to out-perform the United States in technology and scientific achievements, it was the perfect recipe for a call-to-arms.

Appearing on screens less than five months before Eisenhower announced the decision on 29 July 1955 to launch a satellite in a little over two years through the Vanguard programme, Kimball wrote to von Braun excitedly proposing that Disney hail that programme as having tipped the President into a national commitment. Shocked by this, von Braun wrote to Kimball imploring him not to do so, justifiably fearing a backlash against further publicity. Spurred to avoiding restraint on von Braun and seeing in him a significant source of stimulus for its own stunted missile programme, the Army had given the German scientist a generous leash which could easily be drawn in should the ire of the White House descend on the Pentagon. But to this day, some cinematographic historians claim that Eisenhower called Disney to congratulate him and to request a copy of the film so that it could be shown to top officials.

There is great doubt about the veracity of that claim, but it was written as so by Rip Bulkeley, despite officials at the Pentagon and at the Eisenhower library being unable to find any supporting documentation to prove this. But there is a proven record of how the films were taken as acclaimed portents of the future. When the American Rocket Society (forerunner of today's American Astronautical Society) held its regional meeting in Los Angeles, 600 people were given a tour of Disneyland and given a special showing of *Man In Space*. But there is evidence for another influential impact. On 24 September 1955, while speaking on behalf of the Soviet aspiration for launching a satellite, the Russian scientist Leonid Sedov asked Fred Durant, President of the International Astronautical Federation, for a copy of the Disney film, as it would help, he said, 'considerably for the cause of' international cooperation'.

Art played a major part in the presentation of rocketry, of its potential for both war and peace and for simulating humankind into the promise of the possible. Von Braun had realised this in his discussions with political and military leaders in the Third Reich and never forgot the impression

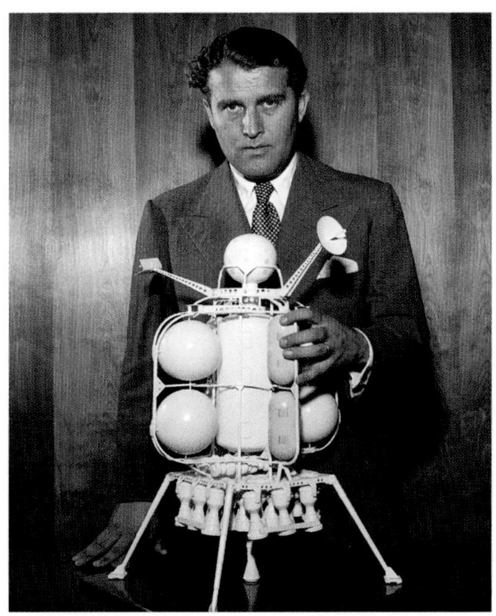

Below: Thanks to his collaboration with Walt Disney, von Braun was able to excite the American public about the coming Space Age. Here, he demonstrates models of his Moon lander (left), a reusable spaceplane (centre) and global telecommunications network. *Author's collection*

made by the artist Gerd De Beek, who once had painted motifs and caricatures on the fins of A-5 and A-4 rockets. Morale and some level of digression was provided by these light-hearted cartoons. Under the opportunities presented to former-Peenemünde personnel for jobs in the United States, De Beek got in to America, as he was qualified in guided missile technology. He soon joined von Braun in Fort Bliss and played a major role in developing a new genre of art suitable for the anticipated age of space exploration.

De Beek arrived in the US on 16 November 1945 from Le Havre aboard the merchant ship *Argentina* and remained with the von Braun team, eventually becoming head of Management Services, Graphic Engineering and Model Studies Branch at the Marshall Space Flight Center. Many of the facility's artworks representing the rockets, spacecraft and concept designs produced during the 1950s and 1960s originated there, either from De Beek himself or through the several colleagues he recruited and worked with. These graphic representations helped project the message about future rocketry and did much to stimulate public fascination. With so close a working relationship with von Braun, De Beek was able to make highly accurate illustrations and provide references for model makers producing scale versions of projected rockets and missiles.

Many of the public relations pamphlets and fold-out posters from the public relations department at Marshall and across NASA were attributed to De Beek or his team of talented artists and these included a large fold-out poster of the Saturn V, among many others. Von Braun frequently visited De Beek's department and exchanged views and ideas about how best to represent the vehicles he knew he wanted to attract funds for, encouraging the artists by informing them how valued was their work in displaying to cynical politicians concepts and configurations they had no imagination to visualise. Highly motivated, the drawings appealed to von Braun, who visibly drooled over the way these art works could fire the imagination, inspiring the great rocket engineer to become an outstanding benefactor for the creative arts in Huntsville. De Beek remained with the Marshall Center for the rest of his working life and until his death on 2 December 1989 at the age of 85 years.

Prelude to Growth

Emphasising a lifelong commitment to space travel, von Braun had disclosed the extent to which he would go to fulfil his passion and aspirations in a letter to Val Cleaver of the British Interplanetary Society on 21 May 1951: 'We should stop bewailing the fact that our beloved space travel idea is being pulled into the capacious maw of the military. It is certainly deplorable that the world is faced with a grave new political crisis, but we should eventually realise that this is beyond our control.' There is a hint here at how von Braun had sought to reconcile personal decisions, which had drawn him inexorably into the clutches of a tyrannical Nazi regime, with a passion for advancing the day of an interplanetary society.

But Cleaver was disturbed at the seeming domination of the military in the development of an impending space capability and expressed this to von Braun. Yet, in a further letter dated 20 June, the German scientist offered a pragmatic realism: 'I am tempted to call "Pax Americana" what maintains peace today. The ethics of the Western World alone, or the general public's desire for peace, just is not enough to keep the Communist tide in check. It is "The Bomb in Being", that does the trick.' Von Braun urged Cleaver to grasp the future through a new generation of young people to cultivate international contacts, a further expression of his support for the International Astronautical Federation (IAF) of which the BIS was a founder member and which is today the representative body for space professionals around the world.

The total commitment to garnering support for his 'paradigm' was undiminished and, convinced that he had a great story to tell, von Braun wrote a short book for Ken Gatland of the BIS in which he described his years at the Peenemünde rocket station. It was never published and there is uncertainty as to whether that still exists. But there were others, jealous of his rising position, who claimed that von Braun was a charlatan, that he did not invent the V2, that he did not develop it and that he used his influence to move people away from central development of that rocket so as to promote himself. Fellow rocket engineer Rudolf Nebel was a particularly persistent thorn in his side, along with Hermann Oberth who was critical of von Braun's work. Oberth is widely considered one of the founding fathers of rocketry, playing a major part during the 1930s and serving with von Braun at Peenemünde before moving to the US

Left: An early Saturn C-1 concept in 1960, with von Braun in front of the first two IBM 7090 computers installed at Huntsville. *Author's collection*

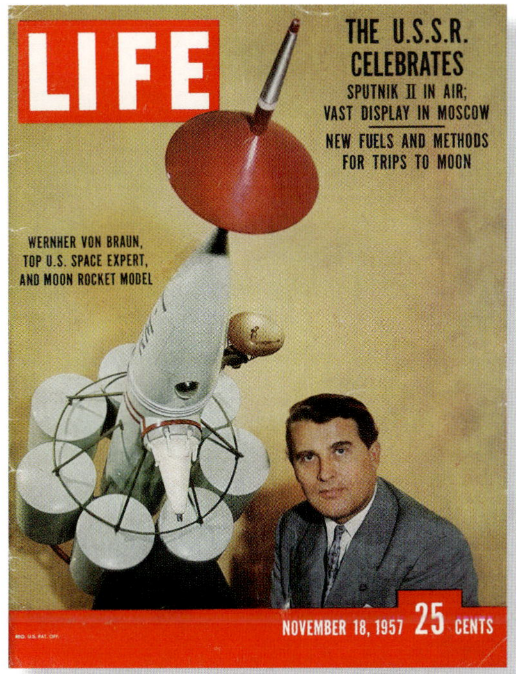

Left: Publicity came naturally to von Braun and he was a favourite with publishers during the late 1950s, not least the prestigious *Life* magazine when Russian successes eclipsed US achievements. *Author's collection*

and working for Convair on the Atlas ICBM.

While von Braun's reputation would later come to speak for itself, in 1952 he was still under judgement as to precisely where his allegiance lay and on 17 September he addressed a prestigious audience in Washington, DC, where he pushed hard for support of a military space station. Listening intently were senior military leaders, power barons from the upper echelons of the aviation and defence industries and legislators from Capitol Hill. He described a military space base in orbit capable of manoeuvring out of the way of intercepting spaceplanes, avoiding shrapnel placed in the path of the facility, loaded with weapons to 'destroy with absolute certainty an enemy spacecraft prior to its launching'. Seeing the station as a powerful deterrent, a buttress to 'pax Americana', he asked for a Manhattan-style commitment – referring there to the atom bomb project of World War 2 – even inferring that a pre-emptive strike on the Soviet Union may have to be a last resort. Perhaps von Braun failed to see that, in his strident invocation for a strong military presence, he was revealing just how fervently he must have argued for the V2 to a reluctant Hitler just 10 years earlier.

There were indeed dire consequences for ignoring the military threats of the day. Within two years the French were under siege by the North Vietnamese leader Ho Chi Minh at a place called Dien Bien Phu and there were deep concerns over the increasingly tight grip exercised by the Chinese communist ruler Mao Zedong. But the rapid acceleration in the Jupiter IRBM programme brought personal reward, raising von Braun to a higher grade, increasing his salary and enhancing the professional grades of several German scientists. And neither was it all about excellent design concepts and high success rates in engineering tests; it was also about embracing a collective organisation and shaping them into a group of people with trust in each other – managerial skills which are as hard to get right today as they ever were.

Back in the late 1940s when the Germans were firing off V2s from White Sands under the Hermes and Bumper programmes, von Braun had asked launch director Kurt Debus to send weekly notes – only one sheet of paper, succinct, brief and to the point. This he found particularly effective and by the end of 1956 he had all his under-managers, below his immediate deputies Eberhard Rees and Arthur Rudolf, to do the same – but on a Monday morning. Throughout the next decade and more, the 'Monday notes', as they were called, became a hallmark of von Braun's oversight across everything that was going on in the laboratories, workshops and test rigs at Redstone Arsenal. During the second half of the 1950s, von

Below left: With the status board hidden for a publicity shot, Don Ostrander, director of programmes on the Saturn project from 1959, and von Braun review a model of the LC-34 complex. Von Braun threatened to resign when Ostrander tried to force Huntsville to use Centaur electronics on the Saturn rocket. Ostrander left in 1961. *ABMA*

Below right: Always attentive to the technical details associated with rocket development, von Braun inspects a Launch Vehicle Digital Computer with colleagues in the MSFC Astrionics Laboratory. *MSFC*

Braun instituted his Development Board, at first exclusively the German scientists and then some invited Americans, all scrupulously arranged around the large, long table, in order of seniority and position within the organisation.

Von Braun's meetings were guided by the agenda but open to discussion and debate around its focus, leaving miniscule adjustments to previous decisions available if they proved productive. These elements of von Braun's management style are crucial to understanding why certain decisions were made in the way they were. There was a degree of consensus, a trust in those responsible for particular systems, subsystems or components. The management style laid open to collective discussion the way decisions were made. This is why von Braun was determined to bring as many of his trusted specialists to the US with him as he could convince the Army to employ. He knew and understood their work on the A-4/V2 and built on this to capitalise on experience.

The critical years from 1956 to the early 1970s defined the way the Saturn launch vehicle programmes progressed and the manner in which they were developed, because it was very different to the way the American institutions, certainly the military, had worked before. Much of the thinking behind concepts and ideas lay within the Development Board, which fed directly into theoretical analysis within the upper echelons of the Development Operations Division and, in this way of working, von Braun differentiated himself and those around him from the goal-driven activities of the armed services, and to a certain extent, to NASA itself when it emerged from the NACA.

But it would be wrong to believe in the 'inflexibility' approach so many analysts have accused the von Braun group of inculcating at Huntsville. The Huntsville team was only too willing to put aside the bigger picture to focus on the imperatives of the present. And in that they found a working formula that brought results through adaptation and a deep sense of commitment to the long-term goal. In those early days of the space programme a lot was learned through experience and the ability of teams to respond quickly to lessons learned. There was little dogma in the management of lower tiers in the decision-chain, managers being encouraged to remain open about new ideas and to foster a sense of purpose within the work staff.

Above: Huntsville honours one of its own as von Braun and his nine-year old son Peter ride a motorcade through the streets of his adopted town on 24 July 1969 in celebration of Apollo 11. *MSFC*

Von Braun is remembered for two things: his 'paradigm', in which space stations would precede Moon landings leading to Mars missions, and for the design of the Saturn I series. While the former is true, the latter is not, as many people, some outside the German group, had a hand in bringing to von Braun the opportunities afforded through clustering and the concept of 'all-up testing' which dramatically reduced the number of qualifying launches prior to full operational capability that could be achieved. Sadly, von Braun is not usually remembered for the engaging inclusivity he brought to the programmes managed at Huntsville but this was the definitive gift he gave to the nascent Space Age. Had it not been for that, it is unlikely that the Saturn series would have been as successful as they were.

Below left: The citizens of Huntsville recognise the achievements of the German rocket scientist. Pictured with von Braun are (left to right), his daughter Iris, wife Maria, Senator John Sparkman and Alabama Governor Albert Brewer, von Braun, his son Peter, and daughter Margrit. *NASA*

Below right: Copious written records are a legacy von Braun has left for historians and writers to mull over for decades to come. *Author's collection*

Appendix 2

Build and Test

Fabrication of the Saturn I Series

The challenges brought about by clustering nine tanks and eight rocket motors in a single stage and assembling on top of that an additional stage with cryogenic propulsion were many. But critical to its success were a series of engineering solutions which were unique at the time and which forged a precedent for future vehicles.

Here, we take a brief look at the processes and procedures for fabricating the S-I, S-IB, S-IV and S-IVB stages, together with the assembly sequence which successfully produced a launch vehicle with tremendous potential and realised growth across two successive generations.

Above: Six S-IB stages during fabrication in Building 4705 at MSFC during early 1966. Tanks for the AS-209 flight lie in the foreground awaiting installation into the stage build-up at extreme left. *NASA*

1 A 25 per cent mock-up of the Saturn C-1 for sizing and fit check at the ABMA, December 1959. *ABMA*

2 Assembly of the centre LOX tank on a circumferential welding machine from cylindrical and bulkhead sections that have been preassembled on other tooling. For joining, tank sections were rotated under a stationary welding head. Tanks were lengthened a section at a time. *MSFC*

3 The circumferential welding machine (left) for the outer tank welds displays the internal back-up bar in a 'serpentine' design which can be collapsed and expanded by pneumatic action. It provided clamping pressure and precision fit of joints – prerequisites for joint quality. *MSFC*

4 Preassembled outriggers and tail barrel are joined on the optical alignment stand. The turntable allowed precision installation of the support frames and mounting brackets for the outer tanks, engines and gimbal actuators. The tail barrel had the same diameter as the centre tank and would be bolted on to it later. *MSFC*

1

2

3

4

5 Completed tail section subassembly in a 7.31m (24ft) diameter tool ring attached to it, hoisted on to the rear cradle support. The centre tank would be moved in and connected to it. *MSFC*

6 The assembly tool at the centre of the final assembly station with fully adjustable controls operated from a central location. Moveable service structures were equipped with hoists and elevators to allow assembly at three levels. *MSFC*

7 Welded by Alcoa, aluminium sheet is prepared in cylindrical segments in Building 4707 for the Saturn S-I-T LOX tank. *ABMA*

8 The centre LOX tank installed to the tail section subassembly (right) and the spider beam (left) with pressurisation spheres installed. Transducer readouts and push-button remote controls were used to maintain precision alignment. The same controls permitted rotation of the entire fixture for assembly of the outer tanks, engines and shrouds. *MSFC*

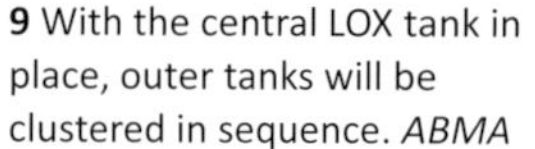

9 With the central LOX tank in place, outer tanks will be clustered in sequence. *ABMA*

10 Inside view of an outer LOX tank partly assembled. Stiffening ring frames were spot welded into cylindrical sections before seam welding. Openings in the rear bulkhead were for tank interconnecting lines and suction lines leading to the engines. *MSFC*

11 Engineers confer at the Fabrication and Engineering Laboratory at the Marshall Space Flight Center with S-I-T stage clustering underway adjacent to Juno II tanks. *MSFC*

12 Four outer LOX tanks are installed and will be followed by installation of the four fuel tanks. The fixture frame stands three stories tall and had a total length of 19.2m (63ft) between tool rings. *MSFC*

13 Viewed from the boattail end, the S-I stage now incorporates the four fuel tanks which are identified by a black coating and the national identity marking. The rear service structure has been moved back to allow 3-point suspension of the fixture and airframe so that the weight and centre of gravity can be determined. Precision load cells are incorporated in the tooling. *MSFC*

14 January 1961 and the SA-1 booster is complete and will be taken for stage testing at the East Tower where first firing will take place on 29 April 1961. *MSFC*

15 A giant in its time, the first SA-1 booster stage is readied for transportation. *MSFC*

16 While still in Building 4705, in February 1961 the S-I stage was temporarily mated with dummy S-IV and S-V stages together with a payload fairing in the horizontal configuration. *MSFC*

17 Building 4705 and the S-I-1 booster is ready for rollout. *MSFC*

18 SA-4 (right) all but complete and ready for shipment to Cape Canaveral, with SA-6 (left) and SA-7 (centre background) in assembly, the latter with its tanks in the foreground awaiting installation. *NASA*

19 H-1 engines on the production line. Some 152 were flown on the 19 Saturn I and IB missions. *Rocketdyne*

20 A view taken during the 1990s of the Cold Calibration Test Stand (CCTS) in the East Test Area at MSFC with the blockhouse in the foreground. *MSFC*

21 An H-1 engine turbopump installed in the CCTS. *MSFC*

22 Rocketdyne's Santa Susana test facility plays host to an H-1 firing. *Rocketdyne*

23 Bearing the sign 'US Army' down the side of the test stand at Redstone Arsenal, Saturn booster (S-I-T) is prepared for a static firing. *AOMC*

24 Showing the thrust structure and engine fixtures for the six RL-10 engines, an early S-IV stage is seen during final assembly. *Douglas Aircraft Co.*

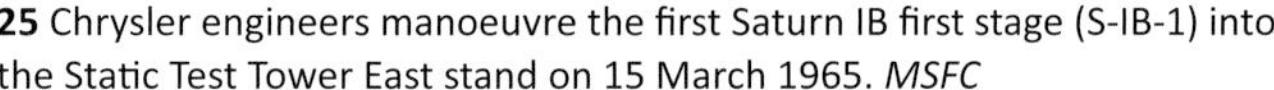

25 Chrysler engineers manoeuvre the first Saturn IB first stage (S-IB-1) into the Static Test Tower East stand on 15 March 1965. *MSFC*

26 S-IV fabrication at Douglas Aircraft with the stage bulkhead welding machine at left and the LOX tank welder at right. *Douglas Aircraft Co.*

27 The S-IV stage thrust structure at left foreground with aft Interstage in background during fabrication at Douglas. *Douglas Aircraft Co.*

28 The S-IV bulkhead dome with the honeycomb structure clearly visible at the 'dollar' manhole centre cover. *Douglas Aircraft Co.*

29 An S-IV stage being made ready for a test at SACTO, with detail around the thrust structure visible. *Douglas Aircraft Co.*

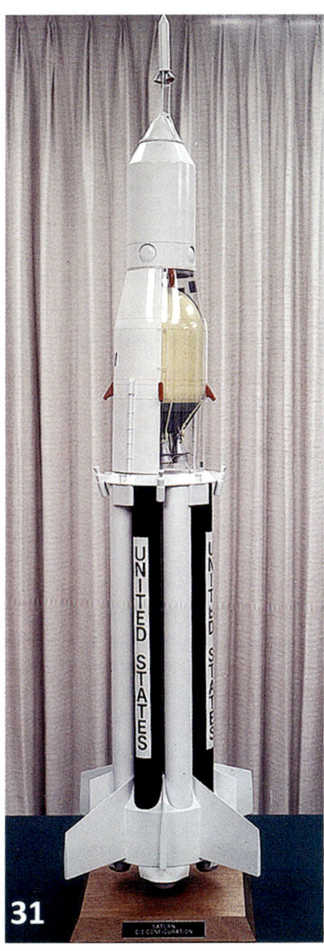

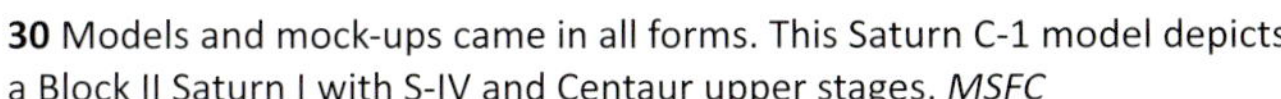

30 Models and mock-ups came in all forms. This Saturn C-1 model depicts a Block II Saturn I with S-IV and Centaur upper stages. *MSFC*

31 Reconfigured for its anticipated role in the Apollo programme, a model of the Block II with Command and Service Modules together with a Launch Escape System. *MSFC*

32 Historian Evelyn Falkowski, who compiled MSFC's *Illustrated History of the Saturn Rocket,* showing visitors William and Joseph Tung the Space Orientation Center at MSFC on 30 December 1964. Pictured are models of Nova (left) and Saturn C-2, neither of which were ever built. *NASA-MSFC*

33 Building 4708 at MSFDC: an H-1 is extracted for alignment checks, February 1960. *ABMA*

34 Engineers at NASA's Lewis Research Center equip a model of the Saturn I with sensors in the hypersonic wind tunnel, July 1961. *MSFC*

Appendix 3

Alternatives to Saturn A, B and C

Could Saturn Have Taken a Different Form?

Above: A selection of US Air Force missiles Juno in the 00m (140ft) high Missile Gallery at Dayton, Ohio. *Ty Greenlees/USAF*

ITEM	"ATLAS" CLUSTER	TITAN CLUSTER PHASE I	TITAN CLUSTER PHASE II	THOR CLUSTER CASE I	THOR CLUSTER CASE II
PAYLOAD					
300 N.MI. ORBIT	31,500	33,500	42,000	29,400	35,900
24 HR. EQ. ORBIT	7,200	8,000	10,800	6,800	8,400
INTERPLANETARY LAUNCH	9,200	10,300	13,700	8,800	10,800
LUNAR LANDING	4,200	4,600	6,800	3,800	5,000
TOTAL VEHICLE					
GROSS WT., LESS P.L.	968,000	932,000	957,500	988,500	1,247,000
MFG. WT., LESS P.L.	55,400	53,600	56,400	71,600	87,900
THRUST (S.L.)	1,200,000	1,440,000	1,440,000	1,155,000	1,485,000
THRUST/WT. (MIN.)	1.20	1.50	1.45	1.21	1.16
FIRST STAGE					
GROSS WEIGHT	675,000	686,000	686,000	742,000	954,000
PROPELLANT WEIGHT	632,460	646,800	646,800	679,000	873,000
BURNOUT WEIGHT	42,540	39,200	39,200	63,000	81,000
EFF. STRUCTURAL FACTOR W_{bo}/W_0	.063	.057	.057	.085	.085
MFG. WEIGHT	39,500	38,000	38,000	56,000	72,000
NUMBER OF ENGINES	6 | 3	8	8	7	9
ENGINE THRUST (S.L.)	170K | 60K	180,000	180000	165,000	165,000
CARRYING CAPACITY	324,000	280,000	313,500	275,900	328,900
NUMBER OF BASIC VEHICLES	3	4	4	7	9

Above: Essentially covering the payload range achieved by the Saturn I/IB, five potential configurations of existing hardware matched the first-stage thrust capabilities of the Juno V/Saturn I designs. *Author's collection*

It was not always clear that the ABMA would get approval for the Juno V/Saturn I proposal and alternative contenders continued to garner consideration. Based on payload performance, cost and mission applicability, existing missiles such as Thor, Atlas and Titan in various clusters were considered as alternatives. There was resistance from the Air Force to the Army's Juno series, which assumed the lead in developing long range rockets until NASA absorbed the Development Operations Division of the ABMA and drew a line under alternative options to the Saturn family.

The representative alternatives presented here are the culmination of several dozen configurations and clusters utilising missile and rocket motors already in existence and which were examined for both engineering and cost solutions. All options here were within the gift of the Air Force, which strenuously sought ways to provide the launch vehicles for the nascent space programme. None of those shown here were economically viable or technically superior to the Juno V/Saturn proposals and none were developed.

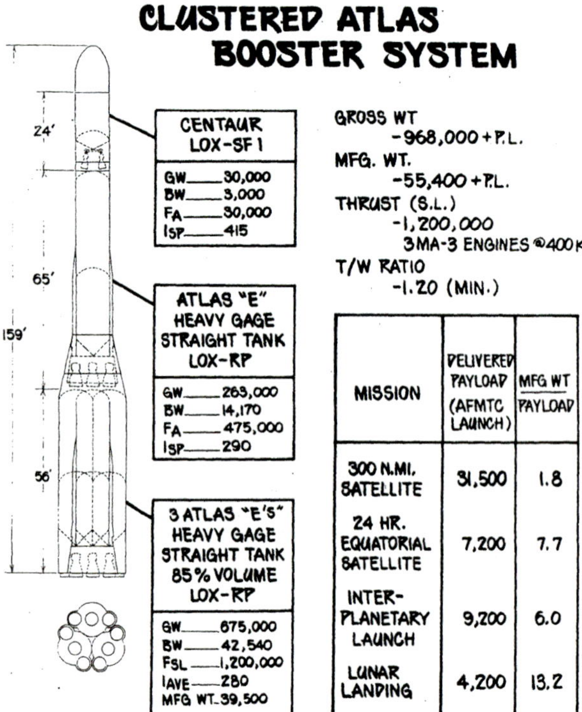

Above: By clustering several Atlas E missiles, adding a fourth as second stage and topped off with a Centaur, the capabilities of a Saturn I could be achieved with the 'Clustered Atlas' concept. *Author's collection*

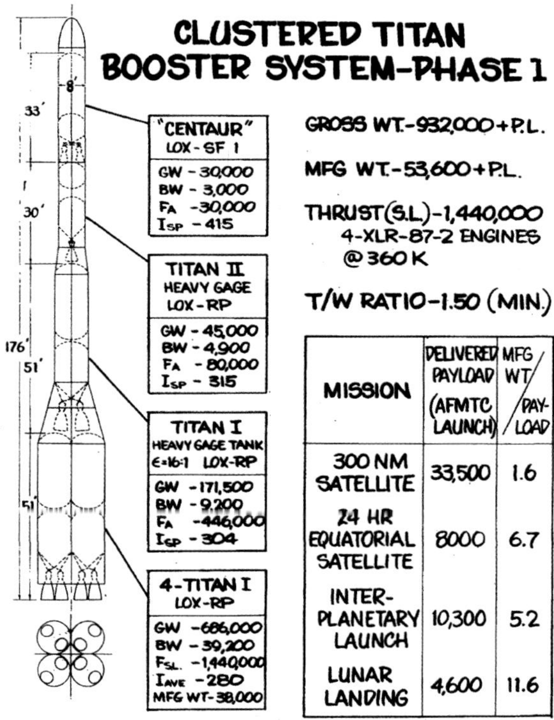

MISSION	DELIVERED PAYLOAD (AFMTC LAUNCH)	MFG WT/PAYLOAD
300 NM SATELLITE	33,500	1.6
24 HR EQUATORIAL SATELLITE	8000	6.7
INTER-PLANETARY LAUNCH	10,300	5.2
LUNAR LANDING	4,600	11.6

Above: Clustered Titan stages, as here, would provide a marginally better performance than the projected Saturn I but the complexity was greater and the mass slightly less than the Atlas concept. *Author's collection*

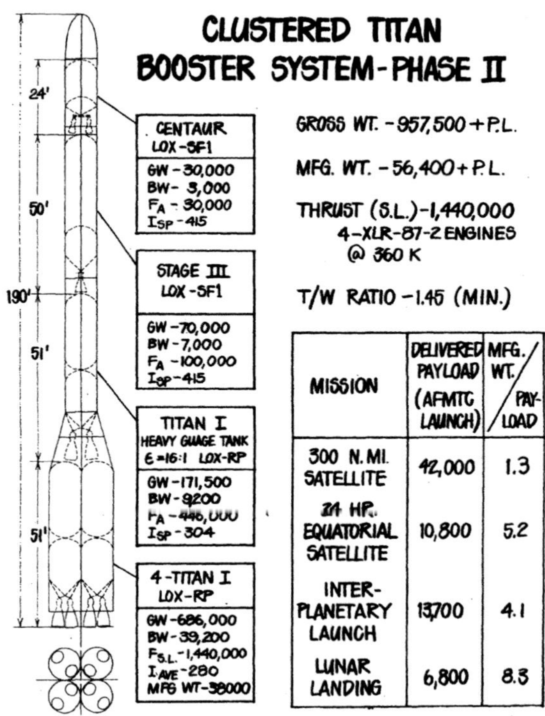

MISSION	DELIVERED PAYLOAD (AFMTC LAUNCH)	MFG. WT./PAYLOAD
300 N. MI. SATELLITE	42,000	1.3
24 HR. EQUATORIAL SATELLITE	10,800	5.2
INTER-PLANETARY LAUNCH	13,700	4.1
LUNAR LANDING	6,800	8.3

Above: Growth with the clustered Titan approach would yield a vehicle equalling the capacity of the Saturn IB but using four Titan I rockets as the first stage and three upper stages as shown. *Author's collection*

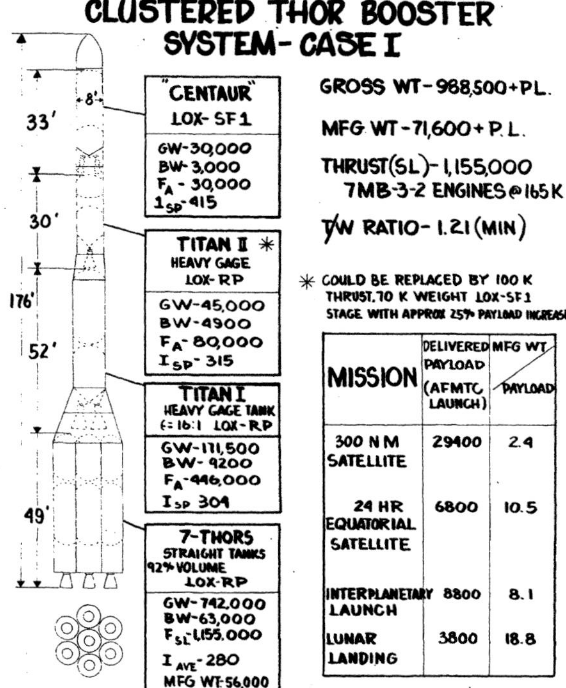

MISSION	DELIVERED PAYLOAD (AFMTC LAUNCH)	MFG WT/PAYLOAD
300 N M SATELLITE	29400	2.4
24 HR EQUATORIAL SATELLITE	6800	10.5
INTERPLANETARY LAUNCH	8800	8.1
LUNAR LANDING	3800	18.8

Above: Basing a clustered system on existing Thor, Titan I, Titan II and Centaur stages, an alternative to Saturn I was up for consideration. But seven Thor rockets would be required for the first stage. *Author's collection*

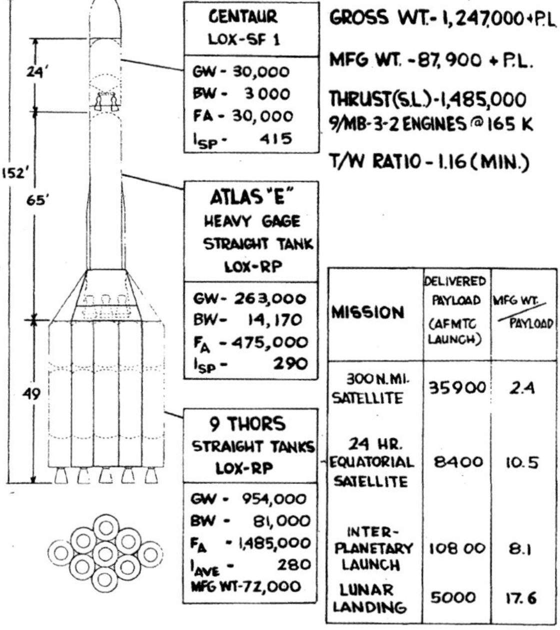

MISSION	DELIVERED PAYLOAD (AFMTC LAUNCH)	MFG WT/PAYLOAD
300 N.MI. SATELLITE	35900	2.4
24 HR. EQUATORIAL SATELLITE	8400	10.5
INTER-PLANETARY LAUNCH	108 00	8.1
LUNAR LANDING	5000	17.6

Above: Simplifying the complexity of a multiple use of upper stages, nine Thor rockets could grow the capability somewhat but not as much as that provided by the Saturn IB with an S-IVB stage. *Author's collection*

Appendix 4

Manned Mission Plans

The Evolving Strategies

By early 1963 the development of the Apollo programme was reaching crisis point, with a protracted sequence of unmanned and manned flights built on a conservative approach to development, test and verification prior to an equally protracted mission evolution phase. With time of the essence, managers were unable to effectively plan a fast-track approach to qualification.

By February 1963 the full extent of the commitment was self-evident and that is shown in these manifests for Saturn I, Saturn IB and Saturn V launch vehicles. These represent the most ambitious use of launchers before the transformation of mission manifests and flight planning later in the year, which introduced a new level of programme management. This successfully saw four NASA astronauts landing on the Moon before the end of the 1960s and a further eight before the end of the series in 1972.

It was made possible through compressed development schedules and the concept of all-up flight testing in which new launchers flew with all stages live rather than sequentially. This broke the policy established by the German rocket engineers at Peenemünde, whereby a steady and progressive test of various systems and elements of a preassembled stack preceded full duration unmanned flights followed by manned missions.

The charts displayed here reflect a less effective approach to the evolution of launcher Saturn development which pertained up to mid-1963. They are included as evidence of the more protracted and conservative approach which, had it been retained, would arguably have prevented NASA fulfilling the goal of landing astronauts on the Moon by the end of the 1960s.

Above: Former Peenemünde guidance and control specialist Dr. Walter Haeussermann (right) discusses the configuration of the Saturn I Instrument Unit with Wernher von Braun, and Clinton Grace (left), operations manager for the builders, IBM, at MSFC. *IBM*

PRELIMINARY

APOLLO/SATURN I FLIGHT MISSION ASSIGNMENT SUMMARY

1. LAUNCH DATE	OCT 61 / APR 62 / NOV 62 / MAR 63	AUG '63	DEC '63	MAR '64	JUN '64	OCT '64	DEC '64	MAR '65	JUN 65	SEP 65	DEC 65	MAR 66	JUN 66
PROPOSED CHANGES	(LAUNCHED)			APR '64									
	◄— RESEARCH & DEVELOPMENT —►							◄— OPERATIONAL —►					
2. LAUNCH VEH.	SA 1 / SA 2 / SA 3 / SA 4	SA-5	SA-6	SA-7	SA-9	SA-8	SA-10	SA-111	SA 112	SA 113	SA 114	SA 115	SA 116
3. MISSION OBJECTIVES (P-PRIMARY, S-SECONDARY)	◄— UNMANNED —►							◄— MANNED —►					
a. LAUNCH VEHICLE (LV) (MSFC Responsibility)	P 1. Structures 2. Propulsion (165K Engines)	P 1. Structures 2. Propulsion (1st 188K Engine) 3. Guidance (Passengers) 4. S-I/S-IV Stg. Separation	P 1. Structures 2. Propulsion 3. Guidance (1st Active Sys) 4. S-I/S-IV Stg. Separation ◄—FLIGHT RESTRICTIONS in Angle of Attack and Probability of Launch—►	P 1. Structures 2. Propulsion 3. Guidance (Active) 4. S-I/S-IV Stg. Separation	P 1. Structures 2. Propulsion 3. Guid. (Active) 4. EDS Full Capability 5. S-I/S-IV Stg. Separation	P 1. Structures 2. Propulsion 3. Guid. (Active) 4. EDS Full Capability 5. S-I/S-IV Stg. Separation	P LV QUALIFICATION 1. Structures (1st to Accommodate 22,500 lb. Payload) 2. Propulsion 3. Guidance (Active) 4. S-I/S-IV Stage Separation 5. EDS Full Capab.	S ◄— Without R&D Instrumentation —►	S	S	S		
b. SPACE CRAFT (SC) (MSC Responsibility)	NONE	NONE	S 1. Launch & Exit Environmental Parameters 2. LES Struct. Character. 3. LES Jettison Character.	S Back-up to SA-6 (if SA-7 Launched in April or Later)	S Study to be Conducted by MSC & MSFC to Determine Possible Additional Mission for Spacecraft	S MSC to Define Alternate Mission in Event SA-8 is not Required for Back-up to Micrometeoroid Experiment	P SC QUALIFICATION 1. Structures 2. SC Systems Opn'l Characteristics 3. SM Prop. (off-load) Includ. Restart 4. CM Re-entry 5. LES Jettison 6. SM-CM Separat. 7. Crew Safety 8. Recovery Sys. 9. Guid. & Navig.	P 1. Manned Orbital Flight 2. Opn. CSM Sys. w/Man. Control 3. Evaluation of Guid. System 4. Evaluate Multiman Inflight Crew Performance 5. Maneuvering Capability of CM RCS During Reentry 6. Complete Recovery Operation	P	P	P		
c. SPACE VEHICLE (SV)	NONE	NONE	S 1. Physical & Flt. Compatibility of LV & SC 2. Compatibility of R&D Communications & Instrumentation Between SV and Ground	S Back-up to SA-6 (if SA-7 Launched in April or Later)	S See Note Under SC Above	S See Note Under SC Above	P SV QUALIFICATION 1. EDS Full Capab. 2. Physical & Flight Compatibility of LV & SC 3. LV-SC Separation 4. Instrumentation, Communications, Tracking	P	P	P	P	SPARE	SPARE
d. OTHER					S Micrometeoroid Experiment	S Back-up to SA-9 Micromet. Exp.							

OMSF NO. G-27 REV. A DATE JAN. 10, '63 REV. DATE FEB. 14, '63 PREPARED BY: SYSTEMS ENGINEERING HUNTSVILLE OFFICE APPROVED: [illegible]

PART 1 OF 2

Above: With 16 Saturn I missions and hardware for those on order, NASA envisaged 12 development and qualification flights with Block II configuration of orbital capability, the last six of which were assigned to operational manned flights between March 1965 and June 1966, each numbered in the 100-series. *Author's collection*

PRELIMINARY

APOLLO/SATURN I CONFIGURATION AND FLIGHT DATA SUMMARY

1. LAUNCH DATE	OCT 61 / APR 62 / NOV 62 / MAR 63	AUG 63	DEC 63	MAR 64	JUN 64	OCT 64	DEC 64	MAR 65	JUN 65	SEP 65	DEC 65	MAR 66	JUN 66
PROPOSED CHANGES	LAUNCHED			APR' 64									
	◄— RESEARCH & DEVELOPMENT —►							◄— OPERATIONAL —►					
2. LAUNCH VEHICLE	SA 1 / SA 2 / SA 3 / SA 4	SA-5	SA-6	SA-7	SA-9	SA-8	SA-10	SA-111	SA-112	SA-113	SA-114	SA-115	SA-116
3. SPACE VEH. CONFIG.												SPARE	SPARE
A. LAUNCH VEH. (LV)													
(1) FIRST STAGE (S-I)	LIVE	LIVE	LIVE	LIVE	LIVE	LIVE	LIVE	OPN' L	OPN' L	OPN' L	OPN' L		
(2) SECOND STAGE (S-IV)	INERT	LIVE (1st)	LIVE	LIVE	LIVE	LIVE	LIVE	OPN' L	OPN' L	OPN' L	OPN' L		
(3) THIRD STAGE (S-VD)	INERT	NONE	NONE	NONE	NONE	NONE	NONE	NONE	NONE	NONE	NONE		
(4) INSTRUMENTATION UNIT (IU)	R & D	R & D	R & D	R & D	PROTOTYPE	PROTOTYPE	PROTOTYPE	OPN' L	OPN' L	OPN' L	OPN' L		
B. SPACE CRAFT (SC)													
(1) LUNAR EXCUR. MODULE (LEM)	NONE	NONE	NONE	NONE	NONE	NONE	NONE	NONE	NONE	NONE	NONE		
(2) SERVICE MODULE (SM)	—	—	◄— BOILERPLATE AFRM External Configuration —►				AFRM (First Compl)	AFRM	AFRM	AFRM	AFRM		
(3) COMMAND MODULE (CM)	◄— INERT JUPITER NOSE CONE —►		◄— BOILERPLATE AFRM External Configuration —►				AFRM (First Compl)	AFRM	AFRM	AFRM	AFRM		
(4) LAUNCH ESCAPE SYS. (LES)	NONE	NONE	◄— PROD Tower Jettison Motor Active Only —►				PROD (First Compl)	PROD	PROD	PROD	PROD		
4. LV PAYLOAD CAP. (LBS)	—	18,500	18,500	18,500	16,600	16,600	22,000+	22,500	22,500	22,500	22,500		
5. SC WEIGHT (LBS)	—	—	12,360 (BALLAST	12,360 TO 18,500)	12,360	12,360	22,000	22,500	22,500	22,500	22,500		
6. FLIGHT DATA													
A. PROFILE	BALLISTIC	ORBITAL	ORBITAL	ORBITAL	ORBITAL	ORBITAL	ORB. OR SUB. ORB.	ORBITAL	ORBITAL	ORBITAL	ORBITAL		
B. APOGEE (NM)		100	100	100	675	675	100	100	100	100	100		
C. PERIGEE (NM)		100	100	100	255	255	100 ~~255~~	100	100	100	100	BHS	
D. FLIGHT AZIMUTH		105°	105°	105°	105°	105°	72°/105°	72°	72°	72°	72°		
7. RECOVERY	NO	NO	NO	NO	NO	NO	YES (WATER)	YES (WATER)	YES (WATER)	YES (LAND)	YES (LAND)		
8. LAUNCH COMPLEX	34	37B	37B	37B	34	37B	34	34	34	34	34		

OMSF NO. G-27 REV. A DATE JAN. 10, '63 REV. DATE FEB. 14, '63 PREPARED BY: SYSTEMS ENGINEERING HUNTSVILLE OFFICE APPROVED: [illegible]

PART 2 OF 2

Above: In allocations on hand in February 1963, the last two Saturn I hardware sets on order were regarded as spares. The schedule for manned Apollo flights with Saturn I overlapped the actual period in which the first seven two-man Gemini spacecraft were launched. Note that rated payload capability was 10,206kg (22,500lb) to low Earth orbit. *Author's collection*

PRELIMINARY
APOLLO/SATURN I-B CONFIGURATION AND FLIGHT DATA SUMMARY

1. LAUNCH DATE	AUG'65	NOV'65	JAN'66	MAR'66	MAY'66	AUG'66	NOV'66	FEB'67	MAY'67	AUG'67	NOV'67	FEB'68
PROPOSED CHANGES												
	← RESEARCH & DEVELOPMENT →				← OPERATIONAL →							
2. LAUNCH VEHICLE	SA-201	SA-202	SA-203	SA-204	SA-205	SA-206	SA-207	SA-208	SA-209	SA-210	SA-211	SA-212
3. SPACE VEH. CONFIG.										SPARE	SPARE	SPARE
A LAUNCH VEH. (LV)												
(1) FIRST STAGE (S-I)	LIVE (Thirteenth)	LIVE	LIVE	LIVE	OPN'L	OPN'L	OPN'L	OPN'L	OPN'L			
(2) SECOND STAGE (S-IVB)	LIVE (First)	LIVE	LIVE	LIVE	OPN'L	OPN'L	OPN'L	OPN'L	OPN'L			
(3) THIRD STAGE	NONE	NONE	NONE	NONE	NONE	NONE	NONE	NONE	NONE			
(4) INSTRUMENTATION UNIT (IU)	R & D	R & D	R & D	PROTOTYPE	OPN'L	OPN'L	OPN'L	OPN'L	OPN'L			
B SPACECRAFT (SC)												
(1) LUNAR EXCUR. MODULE (LEM)	← BOILERPLATE (AFRM LEM Adapter) →			← AFRM (Ascent Stage Only) →								
(2) SERVICE MODULE (SM)	← BOILERPLATE Or (AFRM Structure) →			AFRM	AFRM	AFRM	AFRM	AFRM	AFRM			
(3) COMMAND MODULE (CM)	← BOILERPLATE Or (AFRM Structure) →			AFRM	AFRM	AFRM	AFRM	AFRM	AFRM			
(4) LAUNCH ESCAPE SYS. (LES)	← PROD (Tower Jettison Motor Active Only) →			PROD (1st Compl)	PROD	PROD	PROD	PROD	PROD			
4. LV PAYLOAD CAP. (LBS)	30,000	30,000	30,000	32,000	32,500	32,500	32,500	32,500	32,500			
5. SC WEIGHT (LBS)												
6. FLIGHT DATA												
A PROFILE	ORBITAL	ORBITAL	ORBITAL	ORBITAL	ORBITAL	ORBITAL	ORBITAL	ORBITAL	ORBITAL			
B APOGEE (NM)	105	105	105	105	105	105	105	105	105			
C PERIGEE (NM)	105	105	105	105	105	105	105	105	105			
D FLIGHT AZIMUTH	105°	105°	105°	105°	72°	72°	72°	72°	72°			
7. RECOVERY	NO	NO	NO	YES LAND	YES LAND	YES LAND	YES LAND	YES LAND	YES LAND			
8. LAUNCH COMPLEX	37A	37B	37A	37B	37A	37B	37A	37B	37A			

PART 2 OF 2 — OMSF, NO. 6-28, REV. A, DATE JAN. 10,'63, REV. DATE FEB. 14,'63

Above: The February 1963 schedule for Saturn IB envisaged eight Apollo manned missions following four development and qualification flights. As defined, two Saturn I manned flights were regarded as contingency slots for potential delays to the Saturn IB, the last definitively planned manned flight for that vehicle being December 1965 prior to the first scheduled Saturn IB manned mission launching in May 1966. *Author's collection*

PRELIMINARY
APOLLO/SATURN I-B FLIGHT MISSION ASSIGNMENT SUMMARY

1. LAUNCH DATE	AUG '65	NOV '65	JAN '66	MAR '66	MAY '66	AUG 66	NOV 66	FEB 67	MAY 67	AUG 67	NOV 67	FEB 68
PROPOSED CHANGES												
	← RESEARCH & DEVELOPMENT →				← OPERATIONAL →							
2. LAUNCH VEH.	SA-201	SA-202	SA-203	SA-204	SA-205	SA 206	SA 207	SA 208	SA 209	SA 210	SA 211	SA 212
3. MISSION OBJECTIVES (P-PRIMARY, S-SECONDARY)	← UNMANNED →			Manning of SA-204 to be Considered →	← MANNED →							
a. LAUNCH VEHICLE (LV) (MSFC Responsibility)	P 1. Structures 2. Propulsion 3. Guid. (Sat. I Guid. Sys. Active) 4. S-I/S-IVB Stg. Separation 5. EDS Full Capab.	P 1. Structures 2. Propulsion (Study Inclusion of S-IVB Re-start) 3. Guid. (Sat. I Guid. Sys. Active) 4. S-I/S-IVB Stage Separation 5. EDS Full Capab.	P 1. Structures 2. Propulsion (Study Inclusion of S-IVB Re-start) 3. Guid. (Saturn V Guid. Sys. Active) 4. S-I/S-IVB Stage Separation 5. EDS Full Capab.	P LV QUALIFICATION 1. Structures 2. Propulsion 3. Guid. (Sat. V Guid. Sys. Active) 4. S-I/S-IVB Stg. Separation 5. EDS Full Capab.	S ← Without R & D Instrumentation →	S	S	S	S			
b. SPACE CRAFT (SC) (MSC Responsibility)	S ←	S 1. Launch & Exit. Environmental Parameters 2. LES Structural Characteristics 3. Structural Evaluation 4. LES Jettison	S →	P SC QUALIFICATION 1. Structures 2. SC Systems Opn'l Characteristics 3. SM Prop. (Off-Load) Includ. Re-start 4. CSM/LV Separat'n 5. LEM/LV Separat'n 6. LEM Propulsion 7. Recovery System 8. LES Jettison	P ←	P 1. Manned Orbital Flights 2. Operation of SC Systems 3. Operational Techniques 4. Crew Training 5. Recovery Systems, Complete 6. CM & LEM Rendezvous and Docking 7. LEM Short Excursions, Rendezvous, and Docking 8. LEM Propulsion	P	P	P →			
c. SPACE VEHICLE (SV)	S ←	S 1. Physical & Flight Compatibility of LV and SC 2. Compatibity of R & D Communications, Instrumentation and Tracking Between SV and Ground	S →	P SV QUALIFICATION 1. Physical & Flt. Compatibility of LV and SC. 2. Communications, Instrumentation and Tracking Between SV & Gnd 3. EDS, Full Capab. 4. LV & SC Separat.	P	P	P	P	P	SPARE	SPARE	SPARE
d. OTHER												

PART 1 OF 2 — OMSF, NO. 6-28, REV. A, DATE JAN. 10,'63, REV. DATE FEB. 14,'63

Above: Here, three of the eight planned manned Saturn IB missions were declared as back-up and not as scheduled flights. With the fifth and last scheduled manned Saturn IB flown in May 1967, Apollo missions after that would migrate to the Saturn V. With a calculated maximum 14,742kg (32,500lb) payload capability, the Saturn IB was unable to lift a fully fuelled Block II CSM – but off-loaded for specific missions, it could conduct many of the development and qualification tests for Apollo, including the Lunar Module on some flights. *Author's collection*

Appendix 5

Astronauts of Saturn IB

Men of Apollo – the First and the Last

Eighteen NASA astronauts were assigned to fly on the Saturn IB but three were killed in the Apollo 1 fire at LC-34 on 27 January 1967, several weeks before their flight. The five manned Saturn IB missions included the first and last manned flights of an Apollo spacecraft as well as crew for the world's first reusable space station.

While Russia's Salyut of 1971 is notionally considered the 'first' space station, not until August 1974 did a Russian crew revisit a previously configured orbiting laboratory. On the definition that a space station is one visited by more than one crew, and not merely a habitable vehicle visited once, Skylab was the world's first such facility.

Only four of the 15 astronauts who flew a Saturn IB made another flight: Jack Lousma commanded the third Shuttle flight (STS-3) in March 1982; Paul Weitz commanded the sixth Shuttle flight in April 1983; Owen Garriott flew as a Mission Specialist on the first Shuttle long-duration science mission (STS-9) in November 1983; and Vance Brand flew Shuttle missions STS-5 (November 1982), STS-41B (February 1984) and STS-35 (December 1990).

Above: The Apollo 1 crew who lost their lives on 27 January 1967 in a fire inside CM-012 on top of AS-204, planned for launch on 12 February that year. From left: Edward H White II, Virgil I Grissom, Roger B Chaffee. *NASA*

Above: The Apollo 7 crew who flew the first manned Apollo flight in CSM-101 launched by AS-205 in October 1968. From left: Lunar Module Pilot R Walter Cunningham, Commander Walter M Schirra, Command Module Pilot Donn F Eisele. *NASA*

Above: The SL-2 crew, the first manned occupancy of the Skylab space station, launched aboard CSM-116 by AS-206 in May 1973. From left: Science Pilot Joseph P Kerwin, Commander Charles 'Pete' Conrad Jr, Pilot Paul J Weitz. *NASA*

Above right: The crew of the second manned flight to Skylab, SL-3, launched in CSM-117 by AS-207 in July 1973. From left: Science Pilot Owen K Garriott, Pilot Jack R Lousma, Commander Alan L Bean. *NASA*

Right: Launched aboard CSM-118 by AS-208 in November 1973, SL-4 was the last manned visit to Skylab. From the left: Commander Gerald P Carr, Science Pilot Edward G Gibson, Pilot William R Pogue. *NASA*

Below: The last astronauts to fly an Apollo spacecraft. CSM-111 carried the crew of the Apollo Soyuz Test Project into orbit aboard the last Saturn IB ever launched, AS-210 in July 1975. From left: Docking Module Pilot Donald K 'Deke' Slayton, Command Module Pilot Vance D Brand, Commander Thomas P Stafford. *NASA*

Glossary

A-1–A-10 Sequential project numbers for German rocket designs
ABMA Army Ballistic Missile Agency
ABMC Army Ballistic Missile Committee
Adm Admiral (of the US Navy)
AFBMD Air Force Ballistic Missile Division
ASMTC Air Force Missile Test Center
AFMTR Air Force Missile Test Range
AIAA American Institute of Aeronautics & Astronautics
AIS American Interplanetary Society
AMC Air Materiel Command (of the USAF)
AMR Atlantic Missile Range
AOMC Army Ordnance Missile Command
ARC Atlantic Research Corporation
ARDC Air Research Development Command (of the USAF)
ARPA Advanced Research Projects Agency
ARS American Rocket Society
ASTP Apollo Soyuz Test Project
BIS British Interplanetary Society
BMEWS Ballistic Missile Early Warning Station
BoB US Bureau of the Budget
BP Boilerplate
BSD Ballistic Systems Division (of the USAF)
C Centigrade
CalTech California Institute of Technology
CCAFS Cape Canaveral Air Force Station
cep Circular Error Probability – radius within which 50% of warheads will fall
CIA Central Intelligence Agency
cm Centimetre (length)
DA Direct Ascent (lunar landing mode)
DC Direct Current (or dc)
D-Day Invasion of Nazi occupied Europe on 6 June 1944
Deg Degrees of orbital inclination or orientation (not temperature)
DETA Diethylenertriamine
DMSP Defense Meteorology Satellite Program
DoD Department of Defense (US)
E East in degrees of longitude from the central meridian (Greenwich, England)
EDT Eastern Daylight Time
EOR Earth Orbit Rendezvous (lunar landing mode)
EST Eastern Standard Time
F Fahrenheit
FEBA Forward Edge of the Battle Area
FBW Fly-by-wire
ft Feet (length)
GALCIT Guggenheim Aeronautical Laboratory
gall Gallon
GDP Gross Domestic Product, a measure of national economic size and wealth
GE General Electric
GEM boxes George E. Mueller reporting nodes in management flow
Gen General (military rank)
GIRD Gruppa izucheniya reaktivnogo dvizheniya
gm Gram
GSE Government Supplied Equipment
H_2O_2 Hydrogen peroxide
HSTT High Speed Test Track
HQ Headquarters (usually of NASA)
IBM International Business Machines
ICBM Intercontinental Ballistic Missile
IGY International Geophysical Year (1957-58)
in Inch (length)
IRBM Intermediate Range Ballistic Missile
I_{sp} Specific impulse (thrust integrated over time/unit weight of propellant)
JANBMC Joint Army-Navy Ballistic Missile Committee
JATO Jet Assisted Take Off
JCS Joint Chiefs of Staff
JPL Jet Propulsion Laboratory
kg Kilogramme
km Kilometres
kN Kilonewtons
KT Kilotonne (of equivalent TNT explosive yield)
lb Pounds mass
LC Launch Complex
LEO Low Earth Orbit
LES Launch Escape System
lit Litre
LOR Lunar Orbit Rendezvous (lunar landing mode)
LOX Liquid oxygen
LH_2 Liquid hydrogen
LRPG Long Range Proving Ground
Luftwaffe German Air Force (1935-1945)
LUT Launch Umbilical Tower
m Metres
Mach Multiples of the speed of sound (after Ernst Mach)
MAF Michoud Assembly Facility
MFL Missile Firing Laboratory
min Minute(s)
MISS Man In Space Soonest (USAF programme)
MIT Massachusetts Institute of Technology
mls Statute miles
MSS Mobile Service Structure
MT Megatonne, a yield of thermonuclear weapons in equivalent TNT
MTF Mississippi Test Facility
N_2H_4 Hydrazine
N_2O_4 Nitrogen tetroxide
NAA North American Aviation
NACA National Advisory Committee for Aeronautics
NASA National Aeronautics & Space Administration
NATO North Atlantic Treaty Organization
N Newtons
N Geographical latitude up to 90deg north from the equator
NME National Military Establishment
NOAA National Oceanic & Atmospheric Administration
NRL Naval Research Laboratory
° Degrees of temperature (not orbital inclination)
ORDCIT Ordnance California Institute of Technology
OSSA Office of Space Science & Applications (NASA)
PAA Pan American Airlines
P&W Pratt and Whitney
PCM Pulse Code Modulation
PERT Program Evaluation Review Technique
PSAC President's Science Advisory Committee
Q-ball Housing for sensors measuring aerodynamic pressure
RAND Corp. Think-tank set up by Douglas Aircraft in 1948
RATO Rocket Assisted Take Off
REAP Rocket Engine Enhancement Program
Red Army Soviet Army
RF Radio Frequency
RFNA Red Fuming Nitric Acid
RNII *Reaktivnyy nauchno-issledovatel'skiy institute*
RP-1 Rocket Propellant No 1 (a course-grain kerosene)
rpm Revolutions per minute
S Geographical latitude up to 90deg south from the equator
SI International System of scientific units
S-I First stage of Saturn I
S-IB First stage of Saturn IB
S-IC First stage for Saturn V
S-II Second stage of Saturn V
S-IV Upper stage for Saturn I
S-IVB Upper stage for Saturn IB and Saturn V
S-V Proposed upper stage for Saturn I series
SA *Sturmabteilung* (assault division)
SAC Strategic Air Command (of the USAF)
SACEUR Supreme Allied Commander Europe
SACTO Sacramento Test Operations facility for S-IV ad S-IVB stage tests
SAM Surface to Air Missile
SAMSO Space & Missile Systems Organization
sec Second(s)
SecDef US Secretary of State for Defense
SHAEF Supreme Headquarters Allied Expeditionary Force
SLBM Submarine Launched Ballistic Missile
SMEC Strategic Missiles Evaluation Committee
SRBM Short Range Ballistic Missile
SS *Schutzstaffel* (protective squadron, or unit)
SSD Space Systems Division (of the USAF)
SSFC Santa Susana Field Laboratory
ST Stable platform for guidance system
STL Space Technology Laboratories
TCP Technological Capabilities Panel
Teapot Committee Colloquial name for SMEC chaired by Dr. von Neumann
T-Force Technical Force, an Anglo-American unit for obtaining German war scientists
TNT Trinitrotoluene
Ton Multiples of 2,000lb (1 ton)
Tonne Multiples of a metric ton (2,205lb)
T-Stoff Hydrogen peroxide
UAC United Aircraft Corporation
UDMH Unsymmetrical dimethyl hydrazine
UK United Kingdom
UN United Nations
USAF United States Air Force
USSR Union of Soviet Socialist Republics
UTC Universal Time Clock (time at Greenwich meridian)
V1 'Vengeance Weapon 1' name given to the Fieseler F1 103 flying bomb
V2 'Vengeance Weapon 2' name given to the A-4 ballistic missile
VAB Vehicle Assembly Building
VfR *Verein für Raumschiffahrt*
W West in degrees of longitude from the central meridian (Greenwich, England)
W39 Type of nuclear weapon
WAC Prefix for a version of Corporal rocket, believed to mean 'Without Attitude Control'
WADC Wright Air Development Center
WDD Western Development Division (of the USAF)
WSPG White Sands Proving Ground
Z-Stoff Potassium permanganate

Index

Aircraft

Government Departments, Institutions and Organisations

Manufacturers and Companies

Places and geographic locations

Rockets, Missiles and Spacecraft

Ships

Unattributed entries